21st International Conference on Condensed Matter Nuclear Science 2018

Published in the Journal of Condensed Matter Nuclear Science (JCMNS) Volume 29

Fort Collins, Colorado, USA
3 – 8 June 2018

Editor:

Jean-Paul Biberian

ISBN: 978-1-5108-9472-3

Printed from e-media with permission by:

Curran Associates, Inc.
57 Morehouse Lane
Red Hook, NY 12571

Some format issues inherent in the e-media version may also appear in this print version.

Printed with permission by Curran Associates, Inc. (2020)

For permission requests, please contact
International Society for Condensed Matter Nuclear Science (ISCMNS)
at the address below.

International Society for Condensed Matter Nuclear Science (ISCMNS)
The Willows, Hobro,
Wolverley, Kidderminster
Worcs DY11 5ST
England

www.iscmns.org
webmaster@iscmns.org

Additional copies of this publication are available from:

Curran Associates, Inc.
57 Morehouse Lane
Red Hook, NY 12571 USA
Phone: 845-758-0400
Fax: 845-758-2633
Email: curran@proceedings.com
Web: www.proceedings.com

JOURNAL OF CONDENSED MATTER NUCLEAR SCIENCE

Experiments and Methods in Cold Fusion

Proceedings of the 21st International Conference on Condensed Matter Nuclear Science, Lory Student Center, Colorado State University in Fort Collins, Colorado, USA, June 03–08, 2018

VOLUME 29, August 2019

JOURNAL OF CONDENSED MATTER NUCLEAR SCIENCE

Experiments and Methods in Cold Fusion

JOURNAL OF CONDENSED MATTER NUCLEAR SCIENCE

Volume 29, August 2019

 ISSN 2227-3123

Preface

The 21st International Conference on Condensed Matter Nuclear Science was held during the week of 3–8 June of 2018 in the Lory Student Center on the campus of Colorado State University in Fort Collins, Colorado. ICCF-21 was organized as a single-fee turn-key conference, which incorporated the technical program, housing, meals and all events. The campus setting, where participants shared meals and were lodged in the same village. encouraged the exchange of ideas, and also the development and refreshing of relationships. There were 104 abstracts submitted, and the program accommodated 63 oral presentations. The rest were posters.

The conference web site provides many details on the organization and content of the conference: iccf21.com. Importantly, all of the abstracts and presented graphics are available at that site, as are high-definition videos of all the oral presentations at ICCF-21. Those primary contents of the conference will be available globally for the next decade.

There were two classes of participants in ICCF-21. The first was the normal mix of scientists, technologists, engineers and business people. There were 147 such people at the conference from 22 countries. In addition, there was a set of participants who had won scholarships to attend the meeting without cost other than their travel. There were 50 applications, of which 42 were approved for funding, and 21 attendees. The scholarship participants were students from the U.S. and six other countries, including three high school science teachers. Those who had little prior exposure to LENR were introduced to the field. They also saw how a scientific conference works. The Scholarship Program was funded by the Ralph and Trish Nagel Foundation.

An introductory Short Course was held on the first day of the conference. The organization of that course was very straightforward, with the schedule and topics given in Table 1. After the introduction, there were two talks on the primary means of bringing together either protons or deuterons onto or into metals. That process is called “loading”. Then, there was another pair of talks on the two most important measurements in LENR experiments, namely heat and elemental products. The field has two main challenges, materials and understanding, so a third pair of talks dealt with them. Finally, there was a presentation on the state of commercialization of LENR energy generators, a part of the field that has widespread interest. There were 78 people who attended part or all of the Short Course, substantially more than usual.

Table 1. The schedule, topics and speakers for the Short Course held on 3 June 2018.

Time	Minutes	Topic	Speaker
1000–1040	40	Introduction and Issues	Nagel
1040–1120	40	Electrochemical Loading	McKubre
1120–1200	40	Gas Loading	Biberian
1200–1330	90	Lunch	
1330–1410	40	Calorimetry and Heat Data	Letts
1410–1450	40	Transmutation Data	Srinivasan
1450–1510	20	Break	
1510–1550	40	Materials Challenges	Imam
1550–1630	40	Theoretical Considerations	Hagelstein
1630–1700	30	Commercialization	Seccombe

There were two keynote presentations at the start of ICCF-21. The first gave a larger perspective, and the second was a technical overview of the field.

Thomas Darden, the founder and CEO of Cherokee Investment Partners LLC, also founded Industrial Heat in 2012. He focused his talk on the energy challenges that motivate many of the people trying to understand and advance LENR. Much of Darden's address dealt with social, cultural, scientific, financial and political factors relevant to LENR. He ended with a call for increased openness and collaboration within the LENR field.

Michael McKubre is an electrochemist, who worked at SRI international from 1978 until 2016. He lead a large and productive research program on LENR since 1989. Rather than review those many results, McKubre gave a perspective on what is needed for LENR to progress and realize its promise for the benefit of mankind. He emphasized five "tion" words: verification, replication, correlation, demonstration and utilization. McKubre noted, as he has done in the past, the need for collaboration, cooperation and communication within the field

Near the end of the conference, Dana Seccombe presented an important talk entitled "Experience with Semiconductor Technology Development Potentially Relevant to LENR." His experience in a major and innovative semiconductor development program is relevant to the commercialization of LENR generators. Yield in semiconductor production is conceptually equivalent to success (reproducibility) in LENR experiments. If funding of LENR development were adequate, it would be possible to use the iterative, focused practices from the massive semiconductor industry to speed knowledge, commercialization and exploitation of LENR.

Several of the ICCF series have had unusual sessions for specific purposes. At ICCF-14, there were two sessions devoted to the work of LENR pioneers in order to review and highlight the many contributions to this field by those people, and to honor them for their work and results. A session at ICCF-21 on Friday morning was devoted to talks by three long-time contributors of experimental results about LENR. The intent was again to honor them and their contributions, and to give them a chance to 'teach what they know', which they learned over the decades. Those three presentations were by Edmund Storms, Jean-Paul Biberian and Mitchell Swartz.

An evening meeting on Nuclear Structure and its relation to LENR was organized by Norman Cook and Bob Cook. It attracted about 20 participants. Many of them continued to interact after the conference.

The excursions on one afternoon were unique. Rather than only going to one cultural or natural resource, four visits by groups of conference participants were made to technical organizations near Colorado State University. They were the National Institutes of Standards and Technology, the National Renewable Energy Laboratory, the National Oceanic and Atmospheric Administration and the National Center for Atmospheric Research.

A detailed review of the technical content, and other aspects of ICCF-21, has been published in Infinite Energy magazine. https://www.infinite-energy.com/iemagazine/issue141/index.html.

Sincerely,

Steven B. Katinsky and David J. Nagel
(*August 2019*)

J. Condensed Matter Nucl. Sci. 29 (2019) 1–547
 ISSN 2227-3123

JOURNAL OF CONDENSED MATTER NUCLEAR SCIENCE

Volume 29 2019

CONTENTS

RESEARCH ARTICLES

J. Condensed Matter Nucl. Sci. 29 (2019) 1–11

Research Article

Fabrication and Characterization of Palladium–Boron Alloys Used in LENR Experiments

M. Ashraf Imam* and David J. Nagel
The George Washington University, Washington, DC 20052, USA

Melvin H. Miles
Dixie State University, St. George, UT 84770, USA

Abstract

Most Low Energy Nuclear Reaction (LENR) electrochemical experiments have been performed with commercially pure palladium cathodes. There has also been interest in the use of alloys of palladium, which retain the ability of that element to absorb high fractions of deuterium, but also offer better mechanical properties. Alloying palladium with low levels of boron is a prime example. The fabrication and characterization of Pd–B alloys is described in this paper. Three alloys with nominal composition in weight percent of boron, 0.25, 0.5 and 0.75, were produced by arc melting, followed by annealing. Transmission electron microscopy and X-ray diffraction showed them to consist of two face-centered cubic phases with different lattice parameters, one dispersed as fine particles within the other. It was found that these alloys produce highly reliable LENR results.

Keywords: Annealing, Excess heat, Face-centered cubic, LENR, Palladium–Boron alloys, X-ray diffraction

1. Introduction

Low Energy Nuclear Reactions (LENR) have four remarkable empirical characteristics, which make them very attractive as future energy sources. They (a) do not involve dangerous radiation during operation, (b) do not leave radioactive waste, (c) emit no greenhouse gases and (d), very significantly, provide energy gains. The gain is defined as the ratio of the output thermal energy to the input electrical energy, which initiates and controls the LENR.

The attractive features of LENR are currently counterweighted by two major scientific problems. The first is the lack of understanding of the mechanisms that cause LENR. There have been dozens of theories, but none of them has been adequately developed for quantitative comparison with measurements. The second challenge involves the materials used in LENR experiments. Many experiments have shown that the production of LENR depends on some presently unknown characteristics of materials. It is not known if either the composition or structure, or possibly both

*Corresponding author. E-mail: ashrafimam@gwu.edu.

© 2019 ISCMNS. All rights reserved. ISSN 2227-3123

of those characteristics, determine the efficacy of LENR materials. This causes the irreproducibility issue for energy generation from heavy water electrolysis (the Fleischmann–Pons effect) when palladium metal is used as the cathode.

Palladium has a Face Centered Cubic (FCC) structure with properties akin to gold, i.e., soft, ductile, and resistance to corrosion. It is worth noting that palladium is not tarnished by dry or moist air at room temperature, but at about 600°C a thin oxide film forms in air. Above 800°C, the superficial oxide decomposes, leaving a clean metal surface. Some oxide formation occurs again above 1000°C [1]. The intrinsic hardness and tensile strength of palladium are too low for many applications. The addition of boron to palladium within the solubility limit creates two FCC phases with different lattice parameters, one phase being distributed as fine particles within the other phase.

The Pd–B system has been studied for almost 70 years. A 1996 review of the system contains 19 references [2]. A paper on the Pd-rich part of the Pd–B phase diagram published in 2006 cites 24 papers [3]. The preparation of single-phase alloys made of palladium and other minor elements has been known. For example, various palladium alloys which include boron were reported, such as Weber et al. US Patent No.5,518,556 (a boron-containing surface layer), Hough et al. US Patent No. 4,341,846 (an electroless Pd–B plating material), Smith Jr. et al. US Patent No. 4,396.577 (a brazing alloy containing boron, palladium and other metals) and Prosen US Patent No. 4,046,561 (an alloy for porcelain applications containing palladium, boron and other metals). However, what is lacking in these patents is a palladium and boron composition of sufficient strength to be used as a reactive structure rather than a coating material. Such a bulk material also may be used in thin hydrogen purification membranes or as an electrode in a heat-generating process.

Based on almost 30 years of research, two sources of palladium materials yielding good reproducibility for generation of excess enthalpy effects have been identified: (1) palladium materials prepared by co-deposition method and (2) Pd–B alloys. A common feature for both these methods is that they yield palladium that is relatively free of oxygen as an impurity. A beneficial effect of the added boron is that it minimizes the activity of dissolved oxygen in the palladium by converting it to B_2O_3 during processing. The low density B_2O_3 floats to the surface and is removed during the molten phase of the Pd–B alloy preparation. Further, the creation of two FCC phases makes the material harder and less susceptible to cracking. That is attractive for some applications. In particular, it is the likely explanation for reproducible LENR energy generation.

This paper deals with the fabrication, characterization, and evaluation of Pd–B alloys, which have produced excess enthalpy in nearly every experiment [4,5]. Two US Patents have been granted for these materials [6,7]. Success in producing LENR energy with these materials is described in a companion paper [8]. Plans for the employment of Pd–B alloys in a new program are described in another planned paper [9]. The focus of this paper is on the production and processing of Pd–B alloys, including the effects of annealing for different times at different temperatures and cold deformation. Those activities are described in Section 2. Since the mechanical and many other properties of palladium depend on purity, elemental analyses and X-ray diffraction studies will be presented and discussed in Section 3. The Section 4 summarizes the reported activities and characteristics of the Pd–B alloys, and LENR results obtained with them.

2. Production and Processing of Pd–B Alloys

A copper hearth with the cavity containing the Pd and B, suitable for arc melting and cooling, was employed. Although there are several other forms of heating that would be suitable, an electric arc is preferable. Copper is preferably used as the material for the hearth because of its excellent thermal conductivity to remove useless heat during the arc melting process. The copper hearth has a lower melting point than either boron or palladium. Hence, it was cooled by water flowing within it to prevent the copper from melting into the alloy. The hearth, and the mixture resting upon it, were cooled by the transfer of heat from the arc to the alloy composition, and then through the hearth to the coolant water. Exposure to air cannot be allowed because oxygen or air would oxidize the palladium and the boron, ruining the process. The process was performed in a container that was evacuated and then backfilled with a noble gas, typically high purity argon.

The processes used to produce the Pd–B alloys started with boron in powder form, commonly referred to as five-nines boron (99.999% pure). It was first placed within a cavity in the hearth structure. After the weighed boron powder was dispensed into the structure, the weighed amount of Pd sponge was placed on top of it. The palladium was a pure palladium sponge form, commonly referred to as five-nines palladium (99.999% pure). It was positioned in the compartment in that manner, so that it was overlying the boron powder in the cavity. The denser palladium over-layer prevented splatter or other loss of the boron. Additionally, the lower melting point of palladium allowed it to form a protective shell over the boron before the boron begins mixing with the palladium. This arrangement enabled the metal to melt first and envelop the boron powder, which insured little or no loss of the boron powder during the melting and alloying.

A typical arc melting apparatus having 12 V and 300 A (3.6 kW) was used for the melting. The two electrodes used for the arc melting are a tungsten tip immediately above the sample and the copper hearth. The melting was performed at about 2100°C. The melting point of boron, the higher of the pair, is 2079°C so the melting must be done at least at this temperature. However, the temperature should not exceed about 2200°C or the boron will begin to vaporize. The melting time was about five minutes. A melting time greater than 10 min at the high temperatures would result in vaporization of a portion of the boron.

After the initial melting to mix the loose boron powder into the palladium, the mixture was cooled for approximately 30 min, and then turned over in the circular cavity in the copper hearth with the tungsten tip. Then, the melting, cooling, and turning over steps were repeated as often as necessary to eliminate any boron or palladium pockets, and for thoroughly mixing the material into a solution of a desired homogeneity. The steps were repeated anywhere from about 3 to 10 times. The molten solution was homogenized by a combination of gravity and uncontrolled movements within the melts. Finally, the alloy button was transferred to a slot-shaped cavity in the copper hearth for a final melting to produce an ingot of the alloy about 15 cm long and 1 cm in lateral oval cross section.

After the homogeneous alloys were prepared, the material underwent additional steps of swaging or rolling, if so desired, for reducing the alloy to a fixed diameter or plate. After the material was swaged or rolled, the alloy is annealed to reduce the residual stress. The alloy begins at room temperature and is heated to approximately 650°C. for about two hours. The time and temperature in this step are important because too high a time or temperature would result in a larger grain size of the composition which would detract from hardness, and may render the composition ineffective. After annealing, the alloys were cooled to room temperature. The composition resulting from the above process has boron in an interstitial solid solution within in the palladium, with the alloy having a two-phase structure.

The Pd–B alloy produced contains about 0.1–0.8% by weight boron, and hence, from about 99.2 to 99.9% by weight percent palladium. Three compositions were prepared with nominal 0.25, 0.5 and 0.75 wt.% B compositions. The amount of boron in the mixture appears to be critical. It has been found that the amount of boron must be maintained below 2 wt.% of the mixture for full dissolution of the boron. Anything more will react with the palladium, preventing formation of the two solid solution phases. Hence, the amount of boron placed in the cavity for each case was chosen to be small enough not to form a compound of boron in the palladium. However, it was sufficient to react with impurity oxygen in the palladium, while the boron and palladium were molten. Typically, about 100 g of palladium was used, with amounts of boron consistent with the ranges noted above.

The phases of the two-phase structure have the same crystal structure, but different sets of lattice parameters. It was found, and is shown below, that the crystals of the second phase were larger than the crystals of the first phase. The diameter of the crystallites in the first phase was in the range of 1–10 nm, whereas the diameter of the crystallites in the second phase is much larger. The differing sizes of the crystals of the phases creates a "miscibility gap", as shown in the phase diagram of Fig. 1. This means that the miscibility of the two phases with each other is high because the crystals of the smaller first phase can easily rest in gaps between the larger crystals of the second phase. This filling of the gaps of the larger crystals binds the crystals of both phases together and results in a hardened composition. A more recent phase diagram is shown in Fig. 2. It is little different from the diagram in Fig. 1.

3. Results and Discussion

Three Pd–B alloys were prepared with the aim to achieve 0.25% B, 0.50% B and 0.75% B in accordance with the details given above. The final composition of the three alloys was 0.18% B, 0.38% B and 0.62% B based on Glow-Discharge Mass Spectroscopy (GDMS). Analysis is shown in Table 1. The overall decrease in the boron content after processing, compared to targeted content, was due to the formation of B_2O_3 by reacting with the dissolved oxygen in palladium. Boron acts as scavenger for oxygen during processing. The low density B_2O_3 floats to the surface, and is removed during the molten phase of the Pd–B alloy preparation. Table 1 shows relatively high concentrations of Si, Fe, Rh and Pt in most of the diverse materials, including wire from the Naval Air Warfare Center (NAWC).

To make electrodes from the compositions, the alloys were swaged to 0.4 cm diameter. The swaged rods were cut to length, and machined into usable electrode forms, commonly a few cm in length. The samples were annealed at 650°C for 2 h and furnace cooled to room temperature. A similar procedure was followed for plates or thin foils after rolling instead of swaging. If the annealing is done at different temperatures and different times, the grain size will be affected. An example is shown in Fig. 3. It gives the actual composition of boron as compared to attempted

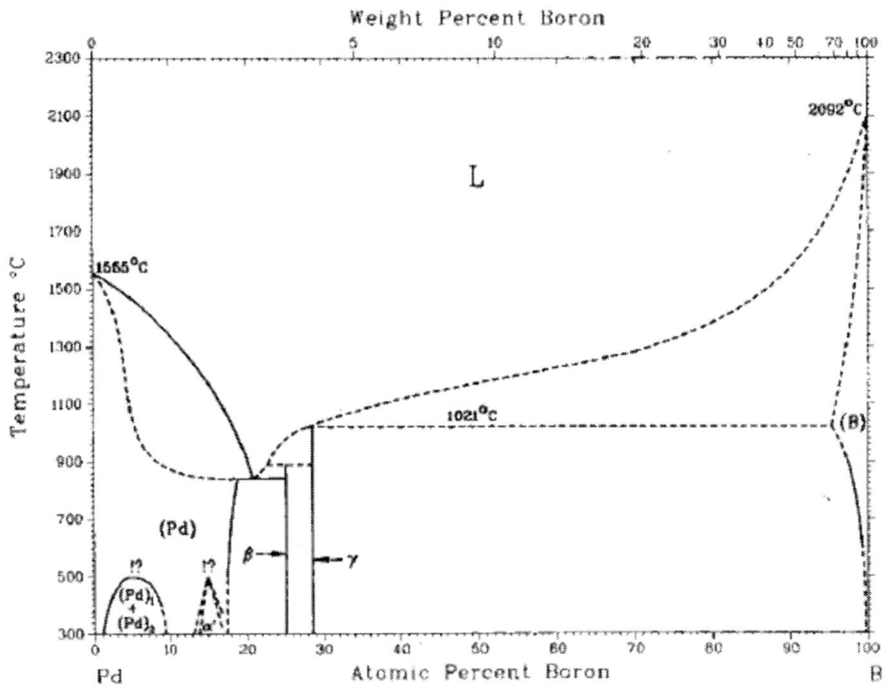

Figure 1. Binary phase diagram of Pd–B system from Binary Alloy Phase Diagrams, Editor-in-Chief: Thaddeus B. Massalski, Volume 1, p. 372, American Society for Metals (August 1987).

composition of 0.25 wt.% B, 0.50 wt.%B and 0.75 wt.% B. Grain size has two major effects: (1) the smaller the grain size, the larger strength and (2) diffusion rates are times higher in grain boundaries, compared to bulk, so that loading with deuterons will be faster in small grain size materials. Detailed analyses of the effects different annealing temperatures and times on grain size are described in [2].

X-ray diffraction studies were carried out to characterize the three compositions of the two-phase Pd–B alloy. The diffraction patterns were obtained with a Phillips diffractometer with generator settings of 50 kV, 30 mA and a copper target. Two distinct phases of the same cubic structure were found in all three compositions of the alloy, as shown in Fig. 4. Lattice parameters for the samples were measured. As can be seen in Fig. 5, the two distinct phases have the same crystal structure but different lattice parameters. The lattice parameter in a first phase remains constant with changes in the boron content of the alloy, whereas the lattice parameter of a second phase increases with an increase in the boron content. As the boron content increases, the fraction of crystals in the second phase increases at the expense of the first phase, as expected.

The 0.62% B sample was studied with a transmission electron microscope. A typical micrograph is shown in Fig. 6. The micrograph shows two phases, one minor (black) in the matrix of major phase (white). Figure 7 shows the Selected Area Diffraction (SAD) pattern of the same as Fig. 6. It shows rings along with the main diffraction spots, indicating no intermetallic compound. Lattice parameters of the two phases measured from X-ray diffraction and SAD are consistent, so they both confirm the production of the Pd–B alloys.

Table 1. Glow-discharge mass spectroscopy (GDMS) analyses of various cathode materials (concentrations in ppm by weight).

Element	NAWC wire	NRL Pd	Pd–0.62B	Pd–0.38B	Pd–0.18B
B	0.007	<0.001	6200	3800	1760
c	<1	0.02	<1	<5	<1
N	<3	0.03	<0.1	<0.1	<5
O	<10	0.45	<10	<10	<20
Mg	0.009	1.2	3.5	2.7	2.9
Al	0.63	0.53	4.1	3.3	1.5
Si	3.5	0.31	15	11	6.8
Ca	0.29	0.58	7.9	2.9	2.4
Cr	0.21	1.2	0.98	1.1	1.1
Mn	0.004	0.75	8.2	5.9	2.6
Fe	2.9	33	56	47	36
Ni	0.03	0.85	1.4	1.7	1.6
Cu	0.76	27	26	26	16
Zn	0.02	1.2	2.3	1.7	1.6
Zr	0.04	0.3	3.9	0.79	0.84
Rh	4.2	11	11	9.6	8.5
Ag	0.45	0.71	0.75	1.4	1.5
W	0.10	3.8	2.2	1.0	0.67
Pt	2.2	30	47	38	18
In	<0.05	<0.05	<0.05	1.9	1.2
Au	1.0	0.17	0.2	0.65	0.22
Ir	1.1	0.4	0.33	0.23	0.18

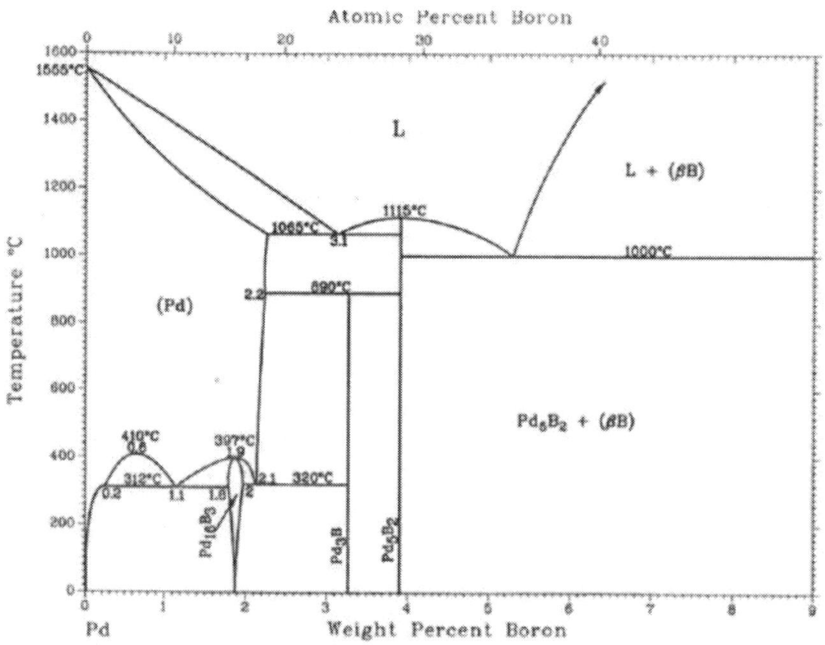

Figure 2. A revised binary phase diagrams of Pd–B system from ASM International Handbook, Volume 3, Alloy Phase Diagram, pp. 2–83 (1999).

4. Summary and Discussion

Three compositions of Pd–B alloys were prepared and characterized. Preparation of the palladium alloys with low boron contents were done in order to keep within the miscibility regime. The compositions had nominal boron concentrations of 0.25, 0.50, 0.75 wt.% B. GDMS analyses of the three as-prepared materials showed 0.62, 0.38 and 0.18 wt.% B. Annealing of the swaged sample at 650°C for 2 h resulted in grain sizes of about 100 μm. The alloying with boron expands the palladium lattice by forming an interstitial alloy with palladium [10]. X-ray diffraction studies showed two distinct phases of the same face centered cubic structure with different lattice parameters. The lattice parameter of one phase remains constant with changes in boron content, whereas the lattice parameter of the other phase increases with the increase in boron content. As the boron content increases, the fraction of one phase decreases at the expense of the other phase.

The alloys described in this paper have proven to be effective in production of energy by LENR. They gave excess powers in seven out of eight experiments in China Lake, California. The materials also produced energy in one study in the New Hydrogen Energy laboratory in Japan, and in another in a private laboratory in California. Three different calorimeters were employed in that set of 10 studies. It was found that excess power appears early in experiments with these Pd–B alloys, in contrast to most other LENR experiments. Details of the past measurements are available [8]. Plans to use these Pd–B materials in future experiments are also documented [9].

Despite all the early work on the Pd–B system, and the LENR-motivated work in the past quarter of a century, there remain several issues regarding the system and its success in LENR experiments. We conclude by noting and commenting briefly on them:

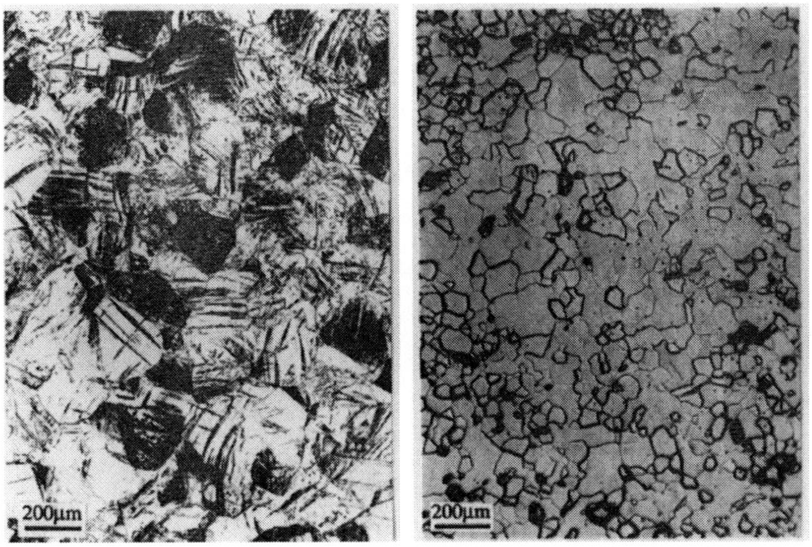

(a)

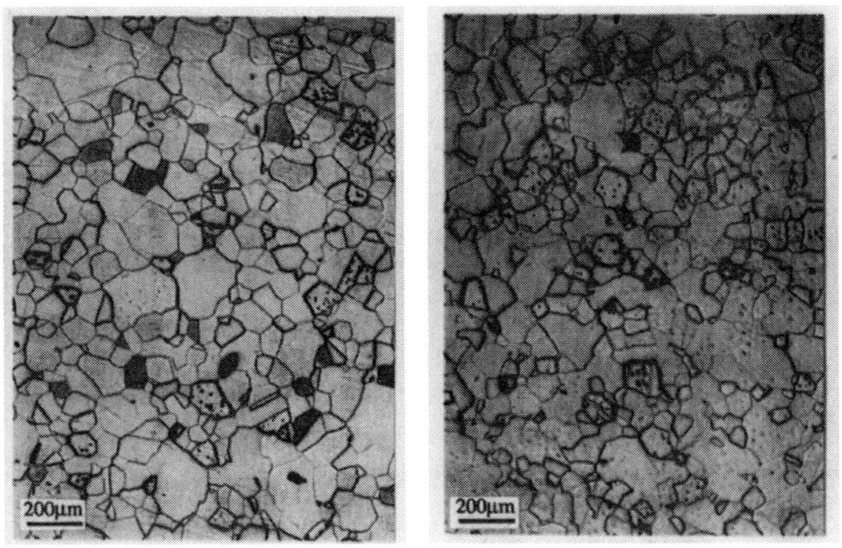

(b)

Figure 3. (a)*Left*: Pd-0.62 wt.% B as swaged. *Right*: Pd-0.62 wt.% B annealed at 650°C for 2 h (grain size about 95 μm). (b) *Left*: Pd-0.38 wt.% B annealed at 800°C for 1 h (grain size about 100 μm). *Right*: Pd-0.18 wt.% B annealed at 800°C for 1 h (grain size about 125 μm), percentages of boron are actual compositions.

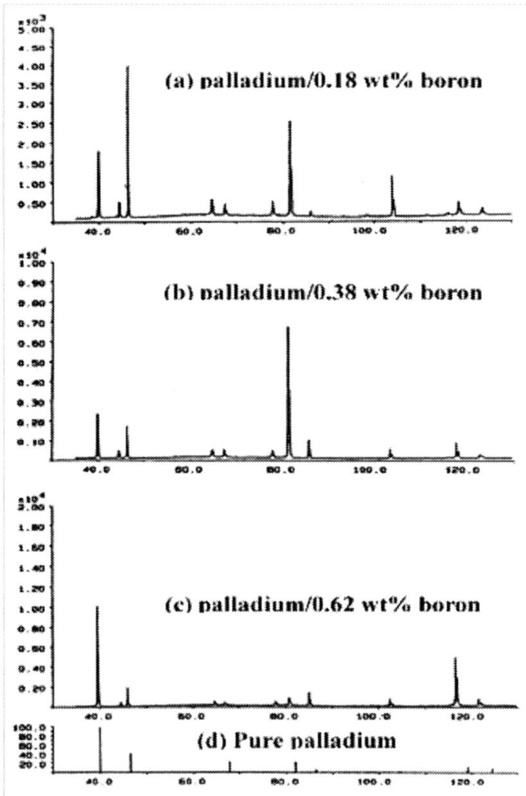

Figure 4. X-ray diffraction patterns of (a) Pd-0.18 wt.% , (b) (b) Pd-0.38 wt.% boron, (c) Pd-0.62 wt.% boron, and (d) pure palladium.

(A) Presence of B as a nuclear reactant. There is ample evidence from the production of helium during LENR experiments, and its correlation with heat production in over a dozen experiments. However, there has been discussion about the possibilities that lithium or boron are also reactants in LENR experiments. This remains an open question.

(B) Oxygen gettering. The possibly deleterious role of oxygen in LENR experiments has made it desirable to eliminate oxygen from materials in such experiments. Sometimes that was done by use of an oxygen-free atmosphere while alloying. The possibility that boron getters residual oxygen in the arc melter used for preparation of our alloys was discussed in this paper. That process is consistent with observations, but not proven.

(C) Increased mechanical strength. The mechanical properties of our Pd–B alloys were not measured. However, it is clear that their strength is much greater compared to pure palladium due to the boron-produced dispersoids and the observed grain structure. The influence of mechanical properties on the generation of LENR is another open question. If there is such an influence, it might be either direct or indirect, the result of a change in materials characteristics causing changes in diffusion or other processes.

(D) Less volumetric expansion? The amount of expansion of cathode materials has been found to be important in

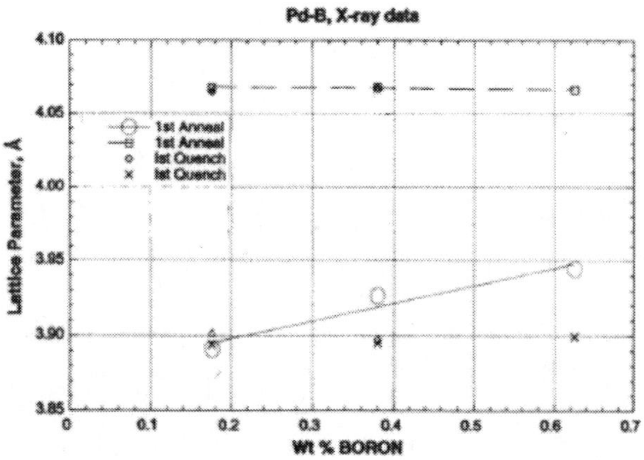

Figure 5. Two distinct phases having same crystal structure, but different lattice parameters, are shown. The lattice parameter in a first phase remains constant with changes in the boron content of the alloy, whereas the lattice parameter of a second phase increases with an increase in the boron content.

some LENR experiments. Maintenance of volume might be related to lesser production of voids and cracks, and the greater retention of deuterium in the metallic lattice.

(E) Asymmetric loading and deloading. It appeared that the rates of loading deuterons into the Pd–B alloys were greater than the rates of deloading after voltages to cells were turned off. This might be due to the mechanical and expansion characteristics just noted. However, it is realized that any alloy changes during loading. That is, the material during loading is different from the material during deloading.

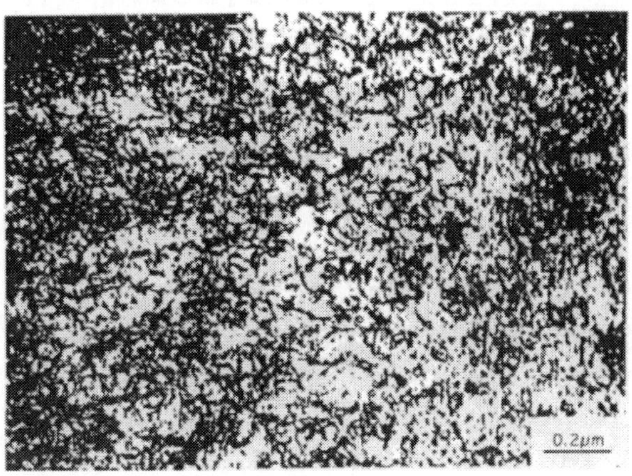

Figure 6. Transmission electron micrograph of Pd–0.62 wt.% B showing two phases. The minor phase is roughly 1–10 nm in diameter.

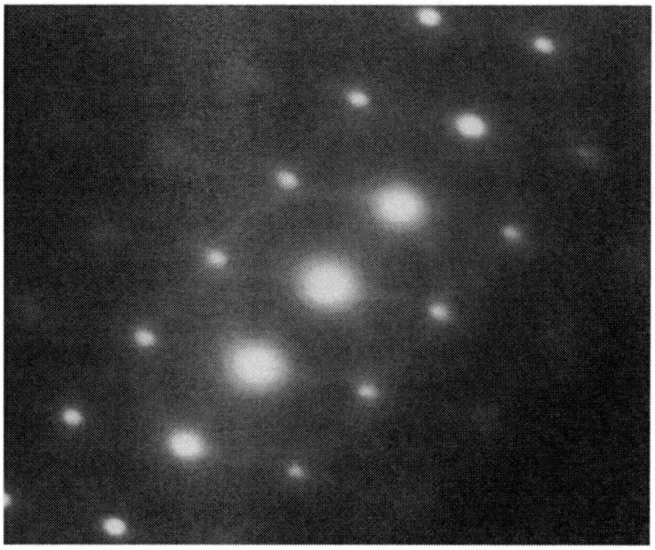

Figure 7. Selected area diffraction pattern of the same material and area as in Fig. 6 showing rings along with the main diffraction spots.

Appendix: Other Applications of Pd–B Alloys

The alloys produced in this work show the same or better strength than pure palladium with much less thickness. This is advantageous for the creation of hydrogen purification membranes because less palladium would be needed to create a membrane and achieve the same results. This is, sturdy membranes of much less thickness are enabled compared to using palladium alone. Put another way, the increased hardness means that a much smaller amount of expensive palladium may be used to provide a membrane of the same capacity compared to costly palladium alone. This would allow much greater membrane capacity through reduced material costs. How much the thickness of the membrane would be able to be decreased with the present composition would depend upon such factors as the geometrical design, gases to be purified, and the extent of purification desired.

The hardened Pd material would also be advantageous for use as electrodes in etching, polishing, electrochemical machining, semiconductor wafer manufacture and other electrochemical processes. Palladium cathodes hardened by alloying, as described in this paper, retain their superior electrical characteristics and resist erosion better than pure palladium.

References

[1] A. S. Khanna, Introduction to high temperature oxidation and corrosion, *ASM Int.*, p. 108, 2002.

[2] P.K. Liao, K.E. Spear and M.E. Schlesinger, The B–Pd (boron–palladium) system, *J. Phase Equilibria* **17** (1996) 340–345.

[3] T. G. Berger, A. Leineweber, E.J. Mittemeijer, C. Sarbu, V. Duppel and P. Fischer, The Pd-rich part of the Pd–B phase diagram, *Zeitschrift fur Metallkunde* **97** (2006) 3–9.

[4] D.D. Dominguez, P.L. Hagans and M.A. Imam, A summary of NRL research on anomalous effects in deuterated palladium electrochemical systems, Naval Research Laboratory Memorandum Report # NRL/MR/6170–96-7803, January 9, 1996.

[5] M.H. Miles, M. Fleischmann and M.A, Imam, Calorimetric analysis of a heavy water electrolysis experiment using a Pd–B cathode, NRL Memorandum Report NRL/MR/6320-01-8526.

[6] Melvin H. Miles and M. Ashraf Imam, US Patent # 6,764,561, June 20, 2004.

[7] Melvin H. Miles and M. Ashraf Imam, US Patent # 7,381,368B2, June 3, 2008.

[8] M.H. Miles and M.A. Imam, Excess power measurements for Pd–B cathodes, *Proc ICCF-21, J. Condensed Matter Nucl Sci.* , to appear in Vol. 29, 2019.

[9] S.B. Katinsky, D.J. Nagel, M.H. Miles and M.A. Imam, LEAP: The LENRIA Experiment and Analysis Program, to be published.

[10] S. Szpak, P.A. Mosier-Boss and J.J. Smith, Reliable procedure for the initiation of the Fleischmann–Pons Effect, *Conf. Proc.*, Vol. 33, *The Science of Cold Fusion,* T. Bressani, E. Del Giudice and G. Preparata (Eds.), Bologna, 1991, pp. 87–91.

J. Condensed Matter Nucl. Sci. 29 (2019) 12–20

Research Article

Excess Power Measurements for Palladium–Boron Cathodes

Melvin H. Miles*

College of Science and Technology, Dixie State University, St. George, UT 84770, USA

M. Ashraf Imam†

School of Engineering and Applied Science, The George Washington University, Washington, DC 20052, USA

Abstract

Palladium–Boron (Pd–B) cathodes prepared at the US Naval Research Laboratory have produced electrochemical excess power effects using D_2O + LiOD electrolytes in nine of ten experiments conducted at three different laboratories and using three different types of calorimeters. The one failure was due to a structural defect in the Pd–B cathodes. An unusual result is the early appearance of the excess power effect for Pd–B cathodes. Three other research groups have also found excess power effects using these Navy Pd–B cathodes. Possible important factors for Pd–B electrodes are the removal of oxygen by the boron during processing, the increased mechanical strength versus pure palladium, less volumetric expansion of these electrodes during loading with deuterium, and the much slower escape of deuterium from this cathode.

Keywords: Boron, China Lake, Cold Fusion, Deuterium, Excess power, NRL, Palladium, US Navy research

1. Introduction

One of the major goals of the US Navy program (Anomalous Effects in Deuterated Materials, 1992–1995) funded by the Office of Naval Research (ONR) was to produce our own palladium cathode materials at the Naval Research Laboratory (NRL). It was obvious at that time (1992) that the major problem in reproducing the Fleischmann–Pons (F–P) excess power effect in the Pd/D system resided within the palladium material. However, none of these Navy palladium metals and alloys were successful in producing significant excess power during the first two years (1992–1993) of this Navy program. This changed in 1994 with the NRL preparation of palladium–boron (Pa–B) alloy cathodes. Seven out of eight experiments using these Pd–B cathodes produced significant excess power in calorimetric experiments at the Navy laboratory at China Lake, California (Naval Air Warfare Center Weapons Division, NAWCWD). A different NRL Pd–B cathode produced excess power in 1997–1998 at the New Hydrogen Energy Laboratory (NHE) in Sapporo, Japan. An experiment in 2017 using this same Pd–B cathode again produced excess power at a Ridgecrest, California

*E-mail: mhmiles1937@gmail.com.
†Corresponding author. E-mail: ashrafimam@email.gwu.edu.

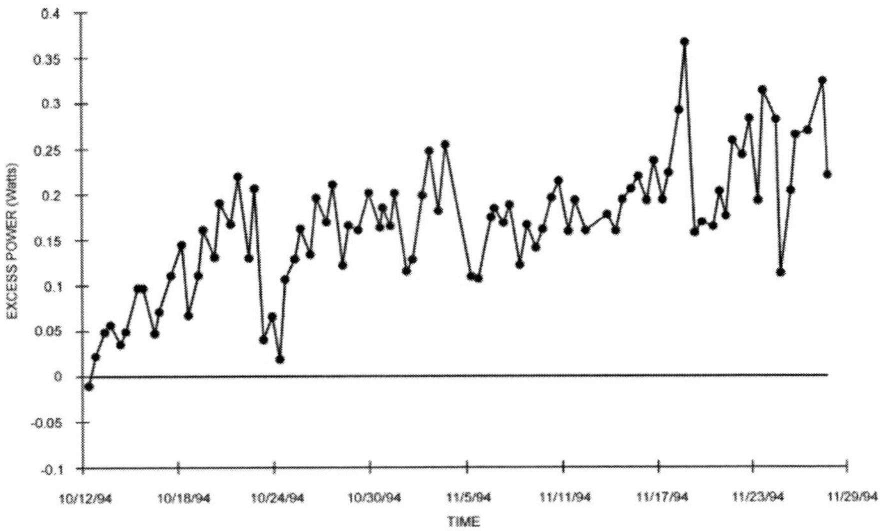

Figure 1. Excess power measurements for an NRL Pd–B rod (0.25 × 2.5 cm, 0.75 wt.% boron).

location. Two US Patents have been granted for this unique NRL Pd–B material [1]. These patents provide details on how these Pd–B materials were prepared.

2. China Lake Results, 1994–1995

These eight experiments at China Lake using the NRL Pd–B cathodes are documented in a Navy Report [2]. Figures for each experiment show the measured excess power for several different times each day. An example for one of the best experiments is shown in Figure 1.

A gradual increase in the excess power was observed over a 45-day time period and reached a maximum of about 350 mW. The total excess energy produced in this experiment was 636 kJ or 5.2 MJ per cubic centimeter of the Pd–B material.

Four isoperibolic calorimeters (A–D) of the same basic design as shown in Fig. 2 were used in these Pd–B experiments. Results for all eight experiments at China Lake are summarized in Table 1.

Table 1. Summary of China Lake Pd–B Experiments (1994-1995).

Start date	Calorimeter	Pd–B dimensions (cm)	Weight % B	Maximum excess power (mW)
5/28/94	B	0.60 × 2.0	0.75	300
10/24/94	B	0.60 × 2.0	0.75	450
10/12/94	C	0.25 × 2.5	0.75	370
10/12/94	D	0.25 × 2.5	0.75	100/0
3/14/95	A	0.40 × 2.0	0.50	140
3/14/95	B	0.40 × 2.0	0.50	240
3/14/95	C	0.40 × 2.0	0.25	220
3/14/95	D	0.40 × 2.0	0.25	90

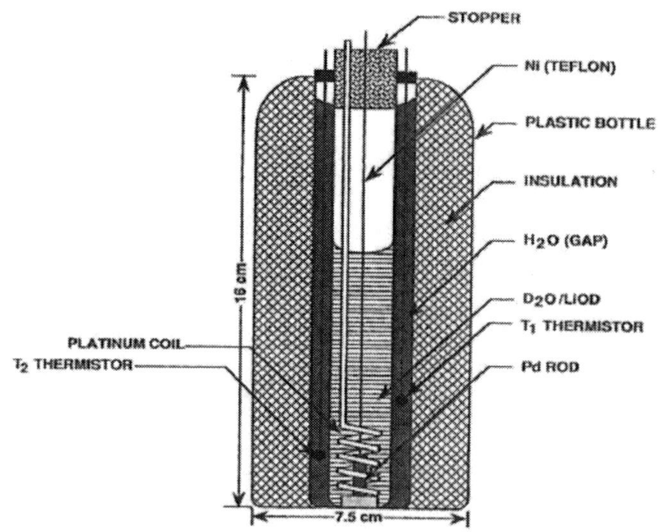

Figure 2. China Lake calorimetric design used for the Pd–B experiments (see [3,4]).

The structural defect for the Cell D (10/12/94) electrode (Table 1) was a folded metal region along almost the full length of the cathode similar to a crack which would be devoid of active electrochemistry. Even this cell showed a maximum of 100 mW in excess power, but most results were nearly evenly distributed between the normal three sigma calorimetric error range of ±30 mW. Cell D (3/14/95) showed a brief 5-day initial period of excess power, but this promising start suddenly ended due to some unknown factor.

The small volume of this Pyrex glass calorimetric cell (18 mL of electrolyte) made it sensitive to temperature changes due to changes in the input power. Thermistors were used to measure the cell and bath temperatures to within 0.01 K [3,4].

3. Japan Pd–B Results, 1997–1998

An appointment at the New Hydrogen Energy Laboratory (NHE) in Sapporo, Japan, for one of us (M.H. Miles) led to another study of a different NRL Pd–B electrode. This material contained 0.50% B by weight (Pd–0.5B) with rod dimensions of 0.47 × 2.01 cm. This new study was conducted in a Fleischmann–Pons Dewar calorimeter labelled ICARUS-1 with inner dimensions of 2.5 × 25.0 cm with the top 8.0 cm silvered. This glass Dewar cell contained 90 mL of 0.1 M LiOD. Calibrated thermistors gave the cell temperatures at two different cell locations to within 0.001 K. Calorimetric measurements were recorded every 300 s throughout this experiment. The analysis of the calorimetric data while in Japan [5] showed the same basic features for excess power versus time as later reported by Fleischmann in his independent and more compete analysis [6]. The peak excess power reached 460 mW (Ref. [6], Fig. A.19). Moreover, during the final cell boiling phase the excess power exceeded 9 W (Ref. [6], Fig. A.22). During the cell boiling, the Pd–0.5 B cathode was obviously the hot spot in the cell based upon the observation of the intense boiling and swirling action centered around this cathode over a three-hour period. This NHE experiment produced a total excess energy of 1.065 MJ over the 67-day period prior to cell boiling or 3.05 MJ per cubic centimeter of the Pd–B

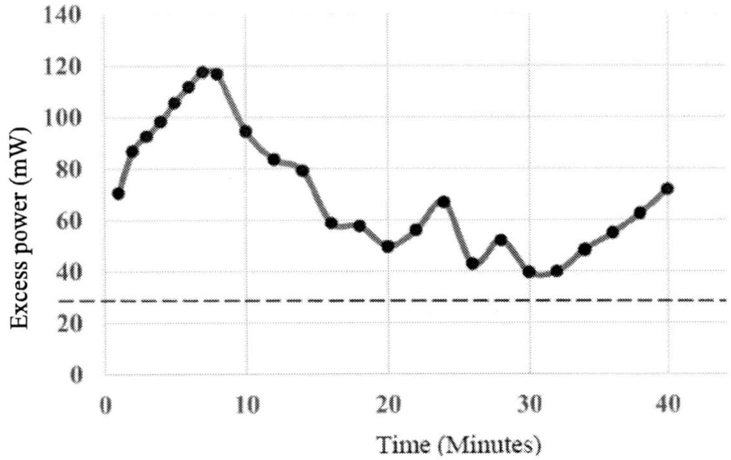

Figure 3. Early excess power for Pd–0.5 B with a peak of 118 mW at 7 min. The dashed line shows the maximum excess power possible due to deuterium loading at 0.150 A.

electrode [6].

An unusual feature of this Pd–0.5B experiment was the very early onset of the excess power effect. This early excess power was even measurable during the initial deuterium loading of the cathode (Ref. [6], Fig. A.21) and was well in excess of the maximum power possible (27 mW) for the exothermic loading of deuterium into palladium (–35.1 kJ/mol D_2) based on the applied current (0.150 A) and formation of $PdD_{0.6}$. This NHE Pd–0.5 B study gave an early mean excess power of 57 mW for the first 55 min and remained above 27 mW for the entire loading period of about four hours [6].

4. Ridgecrest Pd–B Results, 2017

The experiment conducted at Ridgecrest, California used the same Pd–0.5 B cathode as was used in Japan in 1997–1998. However, a different calorimeter was used consisting of two concentric copper tubes separated by insulation material as described elsewhere [7]. The most remarkable feature of this experiment was the very early appearance of excess power as shown in Figure 3. The dashed line in Fig. 3 shows the maximum excess power (27 mW) which could result from the exothermic loading of palladium with deuterium at the experimental cell current of 0.150 A assuming that the total cathodic current is used for the formation of $PdD_{0.6}$. The NHE and Ridgecrest experiments used the same Pd–0.5 B cathode but different calorimeters and gave similar results for the early excess enthalpy measurements. Both experiments gave more than double the excess enthalpy expected for the deuterium loading (–416 J). The measured values were –928 J for the Ridgecrest experiment and –1085 J for the NHE study assuming a loading period of 336 min for each. Any chemical reaction with boron would require about –260 kJ/mol B to explain this extra excess enthalpy. Additional measurements showed a decreasing excess power effect down to 18 mW at 336 min and near zero for the following day (Day 1 in Fig. 4).

The correct explanation for this early excess power using Pd–B cathodes is unknown. Possible explanations include chemical reactions involving boron or perhaps even a deuterium + boron-10 fusion reaction. Boron may somehow be an essential catalyst for the F–P excess heat effect. Normally, long electrolysis times may be required for boron in a Pyrex glass cell to be properly transferred to the palladium for cathodes not containing added boron. For Pd–B

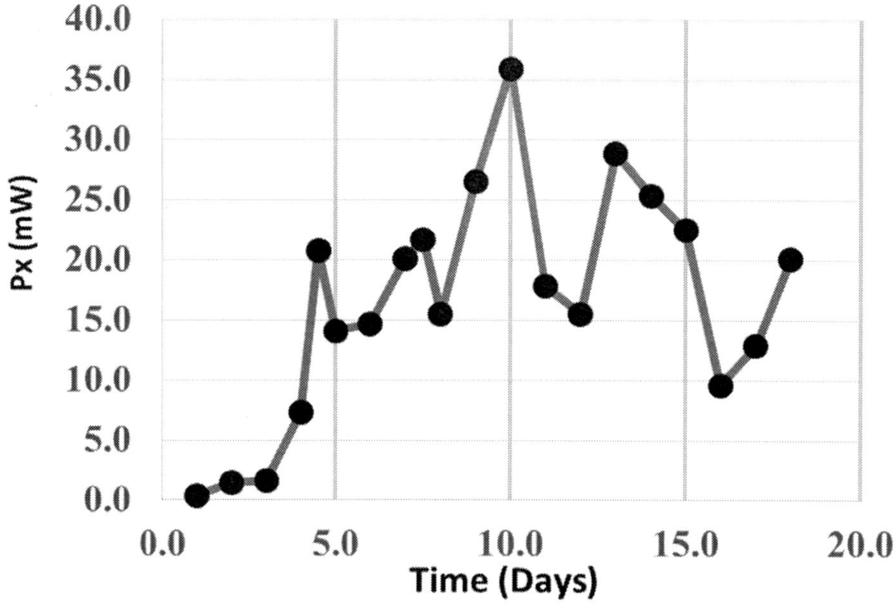

Figure 4. Daily averages for excess power (P_x) in the Pd–0.5 B Ridgecrest experiment of 2017.

electrodes, this catalyst would already be present at the beginning. The most likely explanation of the very early excess power shown in Fig. 3 is presented in the Appendix.

The measured excess power for other days of this Ridgecrest Pd–0.5 B experiment is shown in Fig. 4. The calorimetric results are accurate to within a few milliwatts of excess power ($\pm$ 3 mW), but the effect is much smaller than previously observed in Japan. This could be due to problems with the evolution of gas bubbles. Gas bubbles in this experiment tended to collect at the bottom of the cathode, which was sitting on a Teflon support. The growth and release of these bubbles also caused rather large uncertainties in the cell voltage. These lingering gas bubbles would block the electrochemistry at the bottom of the cathode and act similarly to a crack causing loss of the deuterium loading. This would likely result in smaller excess power effects. Problems with gas bubbles can also vary with the source of the D_2O [2].

5. Related Work

Two Pd–B experiments were also conducted at NRL in January 1995 using a new and accurate Hart Seebeck calorimeter. Both experiments used the NRL Pd–0.75 wt.% B cathodes with dimensions of 0.4 $\times$ 3.5 cm. These experiments were set up by one of us (M.H.M) during a two week visit at NRL. The exothermic deuterium loading for one experiment (Cell D) was measurable at about 7 mW for 20 h and gave –33 $\pm$ 3 kJ/mol D_2 for $PdD_{0.6}$ formation compared to the literature values of –35.1 kJ/mol D_2. However, the other Pd–0.75 B cathode (Cell C) gave evidence of an early excess power effect in addition to the exothermic deuterium loading. The initial excess power continued well beyond the expected loading time frame of about 20 hours for the cell current used. Notebook data recorded by one of us (M.H. Miles) for the first 64 h gave a maximum excess power of 12.8 mW for Cell C at 61.0 h. Even Cell D gave a small maximum excess power of 5.6 mW following the cell current increase to 24 mA/cm^2 (0.060 A). These accurate

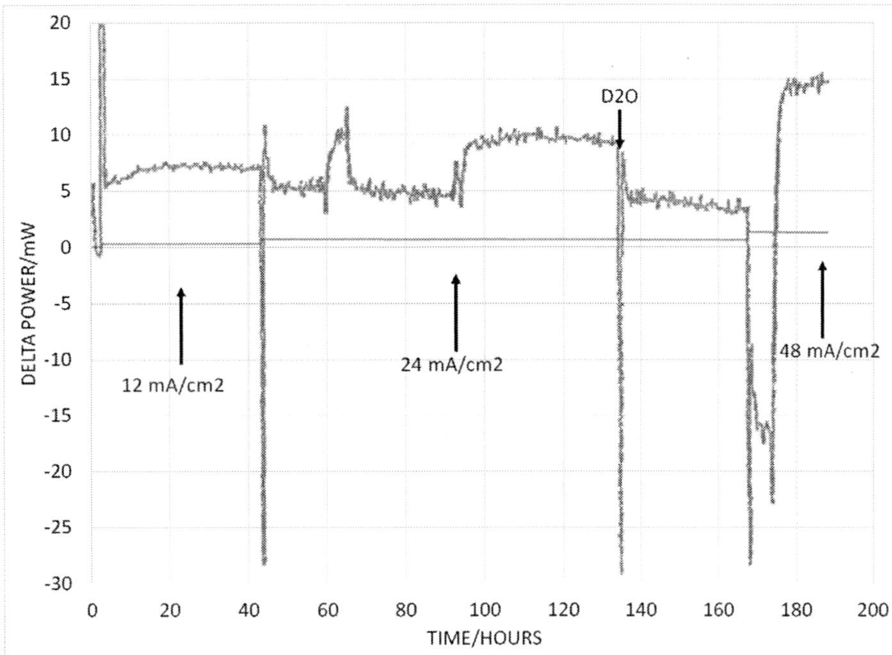

Figure 5. Early excess power for Pd–0.75 B at NRL using a Hart Seebeck calorimeter.

calculations included small additional power terms related to the gas evolution, which were not considered in the NRL calculations [4,11].

This experiment for Cell C is shown in Fig. 5 for the first 200 h [8]. This is additional evidence for early excess power effects in the Pd–B system.

The gas exit lines were too long in this experiment and later became clogged with condensed D_2O vapor. Therefore, these two experiments were terminated, and these results were only reported in an internal NRL-ONR 1995 report [8].

Several other groups have tested this NRL Pd–B material and have found excess power effects, but none of this work has been published. Storms in New Mexico reported excess power for an alloy prepared by one of us (M.A.I) at NRL which contained 0.25 wt.% B with a cylindrical shape and dimensions of 0.47×1.75 cm [9]. This study used 0.008 M LiOD in 99.96% D_2O. The excess power measured by a Seebeck calorimeter reached 80 mW at a current density of 0.5 A/cm^2. Although a paper was prepared by Storms, this Pd–B study was never published. An NRL Pd–B cathode was also tested at SRI in a prismatic Seeback calorimeter by B. Bush and was observed to produce excess power at statistically significant levels [10].

6. Discussion

The major question is why do these NRL Pd–B cathodes produce the F–P excess heat effect while most other palladium materials do not? One possible answer is that the added boron removes oxygen from the palladium by forming B_2O_3 during the melting process. The less dense boron oxide then separates from the molten metal. Other clues for oxygen effects are the successful Johnson–Matthey materials specially produced under a blanket of cracked ammonia

(N_2+H_2). The hydrogen removes oxygen from the metal during the melting process in the form of H_2O vapor. These Johnson–Matthey cathodes also generally produced excess energy in F–P related electrochemical experiments [2,3,11]. A possible third clue is the electrochemical deposition of palladium and deuterium (co-deposition) from $D_2O + PdCl_2$ solutions which provides oxygen-free palladium and reproducible excess power effects (if done correctly) [12].

Another possible important factors for Pd–B cathodes is that the added boron produces a material of much greater mechanical strength than pure palladium [1,6]. There is very little volumetric expansion when Pd–B cathodes are loaded with deuterium. This suggests that Pd–B materials are less likely to crack during the loading process. Another feature is that these Pd–B materials load similarly to palladium cathodes, but the escape of deuterium (de-loading) when the current is removed is at least ten times slower than for pure palladium cathodes based on gravimetric studies. A possible explanation for such large differences in the rate of deuterium loading and de-loading for these Pd–B materials is that Pd–B may load electrochemically across the grains, but when the electrochemical current is removed, most of the deuterium escapes along grain boundaries which may be clogged with the boron atoms. With no applied current, there is no electrochemical potential to drive deuterium into other grains. When the cell current is first turned off for pure palladium cathodes, the escape of deuterium gas is much too rapid to be explained by the simple diffusion of deuterium from palladium grains at the electrode surface. It seems likely that the Pd–B materials are somehow much more restrictive than pure palladium cathodes in allowing deuterium to escape via the grain boundaries.

7. Summary

The NRL Pd–B cathodes have produced excess power in 9 out of 10 experiments conducted at three different laboratories and using three different types of calorimeters. The one exception was due to a structural defect, but even this electrode may have produced small excess power effects. Three other research groups using different calorimeters have also found excess power using NRL Pd–B cathodes in their experiments. There is often an unusual early excess power effect with the use of Pd–B cathodes. These Pd–B results indicate that the prolonged cold fusion controversy is likely related to the dependence of excess power effects on the palladium materials properties.

Acknowledgements

Long term support has been received for one of us (M.H.M.) from an anonymous fund at the Denver Foundation through the Dixie Foundation at Dixie State University. An Adjunct faculty position at the University of LaVerne and a Visiting Professor position at Dixie State University for M.H.M. are also acknowledged. The authors especially thank Dr. David J. Nagel for encouragement and helpful suggestions. The authors also thank Steve Krivit for providing the NRL figure and report [8].

Appendix A. Notes Added in Revision

The immediate excess power for this Pd–B cathode (see Fig. 3) did not seem to be a real effect and some other explanation was needed. It now seems that a likely explanation is that the thermoneutral potential ($E_H = 1.527$ V) generally used in these studies does not apply at the beginning of an experiment when most of the electrogenerated deuterium gas is being loaded into the cathode rather than escaping into the gas phase and leaving the cell. During this period, the main electrochemical reaction would be

$$\mathrm{Pd} + 0.3\,\mathrm{D_2O} \rightarrow \mathrm{PdD_{0.6}} + 0.15\,\mathrm{O_2} \tag{A.1}$$

and not the normal $D_2O \rightarrow D_2 + 0.5\ O_2$ electrolysis reaction where $E_H = 1.527$ V.

The enthalpy change (ΔH°) for Reaction (A.1) is 53.3 kJ/mol, thus the thermoneutral potential is 0.921 V rather than 1.527 V assuming 100% of the deuterium is used for the deuterium loading. This assumption is likely a good approximation for the first 5–10 min of electrolysis but would become less accurate at longer time periods. If 50% of the deuterium was used for loading, then $E_{\rm H}$ would increase to 1.224 V. The use of 1.527 V for $E_{\rm H}$ is valid over most of the experiment except during the time period of deuterium loading.

The change in excess power (P_x) due to the use of $E_{\rm H} = 1.527$ rather than $E_{\rm H} = 0.921$ V at the beginning of an experiment is given by

$$\Delta P_x = (1.527 - 0.921)I = 0.606\,I. \qquad \text{(A.2)}$$

Thus, the initial cell current of $I = 0.150$ A yields $\Delta P_x = 0.0909$ W or about 91 mW. This would account for most of the excess power in Fig. 3 for the first 10 min. If there were no cold fusion effects, then the initial excess power should be zero because the exothermic loading of PdD_x is included in Reaction (A.1). There is still a suggestion of a small excess power effect for Pd–0.5B in Fig. 3 after about 5 min where P_x exceeds 100 mW. The decreasing P_x in Fig. 3 after nine minutes is likely related to the decreasing percent of deuterium used for loading. It is not possible to accurately determine $E_{\rm H}$ during the intermediate loading periods. There were also excess power effects of about 20 mW measured after about 4 h of electrolysis when the loading process became insignificant. Fleischmann also noted a small excess power effect for Pd–0.5 B following the loading process [6]. The very early excess power in Fig. A.21 of [6] was also likely compromised by the use of $E_{\rm H} = 1.527$ V during the initial loading period.

Verification of the effect of Reaction (A.1) was obtained from examination of early data from a previous experiment using a pure palladium cathode. Because of the smaller electrode size, the applied current was only 0.0365 A. Therefore, the expected effect from Eq. (A.2) was calculated as 22 mW. Experimentally, using $E_{\rm H} = 1.527$ gave a similar early excess power effect of 20 mW within the first 10 min of electrolysis. Even the Pd–0.75 B studies at NRL shown in Fig. 5 shows a similar early excess power effect due to the use of $E_{\rm H} = 1.527$ V. This explains the initial short spike of excess power exceeding 20 mW. From the cell current of 54 mA (12 mA/cm^2), the value of ΔP_x from Eq. (A.2) is 33 mW. Regarding Fig. 5, the sharp endothermic behavior when the cell current is increased or when D_2O is added results from the neglect of power that is used to heat the cell contents ($P_{\rm CALOR} = C_{\rm p}M\ \mathrm{d}T/\mathrm{d}t$). The delayed recovery of the excess power at 48 mA/cm^2 is likely related to the gas exit lines becoming clogged with condensed D_2O at the higher cell current [8]. Most researchers use short and vertical glass tubing for the exit of the gases and not long tubing lines that are prone for D_2O condensation.

There is still an early excess power effect for Pd–B cathodes, but it is not the immediate effect inferred by Fig. 3. However, this early excess power does appear near the end of the loading period as shown by Fig. 5 from the NRL Seebeck study as well as in the Japan and Ridgecrest results. Fleischmann commented on this in several places in the NRL Report [6]. An important comment by Fleischmann on p. 25 of [6] states: The major unexpected difference for this Pd–B electrode has been the observation of the development of "positive feedback" at a very early stage of the experiment (Day 3), at a low current density and at a low temperature. It is obviously very important to establish whether this early establishment of 'positive feedback' is a property of Pd–B alloys (the one used in this study) or whether some other factor is involved.

Fleischmann's use of "positive feedback" refers to the increase in excess power during the cell temperature increase with the first application of the in-cell heater. (See Fig. A.15, p. 75 of [6].) For Day 3, this observation of early excess power would not involve any change from the normal $E_{\rm H} = 1.527$ V. Early excess power development late in the loading period or within a few days of electrolysis is an important property of Pd–B alloys.

References

[1] M.H. Miles and M. Ashraf Imam, Palladium–boron alloys and methods for making and using such alloys, US Patent No. 6,764,561B1, July 20, 2004 and 7,381,368B2, June 3, 2008.

[2] M.H. Miles, B.F. Bush and K.B. Johnson, Anomalous effects in deuterated systems, Naval Air Warfare Center Weapons Division Report, NAWCWPNS TP8302, September 1996, 98 pages. See http://lenr-canr.org/acrobat/MilesManomalousea.pdf.

[3] M.H. Miles, K.H. Park and D.E. Stilwell, Electrochemical calorimetric evidence for cold fusion in the palladium–deuterium system, *J. Electroanal. Chem.* **296** (1990) 241–254.

[4] M.H. Miles, B.F. Bush and D.E. Stilwell, Calorimetric principles and problems in measurements of excess power during Pd–D_2O electrolysis, *J. Phys. Chem.* **98 (1994)** 1948–1952.

[5] M.H. Miles, Calorimetric studies of palladium alloy cathodes using Fleischmann-Pons Dewar type cells, in *Proc. 8th Int. Conf. on Cold Fusion*, F. Scaramuzzi (Ed.), Italian Physical Society Conference Proceedings, Vol. 70, 2000.

[6] M.H. Miles, M. Fleischmann and M.A. Imam, Calorimetric analysis of a heavy water electrolysis experiment using a Pd–B alloy cathode, Naval Research Laboratory Report, NRL/MR/6320-01-8526, March 26, 2001, 155 pages. See http://lenr-canr.org/acrobat/MilesMcalorimetrd.pdf.

[7] M.H. Miles and M. Fleischmann, New approaches to isoperibolic calorimetry in condensed matters nuclear science, *ICCF-15*, V. Violante and F. Sarto (Eds.), ENEA, 2009.

[8] D.D. Dominguez and P.L. Hagans, NRL Internal Report, 1995 (unpublished).

[9] E. Storms, Measurements of excess heat produced by a palladium alloy containing 0.25% boron while being electrolyzed in D_2O +LiOD, 2008 (unpublished).

[10] M.C.H. McKubre, Personal Communication, August 2, 2018.

[11] M.H. Miles and M.C.H. McKubre, in *Developments in Electrochemistry: Science Inspired by Martin Fleischmann,* D. Pletcher, Z.-Q. Tian and D.E. Williams (Eds.), Wiley, UK, 2014, pp. 245–260.

[12] S. Szpak, P.A. Mosier-Boss, M.H. Miles and M. Fleischmann, Thermal behavior of polarized Pd/D electrodes prepared by co-deposition, *Thermochimica Acta* **410** (2004) 101–107.

J. Condensed Matter Nucl. Sci. 29 (2019) 21–33

Research Article

Excess Heat from Palladium Deposited on Nickel

Tadahiko Mizuno*

Hydrogen Engineering Application and Development Company, Kita 12, Nishi 4, Kita-ku, Sapporo 001-0012, Japan

Jed Rothwell†

LENR-CANR.org, 1954 Airport Road, Suite 204, Chamblee, GA 30341, USA

Abstract

Two methods of generating excess heat with palladium on nickel are described: an older method, and a newer, faster method. With the older method after sufficient pretreatment, the output heat peaked at 232 W, which was nearly two times input power. However, the pretreatment was complicated and took many weeks or in some cases months before heat appeared. The newer method is to directly apply palladium to nickel. This is simpler and quicker, taking only about a day, but so far it has produced only 10–30 W, and 40 W in one instance.

Keywords: Air flow calorimetry, Deuterium gas, Excess heat, Nickel reactant, Pd coating, Simple method

1. Summary of Old Results

1.1. Old method

A reactor with a cruciform shape was first used in this project (Fig. 1). It weighs 50 kg. Later, 20-kg versions of this reactor were used, as well as cylindrical reactors. All have palladium rods in the center. The rods are 250 mm long, wound with palladium wire (Fig. 2). This is the positive electrode. The negative electrode is a nickel mesh which is fitted against the inside wall of the reactor, and connected to ground (Fig. 3).

The older method is described in detail elsewhere [1]. It includes several rounds of cleaning the electrodes by heating and evacuation to remove impurities, followed by weeks or months of glow discharge to erode the center palladium electrode and sputter palladium onto the nickel mesh.

The highest power observed with the older method was 480 W output with 248 W input, or 232 W excess (Fig. 4). In this example, heating was stopped at 83 ks and the reaction stopped, but when heating is continued, excess heat

*Corresponding author. E-mail: head-mizuno@lake.ocn.ne.jp.

†E-mail: JedRothwell@gmail.com.

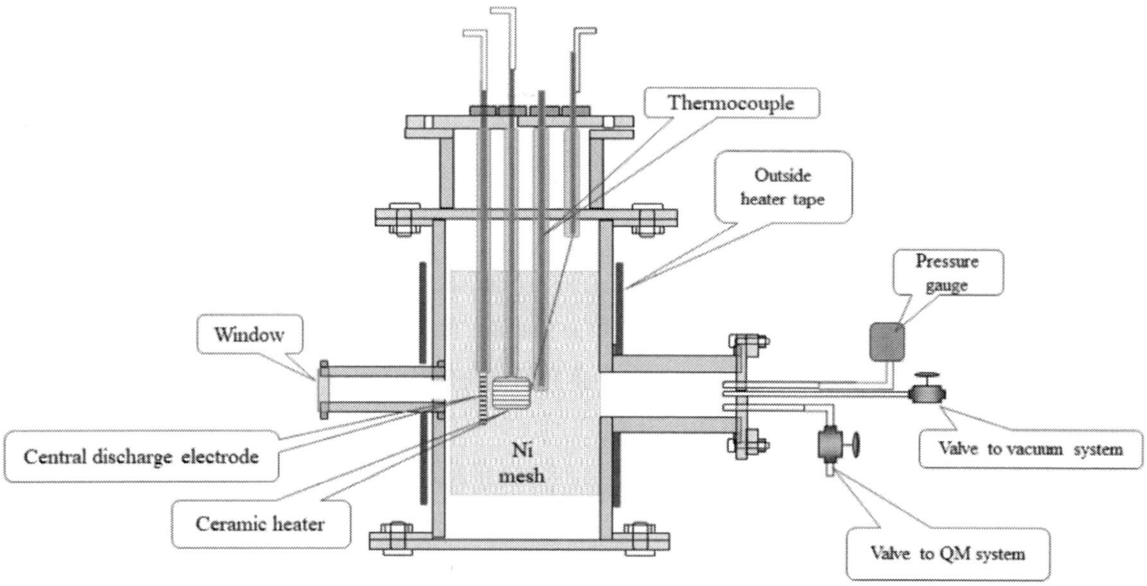

Figure 1. Schematic of the old cruciform reactor (height 500 mm, diameter 213 mm, volume 5.53 L, weight 50.5 kg) A pair of 20 kg versions were used with the new method, one for control, and one for the active cell. The cell contains a central discharge electrode, nickel mesh electrode, and a ceramic heater with palladium wire wound around it.

generation also continues. Unfortunately, this method sometimes took months to turn on. This long delay tied up equipment and hampered the research.

1.2. Old electrode preparation method

Here are the electrode activation procedures for a mesh of 50–100 g as follows.

(1) Place the electrodes in the reactor. The reactant metal is degassed initially at room temperature. This is to avoid the formation of an oxide film or nitride film that would be strengthened by the in-system gas when it is processed at high temperature, so that the subsequent activation treatment becomes difficult.

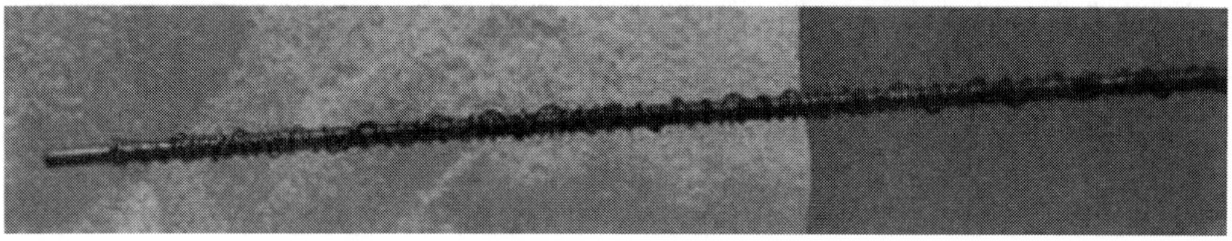

Figure 2. Central discharge electrode. A 250 mm palladium rod, wound with palladium wire.

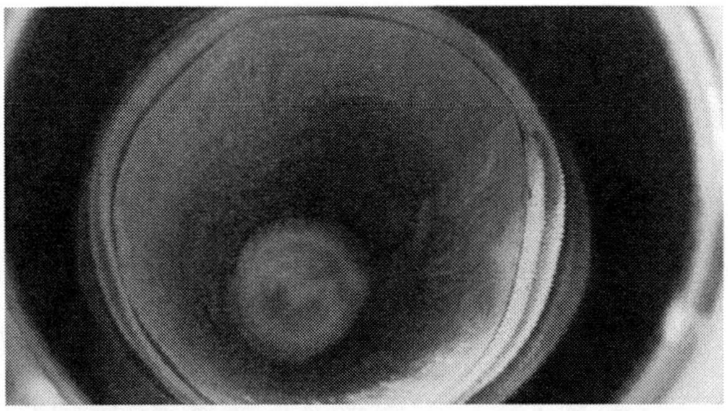

Figure 3. Mesh electrode mounted inside the reactor placed against the reactor wall, connected to ground.

(2) Heat in vacuum to remove the impurity gases generated in the metal film formation process. It is important to begin heating after sufficient removal of impurities at low temperatures (step 1). The reactor gas should be checked with a mass spectrometer each time heating in vacuum is repeated, to confirm that contamination is decreasing. When the reactor is not sufficiently exhausted, H_2O, nitrogen and oxygen gases will remain in it. In a Q-Mass observation, component peaks of H_2O (16,17, and 18) should be lower than the ionic current value of 10^{-9} A.
(3) After the temperature measured on the reactor wall at mid-height reaches 200°C, and the reactor is at uniform temperature, D_2 gas is introduced at 50 Pa. If the mesh was previously used in experiments, and it was sputtered with palladium from glow discharge, it should be successfully re-activated, and excess heat will be generated immediately. If not, steps 2 and 3 may need to be repeated several times.
(4) An unused mesh has to be exposed to glow discharge over many days or weeks, and sputtered with palladium.

After these steps, gas pressure in the reactor is set at 200–1000 Pa. Deuterium absorption/desorption treatment on

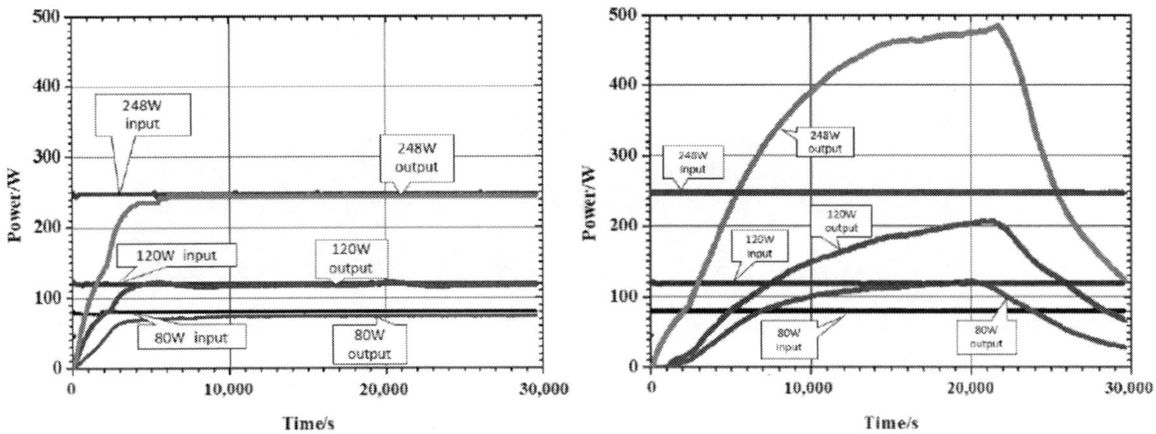

Figure 4. Old results – (*Left*) Calibrations at various power levels. (*Right*) Excess heat at the same power levels.

Figure 5. The calorimeter acrylic box, with the insulation removed to show the equipment placed inside. The active and control reactors are side-by-side in box. These are the 20-kg cruciform reactors.

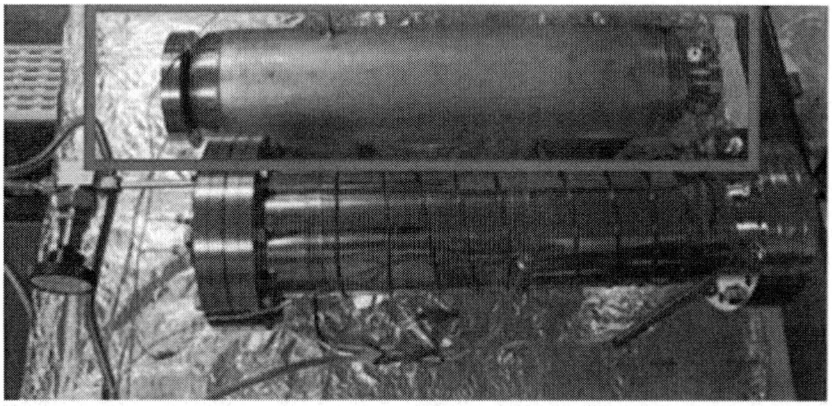

Figure 6. Cylindrical reactors. Length 500 mm, diameter 120 mm, weight 20 kg. A cylindrical reactor and a control reactor (*in red box*) are placed side-by-side in the calorimeter.

the reactant metal surface causes hydrogen embrittlement to atomize the surface metal. This method is to raise the temperature of the reactor wall from room temperature to 300°C, and repeatedly cool it.

The heating time is 10–15 h and the cooling time is 4–6 h, that is, about one cycle per day. This process should be repeated as many times as possible, but as a practical matter it is usually repeated about 5–6 times. This process is essential for large excess heat generation.

Excess heat is governed by gas pressure, reactant weight, and input power. When excess heat is generated, the temperature of the reactant increases by several tens of degrees. The reactant is the nickel mesh that is thinly covered with Pd. The temperature of the nickel mesh cannot be measured directly.

As the reactant temperature rises, the amount of deuterium in the reactant decreases. This causes the exothermic reaction to decrease and the reactant temperature to fall. When the temperature decreases, more deuterium enters the reactant again, and excess heat increases.

1.3. Heat apparently produced by mesh

The nickel mesh apparently produces excess heat, presumably because it was gradually sputtered with palladium nanoparticles during the preparation and subsequent glow discharge. The mesh appears to be where the excess heat reaction occurs, rather than the central palladium rod electrode. Evidence for this includes:

(1) The reaction increases when the reactor vessel itself is heated from the outside. This affects the nickel mesh more than other components.
(2) The reaction can be reliably triggered with palladium deposited on the mesh, with the newer method.

At a power density of 10 W/g, in principle the maximum temperature might reach ~1000°C, depending on the heat resistance and heat losses of the reactor. A temperature apparently as high as 850°C has already been reached.

The temperature is determined by the amount of reactant and the removal rate of heat. In principle, a high temperature can be obtained at any rate, by increasing the insulation of the calorimeter.

1.4. Old method of calorimetry

In the beginning of this project, isoperibolic calorimetry was used. In this method, water was circulated through tubes wrapped around the reactor, and the entire reactor was set to a uniform temperature. Thermal calculation accurately

Figure 7. A layer of this reflective padded aluminum insulation is placed on the inside of the box reducing heat leaks to a minimum. At 100–200 W, the inside of the box is only moderately hotter than the surroundings at ~36°C, so only ~4% of the heat is lost through the walls and insulation. The rest is recovered in the flowing air.

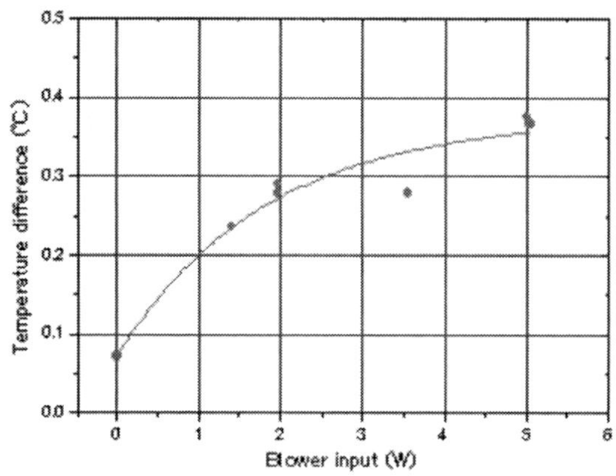

Figure 8. Calibration with blower power only, and no power to the reactors. The temperature rise of ~0.35°C is probably caused by waste heat from the blower motor reaching the outlet RTDs.

measured the heat transfer between three systems: the surrounding environment, the reactor body, and the circulation pump reservoir. This method was repeatedly tested with calibrations and exothermic tests to statistically verify the difference between both tests. Significant excess heat was confirmed by Welch's test [2]. However, although this method of calculating excess heat generation is in principle correct, it was complicated, so it was replaced with airflow calorimetry. Airflow calorimetry is more direct, easier to understand, and it allows the reactor body to reach a higher temperature. It is described in Section 2.

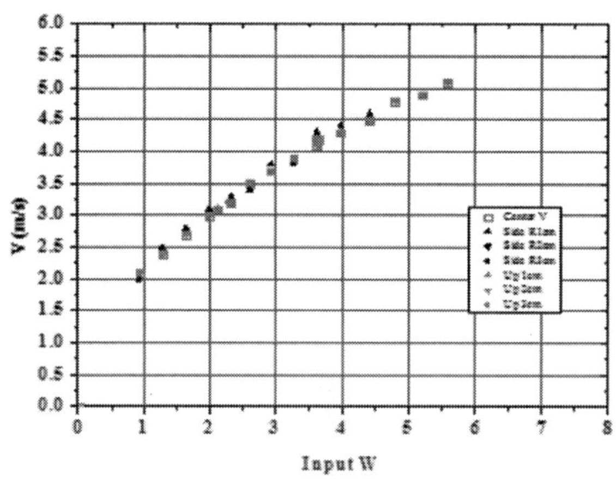

Figure 9. Relationship between blower power and air velocity at the outlet for different locations of the anemometer across the tube section. The velocity profile is almost uniform.

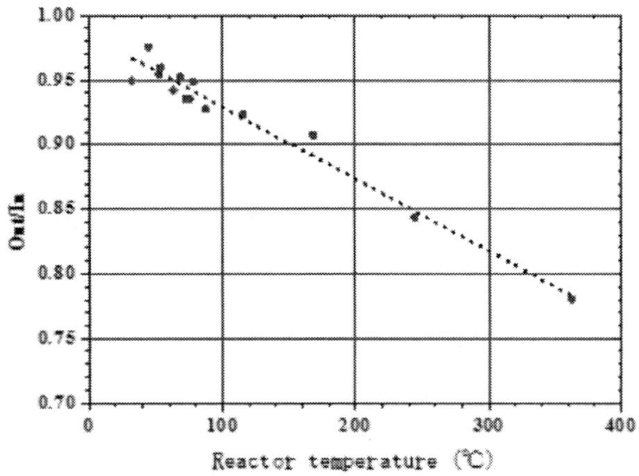

Figure 10. Calibration reactor temperature and heat recovery rate.

2. New Method

2.1. Newer, simplified method of preparing mesh

The newer method is to directly apply palladium to the nickel mesh, either by rubbing a palladium rod onto the mesh, or by electroless plating , rather than waiting for glow discharge to gradually sputter the palladium onto the mesh. This is simpler and quicker, taking only about a day, but so far it has only produced 10–30 W, or 5–15% of input power, and up to 40 W in one instance.

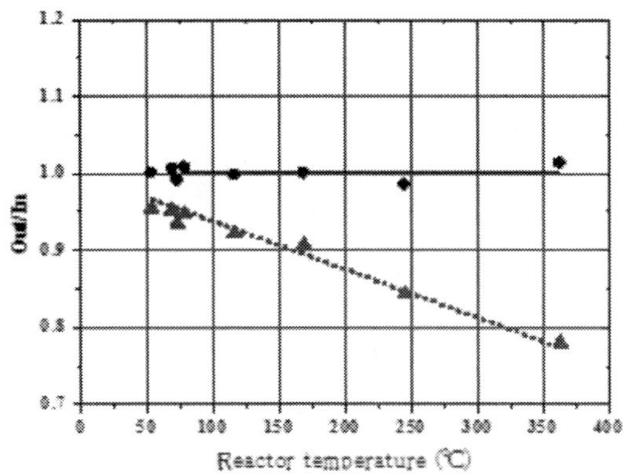

Figure 11. Thermal calibration with and without including heat recovery.

Table 1. Ni mesh surface rubbed with Pd rod

Reactor type	Y/M/D	Gas	Pressure (Pa)	Input (W)	Input (MJ)	Output (MJ)	Out/In ratio
Horizontal reactor	2017-07-31	H_2	1000	100	1.68	1.85	1.10
Cruciform reactor	2017-08-11	H_2	1000	100	1.99	1.99	1.00
Control reactor	**2017-08-11**			**100**	**0.95**	**0.95**	**1.00**
Small reactor	2017-08-16	D_2	4300	100	1.13	1.31	1.16
Control reactor	**2017-08-17**			**100**	**1.10**	**1.05**	**0.96**
Horizontal reactor	2017-08-22	H_2	4400	100	2.48	2.65	1.07
Control reactor	**2017-08-23**			**100**	**1.07**	**1.05**	**0.98**
Control reactor	**2017-08-24**			**100**	**0.72**	**0.70**	**0.96**
Cruciform reactor	2017-09-03	D_2	5	100	0.87	1.08	1.24
Cruciform reactor	2017-09-04	D_2	440	100	2.02	2.18	1.08
Cruciform reactor	2017-09-05	D_2	440	100	4.68	4.73	1.01
Cruciform reactor	2017-09-06	D_2	450	100	2.22	2.49	1.12
Control reactor	**2017-09-22**			**100**	**7.13**	**7.13**	**1.00**
Control reactor	**2017-10-24**			**100**	**5.95**	**5.95**	**1.00**
Cruciform reactor	2017-10-25	D_2	700	100	6.95	6.97	1.00
Cruciform reactor	2017-10-30	D_2	700	100	8.49	8.57	1.01
Cruciform reactor	2017-11-04	D_2	450	200	35.01	34.39	0.98
Cruciform reactor	2017-11-06	D_2	450	200	34.87	34.42	0.99
Cruciform reactor	2017-11-11	D_2	450	100	9.27	9.35	1.01
Cruciform reactor	2017-11-14	D_2	450	400	5.20	5.54	1.07
Control reactor	**2017-11-15**			**400**	**5.19**	**5.12**	**0.99**
Cruciform reactor	2017-11-16	D_2	450	400	8.70	9.16	1.05
Cruciform reactor	2016-11-17	D_2	450	200	4.79	5.21	1.09
Cruciform reactor	2017-11-20	D_2	1000	200	17.18	18.40	1.07
Cruciform reactor	2017-11-21	D_2	750	200	12.28	13.30	1.08
Control reactor	**2017-11-22**			**200**	**31.02**	**31.03**	**1.00**
Control reactor	**2017-11-24**			**200**	**7.86**	**7.84**	**1.00**
Cruciform reactor	2017-11-25	D_2	750	100	15.80	16.98	1.08

Note: Rows in bold face are calibration tests.

All of the tests with the new method used airflow calorimetry. Two different pairs of reactors were used. The first pair were cruciform reactors similar to the one used with the older method, but smaller and lighter, weighing 20 kg instead of 50 kg. They are shown in Fig. 5 placed side-by-side in the calorimeter. The second pair were horizontal pipes (Fig. 6).

The electrode configuration was the same as with the old method, with a central rod electrode, and a mesh electrode placed against the reactor wall and connected to ground. The reaction gas was usually deuterium.

2.2. Airflow calorimetry

An insulated acrylic box is used for airflow calorimetry. It is 400 mm × 750 mm, height 700 mm. During a test, the inside of the plastic box is covered with 1.91 m^2 of reflective padded aluminum insulation (shanetsu.com, Fig. 7). This minimizes losses to radiation. These losses are low in any case, because the cooling air keeps the inside of the box at ~36°C (16°C above ambient). Similar insulation from a US vendor (US Energy Products) has an R-value of 11, which is 1.9 in SI units. The air inlet and outlets are circular, 50 mm in diameter. The inlet is located near the bottom of one side, and the outlet is on the top surface. The outlet is connected to a pipe with an outlet of 66 mm, which makes the airflow more uniform across all parts of the cross section of the outlet, to increase the accuracy of the airflow measurement. The power to the blower is continuously monitored.

The blower is operated at 6.5 W. The outlet air temperature is measured with two RTDs. They are installed in the

Table 2. Nickel mesh with Pd deposited by electroless plating

Reactor type	Y/M/D	Gas	Pressure/Pa	Input/W	Input/MJ	Output/MJ	Out/In ratio
Horizontal reactor	2018-02-07	D_2	730	100	8.36	9.80	1.17
Horizontal reactor	2018-02-09	D_2	730	200	4.83	5.52	1.14
Horizontal reactor	2018-02-11	D_2	670	200	34.74	40.14	1.16
Horizontal reactor	2018-02-19	D_2	670	200	16.38	17.94	1.10
Horizontal reactor	2018-02-21	D_2	670	200	17.66	19.43	1.10
Horizontal reactor	2018-02-23	D_2	670	200	19.99	20.30	1.02
Horizontal reactor	2018-02-26	D_2	670	200	24.71	27.65	1.12
Horizontal reactor	2018-02-27	D_2	670	200	34.05	38.51	1.13
Control reactor	**2018-03-02**			**200**	**12.02**	**12.01**	**1.00**
Control reactor	**2018-03-03**			**200**	**40.06**	**39.53**	**0.99**
Horizontal reactor	2018-03-06			500	9.49	9.23	0.97
Control reactor	**2018-03-07**			**500**	**11.77**	**10.53**	**0.89**
Horizontal reactor	2018-03-16	D_2	670	200	12.65	14.49	1.15
Horizontal reactor	2018-03-29	D_2	670	200	6.94	7.57	1.09
Horizontal reactor	2018-04-04	D_2	590	200	14.99	17.09	1.14
Control reactor	**2018-04-05**			**200**	**6.44**	**6.44**	**1.00**
Horizontal reactor	2018-04-11	D_2	310	200	3.62	3.89	1.08
Horizontal reactor	2018-04-13	D_2	590	200	10.04	10.67	1.06
Control reactor	**2018-04-16**			**200**	**12.10**	**12.13**	**1.00**
Horizontal reactor	2018-04-20	D_2	490	200	11.93	12.57	1.05
Control reactor	**2018-05-16**			**100**	**4.06**	**4.06**	**1.00**
Horizontal reactor	2018-05-16	D_2	500	100	5.06	5.37	1.06
Horizontal reactor	2018-05-18	D_2:500+H_2:6000	6500	100	7.05	7.53	1.07
Control reactor	**2018-05-21**			**100**	**5.45**	**5.45**	**1.00**
Control reactor	**2018-05-23**			**0**	**4**	**–1,547**	
Horizontal reactor	2018-05-23	D_2:500+H_2:6000	6500	200	8.04	8.42	1.05
Horizontal reactor	2018-05-25	D_2:500+H_2:6000	6500	200	14.03	14.18	1.01

Note: Rows in bold face are calibration tests.

center of the pipe, one in the stream of air before it reaches the blower, and one after it, to measure any heat added to the stream of air by the blower motor. The difference between the two is less than 0.1°C. However, there are indications that heat from the blower motor is affecting both of them. A calibration with no input power to the reactors shows that when the blower power is stepped from 1.5 to 5 W, the outlet RTDs are ~0.35°C warmer than inlet (Fig. 8). This is a much larger temperature difference than the moving air in the box alone could produce.

Blowers are inefficient, so most of the input power to the blower converts to waste heat in the motor. It seems likely some of this heat is conducted by the pipe to raise the temperature of the two outlet RTDs.

The airflow rate is measured with a hot wire digital anemometer (CW-60, Custom Co. Ltd.). The flow of air through the box is confirmed by letting smoke from incense into the box, and then timing how long it takes for all of the smoke to clear out.

The air velocity profile across the tube section was checked by placing the hot wire at different locations along two perpendicular diameters. The results are displayed in Fig. 9. The distribution is fairly uniform so that the average air velocity across the tube is well characterized by the velocity measured on the axis.

With the new method, a pair of reactors are installed in the air flow box: an active cell and a control cell. They are placed on insulating bricks to reduce heat losses through the table. They are the same size and design. One is active and the other is the calibration control. The control reactor is kept at the same pressure and input power as the active reactor. The two are periodically swapped. The control becomes the active reactor. The calorimeter is not opened while a series of control and active tests are performed, so conditions in it remain closely similar for both tests.

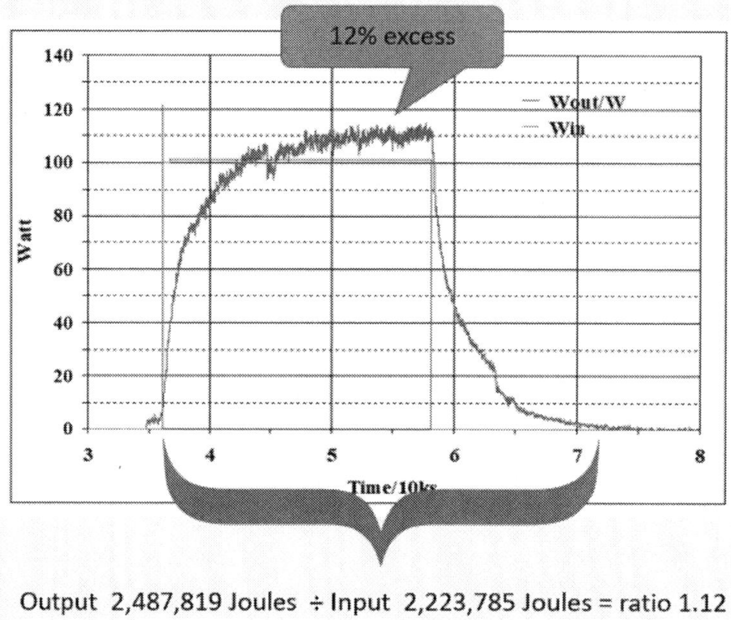

Figure 12. An example of 12% excess heat. Both input and output energy are measured from turn on to cool down.

Two kinds of control tests are done. In the first, the control reactor is heated at the outer wall heater. In the second, glow discharge using ordinary electrodes and hydrogen was performed. Both methods show no excess heat.

In the first approximation, the airflow calorimetry is simple. The number of liters of air that pass through the box per second is computed, based on the size of the duct and the speed of the air leaving the duct. The temperature difference between the inlet and the outlet is measured. Energy is computed as follows:

$$E_{\text{out}} = V\rho C_{\text{p}}\Delta T, \tag{1}$$

where E_{out} is the output energy captured by the airflow (Js^{-1}), V the volumetric air flow rate (m^3s^{-1}), ρ the density of air at the outlet temperature (kgm^{-3}), C_{p} the heat capacity of air at constant pressure ($\text{Jkg}^{-1}\text{K}^{-1}$) and ΔT is the temperature difference between the air box inlet and outlet (K).

The air density is calculated for the temperature at the outlet where the velocity is measured. The variation of the atmospheric pressure is not taken into account because its influence is small. The heat losses of the air box were evaluated. The heat losses increase with the energy input that influences the temperature difference between the air inlet and outlet as well as the reactor wall temperature. The ratio of the heat captured by the airflow to the energy input is called the heat recovery rate.

Figure 10 shows the reactor temperature vs. the heat recovery rate. When there is no input power, and the reactor body temperature is 25°C, the recovery rate should be close to 1. When the reactor body temperature is 100°C, the recovery rate is 0.93; it is 0.82 at 300°C, and 0.78 at 360°C. Naturally this is the same in the test reactor as the calibration reactor; the recovery rate decreases as the temperature of the test reactor rises. The approximate expression showing this ratio is the linear approximation:

$$\text{O/I} = 0.98 - 5.0811 \times 10^{-4} \times t, \tag{2}$$

where t is the reactor temperature (°C).

When the calibration data is recalculated taking into account the heat recovery rates in Fig. 10, the results shown in Fig. 11 are obtained. In this graph, the temperature of the calibration reactor is plotted on the horizontal axis and the output/input ratio is plotted on the vertical axis. In Fig. 11, the marks do not include the heat recovery rate calibration, and the · marks show the data after taking into account the heat recovery rate. If the temperature rise during the calibration of the reactor is computed without taking into account heat recovery the O/I ratio = 0.78 at 363°C for example. These values are extremely close to O/I = 1 after the heat recovery correction according to Eq. (2).

Six parameters (reactor temperatures, air inlet and outlet temperatures, input current and voltage, and blower power) were recorded every 5 s in spreadsheets. The power for each 5-second interval is computed. The record begins when power is turned on. After the power is turned off, the data continues to be recorded until the temperature falls back to room temperature. The total input and output energy for this entire period is added up, and the ratio is computed. Figure 12 shows an example with 12% excess heat.

2.3. Activation process with new method

Palladium is first applied to the nickel screen. This has been done with two methods. The first is to rub the nickel screen with a palladium rod. The second is to deposit a palladium film on the nickel mesh surface by plating it with an electroless plating solution of palladium of Pd-10 (High Purity Chemical Co. Ltd.), Pd concentration 10 g/L. The plating conditions are: 40–60°C, pH 1.5.

The preparation steps are as follows.

(1) Wash with a mild detergent in ordinary tap water, and scrub with a plastic dish washing pad.
(2) Sand with water resistant sandpaper, starting with 250 grit, then 400, then 1000. Wash with mild detergent and rinse with tap water.

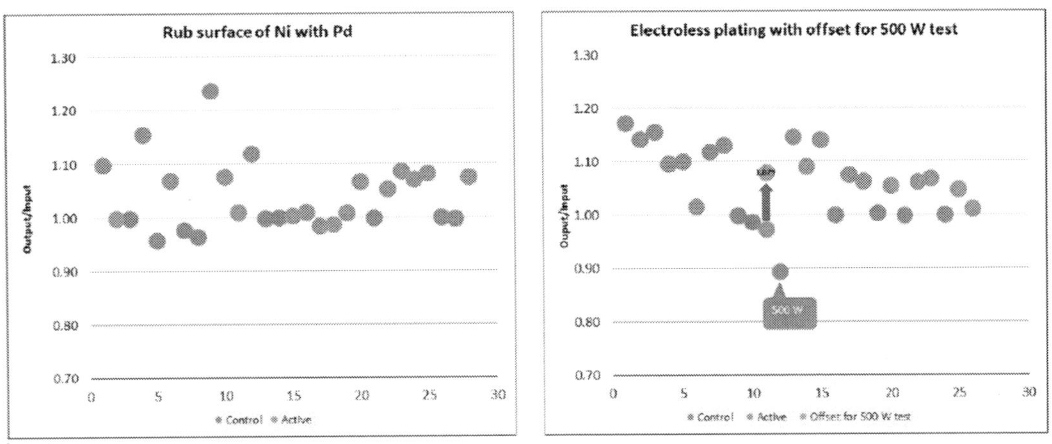

Figure 13. (*Left*) Nickel rubbed with palladium. (*Right*) Electroless plating. The 500 W active test is adjusted to the recovery rate shown by the 500 W calibration (*gray* dot).

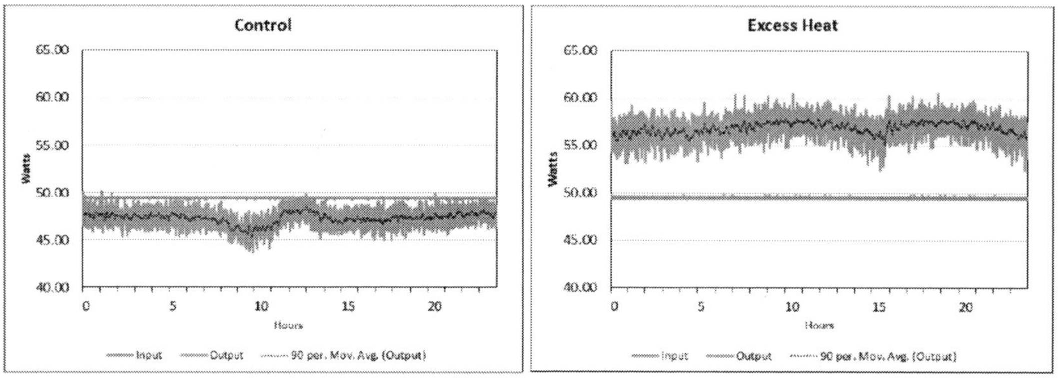

Figure 14. Recent results. 24-hour samples of: (*left*) Control test (2018–08–08) and (*right*) Excess heat test (2018–08–20).

(3) Soak in tap water at about 90°C for a while.
(4) Wash with alcohol.

This process removes oil and roughens the metal surface. After cleaning:

(1) The mesh is placed in reactor. Gas discharge, vacuum evacuation to 10^{-2} Pa at room temperature for about 2 h.
(2) Heat treatment, in which pretreatment (removal of impurities and surface refinement) of the reactant metal surface is carried out. The temperature is 100–120°C duration 5–20 h.
(3) Mesh is removed from reactor. At this stage the surface should be free of contaminants. Coating of palladium on nickel, with rubbing or electroless deposition plating.
(4) Mesh is placed in reactor again. Heated for 1–2 h.
(5) Cool down in the reactor, 1–2 h.

Steps 4 and 5 may need to be repeated a few times.

Although this method is simple and fast compared to the old method, excess heat is generally 10 W, up to 40 W in one instance, and the output/input ratio is small. Tables 1 and 2 and the graphs in Fig. 13 shows the test results of palladium adhered to the nickel mesh by rubbing and electroless plating. Figure 14 shows the results of a particular test as an example.

3. Summary

The previous method was effective for generating excess heat. In the best case, this method generated several hundred watts of excess heat. The reaction sometimes continued for several months. However, the process of fabricating and then activating the nano-particles was complicated and cumbersome, and it took a long time. With the new method of applying palladium to the nickel surface, excess heat is more likely to occur. This method is simpler and much faster. However, the excess is usually only 5–6% of the input power, and the best results so far were $\sim$20% (10–40 W). The future goal is to investigate higher temperature excess heat and reactants.

References

[1] T. Mizuno, Observation of excess heat by activated metal and deuterium gas, *J. Condensed Matter Nucl. Sci.* **25** (2017) 1–25.

[2] T. Mizuno and H Yoshino, Confirmation of excess heat generation during metal–hydrogen reaction, *Proc. JCF16*, Kyoto, Japan, Dec. 11–15, 2015, pp. 16–27.

J. Condensed Matter Nucl. Sci. 29 (2019) 34–40

Research Article

Overview of Pd/D Co-deposition

Pamela A. Mosier-Boss* and Lawrence P. Forsley
Global Energy Corporation, San Diego, CA 92123, USA

Frank E. Gordon†
SPAWAR Systems Center Pacific, San Diego, CA 92122, USA

Abstract

Pd/D co-deposition has been used by a number of researchers to explore the condensed matter nuclear reactions occurring within the Pd lattice. Reported reaction products include heat, transmutation, tritium, energetic charged particles, neutrons, and gamma/X-ray emissions. An overview of these results are discussed in this communication.

Keywords: Co-deposition, Energetic particles, Gamma/X-ray emissions, Heat, Transmutation

1. Introduction

Knowing that long incubation times were needed to achieve high D/Pd loadings in bulk Pd cathodes in order to induce heat production, Stanislaw Szpak, an electrochemist at a Navy laboratory in San Diego, began to explore using Pd/D co-deposition to initiate this effect. In this process, working and counter electrodes are immersed in a solution of palladium chloride and lithium chloride in deuterated water (Fig. 1a). Palladium is then electrochemically reduced onto the surface of the working electrode in the presence of evolving deuterium gas. SEM analysis of electrodes prepared by Pd/D co-deposition exhibit highly expanded surfaces consisting of small spherical nodules (Fig. 1b).

Cyclic voltammetry and galvanostatic pulsing experiments indicate that, by using the co-deposition technique, a high degree of deuterium loading (with an atomic ratio D/Pd>1) is obtained within seconds [1–3]. Because an ever expanding electrode surface is created, non-steady state conditions are assured, the cell geometry is simplified because there is no longer a need for a uniform current distribution on the cathode, and long charging times are eliminated. Using variations of Pd/D co-deposition, researchers have reported on observing excess heat, gamma/X-ray emissions, transmutation, as well as the production of tritium and energetic particles.

*Corresponding author. E-mail: pboss@san.rr.com.
†Retired.

Figure 1. (a) Photograph of the cell showing the cathode, anode, and plating solution. The original plating solution was $PdCl_2$, LiCl, in D_2O, (b) SEM of an Au/Pd cathode showing the cauliflower structure of the deposit.

2. Results and Discussion

2.1. Heat

Szpak et al. used thermocouples in the cathode and solution [4] and infrared imaging [5] to show that the heat source was the cathode and not Joule heating. Infrared imaging of the cathode showed that heat generation was not continuous but occurred in discrete spots on the Pd deposit. Crater formation and evidence of Pd melting under water have also been observed [5]. Depending upon their geometry and diameter, the melting temperature of Pd nanoparticles occurs between 227–1527°C [6]. Nagle [7] estimated that the energy to create a crater was 1.8×10^{-4} J, which agrees well with the value calculated by Kim [8] (6.5×10^{-4} J), who assumed that craters were formed by vaporization. Since the energy of 1 μJ is equivalent to $6.2 \times 10^{5}10$ MeV energy releases, crater formation requires the nearly simultaneous occurrence of many LENR events. This is also inconsistent with the primary hot fusion reactions: D(d,p)T, D(d,n)^{3}He and subsequent D(t,n)α and D(^{3}He,p)α reactions where 2/3 of the 24 MeV released is in fast neutrons. CR-39 neutron rates, as well as results from other LENR researchers, find a dearth of neutrons.

Researchers who have seen evidence of heat production, in Pd/D co-deposition experiments, using calorimetry include Miles [9], Cravens and Letts [10], Letts and Hagelstein [11], DeChairo et al. [12], Swartz and Verner [13], and Dash [14]. Despite the fact that different calorimeters were used as well as different plating solutions, excess heat was measured by these researchers. Miles [9] used the ammonia complex of Pd and an open, isoperibolic calorimeter. He showed that the rates of excess enthalpy generation using electrodes prepared by the Pd/D co-deposition technique were comparable to that obtained when Pd bulk electrodes were used. Miles showed that shuttle reactions were not responsible for the excess heat [15]. Positive feedback [9,10] and heat-after-death effects [4] were also observed with the Pd/D co-deposited electrodes. Letts and Hagelstein [11] used the chloride complex of Pd and a closed, Seebeck calorimeter. They showed that when Pd/D co-deposition was done using low current densities, which improved adherence, no excess heat was produced. However, co-deposition using high current densities did produce excess heat. Like Letts and Hagelstein, DeChairo et al. [12] also used the chloride complex of Pd. They used an open, Seebeck calorimeter. Like Letts and Hagelstein [11], they observed no heat production when Pd/D co-deposition was done using a low current and heat production using high current densities. Swartz and Verner [13] obtained 100,000 J of excess heat doing Pd/D co-deposition in low electrical conductivity D_2O using a spiral wound palladium cathode.

This configuration resulted in asymmetric electrolysis on one side of the cathode which caused deuteron flux inside the metal lattice. They did multi-ring thermal spectroscopy with thermal controls and an open system. Dash and Ambadkar [14] used a closed, Seebeck envelope calorimeter. In their experiment, the Pt was the cathode and Pd was the anode. Pd dissolved from the Pd anode and plated out, in the presence of evolving D_2 gas, on the Pt cathode. The observed excess thermal output was 0.93 ± 0.1 W.

2.2. Tritium

Tritium content during Pd/D co-deposition was measured by researchers using the liquid scintillation technique. Bockris et al. [16] measured tritium in gas and liquid phases during electrolysis. They observed bursts of tritium production when low tritiated D_2O was used and a loss of tritium when highly tritiated D_2O was used. Loss of tritium suggests there is a reaction that consumes thermal tritium. These results have been verified by other researchers. Szpak et al. [17] did a similar experiment and also saw bursts of tritium production when low tritiated D_2O was used. Using highly tritiated D_2O, Miles [18] observed a decrease in tritium in the electrolyte upon termination of Pd/D co-deposition experiments. Lee et al. [19] used a closed system to measure changes in the tritium content in Pd/D co-deposition cells. They observed an increase in the tritium content when low tritiated D_2O was used. However, a decrease in tritium was observed when highly tritiated D_2O was used.

2.3. Gamma/X-ray emissions

Using a Si(Li) detector with a Be window, the cathodically polarized Pd/D cathode was observed to emit X-rays with a broad energy distribution and with the occasional emergence of recognizable peaks attributed to the Pd K_α and Pt L lines [20]. The emission of X-rays was sporadic and of limited duration. Fogging of photographic film was observed after exposure to Pd deposited on a Ag disk cathode upon completion of a co-deposition experiment [21]. A thin Mylar sheet separated the cathode from the film. The circular shape of the cathode could be seen on the film. Also the fogging was inhomogeneous indicating that some sites were more active than others.

Miles et al. [22] used a Geiger-Müller (GM) detector with a thin window to monitor Pd/D co-deposition cells. The co-deposition experiments showed high radiation counts within a few hours of beginning the electrolysis. In contrast, solid Pd rods always required one to two weeks before anomalous radiation was detected.

2.4. Transmutation

As discussed *vide supra*, Dash and Ambadkar's Pd/D co-deposition experiment produced an excess power output of 0.93 ± 0.1 W [14]. Upon termination of the experiment, the Pt/Pd electrode was subjected to SEM/EDS analysis. EDS analysis of several spots showed three peaks at 2.84, 2.99, and 3.18 keV that were attributed to the Pd $L\alpha_1$, $L\beta_1$, and $L\beta_2$ X-ray emissions, respectively. For pure Pd, the $L\beta_1/L\alpha_1$ ratio is 0.4. However other spots showed ratios as high as 0.75. This increase in the ratio indicates that another element is contributing to the peak at 2.99 keV. The $L\alpha_1$ peak of Ag occurs at 2.98 keV. Changes in this ratio can be used to quantitate the amount of Ag present. A ratio of 0.75 indicates that this spot contains about 19% Ag. Dash and Ambadkar attributed the formation of Ag to the following reactions:

$$^{108}\mathrm{Pd} + \mathrm{n} \rightarrow {}^{109}\mathrm{Pd} \rightarrow {}^{109}\mathrm{Ag} + \mathrm{e}^- + \mathrm{energy}.$$

Mosier-Boss et al. conducted a Pd/D co-deposition experiment on a Au foil in the presence of an external magnetic field [23]. Upon termination of the experiment. The Au/Pd cathode was subjected to SEM/EDS analysis. In areas where the deposit exhibited a cauliflower like morphology consisting of aggregates of spherical micro-globules, EDS analysis showed X-ray emissions from Pd. In some areas, the deposit looked sintered. EDS spectra of these areas showed large peaks due to Fe, Cr, Ni, and Al with a very small peak due to Pd. ICP-AES, and ICP-MS analysis of the electrolyte has shown that the presence of these new elements are not due to contamination [12] as the concentration of these new elements on the cathode were several orders of magnitude larger than was found in the electrolyte. The same was true of other cell components. The inhomogeneous distribution of these new elements also supports the notion that the presence of these new elements is not due to contamination. The small size of the Pd X-ray peaks indicates that Pd has been consumed. Since Fe, Cr, and Ni are approximately half the atomic mass of Pd suggests that Pd has been fissioned. The observation of long range alphas ($E_\alpha \geq 12$ MeV) in other Pd/D experiments support fissioning of Pd as these alphas can only form as the result of ternary fission [24–26]. These magnetic field effects were verified by DeChiaro et al. [12], who further showed that the excess heat correlated with the occurrence of these new elements.

3. Energetic Particles

Mosier-Boss et al. [27] used CR-39, a solid state nuclear track detector, to measure the emissions of energetic particles in Pd/D co-deposition experiments. It was found that what substrate was used impacted what was observed. When Pd/D co-deposition was done on a Ni screen cathode, in the absence of an external electric/magnetic (E/B) field, no tracks due to energetic particles were measured. Instead the impression of the Ni screen was observed on the CR-39 detector. With time, the area where the Ni screen was in contact with the detector was observed to swell. Both the swelling and the impression of the screen were observed when a detector, covered with a metal screen, was exposed to a Cs-137 γ source. Consequently, the damage observed for the Ni screen cathode in the absence of a field was attributed to γ/X-rays. When either an external E (with a 6% AC ripple) or B field was applied in a co-deposition experiment on a Ni screen cathode, tracks were observed in the CR-39 detector. When the cathode substrate was either Ag, Au, or Pt wires, tracks were observed in the detector in both the presence and absence of an external E/B field. A series of control experiments demonstrated that the tracks observed in the CR-39 detectors as a result of Pd/D co-deposition were not due to either chemical or mechanical damage or to radioactive contamination. In addition to tracks due to energetic particles, Mosier-Boss et al. [28,29] also reported seeing triple tracks in CR-39 detectors. These triple tracks are diagnostic of the ^{12}C(n,n')3α carbon breakup reaction due to the reaction of a ≥ 9.6 MeV neutron with a carbon atom in the detector and were shown to be indistinguishable from DT neutron generated triple tracks. No triple tracks were observed in CR-39 detectors used in control experiments.

The CR-39 results have been verified by Tanzella et al. [30], NASA Glenn [31], and three groups of UCSD chemical engineering students [32,33]. Tanzella et al. [30] conducted experiments in which the polyethylene covered CR-39 detectors, in contact with the cathode, were immersed in the electrolyte. The polyethylene film covering the detectors was 60 μm thick. Upon termination of the experiment, these detectors underwent microscopic examination, scanning using a TASL (Track Analysis System Ltd.) system, linear energy transfer (LET) analysis of the scanned data, and sequential etching [30,34]. The LET analysis determined that the tracks on the front surface in contact with the cathode were caused by $\geq$ 1.8 MeV protons, $\geq$1.8 MeV alphas, and secondary particles due to recoils from either energetic protons and/or neutrons [34]. On the obverse surface of the detectors, the particles that created the tracks were identified as $\geq$11.8 MeV protons and/or recoils from either energetic protons and/or neutrons. Destructive sequential analysis of the detectors [30] identified tracks due to 3 MeV protons, 12 MeV alphas, and 16 MeV alphas – which was in agreement with the LET spectrum analysis. This analysis also showed 2.2–2.5 MeV neutrons. It was estimated that these neutrons were produced at a rate of 0.3 ± 0.1 n/s. Tanzella et al. [30] conducted additional experiments in which the CR-39 detector was placed outside the cell. In these experiments, a 6 μm thick Mylar film separated the detector from the cathode. These experiments were conducted in both H_2O and D_2O. The detectors were subjected to destructive sequential etching analysis. No neutrons were detected in the H_2O experiment. However, the

D_2O experiment showed recoil tracks due to 2.2–2.5 MeV neutrons. In this experiment, the rate of neutron generation was 0.6 ± 0.1 n/s.

4. Conclusions

The Pd/D co-deposition experiment has offered great flexibility in experimental design. Different Pd plating solutions have been used as well as different cell configurations (e.g., parallel electrodes or concentric electrodes) working electrode surfaces (Au, Ag, Ni, Cu, or Pt), and electrode geometries (wire, sheet, or screen). Both closed and open system have been used. The Pd/D co-deposition process has been used by several researchers to investigate the phenomenon of condensed mater nuclear reactions within the Pd lattice. Reported reaction products include excess heat, gamma/X-ray emissions, transmutation, as well as the production of tritium and energetic particles. These products indicate that several varieties of nuclear reactions are occurring in the system. These include primary and secondary fusion reactions to produce neutrons, protons, tritium, and ≥ 10 MeV protons and neutrons. There is evidence of transmutation as shown by the production of Ag that can arise from either proton (≥ 10 MeV) or neutron capture by Pd. The observation of long range alpha particles indicate the occurrence of possible ternary and quaternary fission of Pd that is supported by the presence of such elements as Fe, Cr, Ni, and Al with a corresponding decrease in Pd.

Acknowledgements

The authors acknowledge the contributions of Stan Szpak for initially developing the Pd/D co-deposition technique. They are grateful for the support of Frank Gordon which allowed Stan Szpak and the authors to conduct their research at the Navy laboratory. They are thankful for the support of Dr. Jay Khim of JWK Corporation. Finally, they are very grateful to all the researchers who have conducted Pd/D co-deposition experiments as their findings not only corroborated our results, but also provided additional data to elucidate the nature of the reactions occurring inside the Pd lattice.

References

[1] S. Szpak, P.A. Mosier-Boss and S.R. Scharber and J.J. Smith, Charging of the Pd/^{n}H system: role of the interphase, *J. Electroanal. Chem.* **337** (1992) 147–163.

[2] S. Szpak, P.A. Mosier-Boss, S.R. Scharber, J.J. Smith, Cyclic voltammetry of Pd+D codeposition, *J. Electroanal. Chem.* **380** (1995) 1–6.

[3] S. Szpak, P.A. Mosier-Boss and J.J. Smith, Deuterium uptake during Pd-D codeposition, *J. Electroanal. Chem.* **379** (1994) 121–127.

[4] S. Szpak, P.A. Mosier-Boss and J.J. Smith, On the behavior of Pd deposited in the presence of evolving deuterium, *J. Electroanal. Chem.* **302** (1991) 255–260.

[5] S. Szpak and P.A. Mosier-Boss, On the behaviour of the cathodically polarized Pd/D system: a response to Vigier's comments, *Phys. Lett.* A **221** (1996) 141–143.

[6] G. Guisbiers, G. Abudukelimu, D. Hourlier and Size-dependent catalytic and melting properties of platinum–palladium nanoparticles, *Nanoscale Res. Lett.* **6** (2011) 396.

[7] D.J. Nagel, Characteristics and energetics of craters in LENR experimental materials, *J. Condensed Matter Nucl. Sci.* **10** (2013) 1–14.

[8] Y.E. Kim, Conventional nuclear theory of low-energy nuclear reactions in metals: alternative approach to clean fusion energy generation, in *Proc. the ICCF17*, 2012.

[9] S. Szpak, P.A. Mosier-Boss, M.H. Miles and M. Fleischmann, Thermal behavior of polarized Pd/D electrodes prepared by co-deposition, *Thermochim. Acta* **410** (2004) 101–107.

[10] D.J. Cravens and D.G. Letts, Practical techniques in CF research – triggering methods, in *Proc. ICCF10*, 2003.

[11] D. Letts and P.L. Hagelstein, Modified Szpak protocol for excess heat, *J. Condensed Matter Nucl. Sci.* **6** (2012) 44–54.

[12] L. Dechario, L.P.G. Forsley, P.A. Mosier-Boss, B.M. Steinetz, R.C. Hendricks, K.J. Long, P. Rayms-Keller, M.L. Shea, S. Barker, T.L. Benyo, D.L. Ellis, I. Locci, D. Johnson and W.D. Jennings, A Multi-laboratory study of anomalous elements and magnetic orientation effects in LENR codeposition experiments, Naval Surface Warfare Center Dahlgren, VA, 2018.

[13] M. Swartz and G. Verner, Excess heat from low electrical conductivity heavy water spiral-wound Pd/D_2O/Pt and Pd/D_2O-$PdCl_2$/Pt devices, in *Proc. ICCF10*, 2003.

[14] J. Dash and A. Ambadkar, Co-deposition of palladium with hydrogen isotopes, in *Proc. ICCF11*, 2004.

[15] M.H. Miles, Investigations of possible shuttle reactions in co-deposition systems, *J. Condensed Matter Nucl. Sci.* **8** (2012) 12–22.

[16] J. O'M. Bockris, C.-C. Chien, D. Hodko and Z. Minevski, Tritium and helium production in palladium electrodes and the fugacity of deuterium therein, in *Proc. ICCF3*, 1992.

[17] S. Szpak, P.A. Mosier-Boss and R.D. Boss, On the behavior of the Pd/D System: evidence for tritium production, *Fusion Technol.* **33** (1998) 38–51.

[18] M.H. Miles, NEDO final report – electrochemical calorimetric studies of palladium and palladium alloys in heavy water, University of La Verne, 2004.

[19] K.-H. Lee, H. Jang and S.-J. Kim, A change of tritium content in D_2O Solutions during Pd/D Co-deposition, *J. Condensed Matter Nucl. Sci.***13** (2014) 294–298.

[20] S. Szpak and P.A. Mosier-Boss, On the behaviour of the cathodically polarized Pd/D system: search for emanating radiation *Phys. Lett. A.* **210** (1996) 382–390.

[21] P.A. Mosier-Boss, J.Y. Dea, F.E. Gordon, L.P.G. Forsley and M.H. Miles, Review of twenty years of LENR research Using Pd/D co-deposition, *J. Condensed Matter Nucl. Sci.* **4** (2011) 173–187.

[22] M.H. Miles, B.F. Bush and K.B, Johnson, Anomalous Effects in Deuterated Systems – Final Report, Naval Air Warfare Center Weapons Division, NAWCWPNS TP 8302, 1996.

[23] P.A. Mosier-Boss, It is Not Low Energy – But it is Nuclear, *J. Condensed Matter Nucl. Sci.* **13** (2014) 432–442.

[24] A.G. Lipson, A.S. Roussetski, G.H. Miley and C.H. Castana, In-situ charged particles and X-ray detection in Pd thin film-cathodes during electrolysis in Li_2SO_4 /H_2O, in *Proc. ICCF9*, 2002.

[25] A. Lipson, I. Chernov, V. Sokhoreva, V. Mironchik, A. Roussetski, A. Tsivadze, Y. Cherdantsev, B. Lyakhov, E. Saunin and M. Melich, Charged particle emissions and surface morphology of Pd/PdO:D_x and TiD_x targets under electron beam excitation, in *Proc. ICCF15*, 2009.

[26] A.G. Lipson, G.H. Miley, A.S. Roussetski and E.I. Saunin, Phenomenon of an energetic charged particle emission from hydrogen/deuterium loaded metals, in *Proc. ICCF10*, 2003.

[27] P.A. Mosier-Boss, S. Szpak, F.E. Gordon and L.P.G. Forsley, Use of CR-39 in Pd/D co-deposition experiments, *Eur. Phys. J. Appl. Phys.* **40** (2007) 293–303.

[28] P.A. Mosier-Boss, S. Szpak, F.E. Gordon and L.P.G. Forsley, Triple tracks in CR-39 as the result of Pd–D co-deposition: evidence of energetic neutrons, *Naturwiss.* **96** (2009) 135–142.

[29] P.A. Mosier-Boss, J.Y. Dea, L.P.G. Forsley, M.S. Morey, J.R. Tinsley, J.P. Hurley and F.E. Gordon, Comparison of Pd/D co-deposition and DT neutron generated triple tracks observed in CR-39 detectors, *Eur. Phys. J. Appl. Phys.* **51** (2010) 20901.

[30] A.S. Roussetski, A.G. Lipson, E.I. Saunin, F. Tanzella and M. McKubre, Detection of high energy particles using CR-39 detectors part 2: results of in-depth destructive etching analysis, *Int. J. Hydrogen Energy* **42** (2017) 429-436.

[31] P. Smith, L.P. Forsley, B. Steinetz, K.P. Prokopius, R. Hendricks and P.A. Mosier-Boss, Energetic nuclear particles observed following the SPAWAR Pd/D co-deposition protocol, NASA John H. Glenn Research Center, Cleveland, OH, 2018.

[32] N. Robertson, H. Saito, J. Yurkovic and S. Zakskorn, Field assisted electroplating, *J. Sci. Explor.* **23** (2009) 452–455.

[33] P.A. Mosier-Boss and L.P. Forsley, Energetic particle emission in Pd/D co-deposition: an undergraduate research project to replicate a new scientific phenomenon, *J. Lab. Chem. Educ.* **6** (2018) 69–76.

[34] P.A. Mosier-Boss, F.E. Gordon, L.P.G. Forsley and D. Zhou, Detection of high energy particles using CR-39 detectors part 1: results of microscopic examination, scanning, and LET analysis, *Int. J. Hydrogen Energy* **42** (2017) 416–428.

J. Condensed Matter Nucl. Sci. 29 (2019) 41–51

Research Article

High-temperature Calorimetric Measurements of Heat for Ni/H_2 Exothermic Reactions

Edward J. Beiting* and Dean Romein

TrusTech, Redondo Beach, CA 90278, USA

Abstract

Instrumentation developed to measure heat power from a high-temperature reactor for experimental trials lasting several weeks is being applied to gas-phase Ni/H_2 LENR. We developed a reactor that can maintain and record temperatures in excess of 1200°C while monitoring pressures exceeding 7 bar. This reactor is inserted into a flowing-fluid calorimeter that allows both temperature rise and flow rate of the cooling fluid to be redundantly measured by different physical principles. A computerized data acquisition system was written to automate the collection of more than 20 physical parameters with simultaneous numerical and dual graphical displays comprising both a strip chart and complete history of key parameters. Initial studies of the absorption of light hydrogen (H_2) by nickel are presented.

Keywords: Anomalous heat, Gas loading, Gas-phase LENR, High-temperature calorimetry, High-temperature cell, High-temperature hydrogen permeation, Hydrogen gas, Nickel-hydrogen absorption, Parkhomov replication

1. Introduction

A laboratory was built to study the heat release from the gas phase manifestation of the Anomalous Heat Effect (AHE aka LENR). We are studying two implementations of this manifestation. The first heats at low temperature (<400°C) nanoparticles of bimetals imbedded in micron-sized particles of refractory. The second heats nickel particles and hydrogen at high temperatures as described by Parkhomov [1,2] in Russia and by Qi et al. [3] in China.

2. Instrumentation

This second task requires the ability to maintain samples at temperatures up to 1300°C while monitoring temperature, pressure, and excess heat continuously for a period of weeks. Furthermore, the calorimeter is capable of accommodating kilowatt thermal power levels with accuracy and precision. The technology developed and initial results of the high temperature and high power measurements are described in this paper.

*Corresponding author. E-mail: ebeiting@ymail.com.

© 2019 ISCMNS. All rights reserved. ISSN 2227-3123

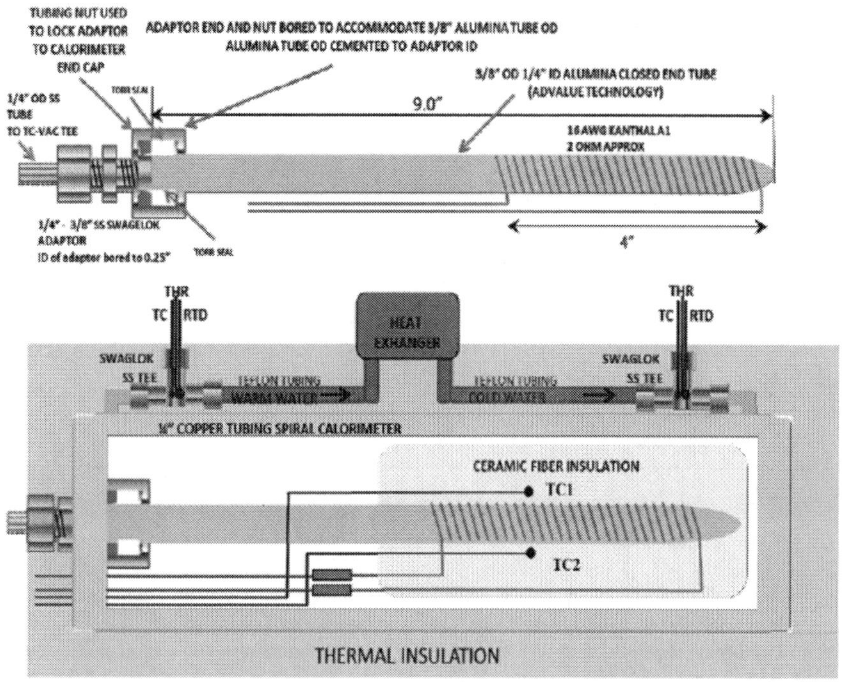

Figure 1. *Top*: Details of the high temperature reactor; *Bottom*: reactor in flowing fluid calorimeter. The heat exchanger circuit comprises a variable rate pump, a flow meter, a fan-forced radiator in a controlled temperature environment, and optionally a copper coil in a temperature-controlled refrigerated bath.

A drawing of the system is shown in Fig. 1. The design philosophy is to redundantly measure critical physical parameters using different physical principles. This calorimeter requires only two measured quantities, the volume flow rate ($\mathrm{d}V/\mathrm{d}t$) and temperature rise (ΔT), to measure the heat power (P), viz,

$$P = \rho c_\mathrm{w} \frac{\mathrm{d}V}{\mathrm{d}t} \Delta T, \tag{1}$$

where ρ and c_w are the density and mass specific heat capacity of the fluid, respectively. Typically this temperature difference is less than 10°C for an insulated reactor at a temperature of 1300°C and a water flow rate of less than 10 ml/s.

The inlet and outlet temperatures are registered using thermocouples (TCs), resistive thermal devices (RTDs), and thermistors. The flow rate is measured first by passing the fluid through a Hall effect flow meter that had an absolute accuracy of 1%. For a redundant flow measurement a second calorimeter can be connected in series; i.e. the flow from first passed through the second that houses the test reactor. The first calorimeter houses a heat source of known power and thus by measuring the temperature difference the flow rate could be calculated by Eq. (1).

A major task was to design a sealed container that heated the sample to high temperature and permitted the pressure in the volume to be monitored continuously. Its development resulted in a cell system that could be held continuously at a measured temperature of 1300°C for a period of weeks. Challenges were thermocouples that remained operational and accurate and heaters that did not fail after a few hundred hours.

The current reactor design is shown in top of Fig. 1 and its integration in the calorimeter is shown in bottom of Fig. 1. The length of the reactor's alumina tube [4] is chosen so that its unheated end remains below 100°C when the interior volume of the heated end is 1300°C. Empirical testing found that a 16 AWG Kanthal A1 wire spiral wrapped on the alumina tube is durable for the long trial periods. The method shown of sealing the ceramic tube to the stainless steel fitting was tested to a pressure of 8 bar.

Two external type-N thermocouples are to monitor the interior temperature of the reactor. Using external thermocouples have a number of advantages: there is no opportunity for the thermocouple metals to react with the reactant or their products; the thermocouple temperature is kept at a lower temperature (usually <1000°C) than that of the interior of the reactor thus increasing the thermocouple's life and accuracy; no high pressure/vacuum feed through is required; no high temperature electrical insulation isolating the thermocouple wire is required. Type-N thermocouples are more durable and accurate than type-K thermocouples at high temperatures.

The external thermocouples must be correlated with the interior temperatures of the reactor. This is done during a calibration run described below. The interior temperatures vary along the length of the tube under the spiral heater. The

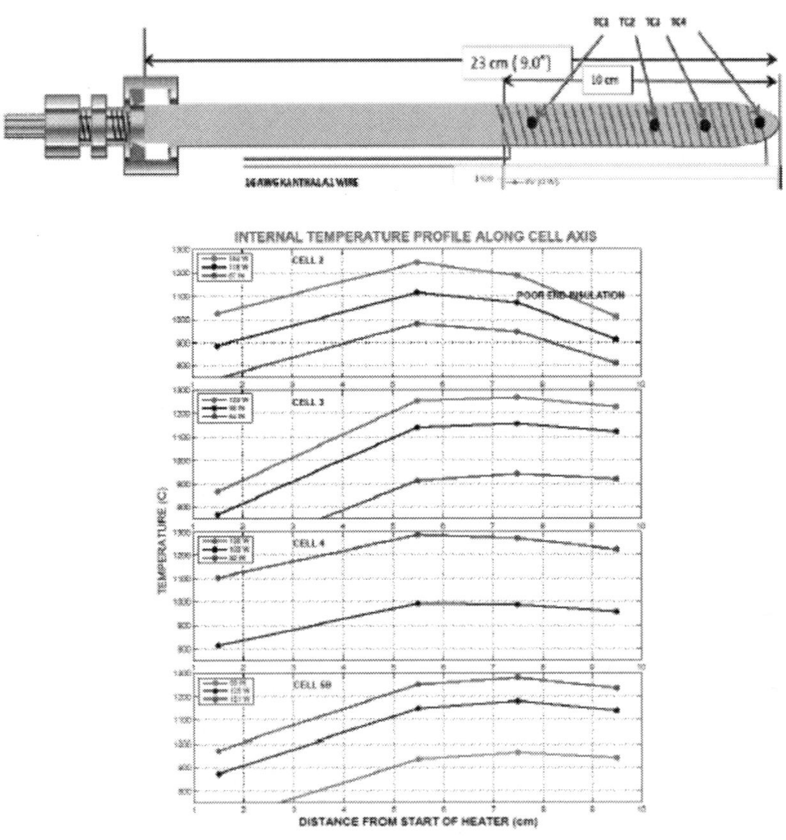

Figure 2. Temperature variation along the axis of cells

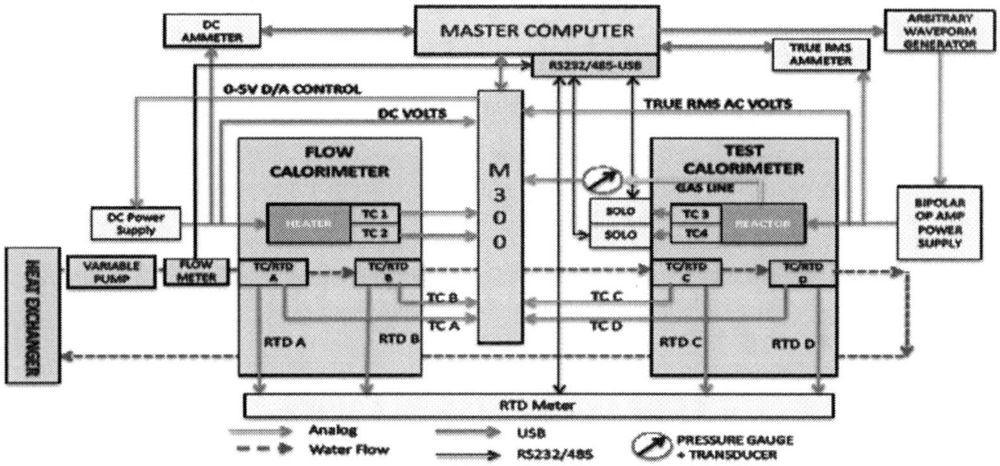

Figure 3. Functional diagram of the data acquisition system (DAQ).

variation is dependent on the insulation wrapped around the reactor. Figure 2 shows the temperature variation along the axis of four different cells taken with a four-thermocouple probe. The locations of the thermocouples are shown in the drawing at the top of the figure. If the end of the cell is poorly insulated, the temperature decreases significantly at the end as shown in the top plot.

Typically, the sample occupies about 3 cm of the 10 cm heated volume allowing it to be placed in the zone with the most uniform temperature using ceramic fiber spacers.

Figure 3 shows a functional diagram of an instrument with the optional calorimeter for the redundant flow rate measurement integrated into the data acquisition system (DAQ). A computer controls the experiment through USB ports using MATLAB® and its Instrument Control Toolbox software. Most measurements are made using a Rigol M300 DAQ system with 6.5 digit precision. The Rigol DM 3058E used as ammeters have 5.5 digit precision. SOLO 4824 temperature controllers monitor the reactor temperature for the type-N thermocouples. The output of a Rigol DG 1022 Arbitrary Waveform Generator is amplified by a Kepco BiPolar Operation Amplifier Power Supply when AC power heats the reactor. The calorimeter flow rate is measured using an Omega FMG81 flow meter with Laurel readout. A Define meter reads the RTDs. The water flow is controlled with a TCS MG1000S or MG2000S pulseless pump. Pt100 RTDs and thermocouples are purchased from Omega Engineering. Data were appended to a file every acquisition cycle with a maximum rate of 3 cycles/min, assuring that no data is lost if there were a system failure.

The AHE software was built for a class of experiments that start with initializing the software and then proceeds to collect experimental data over a long period of time. The experimental data collection is performed by a data acquisition loop that collects data from a suite of sensors connected to instruments. Every time through the acquisition loop, data are displayed to the operator using numeric displays and graphs. Some processing is performed to display values computed from the raw data. This processed data can also be displayed to the operator using numeric displays and graphs. Finally, the software allows experimental conditions, such as voltage to a power supply, to be changed either automatically by software or to be changed by operator actions.

The software has been built using object oriented programming principles that define software objects that are connected into a network of processing objects. The goal has been to build software that is maintainable and may be ported from one experiment platform to another experiment platform. Its architecture is shown in Fig. 4.

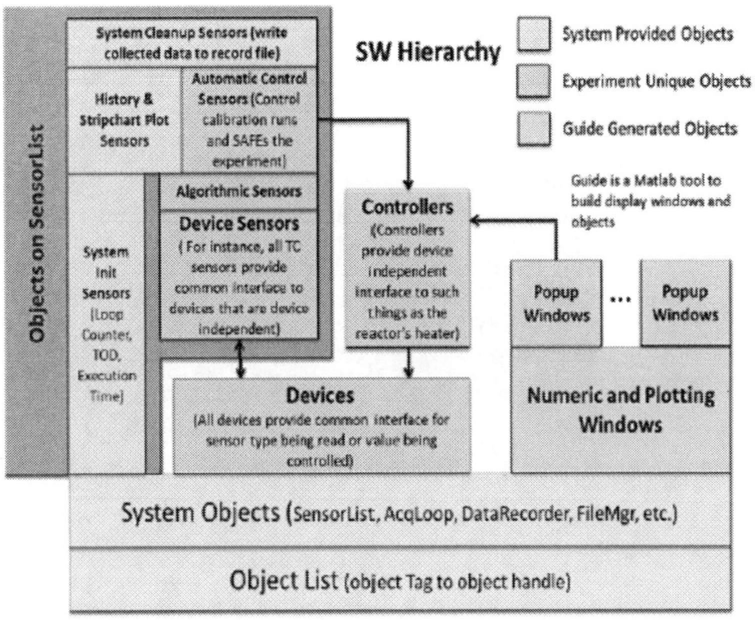

Figure 4. The software hierarchy.

The program runs in two phases. The first initializes all of the software objects and connects one object to another as needed. The second phase executes the data acquisition loop. The goal has been to build the software objects so that they are debugged once and then used as boxes that have certain properties that need to be defined during initialization. It should be possible to change no code except in the initialization functions that have been built for each set of software objects.

There are two types of objects used by the AHE software. The first displays objects provided by MATLAB. These objects include MATLAB figure, axes, and uicontrol objects (user interface controls), such as text or edit widgets. Each of these displays objects has the Tag property. The second are the user objects developed as part of the AHE software. These objects also have the Tag property. When a display object and a User object have the same tag, the AHE software automatically pairs the two during initialization phase so that the User object's data can be displayed during the acquisition loop. The MATLAB source code for this program is well documented and user manual is available. Its source code will be made available to interested parties.

The interactive two screen graphical inter-face is shown in Fig. 5. The axes of any of the plots are adjustable and operational parameters can be changed with push button controls.

The software automates the acquisition of the calibration data. An example of the data from a calibration sequence is shown in Fig. 6. This dense figure shows seven plots. From the top these are: (1) pressure (3×10^{-3} Torr here), (2) two interior and exterior temperatures in automated steps, (3) coolant flow rate, (4) the calorimeter entry temperatures measured with a thermocouple, RTD, and thermistor, (5) the triply redundant changes in calorimeter fluid temperatures, (6) the input power and triply redundant inferred powers, and (7) the measured excess power. The bottom plot showing the measured excess power requires explanation. The excess power during this trial is 0 W. Whenever the input power to the heater is changed, the temperature of the fluid (water in this case) flowing through the calorimeter begins to

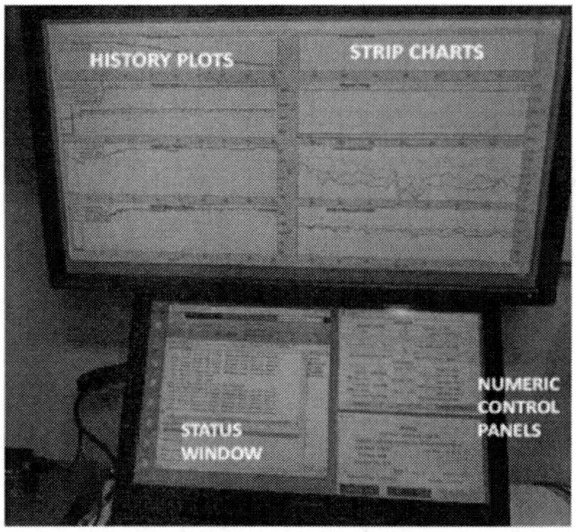

Figure 5. The two display monitors of the operating software.

change until it equilibrates. For the flow rate in this run (7.5 ml/s shown in the third plot), the equilibrium time is about 30 min. An increase in power results in a negative excursion (reading too low a power) and a decrease in power gives the positive excursion shown. A calibration parameter is adjusted for each temperature sensor (two are required for the thermistor) to give the accuracy shown in the bottom plot. The accuracy shown is about ± 0.5 W, which is better than 0.5% for the maximum input power of about 140 W.

The second plot from the top shows interior temperature registered by two of the interior thermocouples (red lines) and the temperatures registered by the two external thermocouples (blue lines). Third order polynomial fits of interior to exterior temperatures are used to infer the interior temperatures from measured external temperatures during a test run. This creates four inferred temperatures and the highest and lowest inferred values are display on the numeric panel in Fig. 5.

3. Initial Experimental Studies

Initial studies focused on replication of work of Parkhomov et al. [1,2]. We obtained three samples of Vale nickel powder labeled Type 123, Type SNP, and Type 255 [5]. Electron micrographs of these samples are shown in Fig. 7. Type 123 was chosen as most closely resembling that used by Parkhomov.

The sample tested was 1 g of Type 123 nickel powder and 0.1 g of $LiAlH_4$. Of several test trials, only one showed the characteristic pressure profile described by Parkhomov et al. These data are shown in Fig. 8. The first pressure rise occurred at below 200°C and the first pressure decrease at 250°C. Two additional steps to 550°C and 950°C induced the final pressure increases and taking the temperature to 1250°+C inducing the long pressure drop to sub-atmospheric at 65 h. No excess power was observed as seen in the bottom plot.

A similar inability to observe subsequent pressure drops after initial success was observed by Qi et al. [3] who did observe excess heat on the initial run. They attributed this behavior to nickel reaction with the atmosphere after their initially sealed container was left open. Our container of nickel was never sealed and was always exposed to the atmosphere.

In order to study the lack of pressure decrease (assumed to be nickel absorption) in subsequent trials, it was decided

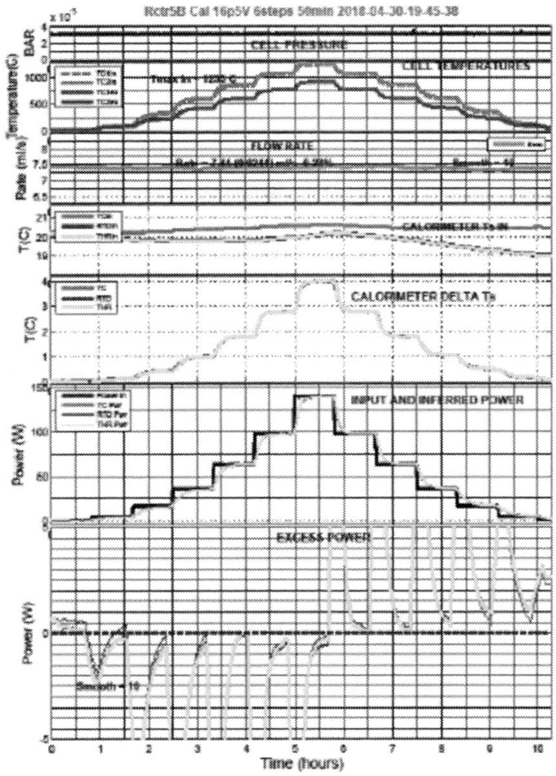

Figure 6. Example of calibration data.

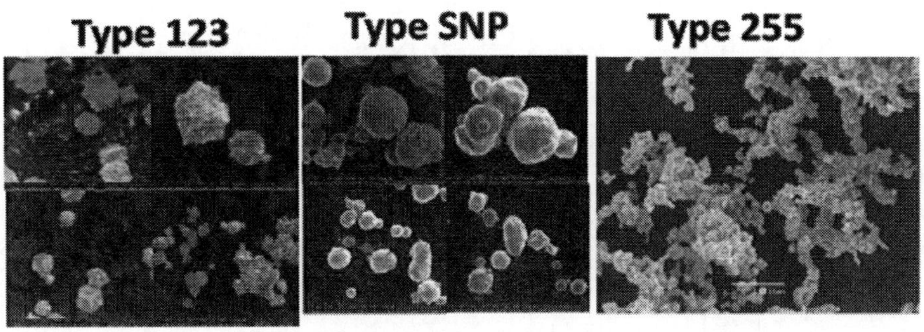

Figure 7. Electron micrographs of three types of Vale nickel powder.

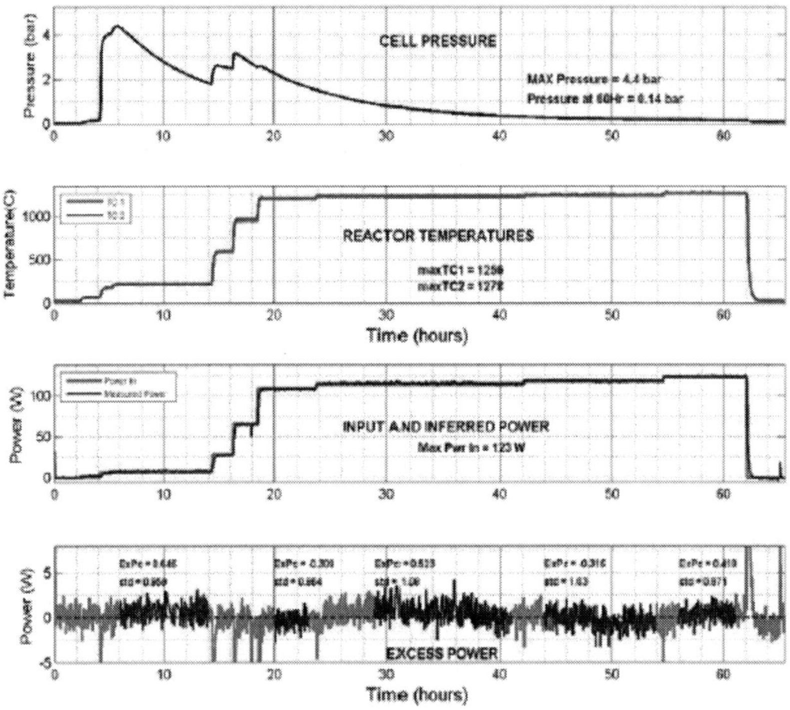

Figure 8. One-time pressure profile of a mixture of nickel powder and $LiAlH_4$.

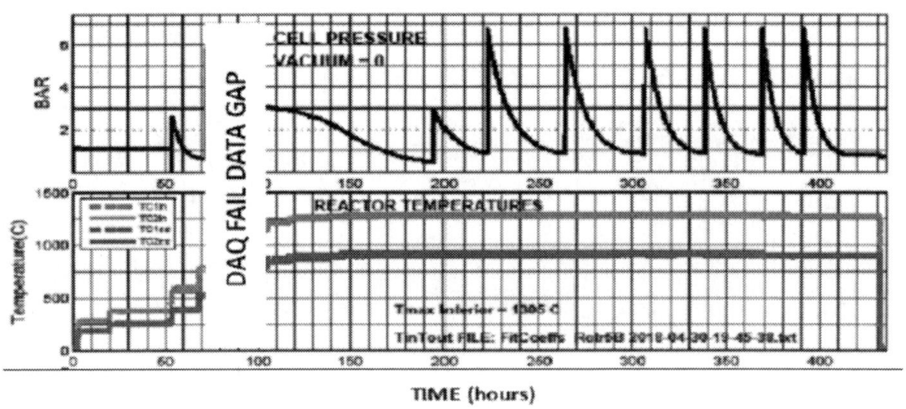

Figure 9. Data run interrupted by ICCF-21 conference.

Table 1. History of pressure changes in cell.

Moles Ni=0.034075			
Press No.	ΔP(bar)	H moles	H mole frac.
1	1.5	0.0028	0.0811
2	2.3	0.0042	0.1243
3	2	0.0037	0.1081
4	5.75	0.0106	0.3108
5	5.75	0.0106	0.3108
6	5.75	0.0106	0.3108
7	5.75	0.0106	0.3108
8	5.75	0.0106	0.3108
9	5.75	0.0106	0.3108
	Total=	0.0742	2.1781

to load the cell only with Vale 123 nickel and pressurize with pure hydrogen. The data from a 460 h (19 days) run shown in Fig. 9 were presented in the poster paper at ICCF-21. Subsequently, the cell was cooled and remained sealed at room temperature for three weeks when work was interrupted to attend the conference. After returning from the conference, the cell was reheated and re-pressurized to 5.75 bar and it again decreased to a sub-atmospheric pressure in a similar manner to that seen in the last six pressurization shown in Fig. 9. The total time this nickel sample was heated in the cell was 520 h (22 days) with 420 h (17.5 days) at a temperature above 1200°C. During these periods the cell was repeatedly pressurized with hydrogen as shown in these figures.

If one assumes that all the hydrogen is absorbed by the nickel in Fig. 9, the repeated pressurizations in Fig. 9 lead to the values presented in Table 1. This table, also presented at ICCF-21, shows an atomic H to atomic Ni mole fraction of a little less than 2.2 for the cumulative absorptions. Adding the additional post conference pressurization increases this mole fraction ratio to 2.5. Note that the pressure value in each pressurization is allowed to decrease to a sub-atmospheric value before re-pressurization in order to assure that there were no leaks to atmosphere in the system.

From Fig. 9 (and the subsequent pressurization) it is seen that the rate of hydrogen absorption increases as the assumed ratio of atomic H-to-nickel atom ratio reaches 2.5. Either nickel can absorb copious hydrogen at high temperature or hydrogen is lost via another mechanism, possibly permeating into or through the walls of the system.

To check this conjecture, a previously used cell that had been held at elevated temperatures for 150 h and temperatures above 1200°C for 70 h was reconfigured for a test. A new heater coil and insulation were installed and the empty cell (except for 3 cm of fiber insulation [6]) was installed in the calorimeter. This cell was then heated to an internal temperature of >1200°C for more than five days. During this period the cell was pressurized twice with 84 psig (6.7 bar) of H_2. The results of this test are shown in Fig. 10.

It is clear from the pressure curve (second from the top) in this figure that nickel is not required to give the pressure drops see in Fig. 9. The first pressurization was allowed to fall to 2 bar before re-pressurization but the second was allowed to fall well below atmospheric pressure (0.6 bar) before allowing the temperature to drop to ambient, reaching a minimum pressure of 0.56 bar before slowly increasing for the next 15 h. Subsequently, the cell was opened to atmosphere to check the calibration of the pressure transducer, which read 1.0089 bar for an accuracy of better than 1%.

The bottom two curves in Fig. 10 show the rates of the pressure drops. Note from the third curve the maximum depressurization rate of the first pressurization is -0.4 bar/h and the second is about –0.25 bar/h. From the bottom curve we note that the depressurization rate is approximately proportional to the pressure.

Because hydrogen is known to have a high permeation rate at high temperatures, the permeation rate was explored for both the stainless steel tubing and alumina of the system. Noting that 304 and 316 stainless steel have nearly the same permeation rate [7,8], we use the values given by [8] to calculate the permeation rate for the gas system of 25

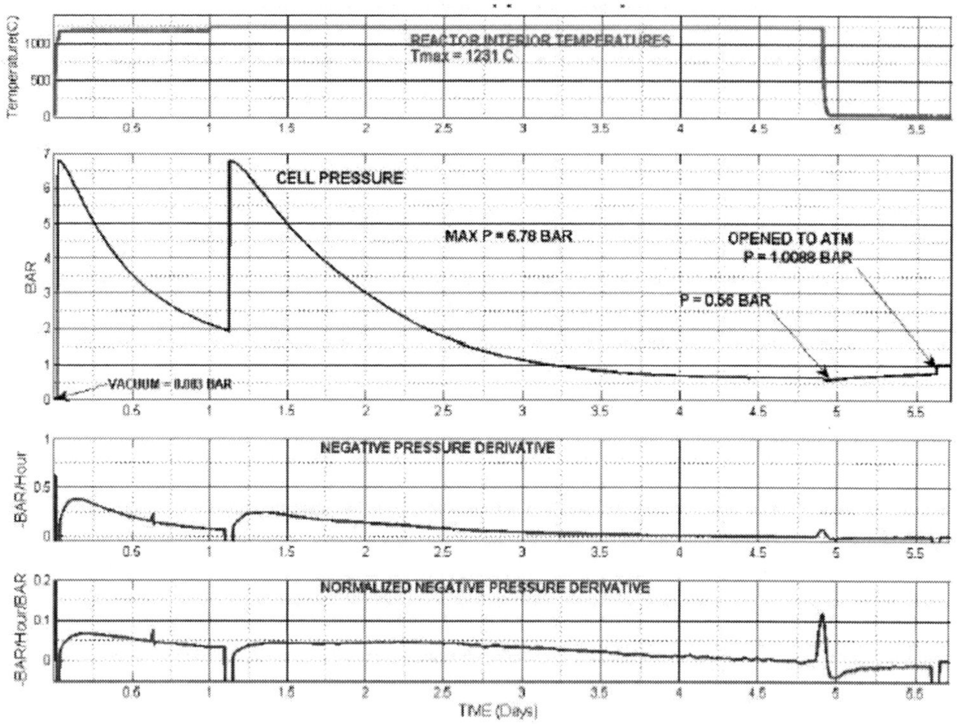

Figure 10. Pressurization of an empty cell.

cm of (0.635 cm OD) stainless tubing at room temperature. This is found to be 3.7×10^{-7} bar/h and thus is not a factor. Alumina permeation by hydrogen has been measured [9]. Using the alumina surface area, wall thickness, and pressures of the experiment and a temperature of 1250°C, a permeation rate of 0.004 bar/h is calculated. This is less than 1% of the observed rate.

For completeness, it is noted that the alumina tube is sealed to a stainless steel fitting using a sealant (Torr Seal) designed for vacuum systems. After a cell is fabricated but before it is used, it is pressurized with 100 psig (7.8 bar) of hydrogen, sealed with a valve, and monitored for a minimum of 24 h at room temperature. If no pressure drop is observed within the resolution of the Bourdon gauge (0.5 psi), the cell is considered leak free.

Because a high temperature could compromise the integrity of the stainless steel-alumina seal, the maximum temperature at the end of the alumina tube sealed with Torr Seal was measured. It was found to be 80°C when the interior temperature of the tube at the location of the nickel is 1250°C. This temperature was measured with the copper end-cap of the calorimeter that holds the stainless steel-alumina seal NOT in contact with the calorimeter (due to the leads of the thermocouples). Thus this end cap reached a considerably higher temperature than when it was used in the previous measurements and consequently it is unlikely that the Torr Seal reached 80° C during the measurements shown in Figs. 9 and 10. Torr Seal can be used "at temperatures from –45°C to 120°C (bakeable temperature)" [10]. Thus the sealant was significantly below the maximum operating temperature recommended by the manufacturer. One possibility not yet explored is absorption by the ceramic fiber in the end of the tube. Its composition is "high purity alumina, zirconia and silica spun ceramic fibers" [6].

Although the data cited above [9] indicates that absorption into the tube is not the cause of these pressure drops, extended periods at high temperatures of the data in Figs. 9 and 10 or a difference alumina grades could make this data comparison invalid. At this writing this pressure drop is a topic of investigation.

Acknowledgements

We thank Drs. Jacques Ruer and Michael R. Staker for enlightening discussions.

References

[1] A.G. Parkhomov and E.O. Belousova, Research into heat generators similar to high-temperature Rossi reactor, *J. Condensed Matter Nucl. Sci.* **19** (2016) 244–256.

[2] An English translation of Parkhomov results using a flowing calorimeter can be found at https://drive.google.com/file/d/0Bz7lTfqkED9Wdm1NeEtxMFJLTmM/view.

[3] B.-J. Qi, M. He, S.-Y. Wu, Q.-Z. Zhao, X.-M. Wang, Y.-J. Pang, X.-L. Yang and S.-S. Jiang (jiang@ihep.ac.cn), Anomalous heat production in hydrogen-loaded metals: possible nuclear reactions occurring at normal temperature, China Institute of Atomic Energy, P.O. 275 (49), Beijing 102413.

[4] AdValue Technology Alumina Tube–Closed One End; 3/8-1/4-12 AL-T-N3/8-N1/4-12-COE; http://www.advaluetech.com/index.html.

[5] Novamet Specialty Products Corp., 1420 Toshiba Drive, Suite E, Lebanon, TN 37087, USA.

[6] Ceramic Blanket 2600, Thermal Products Co. Inc., 4520 S. Berkeley Lake Road, Norcross, GA 30071.

[7] Sun Xiukui, Xu Jian and Li Yiyi, Hydrogen permeation behaviour in austenitic stainless steels, *Matls Sci. Eng.* A **114** (1989) 179–187.

[8] E. Hashimoto and T. Kino, hydrogen permeation through Type 316 stainless steels and ferritic steel for a fusion reactor, *J. Nucl.Matls.* **133,134** (1985) 289–291.

[9] R.M. Roberts, T.S. Elleman, H. Palmour III and K. Verghese, Hydrogen permeability of sintered aluminum oxide, *J. Am. Ceramic Soc.* **62** (1979) 495–499.

[10] Agilent Torr Seal Low Vapor Pressure Resin Sealant Data sheet, Agilent Technologies, 121 Hartwell Avenue, Lexington, MA 02421, USA, 2014.

J. Condensed Matter Nucl. Sci. 29 (2019) 52–74

Research Article

Steps to Identify Main Parameters for AHE Generation in Sub-micrometric Materials: Measurements by Isoperibolic and Air-flow Calorimetry

Francesco Celani*, B. Ortenzi and A. Spallone

INFN-LNF, Via E. Fermi 40, 00044 Frascati (RM), Italy

C. Lorenzetti, E. Purchi, S. Fiorilla, S. Cupellini, M. Nakamura, P. Boccanera and L. Notargiacomo

International Society for Condensed Matter Nuclear Science (ISCMNS-L1), via Cavour 26, 03013 Ferentino (FR), Italy

G. Vassallo

Department of Industrial and Digital Innovation, University of Palermo, Viale delle Scienze, 90128 Palermo (PA), Italy

R. Burri

IETC Laboratories (www.ietclab.org), 6827 Brusino Arsizio-CH, Italy

Abstract

In 2011, we introduced the use of constantan alloy in LENR, in the form of long and thin wires as a hydrogen dissociation promoter. We disclosed for the first time the reason for the choice of such material at IWAHLM-12 Workshop (2017), hypothesizing it was the initiator of the reaction in Andrea Rossi's experiment. We developed a specific treatment to increase the dimensionality of wire surface through the application of high peak power pulses. The wire is inserted in fiberglass ... (continued in the next page)

Keywords: Anomalous Heat Excess (AHE), Electro-migration phenomena, H_2 and/or D_2 absorption at high temperatures into constantan, $H_2 \rightarrow 2H$ dissociation by constantan, Isoperibolic and air-flow calorimetry, Low work function materials, Noble gases, Spontaneous voltage generation along hydrogen-absorbing wires, Surface-modified Cu–Ni–Mn alloy, Thermionic effect

*Corresponding author. E-mail: franzcelani@libero.it.

© 2019 ISCMNS. All rights reserved. ISSN 2227-3123

(continued from the title page)

sheaths, made up of micrometric fibers, impregnated with a solution of an electron-emitter element (Sr). Later, we added Fe and K to the wire surface and the sheaths and adopted the procedure of making equally spaced knots along the wire to produce thermal and magnetic gradients. We also pointed out that the addition of noble gases with low thermal conductivity, and in particular xenon, to the H_2/D_2 atmosphere, produces a considerable rise of temperature in the reactor, maybe because those gases acting as catalyzers in the generation of excess power. Measurements were always performed with isoperibolic calorimetry, which has the advantage of producing non-equilibrium conditions that favor the generation of anomalous heat excess (AHE). With this procedure, we reached a gain of almost a factor 2 at the highest temperature, although with limited stability over time. In this paper, we present SEM observations and EDX analyses of the wire before and after applying the treatment. We have also conducted a series of experiments using air-flow calorimetry. The calorimeter consists of an insulating Styrofoam box whose internal walls are covered with a thick foil of aluminum; the external wall of the reactor was covered with a double layer of black and thick aluminum foil to homogenize temperature. The calorimeter contains the reactor and a halogen tungsten lamp inside a dummy reactor used for calibrations. Even with the air-flow calorimetry approach, which does not produce the most appropriate conditions for AHE, we have obtained excess power, although in quite lower amounts. The best results are: (a) with 100-μm diameter wire, D_2 at 1 bar, input power 90 W, the AHE was over 12 $\pm$2 W, but after 1 day the wire broke and (b) with 200-μm diameter wire, Xe–D_2 mixture each at 0.1 bar and input power of 120 W, AHE was 6–7 W stably for weeks.

1. Introduction: The Choice of Constantan

In 2011, we initiated a research program aimed to increase the magnitude and reproducibility of the so-called *Anomalous Heat Effects* (AHE) as the main result of the occurrence of *Low Energy Nuclear Reactions* (LENR), preferably using *low-cost materials*.

For this purpose, we focused the program from the very beginning on the study of constantan, an alloy of nickel and copper ($Cu_{55}Ni_{44}Mn_1$, abbreviated as CNM or Cst) developed by E. Weston in 1897, instead of other metals that have been extensively studied over the years, such as:

- Palladium (F. Paneth as Pd compounds, and in particular in the form of asbestos in 1926 [1], M. Fleischmann and S. Pons using electrolytic loading in 1989 [2]).
- Ti (S. Jones [3] F. Scaramuzzi since 1989 [4]).
- Ni (S. Focardi and F. Piantelli as bulk-shaped rods since 1991 [5], A. Rossi and B. Ahern as powders in 2008).

At that time, CNM alloy was quite unusual in the field of LENR research and we selected it as a possible substitute for the previously used metals because of the disclosure of Rossi's invention during the demonstration held in Bologna in January 2011 (which was not considered conclusive by the LENR community). We supposed the presence of a *hidden factor* behind the successful AHE generation in early Rossi's experiments that, likely, was not immediately recognized. To our best knowledge, such early experiments took place in 2007 while A. Rossi was collaborating with B. Ahern in the United States.

We would like to emphasize that the real reason for the choice of constantan in our research program was presented for the first time in public and discussed in detail at IWAHLM-12 (5–9 June 2017, Asti-Italy, an International Workshop organized since 1996 by B. Collis). Later on, several scientific meetings on LENR field were devoted to analyzing and discussing such arguments. Among others, the most important were held at the Aeronautic section of the Italian Army (end of June 2017, Rome), and at the 70th Anniversary of ANDI (Associazione Italiana Degli Inventori – Italian Association of Inventors, November 2017, Rome).

After deeply considering Rossi's experimental set-up, we came up with the hypothesis that the *real and main catalyst or initiator* of the reaction with gaseous hydrogen was not the nickel itself, but the *thermocouple* inserted in the reactor with Ni nano-powders. In particular, we refer to J-type (Fe–Constantan) thermocouple, suitable for the temperature range of most experiments (<750°C). This type of thermocouple is especially convenient because of its low cost and high sensitivity (about 50 μV/°C). Furthermore, we believed, based on our experience, that *this thermocouple could be partially damaged after several tests, leading to the activation of its surface* and that the insulating and protective material covering the thermocouple – a kind of glass – may contribute as well to the observed phenomena.

Later, in June 2011, we found a paper of S. Romanowski and collaborators [6] supporting our empirical considerations. On the basis of computer simulations, they show that Ni–Cu alloy, such as constantan, may provide an extremely large amount of energy (up to 2–3 eV for a wide range of Ni–Cu ratio) for the catalytic dissociation of hydrogen molecules to the atomic state ($H_2 \rightarrow 2H$) (Table 1).

Similar catalytic properties can also be expected with deuterium gas and they actually have been retroactively noticed in electrolytic and gaseous experiments carried out by our group since 1989. At that time results obtained in comparative tests of Pd and Ni in electrolytic environmental conditions did not have a clear explanation. After repeated cycles of cathodic/anodic regimes and high/low current densities, we observed that Cu (and Ag) could occasionally show-up in our electrolytic cell, because of contamination coming from the electrical connections (wires of Cu-Ag covered by PTFE) due to the corrosion mediated by LiOD–D_2O vapors and be deposited on Pd surfaces and even more on Ni. The measured excess heat was, some times, larger in the case of the less "noble" Ni than with Pd.

Another fact that made us more confident about the adoption of such material was the discovery of a patent filed in 1993 [7] – brought to our attention by our coworker G. Vassallo in summer 2012 – where B. Ahern appears as the first inventor. In this patent, Cu–Ni is reported as a good candidate for obtaining AHE, especially when at *nanometric dimensions* or under the form of *alternating layers* of copper and nickel. Such indication encouraged us to use and further develop our technique for the preparation of a multilayered and nanostructured texture by means of electrical short-duration pulses of very high peak power.

Table 1. Predicted value of dissociation energy for H_2 dissociation reaction with different materials [6]. The best catalytic material is $Ni_{0.375}Cu_{0.625}$ with more than 3 eV, the worst is Ag.

Material composition	DE (eV) for hydrogen dissociation ($H_2 \rightarrow 2H$)
$Ni_{0.375}Cu_{0.625}$	3.164
$Ni_{0.625}Cu_{0.375}$	2.861
$Ni_{0.8125}Cu_{0.1875}$	2.096
Ni	1.736
$Ni_{0.1825}Cu_{0.8175}$	1.568
$Ag_{0.8125}Pd_{0.1875}$	0.572
$Ag_{0.625}Pd_{0.375}$	0.560
$Ag_{0.325}Pd_{0.675}$	0.509
$Ag_{0.1875}Pd_{0.8125}$	0.509
Pd	0.424
Cu	–1.110
Ag	–1.416

2. Wire Treatment

In order to increase the surface area of CNM wires, we apply to them several hundred electric pulses, typically with a duration of 50 ms (recently up to 70 ms), with a rather large peak power (15–20 kVA/g). In this way, we reach a surface temperature of 700–1000°C in air, that point at which constantan starts to oxidize showing the formation of a spongy sub-micrometric texture. We believe that such behavior is due to the rapid rise in temperature followed by rapid cooling (quenching), which leads to some phase separation with the formation of separated islands of nickel-rich or copper-rich phases, a supposition *confirmed by EDX analysis*.

It is likely that a skin effect (pulse rise-time 1 μs) *concentrates* most of the current, and therefore the power, at the surface of the wire, with predictable effects on the increase of temperature and its gradient.

When we first looked at the treated materials with SEM–EDX after pulsed oxidation, we were surprised to note that the wire surface displayed the presence of mixed oxides of copper and nickel (Cu_xO_y, Ni_wO_z) arranged in a multilayer and sub-micrometric structure.

We would like to highlight that the material prepared per this procedure shows a reduced tendency to self-sintering. Consequently, the inert material that is often added to reduce sintering problems (e.g. ZrO_2 chosen by Yoshiaki Arata since 2002 for nano-Pd) is now replaced with structures that can possibly absorb some amounts of hydrogen and take part in the exothermic reactions. Finally, the pristine CNM is the substrate where the sub-micrometric materials of various composition are supported.

For the sake of clarity, we point out that we do not have precise control of nanoparticle dimensionality as in Arata's procedure with the use of melt-spinning and quenching processes, by which he obtains 2–15 nm nanoparticles with Pd_35% over ZrO_2_65%.

We occasionally observed that the oxidized and partially reduced structures on constantan wires may detach and fall off, affecting negatively the outcome of the experiment. Such adverse effects increase after H/D absorption and several thermal cycles (e.g. 20→400 – 700→20°C).

In our experimental set-up, constantan has been used in the form of long ($l = 1$ m) and thin wires ($\Phi = 200\,\mu$m; weight per unit length 0.28 g/m), whose surface is very smooth before applying our procedure.

3. Importance of Non-equilibrium Conditions

In almost all of our experiments, we realized that having non-equilibrium conditions is a key factor to induce, and possibly increase, any anomaly in the H/Metal systems. Over many years, we noticed that all kinds of *local non-equilibrium conditions* (thermal, electric, H/D concentration, radiation, etc.) play a role in triggering excess power release. Based on this conclusion, in 2015 we introduced a new type of wire geometry, aiming to increase the local thermal gradients.

We already took advantage of fiberglass sheaths, which represent a step-discontinuity to heat transfer from the wire to the local gaseous environment. Such effect further increases if the wire has a current flowing in it, as usual in our experiment; the result is also a large voltage drop along the wire, originating Non-Faradaic Electrochemical Modification of Catalytic Activity (NEMCA) Effect [8] (later independently rediscovered by G. Preparata and E. Del Giudice) and increasing the catalytic activity of constantan wire. After various attempts, we realized that the simplest approach to strengthen the thermal gradients was to introduce several knots in the CNM thin wire.

The steps for the reactor assembly are detailed in Fig. 1. The most critical part is the tightening of the knots, during which the fragile wire can be subjected to damage because of the mechanical stress. Once current I is applied, the knot becomes a current loop that gives rise to a magnetic field whose lines go through the hole and whose strength along the z-axis – directed along the centerline of the loop – is given by:

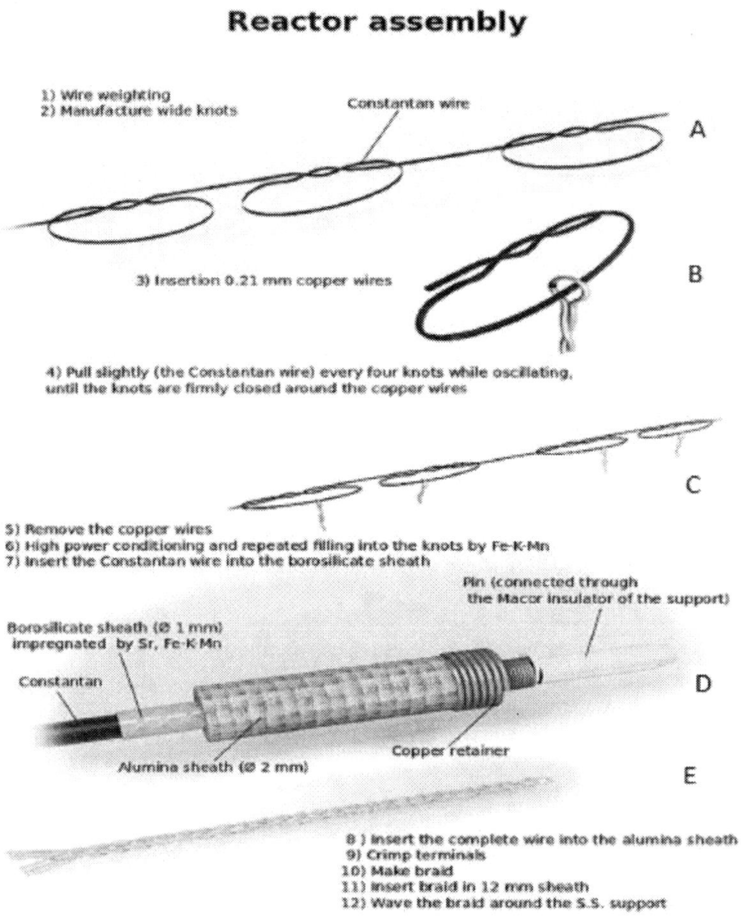

Figure 1. Reactor assembly: (A) Preparation of the knots in the constantan wire. (B) Insertion of the copper wires in order not to completely tighten the knot. (C) Critical point of tightening the node due to possible stress of the constantan wire. (D) After conditioning, insertion of the wires inside the several glass sheaths. (E) Braid of the three wires to be waved around the SS support.

$$B_z = \frac{\mu_0}{4\pi}\frac{2\pi R^2 I}{(z^2+R^2)^{3/2}}, \tag{1}$$

where R is the radius of the loop and z the distance along the z-axis from the loop. Finally, Fig. 2 presents an overview of the reactor once assembled. Note that the core is at the end inserted in a thick-wall borosilicate glass tube. Temperatures are measured at the external wall of the glass tube (T_c) and at the SS central support (T_{ss}) along which the wires are braided inside the reactor.

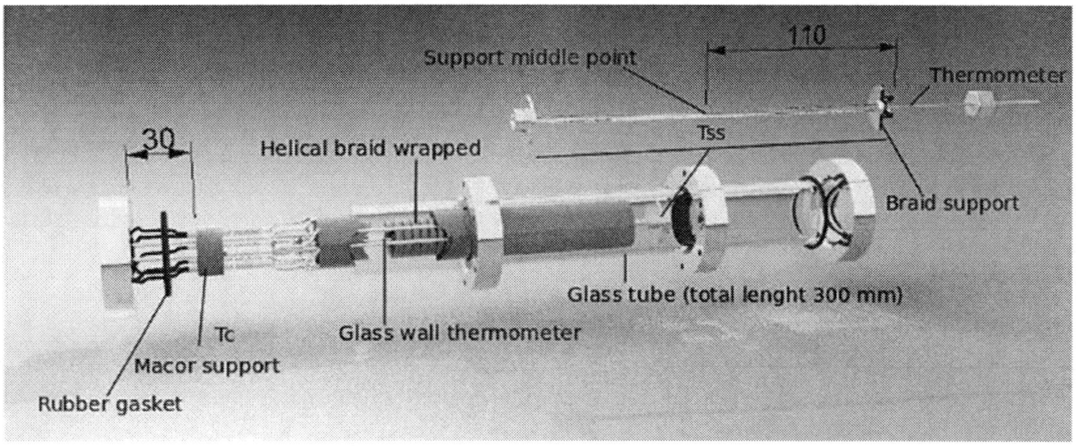

Figure 2. Schematic of the reactor once assembled.

The procedure worked very well and the first results were presented at ICCF20 (Sendai-J, October 2016). We compared two 200-μm constantan wires having 41 and 71 knots in terms of their capability of generating AHE and noticed that the excess power is positively correlated with the number of knots. At the beginning, the knots had an

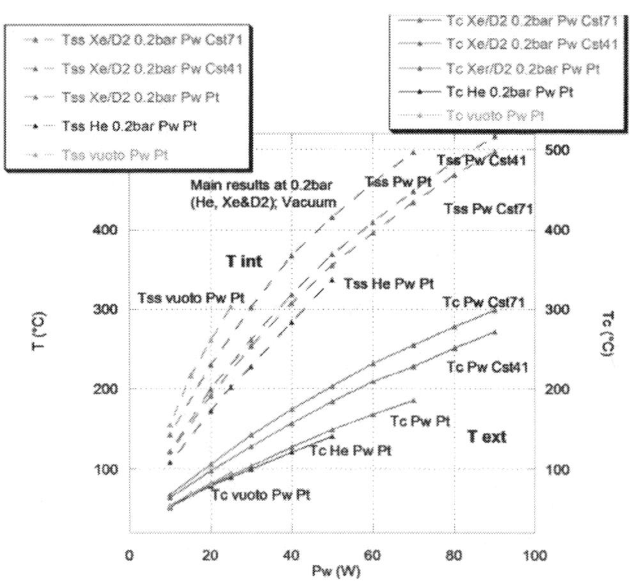

Figure 3. Temperature versus Input Power applied to the reactor in vacuum (P_w at Pt), He (P_w at Pt), Xe+D_2 mixture (P_w at Pt, Cst41, Cst71) atmosphere. A large increase of external temperature T_c is observed with respect to Pt case for P_w applied to Cst41 and Cst71.

internal diameter in the range of 0.2–0.6 mm; later, it was decreased to 0.15–0.20 mm.

In our experiment, we also had mild excitation thanks to gamma ray arising from commercial Thoriated Tungsten rods (used for TIG welding) positioned outside the glass reactor and inside an airtight small SS tube.

A summary of the results achieved with the above-described set-up is shown in Fig. 3. After calibration in vacuum and in helium atmosphere, the experiments were carried out with a D_2–Xe (50% ratio) gas mixture at 0.2 bar. The choice of xenon is motivated by its low thermal conductivity, in order to increase as much as possible the temperature in the reactor at constant input power. We display the behavior of temperature T_{ss} and T_c as a function of input power, for power applied respectively to Pt, 41-knots Cst and 71-knots Cst wire. One can observe the beneficial effect of non-equilibrium conditions, growing T_c with the number of knots in the wire.

4. High Peak Power Pulses

Figure 4 shows a sequence of pictures taken with a time frame of about 20 ms when high peak power (HPP) pulses were applied to a constantan wire of 200 μm diameter with knots whose hole has a diameter smaller than 200 μm and at a distance of ~2 cm from each other.

We can observe that the area of the knots is brighter in comparison to the normal part of the wire, meaning that in those sites, higher temperatures are reached, and local thermal non-equilibrium achieved. Such conditions take place also in DC operation. The HPP pulser is based on the well-known capacitive discharge method. Among its performance characteristics are the quite large value of (multiple) capacitor bank used (6600 μF, parallel of ceramics, polyesters, pulse-operated electrolytic), the relatively large repetition rate (0.1 Hz), the fast rise (<1 μs) and fall time. The electronic switch is based on four Power MOS arranged in parallel, with up to 100 A of current capability and 450 V of voltage limit. The HPP pulser has to provide the wire enough energy to account for:

(A) The thermal inertia of the wire, which opposes to temperature changes. The heat to be transferred to the wire is given by

$$Q = mc_p\Delta T,$$

where m is the mass (g), c_p is the specific heat (J g^{-1}°C^{-1}), ΔT is the temperature variation (°C).

(B) The power lost by irradiation, given by Stefan–Boltzmann law

$$P = \sigma T^4 S,$$

where $\sigma = 5.67\times 10^{-8}$ W m^{-2} K^{-4}, S the surface of the wire.

(C) The power transferred by convection to the air, given by

$$P = hS\Delta T,$$

where the heat transfer coefficient $h \approx 10$ W m^{-2} K^{-1}.

We can therefore estimate the energy that a pulse must have for the wire to increase its temperature to 1000°C. Considering our situation, in the case of the wire of diameter of 200 μm, we have:

$m = 1$ g, $\rho_{CMN} = 8.90$ g/cm^3, $c_p = 0.41$ J g^{-1} K^{-1}, $S = 2\,\pi$ r l $= 6.28$ cm^2, $T = 1273$ K, $\Delta T = 1000$ K.

Hence, the three contributions become:

(A) $E = Q = 410$ J,
(B) $P = 93.6$ W $\rightarrow E = P$ 0.07 s = 6.55 J,
(C) P = 6.28 W $\rightarrow E = P$ 0.07 s = 0.44 J.

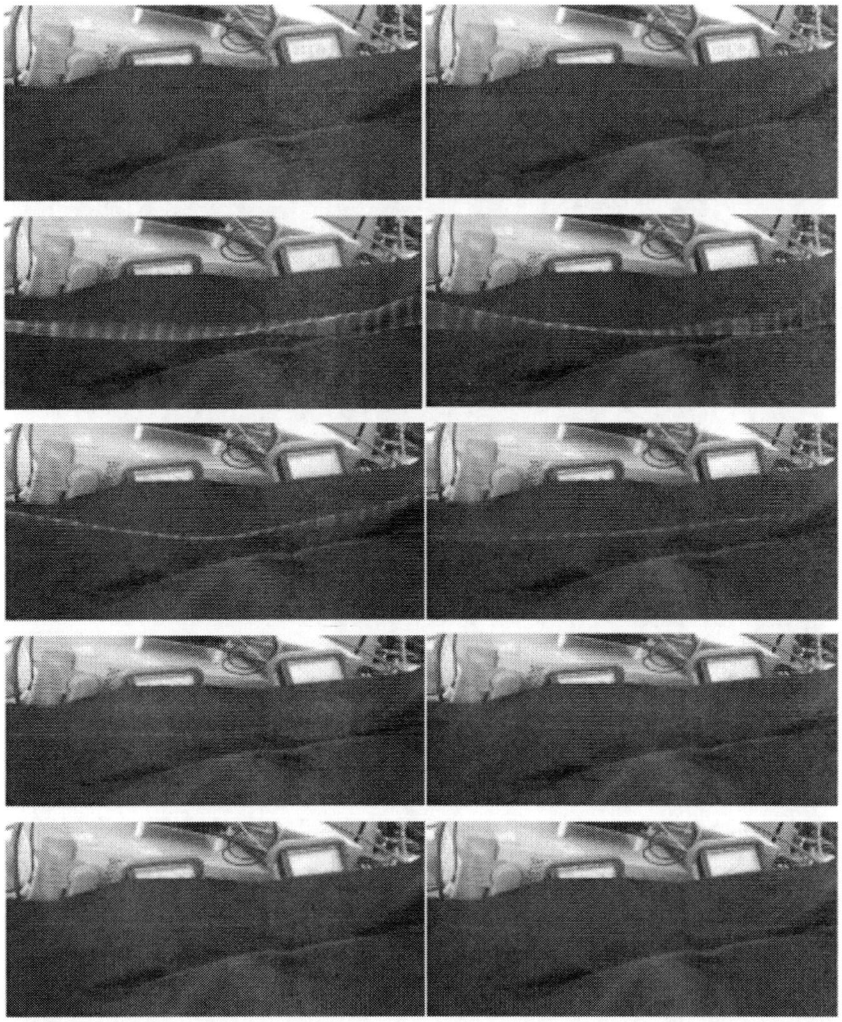

Figure 4. Photographic sequence of a 200-μm-diameter constantan wire subjected to High Peak Power Pulses: the time frame is about 20 ms and pulse duration is 70 ms. The brighter dots along the wire correspond to the knot sites, which reach higher temperatures than the other parts of the wire. In this way, local thermal non-equilibrium conditions are achieved.

The sum is $E = 417$ J, where the leading term is the one associated to the thermal capacity of the wire. The estimated mean pulse energy density is up to 800 J/g, an amount more than enough for our purposes.

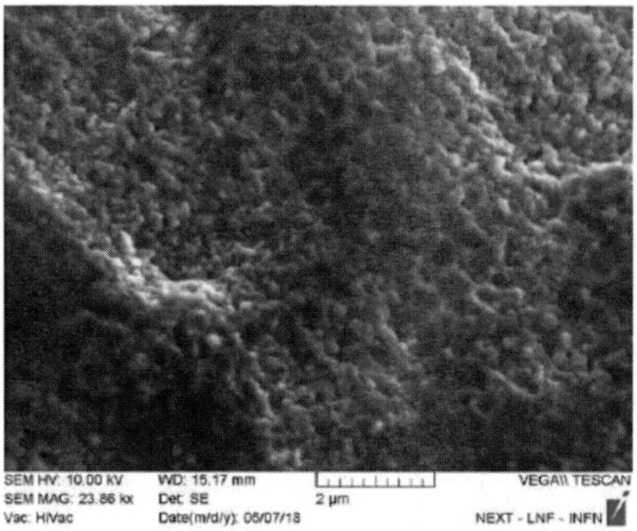

(a)

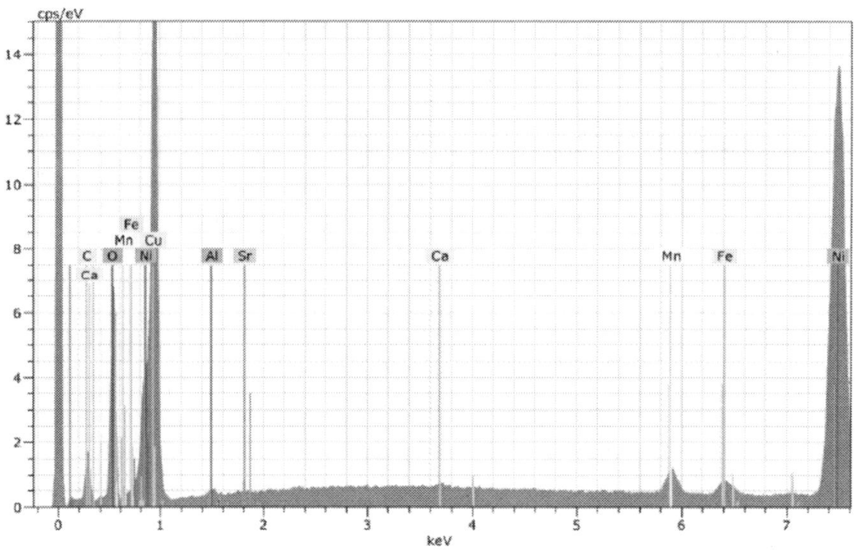

(b)

Figure 5. (a) SEM observation of the final part of the wire ($\Phi = 200\,\mu$m) taken as reference, just heated in air at 600°C for 300 s. (b) Corresponding EDX spectrum: counts per energy channel as a function of the incident X-ray energy. The highest count peaks correspond to Copper and Nickel.

Table 2. EDX spectrum analysis of the final part of the reference wire. Copper is present at 43.5% in weight, nickel at 31.6%.

Element	Z	Series	Count	C unnorm. (wt.%)	C norm. (wt.%)	C atom. (at.%)	C err. (1σ) (wt.%)
Cu	**29**	**K-series**	**167934**	**42.65**	**43.53**	**23.71**	**1.07**
Ni	**28**	**K-series**	**149 919**	**30.99**	**31.64**	**18.65**	**0.79**
O	8	K-series	22943	12.95	13.22	28.59	1.66
C	6	K-series	5231	9.39	9.58	27.61	1.48
Mn	25	K-series	7629	0.97	0.99	0.63	0.05
Fe	26	K-series	4387	0.56	0.57	0.35	0.04
Al	13	K-series	723	0.24	0.24	0.31	0.04
Ca	20	K-series	899	0.12	0.12	0.10	0.03
Sr	38	K-series	71	0.10	0.11	0.04	0.04

5. SEM Observations and EDX Analysis

We performed an analysis of the elemental content of the wire before and after the application of HPP procedure. For a different part of the wire we report SEM observation and EDX analysis. The maximum energy of electrons used in the spectroscopy is 30 keV, corresponding to a resolution of about 2 μm in depth. We have taken an untreated constantan wire, just heated at 600°C for 300 s in air, as reference for the comparison with a treated one. Figure 5 shows the SEM observation of the final part of such wire, while Table 2 reports the corresponding EDX analysis: the untreated wire is 43.5% Cu (by weight) and 31.6% of Ni.

After the wire is subjected to hundreds of HPP pulses, the external surface of the knot is treated with a water solution of mixed nitrates of Sr–Fe–K–Mn.

This treatment is executed by hand and repeated several times. Nitrates are subsequently decomposed to oxides by heating (up to 700–1000°C). Figure 6 and Table 3 show the composition on the surface of the knots. Here, Cu content is larger (58.40 wt.%) for the partial decomposition of CNM alloy – Cu has a lower melting temperature with respect to Ni – and the diffusion of the elemental Cu toward the surface. Note the increased amount of Fe (4.98 wt.%) and Sr (3.58 wt.%). Finally, the area between two knots, after the treatment, is shown in Fig. 7 and Table 4.

We can conclude that the relative composition of main materials (Cu, Ni) is *strongly* dependent on the position along the wire and varies largely according to the performed treatments and the geometrical position.

Table 3. EDX spectrum analysis of the area at the top of a knot after applying HPP procedure. Cu content has raised to 58.4 wt.%, due to its diffusion toward wire surface, while Ni is at 5.50%.

Element	Z	Series	Count	C unnorm. (wt.%)	C norm. (wt.%)	C atom. (at.%)	C err. (1σ) (wt.%)
Cu	**29**	**K-series**	**126 198**	**58.16**	**58.40**	**30.91**	**1.46**
O	8	K-series	11977	13.56	13.61	28.61	1.88
C	6	K-series	3187	10.58	10.62	29.74	1.83
Ni	**28**	**K-series**	**15 276**	**5.47**	**5.50**	**3.15**	**0.17**
Fe	26	K-series	19339	4.96	4.98	3.00	0.16
Sr	38	K-series	1359	3.57	3.58	1.38	0.16
Ca	20	K-series	4517	1.07	1.08	0.90	0.06
Mn	25	K-series	3598	0.88	0.89	0.54	0.05
Mg	12	K-series	552	0.70	0.70	0.97	0.08
Al	13	K-series	820	0.64	0.64	0.80	0.07

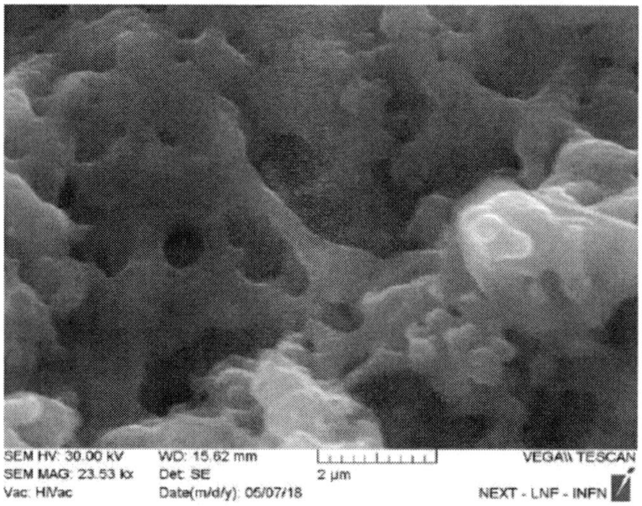

(a)

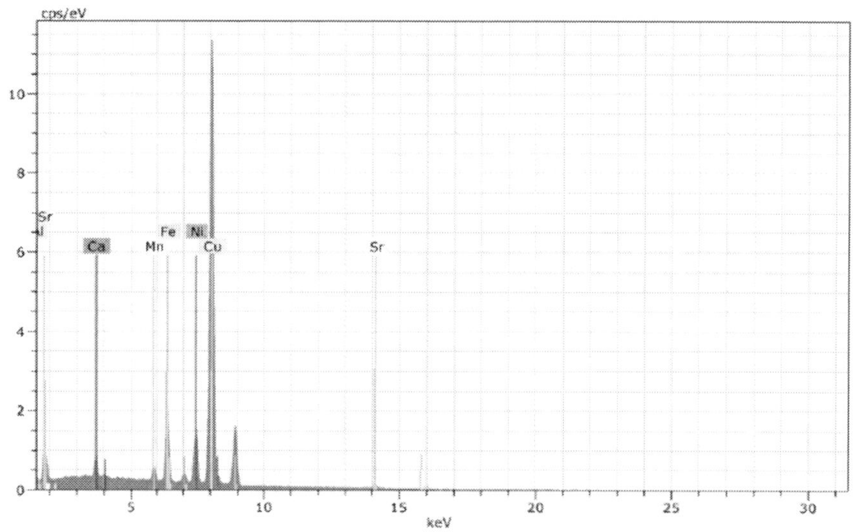

(b)

Figure 6. (a) SEM observation of a typical area at the top of the knots (wire with Φ =200 μm). (b) Corresponding EDX spectrum. Surface element composition changes remarkably after the application of HPP procedure, particularly as regards Nickel.

6. The Introduction of Fiberglass Sheaths

Since 2013, we have systematically adopted the procedure of putting the wires, immediately after their preparation, within a fiberglass sheath to prevent or minimize the separation of active surface layers from the wire core.

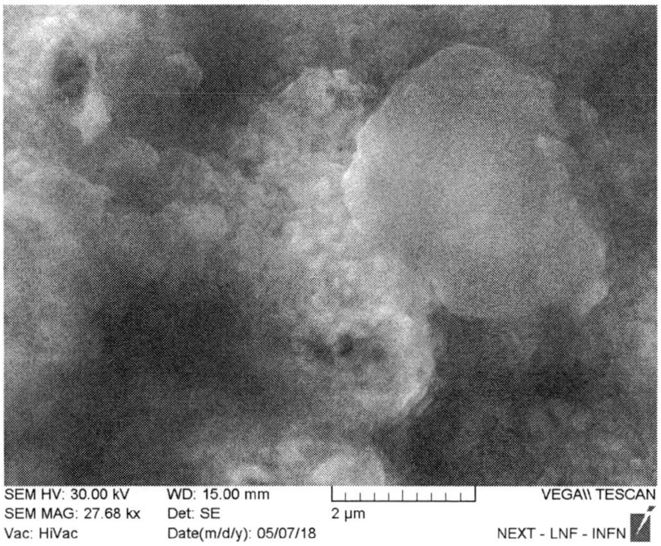

(a)

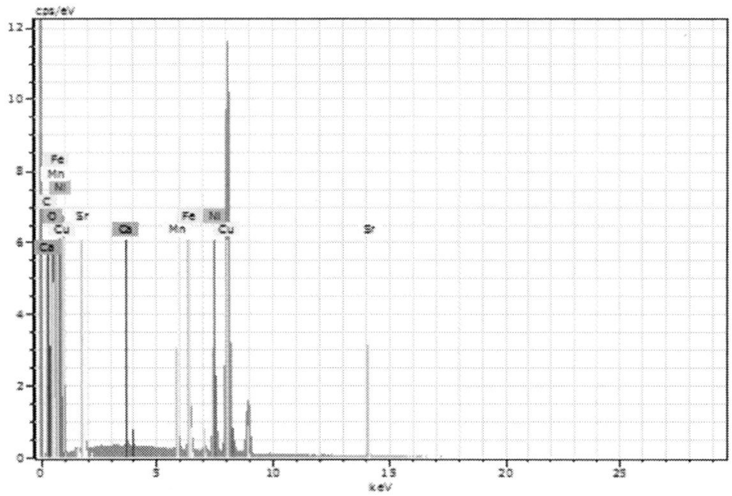

(b)

Figure 7. (a) SEM observation of the area between two adjacent knots (wire with $\Phi = 200\ \mu$m). (b) Corresponding EDX spectrum.

We have reason to believe that this glass sheath may contribute to the generation of AHE under certain circumstances. We also observed this phenomenon in a previous set of experiments with palladium wires in 2008, where we employed them just for electrical insulation purposes. At that time, we tested several types of fiberglass sheaths and realized that Alumina (Al_2O_3)-based materials, although able to withstand quite large temperatures (T>1200°C), never produced significant excess heat effects. Only more common glass types such as E Glass, i.e. boro-silico-

Table 4. EDX spectrum analysis of the area between two adjacent knots. Cu is present at 62.5% in weight, Ni at 9.77%.

Element	Z	Series	Count	C unnorm. (wt.%)	C norm. (wt.%)	C atom. (at.%)	C err. (1σ) (wt.%)
Cu	**29**	**K-series**	**130 832**	**62.34**	**62.50**	**36.21**	**1.56**
O	8	K-series	11531	12.27	12.30	28.31	1.71
Ni	**28**	**K-series**	**26 135**	**9.75**	**9.77**	**6.13**	**0.27**
C	6	K-series	2364	8.17	8.19	25.11	1.51
Fe	26	K-series	13484	3.44	3.44	2.27	0.12
Sr	38	K-series	938	2.45	2.46	1.03	0.13
Mn	20	K-series	4477	1.11	1.11	0.74	0.06
Ca	25	K-series	928	0.22	0.22	0.20	

alumino-calcic glass, showed synergistic effects concerning AHE phenomena, although presenting limits on maximum continuous temperature of 550°C. In particular, we utilized the glass sheaths made by SIGI-Favier (IT-F), a derivative of borosilicate glass with further addition of TiO_2, Fe_2O_3 and Fe. We first thought that the glass itself could play an important role but we did not have any justification for our experimental observations yet.

Having in mind the idea of a "vessel" or containment made of micrometric fibers (diameter 4–6 μm, almost porous), we started modifying the sheaths by impregnation. This was done by dipping the sheaths into a solution of $Sr(NO_3)_2$ in H_2O or D_2O that was later decomposed to SrO at high temperature. Finally, we inserted the CNM wires into the impregnated sheaths.

We note that SrO has the feature of a low work function for electron emission, together with materials such as CaO and Y_2O_3. The use of these materials is in accordance with Yasuhiro Iwamura's procedures (1999) in the field of transmutations (Sr→Mo, Cs→Pr) induced by deuterium flux on multilayered (CaO, Pd) structures [9].

The work function W of a material is the main factor for electron emission at high temperatures. Such phenomenon is known as thermoelectric effect and is described by Richardson's law:

$$J = A_g T^2 \exp\left(-\frac{W}{k_B T}\right), \tag{2}$$

where J is the emission current density (A/m^2), $A_g = \lambda_R A_0$ is a constant factor (λ_R is a correction factor depending on the material, $A_0 = (q_e m_e k_B^2)/(2\pi^2\hbar^3)$ =120.173 A cm^{-2} K^{-2}, with electron charge q_e =1.60× 10^{-19} C and electron mass m_e =5.11× 10^5 eV), T is temperature (in K), k_B = 8.617× 10^{-5} eV/K is the Boltzmann constant. At T constant, the reduction of only 1 eV of the work function implies an increase of the current density of a factor $\exp(1/(k_B T))$. As an example, at T = 1000 K such factor is around 1.1× 10^5. In other words, a reduction from 2 to 1 eV increases the current density of a factor of 10^5! In Fig 8 Richardson's law is plotted for different values of the work function and one can clearly see how strongly dependent on it the current density is. Recent data show a value as low as 1.2 eV for SrO at low dimensionality with respect to 2.3 eV in bulk shape [10]. We produce thin layers of SrO at the surface of the glassy sheaths and mainly at CNM surface.

The atomic hydrogen produced by the dissociation of molecular hydrogen through the catalytic action of our sub-micrometric structured CNM wires is largely adsorbed onto the surface of the micrometric glass fibers. It is important to mention that the properties of some specific glasses with respect to their interaction with atomic hydrogen were discovered by Irving Langmuir around or before 1927. Langmuir selected the optimal type of glass for the construction of incandescent lamps, i.e. able to withstand repeated (>1000 times) fast cycling of temperatures (20→200→20°C) without self-damaging. We believe that the impregnation steps with SrO, Fe_xO_y and K–Mn, as a whole, further enhance the intrinsic tendency of borosilicate glass fibers to absorb or interact with atomic hydrogen. Moreover,

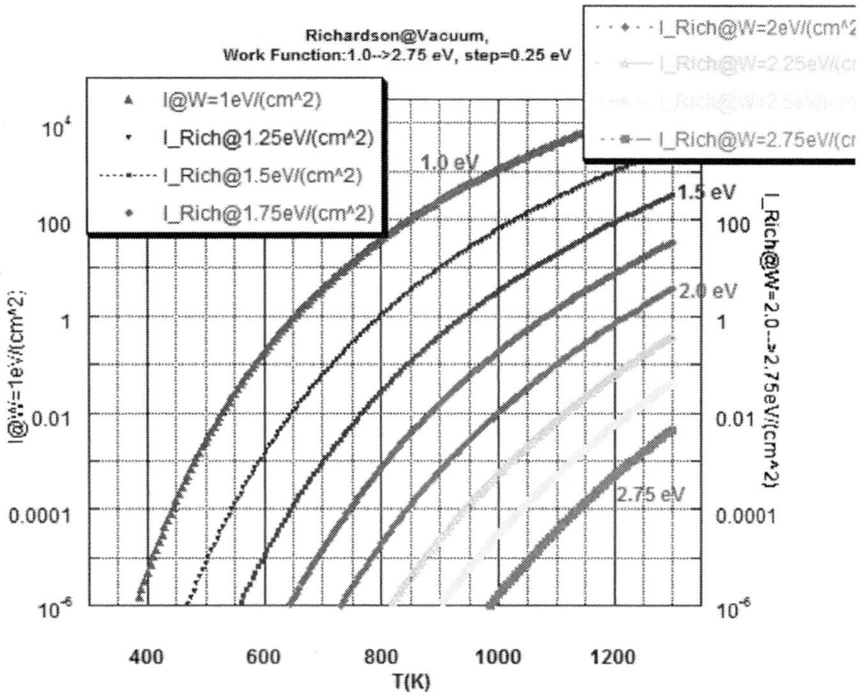

Figure 8. Plot of Richardson's law: current density versus temperature for values of work function ranging from 1 to 2.75 eV with a step of 0.25 eV. The scale is semilogarithmic.

the "chemically aggressive" alkaline elements used for impregnation and drying by high temperatures increase the surface area of glassy sheaths. The atoms being much closer to each other compared to atoms in free motion in a gas, the formation of a thin film of atomic hydrogen adhering to glass surface increases the probability of hydrogen recombination in its molecular form, an exothermic reaction with the release of 4.52 eV. We think that glass adsorption is therefore a cofactor in the generation of excess heat together with the main LENR process and we cannot exclude the possibility that the H film act as a reservoir for the diffusion process of atomic hydrogen into the metallic wires [11].

We presented the main results of our experiments at "2014 MIT Colloquium on Cold Fusion effect" (21–23 March 2014;IT Cambridge, USA). The reactor consisted of two constantan wires and one Pt wire, inserted inside glass sheaths, braided around the central stainless steel (SS) support and, finally, this latter covered with a fiberglass sleeve. The whole reactor is inserted and thermally sealed in a thick borosilicate glass tube. In order to definitively clarify the role of the glass in heat generation, we repeated the experiment with two additional sleeves on the steel support and compared the results. Temperature data in Fig. 9 clearly show the positive role of glass in heat generation and present behavior growing at low pressure, in accordance with Langmuir's observations about hydrogen recombination. Hydrogen recombination reaction is highly exotermic: in principle, the local temperature can be as high as 36 000 K at very low pressures.

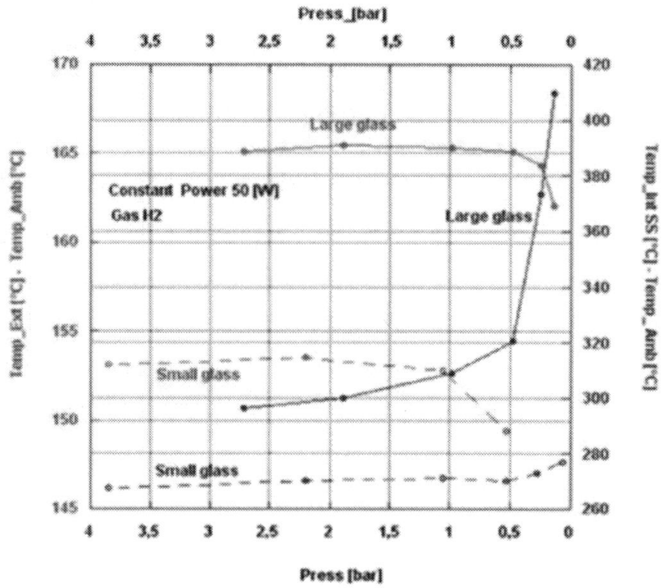

Figure 9. Data comparison for tests conducted with a small and a large amount of glass enveloping the SS support in the reactor, in H_2 atmosphere and at constant input power of 50 W. Temperature difference with ambient at the external glass tube surface (*red*) and at the central SS tube (*blue*) is plotted against pressure in the reactor.

7. The Introduction of Iron and Potassium

After consuming the first batch of constantan wires (excess heat up to 5–20% of the input power at 50 W input), the use of new batches resulted in a lower AHE, although using the same reactor operated in the isoperibolic regime. We later realized that the first batch of CNM that we used was prepared before 1970. We measured that the Fe content, either as an impurity or intentionally added, was as large as 0.5–1% on average, while in certain spots it could reach 3–5% concentration, especially at the surface. In addition to the lower purity of starting materials (i.e. Cu, Ni, and Mn), we have considered possible contamination of the surface during the swaging process and/or inappropriate storage in a rusted iron cupboard. Moreover, if we consider that the aforementioned J-type thermocouples also contain iron, the case for its synergistic effect on AHE is strongly reinforced.

It is worth mentioning also that iron has the unusual property of absorbing H_2 at high temperatures ($T > 600°C$) and releasing it at lower temperatures. Aiming to further increase AHE, we eventually decided to add Fe together with K, as mixed oxides, on the surface of the wires and sheaths as a dissociation promoter of $H_2 \rightarrow 2H$ reaction. We were inspired by Fischer–Tropsch and Bosch–Haber type catalysts for the synthesis of gasoline and ammonia. The role of K was brought to our attention by C. Lorenzetti (co-author) in 2014. The addition of iron was made before treating the wires with the HPP procedure. Later, in May 2016, we added also Mn in order to stabilize the observed effects, thanks to Mn ability of reducing the evaporation of K at high temperatures [12,13]. Just as a similarity, the effect of Mn resembles that of Iodine to extend the useful life span of the well-known incandescent (halogen tungsten) lamps.

Table 5. Effect of argon, xenon and heavy water (D_2O) addition on temperatures measured in the reactor for Cst71 wire at 50 W input power.

Gas type	D_2	Xe / D_2	Xe / D_2	Ar / D_2	He	H_2	D_2O (2 cm^3) Air (10 cm^3)	D_2O (2 cm^3) Air (10 cm^3) Xe (45 cm^3)
P at reactor off (bar)	1	1 / 0.8	1 / 0.8	1 / 1	1	1		
P at reactor on (bar)	1.41	2.63	2.64	2.71	1.346	1.32	0.20	0.51
P_w (W)	50.18	50.13	50.38	49.91	49.75	50.25	50.11	50.30
T_c (°C)	197.2	208.3	208.6	205.5	191.03	192.8	203.73	212.42
T_{ss} (°C)	258	342	342.9	308.9	253.5	243.7	367.8	400.08
T_{Pt} (°C)	257	318	319.4	292.26	252.1	244.6	344.69	376.14
T_{amb} (°C)	20.7	20.4	19.98	18.35	17.04	17.8	20.53	20.57

8. Introduction of Noble Gases with Low Thermal Conductivity

Since 2008, we have observed that the addition of argon and other low-conductivity noble gases to H_2 or D_2 increases the likelihood of AHE occurrence. This was noticed even in experiments conducted at low pressure, e.g. at 0.1 bar.

Our motivation for inert gas addition was to reduce the applied input electric power thanks to lower thermal conductivity losses, allowing, for instance, the enhancement of the detection of very small AHE. Such effect was further magnified when Ar was replaced with Xe, the gas with the lowest value of thermal conductivity among the non-radioactive, $\kappa = 5.5$ mWm^{-1}K^{-1} at 300 K and 1 bar.

For xenon in particular, we were able to measure an AHE magnitude much larger than expected.

At the end of November 2016, while still trying to interpret the effect of xenon on AHE, we found with great surprise a 1993 German patent (DE4300016) claiming the addition of xenon to deuterium to enhance the reaction $2D^+ \rightarrow He^4$ [14]. We also noticed similarities in the experimental apparatus where this reaction is supposed to occur, a thin glass tube containing ionized Xe and D.

In Table 5 we compare temperature values of measurements performed at 50 W input power on Cst71 wire (Φ = 200 μm) for various combinations of gas mixtures. It is noticeable the rise in temperature when noble gases are added to reactor atmosphere, a phenomenon more pronounced in the case of Xe, showing that such gases play a role in the generation of AHE, may be acting as catalyzers.

9. Air Flow Calorimetry

In our experiments, we have always used isoperibolic calorimetry to measure the heat generated by the reactor and exchanged with the ambient air. Thermometers are placed in proper sites of the reactor, such as at the SS central tube and at the external wall of the borosilicate glass tube, and temperatures constantly monitored. In this way, we are able to estimate the heat radiated by the reactor wall (Stefan–Boltzmann law) and the heat lost by convection to ambient (Newton's law of cooling). The sum of these contributions gives the calculated heat produced by the reactor. Isoperibolic calorimetry has revealed to fit very adequately the requirements for LENR experiments, as it allows working in non-equilibrium conditions, fundamental to get and maximize AHE. We measured gains near a factor 2 in the experiments at the highest temperature, although with limited stability over time.

Over the years, the scientific debate produced evidence that local measures taken at specific spots may lead to an overestimate of the actual total amount of heat, and therefore of energy, outgoing from the reactor. Mass-flow calorimetry is considered a more reliable type of measurement, as the heat produced is exchanged with a substance flowing in a controlled environment where temperature tends to be homogeneous. When applied to our experiments,

Table 6. Values of density, specific heat and maximum water content of air (temperature range 0–100°C, 1 atm).

T (°C)	Density (kg/m^3)	C_p (kJ/kg K)	Max H_2O content (g H_20/m^3 air)
0	1.292	1.005	4.84
5	1.268	1.005	6.78
10	1.246	1.005	9.37
15	1.225	1.006	12.77
20	1.204	1.006	17.19
25	1.184	1.006	22.88
30	1.164	1.006	30.1
35	1.146	1.006	39.2
40	1.127	1.007	50.60
45	1.11	1.007	64.63
50	1.093	1.007	81.85
55	1.076	1.007	102.82
60	1.060	1.008	128.1
70	1.029	1.009	194.6
80	1.000	1.009	194.6
90	0.973	1.010	414.7
100	0.946	1.011	584.5

however, the reactor does not operate under the optimal conditions for obtaining AHE and performances are expected to be lower.

In the light of such considerations, we have decided to replicate the experiments in air-flow calorimetry, aiming to prove in a more rigorous way the excess power generated by the active reactor.

The air flow calorimeter operates according to the simple formula:

$$Q = mc_p\Delta T, \tag{3}$$

where Q is the heat exchanged, m air mass (kg), c_p the specific heat (kJ/kg K), ΔT is the temperature difference between outcoming and incoming air (K). Difficulties arise because, being a gas, air density decreases a great deal with increasing temperature, while c_p increases only slightly. Moreover, air water content increases a great deal with temperature. The energy needed inside the measuring chamber to increase water temperature (4.184 J/g) or even evaporate it (i.e., phase transition) can be large. Such water is a negative term in the system. Table 6 shows air density, specific heat and maximum water content for temperature ranging from 0 to 100°C, used in our calculations. We made a consistent effort during the measurements to keep room temperature constant and room humidity in the chosen range. We controlled such quantities by means of an air-conditioner working also in dehumidification mode and verified with an hygrometer that relative humidity was maintened between 45 and 55%.

The calorimeter consists of a large insulating Styrofoam box whose internal walls are covered with a layer of thick aluminum foil in order to make the temperature uniform. During the experiments, the external wall of the glass reactor is covered with a double layer of black and thick aluminum foil to further homogenize the internal temperature. For calibrations, we have used a halogen tungsten lamp, with maximal power 230 W, but underpowered during operations in order to reduce temperature. We have put the lamp inside a borosilicate glass tube with the same composition and dimension as that of the reactor (L = 29 cm, Φ = 34–40 mm), as shown in Fig. 10. Calibrations have been performed powering the lamp in a whole cycle 0 $\rightarrow$120 W $\rightarrow$ 0, with steps of 20 W.

Figure 11 summarizes the results of the experiments conducted in air-flow calorimetry. The variation of temperature ΔT between outcoming and incoming air, divided by input power (°C/W), is plotted against the given input

Figure 10. Interior of the air-flux calorimeter with the reactor (*above*) and the Halogen Tungsten lamp used for calibration (*below*, max power 230 W). The lamp is inserted inside a borosilicate glass tube of the same dimensions of that used for the reactor. The internal walls of the insulating box are covered with an aluminum foil in order to make temperature as uniform as possible inside the calorimeter.

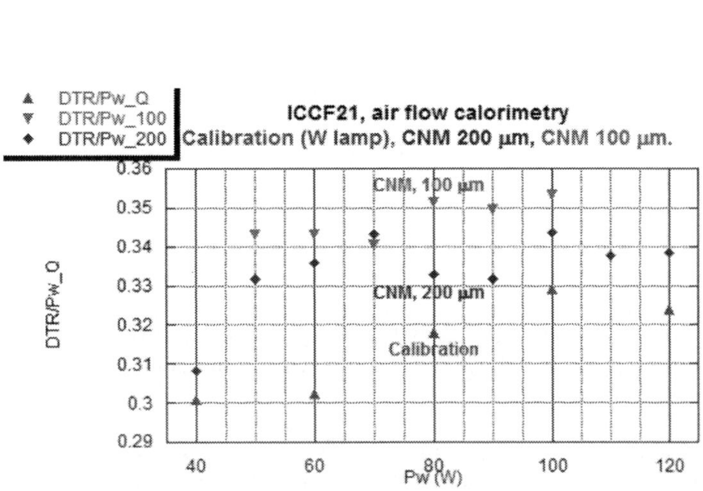

Figure 11. Comparison of the results obtained with experiments conducted with air-flow calorimetry. The change in temperature between outgoing and incoming air divided by the input power ($\Delta T/P$), is plotted against the input power (*P*). The first series of measurements (*in green*) are performed giving power to the calibration heater, a halogen tungsten lamp. Red points correspond to the measurements taken with the reactor operating with 100-μm diameter CNM wire, and they are interrupted at 100 W because of the breaking of the wire. In blue, the points referring to the reactor with 200-μm diameter CNM wire. The wires are in an atmosphere of deuterium D_2 at 1.2 bar at room temperature. Data above the calibration curve indicate the production of AHE.

Figure 12. Picture of the borosilicate glass reactor after last experiment with air-flow calorimetry: it is clearly visible, at center, the imprint of the sheaths caused by partial melting of the internal wall due to local overheating.

power (W). The green points define the calibration curve obtained with the tungsten lamp. The red points correspond to measurements using a CNM wire of 0.1 mm of diameter, which broke after only 840 s at 100 W of input power. Finally, the blue points show the measurement with a CNM wire of 0.2 mm diameter. The reactor operated in a D_2 atmosphere, with pressure of 1.2 bar at room temperature. All the data shown are raw; i.e. without corrections for air density inside the thermal-insulating calorimeter box. Only external room temperature and humidity are kept almost constant. Moreover, with increasing temperature the heat losses increase by external air convection from the large insulating box to ambient. We made some cross checks of the results from time to time, by calculations using the value of Table 6 and by measuring the temperature of the external wall of the insulating box through a high sensitivity (0.1°C) IR video-camera (Nikon). We can clearly observe that the points corresponding to the active reactor are above the threshold given by the calibration curve, indicating therefore the production of excess heat. We note that the data are very conservative, as we were very careful not to overestimate any contribution.

The best results obtained are the following:

(a) With a CNM wire with diameter 100 μm, D_2 atmosphere at 1 bar, internal reactor temperature 500°C, input power 90 W, the AHE was over 12±2 W, i.e. over 150 W/g, but after 1 day the wire broke. In details, when supplied with 100 W, the wire broke only after 840 s; we estimated from the behavior an excess power of about 30 W, but, of course, it is just a guess because of the lack of satisfactory data points.

(b) With a CNM wire with a diameter 200 μm, it was necessary to use a Xe–D_2 mixture (each 0.1 bar) and input power of 120 W in order to obtain an AHE of 6–7 W stably for weeks.

As proof of the high temperatures reached by the wire, after the reactor was switched off and opened, we observed that some overheated segments of the fiberglass sheaths partially melted the internal wall of the borosilicate glass tube and left their imprint on it (Fig. 12). According to the specifics of the Schott Duran glass used, the softening temperature is 825°C.

10. Addendum in Response to Private Questions Received during and after the Presentation: toward a New Reactor Design and Similarities with a Thermionic Diode

In 2014, a second independent constantan wire was introduced in the reactor. When the first treated wire was heated passing a current into it, we observed a weak current flowing in the second wire as well as the rise of a voltage with respect to the first. Moreover, the phenomenon is clearly driven by a temperature difference of several hundred degrees Celsius between the cathode (heated via direct current) and the anode. (It is important to note that the wires comprise a surface with a low work function due to the presence of oxides of alkaline metals, typically between 1 and 2.5 eV, depending on surface morphology.) The spontaneous current and voltage were also observed independently by Mathieu Valat and Bob Greenyer using a measurement set-up arranged by them completely independently of our work

This voltage turned out to be the consequence of the thermionic emission from the first wire that behaved as a cathode while the second wire as an anode. Given the relatively high pressure at which the phenomenon was observed, the exchange of charge between the first wire (cathode) and the second wire (anode) is supposed to be mediated by ions. After much work to stabilize the phenomenon, we recognized that the current density produced in the non-connected wire, according to preliminary data analyses, is quite well described by Child–Langmuir law

$$J = K\,\frac{V^{3/2}}{d^2}\,, \tag{4}$$

where the current density J is proportional to the three-halves power of the differential of potential between cathode and anode and inversely proportional to the square of their distance. Consequently, the amount of spontaneous current was dependent on the sign of the current, i.e. the potential, applied to the cathode wire.

Interestingly, the voltage arising between the two wires did show a strong correlation with AHE. In a set of still on-going experiments, an external voltage (bias) was applied between the cathode (the treated wire heated by direct current) and the anode through an external power supply, leading to an AHE increase sufficient to melt the cathode. These observations draw attention to the stunning similarities between the reactor and a thermionic diode. Furthermore, we can elaborate on the possibility of converting the AHE generated on the cathode into electricity using its thermionic emission. In other words, the on-going work is exploring a highpressure thermionic converter where the cathode is heated by an AHE occurring on it, or in close proximity [15]. In this reactor configuration, AHE will be driven by the voltage between the two electrodes and by their temperature. While part of the electrical output of this converter will be used to control the reactor (i.e. maintaining a certain temperature of the cathode), the remaining part could be used for an external load.

11. Conclusion

With this last set of experiments, we have observed that the whole of procedures adopted, such as:

- The use of constantan – our main original contribution in LENR research – as a catalytic material for molecular hydrogen dissociation, able to reduce its resistance in correspondence with atomic hydrogen/deuterium absorption [16].
- The pulsed oxidation of constantan wires through the High Peak Power procedure, to increase surface area and form layered nanostructures on wire surface [16].

- The addition of potassium modified iron oxides as a hydrogen dissociation catalyst [11].
- The use of fiberglass sheaths impregnated with solutions of strontium nitrate – an electron emitter – and iron nitrate/potassium permanganate in D_2O [11,17] (decomposed to oxides).
- The use of manganese to stabilize the system over time, reducing potassium evaporation [11,17].
- The introduction of knots with small-diameter holes along the wire, to enhance non-equilibrium conditions, being the knots location of thermal and magnetic gradients [11].

allows the production of AHE – although at levels quite lower than with isoperibolic geometry – even when using flow-calorimetry measurement techniques, which typically *minimize* the non-equilibrium situations. Wire temperature appears to be a key factor and has to be as high as possible, while avoiding the sintering of sub-micrometric materials.

The amount of AHE is inversely proportional to the wire diameter, because, for a given power, a thinner wire reaches higher temperatures. On two occasions we have observed that the 100-μm diameter constantan wire is *not* able to sustain, for a long enough time, an input power close to 100 W in our experimental conditions (length about 130 cm, 115 cm after tying the knots). In this regime, the AHE values grow too fast, perhaps in a positive feedback process.

We also have indications that the voltage drop along the wire (i.e. NEMCA effect [8]) plays an important role in AHE generation. Further systematic work is needed to keep under full control such extremely interesting results and operating regimes.

From the point of view of futuristic technological applications, we note that the maximum excess power density, with 100-μm diameter wire, was larger than 100 W/g, supposing that also the bulk of the wire, not only surface, is operative. On the other hand, if we suppose that the glassy sheaths, wetted with the usual mixture of Sr–Fe–K–Mn (total weight of about 3 g), played a special role for AHE generation, the power density drops largely. Anyway, the industrial cost of glassy sheath is quite low, a few Euros per gram.

Such values are qualitatively similar to our previous experiments (up to 2008), when we used mainly long Pd wires (diameter of 50 and 100 μm) treated with several cycles of deposition of specific materials (Th-nitrate, Sr-nitrate, liquid glass home-made) and inserted in glassy sheaths. Sadly, almost all relevant logbooks, where all the details and know-how were reported, were scrapped by people working at LNF in February 2015.

For the last 10 days before the ICCF21, a new reactor has been operating, with minor changes in the overall geometry. Our aim is to reconfirm the previous data and increase, step by step, with the input power controlled to avoid uncontrolled self-breaking/burning of the wire. The first results at low power are similar to the previous experiment just reported. With regard to the measurement devices, the realization of an air-flow calorimeter with higher accuracy is an on-going activity parallel to the development of the reactor.

Acknowledgments

The experimental work described in this paper was carried out mainly at INFN-LNF; some key trials were conducted at the premises of a LENR Laboratory located in a Metallurgical Company of North Eastern-Italy, which has also helped financially since 2011. Società Italiana di Guaine Isolanti (SIGI) designed and produced innovative/unconventional types of glassy sheaths among a joint collaboration with our Group, the Favier Company (France side of SIGI) and the aforementioned Metallurgical Company. The replication of previously selected experiments was also attempted within Martin Fleischmann Memorial Project by Mathieu Valat and Bob Greeyner with the Live Open Science approach. We are indebted to Prof. Brian Josephson, Nobel Laureate in Physics (who has long been involved in the LENR field) for his "Open Letter" addressed also to the Scientific Coordinators of INFN, where he asked for the continuation of F. Celani's LENR Research in Frascati even after the retirement age limit. We heartily thank Laura Vantini for her promotion of a petition on the website platform Change.org asking for the conservation of our laboratory at INFN-LNF and its availability to F. Celani after his retirement. The petition was addressed to the President of the Italian

Republic to the Prime Minister of the Government and to the Minister of Education, Universities and Research and it was presented to them on 21 October 2017 [18]. In September 2017, F. Celani was thoroughly interviewed by Pandora TV, a Web TV whose responsible Giulietto Chiesa, a former member of European Parliament, is very interested in energetic and environmental problems [19]. An interview with F. Celani by the journalist Alessandro Milan was published on the newspaper "La Verità" in September 2017, in agreement with the deputy editor Massimo De Manzoni [20]. Both people are very involved, in their professional activity, in issues concerning sustainable development and low environmental impact technologies Starting from October 2017, an important political Group in Italy (North League) pushed the President of INFN to allow the continuation of F. Celani's experiments in the Frascati Laboratory, although he had reached the age-limit of retirement. The main people involved were Francesco Malagoli, Filippo Panini, Paolo Varini (from Modena City). All of them have followed LENR studies since 2010, among their key political framework/projects on environmental conservation and pollution reduction.

Antonino Cataldo and Stefano Bellucci (NEXT collaboration) performed SEM and EDX analysis at INFN-LNF. Expenses at ICCF21 were partially supported by IFA organization (main topics are "Water and Energy"-Italy. We are indebted to both Jed Rotwell (USA) and Jean Paul Biberian for their invaluable work to improve the paper about our (several) misprintings and Italian–English neologisms.

References

[1] F. Paneth and K. Peters, Über die Verwandlung von Wasserstoff in Helium, *Naturwissenschaften* **14** (43) (1926) 956–962.

[2] M. Fleischmann, S. Pons and M. Hawkins, Electrochemically induced nuclear fusion of deuterium, *J. Electroanal. Chem.* **261** (1989) 301.

[3] S. E. Jones, E. P. Palmer, J. B. Czirr, D. L. Decker, G. L. Jensen, J. M. Thorne, S. F. Taylor and J. Rafelski, Observation of cold nuclear fusion in condensed matter, *Nature* **338** (1989) 737–740.

[4] A. De Ninno, A. Frattolillo, G. Lollobattista, L. Martinis, M. Martone, L. Mori, S. Padda and F. Scaramuzzi, Emission of neutrons as a consequence of titanium–deuterium, *Il Nuovo Cimento* A **101** (5) (1989) 841–844.

[5] S. Focardi, R. Habel and F. Piantelli, (Anomalous Heat Production in Ni-H Systems,(*Il Nuovo Cimento A,* vol. 107, pp. 163-167, 1994.

[6] S. Romanowski, W. M. Bartczak and R. Wesołkowski, Density functional calculations of the hydrogen adsorption on transition metals and their alloys. An application to catalysis, *Langmuir* **15** (18) (1999) 5773–5780.

[7] B.S. Ahern, K.H. Johnson and H.R. Clark Jr., Method of maximizing anharmonic oscillations in deuterated alloys, Patent US5770036, US5411654, US5674632, 23 June 1998.

[8] C.G. Vayenas, S. Bebelis and S. Neophytides, Non-Faradaic electrochemical modification of catalytic activity, *J. Phys. Chem.* **92** (1988) 5083–5085.

[9] Y. Iwamura, T. Itoh and M. Sakano, Nuclear products and their time dependence induced by continuous diffusion of deuterium through multi-layer palladium containing low work function, *Proc. of ICCF8,* 2000, pp. 141–146.

[10] T.J. Sommerer, Barium-free electrode materials for electric lamps and methods of manufacture thereof, Patent WO2007064570 A2, 7 June 2007.

[11] F. Celani, A. Spallone, B. Ortenzi, S. Pella, E. Purchi, F. Santandrea, S. Fiorilla, A. Nuvoli, M. Nakamura, P. Cirilli, P. Boccanera and L. Notargiacomo, Observation of macroscopic current and thermal anomalies, at high temperatures, by hetero-structures in thin and long constantan wires under H_2 gas, *J. Condensed Matter Nucl. Sci.* **19** (2016) 29–45.

[12] A. Miyakoshi, A. Ueno and M. Ichikawa, Mn-substituted Fe–K mixed oxide catalysts for dehydrogenation of ethylbenzene towards styrene, *Appl. Catalysis* A: *General* **216** (2001) 137–146.

[13] A. Kotarba, I. Kruk and Z. Sojka, How the iron oxide catalyst for EBDH is stabilized via Mn addition, *J. Catalysis* **221** (2004) 650–652.

[14] H. Preusker, Design of a helium fusion reactor, Patent DE4300016, 02 01 1993.

[15] K.A.A. Khalid, T.J. Leong and K. Mohamed, Review on thermionic energy converters, *IEEE Trans. Electron Devices* **63** (2016) 2231–2239.

[16] F. Celani et al., Experimental results on sub-micro structured Cu–Ni alloys under high temperatures hydrogen/deuterium interactions, *Chem. Mater. Res.* **3**(3) (2013) 25–76.

[17] F. Celani, G. Vassallo, E. Purchi, S. Fiorilla, L. Notargiacomo, C. Lorenzetti, A. Calaon, B. Ortenzi, A. Spallone, M. Nakamura, A. Nuvoli, P. Cirilli, P. Boccanera and S. Pella, Improved stability and performance of surface-modified Constantan wires, by chemical additions and unconventional geometrical structures, *J. Condensed Matter Nucl. Sci.* . **27** (2018) 1–13.

[18] L. Vantini, No alla distruzione del laboratorio di Francesco Celani, 21 Oct 2017 Online. Available: https://www.change.org/p/sergio-mattarella-no-alla-distruzione-del-laboratorio-di-francesco-celani.

[19] Pandora TV, Francesco Celani: Una scienza, dogmatica impedisce la ricerca e danneggia il paese , (27 Sept. 2017. Online. Available: https://www.pandoratv.it/francesco-celani-una-scienza-dogmatica-impedisce-la-ricerca-e-danneggia-il-paese/.

[20] A. Milan, Uno scienziato contro il Monopolio, 23 Sept. 2017. Online, Available: http://edicola.laverita.info/laverita/books/laverita/2017/20170923laverita/#/1/.

J. Condensed Matter Nucl. Sci. 29 (2019) 75–84

Research Article

Cavitation Effects on Various Metals in D_2O

Thomas N. Claytor*
High Mesa Technology, Los Alamos, NM, USA

Roger S. Stringham†
Firstgate Energy, Kauai, Hi, USA

Malcolm M. Fowler‡
McFarland Instrumentation Services Inc., ESPANOLA, NM 87532- 3502, USA

Abstract

Cavitation at high frequency in close proximity to metal surfaces was investigated in D_2O. A new miniaturized cavitation system was operated at 1.7 MHz and was small enough to fit into a sensitive Seebeck calorimeter. The 1.7 MHz reactor was designed to hold 20 g of D_2O or H_2O. The cells were operated in a pulsed mode for 1 or 2 min and then allowed to cool in the calorimeter. Target foils were placed in close proximity to the transducer and different foils show a maximum of about 13% change in measured excess heat output. Tritium measurements were made on the resulting liquid and showed a small initial increase. However, longer runs did not show a linear increase in tritium as the cavitation times were increased. No detectable signals were measured on a pancake gamma detector placed below the cavitation cells. Many different target foils (TF) (7 × 19 × 0.1 mm) were run in the system. A few showed some interesting surface features.

Keywords: Bubble collapse, Cavitation, Cavitation damage, LENR, Particulate generation, Ultrasound

1. Introduction

Many have speculated that cavitation with either single or multiple bubble could result in dd fusion in the presence of D_2O or other deuterided liquids [1–3]. Here, we investigate cavitation not only in the liquid but primarily as it impinges on different metal surfaces. Cavitation damage on metals in H_2O is well known [4]. Less studied is cavitation damage in metals immersed in D_2O [5–10]. In particular, we investigated the assertion by Stringham [6–10] that very anomalous effects occur when metal foils are immersed in a cavitation field at fairly high frequencies.

*Corresponding author. E-mail: Claytor@att.net.
†E-mail: Rogerssbiz@gmail.com.
‡E-mail: Malcolm-fowler@zianet.com.

© 2019 ISCMNS. All rights reserved. ISSN 2227-3123

Cavitation damage is typically seen at low frequencies (30 kHz) where the bubble can grow to microscopic dimensions (>100 μm) and with about 90 nJ energy. At the high frequencies ($\sim$1.7 MHz) of this study the energy contained in a single bubble is small, but there may be millions of bubbles. Little work has been done to characterize the damage or the damage products from the metal surface-cavitation interaction at these frequencies because such low energies are not expected to produce much surface damage [4]. The cavitation threshold in an ideal fluid is thought to be very high (>800 W/cm^2) at the frequency used in this study. However, our liquid is less than ideal because the target metal and transducer will shed oxide particles and, in addition, there may be nano-cavities [4] in the liquid formed during the cavitation. Thus, the cavitation threshold is lowered substantially ($\sim$20 W/cm^2). This particular experiment has many parameters that are difficult to control and will impact the reproducibility of the experiments. With further refinements, we expect the reproducibility to improve. The basic purpose of this research is to determine if the "Stringham effect" can be reproduced and in what metals is it most evident. By the term "Stringham Effect" we are referring to anomalous behavior of cavitation impinging on a metal. This could be the appearance of excess heat, helium or tritium in the liquid or metal, or other elements that were not initially present in the liquid or metal.

2. Procedure

Tritium was measured with a Beckman LS6500. A unique feature of the scintillation counter is that it is totally enclosed in a sealed volume to reduce radon prodigy infiltration and has a real time radon monitor in the enclosure. This allows for a very consistent and low background. Ultima Gold (Perkin Elmer scintillation fluid) was used for all measurements in low background poly vials. Vials were typically counted four times, no less than a day apart, to ensure no short half lived isotopes were present.

Radiation detection directly from the cavitation cell or foil (post run) was done with an RM-80 pancake detector (Aware Electronics) and Silicon Surface Barrier Detectors from ORTEC and an AMPTEK 8000A multichannel analyzer.

Excess heat measurements were made in a custom calorimeter that was obtained from Coolescence (Fig. 1). The calorimeter outer chamber is temperature controlled to 0.1°C (via a heat exchanger and water chiller) and the interior envelope is 86.3 $\times$ 38.1 $\times$ 29.2 cm (l,w,h). Located in the controlled environment is the inner calorimeter chamber that has an interior volume of 18.0 $\times$ 11.5 $\times$ 13.0 cm (l,w,h). The entire top of the inner chamber consists of a Peltier sensing element. Typically, the outer chamber was held at 19–20°C and, at this temperature, the sensitivity of the calorimeter was 27.1 mV/W at a 22 W input.

SEM measurements were made with a Tescan Vega with a Bruker EDX. ICPMS analysis was done by A&B Labs. Houston TX. using an Agilent 7700 and 7900.

3. Cavitation Cells

Several iterations of the cavitation cells were investigated for this study. Initially an all acrylic cell was used, but it was found that the heat from the transducer and the heat from the absorbed ultrasonic beam would eventually distort the acrylic body resulting in leaks. Acrylic bodies were used exclusively for long runs to collect nanoparticles for analysis in the ICPMS. The final design of the cells for use in the calorimeter was made with an aluminum transducer holder and a glass tube with an acrylic top and O-ring seal. This is shown in Fig. 2. To minimize pressure buildup in the cell, the units were only run for a maximum of 2 min. The transducers were 1.7 MHz PZT disks 20 mm diameter by 1.3 mm thick and were glued into the aluminum holder with thermosetting glue. The transducers were from various sources (but all PZT) having different working surfaces. In some cases the surface electrode was silver, or stainless steel or even Ni with a TiN coating. Since all the transducers had different efficiencies and even different units of the same model varied by 20%, it was important that during the calorimeter tests of different foils the same transducer

Figure 1. Coolescence calorimeter, outer box (*left*), and inner chamber with Peltier elements (*right*).

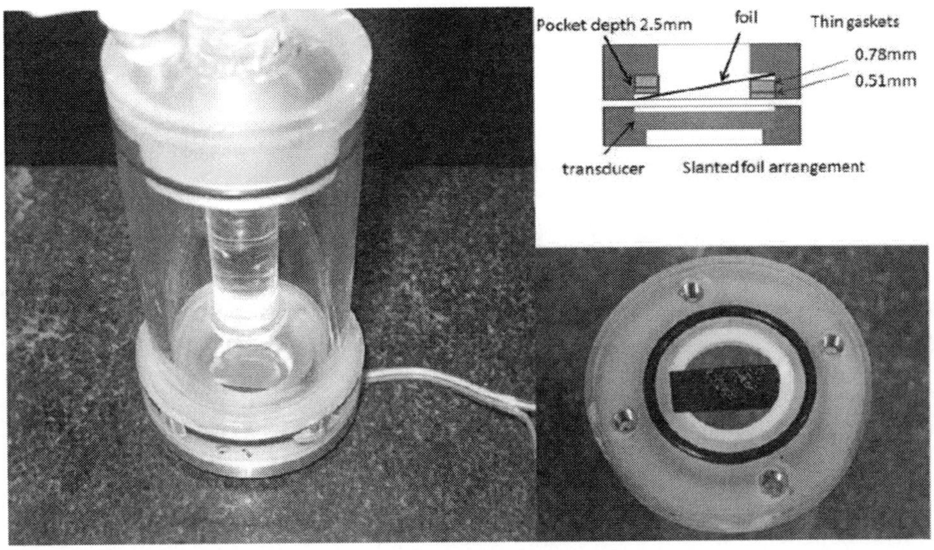

Figure 2. Cavitation cell with reflector and copper sample (*left*), arrangement of foil in cell and a Ni foil that was painted to show cavitation streaks and pits after a short run (*right*).

was used. For electrical input powers of greater than 30 W the transducers would eventually fail (due to the electrical contacts or surface foil delamination). Thus, for all the data reported here for calorimetric measurements, the power input to the transducer was kept to about 18–24 W.

4. Materials

The foils (technical grade) used in this study were 19 mm in length and 7 mm wide by 250 μm or 125 μm thick. The D_2O was from Cambridge Isotopes (99.9%) and the H_2O was distilled water. The Cambridge Isotopes D_2O contained 90 or 7 dpm per ml of tritium depending on the batch. The water did not contain any detectable tritium (<1.5 dpm). Prior to the experiments, the D_2O or H_2O was degassed in a vacuum chamber for a few minutes. After the vacuum process, Argon was introduced into the chamber. Most runs were performed with an Argon cover gas.

5. Experimental

Initially it was uncertain if strong cavitation could be generated at the high frequencies used in this study because of the low energy contained in the bubble. To determine that indeed there was cavitation, several foils were painted with a thin layer of black paint. After exposure in the cell, these foils would often show small round <30 μm features where the paint had been removed and the specular reflection from the metal surface sometimes showed a frosted texture. However, the density of these features varied considerably from run to run, sometimes none were seen and other times there was a high density. This was attributed to the fact that in the near field of the transducer the intensity varies rapidly with distance from the transducer surface and also with the liquid temperature. Since these parameters could vary from run to run, it was decided to slant the foil slightly to present at least some of the foil surface at an intensity maxima. The microscopic pits could be in lines or in a circular pattern depending on the standing wave pattern in the liquid or on the foil.

Cavitation was also detected using a thin (0.02 mm) piezoelectric transducer attached to the side of the cell. When the system was run at 1.7 MHz, there was a broad peak in sound intensity (as measured by the probe transducer) centered near 200 kHz suggesting the maximum bubble size was 10 μm. It was found that in clean, cold, degassed D_2O or H_2O the intensity of the 200 kHz spectra dropped 10 dB within the first 20 s of operation, indicating that perhaps the bubbles became gassy or the particulates were nucleating the cavitation.

It should be noted that a thermocouple on the backside (side not in contact with the liquid) of the piezo driver would often rise to temperatures >100°C within 10 s after the power was applied. Because the temperature is constantly changing during a run, it was difficult to maintain the resonant condition for very long, if just a driver amplifier was used at a constant frequency. So, for this study we used a custom oscillator that maintained the resonant condition for the transducer as the temperature of the piezoelectric transducer and water increased. Our goal was to hold the power constant to the transducer to within 5–10% during a 1 or 2 min run. For each material used, the power delivered to the transducer would be slightly different. Therefore, the drive was adjusted to maintain about 22 W to the transducer. The reason for this was to present to the calorimeter and the electronic power measurement circuit the same power for each foil. Our objective was to determine if there is a difference in heat output between metals under the influence of the cavitation. The weight of the cell was measured prior to the run and after to determine if any water or D_2O had been lost. If the seal obviously leaked or more than 10 mg of water was lost, the run was discarded.

6. Power Measurements

To measure the RF power to the transducer a Tektronix CT-2 current transformer was used for the current measurement, and a Tektronix 2220 voltage probe measured the voltage across the transducer leads. Because the current transformer is nominally accurate to within 2% and the voltage probe is accurate to within 2% it is not expected that our AC measurement compared to a DC measurement of a purely resistive load would give agreement better than 5%. Thus, we restrict ourselves to presenting comparative outputs from each different foil. A Tektronix TDS 2024B scope (200 MHz, 2 Gs/s) was used to make the measurement of $V \times I$ for each point on the voltage and current waveforms. Then the resultant waveform was summed over an integral number of cycles to obtain the power measurement. All measurements were made without changing any of the gain ranges on the scope between runs with different foils. It was found that any gain changes on the scope introduced an unacceptable change in apparent power. PC data analysis and acquisition software (Keysight VEE) was optimized to acquire a power reading every 250 ms resulting in 240 power samples for a 1 min run. The Seebeck calorimeter output (voltage) was recorded every 2 s along with room, interior box and Peltier fin temperature using K-type thermocouples. These measurements were made with a Keithley 2700 multichannel scanner connected to a PC.

7. Results

Shown in Fig. 3 are SEM images of a PdAg 2% foil after several 2 min runs in D_2O at 30 W (1.7 MHz) input power to the transducer. The left image shows a large (>100 μm) feature that was formed in the center of the plate. The material was found to contain oxides of Na, K, Al, Ca, Mg, Si, and Cl, S and P by EDX. After the material was removed a rather large hole in the surface was revealed. EDX focused on the interior of the large hole and small hole clusters showed the same EDX signature as the large feature.

Figure 4 is an SEM image of an Ag foil that had been run under similar conditions to the PdAg 2% foil. It can be seen that a large consolidated clump of material is in the center of the foil where the acoustic intensity is expected to be highest. It should be noted that under the large clump in the right image a hole was found suggesting that the material on the surface originated from the hole. Since no SEM images were made of the foils prior to the sonic irradiation it

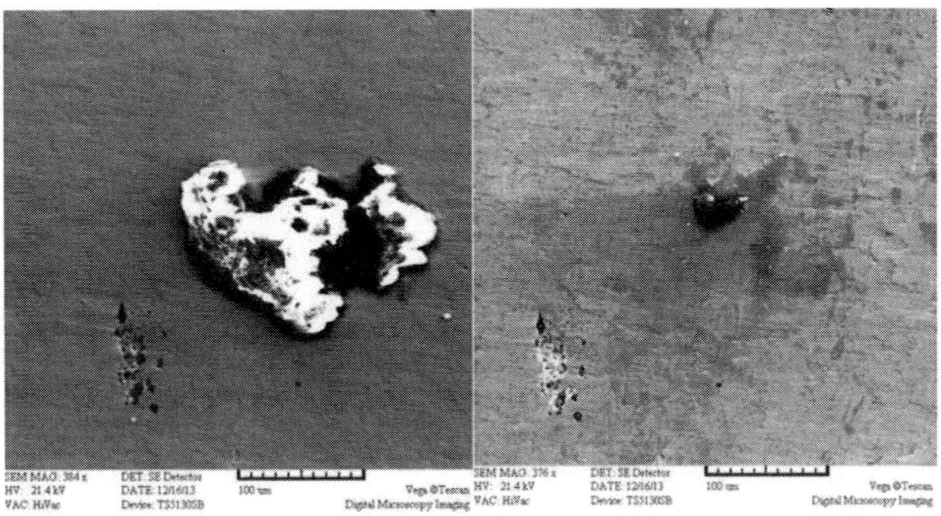

Figure 3. Material found on a PdAg 2% foil after runs at 1.7 MHz. Small holes are to the lower left of the large feature (*left*). After the large feature was removed a large hole was revealed (*right*).

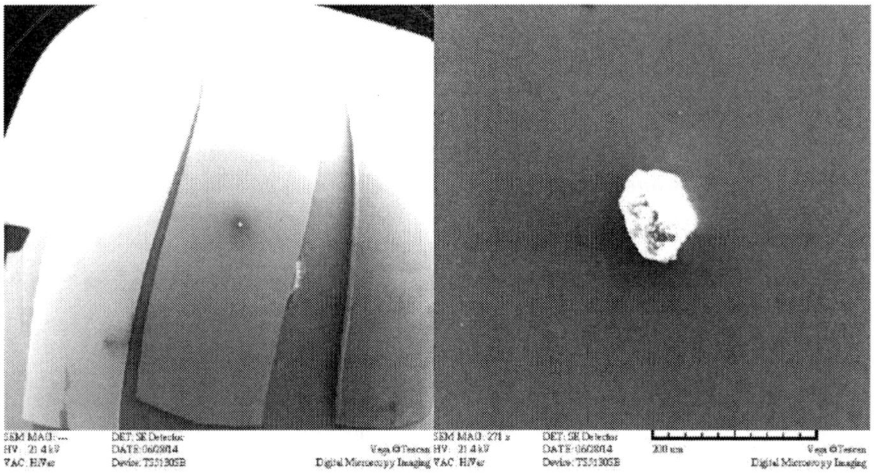

Figure 4. Ag foil (left low mag, right high mag) showing clump of material similar to that found on the PdAg 2% foil. When the material on the surface was removed a hole similar to the one found on the PdAg 2% foil was revealed.

cannot be ruled out that the holes were not initially present. However, each foil was initially visually inspected under $\times 10$ magnification and the holes were not noticed.

Shown in Fig. 5 is a picture is of a pure Ni plate that had a stripe of round Ni particles deposited on the surface facing the transducer. These were not present prior to the cavitation run as determined by visual inspection. None of

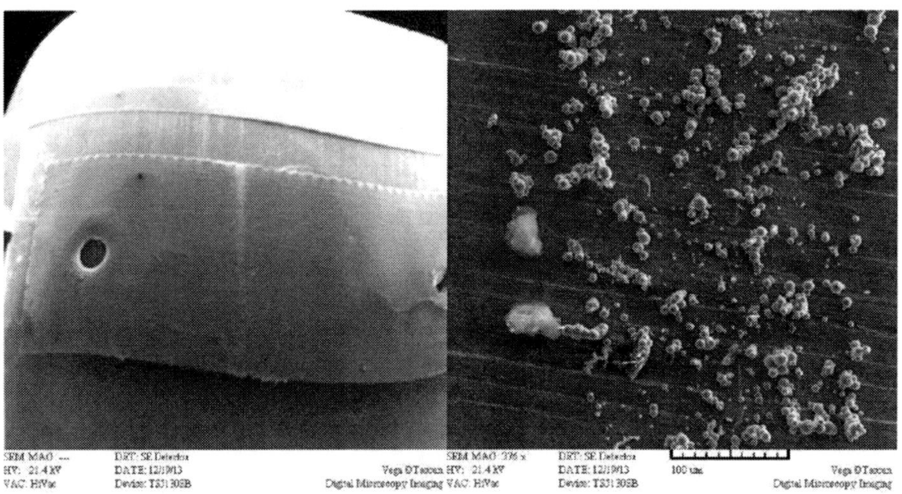

Figure 5. Ni foil 7 mm wide showing stripe of Ni particles (*left*), close up of Ni balls in stripe (*right*).

Table 1. Results of ICPMS (elements not listed in the table were below detection limits).

Run	Ni, D_2O 7.6 h	PdCu, D_2O 3.5 h	PdRhCoB, D_2O 9.3 h	PdB, D_2O 10.5 h	PdAg, D_2O 11.75 h	Cambridge isotopes D_2O unused material	Detection limit mg/l
Al mg/l	0.0071	0.0057	0.024	0.0068		0.0048	0.002
Sb	0.012	0.012	0.023	0.014	0.010	0.012	0.0012
Cr	0.0016			0.0014			0.0012
Cu	0.0042	0.0076	0.0037	0.0066	0.011	0.0014	0.0012
Pb	0.0048	0.002		0.0031			0.0012
Ni	0.0036					0.0014	0.0012
Ag		0.003		0.0086	0.0083		00012
Na	3.3	1.12	4.05	3.8	4.27	2.4	0.25
Zn	0.0092	0.015	0.018	0.0088	0.025		0.005

the foils showed any radiation over background when placed on the pancake Geiger detector (RM-80). However, this measurement was not done until several hours after the runs.

Given the apparent fact that certain elements not initially present in the foil were ejected from the foil, we ran several foils for an extended time ($\sim$10 h) (but at lower power levels of 20 W) to attempt to build up particulates in the D_2O. Light scattering (405 nm) revealed that in PdB, PdCu, PdAg 2% and Ni foils nano-particulates were present. In addition, some of the particulates strongly fluoresced in the infrared. ICPMS was performed on 10–20 g of the D_2O from these runs as well as a blank D_2O sample. Because the samples were relatively small, the ICPMS analysis was not as sensitive as desired, in particular Na, K, Mg, Ca, Fe were only able to be detected at the 250 μg/l level, the other 20 elements (Al, Sb, As, Ba, Be, Cd, Cr, Co, Cu, Pb, Mn, Mo, Ni, Se, Ag, Tl, Th, Ti, V, and Zn) were detectable at the 5–1 μg/l level. The results of heavy water analysis are shown in Table 1.

The pure CI D_2O is seen to have Al, Sb, and Na at values well over the detection limit, the other elements in the D_2O, Cu and Ni are at the detection limit for the instrument. Zn is found in all the post processing samples, but ZnO is a ubiquitous material often found in ceiling tiles. Interestingly only a modest amount (2.5$\times$ background) of Ni was found in the Ni sample. One might expect to find Pd in the analysis, but this was not detected, due to the fact that its detection required a more complicated digestion. Co was also not detected in the PdRhCoB sample. This along with the paucity of Ni in the Ni sample leads one to believe that the cavitation erosion of the parent metal was not a vigorous as one might have expected.

The highest sodium level is obtained from the sample that was run the longest and Cu is also present at the highest concentration in the sample that was longest run. Interestingly, the face of the transducer in these runs was stainless steel (Fe 66.4%, Cr 16.6%, Ni 13.7%, Mo 1.5%, Mn 1.28%, and Cu 0.41%). Thus, one would expect to see elevated amounts of Fe, Cr, and Ni in each sample. The data is suggestive that some of the material seen in the eruptions (in Figs. 3 and 4) has been detected as nano-particulates in the liquid especially after longer runs. Given the fact that others have found similar elements in their systems, more work should be done to confirm these results [12].

Not all of the material generated in these cavitation runs ends up in the liquid. Some is deposited on the transducer and foil. Very small amounts of material can be detected in the SEM and analyzed by the EDX if it occurs as particles on the foil surface. For instance, the PdB sample had Cu, Pt, and Pd particulates present on the foil, the PdRhCoB had a cluster of Fe, Mn, and Co particulates. Similarly, a Pure Pd sample showed a cluster of Si, Ni, Cu, Zr, S, Al, Mn, and Fe as well as Pd particulates. Because of the micro-nature of the particulates (usually <5 μm) and the fact that they are sparse, they would not be detectable by ICPMS even if dissolved in the liquid. Even the large objects on the surfaces in Figs. 3 and 4 would not have been detected in the ICPMS analysis, even if the material was completely dissolved because of the number of elements present.

8. Calorimetry

The Coolescence calorimeter was slightly modified to make it more efficient. Aluminum foil (250 μm thick) was added to all the walls of the calorimeter, except the upper surface where the Peltier elements were located. The aluminum plate that held the Peltier elements was painted black on the interior to more effectively absorb the heat. A fixture in the calorimeter held the cavitation cell in the same location for each run and because the transducer gets hot, during a run, a fan was placed in the fixture that blew directly on the transducer. Since the transducers could not be run for more than 2 min at a time without severe heating, pulsed calorimetry was used for the tests. In the pulsed mode, RF was applied for 1 or 2 min, the calorimeter responded promptly and then the response decayed over a period of 2–3 h.

To compare each metal against the others in D_2O and with the no-sample runs in H_2O and D_2O, the peak response and total integrated heat was measured during a 3 h period and subtracted from the baseline. Repeated measurements using a resistor as a calibration show that the repeatability was $\pm 1.7\%$ for the 1 min runs and $\pm 0.5\%$ for the 2 min runs. The largest non-random error in the calorimeter output is due to the temperature control in the lab room (and thus the outer calorimeter chamber) that can vary depending on the state of the AC or heating system. The peak value of the heat output was less sensitive to the ambient changes since that measurement could be completed in 10 min rather than 3 h. About 100 calorimeter runs were made to assess apparent heat from various foils and background runs.

Figure 6 summarizes the peak heat data obtained for various foils and no foil. The conditions for the calorimeter runs were that the sample and inner chamber of the calorimeter was 20°C and the power applied (21–24 W) to the transducer was on for 60 s and normalized to 22 W. Twenty grams of low tritium D_2O was used for all runs. Several runs on the same sample were performed and the results in Fig. 6 are the average of those runs. The difference between the heat from the highest sample and the no sample run was 13.8% with a maximum error of 3.4%, thus the data has a significance of about 4σ.

Interesting, among the top four alloys are PdB and PdRhCoB alloys that are implicated in electrochemical excess heat experiments [13] and tritium production experiments [14]. Excess heat from Cu and Nb foils is unexpected since there is no prior indication that these materials would be active in excess heat experiments. But given the large uncertainty in the experiment, these data should be revisited. Improvements to the calorimeter are being made and further work will utilize 2 min runs and more stable temperature regulation of the outside box.

Low Tritium D2O (~ 20C)
60s on at 22W

	mV/W x 10^6	
PdRhCoB	816	
Cu 4 mil	817	
PdB	782	
Nb	785	
PdLi	780	13.8%
PdHg	775	
Zr 4mil	774	
Ta 4mil	769	
PdCu	761	
PdAg2%	751	
PdAg2% D2	727	
No sample	717	

Figure 6. Ranking of foils.

The excess power measured here is a small fraction of the total power supplied to the system. For instance, for 22 W delivered to the transducer, 32 W has to be input to the resonator board. The maximum excess power is about 2.8 W for the data in Fig. 6. However, this is consistent with the excess power often reported by Stringham for operation at this frequency.

9. Tritium Results

Active material often results in the appearance of tritium in the D_2O. In many instances the tritium will increase initially, then either level off or decrease slightly. Here we adopted multiple foil experiments to determine if any one of many foils produced excess tritium. In other words, the same low tritium D_2O was used for each successive foil. The resulting low tritium D_2O was filtered (0.1 μm filter) or distilled prior to scintillation counting. Neither process removed all the nano-particulates. After running multiple samples for 76 min (1 min on and 5–10 min off to cool down), the excess tritium generation rate confidence was 2σ (79.5 pCi/h $\pm$ 42) over the background. This rate is significantly higher than most of the rates obtained from gas discharge systems [14]. However, the uncertainty is larger than in those experiments. PdRhCoB and PdCu were run independently (for >3 h) and gave low rates of tritium generation (5.5 pCi/h and 10 pCi/h) that also did not exceed the 2σ confidence level. The fact that longer runs yielded more tritium is encouraging, however, the longer the run, the lower the rate.

10. Conclusion

The data presented here indicates that cavitation on metals in D_2O and perhaps H_2O can produce anomalous results that are defined as the "Stringham effect" after Roger Stringham, who originally investigated these phenomena. This includes the evidence for the appearance of elements not originally present or elements in unusual forms in the system after cavitation. While these supposed transmutation effects occur also in electrochemical cells, the electrochemical cells contain a soup of chemicals that often are claimed to be the source of the impurities. In the present study, the materials are clean foils and pure D_2O or H_2O with no additives other than perhaps an Argon cover gas.

Tritium in small but consistent amounts has been found in the liquid after D_2O runs. No mixed D_2O/H_2O runs have been done, in this study, but they would be the next step since a D_2O/H_2O ratio of 20% seemed to be more effective than pure D_2O in producing tritium in catalytic systems based on our prior unpublished work.

The lack of X-rays or gammas is the most obvious problem when trying to explain that this process is nuclear. However, if the energy is less than 10 keV then it is doubtful that the X-rays would escape from the cells. Some foils did pick up radon prodigy (as determined by the SSD detector) after removal from the cell, but those can be easily discriminated from longer term emissions from the foils. Nevertheless, the preponderance of data presented here definitely supports Stringham's claims of anomalous behavior in these systems.

Acknowledgements

We would like to acknowledge the generous support of Dewey Weaver and J.T. Vaughn of Industrial Heat LLC, and Rick Cantwell and Matt McConnell of Coolescence LLC. Edmund Storms provided very useful discussions/suggestions and access to his SEM and EDX for this work.

References

[1] W.C. Moss, D.B. Clarke, J.W. White and D.A. Young, Sonoluminescence and the prospects for table-top micro-thermonuclear fusion, *Phys. Lett.* A **211** (1996) 69.

[2] R.P. Taleyarkhan et al., Evidence for nuclear emissions during acoustic cavitation, *Science* **295** (2002) 1868.

[3] Nanospire Inc.

[4] D.G. Shchukin, E. Skorb, V. Belova and H. Mohwld, Ultrasonic cavitation at solid surfaces, *Adv. Materials* **23**(17) (2011) 1922–1934.

[5] J. Jorne, Ultrasonic irradiation of deuterium-loaded palladium particles suspended in heavy water, *Fusion Technol.* **29** (1996) 83.

[6] R.S. Stringham, When bubble cavitation becomes sonofusion, in *237rd ACS National Meeting*, Salt Lake City, 2009. http://citeseerx.ist.psu.edu/viewdoc/download?doi=10.1.1.380.6715&rep=rep1&type=pdf.

[7] R.S. Stringham, When bubble cavitation becomes sonofusion, *J. Condensed Matter Nucl. Sci.* **6** (2012) 1–12.

[8] R.S. Stringham, Sonofusion produces tritium that decays to helium three, *ICCF 15*, V. Violante and F. Sarto (Eds.), Oct. 5–9, 2009, p. 57

[9] R.S. Stringham, Deuteron plasmas driven to neutrality and ^{4}He, *ICCF 21*, S. Katinsky and D. Nagle (Eds.), University of Colorado State, Fort Collins CO, June 3–8, 2018, to be published in *J. Condensed Matter Nucl. Sci.*

[10] R. Stringham, Helium measurements from target foils at LANL and PNNL, *J. Condensed Matter Nucl. Sci.* **24** (2017) 284–295.

[11] M. Alheshibri, J. Qian, M. Jéhannin, and V.S. J. Craig, A history of nanobubbles, *Langmuir* **32** (2016) 11086–11100.

[12] Arik El-Boher, University of Missouri, private communication.

[13] M. Miles, K.B. Johnson and M.A. Imam, Electrochemical loading of hydrogen and deuterium into palladium and palladium–boron alloys, in *Sixth Int. Conf. on Cold Fusion, Progress in New Hydrogen Energy*, 1996, p. 208.

[14] T.N. Claytor et al., Tritium evolution from various alloys, *LAUR* 98-2697.

J. Condensed Matter Nucl. Sci. 29 (2019) 85–94

Research Article

Temperature Dependence of Excess Power in Both Electrolysis and Gas-loading Experiments

Zhan M. Dong*, Chang L. Liang and Xing Z. Li†

Department of Physics, Tsinghua University, Beijing, China

Shu X. Zheng

Department of Engineering-Physics, Tsinghua University, Beijing, China

Abstract

The earlier data from a "Heat after Death" electrolysis experiment and from a Tsinghua University gas-loading experiment are reviewed to show that temperature dependence of excess heat in both electrolysis and gas-loading experiments supports the straight-line behaviour in the semi-logarithmic plot discovered by Storms. Additional gas-loading data in seven Pd-tubes show that excess heat is correlated to a deuterium flux as a result of the diffusion process which is implied in this temperature dependence of excess heat.

Keywords: Deuterium flux, Gas-loading experiment, NAE, NAZ, Pumping effect, Resonant surface capture model, Temperature dependence of excess heat

1. Introduction

We presented two papers about the temperature dependence of excess heat at ICCF-21: a theoretical study and an experimental study. The theoretical study showed that this temperature dependence of excess heat [1] might reveal the resonant mechanism of excess heat: *resonant surface capture reaction* [2]. In this paper we further show that this temperature dependence of excess heat is confirmed not only by earlier Fleischmann–Pons' electrolysis experiments (the famous "heat after death" experiment (ICCF-3) [3]), but also by the earlier gas-loading experiments at Tsinghua University in three aspects: (i) "pumping effect" in a long-thin palladium wire (1999 Asti Meeting and ICCF-9) [4], (ii) the comparison of cooling curves between Pd/D and Pd/H systems (ICCF-6) [5], and (iii) the temperature cycling effect on a long-thin palladium wire (ICCF-7) [6]). Moreover, additional experiments at Tsinghua University (ICCF-

*E-mail: dongzm@tsinghua.edu.cn.

†Corresponding author. E-mail: lxz-dmp@tsinghua.edu.cn.

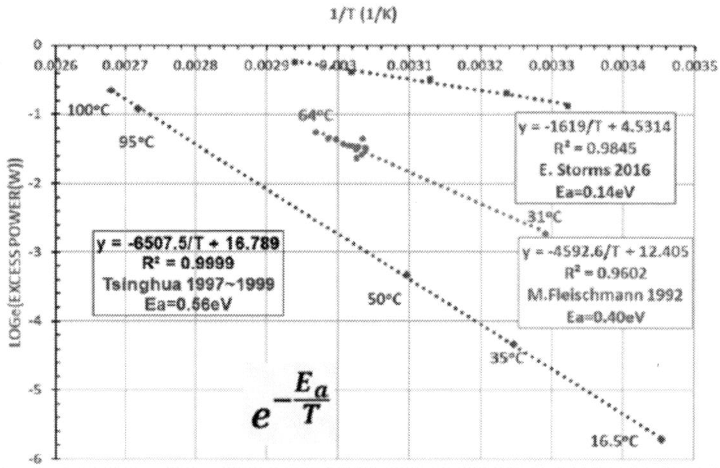

Figure 1. Three straight lines in semi-logarithmic scale plot.

18) [7] have directly shown that this excess heat effect in gas-loading experiments is related to the deuterium flux diffusion through the palladium thin wall of the Pd-tubes.

2. The Temperature Dependence of Excess Heat before Boiling Period of Electrolyte

In 1992 (ICCF-3), Fleischmann and Pons first presented their most compelling "heat after death" experiments. After a period of incubation at low current density, the electrolytic current was raised to trigger a positive feed-back effect in temperature. In their plot the excess heat power was given along with the temperature of the electrolyte [3]. Twelve pairs of data points are available directly from their plot, and are plotted in Fig. 1 (*red circles*). It is interesting to note that these data points are on a straight line again in the semi-logarithmic plot with ($1/T$) as abscissa as first discovered by Storms 2016 [1]. The data point before triggering has been very important to confirm this straightline behavior, because the temperature of electrolyte after triggering showed somehow a little scattering This implies that although the excess heat power level was low before triggering it was still a real effect. It might be a hint that the so-called threshold effect for current density on the surface of Pd cathode might not be real. It was just the onset point of positive feed-back due to the temperature dependence of excess heat power

3. Confirmation in Gas-loading Experiments

There are three sets of data points from gas-loading experiments with the same long-thin palladium wire in D_2 gas, and a comparable data set in H gas from Tsinghua University in 1997–1999.

1. *"Pumping effect" at 100, 95, and 50°C*

A long-thin palladium wire ($\sim$250 cm$\times\phi$34 mm) was wound on a Quartz frame, and was sealed in a stainless vacuum Dewar filled with deuterium gas (Fig. 2). A tungsten wire (ϕ0.1 mm) was used to dissociate the deuterium molecule, D_2, into atoms in order to increase the chemical potential of D atoms which was believed to be essential for loading the D atoms into palladium wire [8]. A Pt-100 thermistor monitored the temperature at the middle of the

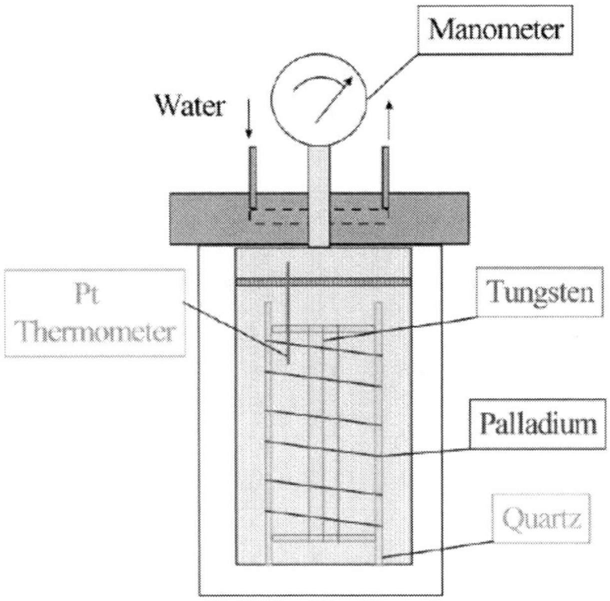

Figure 2. The early gas-loading apparatus in 1997–1999 at Tsinghua University.

palladium wire (T_{Pd}) An intelligent DC power supply was used to heat the palladium wire to specified temperatures (100, 95, and 50°C) When we pumped out the deuterium gas after loading, the necessary power to keep the specified temperature dropped. This power reduction was a good measurement of any excess heat effect in the Pd–D system.

Figure 3 shows the necessary power (red line in upper plot) and the specified temperature of Pd wire (purple line in the lower plot). At first the temperature was kept as 95 ± 0.2°C for more than half a day with 2.25 ± 0.05 W DC power. When the pumping started, the power needed to maintain this temperature dropped instantly. We might expect to see the power jump up, because the degassing from Pd wire is supposed to be an endothermic process. We did see such a power jump as a short spike in the red line, but it was quickly followed by an exothermic effect–the DC power decreased until it reached a steady value. The DC power reduction is about 2.25–1.85=0.4 W for 95°C. Figure 3 also shows that this pumping effect is reproducible after one day (near 220 000 s). At 250 000 s, we changed the set temperature from 95°C to 100°C, and the necessary DC power needed to maintain this higher temperature increased to 1.96 W while pumping was continuing. The sudden stop of "pumping" near 300 000 s increased the necessary power to 2.48 W. The power reduction of "pumping effect" at 100°C is about 0.52 W.

In order to exclude the possible effect of the reduction in the heat transfer coefficient, we note that Fig. 4 showed the gas pressure in the Dewar (purple line in the lower plot) while the T_{Pd} is fixed at 50°C for 1 day The "pumping" was done in two phases: weak pumping and strong pumping. At 40 000 s we started weak pumping with a controllable valve. In the period of 40 000–48 000 s, the gas pressure was increasing slowly; however, the necessary heating power (red line in Fig. 4 upper plot) was decreasing. Evidently, this power reduction was due to some exothermic effect inside the Pd–D system, and not by the reduction in the heat transfer coefficient. When the strong pumping started near 48 000 s, the necessary heating power dropped quickly at first; this was then followed by a slow reduction. However, the gas pressure showed totally different time-behavior: a step-wise reduction at 48 000–63 000 s. When we stopped

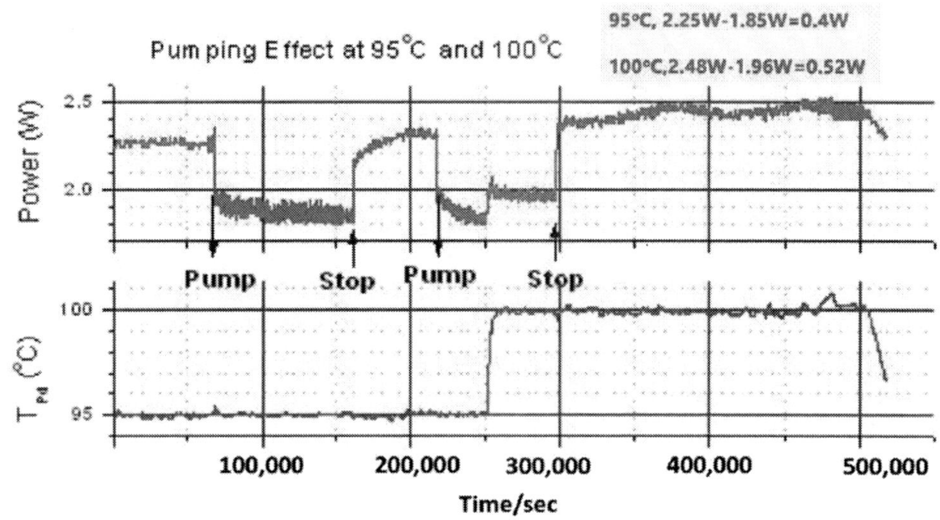

Figure 3. "Pumping effect" in gas-loading experiments near 95 and 100°C.

strong pumping near 63 000 s, the time-behavior of power and gas pressure are so different there is no way to attribute this power variation to the variation in gas heat conductivity. This exothermic effect at 50°C is about 0.24–0.204 = 0.036 W.

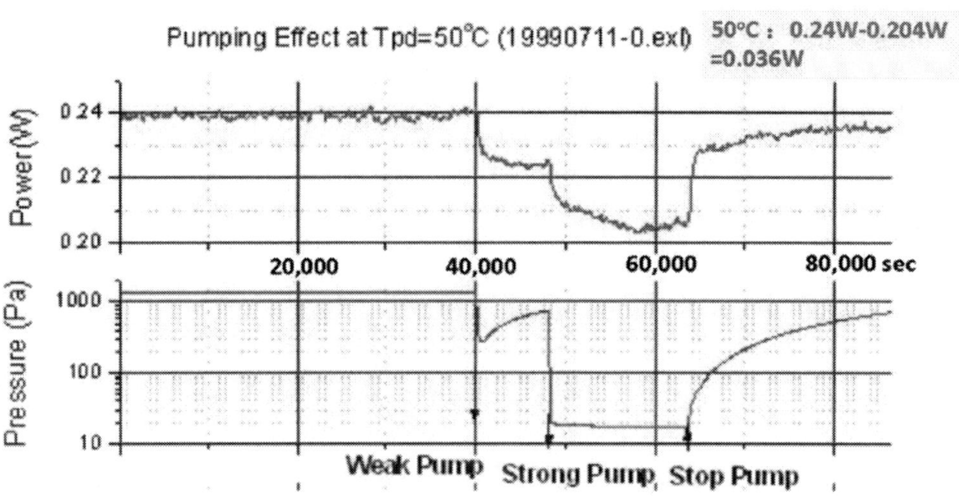

Figure 4. "Pumping effect" in gas-loading experiment at 50°C.

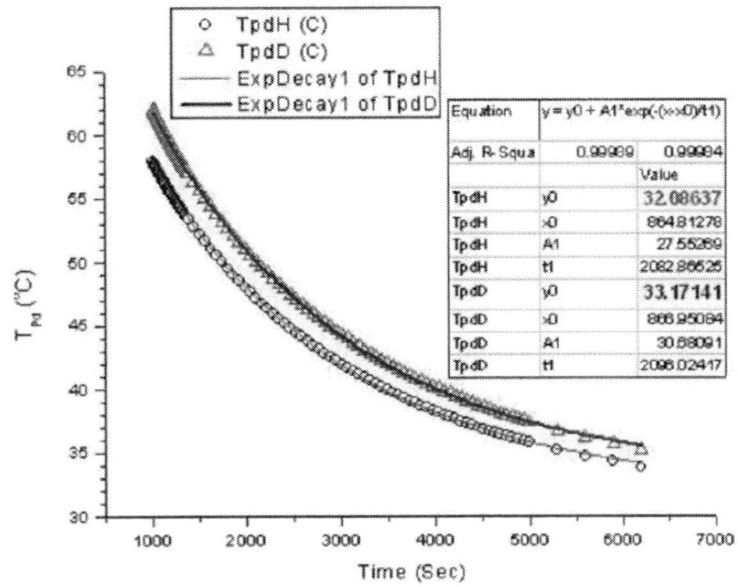

Figure 5. Comparison of cooling curves of D/Pd and H/Pd systems.

2. *Comparison of cooling curves between Pd/D and Pd/H systems near 35°C*

Although we do not know exactly at which temperature the resonant exothermic reaction occurs, we can still observe this exothermic effect during the cooling process. Particularly, we can compare the difference between the Pd/D and Pd/H system, and confirm this exothermic effect. We built two gas loading Dewars as shown in Fig. 2. The twin systems are almost the same (see Table 1)

Table 1. Parameters of twin systems for gas-loading.

	D/Pd	H/Pd
Weight of Pd wire (g)	2.846	2.844
Resistance of tungsten wire (Ω)	4.1	4.3
Vacuum pressure before filling (Pa)	10^{-3}	10^{-3}
Filling pressure(mmHg)	600(D)	600(H)

except they were filled with different gases (deuterium and hydrogen). After they were heated to similar temperatures, the heating power was shut off, and the twin systems approached different constant terminal temperatures: the temperature in the D/Pd bottle was always a little higher than that in H/Pd bottle. In Fig. 5, the red triangles show the temperature in the D/Pd bottle, and the black circles show the temperature in H/Pd bottle. They are fitted to the cooling curve very well with:

$$T_{\mathrm{Pd}} = T_{\mathrm{Pd0}} + A_1 \mathrm{e}^{-\frac{(t-t_o)}{t_1}}. \tag{1}$$

Here, t is the time, and t_o is the starting point of their cooling curves. It is evident that when $t \to \infty$, $T_{\mathrm{Pd}} \to T_{\mathrm{Pd0}}$.

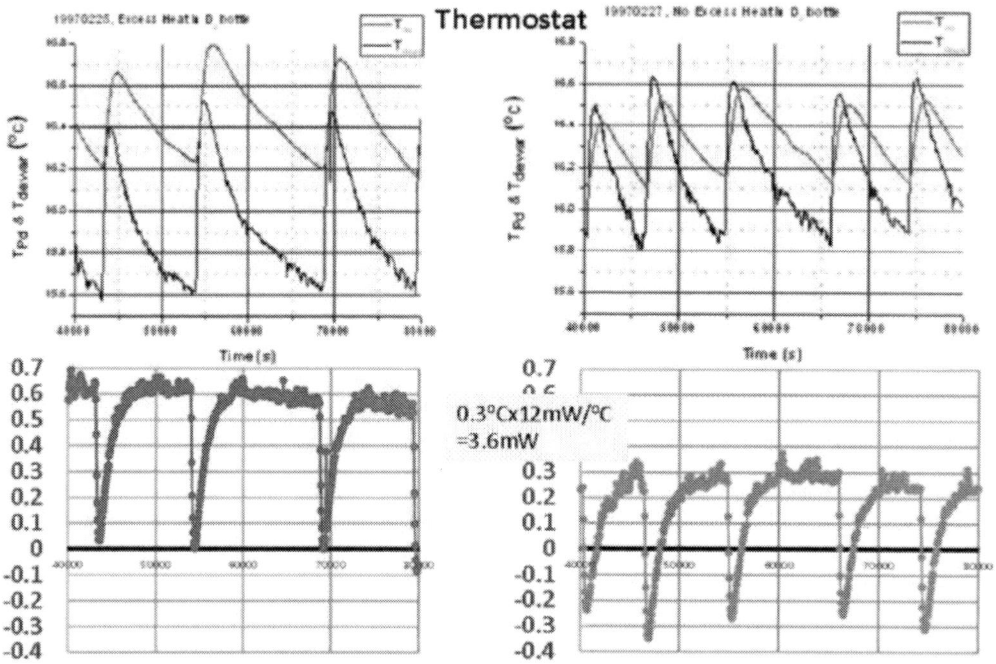

Figure 6. Temperature cycling in a thermostat for D/Pd system in different days.

However, the D/Pd and H/Pd systems approach different levels: 33.17 and 32.08°C, respectively, although they were cooling in the same room temperature. The D/Pd system is 1.09°C warmer than the H/Pd system which implies an exothermic reaction is occurring inside the D/Pd bottle even when the temperature approaches room temperature. We determined that the calorimeter constant of the H/Pd bottle was 0.012 W/°C in this temperature range; hence, the excess power was 12 mW near 35°C.

3. *The temperature cycling effect in D/Pd system around 16.5°C*

In winter, when the room temperature in our laboratory is much lower, we put the D/Pd system in a thermostated water bath. The thermostat kept the temperature of the water bath oscillating around $T_{\rm water} = 16.0°C$ (black line in Fig. 6 upper-left plot). It is evident that the temperature of the Pd wire, $T_{\rm Pd}$ (red line), is always a little higher than the $T_{\rm water}$. The lower-left plot in Fig. 6 shows that the difference ($T_{\rm Pd} - T_{\rm water}$) is almost always greater than zero. It shows again the exothermic effect inside the D/Pd bottle. This difference is about 0.6°C most of the time. One might worry about the accuracy of these Pt-100 thermistors. The right plots of Fig. 6 shows the accuracy of the Pt-100 thermistors. On the next day, the colder weather triggered the thermostat more often; hence, the bath temperature was oscillating faster. It oscillated five times in 40 000 s, which was two times more than that of the previous day. According to the resonant theory of energy bands (ICCF-6) [9], if the cooling speed was too fast, the D/Pd system might not have enough chance to produce resonant reactions before it cooled down through and below the resonance energy band. Therefore, on the second day, the temperature of D/Pd system was no longer kept higher than the water bath temperature (lower right-hand side plot of Fig. 6). When $T_{\rm water} > T_{\rm Pd}$, the D/Pd system was heated by the water bath, the $T_{\rm Pd}$ was increasing until $T_{\rm water} = T_{\rm Pd}$ When $T_{\rm water} < T_{\rm Pd}$, the D/Pd system was cooled by water bath, and $T_{\rm Pd}$ was decreasing until $T_{\rm water} = T_{\rm Pd}$. As a result, while the $T_{\rm water}$ was oscillating, $T_{\rm water}$ always intercepted the $T_{\rm Pd}$ at maximum $T_{\rm Pd}$ or minimum $T_{\rm Pd}$. The upper-right-hand side plot of Fig. 6 clearly shows that our Pt-100 thermistors were working properly. Both $T_{\rm Pd}$ and $T_{\rm water}$ are accurate enough to show this tiny exothermic effect. The lower-right-hand side plot of Fig. 6 shows that the difference $(T_{\rm Pd}–T_{\rm water}) = 0.3°{\rm C}$ is enough to drive the D/Pd system to follow the oscillating $T_{\rm water.}$ Having compared left and right-hand side plots of Fig. 6, we might conclude that there was an exothermic effect on the first day to keep the D/Pd system 0.3°C higher than $T_{\rm water}$. Using the same calorimeter constant, 12 mW/°C, we have 3.6 mW excess heat power near 16.5°C It was a big surprise when we put these five pairs of points on the semi-logarithmic plot: they were on a straight line again (purple diamond points in Fig. 1). These experiments at low temperatures in early 1997 were designed to show an exothermic effect qualitatively only. We did not expect such a good quantitative result. Now the question is why the slope of this straight line is different from that of Storms' and Fleischmann–Pons' lines. In our theory paper [2], we showed that the slope of Fleischmann–Pons' straight line is probably related to the activation energy of deuteron in lithium deuteride. The slope of the straight line here for the gas-loading experiment might also be related to the activation energy of deuterons in tungsten deuteride. Although we have some evidence for this assumption, we would prefer to discuss this possibility when more experimental data are available. We will first show that this excess heat is related to a diffusion flux of deuterium though palladium.

4. Excess Heat is Correlated with Deuterium Flux but Not "Pumping"

To further exclude the effect of heat conductivity induced by "pumping", and show the correlation between excess heat and deuteron flux, the gas-loading experiments were continued at Tsinghua University using a Pd-tube instead of Pd-wire after 1999. The first experiment used a Seebeck calorimeter (C-80D) which is supposed to be independent of heat conductivity with the accuracy of 1 μW. The "pumping effect" has been confirmed again in a short thin wall

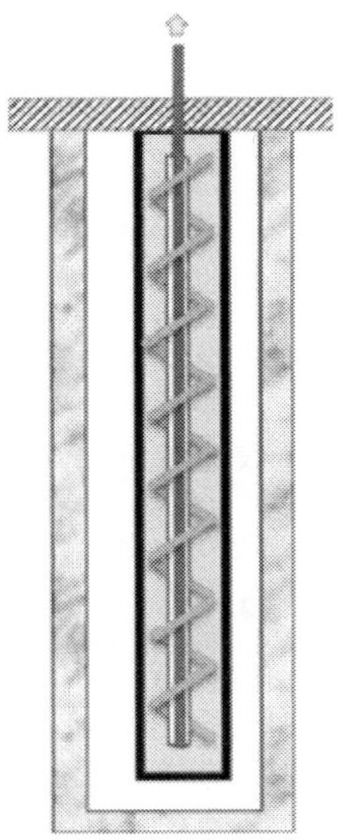

Figure 7. A new gas-loading system with a bunch of seven long-thin Pd tubes.

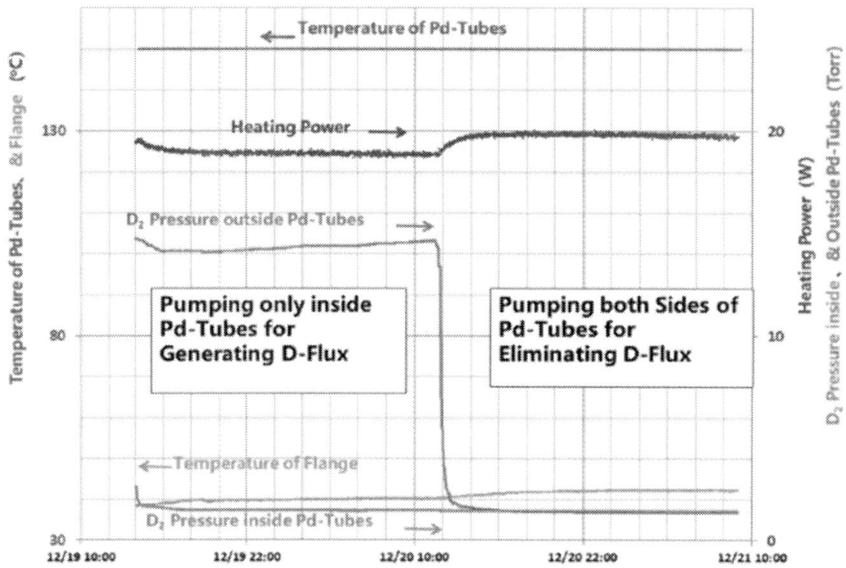

Figure 8. Excess heat is correlated with the deuterium flux but not "pumping effect".

Pd-tube (26 mm × 0.1 mm × ϕ 4 mm) [10]. A peak of excess power was found near 140–150°C which was correlated with a peak of deuterium flux through the thin wall of the Pd-tube. In order to further scale-up this deuterium flux effect, a new Pd-D system was built using seven long-thin Pd-tubes (40 cm × 80 μm×ϕ 3 mm) (Fig. 7) A bunch of seven long-thin Pd-tubes were sealed at their lower ends and the upper ends were connected to a mechanical pump to pump out the gas inside the Pd-tubes. The deuterium gas was filled outside the Pd-tubes in a stainless steel cylinder. An electrical heater was wound around the Pd-tubes, and the whole cylinder was put into a glass Dewar for heat insulation. The major heat conducting path was through the stainless steel flange on the top of the Dewar. An intelligent power supply was applied to keep the Pd-tubes at a specified temperature (150°C). When we pumped inside of Pd-tubes only a deuterium flux was generated from outside into the inside of the Pd-tubes. If we pumped both sides of Pd-tubes the "pumping effect" was supposed to be stronger, but the deuterium flux was reduced greatly. We might compare these two cases to judge if the heat effect was correlated with deuterium flux or correlated with "pumping". Figure 8 shows the experimental result over two days. The red line on the top shows the temperature was kept at 150°C by an intelligent power supply very closely. The green line at the bottom shows that the gas pressure inside the Pd-tubes was kept at less than 2 torr by a mechanical pump. The central pink line shows that the D_2 gas pressure outside the Pd-tubes was about 14 torr at first, and was pumped down at 12:00 AM of the second day. The necessary heating power of the intelligent power supply is shown by the brown thick line above the pink line. A very clear power jump was correlated with pumping both sides of Pd-tubes. The necessary heating power jumped from 19 to 20 W to keep the temperature at 150°C when the "pumping effect" was stronger, and heat conductivity was supposed to be smaller. Therefore, we have to attribute the DC power reduction on the first day to the deuterium flux generated by pumping inside of Pd-tubes only. Indeed the heat effect correlated with deuterium flux is more than 1 W if we consider the reduction of heat conductivity.

5. Conclusion.

Three conclusive remarks might be tentatively draw from our two presentations at ICCF-21:

(1) A straight line in semi-logarithmic scale plot is confirmed by three sets of data from gas-loading and from electrolysis experiments in different laboratories which were published well before the publication of Storms' discovery

(2) A two-step model is implied in this straightline behavior. An elastic diffusive process creates a mother state for the next inelastic nuclear transition to daughter state. Indeed this might be just the model we need for understanding the nuclear active environment (NAE) or nuclear active zone (NAZ).

(3) The additives in anomalous heat experiments might play a more important role than we thought previously. The lithium (LiOD) in electrolysis experiments was only thought to be an additive to increase conductivity of electrolyte. The lithium ($LiAlH_4$) in Ni–H system was considered only a catalyst in reactions. The tungsten wire was considered only a tool to dissociate the molecule D_2. Now we understand that: the slope of Fleischmann–Pons' straight line is very close to the activation energy of deuterium diffusion coefficient in lithium–deuteride; the slope of Storms' straight line is very close to the activation energy of deuterium diffusion coefficient in palladium–deuteride; and the slope of Tsinghua University's straight line might be close to the barrier energy of the deuterium diffusion coefficient in tungsten deuteride. Based our resonant surface capture model we might ask: What was the major fuel in these anomalous heat effects?

Acknowledgements

Grateful thanks to Jed Rothwell for his unremitting effort to keep a complete literature available to our CMNS community, which makes it possible to retrieve the old data to make new discoveries. This work is supported by The Ministry of Education (#20091770437), The Ministry of Science and Technology (Fundamental Research Division, #2009CB226113), Natural Science Foundation of China (#10475045 & #21153003) and Tsinghua University (Basic Research Fund (985-III)).

References

[1] E. Storms, *J. Condensed Matter Nucl. Sci.* **20** (2016) 100–138.

[2] Xing. Z. Li, Zhan M. Dong, Chang L. Liang and Gui S. Huang, Resonant surface capture model, *Proc. ICCF-21*.

[3] M. Fleischmann and S. Pons, Calorimetry of the PD–D_2O system: from simplicity via complications to simplicity, In the *Proc. Third Int. Conf. on Cold Fusion*, 1992, Nagoya, Japan, Universal Academy Press, Tokyo, p. 47.

[4] Xing Z. Li, Bin Liu, Xian Z. Ren, Jian Tian, Wei Z. Yu, Dong X. Cao, Shi Chen, Guan H. Pan and Shu X. Zheng, "Pumping effect" – Reproducible excess heat in a gas-loading D/Pd system– in the *Proc 9th Int Conf. on Cold Fusion, Condensed Matter Nucl Sci* 2002 Tsinghua Univ., Beijing, China Tsinghua Univ. Press pp. 197–201.

[5] Xing Z. Li, Wei Z. Yue, Gui S. Huang, Hang Shi, Lan Gao, Meng L. Liu and Feng S. Bu, "Excess heat" measurement in gas-loading D/Pd system, in *Proc. The 6th Int. Conf. on Cold Fusion, Progress in New Hydrogen Energy,* M. Okamoto (Ed.) Vol. II, Toya, Hokkaido, Japan, NEDO Press, Oct. 13–18, 1996, pp. 455–462.

[6] Xing Z. Li, Shu X. Zheng, Hai F. Huang, Gui S. Huang and Wei Z. Yu, New measurements of excess heat in a gas loaded D–Pd system, in *Proc. The 7th Int Conf on Cold Fusion on Cold Fusion*, pp197–201, Vancouver, Canada, April 19–24, 1998, F. Jaeger (ENECO).

[7] Zhan M. Dong, Shu X. Zheng, Qiang Du, Chang L. Liang and Xing Z. Li, Anomalous heat induced by deuterium flux in a bunch of long-thin palladium tubes using PID method for calorimetry, Poster Presentation in *The Proc. The 18th Int Conf on Cold Fusion*, Univ. of Missouri, Columbia, USA, July 21–27, 2018, Bob Duncan.

[8] W. A. Flanagan and T. B. Oates, Formation of nearly stoichiometric palladium-hydrogen systems, Nature Physical Science, 231, 19 (1971).

[9] Xing Z. Li, Hai F. Wang, and Zhi G. Bian, "Fine tuning" mechanism for resonant tunneling in Pd/D systems, in *Proc. The 6th Int. Conf. on Cold Fusion, Progress in New Hydrogen Energy,* M. Okamoto (Ed.) Vol. II, Toya, Hokkaido, Japan, NEDO Press, Oct. 13–18, 1996, pp. 507–511.

[10] X.Z. Li, B. Liu, J. Tian, Qing M. Wei, Rui Zhou and Zhi W. Yu, Correlation between abnormal deuterium flux and heat flow in a D/Pd system, *J. Phys.* D*: Appl. Phys.* **36** (2003) 3095–3097.

J. Condensed Matter Nucl. Sci. 29 (2019) 95–118

Research Article

Space Application of the GeNIE Hybrid^{TM} Fusion–Fission Generator

Lawrence P. Forsley* and Pamela A. Mosier-Boss

Global Energy Corporation, San Diego, CA 92123, USA

Abstract

JWK Corporation and Global Energy Corporation (GEC) have spent the past two decades understanding and applying nuclear reactions in condensed matter with the US Navy and NASA. The Navy cooperation resulted in US Patent, 8,419,919, System and Method for Generating Particles. The use of this patent to fission uranium is described in a companion paper, Uranium Fission Using Pd/D Co-deposition. GEC is applying this technology as a non-fissile reactor core suitable for deep-space power under its second NASA Space Act Agreement. This paper discusses the need for space-based nuclear power, the alternatives, the hybrid fusion-fast-reactor and the spaceflight readiness testing facilities.

Keywords: Fast fission, Fusion, LANR, NASA, Space power

1. Overview

NASA has used solar power for 50 years and nuclear power beginning three years later. Solar powered spacecraft are generally limited to the inner Solar System out to Mars, with the exception of the 60 foot solar panel span of the JUNO Jupiter orbiter. Other than the US SNAP-10 fission reactor, each of nearly 40 missions, including New Horizons to Pluto, were powered by plutonium (^{238}Pu) radioactive thermoelectric generators (RTG). Although run for decades as seen with the now 41 years extended missions of the two Voyager spacecraft, RTGs provide less than 1 kW of electrical power (kWe). Meanwhile the Soviets flew 31 fission reactors in low-earth orbit (LEO) each producing up to 10 kWe. Unfortunately, the Kosmos-954 satellite came down over Northern Canada in 1978 and contaminated 124,000 km^2 of territory. Hence, there's reluctance to fly fissile material and non-fissile RTGs as used on the Jupiter Galileo, Saturn Cassini, Pluto New Horizons and Mars Curiosity spacecraft as well as the earlier Voyager and Pioneer Missions

Another need is to develop high Specific Impulse (Isp) propulsion exceeding chemical rocket efficiencies. Various Hall Effect and Ion Drive systems have flown using solar power. Rather than expel oxidized propellant like a chemical rocket, these systems ionize and exhaust heavy ions, like xenon. The ions can exit at 40 km/s vs. a chemical rocket maximum exhaust of 7 km/s. However, it has long been recognized that neither solar nor RTGs can provide sufficient

*Corresponding author. E-mail: lawrence.p.forsley@nasa.gov.

power for driving larger ion engines for propelling voyages past the asteroid belt. Consequently, NASA has considered higher power fission reactors for decades including the cancelled 500 kWt (>100 kWe) Project Prometheus [1] during the mid-2000s and the current 43 kWt (10 kWe) Kilopower Program. The goal is to produce long duration, multi-kW to multi-MW reactors for planetary probes, planetary surface power and manned nuclear electric propulsion as in the movie, *The Martian* [4]. High power space nuclear reactors for thermal nuclear propulsion have also been tested but not launched.

GEC has had two Space Act Agreements with NASA. The first was a prelude to the NASA Advanced Energy Conversion Project (AEC) [2,3] under the Radioisotope Power System Program (RPS). The second GEC Space Act Agreement is to develop a space-ready, non-fissile nuclear generator using thorium. Both Space Acts have been conducted at NASA Glenn Research Center (GRC), near Cleveland, OH, and at the Plum Brook Station, 45 miles away outside Sandusky, Ohio with related work at JWK facilities in San Diego, CA and at the University of Texas, Austin, Nuclear Engineering Teaching Laboratory. GRC developed ion engines, heat pipe thermal transport, advanced Sterling Engines for power conversion and the KRUSTY fission reactor. Plum Brook provides space flight qualification facilities for both launch and space conditions. Each of these facilities provides staff, equipment and facilities for GEC to develop and test a non-fissile, deep space power generator suitable for long duration power and nuclear electric propulsion.

2. Nuclear Deep Space Power Needs

There are a mix of US RTGs (e.g. Viking, Curiosity) and solar powered (e.g. Phoenix, Spirit, Opportunity) spacecraft on the Martian surface. As has been found several times, and most recently in 2018, Martian dust reduces solar insolation and consequently solar cell output. While solar powered Opportunity eventually succumbed to a planet wide dust storm the nuclear powered Curiosity mission continued. RTG's provide both electricity and warmth through electric heaters and radioactive decay. Another consideration on Mars and our Moon is the length of night. The Martian day is 24 h and 37 min requiring batteries to cover half that time at the equator and longer at higher latitudes, as on Earth. Polar regions are particularly problematic. Worse, our Moon has a 14 day-night, negating chemical battery backup. Consequently, the Apollo lunar missions left RTG powered instruments on its surface.

But the RTG limit of <1 kWe is barely sufficient for planetary probes far from the sun. Planetary probe science packages including communications would prefer 10 kWe. Planetary power for humans requires at least 40 kWe. A nuclear electric ion drive for the Hermes (Fig. 1) requires on the order of 2–5 MWe.

Human travel beyond the Earth-Moon system will require nuclear electric propulsion if astronauts are to arrive healthy. For example, astronauts on the International Space Station in Low Earth Orbit (LEO) are subject to cosmic rays averaging 300 MeV, with GeV and TeV protons and higher Z nuclei up to iron. Solar MeV protons, spallation neutrons and nuclear fragments from interactions with the Space Station complete the dosimetry problem. Some of these particles are identified in Fig. 2 of an etched CR-39 microphotograph [5]. If sufficient power were available, nuclear electric propulsion could reduce the one way Mars trip time to 2 months [6] from several months.

3. Proposed Space-based Fission Systems

Recognizing both the need for more power for space probes and either nuclear thermal or electric propulsion, especially for manned crews, NASA began the short lived Prometheus Project in 2003. The goal was a 200+ kWe fission reactor for a proposed, unmanned, Jovian three moon encounter[a]. Rather than nuclear thermal propulsion, like the Nuclear Engine for Rocket Vehicle Application (NERVA) [7], whereby hydrogen or other gas is heated by passing

[a]Unlike the movie, *2001: A Space Odyssey*, which had a crew.

Figure 1. Hermes spacecraft [4].

through a fission core and expelled as exhaust [8], Nuclear Electric Propulsion (NEP) expels a highly ionized gas, like xenon. Most geosynchronous satellites [9] use related Hall effect thrusters for station-keeping (staying in orbit) and the asteroid encountering Deep Space 1 probe uses an ion engine [10]. All of these are solar powered.

Figure 3 shows an equivalent sized 300 kWe Nuclear Electric Power system (NEP) as compared to scale with a similar powered Solar Electric Power system (SEP) [11].The large, wing-like, solar panels are far larger than the heat dissipation panels of the NEP where both mass and volume drive launch costs. NASA revisited space nuclear

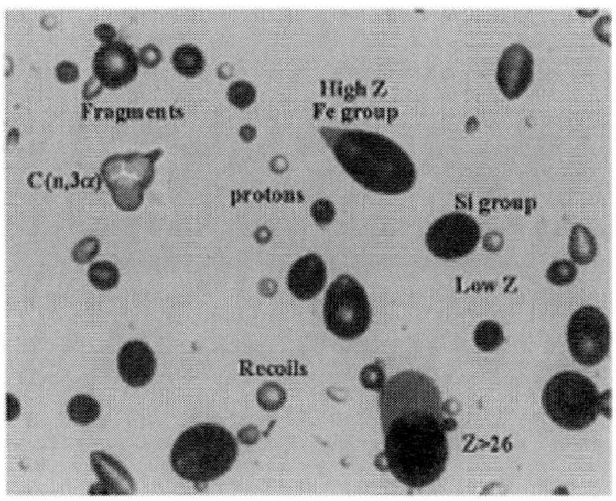

Figure 2. CR-39 particle detector.

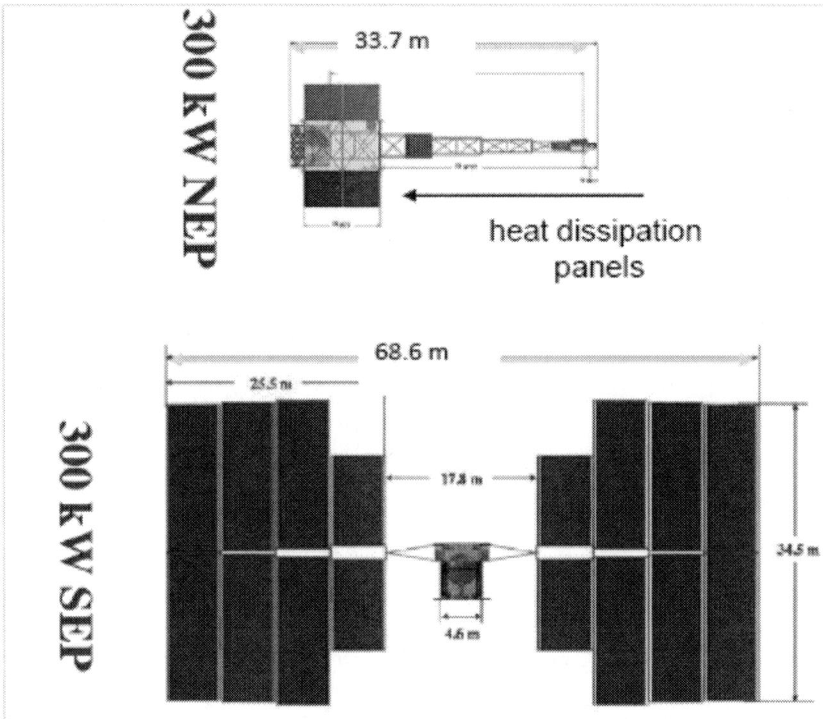

Figure 3. Size comparison.

fission reactors after completing successful tests of the Kilopower KRUSTY 93% ^{235}U reactor with heat pipes and an advanced Stirling Engine for power conversion. This was presented at the American Nuclear Society, *Nuclear Emerging Technologies for Space* conference held in Las Vegas in February, 2018 [12]. The meeting was within driving distance of the DoE NNSA site as the KRUSTY test was conducted.

The goal is to develop a 10 kWe, space-qualified system to run for a decade on our moon, or, with four units, support human activity on Mars.

GEC is able to build upon the power conversion technologies, like the Advanced Stirling Engine, and the various subsystems developed for the Kilopower Program at NASA GRC [13]. For example, Mason [11] has calculated the mass budget of a 2.5 MWe nuclear reactor shown in Table 1. The reactor occupies a small mass percentage, 13%, whereas radiation shielding, heat rejection, power conversion and power management use 87%. This is due to three factors: thermal to electric conversion efficiency, reactor radiation shielding and heat dissipation. Running at higher temperature with more efficient power conversion reduces the amount of waste heat. Direct conversion of charged particles to electricity, not easily realized with a fission reactor, would also increase efficiency.

Table 1. 2.5 MWe nuclear reactor mass budget.

Sub-system	Mass (kg)	Percent (%)
Reactor	3499	13.0
Radiation shield	6734	25.0
Power conversion	2713	10.0
Heat rejection	7930	29.4
Power management	6080	22.6
Total	2695	100.0

4. Proposed Space-based Fusion Systems

Fusion reactors on the ground or in space have problems. A 10^8°C plasma requires multi-tesla magnetic fields while an Inertial Confinement Fusion (ICF) plasma disassembles in nanoseconds. Magnetic tokamaks and Mirror Fusion reactors have been built, some with superconducting magnets, but as yet with no net power production on the ground, *let alone in space*. Furthermore, the size, operating power and mass requirements for both magnetically and inertially confined systems preclude them from being considered for space power. One of the largest fusion spacecraft, employing multiple laser "cannons" and an inertial confinement target delivery system was proposed by Orth in 2000 [14] in Fig. 4. Neither system provides sufficient thrust to launch to orbit, but would need to be ferried up piecemeal by chemical rockets.

Whereas a Mirror Fusion reactor is a solenoid holding a fusion plasma with electrostatic and magnetic mirrors on each end to contain the plasma, Chang-Diaz omitted a "mirror" *allowing the plasma to escape at high velocity.* This high current, ion propulsion system is the VASIMR, or Variable Specific Impulse (Isp) Magnetoplasma Rocket (Fig. 5) [15]. A variable Isp allows a tradeoff between delta-V and thrust. However, using VASIMR for the International Space Station (ISS) to maintain orbit requires 250 kWe. Far higher power is required for trans-lunar propulsion. Yet, this is one of the most efficient rocket engines available with a variable ion exit velocity from 3 to 120 km/s.

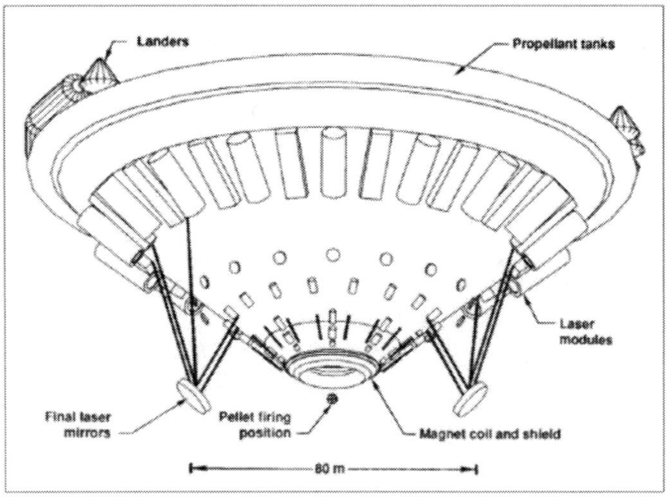

Figure 4. Laser fusion propulsion.

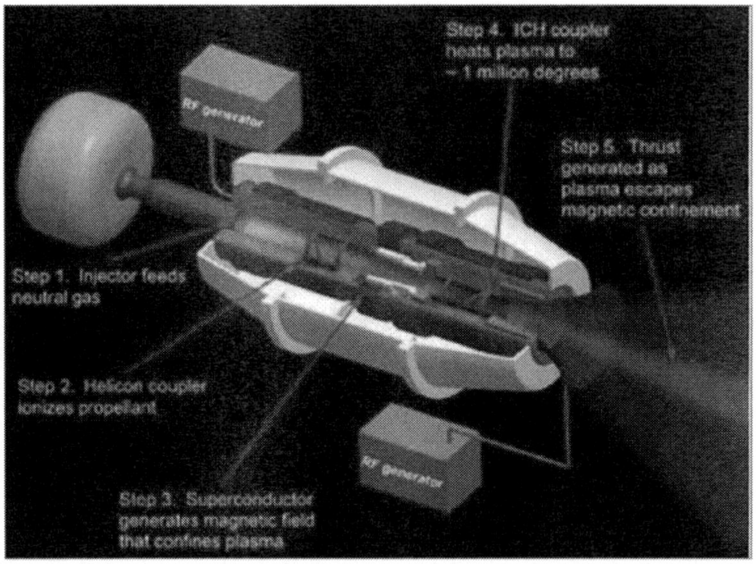

Figure 5. VASIMR propulsion.

5. Proposed Space-based Hybrid Fusion–Fission System

Hybrid Fusion–Fission systems have been considered for decades, largely because 2/3 of the deuteron-deuteron (DD) and deuteron-triton (DT) fusion energy leaves as energetic neutrons. Bethe [16] suggested a fusion hybrid in 1979 as a means to make use of the 2.5 and 14.1 MeV neutrons from the most easily ignited DT plasma. These neutrons would convert fertile isotopes to fissile, fission them, and breed additional tritium from lithium.

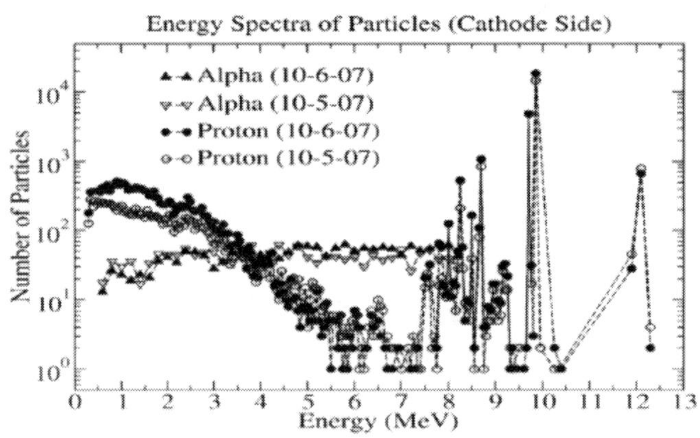

Figure 6. Pd/D Charged particle spectra 1–13 MeV [17].

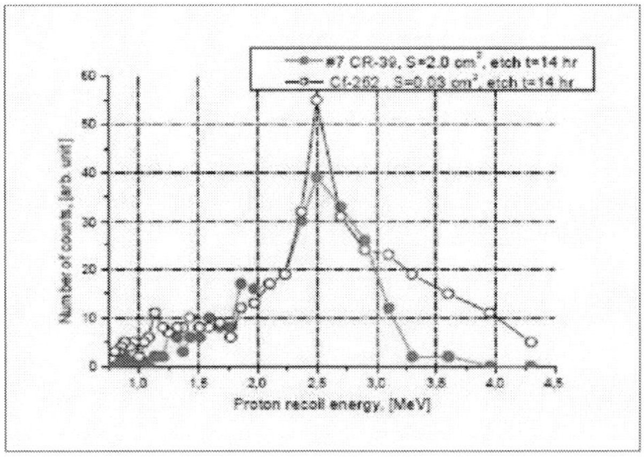

Figure 7. Pd/D Neutron spectra 1–4.5 MeV [18].

Alternatively, the patented (US 8,419,919) Pd/D co-deposition system generates a variety of fast protons, tritons, alphas (^{4}He), helion (^{3}He) [17] (Fig. 6) 2.5 MeV DD (Fig. 7) [18] and 14.1 MeV fusion neutrons [19] These energetic particles fission both natural uranium and thorium with an average neutron energy of $\approx$6.4 MeV [19]. We refer to this as the GeNIE (Green Nuclear Interstitial Energy) Hybrid^{TM}. Given the costly safety and security restrictions on launching either a fissile uranium reactor, or an RTG running on plutonium or americium, a hybrid fusion-fast fission reactor using either natural uranium or thorium has significant advantages. However, present UN treaties [20] for space-based fission reactors specify highly enriched ^{235}U. This requirement has been under UN discussion for years.

5.1. Hybrid reactor

GEC is developing a space-rated, hybrid, fusion-fast-fission, thorium reactor. This has different, and in some ways more stringent, requirements than a terrestrial reactor. For example, it needs a mean-time-to failure exceeding 50,000 h (5.7 years), to enable most missions of interest, as repair is usually impossible [21]. Initially, it will be mated to the NASA Glenn Advanced Stirling Engine that is used with the KRUSTY reactor. This sets specific mass, volume and temperature requirements.

Like KRUSTY, the goal has been to move in steps from tens of watts to tens of kilowatts. The first Kilopower Program demonstration, DUFF, produced 24 We using a Stirling Engine with a heat pipe. These Stirling engines have a conversion efficiency of 10–30% thermal to electric depending upon the temperature difference, ΔT, between the operating temperature and the heat dump. Despite space being cold it is also a well-insulating vacuum.

The Hybrid Reactor upper temperature limit is controlled by materials. But, this has an upside. For example, the hydrided enriched uranium metal fuel rods used in General Atomic TRIGA thermal fission reactors are self-moderating with a rapid, negative temperature co-efficient. *TRIGA reactors are considered inherently safe.*

5.2. ^{238}U and ^{232}Th fission cross-sections

The following figures show the neutron [22] and proton [23] actinide fission cross-sections in barns (1 b=10^{-24} cm^2) and incident particle kinetic energy in meV (10^{-3} eV) to MeV (10^6 eV) units. Colored arrows indicate neutron and proton kinetic energies observed in condensed matter reactions estimating scattering losses through both the co-

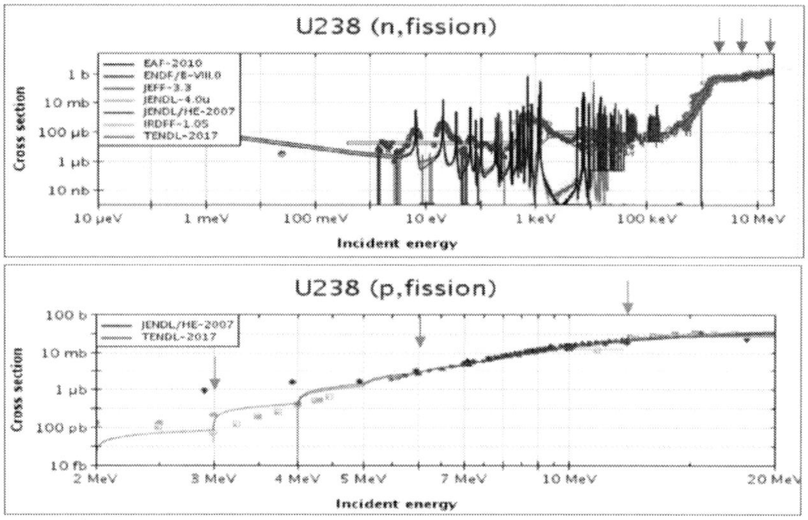

Figure 8. ^{238}U Neutron and proton fission cross-sections.

deposition layer and electrolyte between the active surface and the CR-39. Note that both the log energy and cross-section scales vary by figure. Neutral neutron interactions have a higher cross-section than charged protons due to the lack of a Coulomb Barrier. But, fast protons *will* fission actinides. What is not shown are competing reactions like capture and spallation reactions (n,n′), (n,2n), (n,p), (p,n), etc. These reactions create excited nuclei that do not directly fission.

Sustained thermal neutron fission requires >3%[b] of odd-numbered actinides, like ^{233}U, ^{235}U, and ^{239}Pu that have high thermal neutron (0.025 eV) fission cross-sections, σ_t, of 500–600 b. Most fission reactors use water or graphite to moderate, or thermalize and slow, the 1+ MeV fission neutrons to thermal energy. A fast fission reactor uses unmoderated neutrons but requires nearly 20% enrichment of the odd-numbered, fissile nuclei since the fast neutron fission cross-section, σ_f drops to ≈1 b. Fast and thermal reactors can convert, or breed, even numbered (fertile) nuclei into odd-numbered (fissile) nuclei by neutron capture. Both reactors depend upon a neutron chain reaction producing >2 neutrons/fission. Reactor criticality is maintained by a neutron economy controlling how many neutrons escape (geometry) are captured (fission poisons, control rods and breeding fissile fuel) or are delayed (fission product neutron decay).

5.3. Fusion fast fission reactions

By comparison, our Hybrid reactor is sub-critical relying upon neither fissile fuel nor a fissile chain reaction. It is a fast reactor, fissioning both fissile and fertile nuclei. The fusion-fast-fission reactor is based upon previous work described in "Investigation of Nano-nuclear Reactions in Condensed Matter: Final Report" [24] and discussed in, "Uranium Fission Using Pd/D Co-deposition" [25]. As noted, fast neutron energies of 6.3–6.83 MeV have been measured with average fluxes exceeding 10^6 n/s. The instantaneous flux exceeded this rate.

[b]CANDU reactors can use natural uranium with D_2O which has a reduced n capture cross-section compared to H_2O.

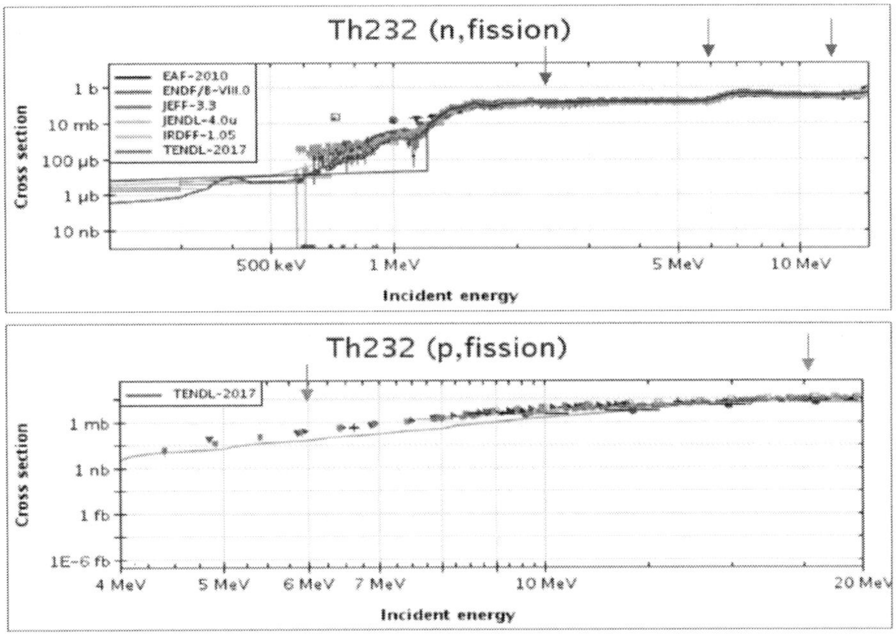

Figure 9. ^{232}Th Neutron and proton fission cross-sections.

Figures 6 and 7 show the measured charged particle and neutron energies and we have observed DT fusion 14.1 MeV neutrons [19]. Figures 8 and 9 indicate these energetic particles will fission both fertile actinides, ^{232}Th and ^{238}U with $\sigma_f \approx 100\ \mu$b - 1 b [22,23]. The higher cross-sections are comparable to the fast fission cross-sections of fissile actinides.

Conventional hot deuteron fusion reaction channels where "D" and "d" are deuterons, "n" neutron, "p" proton, "T" and "t" triton, ^{3}He or helion and α is an alpha particle, ^{4}He or helium ion:

Figure 10 shows the DT Fusion-Fast-Fission reaction. DD fusion-fast-fission is similar. *Both primary and sec-*

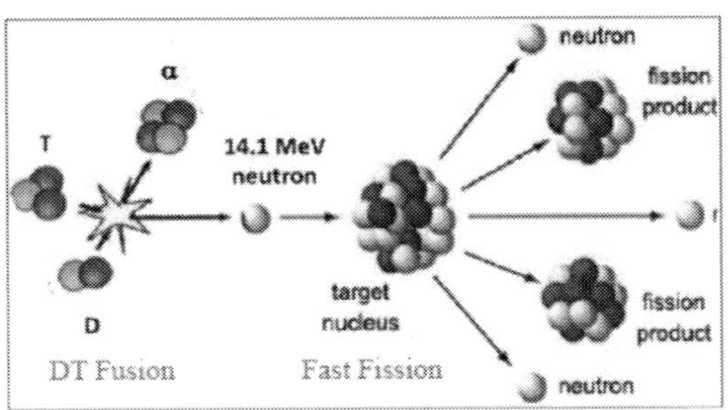

Reaction	Energy (MeV)	Occurrence
D(d,n)^{3}He	4.07	Primary ≈50%
D(d,p)T	3.25	Primary ≈50%
D(^{3}He,p)α	18.3	Secondary
D(t,n)α	17.6	Secondary
T(t,α)2n	11.3	Low probability
^{3}He(^{3}He,α)2p	12.86	Low probability

Figure 10. DT Fusion-fast-fission reaction.

Figure 11. NASA Calorimeter.

ondary fusion and induced fast fission reactions were generated using the patented protocol [24].

Combining the most probable primary and secondary fusion reactions result in $\approx$8 MeV kinetic energy in fast proton, helium and alpha particles with $\approx$16 MeV as neutron kinetic energy. By comparison, actinide fission produces $\approx$170 MeV in charged fission fragments and $\approx$30 MeV in gamma and neutron kinetic energies with $\approx 3 \times 10^{10}$ fissions/watt-thermal Although the fusion neutron energy drives the fission reactions the overriding thermal power is from fission products.

The NASA version of the Hybrid fusion-fast-fission reactor will be tested in a series of stages analyzing neutron flux, stability and pressurized gain with high temperature aqueous operation at <150°C and <4 bar pressure. Low energy X-ray, γ and visible light diagnostics require a 250 ml glass pressure vessel (Fig. 12). Higher temperatures, pressures and volumes require Hastelloy and stainless steel vessels. The reactor is housed in a calorimeter (Fig. 11) that was recently calibrated to 200 mW or better sensitivity with an $\approx$ 40 W upper limit. All of the materials, containment vessels and previous operating procedures have been subject to NASA GRC and Plum Brook Health Physics and Safety reviews as will modified experiment protocols.

5.4. Hybrid fuel rod

Our earliest hybrid fuel rod was a natural uranium wire, (99.3% ^{238}U, 0.7% ^{235}U) 0.05 cm diameter $\times$1 cm long, with a volume of 1.95×10^{-3} cm^3 and a mass of 38 mg. Uranium density is 19.1 g/cm^3. Our previous co-deposition research indicates nuclear reactions occur within a few microns of the surface rather than in the bulk. Consequently, the active region is <1% of the volume and mass, with a 2 μm deep surface. The active cylindrical volume is 3 $\times$

Figure 12. Pressure vessel.

10^{-5} cm^3 with 0.38 mg of natural uranium.

5.5. Hybrid vs. KRUSTY fission power density

Both the Hybrid and the Kilopower KRUSTY are fast fission reactors. The first Hybrid used 38 mg of 99.3% ^{238}U (0.7% ^{235}U) whereas KRUSTY used 28 kg of 93% enriched ^{235}U. KRUSTY ran 28 h and the Hybrid for 33.5 h. One thermal watt requires 3×10^{10} fissions/s. KRUSTY was producing 3 W/cm^3 or 9×10^{10} fissions/cm^3/s. The *unoptimized* Hybrid reactor produced 10^6 n/s from a volume of 3×10^{-5}cm^3 or the rough equivalent of 3.4×10^{10} fissions/cm^3/s: *or 38% of the KRUSTY power density.*

5.6. Modelling

We have experimentally measured and modeled various neutron reflector and moderator materials at the University of Texas, Austin, Nuclear Engineering Teaching Laboratory (NETL) [26], with the review of the health physicist. The experiments were conducted using electrolytic co-deposition and a Thermo-Fisher DT fusion neutron generator within a graphite neutron moderator/reflector (Fig. 14).

Both the DT fusion generator and graphite reflector were modeled using the Los Alamos Monte-Carlo N Particle (MCNP) code [27].

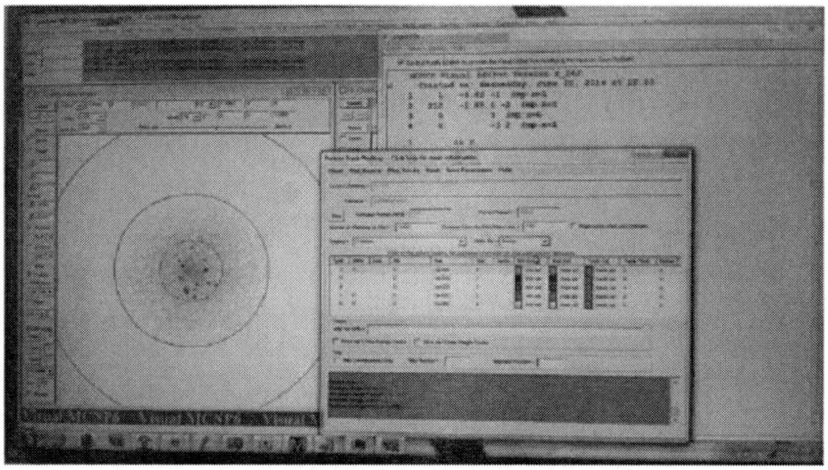

Figure 13. MCNP-6.1 Modeling with Vised Visual Editor.

Although nuclear reactions in condensed matter may appear to side step conventional physics[c] the reaction products follow the Standard Model for particle physics. Consequently, we have modeled the elastic and inelastic nuclear interaction of both neutrons and charged particles as a function of density, nuclear charge, cross-section and the resulting particle mean-free paths.

The Monte-Carlo code, MCNP-6.1 and the University of Michigan developed PoliMi code used with MCNP-2.27 have been used for neutron spectroscopy scintillator response functions. MCNP-6.1 with the visual editor, Vised, [28] was used to model bremsstrahlung triggering of deuterated materials (Fig. 13). Charged particle scattering has been modeled using SRIM/TRIM [29]. Unfortunately, SRIM/TRIM does not handle nuclear interactions and neither MCNP nor the CERN GEANT-4 [30] codes properly model low Z, low energy interactions. Hence, neither code properly models multi-keV energy hydrogen isotope fusion or its products. Previously, we used the known DD and DT fusion neutron energies to model shielding. But, MCNP-6 can be modified to incorporate a subroutine [31] developed in Italy and the UK for DD and DT fusion generators. This code handles deuteron energies between 10 and 50 keV by incorporating SRIM/TRIM scattering tables. With the addition of a screening electron shifted Gamow factor the code may model low energy nuclear reactions in condensed matter.

Nonetheless, both MCNP and GEANT4 have been used to simulate complex reactor configurations, housings, neutron shielding, detector response, as well as neutron moderation, reflection and synthetic HPGe spectra from uranium fission neutron activation and spallation products (Fig. 15). These modeling codes have been instrumental in developing the hybrid fusion-fast-fission reactor.

5.7. Diagnostics

In order to optimize hybrid reactor operation, real-time and post-run diagnostics have been used. The NASA GRC Advanced Energy Conversion Project replicated diagnostics used in JWK and GEC laboratories and developed additional

[c]Nuclear reactions in condensed matter are conventional nuclear reactions in an unconventional place assisted by electron screening and other factors.

Figure 14. Graphite neutron reflector.

ones to study LENR triggering for the NASA GRC Advanced Energy Conversion Project. The nuclear diagnostics include gamma ray, neutron, and charged particle detection as well as witness materials that can be observed via gamma rays *in situ* or counted afterwards. Gamma ray detectors include NaI(Tl), $LaBr_3$, and HPGe, one with a Be window to observe X-rays. Charged particles have been identified afterwords via alpha/beta spectroscopy or can be monitored in real-time with *in situ* fluorescers. CR-39 Solid State Nuclear Track Detectors have been used for both charged particle [32] and neutron spectroscopy [33], and scanned with a TASL 3D scanner [34]. Neutron Time-of-Flight (TOF) is conducted with EJ-200 panels and a CAEN nanosecond timing system (Fig. 16) Real-time neutron spectroscopy uses EJ-309 (Fig. 17), Bicron 501A and Stilbene proton-recoil scintillators with pulse shape discrimination (PSD) [35] These two complimentary techniques give accurate neutron energies from 400 keV to $>$15 MeV. The CERN ROOT [36] system is used for acquisition and data analysis.

6. NASA GRC Plum Brook Station

The NASA Glen Research Center, GRC, includes the Plum Brook Station located 45 miles west of Lewis Field and south of the city of Sandusky, OH, near Lake Erie (Fig. 18) The facility [37] is relatively isolated and built on land reclaimed from a World War II munitions facility. That isolation allowed over 100 live rocket engine tests with many supporting the Apollo moon program. Plum Brook Station was instrumental in developing space nuclear propulsion. The Hydrogen Heat Transfer Facility (HTTF) was built to test NERVA hydrogen exhaust nozzles at high temperature

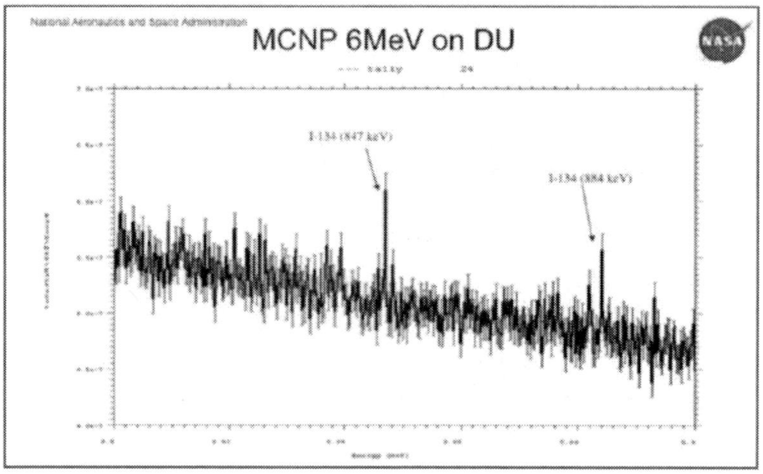

Figure 15. MCNP ^{134}I synthetic HPGe spectra from ^{238}U(γ,f).

and pressure. Later, it became the Hypersonic Test Facility, and housed the Advanced Energy Conversion Project laboratory facilities until the Chemistry/Nuclear Laboratory was setup.

A 60 MWt research reactor was built on site and operated from 1961 to 1973 to test components then fully decom-

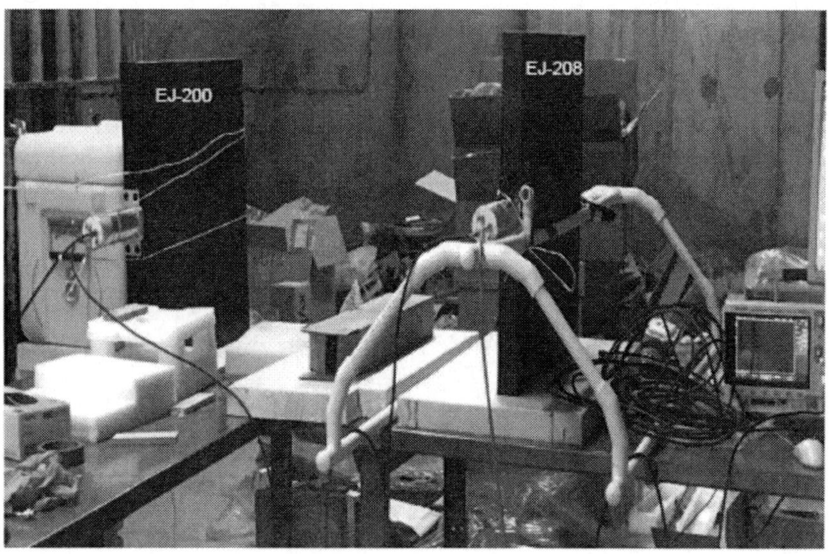

Figure 16. AmBe neutron TOF calibration in Igloo.

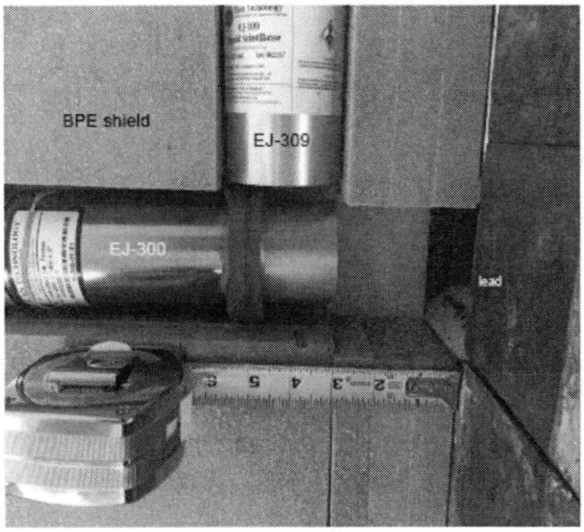

Figure 17. Shielded scintillators.

missioned in the mid-2000s. Although these activities ceased by the 1970s, NASA's need for nuclear power in space, beyond RTG capabilities, for propulsion and planetary power never ceased. This was noted in a 2005 space power review [38] during Project Prometheus and again in a 2015 Johns Hopkins Applied Physics Laboratory (APL) review [39].

As Fig. 19 shows Plum Brook consists of several facilities. The Chemistry/Nuclear Laboratory (Fig. 20) has been used for test sample/vessel construction, post-operation materials assay and will be used for remote reactor operation.

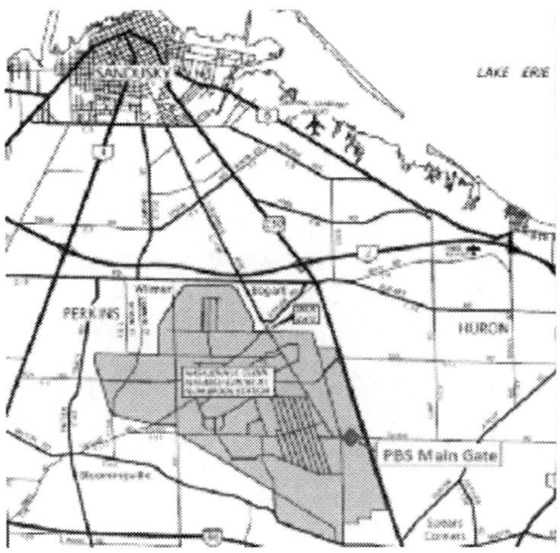

Figure 18. Plum Brook station map.

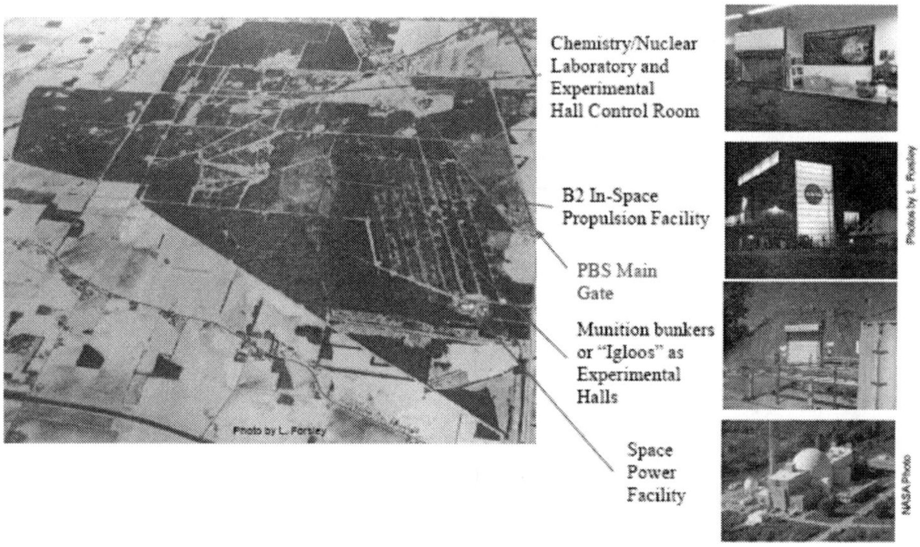

Figure 19. Plum Brook station (PBS) facilities.

Tests have been conducted in the igloos, or experimental halls. The one shown in Fig. 21 has HVAC equipment to dissipate up to 100 kW thermal heat from reactor operation. There are several other igloos available.

The B2 facility is designed for rocket launches and large scale testing under space vacuum at cryogenic temperatures with simulated full spectrum solar irradiation. Complete second stage Centaur rockets were test fired *indoors at altitude* at B2 with the exhaust plume removed. Previous rocket test stands, B1, for testing high energy rockets, and B3, for NERVA nuclear rocket components, predated the Apollo program and were decommissioned.

Figure 20. Chemistry/nuclear laboratory.

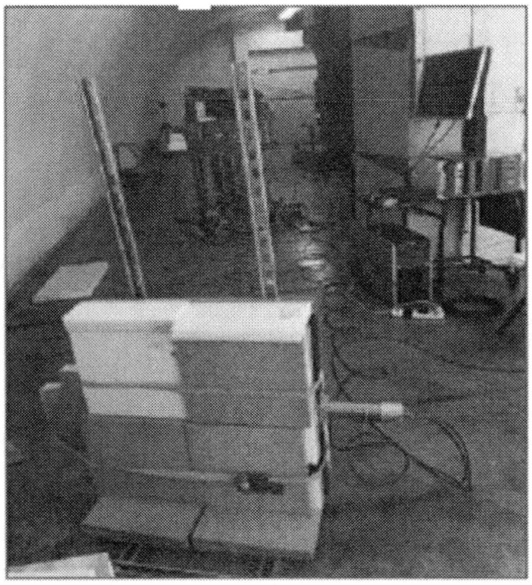

Figure 21. Igloo TOF run.

Figure 22. SPF vacuum.

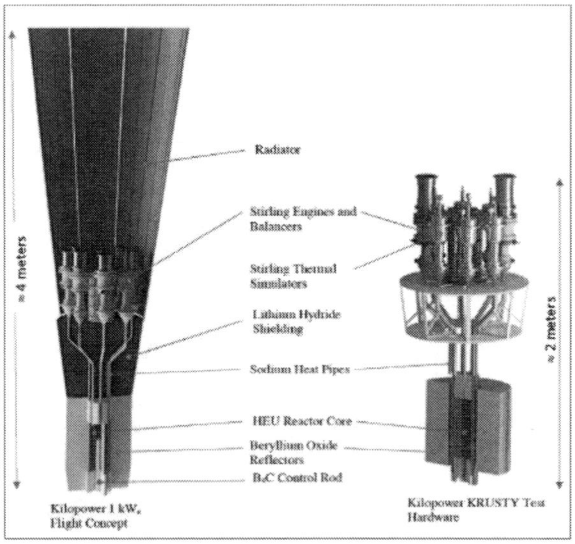

Figure 23. 1 kWe fission reactor flight concept [40]. 1 kWe hybrid generator replaces only the 93% ^{235}U HEU core and B_4C control rod.

The experimental halls, or igloos are used for scale-up and long duration testing. They are operated from a control room adjacent to the igloo or from the Chemistry/Nuclear Laboratory. Figure 21 shows borated polyethylene shielded neutron TOF panels and a lead shielded gamma ray diagnostic with fast timing data collection operating in the igloo. Video, data and control are remotely accessible over a fiber optic network to both the Chemistry/Nuclear Laboratory and the B2 Control Room.

The Space Power Facility (SPF) has the world's largest space vacuum chamber (Fig. 22) along with a shaker table and an acoustic chamber for simulating launch vibration and acoustics. SPF has been used for a variety of US and foreign spacecraft readiness testing including Space-X and ESA modules. The air bag systems used to deliver various Mars rovers were tested under Martian atmospheric pressure. Originally, the facility was designed for, but never used with, nuclear propulsion testing. B2 has a smaller, high vacuum facility, but can simulate long duration, Low Earth Orbit or Martian surface conditions including solar irradiation from UV through Infrared. Smaller space vacuum chambers are available.

7. Hybrid Space Flight Configuration

The goal of the Space Act is to develop a 1 kWe flight ready power system using Hybrid Fusion-Fast-Fission as an alternative to the highly enriched uranium core used in the KRUSTY demonstration [40]. During that test the Stirling engines' efficiency was 30–34% at 50% of Carnot at 800°C. The Hybrid reactor would use all of the power conversion, heat dissipation, and shielding sub-systems developed for the Kilopower Program with a similar form factor, mass and volume as the HEU Reactor Core shown in Fig. 23, replacing only the fissile core. Previously, the non-optimized Hybrid maintained an average nuclear fusion–fission rate/cm^3 that was 38% of the KRUSTY rate. However, since the fusion-fast-fission reactions occur near surface the power scales by surface area, not by mass. The Hybrid reactor components can be folded to occupy less volume to increase the overall fusion-fission-rate and the power output.

8. Conclusion

GEC is developing a high power space-rated generator in conjunction with NASA under an Umbrella Space Act Agreement [42]. Demonstrating a space-flight ready, 1 kWe, (5 kWt) hybrid fusion-fast-fission reactor would meet several existing NASA needs. Scaled up, it would allow deep space nuclear ion propulsion and meet manned planetary power requirements. It is fitting that the GEC Hybrid Reactor technology is being adapted for deep space power missions at Plum Brook Station given two decades of nuclear power in space research there.

The successful KRUSTY Kilopower Program demonstration [40] using highly enriched uranium showed that space-based nuclear fission power needn't be a billion dollar program as KRUSTY was built and tested for $20M. But, by removing the need to launch fissile material or plutonium heat sources, the concerns and costs for production, safety and security are drastically reduced using the GEC Hybrid Reactor. Both the KRUSTY demonstration [41] and the GEC Hybrid Reactor [13] pave the way for high power nuclear reactors in space.

If it is safe enough to launch from Florida, it is safe enough to use in Florida.

Acknowledgements

This research, development and application is a direct result of nearly 30 years of activities beginning at the University of Rochester in the Department of Chemical Engineering under the baleful gaze of Dr. John Huizenga transitioned to JWK facilities and the University of Texas, Austin Nuclear Engineering Teaching Laboratory and under various NCRADAs with the US Naval Research Laboratory (NRL) SPAWAR-Pacific, in conjunction with the Los Alamos National Laboratory (LANL) and the DoE NNSA Special Technology Laboratory (STL), the Naval Air Weapons Station (NAWS), China Lake and the Naval Surface Warfare Center (NSWC), Dahlgren Division. Complimentary research and development continued under contracts and Space Act Agreements with the NASA Glenn Research Center. Related research funding and support was provided by ENECO, JWK, SPAWAR-Pacific ILIR, ONR, DoE, DTRA and NASA.

We specifically thank the following.

JWK:

Dr. Jay Khim, Mr. Scott Phillips, Mr. Merc Joseph, Ms. Adrean Kirk, Ms. Amy Rankin, Ms. Lori Bulat, Ms. Sierra Treanor and Ms. Victoria Leist.

US Navy:

NRL: Dr. David Nagel (*retired*), Dr. Robert August (*retired*), Dr. Gary Phillips (*retired*), and Georgetown University (*retired*) and Dr. Scott Chubb (*deceased*).

SPAWAR-Pacific: Dr. Frank Gordon (*retired*), Dr. Stanislaw Szpak, SPAWAR-Pacific (*deceased*) and Mr. Ryan Friedl, Esq.

NAWS, China Lake: Dr. Melvin Miles (*retired*) and Mr. Donald Thompson (*retired.*)

NSWC, Dahlgren Division: Ms. Stacy Barker, Dr. Louis DeChiaro, Dr. Pearl Rayms-Keller and Ms. Karen Long.

The University of Texas, Austin, Nuclear Engineering Teaching Laboratory:

Dr. Sheldon Landsburger Mr. Tracy Tipping Ms. Rebecca Norris and Dr. Alex Brand.

The University of California, Berkeley:

Dr. Buford Price (Emeritus) and Dr. Winthrop Williams.

The University of California, San Diego:

Dr. Jan Talbott, Mr. Neil Robertson, Mr. Hiroaki Saito, Ms. Julie Yurkovic and Ms. Stefanie Zakskorn.

SRI:

Dr. Fran Tanzella (*retired*) Dr. Michael McKubre (*retired*) and Mr. Ben Earle, Stanford University.

DoE:

Sandia National Laboratory: Dr. Patrick McDaniel (Retired), University of New Mexico.
Special Technologies Laboratory: Mr. Mark Morey, Dr. Jim Tinsley and Mr. Paul Hurley.
LANL: Dr. Thomas Claytor and Dr. Douglas Tasker.
LLNL: Dr. Johan Frenje, also MIT and Dr. Ed Moses (*retired*).
NASA:
GRC AEC: Dr. Theresa Benyo, Dr. Arnon Chait, Dr. David Ellis, Mr. Gustave Fraelick, Mr. Robert Hendricks, Mr. Carl E. Sandifer II, Mr. Philip Smith and Dr. Bruce Steinetz.
GRC: Mr. Chris Blasio, Mr. John Hamley, Dr. Janet Kavandi and Dr. Lee Mason.
GRC Plum Brook Station: Mr. Gerald Hill (*retired*) and Mr. David Stringer.
GRC AEC affiliates: Dr. Bayar Baramsai, Mr. Michael Becks, Dr. Wayne Jennings, Mr. Rich Martin, Mr Nicholas Penney, Dr. Vladimir Pines, Dr. Marianna Pines and Dr. Phillip Ugorowski.
HQ: Mr. Leonard Dudzinski and Dr. Mathew Forsbacka.
JSFC affiliate: Dr Dazhung Zhou (*retired*).
Also several others

Dr. Eugene Malove, Infinite Energy (*deceased*), Dr. Jim Patterson, CETI Inc. (*deceased*), Dr. Andrei Lipson, Institute of Physical Chemistry, Russian Academy of Science (*deceased*), Dr. Alexi Roussetski, Lebedev Institute, Dr. Thomas Passel, EPRI (*retired*), Dr. Brian Ahern, USAFRL (*retired*), Dr. Jenny Vinko, H.E.R.A. and Dr. Paolo Tripodi, H.E.R.A. (*retired*), Dr. John Dash, Portland State University (*deceased*), Dr. Peter Hagelstein, MIT, Dr. Alexander Karabut (*deceased*), Mr. David French, Esq. (*deceased*), Dr. Yasuhiro Iwamura, Mitsubishi, (*retired*), and Tohoku University, Dr. Chino Srinivasan, Bhabha Atomic Research Centre (*retired*), Dr. James Miller, SABIA Inc., Mr. Curt Brown, Point Source Inc., Mr. Sam Tung (*deceased*), Ms. Emily Tung, Mr. Dennis Letts, Letts Laboratory, Dr. Chuck Hurlbut and Mr. James Klecker, Eljen Technologies and Mr. George Murray, Inrad Optics.

Special thanks to: Dr. Martin Fleischmann (*deceased*) and Dr. Stanley Pons, University of Utah (*retired*), Dr. Jacob Jorne, University of Rochester, Fred Jaeger, ENECO (*retired*), Dr. George Miley, University of Illinois (*retired*), Dr. William Wilson, DTRA (*retired*), Mr. Paul Westmeyer, NASA (*retired*), and Dr. Roger Boss, SPAWAR-Pacific (*retired*).

In Memoriam

- Dr. Yoshiaki Arata, *Japan*,
  文化勲章 *Order of Culture*
- Dr. Robert Bass, *US*
- Dr. Yuri Bazhutov, *Russia*
- Dr. John O'M. Bockris, *US*,
  Faraday Medal
- Dr. Scott Chubb, *US*
- Dr. Talbott Chubb, *US*,
  Navy Distinguished Civilian Service Award
- Dr. Norman Cook, *US* and *Japan*
- Dr. John Dash, *US*
- Dr. John Fisher, *US*,
  National Academy of Engineering
- Dr. Martin Fleischmann, *UK*,
  Fellow of the Royal Society
- Dr. Sergio Focardi, *Italy*

- Mr. Hal Fox, *US*
- Mr. David French, Esq., *Canada*
- Dr. John Huizenga, *US,*
  AEC E.O. Lawrence Award
- Dr. P.K. Iyengar, *India,*
  Shanti Swarup Bhatnagar Prize, Chairman Indian Atomic Energy Commission
- Dr. Alexander Karabut, *Russia,*
  Preparata Medal
- Dr. Yan Kucherov, *Russia and US,*
  First Truffle Prize (Preparata Medal precursor)
- Dr. Sven Kullander, *Sweden*
- Dr. Andre Lipson, *Russia*
- Dr. Eugene Mallove, *US*
- Dr. Douglas R. O. Morrison, *UK*
- Dr. Richard Oriani, *US,*
  Alexander Von Humboldt Prize
- Dr. James Patterson, *US*
- Dr. Giuliano Preparata, *Italy*
- Mr. Evan Ragland, *US*
- Mr. James Reding, *US*
- Dr. Andrew Riley, *US*
- Dr. Julian Schwinger, *US,*
  Physics Nobel Laureate
- Mr. Kenneth Shoulders, *US*
- Dr. Stanislaus Szpak, *US*
- Dr. Kevin Wolf, *US*

Losses by Country	
Austria	1
Bahamas	1
Canada	1
France	2
India	1
Italy	8
Japan	4
Netherlands	1
Russia	9
Sweden	1
United Kingdom	2
United States	26

I have added two of the more vociferous voices who have left the playing field: Huizenga and Morrison, both of whom

I knew personally. Huizenga shut down my AMS analysis of a Pd cathode while Morrison and I discussed the wisdom of eating Wimpy hamburgers in the UK in light of Creutzfeldt-Jakob disease.

References

[1] R. Taylor, Prometheus Project: Final Report, *Jet Propulsion Laboratory,* 982-R120461 (2005).

[2] T.L. Benyo, B.M. Steinetz, R.C. Hendricks, R.E. Martin, L.P. Forsley, C.C. Daniels, A. Chait, V. Pines, M. Pines, N. Penney, T.R. Kamm and M.D. Becks, Investigation of deuterium loaded materials subject to X-ray exposure, NASA/TM–2015-218491/REV1, arXiv:1704.01183 (2015).

[3] B.M. Steinetz, T.L. Benyo, V. Pines, M. Pines, L.P. Forsley, P.A. Westmeyer, A. Chait, D. Becks, R.E. Martin, R.C. Hendricks, N. Penney, A.M. Marsolais and T.R. Kamm, Experimental observations of nuclear activity in deuterated materials subjected to a low-energy photon beam, NASA/TM–2017-218963, arXiv:1704.00694 (2017).

[4] The Martian, 20th Century Fox.

[5] J.K. Pálfalvi, Yu. Akatov, J. Szabó, L. Sajó-Bohus, I. Eördögh, International space station, *Radiiat. Prot. Dos* **110** (2004) 393.

[6] J.W. Dankanich, B. Vondra and A.V. Ilin, Fast transits to mars using electric propulsion, *46th AIAA/ASME/SAE/ASEE Joint Propulsion Conference and Exhibit*, Nashville, TN, 25–28 July 2010.

[7] https://fas.org/nuke/space/nerva-spec.pdf.

[8] The NASA Plum Brook facility tested thermal nuclear components in the hydrogen heat transfer facility (HTTF) later repurposed as the hypersonic test facility (HTF) and then an AEC facility.

[9] M. Dudeck, F. Doveil, N. Arcis and S. Zurbach, Plasma propulsion for geostationary satellites and interplanetary spacecraft, *Rom. J. Phys* **56** (Supplement) (2011) 3–14.

[10] https://www.jpl.nasa.gov/news/fact_sheets/ds1.pdf.

[11] L. Mason, High power NEP power concepts, *NASA GRC*, 2017.

[12] D. Poston, M. Gibson, T. Godfroy and P. McClure, Design of the Krusty reactor, *ANS NETS 2018 – Nuclear and Emerging Technologies for Space,* American Nuclear Society, 2018.

[13] L. Forsley, Space power: the genie fast-fission sub-critical core, *ANS NETS 2018 – Nuclear and Emerging Technologies for Space,* American Nuclear Society, 2018.

[14] C.D. Orth, Overview of the VISTA Spacecraft Concept Powered by Inertial Confinement Fusion, *UCRL-JC-141514*, 2000.

[15] http://www.adastrarocket.com/aarc/.

[16] H.A. Bethe, The fusion hybrid, *Phys Today* (May, 1979) 44–51.

[17] P.A. Mosier-Boss, F.E. Gordon, L.P. Forsley and D. Zhou, Detection of high energy particles using CR-39 detectors Part 1: results of microscopic examination, scanning, and LET analysis *Int. J. Hydrogen Energy* **42** (1) (2017) 416–428. D. Zhou, updated results obtained from in-depth LET analysis, personal communication (April, 2019).

[18] A.S. Roussetski, A.G. Lipson, E.I. Saunin, F. Tanzellaand and M. McKubre, Detection of high energy particles using CR-39 detectors Part 2: results of in-depth destructive etching analysis *Int. J. Hydrogen Energy* **42**(1) (2017) 429–436.

[19] P.A. Mosier-Boss, S. Szpak, F.E. Gordon and L.P.G. Forsley, Triple tracks in CR-39 as the result of Pd/D co-deposition: evidence of energetic neutrons, *Naturwissenchaften* **96**(1) (2009) 135–142.

[20] http://www.unoosa.org/pdf/publications/ST_SPACE_061Rev01E.pdf, pp 44–45.

[21] TRW built both Voyager spacecraft with 1 year repair warranties if returned to their facility. However, within a year of launch the spacecraft were half-way to Jupiter!

[22] http://www.oecd-nea.org/janisweb/book/neutrons/.

[23] http://www.oecd-nea.org/janisweb/book/protons/.

[24] P.A. Mosier-Boss, L.P. Forsley and P. McDaniel, Investigation of nano-nuclear reactions in condensed matter: final report, *Defense Threat Reduction Agency* 2016, 41–49.

[25] P.A. Mosier-Boss, L.P. Forsley and P. McDaniel, Uranium fission using Pd/D Co-deposition *J. Condensed Matter Nucl. Sci.*, accepted.

[26] A. Brand and R. Bost, Report on MCNP work for neutron generator for project GeNIE, *NETL* and *JWK* (November, 2011).

[27] https://mcnp.lanl.gov/mcnp_how_to_get_to_mcnp.shtml.

[28] http://www.mcnpvised.com.
[29] http://www.srim.org.
[30] https://geant4.web.cern.ch.
[31] A. Milocco, A. Trkov and M. Pillon, A Monte Carlo model for low energy D–D neutron generators, *Nucl Instrum. Methods Phys Res* B, **271** (2012) 6–12.
[32] P.A. Mosier-Boss, S. Szpak, F.E. Gordon and L.P.G. Forsley, Characterization of tracks in CR-39 detectors obtained as a result of Pd/D co-deposition, *Eur. Phys. J. Appl. Phys.*,**46** (2009) 30901.
[33] P.A. Mosier-Boss, S. Szpak, F.E. Gordon and L.P.G. Forsley, Triple tracks in CR-39 as the result of Pd/D co-deposition: evidence of energetic neutrons *Naturwissenschaften* **96** (2009) 135–142.
[34] http://www.tasl.co.uk.
[35] B.M. Steinetz, T.L. Benyo, A. Chait, R.C. Hendricks, L. Forsley, B. Baramsai, P.B. Ugorowski, M.D. Becks, V. Pines, M. Pines, R.E. Martin, N. Penney, G.C. Fralick and C.E. Sandifer II, Novel nuclear reactions observed in bremsstrahlung irradiated deuterated metals, in review.
[36] https://root.cern.ch.
[37] M.D. Bowles, *Science in Flux: NASA's Nuclear Program at Plum Brook Station, 1955–2005* NASA/SP–2006-4317, 2006.
[38] Space Studies Board, NRC, Priorities in space science enabled by nuclear power and propulsion, *National Academies Press*, http://nap.edu/11432 (2006).
[39] R.L. McNutt (Study Chair), *Nuclear Power Assessment Study Final Report*, The Johns Hopkins University Applied Physics Laboratory (February 4, 2015.)
[40] M. Gibson, D. Poston, P. McClure, T. Godfroy, M. Briggs and J. Sanzi, The Kilopower reactor using Stirling technology (KRUSTY) nuclear ground test results and lessons learned (2018). https://ntrs.nasa.gov/archive/nasa/casi.ntrs.nasa.gov/20180005435.pdf.
[41] https://www.nasa.gov/press-release/demonstration-proves-nuclear-fission-system-can-provide-space-exploration-power.
[42] Umbrella Space Act Agreement, SAA3-1529, and Annex No. 1, SAA3-1529-1, NASA Glenn Research Center and GEC for development and testing of a high power space generator.
[43] Kilowatt Brewery, San Diego, logo used with permission. *It's not just beer!*.

The NASA Kilopower Program 2012 NASALANL/NNSA Demonstration Using Flattop Fission, (DUFF), experiment had *Homer Simpson's favorite beer...*

. . . GeNIE Hybrid^{TM} has *Kilowatt*!

J. Condensed Matter Nucl. Sci. 29 (2019) 119–128

Research Article

Anomalous Heat Effects Induced by Metal Nano-composites and Hydrogen Gas

Yasuhiro Iwamura*, Takehiko Itoh† and Jirohta Kasagi

Research Center for Electron Photon Science, Tohoku University, Sendai, Miyagi 982-0826, Japan

Akira Kitamura‡, Akito Takahashi, Koh Takahashi, Reiko Seto and Takeshi Hatano

Technova Inc., Tokyo 100-0011, Japan

Tatsumi Hioki and Tomoyoshi Motohiro

Green Mobility Research Institute, Institutes of Innovation for Future Society, Nagoya University, Nagoya, Aichi 464-8603, Japan

Masanori Nakamura, Masanobu Uchimura, Hidekazu Takahashi and Shunsuke Sumitomo

Research Division, Nissan Motor Co. Ltd., Yokosuka, Kanagawa 237-8523, Japan

Yuichi Furuyama

Graduate School of Maritime Sciences, Kobe University, Kobe, Hyogo 658-0022, Japan

Masahiro Kishida and Hideki Matsune

Graduate School of Engineering, Kyushu University, Fukuoka 819-0395, Japan

---

Abstract

Collaborative research among Technova Inc., Nissan Motor Co. Ltd., Kobe Univ., Kyushu Univ., Nagoya Univ. and Tohoku Univ. was done from Oct. 2015 to Oct. 2017. For this collaborative work, a new accurate oil mass-flow... (continued in the next page)

Keywords: Anomalous heat, Deuterium gas, Excess heat, Gas loading, Heat burst, Hydrogen gas, Metal nano-composite, Nano-material, Replication

---

*Corresponding author. E-mail: iwamura@lns.tohoku.ac.jp.
†Also at: CLEAN PLANET Inc., 105-0022 Japan.
‡Also at: Graduate School of Maritime Sciences, Kobe University, Kobe 658-0022 Japan.

© 2019 ISCMNS. All rights reserved. ISSN 2227-3123

(continued from the title page)
calorimetry system was developed at Tohoku University to replicate anomalous heat generation experiments reported by Technova and the Kobe Univ. Group. In this paper, we present evidence of anomalous excess heat effects obtained from experiments at our laboratory at Tohoku University. Excess energy experiments were done using nano-sized metal composites with H_2 or D_2 gas. Anomalous excess heat generation were observed for all the samples at elevated temperature (150–350°C) except for the palladium nanoparticles embedded in mesoporous SiO_2 (PSn1). The amount of anomalous heat generation per hydrogen atom ranged from 15 eV/H or D to 2.1 keV/H or D, which is too much to be explained by any known chemical process. Coincident burst events of pressure and gas temperature were observed for all the experiments using the $CuNi_7Zr_{15}$–O_x with H_2 gas, which suggested sudden energy releases in the reaction chamber. These observations suggest large local energy bursts. Excess heat experiments using the same material at Kobe and Tohoku Universities showed similar experimental results. Qualitative reproducibility between the Kobe and Tohoku experiments was good.

1. Introduction

Akira Kitamura and Akito Takahashi team of Technova Inc. and Kobe University have been studying anomalous heat effects by the interaction of metal nanoparticles and hydrogen isotope gas for several years [1–3]. Based on their results, a new research project started on October 2015 through the collaboration of six Japanese organizations: Technova Inc., Nissan Motor Co. Ltd., Kyushu University, Kobe University, Nagoya University and Tohoku University.

The objective of the collaborative research is to clarify the existence of the anomalous heat generation phenomena and to confirm the reproducibility of the phenomena. For these purposes, anomalous heat experiments at Kobe and Tohoku Universities and sample preparation and analyses at Nissan, Kyushu and Kobe Universities have been performed. Replication experiments were performed at Tohoku University using a high-quality heat measurement system similar to the apparatus at Kobe University.

The Research Center for Electron Photon Science of Tohoku University and CLEAN PLANET Inc. established a collaborative research division in 2015 – the Condensed Matter Nuclear Reaction Division [4] – and research on anomalous heat generation was started. Replication efforts have been made on two types of experiments as a first step at our division in Tohoku University; one is the present collaborative work [1–3], and the other is the experiment using nano-Pd/Ni fabricated by glow discharge with D_2 gas [5].

A summary of the collaborative research is shown in Table 1 and in Fig. 1. Excess energy experiments were done using nano-sized metal composites with H_2 or D_2 gas. The nano-sized metal composite samples are composed of nickel, palladium or copper nanoparticles embedded in ZrO_2 or SiO_2 particles with diameter of several microns to tens of micronsμ [1–3]. Experiments using CNZ($Cu_1Ni_7Zr_{15}$–O_x) with H_2, PNZ($Pd_1Ni_7Zr_{15}$–O_x) with D_2, CNS(Cu_1Ni_{10}/SiO_2) with H_2 and PSn1(Pd/SiO_2) with D_2 were performed. Anomalous excess heat generation was observed for all the samples at elevated temperatures (150–350°C), except for the palladium nanoparticles embedded in mesoporous SiO_2 (PSn1). The amount of anomalous heat generation per hydrogen atom ranged from 10 eV/H or D to 100 eV/H or D, which could not be explained by any known chemical process (Fig. 1). Experiments, Nos. 15 and 16, were performed to demonstrate the reproducibility of this excess heat effect.

2. Experimental

A schematic of the experimental apparatus is shown in Fig. 2; it is based on the instrument described previously in [1] with some improvements. The experimental set-up was described previously in [6,7].

The reaction chamber (RC) that contains nickel-based binary nanocomposites and hydrogen isotope gas is located at the center of Fig. 2. Heat generation from the RC is estimated by mass flow calorimetry. The inlet oil temperature

Table 1. Summary of collaborative research

No.	Place	Sample name	Composition	Gas	Temp. (°C)	Max. power (W)	Released enthalpy
1	Kobe	PS3	Pd/SiO_2	D	200–300	~ 0	~ 0
2	Kobe	PNZ3	$PdNi_7Zr_{15}–O_x$	D	200–300	10	7.7 MJ/mol-D, 80 eV/D
3	Kobe	PNZ3r	PNZ3- re oxidized	H	200–300	8.0	2.0 MJ/mol-H, 21 eV/H
4	Kobe	CNZ5	$CuNi_7Zr_{15}–O_x$	H	200–300	3.3	3.6 MJ/mol-H, 37 eV/H
5	Tohoku	PNZ4s	$PdNi_7Zr_{15}–O_x$	D	160–300	3.3	1.4 MJ/mol-D, 15 eV/D
6	Tohoku	CNZ5s	$CuNi_7Zr_{15}–O_x$	H	160–250	5.0	6.5 MJ/mol-H, 68 eV/D
7	Kobe	PSf1	$Pd/SiO_{2-covered}$	D	200–300	~ 0	~ 0
8	Tohoku	PSn1	$Pd/meso–SiO_2$	D	200–300	~ 0	~ 0
9	Kobe	CNS3	$CuNi_{10}/SiO_2$	H	200–400	4.4	67 MJ/mol-H, 700 eV/H
10	Tohoku	CNS3s	$CuNi_{10}/SiO_2$	H	15–300	4.2	11 MJ/mol-H, 120 eV/H
11	Kobe	PNZ5	$PdN_{i7}Zr_{15}–O_x$	D	250–350	4.2	7.6 MJ/mol-D, 70 eV/D
12	Tohoku	CNZ6s	$CuNi_7Zr_{15}–O_x$	H	150–300	2.5	5.3 MJ/mol-H, 55 eV/H
13	Kyushu	PNZ	$PdNi_7Zr_{15}–O_x$	H	23-450	–	–
14	Kobe	PNZ6	$PdNi_{10}Zr_{20}–O_x$	D	250–350	25	200 MJ/mol-D, 2.1 keV/D
15	Kobe	PNZ7k	$PdNi_7Zr_{15}–O_x$	D	250–350	5.0	3.4 MJ/mol-D, 35 eV/D
16	Tohoku	PNZ7s	$PdNi_7Zr_{15}–O_x$	D	250–350	4.0	3.0 MJ/mol-D, 31 eV/D

(T_{in}) is measured by three independent thermocouples and the outlet oil temperature (T_{out}) is also measured by three thermocouples. A oil coolant (BT400; Matsumura Oil Co. Ltd.) enabled the use of the flow-calorimetry method at elevated temperatures. The coolant is driven by a digital liquid tubing pump. A 1 kW sheath heater (W_1) is spirally wound on the outer surface of the RC and a 200 W cartridge heater (W_2) is located at the central axis of the RC to

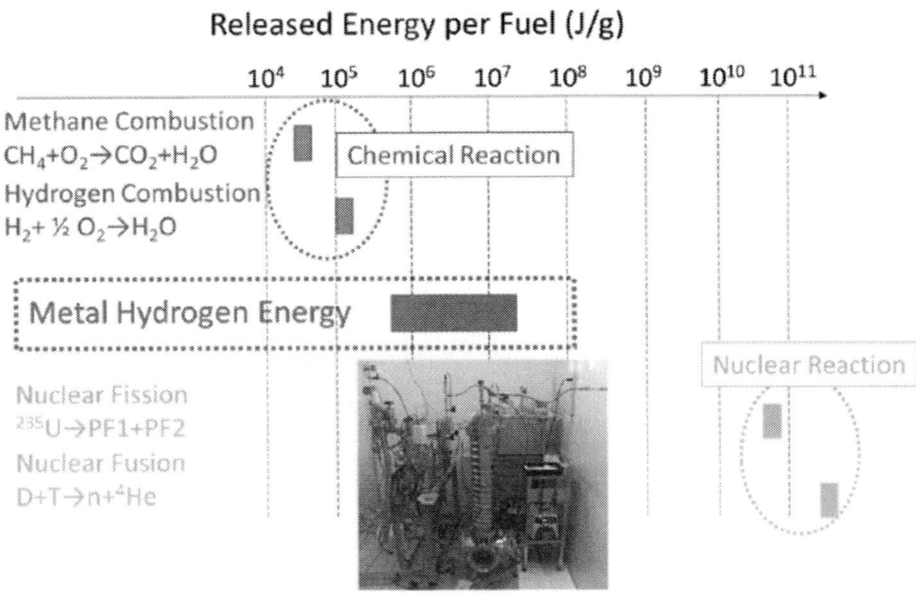

Figure 1. Released energy per gram of fuel; summary of collaborative research.

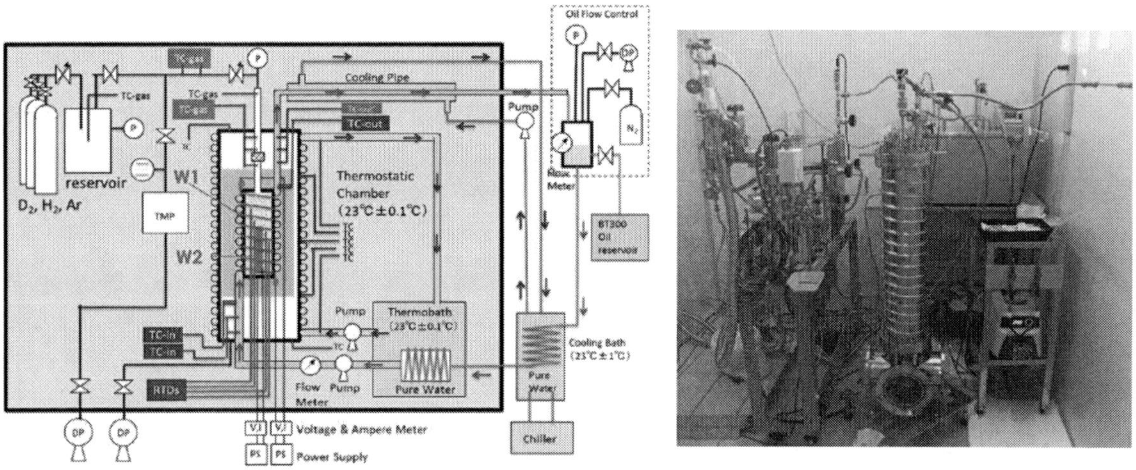

Figure 2. Experimental set-up.

heat the sample in the RC. The power to the heater is supplied by a precision regulated DC power supply. The input electrical power for each heater is monitored by two independent voltage and current meters. H_2 or D_2 gas is fed from a reservoir through a super needle flow regulator to the sample in the RC. Pressures in the RC and the reservoir are monitored continuously. Temperature distribution in the RC is measured by four Resistant Temperature Detectors (RTDs) and temperatures along the oil coolant pipe and the stainless-steel pipe for gas introduction are monitored by thermocouples. Having several temperature measurement points enables us to judge the accuracy of an observed excess heat measurement, although the heat recovery rate of this system rate is low. All the components in this thermostatic chamber are controlled at $23 \pm 0.1°C$ to avoid the influence of outside temperature fluctuations.

Sample preparation and experimental procedures were given in [6,7]. At first, the RC was evacuated by a turbo molecular pump and then heated to 200–300°C remove H_2O or other impurity gases such as hydrocarbons. After baking, the RC is cooled to room temperature. H_2 or D_2 gas were stored in reservoir chambers (1 and 2 L, respectively) at a pressure of about 1 MPa. H_2 or D_2 gas was introduced into the RC by opening the super needle valve.

Initially, we observed H_2 or D_2 gas absorption and heat generation at room temperature due to the presence of palladium. Subsequently, we applied electric power to the heaters located at the inside and outer-surface of the RC to increase the sample temperature. Values of temperature, pressure, voltage, current and a flow rate were logged during experiments. From these measurements, we estimated H or D absorption rates and excess heat generation from the samples. Blank runs to obtain the heat recovery rate of the experimental apparatus were performed using 1 mm-diameter ZrO_2 beads of total mass 1300 g in the RC before or/and after foreground runs. Typical masses of the PNZ and CNZ samples were 100 g.

3. Results and Discussion

3.1. Heat analysis and its error estimation

Heat analysis of this system is based on the equation:

$$\eta Q = F_{\mathrm{R}} \rho (T_{\mathrm{ave}}) C T_{\mathrm{ave}} (T_{\mathrm{out}} - T_{\mathrm{in}}), \tag{1}$$

where η is the heat recovery rate, Q is the heat release rate, F_{R} is oil flow rate, $\rho(T_{\mathrm{ave}})$ is the oil density as a function of temperature, $C(T_{\mathrm{ave}})$ is heat capacity, T_{out} and T_{in} is the outlet and inlet temperatures of the coolant oil, respectively. Physical data of $\rho(T)$ and $C(T)$ of the coolant oil are already known. As temperature dependence of $\rho(T)$ and $C(T)$ is linear, we can postulate that T_{ave} is equal to $(T_{\mathrm{out}} + T_{\mathrm{in}})/2$.

The value of Q is expressed as

$$Q = W_1 + W_2 + H_{\mathrm{EX}}, \tag{2}$$

where W_1, W_2 and H_{EX} are the input power of heater 1, the input power of heater 2 and the excess heat power from the RC.

Based on these equations, η is determined as a function of $(W_1 + W_2)$ by a blank run, because Q, F_{R}, $\rho(T_{\mathrm{ave}})$, $C(T_{\mathrm{ave}})$ and $(T_{\mathrm{out}}-T_{\mathrm{in}})$ are obtained by experimental data. H_{EX} is calculated by a foreground run using the determined η. We simplify Eqs. (1) and (2) to estimate experimental error.

$$H_{\mathrm{EX}} = \frac{F_{\mathrm{R}}\rho C}{\eta}\Delta T - W, \qquad \Delta T = T_{\mathrm{out}} - T_{\mathrm{in}}, \qquad W = W_1 + W_2.$$

Considering that experimental variables are F_{R}, ΔT and W, we can assume that error range of the calculated excess heat is the sum of fluctuations of oil flow rate, temperature difference and input electrical power.

$$\delta(H_{\mathrm{EX}}) \approx |\delta(F_{\mathrm{R}})|\frac{\rho C \Delta T}{\eta} + |\delta(\Delta T)|\frac{f_{\mathrm{R}}\rho C}{\eta} + |\delta(W)|. \tag{3}$$

Experimental data show that the largest contribution to the error of H_{EX} is the F_{R} term; W is the most stable parameter.

3.2. Burst-like coincident increase events of chamber pressure and gas temperature

During the collaborative work, we observed interesting coincident burst events of the pressure in the reaction chamber and gas temperature for CNZ5s and CNZ6s experiments in Table 1, which suggested sudden energy releases in the reaction chamber. After the period of collaborative research, we set out to replicate the phenomena using a re-oxidized CNZ5s sample. As shown in the last column of Table 2, we were able to replicate this. A summary of these experiments is shown in Table 2. These samples contain the same compositions although sample treatment was different for the No. 3 sample. The CNZ5sR sample (No. 3) was prepared by oxidizing the used CNZ5s sample (No. 1) in air for 180 h at 450°C.

Released excess enthalpy for these CNZ samples was 3.3– 6.5 MJ/mol-H, which is too much to be explained by any known chemical reactions. Furthermore, coincident burst events of chamber pressure and gas temperature were

Table 2. Summary of CNZ experiments at Tohoku university.

No.	Samples	Date	Temp. (°C)	Max. power (W)	Released excess enthalpy	Coincident increase of Pr and E2
1	CNZ5s ($CuNi_7Zr_{15}$–O_x)	05–19 Aug. 2016	160–250	5.0	6.5 MJ/mol-H 68 eV/H	Yes; $W_1 = 134$ W, $W_2 = 0$
2	CNZ6s ($CuNi_7Zr_{15}$–O_x)	1 Mar. 1 to 18 April, 2017	150–350	2.5	5.3 MJ/mol-H 55 eV/H	Yes; $W_1 = 80$ W, $W_2 = 0$
3	CNZ5sR ($CuNi_7Zr_{15}$–O_x (Re-oxidized)	11–30 May, 2018	140–260	2.6	3.3 MJ/mol-H 34 eV/H	Yes; $W_1 = 134$ W, $W_2 = 0$

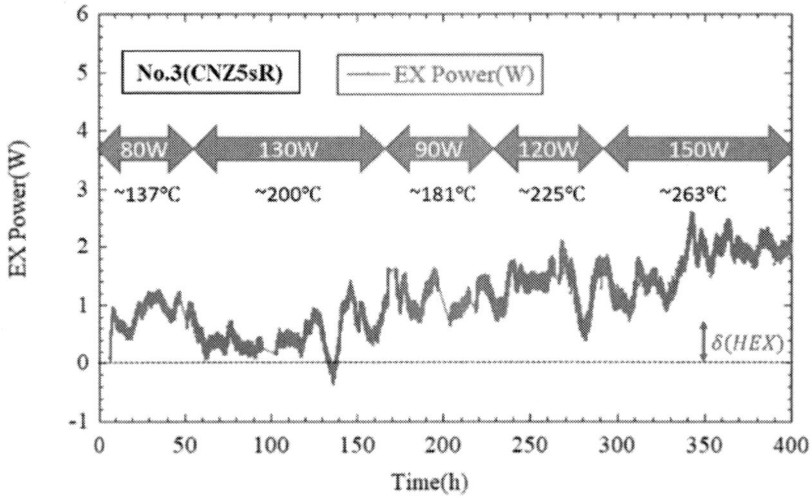

Figure 3. An example of excess heat generation; CNZ5sR with H_2.

observed in all cases. Excess heat data for CNZ5sR is shown as an example in Fig. 3. Total input power $(W_1 + W_2)$ and rough temperature in the RC is shown. Although the measured excess power during 130 W input power was not obvious, excess power continued for more than 200 h. The error range of the excess power for 150 W input based on Eq. (3) is shown in Fig. 3. The error ranges for other input power levels are smaller than the range for 150 W input.

Figure 4 shows blank run data for the pressure of RC (Pr) and upper pipe temperature (E2). The pipe is connected

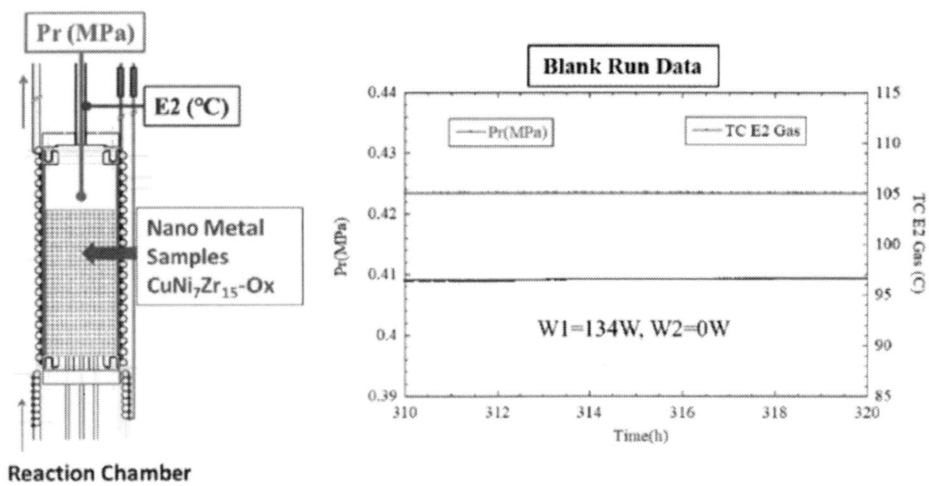

Figure 4. Blank run data for reaction chamber pressure (Pr) and upper pipe temperature (E2).

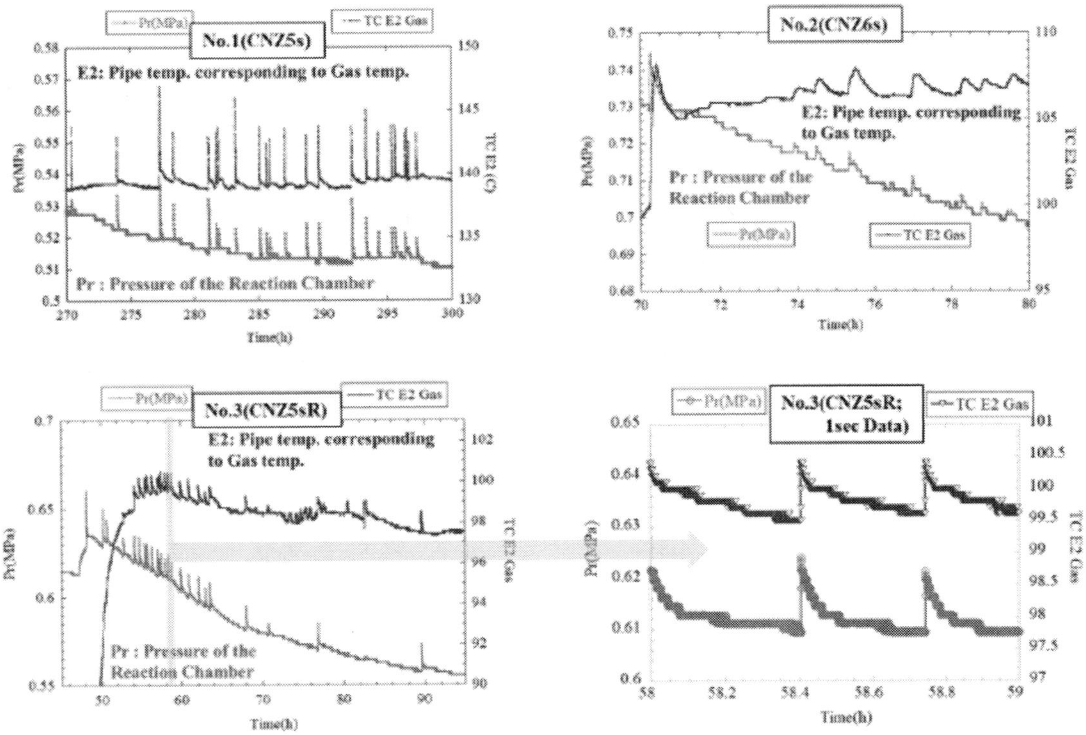

Figure 5. Burst-like coincident increase events of chamber pressure (Pr) and upper side temperature (E2).

to the hydrogen gas reservoir through a super needle flow regulator as shown in Fig. 2. A pressure gauge is connected to the pipe to measure the pressure of RC (Pr) and the surface temperature is monitored by thermocouple (E2). Pr and E2 are kept nearly constant under constant input power and oil flow rate conditions as shown in Fig. 4.

During certain test runs, coincident burst events of Pr and E2 were observed. These events were observed when we used the $CuNi_7Zr_{15}$–O_x sample with H_2 gas when the cartridge heater (W_2) power was set to zero. In other words, the coincident events occurred when the input power was supplied by the sheath heater (W_1), which was wound on the outer surface of the RC when it contained $CuNi_7Zr_{15}$–O_x nano-particles.

These coincident burst events occurred at random for the three experiments in Table 2; the data for these events are given in Fig. 5. The pressure and temperature increases were significant and cannot be explained by noise. For experiment No. 3, an expanded time scale is provided. For Pr and E2 the data acquisition interval was 1 s. Examining these two coincident events closely, we see that the pressure in the RC increased first, followed by a temperature increase of the pipe surface few seconds later. Noting that the time response of the pressure gauge is faster than the thermocouple attached to the pipe, it is reasonable to assume that high temperature H_2 gas was emitted intermittently.

Although it is difficult to interpret the data at the upper end of the RC, coincident temperature and pressure rises suggest burst energy releases from the RC. The temperature distribution in the RC and the temperature gradient in the metal nanocomposites must play some role in inducing these events, as these events were observed only for $W_2 = 0$. An important observation is that these coincident burst events were replicated qualitatively.

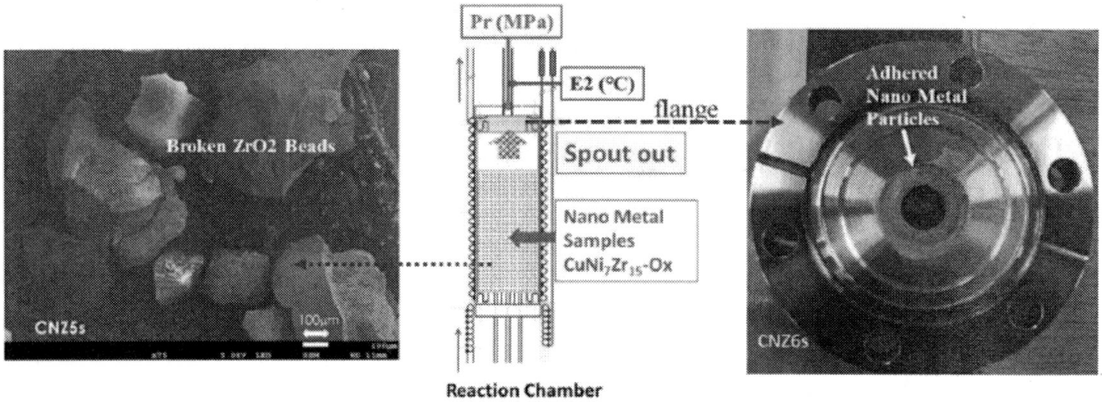

Figure 6. Examples of broken ZrO_2 beads and adhered nano-material on the upper flange after excess heat generation.

Other experimental observations supporting the assumption of burst energy releases are shown in Figs. 6 and 7. After completing these CNZ experiments, the metal nanocomposite samples were sieved out to separate them from the 1-mm diameter ZrO_2 beads. We found that broken parts of ZrO_2 beads were mixed with the sample as shown in the left photo of Fig. 6. ZrO_2 beads are very hard and difficult to break; they are used to grind materials into a powder. This suggests that large local stresses, presumably due to heat bursts, shattered the ZrO_2 beads.

Other evidence that supports the assumption of burst energy release is shown in the right photo of Fig. 6. This photograph shows the lower side surface of the upper flange at the conclusion of the CNZ6s experiment. Material adhered to the surface around the pipe of the upper flange. This white substance was dissolved by HNO_3 and subsequently analyzed by ICP-MS. Figure 7 shows the results of ICP-MS analysis of this material. Reference data for HNO_3 solutions are plotted in black and blue lines and test material data are plotted in red. This analysis shows large clear peaks for Cu, Ni and Zr. As the flange is made of stainless steel, this analysis suggests that $CuNi_7Zr_{15}$–O_x particles were embedded in the flange. This observation is consistent with the assumption of burst energy release and suggests that

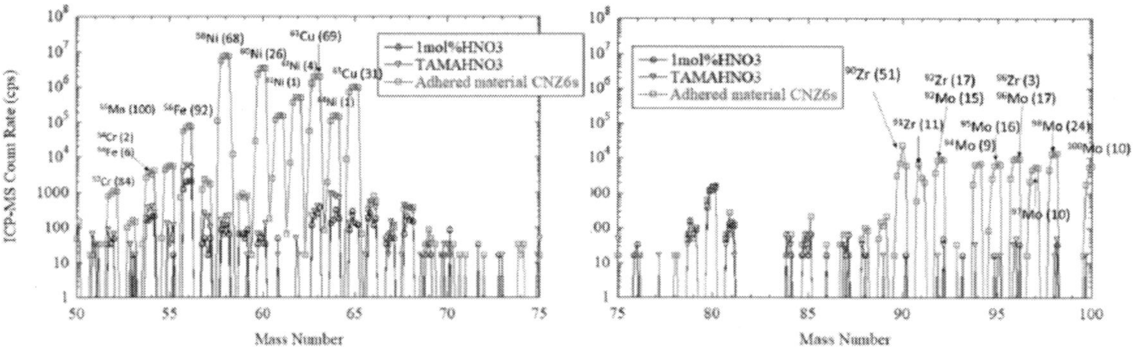

Figure 7. ICP-mass analysis for the adhered nano-metal particles on the upper flange.

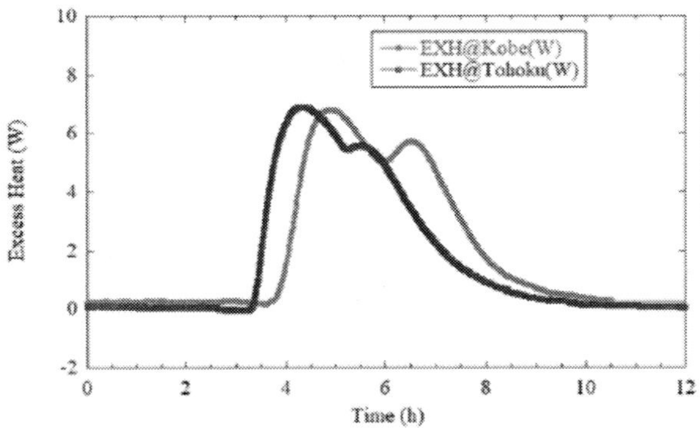

	Kobe	Tohoku
Released Enthalpy (KJ/mol-D)	53.7	50.9
Released Enthalpy (eV/D)	0.56	0.53

Figure 8. Comparison of released heat power between Kobe and Tohoku at room temperature..

certain condensed matter nuclear reactions occur intermittently in or around the metal nanocomposites with hydrogen.

3.3. Experimental reproducibility at different site using same metal nano-composites

As shown in Table 1, PNZ, CNZ and CNS metal nano-composite samples showed anomalous energy generation that cannot be explained by known chemical reactions. However, palladium nano-particles expressed as PSf1 and PSn1 did not show any anomalous phenomena. These experimental results indicate rough experimental reproducibility. To better test reproducibility, we performed experiments at Kobe and Tohoku University using the same $PdNi_7Zr_{15}–O_x$ with D_2 gas.

Figure 8 shows a comparison of released heat between Kobe and Tohoku at room temperature. The released heat power was calculated based on the difference between inlet and outlet coolant temperatures as was done for the

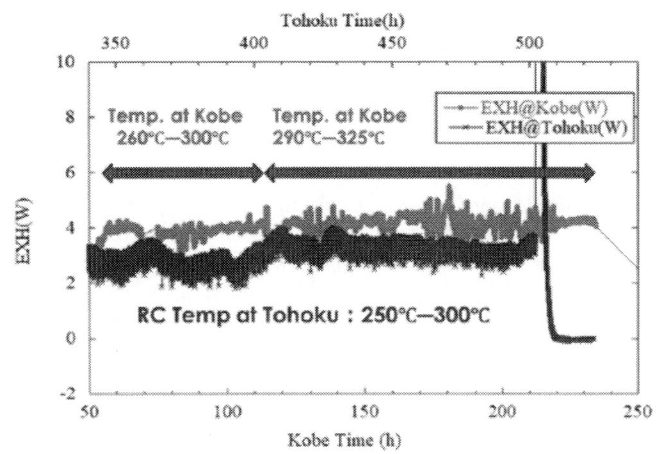

Figure 9. Comparison of excess heat power between Kobe and Tohoku at elevated temperature.

experiments at elevated temperatures. The heat power has two peaks, because D_2 gas was re-pressurized to the RC after the initial absorption of D_2 gas. After the resupply, the pressure of D_2 the absorption rate of D_2 into the $PdNi_7Zr_{15}$–O_x sample increased and enhanced the rate of heat release associated with absorption of D_2. Released excess enthalpies obtained at Kobe and Tohoku Universities agree.

Figure 9 shows a comparison of excess heat power between the experiments at elevated temperatures performed at Kobe and Tohoku Universities. As the excess power obtained by the present method has strong correlation with the temperature in the RC, we compared excess heat at similar RC temperatures. The same level of excess power were observed for similar RC temperatures. These experimental results demonstrate that a similar level of excess power can be obtained if we use the same metal nanocomposites and similar experimental setups.

4. Concluding Remarks

The Collaborative Research Project between six parties on anomalous heat effects was done from Oct. 2015 to Oct. 2017. Anomalous heat generation was observed for all the samples at elevated temperature, except for the palladium nanoparticles. Integrated excess heat reached 1.4–200 MJ/mol-H(D), which is too much to be explained by any known chemical process. Coincident burst events of the reaction chamber pressure and gas temperature were observed many times using $CuNi_7Zr_{15}$–O_x with H_2 gas. This suggested burst energy releases in the reaction chamber. These burst-like events were observed three times and suggest large local heat releases. Qualitative reproducibility was good between the Kobe and Tohoku experiments.

Acknowledgements

The authors thank Mr. Y. Shibasaki (Tohoku University) for support in establishing the experimental system. The authors express appreciation to Technova Inc. for support in building the experimental apparatus. We also acknowledge Mr. H. Yoshino, Mr. S. Hirano, Mr. M. Ise, Mr. S. Murakami and Mr. M. Hattori, who are the members of CLEAN PLANET Inc. for their significant assistance.

References

[1] A. Kitamura, A. Takahashi, R. Seto, Y. Fujita, A. Taniike and Y. Furuyama, Brief summary of latest experimental results with a mass-flow calorimetry system for anomalous heat effect of nano-composite metals under D(H)-gas charging, *Current Science* **108** (4) (2015) 589–593.

[2] A. Kitamura, A. Takahashi, R. Seto, Y. Fujita, A. Taniike and Y. Furuyama, Effect of minority atoms of binary ni-based nano-composites on anomalous heat evolution under hydrogen absorption, *J. Condensed Matter Nucl. Sci.* **19** (2016) 1–10.

[3] A. Kitamura, E. F. Marano, A. Takahashi, R. Seto, T. Yokose, A. Taniike and Y. Furuyama, Heat evolution from zirconia-supported Ni-based nano-composite samples under exposure to hydrogen isotope gas, *Proc. JCF16* 2016 pp. 1–16.

[4] Y. Iwamura, J. Kasagi, H. Kikunaga, H. Yoshino, T. Itoh, M. Hattori and T. Mizuno, The Launch of a new plan on condensed matter nuclear science at Tohoku university, *J. Condensed Matter Nucl. Sci.* **19** (2016) 119–126.

[5] T. Itoh, Y. Iwamura, J. Kasagi and H. Shishido, Anomalous excess heat generated by the interaction between nano-structured Pd/Ni surface and D_2/H_2 gas, *J. Condensed Matter Nucl. Sci.* **24** (2017) 179–190.

[6] A. Kitamura, A. Takahashi, K. Takahashi, R. Seto, T. Hatano, Y. Iwamura, T. Itoh, J. Kasagi, M. Nakamura, M. Uchimura, H. Takahashi, S. Sumitomo, T. Hioki, T. Motohiro, Y. Furuyama, M. Kishida and H. Matsune, Excess heat evolution from nanocomposite samples under exposure to hydrogen isotope gases, *Int. J. Hydrogen Energy* **43** (2018) 16187–16200.

[7] Y. Iwamura, T. Itoh, J. Kasagi, A. Kitamura, A. Takahashi and K. Takahashi, Replication experiments at Tohoku university on anomalous heat generation using nickel-based binary nanocomposites and hydrogen isotope gas, *J. Condensed Matter Nucl. Sci.* **24** (2017) 191–201.

J. Condensed Matter Nucl. Sci. 29 (2019) 129–168

Research Article

Coupled Calorimetry and Resistivity Measurements, in Conjunction with an Emended and More Complete Phase Diagram of the Palladium–Isotopic Hydrogen System

M.R. Staker*

Department of Engineering, Loyola University Maryland, 4501 North Charles St, Baltimore, MD 21210, USA

Abstract

Results of a calorimetric study established the energy produced, over and above input energy, from electrolytic loading of deuterium into Pd was 150 MJ/cm^3 of Pd (14 000 eV/Pd atom) for a 46 day period. High fugacity of deuterium was developed in unalloyed palladium via electrolysis (0.5 mol electrolyte of lithium deuteroxide, LiOD) with the use of an independent electromigration current. In situ resistivity measurements of Pd were used to assay activity of D in the Pd lattice (ratio of D/Pd) and employed as an indicator of phase changes. During this period, two run-away events were triggered by suddenly increasing current density resulting in 100% excess power (2.4 W output with 1.2 W input) and necessitating temporary cut back in electrolysis current. The average excess power (excluding run-away) ranged from 4.7 $\pm$ 0.15 to 9.6 $\pm$ 0.30% of input power while input power ranged from 2.000 to 3.450 W, confirming the Fleischmann–Pons effect. The precision was: Power In = $\pm$ 0.0005 W; $\Delta T = \pm 0.05°$C; Power Out = ± 0.015 W for an overall precision of $\pm 0.5\%$. High fugacity was required for these results, and the triggered run-away events required even higher fugacity. Using thermodynamic energy balance, it was found that the energy release was of such magnitude that the source of the energy is from a nuclear source; however, the exact reaction was not determined in this work. X-ray diffraction results from the recent literature, rules for phase diagram construction, and thermodynamic stability requirements necessitate revisions of the phase diagram, with addition of three *thermodynamically stable* phases of the superabundant vacancy (SAV) type. These phases, each requiring high fugacity, are: γ (Pd_7 $VacD_{6-8}$), δ (Pd_3VacD_4- octahedral), δ' (Pd_3 $VacD_4$- tetrahedral). The emended palladium–isotopic hydrogen phase diagram is presented. The excess heat condition supports portions of the cathode being in the ordered δ-phase (Pd_3VacD_4- octahedral), while a drop in resistance of the Pd cathode during increasing temperature and excess heat production strongly indicates portions of the cathode also transformed . . . (continued in the next page)

Keywords: Calorimetry, Electrolysis, Nuclear energy, Palladium–isotopic hydrogen phase diagram, Resistivity, Superabundant vacancy structures

*E-mail: m.r.staker@alum.mit.edu (best) or mstaker@loyola.edu; Tel.: 410 617 5188.

© 2019 ISCMNS. All rights reserved. ISSN 2227-3123

(continued from title page)... to the ordered δ'-phase (Pd_3VacD_4 - tetrahedral). A dislocation mechanism is presented for creation of vacancies and mobilizing them by electromigration because of their attraction to D^+ ions which aids the formation of SAV-phases. Extending SAV unit cells to the periodic lattice epiphanates δ as the nuclear active state. The lattice of the decreased resistance phase, δ', reveals extensive pathways of low resistance and a potential connection to the superconductivity phase of PdH/PdD.

1. Introduction

Modifications of properties in metals and alloys, apart from hydrogen embrittlement and degradation (reviewed by Robertson et al. [1]), by introducing hydrogen to high activity include: increased and decreased resistivity [2], induced ferromagnetism [3], optical property changes [4,5], increased lattice atom mobility [6,7], induced ordering [8,9], increased levels of vacancies [10,11], and even vacancies at concentrations near 25% [12–22], called superabundant vacancies (SAV). SAV formation in face centered cubic (FCC) metals changes the unit cells from FCC to simple cubic (SC) with vacancies (Vac) at all corner atoms of the FCC unit cell. This Vac ordering is similar to the gold (Au) ordering in copper–gold (Cu_3Au). In palladium (Pd), ordered SAV structures are: $Pd_3Vac_1D_x$ (δ or δ'-phases) [15,21] where x is between 4 and 8, and $Pd_7Vac_1D_{6-8}$ (γ-phase) [22]. Isotopic hydrogen atoms (protium (H), deuterium (D), or tritium (T)) occupy the octahedral interstitial sites (δ-phase) singly or as a pair of closely spaced atoms in $Pd_3Vac_1D_x$ [15,21] and/or occupy tetrahedral sites (δ'-phase) [23,24]. (Naming here follows the convention of phase diagram construction of phases left to right in order of the Greek alphabet.) SAV are observed in other metals/alloys beside Pd and nickel Ni and include: Fe, Mn, Ti, Zr, Nb, Al, Cu, Mo, Cr, Co, Ag, Au, Rh, Pt, Ir, Pu, Pd–Rh alloys, Pd–Ag alloys, and Cu–Ni alloys. SAV have been produced by the following methods: wet electrolysis, high-temperature with high-pressure gas via anvil compression, co-deposited electrolysis, solid state electrolysis (dry electrolyte), ion beam implantation, and plasma-injection. Eliaz et al. [25] have reviewed hydrogen-assisted processing of materials. Links between processing, structure and properties is continuously sought by metallurgists and material scientists. Does increased space between atoms along unit cell edges change conductivity, electron mobility, and redistribution of electron density (Schrodinger equation), and thereby enable nuclear reactions inside a lattice along these edges? The first purpose of this investigation is to position the new phases appropriately on the Pd–isotopic hydrogen equilibrium phase diagram. The second purpose is to investigate if electrolytically loading of D into Pd produces excess heat (more heat out than in) consistently, benefiting from electromigration and high dislocation density from plastic deformation with an associated increase in vacancies. A related purpose is to explore if SAV favor nuclear reactions at high fugacity because of unusual crystallography (open tube lattice) compared to the traditional PdD unit cell (β-phase with its usual electron distribution). Traditional phases of metal hydrides (α and β) might not be unusual enough in structure, and electron distribution to support nuclear reactions; but SAV phases, distinct from β-phase and having open tubes and unfamiliar electron-proton (or deuteron) interaction, are insufficiently explored. Zhang and Alavi [26] have used density-function theory to show electronic structure is more important than entropy effect in forming SAV.

2. Analysis

The purposes of this section are to show: (1) these new phases, γ, δ, and δ', are equilibrium phases, (2) near room temperature, they require *creation* of vacancies by a mechanism other than diffusion (dragging of jogs by moving screw dislocations) and *relocation* of vacancies (aided by attraction to electromigrating D^+ ions), and lastly (3) apt incorporation into the Pd–isotopic hydrogen equilibrium phase diagram.

Although evidence for high vacancy content in SAV phases was originally obtained by unit cell dimensional changes [12], strongest evidence [15,20–24] for these three phases, with distinct crystal structures, comes from X-ray diffraction (XRD). It is also supported by thermal decomposition spectra [15,27,28]. In thermal desorption data for pure Cu [27] and Ni [28], the spectra is the same for samples prepared via high-pressure/high-temperature to those created with electrodeposition (co-deposition of H(D) and Pd during electrolysis) at room temperature. In the former, the kinetics for formation is aided by high-pressure/high-temperature (high fugacity) while in the latter the structure is created atom by atom, so kinetics is bypassed, evolving directly into the lowest energy state, SAV. Only with subsequent thermal activation can hydrogen be coaxed into egressing (desorbed). The *kinetics* and signature of *desorption* is the same regardless of how the SAV state was arrived at.

A distinct unit cell constitutes a separate phase. It is shown from density functional perturbation theory (DFT) [15,29–36] that these new phases are equilibrium phases (lowest free energy), and as such, necessitate they be added to the Pd-D equilibrium phase diagram. Resistivity measurements (Section 4) link these phases, and a phase transition of δ to δ', to measured excess heat. These phases, occurring at high D/Pd ratios, offer unique pathways of open structure (vacancy tubes or channels, Section 5) with low resistance to electron and proton (and deuteron) transport.

SAV phases result from hydrogen-induced vacancy formation [15,20–24,29–36]. Vacancies have higher mobility (validated by DFT calculations [29–36]) from electromigrating D^+ which drag them for building SAV structures, (mechanism in Appendixes A and B). Higher numbers of vacancies are promoted by high dislocation density from plastic deformation, as outlined in Appendix B. These two steps in the formation of room temperature SAV would certainly limit nucleation and growth of δ and δ', but Sections 4 and 5 show that the volume fraction of SAV phases needed to support excess heat is an extremely small fraction.

On SAV Fukai [15] astutely recognized and stated:

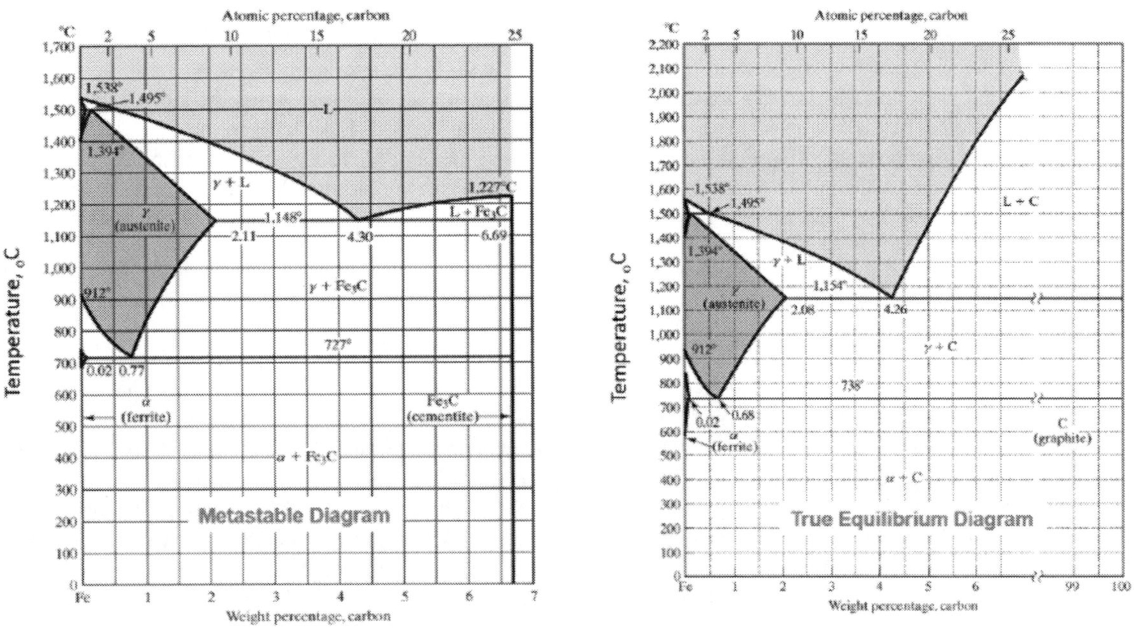

Figure 1. The Fe–Fe$_3$C (*left*) and Fe–C (*right*) phase diagram. The composition axis on the metastable diagram is weight % C (*left*) even though Fe$_3$C is a component (as opposed to C), whereas on the true equilibrium phase diagram, C is properly both composition axis and component; Fe$_3$C does not exist, adapted after [38].

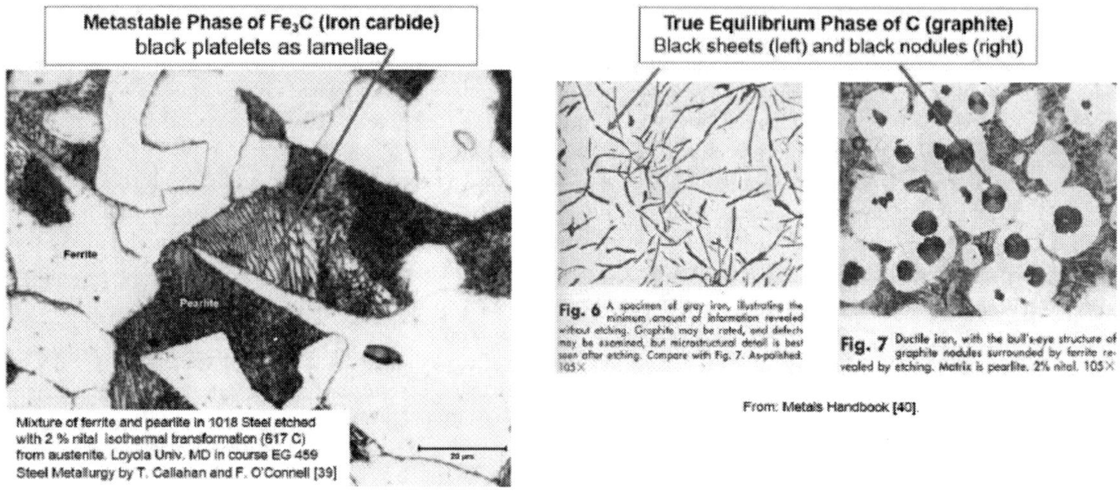

Figure 2. Micro-constituents for the diagrams in Fig. 1: those on the left (from [39]) are for the metastable Fe—Fe_3C phase diagram, those on the right (from [40]) are for the true equilibrium Fe—C diagram, showing that both types of diagrams are useful for steel and cast iron.

> "... most important implication in the physics of SAV is that the most stable structure of all M–H alloys is in fact the defect structure containing a large number of M-atom vacancies. All M–H alloys should tend to assume such defect structures, ordered or disordered depending on the temperature, as long as the kinetics allows. In practice, however, M–H alloys are in most cases prepared under conditions where M-atom vacancies cannot be introduced. Thus it can be said that most (all) phase diagrams of M–H systems reported to date are metastable ones. These metastable diagrams are certainly useful as such, but the recognition that they are metastable ones is of basic importance. The real equilibrium phase diagrams including M-atom(s) vacancies have not been obtained so far."

Emending the phase diagram hinges on distinguishing between metastable and true equilibrium phase diagrams, as well as, the rules for possible and impossible phase diagrams. Okamoto and Massalski [37] express the relevant phase sequence rule: *There should be a two-phase field between two single-phase fields. Two single phases cannot touch except at a point.* Phase diagrams require 'necessary', but not 'sufficient' conditions, for presence of phases. The 'necessary' condition is the change in free energy must be negative ($\Delta F < 0$) for new phases to form. The 'sufficient' condition is from kinetics. In steel, both metastable and true equilibrium diagrams are useful since many heat treatments preclude equilibrium. The iron–iron carbide (metastable) and the true equilibrium phase diagram of iron–carbon are compared in Fig. 1 along with micro-constituents from each (Fig. 2). For Pd–isotopic hydrogen, the presently accepted diagram is metastable since some of the equilibrium SAV phases are absent because of kinetics. Traditional (historical) metastable diagrams of Pd–D(H) omitted SAV phases because they were only recently discovered. The equilibrium diagram with all phases of lowest free energy is presented below. Kinetics may also limit the size (volume percent) of phases in microconstituents. Kinetics for creation, mobilization and conglomeration of vacancies undoubtedly explain the incubation period to initiate excess heat in many low energy nuclear reaction (LENR) experiments

Figure 3 shows four isotherms (red) illustrating the phase sequence rule. In Fig. 4, from [36], one can see compliance and violation of the sequence rule in the V–H system. Figure 5 shows a violation of the phase sequence rule

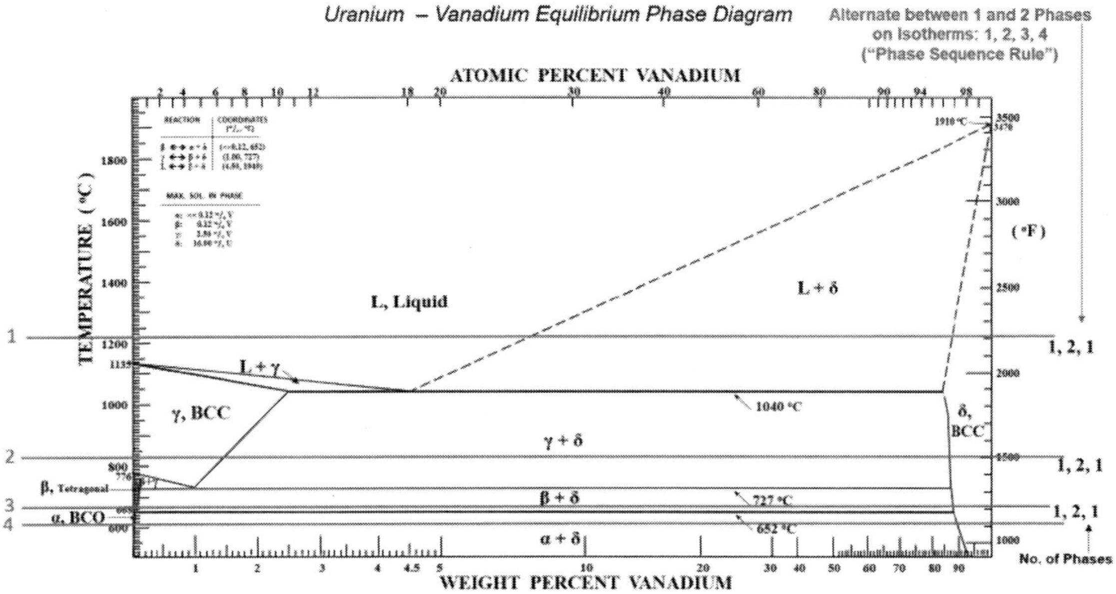

Figure 3. Phase diagram for vanadium–uranium system showing the phase sequence rule for four isotherms from [41].

in Pd–H [36]: the upward sloping phase boundary from D/Pd ratio of 0.66– ∼0.9 separating β from β'. A similar violation is in Fig. 5(B) from Araki et al. [42], but this can be corrected by interpreting their data as in Fig. 6 (red) to comply with the sequence rule, but necessitates a phase boundary at H/Pd ratio of 0.76 and another at 0.85 separating β from $\beta + \gamma$. In addition upward sloping red lines to the left of 0.76 and to the right of 0.85 have to be two-phase regions (curves with a nested two-phase field) as is shown below.

Figure 7 shows the unit cell for γ-phase (Pd_7VacD_{6-8}) from [22]. Delineation is revealed from two "unit FCC cells" (dark outline). From these, one sees the apposite true unit cell and superlattice structure of Pd_7VacD_{6-8}. D shifts slightly toward the corners allowing them to bind more to each vacancy. This is true for all of the D except the one in the central octahedral site not bound or trapped to any particular vacancy. Depending if this site is occupied, stoichiometric ratio of D to Pd is between 6 and 8 for 7 Pd atoms, giving D/Pd ratio between 0.857 and 1.143: γ-phase has mid-point stoichiometry D/Pd = 1 (subscript for D = 7).

Figure 8 combines the metastable diagram with SAV data to yield an equilibrium phase diagram. It has δ-phase Pd_3VacD_4, with D/Pd ratio 1.333 determined (see Appendixes C and D) from XRD [15–17,19,20,28,30,33,35] since there are 4 D for every 3 Pd at strict stoichiometry. These two phases (γ and δ) must, by the sequence rule, be separated by a two-phase field of ($\gamma + \delta$). The size of each phase field is determined as follows. The temperature extent has some uncertainty (dotted). The width of γ is based on the central interstitial site filling: empty in both half cells, filled in one of the two, or filled in both. For off-stoichiometry, the width of δ, from this analysis, is 1.333 ± the same width as the two-phase fields on either side of γ (.095 from Fig 8). This gives D/Pd_{Min} = 1.33–0.095 = 1.24 and D/Pd_{Max} = 1.33 +0.095 = 1.43. This construction follows Araki's et al. [42] data, indicating start and end of the two-phase region ($\beta + \gamma$) at 0.76 and 0.85 respectively. It is suggested that the two-phase region on the right of γ has the same width (0.095) from symmetry and a lack of data to support another value. This layout is qualitatively consistent with Fukai and Sugimoto [30,31] who specify two phases of different vacancy concentrations (named here γ and δ) and [20] who

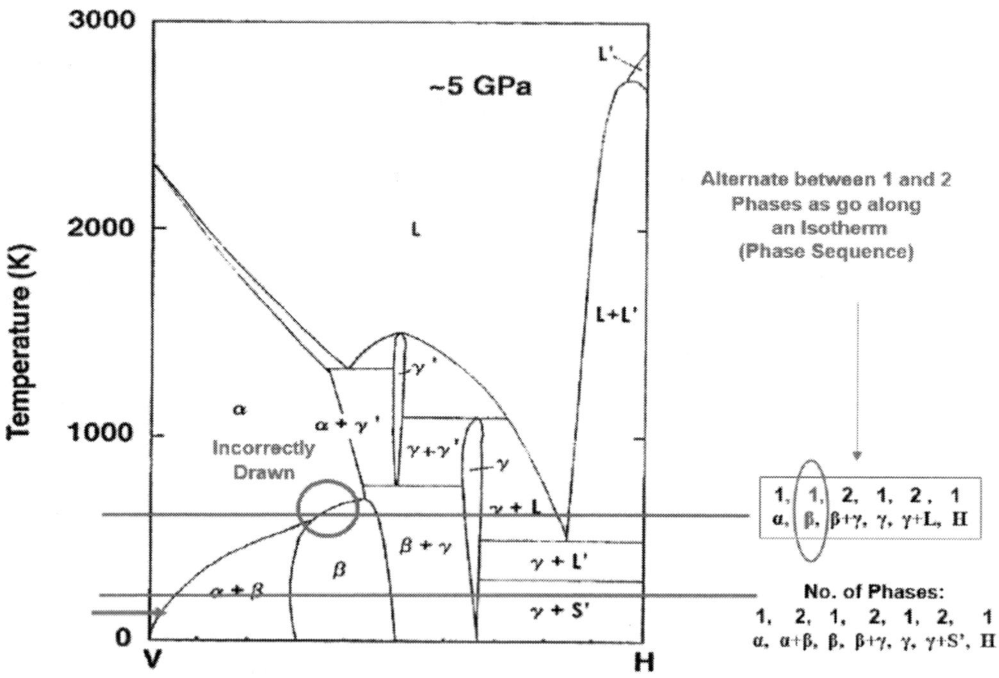

Figure 4. Phase diagram for the vanadium–hydrogen system at 5 Gpa where the phase sequence rule is violated in the top isotherm but upheld in the lower isotherm, adapted after [36].

also show XRD evidence of two concentrations (12 and 20%). In the XRD work of Fukada et al. [22], these two phases were labeled moderate (0.86–1.14) and rich (1.24–1.43) vacancy concentrations.

There is also a δ' field, at D/Pd = 1.333 (Fig. 8). The difference between δ and δ' is D occupies octahedral sites in δ, while D occupies tetrahedral sites in δ'. The δ' appears below a temperature of 375 K based on resistivity data (Section 5), and is supported by tetrahedral occupancy by D from [23,24]. From DTF, Isaeva et al. [29] found, at lower temperatures, tetrahedral site occupancy by H (D) stabilizes SAV more than octahedral site occupancy. Neutron diffraction data of Ferguson et al. [24] and Pitt and Gray [23] show H migrates from octahedral to tetrahedral sites at lower temperatures.

The δ and δ'-phases are of interest to LENR. The δ''-phase has been in superconductivity literature and will not be detailed other than noting its existence and approximate position on the phase diagram. In addition, ε is not speculated on here other than its link to superconductivity at D/Pd ratio of 1.6 as Tripodi et al. [43] have predicted. In summary, SAV phases (γ, δ, and δ') are equilibrium phases, require creation and mobilization of vacancies, and are incorporated into the Pd–isotopic hydrogen equilibrium phase diagram.

3. Materials and Experimental Procedure

Unalloyed Pd from Johnson–Matthey was used as 0.5 mm diameter rod of length 25.4 mm in an electrolytic cell with Cu leads connected to the current-controlled DC power supply. The specimen diameter was chosen consistent with results of electromigration analysis for D^+ (Appendix A). Since Pd has a much higher melting point than Cu,

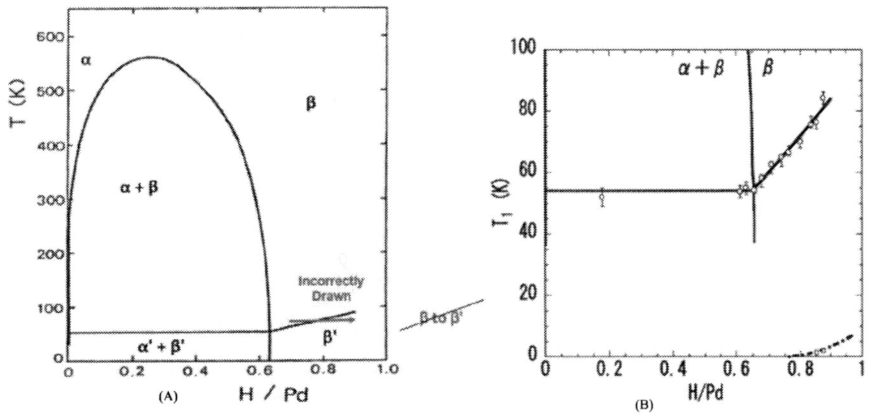

Figure 5. (A) The Pd–H phase diagram with the phase sequence rule violated. This diagram is a metastable diagram lacking equilibrium phases (γ,δ, and δ'), adapted after [36]. (B) A portion of the Pd–H phase diagram of Araki et al. [42] with the phase sequence rule violated. Open circles and open squares are from measurements in their work. This diagram is a metastable diagram lacking equilibrium phases (γ,δ, and δ'), after [42].

attachment was completed as in Fig. 9 by heating tips of 5.19 mm diameter (#4 gauge wire) copper leads with a torch to form a molten bead, inserting the end of the cold Pd specimen into the droplet, while withdrawing the flame,

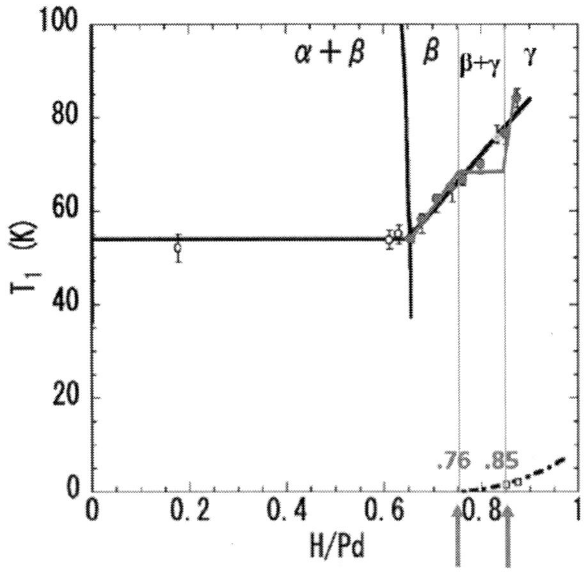

Figure 6. A portion of the Pd–H phase diagram of Araki et al. [42] with red lines being another interpretation of phase boundaries. Open circles and open squares are from measurements in their work. This diagram is also a metastable diagram lacking equilibrium phases (δ and δ'). Adapted after [42].

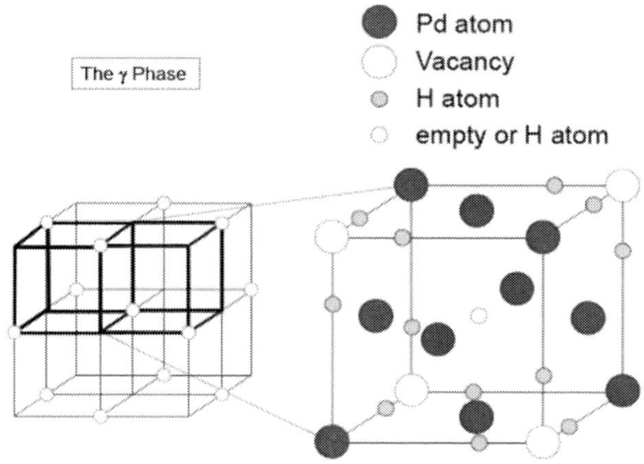

Figure 7. Superlattice structure of Pd_7VacD_{6-8}. *Left*: super-cell lattice showing only vacancies. *Right*: half-cell structure magnified from the super-cell with heavy lines. The H-atom at the body center does not bind to any vacancy, thus the subscript for hydrogen varies within a range from 6 to 8. Adapted from [22].

causing instant solidification of Cu, braze-bonded around the end of the Pd specimen. The Pd remained solid due to higher melting point. This method was used on both ends of the Pd forming a good electrical connection for the cell. It allowed for current to be driven through the Pd independent of electrolysis current from platinum (Pt) anode to Pd cathode. It was also used for Pt specimens in Pt/H_2O control cells. Expansion of Pd during absorption of D was accommodated by high ductility Cu preserving the bond for the lifetime of the experiment. Contact of electrolyte with Cu leads, was avoided by heat shrink electrical insulation tubes and then sealing Pd ends (Fig. 10) with miccrostop

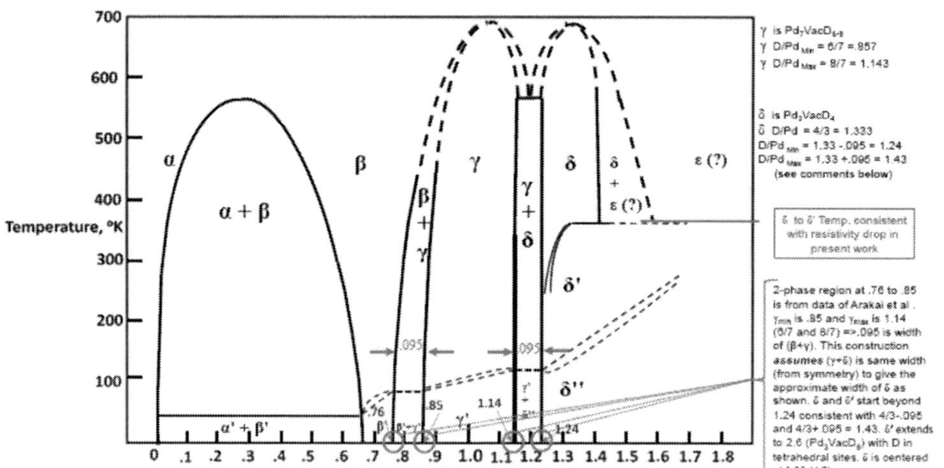

Figure 8. Equilibrium phase diagram for isotopic hydrogen–palladium.

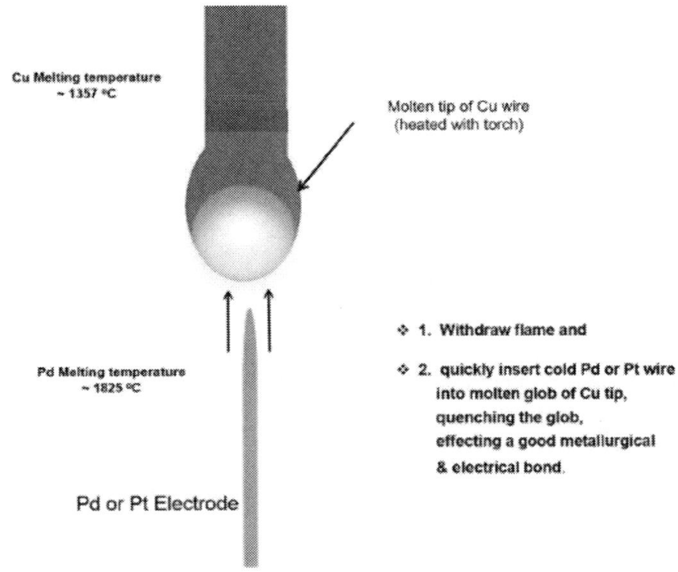

Figure 9. Method of preparing and attaching Cu leads to Pd or Pt cathodes.

stop-off lacquer (such as MICCROSHIELD Stop-Off Lacquer from Tolber Division Miccro-Products).

The electrolyte was 18.0 ml of 0.5 mol LiOD in 99.8% deuterium oxide D_2O from Cambridge Isotope Laboratory. The cells were constructed by nesting four slightly different size Pyrex test tubes, each separated by two O-rings and a thin air space. A Teflon top, sealed with O-ring, excluded ambient atmosphere by allowing the positive pressure of O_2 and D_2 gas to exit through a capillary tube into a reservoir of vacuum pump oil. This arrangement is shown in Figs. 11–13. Fleischmann and Miles [44] showed recombination is either zero or too small to be a source of heat. There was visual monitoring of cell electrolyte level and exit gasses. This configuration of four-nested tubes reduced overall heat transfer from the inner test tube and increased sensitivity (producing a larger delta Temperature, ΔT, for each input watt). Cells were calibrated by measuring the power in and ΔT, the difference between cell temperature and surrounding air temperature inside the calorimeter. Power from electromigration current (current-control mode)

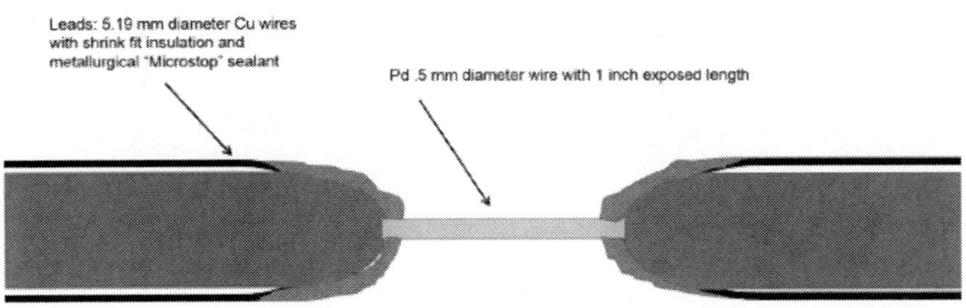

Figure 10. Pd or Pt (control) cathodes with brazed electrical leads of Cu isolated from the electrolyte in the cells by heat shrink tubes and Microstop lacquer sealant.

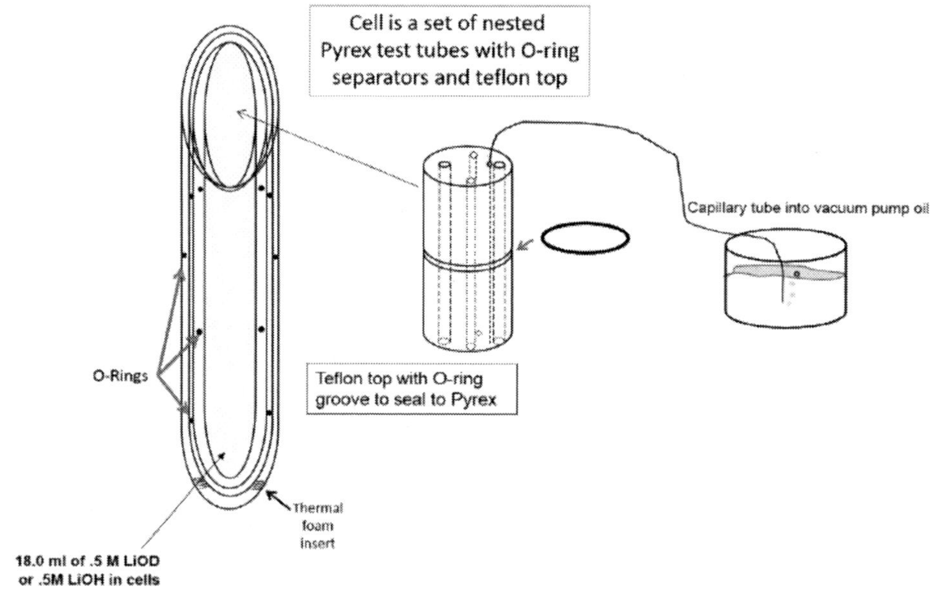

Figure 11. Electrolytic cells constructed by nesting four Pyrex test tubes (with O-rings) and Teflon top, through which cathode, two cathode leads (Cu), anode lead (Pt), two small Pyrex tubes (two) for thermocouples, make-up water from syringe, and exit gas tube come through. Top of cell is distorted in drawing which is not drawn to scale.

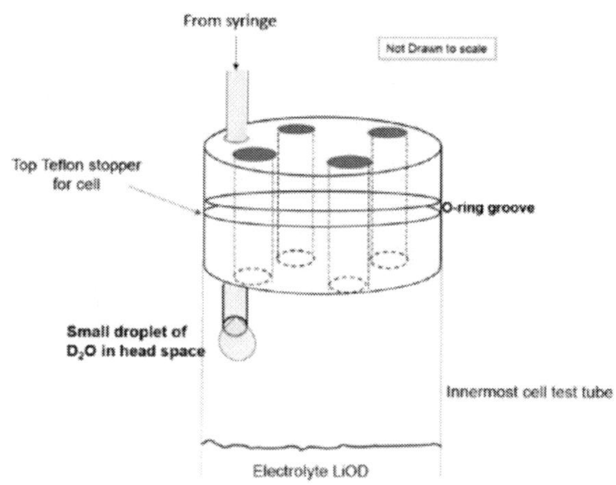

Figure 12. Suspension of D_2O or H_2O droplet from make-up water syringe tube in the head space above electrolyte allowed growth and thermal equilibration before dropping into the electrolyte. Typically this took about 10 min, depending on electrolysis current to which the delivery rate was matched.

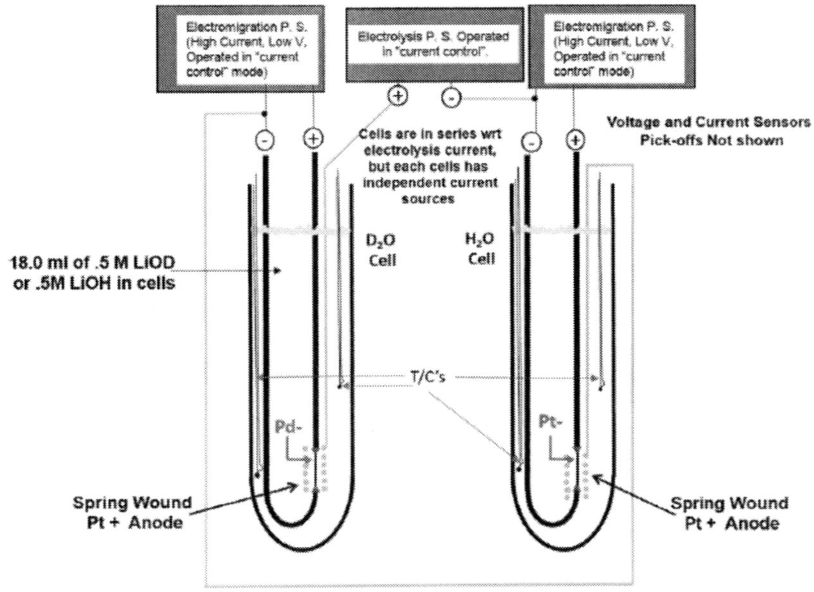

Figure 13. Electrolytic cells of four nested (not shown) Pyrex test tubes, sealed with Teflon tops (not shown). Both cells are in series on one current controlled power supply (PS). Independent electromigration currents from PS's for each cell were also operated in current control mode.

through Pd specimens was the product of current times voltage drop along the Pd specimen. This voltage drop was measured outside the cell on the Cu leads carrying electromigration current since 98.6% of the drop occurred in the PdD portion because its resistivity is higher than Cu (5.9–11.8 times higher depending on D loading), but most importantly because of its greatly reduced cross-section (ratio of Cu area to PdD area = $d_{Cu}^2/d_{Pd}^2 = 108$). The net effect of these two factors, along with the ratio of Cu to Pd lengths, is the ratio of voltage drop in PdD to that of the Cu leads. It is 69.4 to 1. This assessment includes the length of Cu (into and out of the cell) of 35 cm and the average of least and most resistivity values. Power-in for the electrolysis portion (current-control mode) was electrolytic current times electrolytic voltage minus thermoneutral potential (conservative values of 1.52 V for D_2O and 1.48 V for H_2O were used). Cell temperature was obtained from thermocouples placed inside two capillary tubes in each cell (Fig. 13). These were averaged with each other and over time (collected every 0.01 s yielding 5000 measurements, and were averaged for each temperature or voltage collected, giving an output data set every 15 min). Thermocouples were welded and their output voltage compared to a standard thermocouple calibrated periodically at the National Institute of Standards and Technology, assuring absolute accuracy within 0.05°C for Temperature readings after extensive averaging (5000 readings). It should be noted that precision for any given thermocouple was better than 0.05°C since repeatability on the same thermocouple did not include an offset (systematic error) normally involved when measuring absolute accuracy.

A calibration curve of ΔT versus power-in was run for cells with 1–4 nested tubes in an effort to assess sensitivity by adding tubes. This was motivated by the small volume of Pd electrode, anticipated power density releases (W/cm^3), and a desire to distinguish between chemical energy (a few eV/atom) and nuclear energy (between keV/atom and MeV/atom, depending on the fraction of atoms participating, see Section 4). The sensitivity increased with each nesting Pyrex tube and its associated dead air space.

Maintaining electrolyte level in cells used a Harvard Apparatus Model 22 Digital Syringe Pump outfitted with two syringes, one with heavy water from Cambridge Isotope with 99.8% D and one with double distilled light water. The selectable rate of delivery (between 0.002 μl/h and 55 ml/min) was established by developing a table of instrument input settings for each electrolysis current. Make-up heavy and light water in the form of small droplets dangled in the cell head space (lasting about 10 min) before dropping into the electrolyte (Fig. 12) and allowed for thermal equilibration. Use of the syringe pump (slow but constant rate of delivery) eliminated thermal shock when replenishing D_2O and H_2O. Thermal shocks (negative temperature spikes) had been a shortcoming in this field. Sudden large additions had additional surface area for heat transfer and cooler electrolyte mass for a given power level and upset excess heat reactions. The effect of over or under filling of electrolyte (eliminated with correct syringe pump settings) is shown in Section 4. The time constant was determined for the heavy water cell to be about 34.5 min by the standard method of determining a time constant (Fig. 14). It is compared to one of about 116 h for the entire calorimeter chamber (202 times larger). This insured a small but distinctive power spike in the cell could be measured since chamber temperature was constant and capable of absorbing the power of the cells without any change in overall calorimeter temperature (the heat capacity of the calorimeter was 2500 times higher than that of the cell). With the ratio of thermal diffusivity ($\alpha = k/(\rho\ c_p)$ of air to water at 150 ($\alpha_{air}/\alpha_{water} \approx 150$), the temperature inside the chamber was equilibrated and constant more consistently than using a commonly employed water bath. Here k is thermal conductivity, ρ density and c_p is constant pressure heat capacity. There was no forced air velocity inside the

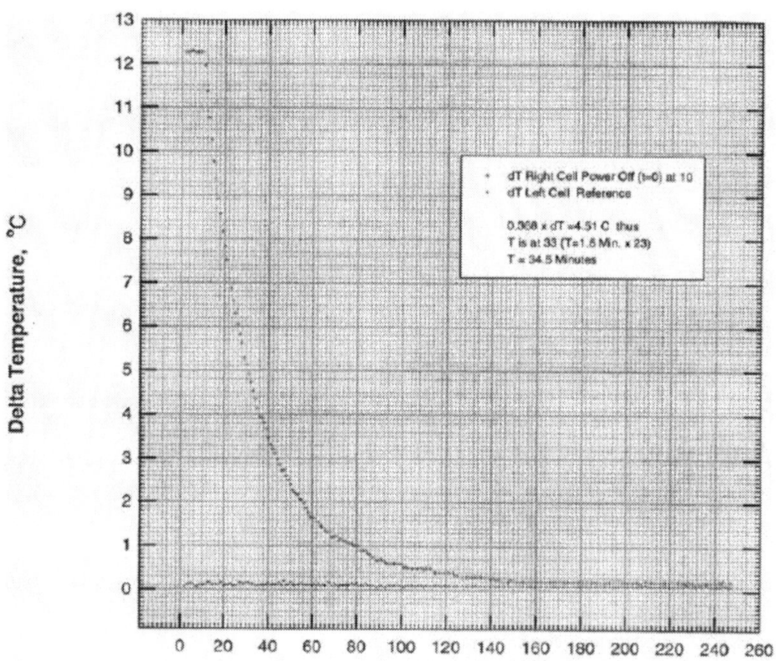

Figure 14. Time constant determined for right and left cell (*right shown*) in the constant temperature chamber to be 34.5 min. The constant temperature chamber was measured at approximately 116 h (about 202 times longer while the heat capacity was 2500 × higher). The time constant τ represents the time for it takes the cell's step response to reach 1–1/*e* ($\approx$ 63.2%) of its final (*asymptotic*) value.

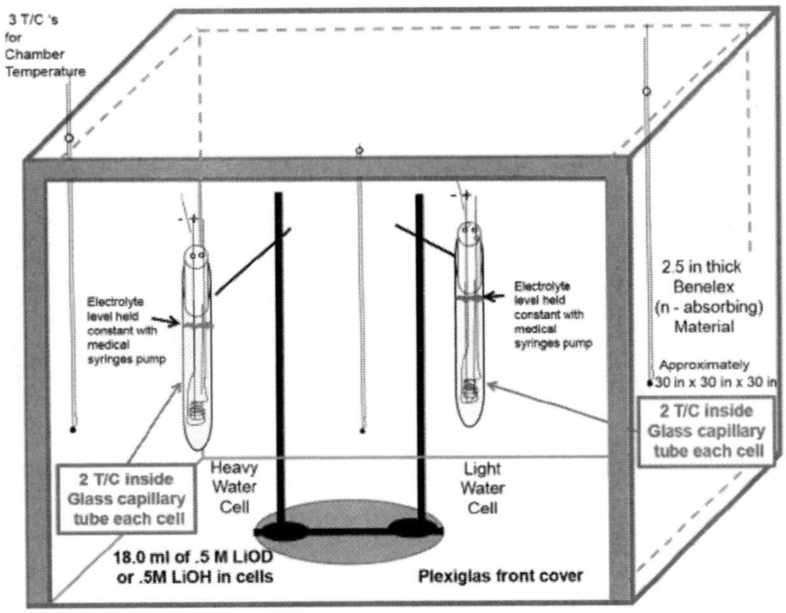

Figure 15. Constant temperature chamber with thermal time constants: τ chamber $\gg$ τ cells.

calorimeter, only natural convection.

The calorimeter itself (Fig. 15) consisted of a chamber approximately 75 cm $\times$ 75 cm $\times$ 75 cm with wall thickness of 6.4 cm of Benelex (pressed fibers board) surrounding cells on all sides except the front, which had Plexiglas for visual monitoring the cells, especially electrolyte level. Three thermocouples, inside the chamber were averaged for calorimeter media temperature. There was never any difference in these three readings (within precision), however, averaging improved reliability and confirmed lack of forced convection.

A data acquisition processor card 2400a by Microstar Laboratories Inc. was used for all data acquisition of voltages, and currents (measured as a voltage drop across a precision resistor) for power input, and for thermocouple readings. Precision, using extensive averaging, for the power-in was $\pm$ 0.0005 W, while that of the ΔT was $\pm$ 0.05°C (or better as explained above) and that of power-out of $\pm$ 0.015 W (based on the limiting value of ΔT of $\pm$ 0.05°C). This gives an overall precision for the experiment of $\pm$ 0.5%, enabled by: (1) Use of the data acquisition card, eliminating human bias and giving large averaged data sets. (2) Use of the calorimeter chamber, with time constant over 200 times larger than that of the cell. (3) Use of the syringe pump eliminating shock and maintaining constant level. Other factors that contributed to successful excess heat production were preconditioning the Pd to a high vacancy and dislocation density by anodically unloading and cathodically re-loading (back over resistive hump and over again) four times at low current density (<50 mA/cm^2) before measuring heat (Appendix B), and selecting the diameter of specimen to promote electromigration of D^+ without adding significant resistive parasitic power (Appendix A). The D/Pd ratio increased with each loading cycle: four cycles were found to be an optimal number. Electromigration promotes concentration enhancement at the bottom end of the Pd, and causes redistribution of vacancies. Using plastic deformation, with a high dislocation density, is suggested by Sugeno and Kowaka [45] and Lewis [46] in Fig. 16. The final hardness (and dislocation density), from hydrogen insertion and removal into well-annealed Pd, is the same as

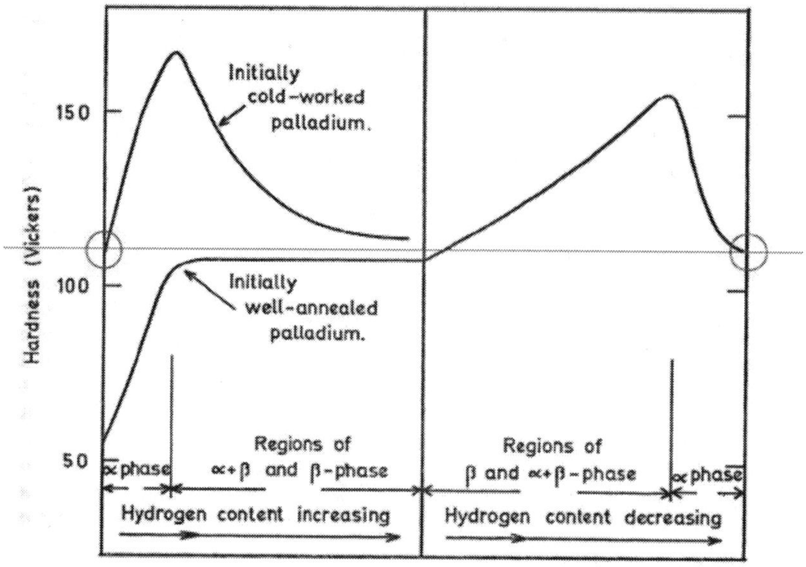

Figure 16. Hardness as a function of hydrogen content during insertion and removal of hydrogen at 25°C after [45,46]. The final hardness (and flow stress τ) and dislocation density ρ from insertion and removal of hydrogen into well-annealed Pd have the same starting values as cold worked Pd. Hardness (flow stress) are related to dislocation cell size d and dislocation density ρ as shown by Holt [47] and Staker and Holt [48]: $\tau = \mathrm{K}d^{-1}$, $d = 16\rho^{-1/2}$, $\tau = 0.57\ \mathrm{G}b\rho^{1/2}$, where G is shear modulus, b Burgers vector.

the starting value of cold worked Pd. Dislocation intersections create dislocation jogs and enables vacancy creation in bulk without need for diffusion from surface or grain boundaries. Hardness and flow stress are measures of dislocation density and dislocation cell size, as shown by Holt [47], Staker and Holt [48], and Kuhlmann-Wilsdorf [49]. Deutges et al. [50] have shown, by transmission electron microscopy that cell formation, common to most metals, also occurs in PdH and results in higher diffusivity of H, consistent with higher vacancy concentration (Appendix B).

This method utilized the Fleischmann–Pons cell except for four-wire leads and three power supplies. In-situ resistance monitoring of D to Pd ratio after Burger et al. [51] and McKubre et al. [52] ensured loading over the resistive hump (D/Pd $= 0.74$) to higher values of D/Pd (0.93), and monitored changes in resistance after excess heat initiated. Their analysis of variation of resistivity ratio $\rho_\beta/\rho_{\mathrm{Pd}}$ with r_{x} (the D/Pd ratio) showed ρ_β follows the relationship: $\rho_\beta/\rho_{\mathrm{Pd}} = 0.97869 + 3.0001(r_{\mathrm{x}}) - 15.090\,(r_{\mathrm{x}})^2 + 44.155(r_{\mathrm{x}})^3 - 49.119\,(r_{\mathrm{x}})^4 + 17.577(r_{\mathrm{x}})^5$, with $R^2 = 0.999$. Thus by measuring the initial (ρ_{Pd}) and in situ (ρ_β) resistances D/Pd ratio as r_{x} in this equation was determined and constantly monitored at all times. Once crossing the maximum (hump) at D/Pd $= 0.74$, one could be confident of the loading ratio as long as resistivity ratio ($\rho_\beta/\rho_{\mathrm{Pd}}$) never rose above the maximum value (2.004) during a run (never crossed back over the hump by de-loading). Rising and falling resistivity values were observed and monitored during electrolysis potential reversals while crossing the miscibility gap four times.

Previous LENR research bid to avoid loading hydrogen and its isotopes through the α to β miscibility gap. This scheme was initiated by the industry using Pd and Pd–Ag alloys for purification of hydrogen and isotope separation. The main concern was to avoid damage to Pd or Pd–Ag alloys, anxious about structural integrity (resistance to fracture). This damage is due to deformation and cracking accompanying large volume change associated with α to β-phase change. It progresses from sever plastic deformation (high dislocation density but unfractured) to eventually exhausting the limit of ductility culminating in cracks. Whether plastic deformation leads to cracking or continues to even larger amounts of deformation depends on the state of stress (a combination of tensile and shear versus compressive and shear). Motion of dislocations (plastic deformation) is attained only by shear (not compressive or tensile) stress, whereas fracture is initiated by tensile stress. The motivation for the use of the protocol of loading and unload hydrogen four times by reversing the electrolysis current is to create a high dislocation density with copious intersections with screw dislocations in dislocation cell walls. These generate vacancies (by the mechanism in Appendix B) and are caused to migrate to unit cells that become ordered SAV unit cells (δ and δ' phases by the mechanism in Appendix A). These unit cells grow to form small islands of multiple unit cells (see Sections 5 and Appendix B) that constitute a volume fraction near 0.03 % (for δ-phase). The way to avoid fracture is to assure that the state of stress minimizes tensile stress. This state of stress, displayed on Mohr's circle, would be shifted as far left (into negative compressive stress) as possible but with a large diameter (high shear stress). In this research, creation of dislocations and their intersections by loading and unloading four times created vacancies with the beneficial side effect of increasing the diffusion rate for hydrogen (of course even much higher diffusion rates are provided within SAV unit cells as they grow in size). If fracture occurs, high loading rates are not possible, so loading slowly in the beginning effects the state

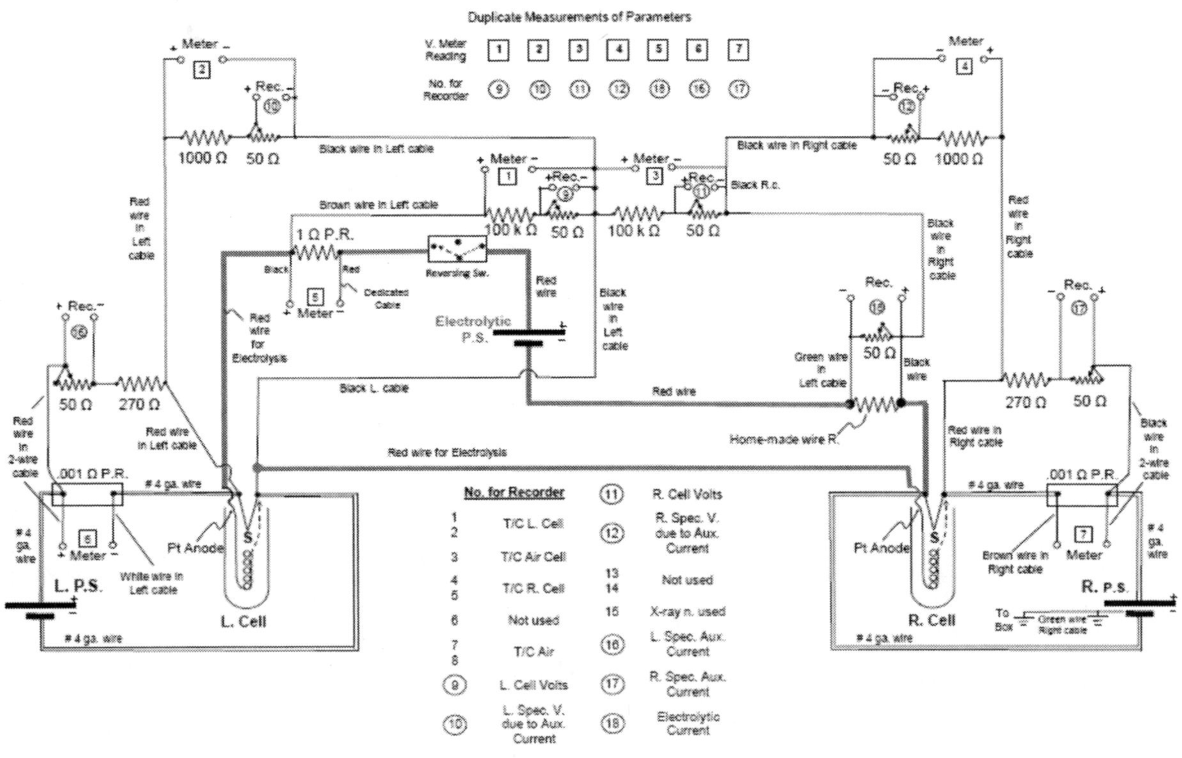

Figure 17. The wiring diagram using three power supplies (PS). The numbers in boxs or circles are locations of readouts (meter and recorder are redundant for data acquisition processor card). Pd and Pt cathodes, S, are connected to 5.19 mm diameter Cu leads, labeled #4 ga. wire, and the electromigration current there in is labeled Aux. Current. The Pd/D_2O and Pt/H_2O are labeled L. cell and R. cell, respectively. Three types of wires are indicated: double line (electromigration circuit), red line (PS for electrolysis with cells in series), and light black lines to data read-outs.

of stress minimizing tensile stresses between regions highly loaded (expanded) and those not yet loaded, as indicated by Cravens [53] and leading to the optimal number of loading/de-loading/reloading cycles of four.

Electromigration was incorporated for monitoring resistance (measuring the voltage drop along the length of Pd) and D/Pd ratio, and to force movement of D^+ ions toward the bottom of the specimen. Because of attraction of vacancies for D^+ ions, it also promoted concentration of vacancies, easing activation energy of SAV. It also balanced the power with a small amount of resistive heating in either light or heavy water cells. They were in series for electrolysis, but electromigration currents were controlled independently.

Figure 13 shows heavy water Pd and light water Pt cells wired in series (same current). The Pt anode was wound as tightly as possible to the cathode, lowering electrolysis voltage and power-in. Use of 0.5 M LiOD and 0.5 M LiOH also reduced power, providing a more sensitive calorimeter. Two thermocouples inside small capillary tubes monitored cell temperature. They were averaged for each cell. Complete wiring diagram for the electrolysis, electromigration, power supplies, reversing switch, precision resisters for currents, other voltage divider resistors for readouts, and locations for the data acquisition processor card tapping are shown in Fig. 17. The numbers in a box or circle are locations of these readouts. The Pd/D_2O cell and Pt/H_2O cell cathodes, marked S, are directly connected to the 5.19 mm diameter Cu leads (labeled #4ga. wire) and to the power supply (PS). The electromigration current is labeled Aux. Current in Fig. 17.

In Fig. 18 there are three hypothetical data sets of ΔT versus power-in, where the *accuracy* (red line) can be improved by adding more data. However the *precision* cannot be improved unless methods are taken to reduce the sources of error in ΔT as described above. This kind of scatter of green data points from the red line is too great in the left figure to provide confidence that any given green point is a result of excess heat. The second hypothetical data set (center) has improved precision but with same accuracy. The right data set has precision and accuracy to measure excess heat for a given specimen size (more excess heat for a larger size).

In developing calibration curves for light and heavy water cells, data were taken either before loading to high ratios or by using a Pt cathode and Pt anode. Data (with 5000 readings) were averaged by the data acquisition card and associated computer software at 15 min intervals from all thermocouples, voltages and currents, filled a file without human intervention, and then plotted as small dots. By changing electrolysis and/or electromigration current slightly

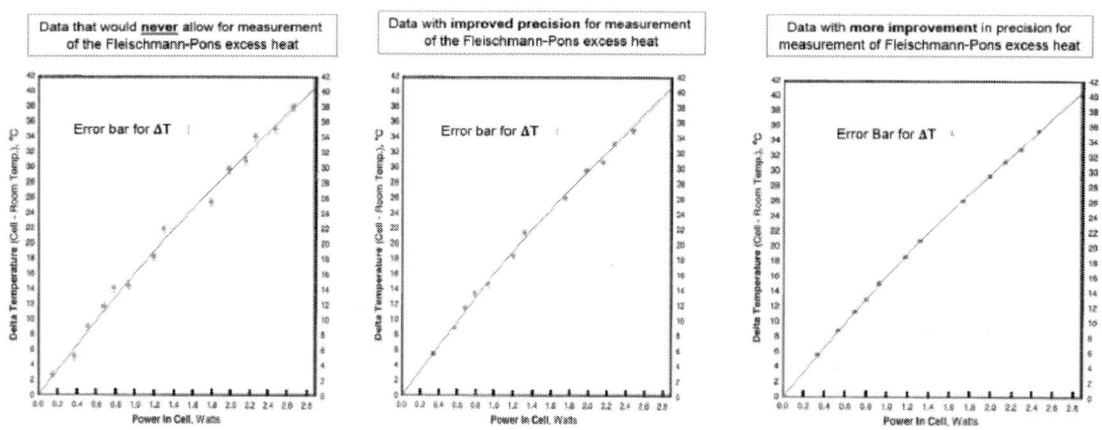

Figure 18. Hypothetical data of ΔT versus power-in, where the *accuracy* (*red line*) can be improved by adding more data to the set, but improvement in *precision* requires other means.

to new settings, data was acquired at new positions on the graph (about every 45 min): this permitted closely spaced values which filled in the entire line (*curve of dots*).

4. Results

Figure 19 shows a calibration curve for ΔT versus power-in with ΔT measured in degrees Celsius and power-in measured in watts obtained as just described (Section 3). It uses a small dot as the data character symbol: there is no additional curve through the data. This demonstrates higher precision than that reported above of ΔT of $\pm$ 0.05°C. Figure 20 shows the same data with a calibration check (data characters of crosses, +) added after 46 days. These crosses are plotted with larger font size to distinguish them from the dots, nonetheless, the centers of the crosses fall on top of the original calibration, indicating no shift in calibration.

Figure 21 shows the calibration curve drawn with a line (replacing the dots) and extended to higher power-in levels. The ΔT and power-in are during operation of the Pd/D_2O cell at various times during the 46 days of operation. A straight line at higher levels of power is a conservative estimate of the calibration curve since it is slightly concave down (slightly lower than shown). There are rare scattered data below and above the calibration curve. All the data below calibration were taken by the data acquisition processor card under non-equilibrium conditions: either by a deliberate attempt to document the effect of cell overfilling (raising the electrolyte level above equilibrium operating level of 18.0 ml) or when electrolysis was deliberately stopped briefly, allowing the cell to cool below equilibrium calibration conditions. These periods are not part of calibration or equilibrium conditions (normal operation of the cell). The

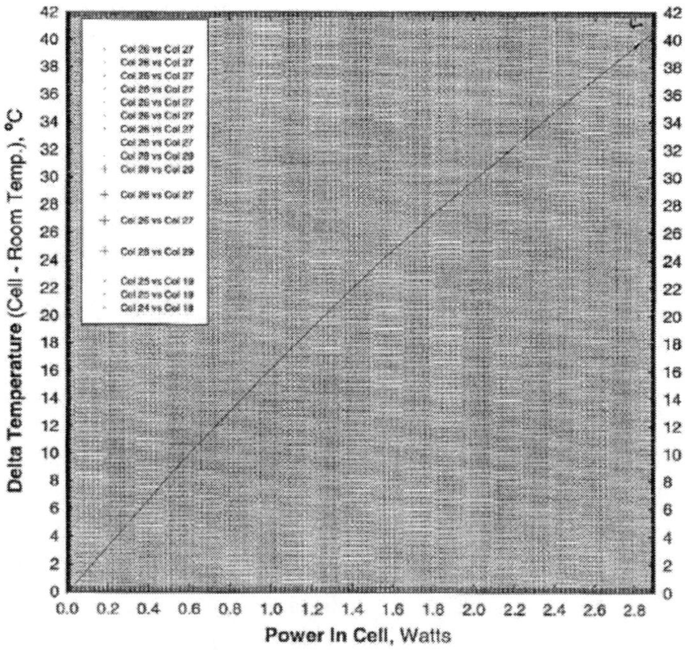

Figure 19. Calibration curve of right cell (Pt/Pt) and left cell (Pd/Pt). The line is a series of small dots closely spaced determined from data acquisition card and computer data readings from Fig. 17, for plotting without human intervention. Precision is as follows: Power In = $\pm$ 0.0005 W; $\Delta T = \pm$ 0.05°C; Power Out = $\pm$ 0.015 W; Over all Precision = $\pm$ 0.5%.

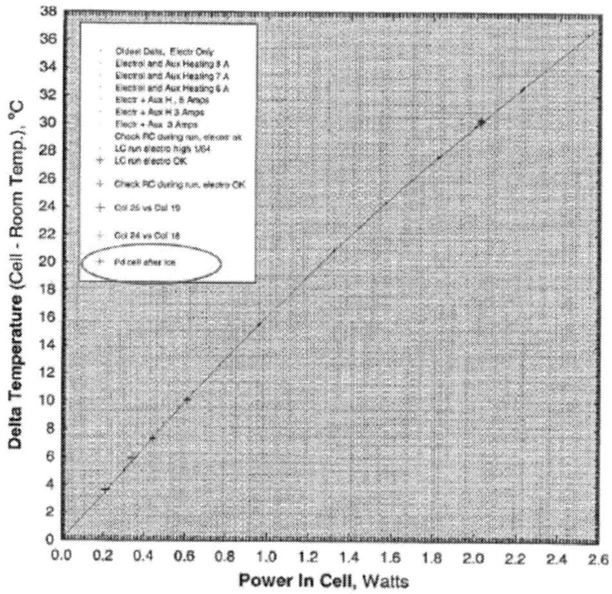

Figure 20. Calibration curve of right cell (Pt/Pt) and left cell (Pd/Pt) *before* electrolytically loading D, dots (·) and *after* ice-quenching the Pd (*left*) cell (*crosses*, +), stopping excess heat 46 days later. Small dots forming line are from voltages, currents and thermocouples inside cells and outside cells in calorimeter chamber. Crosses and dots on same curve proves no shift in the calibration. Data acquired and plotted without human intervention. Precision is as in Fig. 19.

clustered data set at the left and below the line near power-in of 2.85 W is the result of deliberately overfilling the cell by about 1.5 mm from the equilibrium calibration mark, then allowing the electrolysis to run without the syringe pump filling (pump off). At the point where the data characters (small crosses) meet the calibration line, the electrolyte level met the equilibrium level mark of 18.0 ml. This behavior was tested again at 3.29 W (tight cluster) with about 1 mm overfill and then allowing electrolysis to run until a 1 mm under-fill condition (data above the line).

All other data (crosses) above calibration line are a result of excess power, ranging from 0.10 W (about 4.7% of power-in, depending on power-in) to 0.32 W (9.6%). Excess power stayed in this range most of the 46 days, except for two events of run-away power that required cutting back power-in to prevent boiling of the cell. They are shown in Fig. 22. Excess power during run-away was triggered by an increase in electrolysis current. Before run-away, the light water and heavy water cell were matched in input power by trimming the electromigration current. The fact that the heavy water cell was higher in temperature by about 2.5°C shows it was producing excess power. When electrolysis current was increased from 444 to 535 mA and the control cell (light water) was again matched in power, the temperature of the heavy water cell started to run-away necessitating a cut back in current to 435 mA; but excess power continued. With only a modest increase in current (from 435 to 442 mA), temperature ramped into run-away again, but stabilized in the 82°C range, considerable higher than 67°C, prior. This second run-away event produced 2.4 W with 1.2 W input. Total excess *heat* (excess energy) over a 46 day period of 0.775 MJ or 150 MJ/cm^3 of Pd or 14 000 eV/Pd atom was from integrating the suitable power values over this period. This is of such a magnitude that it must be a nuclear reaction; but there is no evidence in this study which nuclear reaction.

In this study seven Pd specimens were run. Six specimens gave excess *power*. Two gave extensive excess *heat*

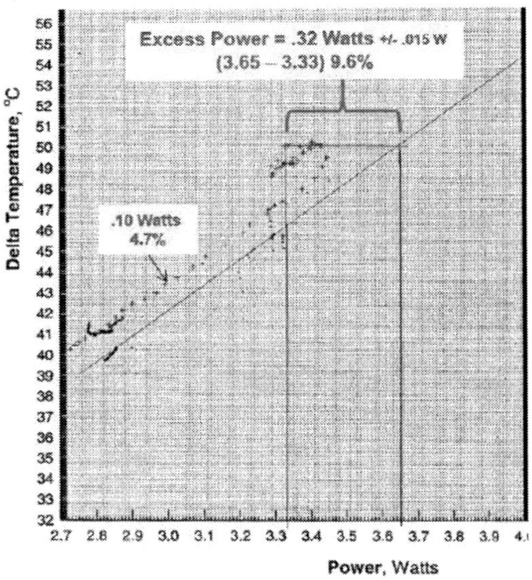

Figure 21. Left cell (Pd/Pt) in D_2O exhibiting excess power (*crosses*, +). The line is drawn through the calibration curves of Figs. 18 and 19 (without excess heat). Watts of excess power are above the calibration curve but varied for the entire 46 day period as shown by periods of 0.10 and 0.32 W excess power.

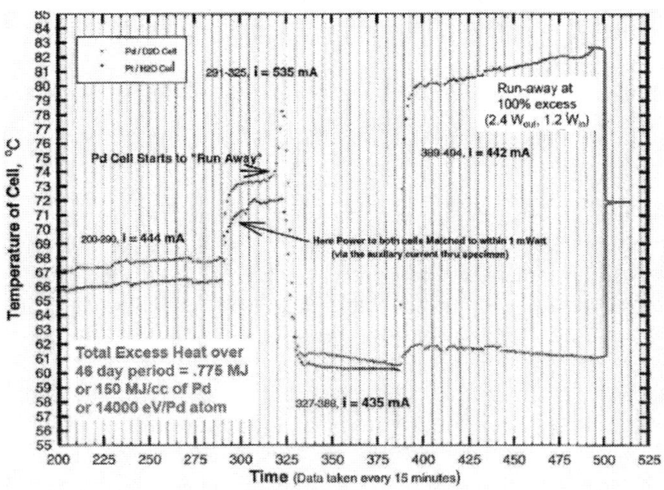

Figure 22. Left cell (Pd/Pt) in D_2O exhibiting excess power (+) and responding to sudden changes in electrolytic current. The temperature of both heavy and light water cells are shown as a function of time (data acquired every 15 min). Before initiation of run-away events, power to both cells was matched to within 1 mW.

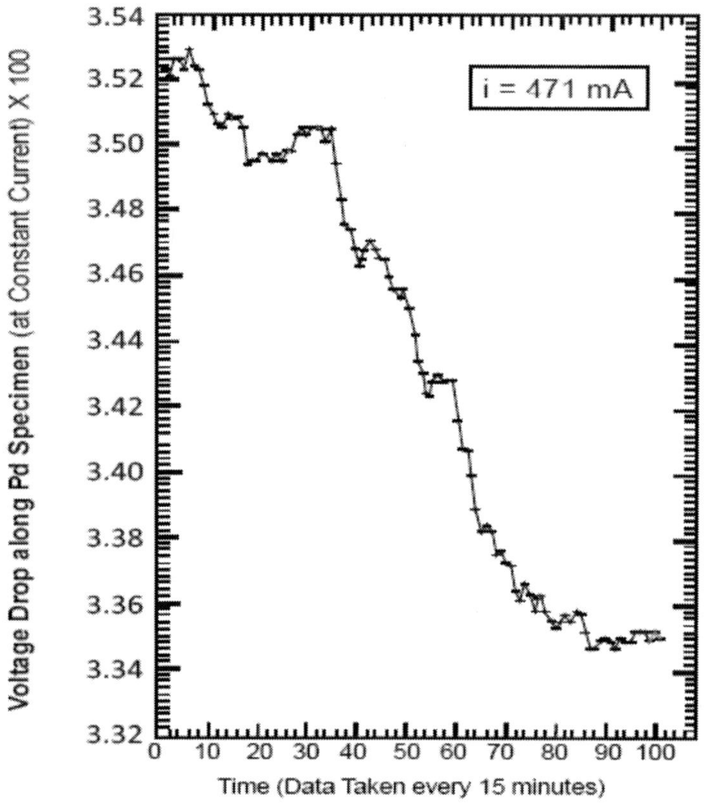

Figure 23. Left cell (Pd/Pt) in D_2O exhibiting excess power (+). The resistivity of the PdD had gone over the hump in resistance (near D/Pd = 0.74) at the beginning of the run assuring specimen is in the range above an average D/Pd = 0.93 for the events featured here. The drop in voltage along specimen is from a change in its resistivity since electromigration current is constant and the temperature of the cell is increasing.

similar to that reported. The ones that gave excess power (but not excess heat) were deliberately stopped for various reasons not associated with excess heat–not run long enough to determine excess heat that exceeds the limits of chemistry. It is reasonable that they would have given excess heat. Another specimen (seventh) had a visible crack, no excess heat, and did not load properly. It was run however to verify the importance of loading on excess heat.

Another important observation was the change in resistivity during excess heat in Fig. 23. Resistivity is measured as a drop in voltage along the Pd with constant current. Each number on the time scale is 12 min (data taken every 15 min). From approximately 5–87 units on time scale, there is an irregular periodic drop in resistivity, interrupted only by a slight diffidence that vanishes quickly, followed by resumed drop. Resistivity of PdD had gone over the hump (near D/Pd = 0.74) at the beginning of the run. This assured the specimen was above the range of average D/Pd = 0.93 for the events featured here and all events of excess power and heat. What makes this drop particularly significant is the fact that temperature of the cell is increasing all the time resistivity is changing. This is shown in Fig. 24 along with the temperature increase of the cell during resistivity drop. These events could be triggered by a sudden increase in current density, but they also happened, most often, spontaneously at constant current density.

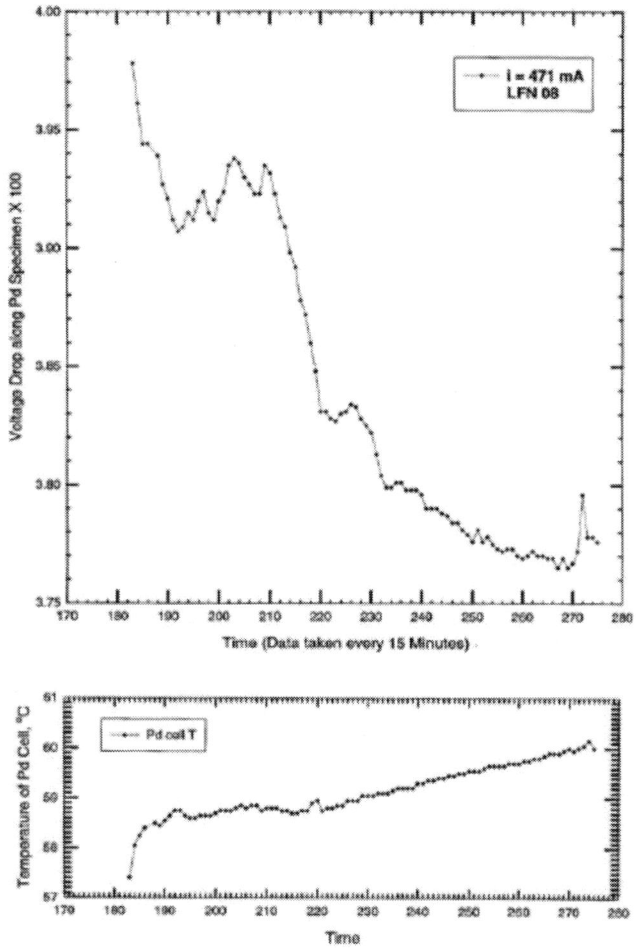

Figure 24. Left cell (Pd/Pt) in D_2O exhibiting excess power with the specimen above an average D/Pd $= 0.93$. The drop in voltage is from a change in resistivity since electromigration current is constant and the temperature of the cell is increasing and electrolysis current density is constant.

5. Discussion

The magnitude of excess heat (Figs. 21 and 22) confirms Fleischmann–Pons heat effect from nuclear origin. The amount of excess heat per cubic centimeter of Pd (150 MJ/cm^3) or per Pd atom (14 000 eV/atom) is too large for a chemical reaction, which produces energy per atom less than 2 eV/atom. The drop in resistivity while temperature increases, is not expected behavior of PdD. Most metals and metal hydrides (or deuterides) show increasing resistivity with temperature [54–56], as in Fig. 25. Nucleation of a new phase, other than beta (β) or gamma (γ), with lower resistance is likely occurring in Figs. 23 and 24. If excess heat is from δ (Pd_3VacD_4 with D in octahedral sites), then formation of δ' (Pd_3VacD_4 with D in tetrahedral sites) (Fig. 26) enables extensive pathways of low resistance for

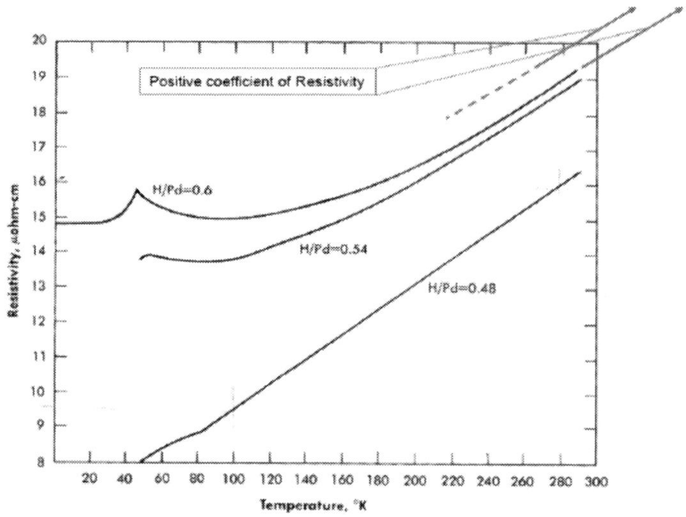

Figure 25. Resistivity versus Temperature for Pd–H samples from low temperature to room temperature and extrapolated to temperatures above room temperature with a positive coefficient of resistivity, from [54,55].

electron transport along tubes, which are vacancy channels free of atoms along edges of unit cells. These extend from one unit cell to the next and intersect at all unit cell corners, as shown in Fig. 27. The solubility of D(H) decreases with increasing temperature in Pd–H [57–59] and decreases in Pd–Ag alloys, as shown by Paolone et al. [60]: they are exothermic [57]. With current density constant, fugacity is constant. Phase change to tetrahedral site occupancy is a change with more order as Isaeva et al. [29] indicate: tetrahedral site occupancy is favored as a more ordered phase. Resistivity, in general, is larger in a disordered state than in an ordered state, as pointed out by Fukai [61]. Therefore, the specimen is unlikely to be further loading itself with D (to lower resistivity), but rather more likely to be undergoing phase change from δ to δ'. As there is insufficient time for δ or δ' to form at room temperature by vacancy diffusion this is not the formation mechanism of SAV ordered phases. Appendices A and B are an alternative of supplying vacancies at small SAV volume fractions for a mixed phase composite microstructure in the bulk (β with small islands of δ or δ'). The δ is estimated below to be 0.03%: thus not many vacancies are required. The co-deposit work that results in SAV at room temperature show: SAV phases form and are stable at room temperature (equilibrium phases, not meta-stable phases). If SAV phases are merely formed at elevated phases and then quenched-in at room temperature then they would be meta-stable and would decompose to an equilibrium form (another that is not an SAV phase) but they do not.

Sites for nucleation of δ' would be less than the total volume fraction f_v of δ-phase in the cathode. This fraction is the active atoms divided by total atoms, determined as follows. The number of Pd atoms in specimens here is = 3.4×10^{20} atoms (size of Pd, Section 3), while nuclear reactions could produces 23 800 000 eV per reaction. The actual energy produced is 14 000 eV/Pd atom or 7000 eV/(D atom pair) over 46 days. Thus $f_v = 7000/23\,800\,000 = 0.0003 = 0.03\%$. The total number of D pairs participating $= 1.0 \times 10^{17}$ pairs out of 3.4×10^{20} atoms. If all of the δ-phase is active in giving heat, then f_v is too small to detect δ-phase from metallurgical microscopy. The δ'-phase is an even smaller fraction, yet it produces a macroscopic effect (measurable lowering of overall resistivity of the bulk specimen). This implies the inherent resistivity of δ' is very low (possibly even zero) since the total resistance of the cathode must obey the law of mixtures, combining resistivity of δ' times its small volume fraction with resistivity of

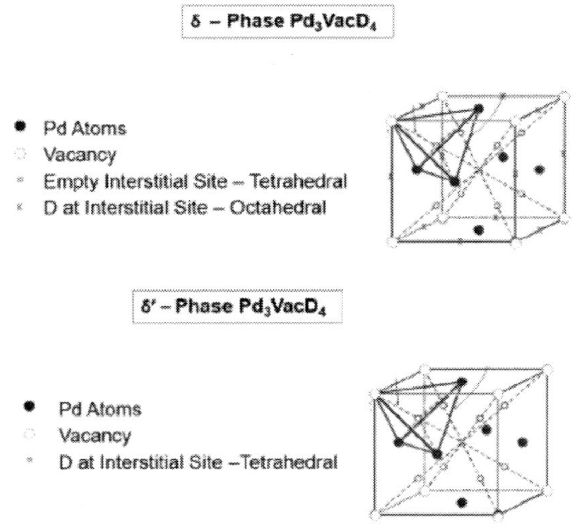

Figure 26. The ordered unit cells of the delta (δ), Pd_3VacD_4 and delta prime (δ'), Pd_3VacD_4-phases. The main difference is that D occupies octahedral sites in δ and tetrahedral sites in δ'. Edges of the unit cell in δ' are straight paths of open tunnels (*or tubes*) because of vacant Pd atoms. In δ, the only atoms in these tubes are D^+ ions.

beta β times its volume fraction (complement fraction). Thus, there is significant variation in resistivity from location to location within the bulk. This is consistent with observed local hot spots for production of tritium observed by Will et al. [62] and Srinivasan et al. [63]. It is also consistent with small and scarce local explosive reactions in the lattice in near-surface region from volcanic-like eruptions observed in optical and scanning electron microscopy of the surface after excess heat [64].

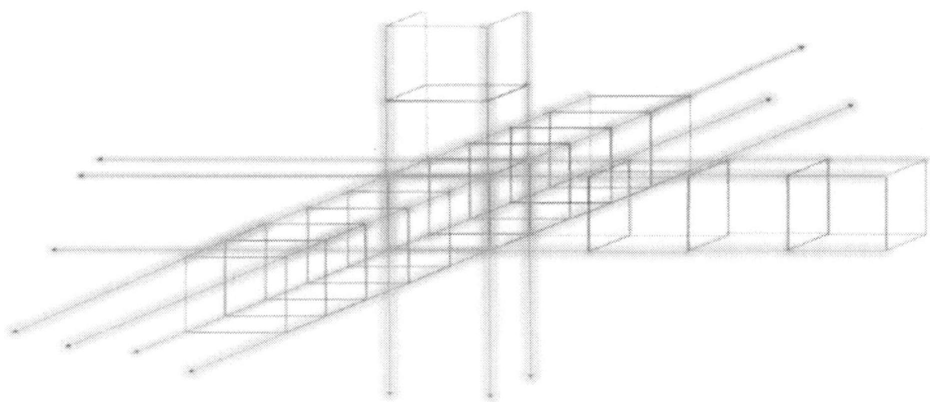

Figure 27. Tubes (*green*) on all edges of unit cells of delta (δ), and delta prime (δ')-phases are a network of intersecting tunnels producing a separate open tube lattice of its own.

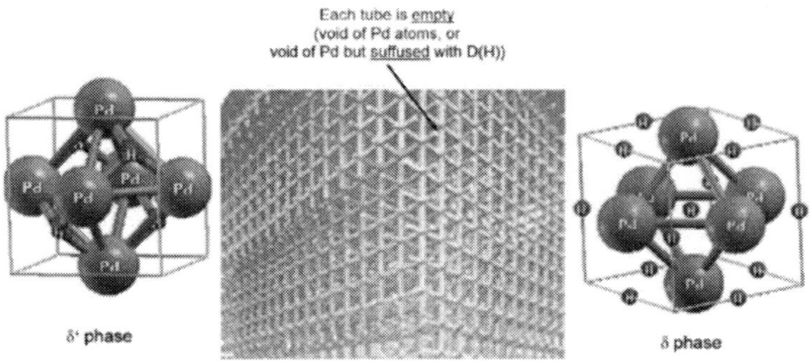

Figure 28. Tubes for each unit cells of either δ or δ'-phases. These phases form a 3-D vacancy tube lattice or network of intersecting tunnels. The tube lattice (*green*) has Pd and isotopic hydrogen in the space between tubes in δ' (left image = Pd_3VacD_4 - T), or has only Pd atoms (right image = Pd_3VacD_4 - O) in the space with D^+ *inside* the tubes in δ. Unit cell images (*blue and red*) after [29].

The tube lattice (Fig. 28) has Pd atoms nested between tube *intersections*: either Pd with D(H) or only Pd, depending on whether the phase is δ' or δ (with D still inside the tubes). If D is still inside tubes (δ) as an ion, then this is a variant of Storms' model [65]: an electron in between each D^+ ion (Fig. 29). Storms' model might be improved by replacing a two-dimensional crack with a one-dimensional tube of diameter equal to about one Pd atom. The tube would maintain alignment and avoid the buckling problem inherent in two-dimensional crack space. Electron shielding, in two-dimensions, needs to be kept aligned to avoid instability (D(H) ion and electron pop sideways). The tube would keep the shielding aligned and avoid elastic buckling instability as in axially loaded beams in compression, Euler Buckling. In addition real metal cracks may be too wide (not sharp) on an atomic scale to align a string of alternating charges of ions as proposed by Storms [65]. The size and geometry of real cracks are shown in Fig. 30

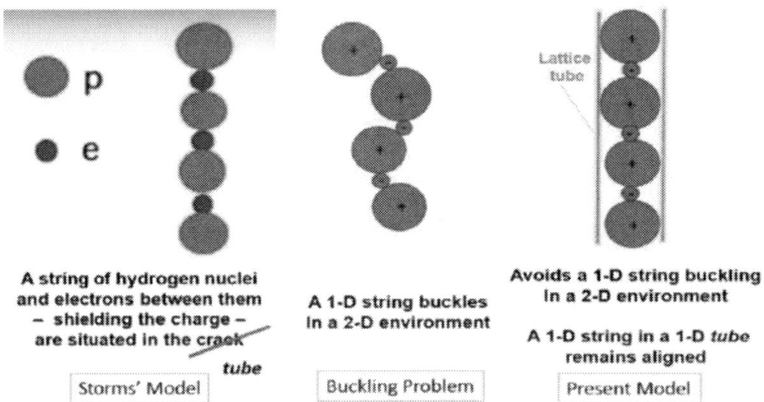

Figure 29. Storms [65] has modeled electron shielding in a two-dimensional crack shown on left and available online: https://www.youtube.com/watch?v=SNodilc6su0. The center shows that a string of alternating electrons and deuterons (*protons*) will buckle when left in a two-dimension crack with a third dimension of width of 1 atom. The present view of SAV in Figs. 26 and 27 corrects the buckling problem since the lattice tube is 1 atom in diameter and maintains alignment when compressed axially.

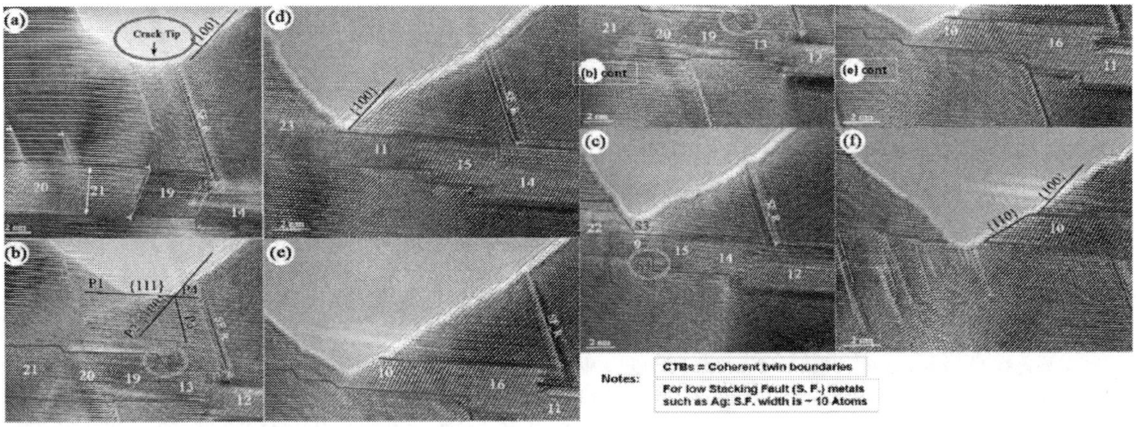

Figure 30. High resolution transmission electron microscopy, HRTEM, images in silver (Ag), a low stacking fault metal: (a)– (c) in situ HRTEM images during the crack propagation in the matric. (d)– (f) in situ HRTEM images during the penetration of the crack across the CTBs. The beam direction is parallel to ⟨110⟩. The CTBs are outlined in red lines and the corresponding twin thickness is labeled in the unit layers. A stacking faults marked as SF_R is chosen as the reference. The crack surface changes from { 100} to {110} after the crack penetration across the CTBs. P1–P4 in (b) indicate slip planes. Crack is too large to support an aligned string of alternating deuterons and electrons. adapted from [66].

adapted from [66] who used high resolution transmission electron microscopy HRTEM to document images at tips of cracks in silver (Ag), a low stacking fault metal. Pd is a higher staking fault metal (approximately 10 times higher).

This model of a regular lattice of tubes filled with an alternating pattern of deuterons and electrons is very different from previous models of multiple deuterons in a monovacancy such as the work of Nordlander et al. [67]. In addition, the local source of vacancies, from severe plastic deformation (SPD) by intersecting dislocations, and the effects of electromigration for minor relocation effects may present additional issues from those discussed by McLellan and Zang [68] related to the kinetics of vacancy migration. Additionally, these issues are directly related to kinetics (rate of production) rather than thermodynamics (stable phases). The kinetic problems were by passed in the work of Szpak et al. [69] and fellow researchers Letts and Hagelstein [70] who obtained excess heat by purposefully creating vacancies directly with co-deposition at high current density.

6. Summary and Conclusions

(1) The energy produced, over and above input energy, was 150 MJ/cm^3 of Pd (14 000 eV/Pd atom) for 46 days. The average excess power ranged from 4.7$\pm$ 0.15 to 9.6$\pm$ 0.30 % of input power while input power ranged from 2.000 to 3.450 W, confirming the Fleischmann–Pons effect. The energy release was of such magnitude that the source is nuclear. The exact nuclear reaction was not determined in this work.

(2) Two run-away events were triggered by suddenly increasing current density resulting in 100 % excess power (2.4 W output with 1.2 W input) necessitating a temporary cut back in electrolysis current.

(3) The emended palladium–isotopic hydrogen phase diagram is presented: Three new phases, from XRD results from recent literature, are shown on the phase diagram as superabundant vacancies (SAV) phases and are: γ-phase (Pd_7VacD_{6-8}), δ-phase (Pd_3VacD_4-octahedral), δ'-phase (Pd_3VacD_4-tetrahedral). These phases are the lowest free energy phases at their respective compositions.

(4) Resistivity of Pd was used to assay D activity in the Pd lattice (ratio of D/Pd) and employed as an indicator of phase changes. The excess heat supports portions of the cathode being in the ordered δ-phase (Pd_3VacD_4-

octahedral), while the drop in resistance of the Pd cathode during increasing temperature and excess heat indicates portions of the cathode transformed to the ordered δ'-phase (Pd_3VacD_4-tetrahedral).

(5) The structure of δ-phase (Pd_3VacD_4-octahedral) and δ'-phase (Pd_3VacD_4-tetrahedral) show a network or lattice arrangement of empty tubes (δ') or tubes filled with isotopic hydrogen (δ). These empty tubes provide extensive pathways of ultra-high mobility of hydrogen (δ) or electrons (δ') or both. It is proposed these tubes provide a pre-condition of nuclear activity.

(6) A model of electromigration is presented where these phases were encouraged by electromigration current, causing D^+ ions (trapped to vacancies) to pull vacancies along and aid the formation of SAV phases. The model of electromigration indicates considerable enhancement of D^+ ions (higher D/Pd) at one end of the specimen raising the likelihood of SAV phases and nuclear activity.

(7) A plastic deformation based model offers a mechanism for vacancy production in the bulk lattice. Vacancies are created by dragging of jogs connected between screw dislocations. Jogs are created by intersecting dislocations. The creation and mobilization of these vacancies raise the likelihood of SAV phases and nuclear activity by mitigating the necessity for bulk diffusion from the surface or grain boundaries. It shows the importance of plastic deformation (by cold work or by a loading/unloading/reloading sequence) in preparing Pd (or Ni) specimens for LENR.

Acknowledgements

Special thanks are expressed for the encouragement of Drs. John J. Antal and David J. Nagel, for many very helpful suggestions by one of the reviewers, and for the advanced copy review efforts of Jed Rothwell.

Appendix A. Effects of Electromigration on Enhancing the Concentration

Early work on isotopic hydrogen ion migration from an electric field in metal lattices began with Coehn [71], co-workers [72–74] and others [75,76]. It was applied to LENR promptly by several groups [77–84]. Bartolomeo et al. [79] predicted in 1993 about the time SAV phases were being discovered that: It is certainly possible that a further phase transformation takes place at high D:Pd ratios leading to the formation of a γ-phase and that it is the properties of this phase which will explain the nuclear processes induced in Pd and Pd-alloy host lattices. This prediction seems to be born out for δ rather than γ, as in this work. The Preparata-Effect [84] uses a very long thin wire with as high a total voltage drop as possible along the wire (axial voltage drop) but with as low an axial current as possible for electromigration. Here the opposite (low voltage drop with high current) is sought. Low voltage drop minimizes parasitic joule heating (for a given current), and high current causes electromigration, but also contributes to joule heating, which must be balanced against electromigration.

Specimens were supported in cells between heavy gauge Cu wire leads. D^+ electromigration was effected by passing current for enhancement of D concentration along the length as in Fig. 31. Fluxes in opposite directions, $J_{D+}^{\text{electricfield}}$ and $J_{D}^{\text{concentration}}$, balance at steady state (after transient period). Flux $J_{D+}^{\text{electricfield}}$ due to electric field ($E = dV/dx$) is $nq\mu E$, where n is number of charges, q charge assigned to D^+ ($\sim$-e), and μ is mobility ($= D/k_BT$). Electrons and protons (deuterons) have different mobilities: for electrons $\mu = 1/nq\rho$, where ρ is resistivity in cm-Ω, but deuteron mobility, from Nernst–Einstein equation, is $\mu = D/k_BT$, where k_B is the Boltzmann constant and D is diffusion coefficient. With Fick's first law, $D(dC/dx) = nq\mu E$, a first order differential equation. If dC/dx is C' and dV/dx is V', and since $C = n$, the equation is $C' = C\,S\,V'$, where S is defined as q/k_BT. The solution is: $C(x) = Be^{SV}$, where B is a constant from boundary condition: $C(x=0) = C_o$ and $V(x=0)$ can be defined as V_o. With these and the boundary condition: $C(0) = Be^{SV_o}$ with $B = C_o e^{-SV_o}$. The solution is: $C(x)/C_o = \exp(q[V(x) - V_o]/k_BT)$, or if Z is defined as (q/e^-), then $C(x)/C_o = \exp\,(Ze[V(x) - V_o]/k_BT)$. This equation

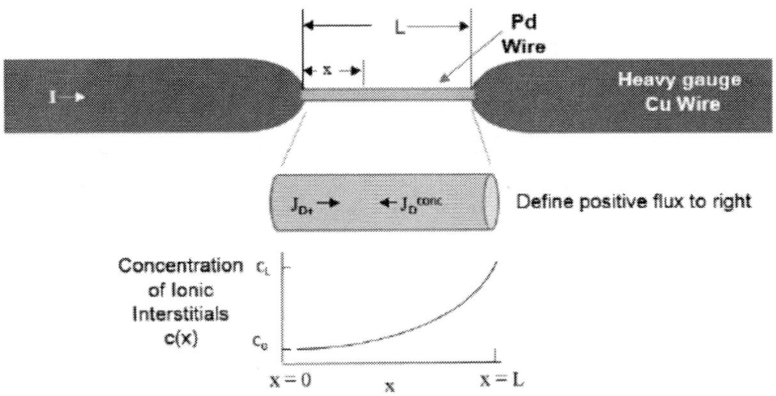

Figure 31. Pd between two Cu leads, the two fluxes that develop and the resulting concentration profile.

was independently published before the discovery of LENR [85], but, as far as the author knows, was first applied to LENR in 1991 [77,78] and shortly thereafter [79–84]. For rod of uniform section A ($= \pi d^2/4$) and resistivity ρ, $E = [V(L) - V(0)]/L$ and $E = I\rho/A$, thus $[V(L) - V(0)] = I\rho L/A$. Then $C_\mathrm{L}/C_\mathrm{o} = \exp(4ZeI\rho L/\pi d^2 k_\mathrm{B}T)$. Evaluating with numbers gives: $C_\mathrm{L}/C_\mathrm{o} = \exp(0.0003723IL/d^2)$ for $z = 0.7$ and $C_\mathrm{L}/C_\mathrm{o} = \exp(0.0005319IL/d^2)$ for $z = 1$. Table 1 shows electromigration effects are significant, in terms of concentration at one end versus the other end (i.e., $C_\mathrm{L}/C_\mathrm{o}$). For values of currents used (typically 3 A, but as high as 10 A), the enhancement is from 506 to 22 200%. If actual enhancement were only 40% it would be significant. It should be pointed out, the parasitic heat added could be significant if specimen size is imprudently selected, but for the conditions here (3 A) it was about 0.1 W.

Table 1. Composition enhancement from one end to the other by electromigration in Pd[a].

Total current, I (A)	Diameter of specimen, d (cm)	Power density by Joule heating , P (W/cm^3)	Total power (= $P \times$ volume of specimen) (W)	Composition ratio (enhancement) C_L/C_o (= C(L)/C(0))	
				Z = 0.7	Z = 1
1	0.10	0.175	0.0035	1.10	1.14
3	0.10	1.58	0.0314	1.33	1.50
10	0.10	17.5	0.349	2.57	3.86
20	0.10	70	1.40	6.60	14.91
1	0.05	2.80	0.014	1.46	1.72
3	0.05	25.21	0.126	3.10	5.06
10	0.05	280	1.40	44.0	222.0
20	0.05	1121	5.59	1902.0	49 425.0

[a]*Note*: Using $\rho = 10.8 \times 10^{-6}\Omega$-cm, $L = 2.54$ cm, temperature (300 K) and constant diffusion coefficient, electric wind force at zero, steady state is assumed (transient period is proportional to length^2/diffusion coefficient, L^2/D) and no leakage back into electrolyte along the specimen surface (the highest potential). Thus, $C/C_\mathrm{o} = \exp(0.0005319IL/d^2)$, for $Z = 1$. Joule heating (total power = $P\times$ volume of specimen) uses $P = 0.000017508I^2/d^4$ in W/cm^3.

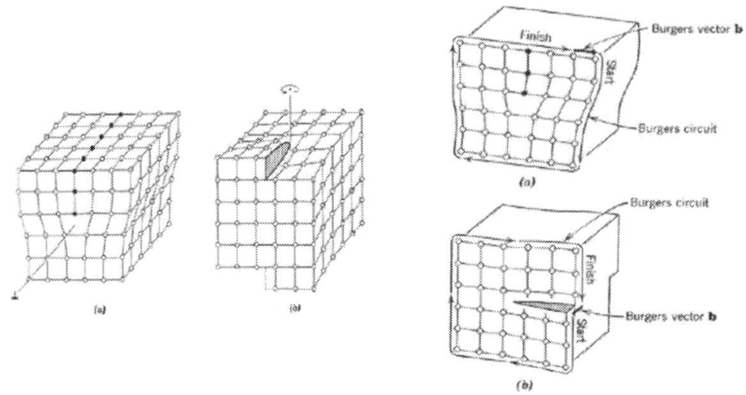

Figure 32. Geometry (*Left*) and Burgers circuits (*Right*) around Edge (a) and Screw (b) Dislocations. After [87].

Appendix B. A Plastic Deformation Mechanism for Creating Apposite Interior Vacancies

A well-known mechanism [86–88] to create vacancies through screw dislocation intersections that generate jogs evidently plays an important role in pre-processing Pd for LENR. Loading and unloading Pd with D four times (three times unloading, seven times through the miscibility gap) encouraged a well-developed dislocation cell structure and its associated excess of vacancies. The Burgers vector for edge dislocations is always perpendicular to the dislocation line and screw dislocations have Burgers vectors parallel to the dislocation line, Fig. 32. Jog-components of dislocations in the slip plane (called kinks) can be either screw or edge (Fig. 33). When dislocations intersect, they always create jogs (Fig. 34). For purposes here, the most important type of intersections are intersections with screw dislocations. Dislocations intersect during heavy cold work. Wen et al. [89] have shown hydrogen enhances homogenous dislocation nucleation, promotes dislocation emission, induces slip planarity, and localizes dislocation activity significantly, leading to locally enhanced vacancy formation from dislocations. In Fig. 34, both screws (AB moving along the slip plane, ABCD) and edge (BC moving left to right on slip plane) components of curved dislocations can create jogs. Unlike kinks, a jog is normal to the main slip plane and resists being dragged by a screw dislocation because it either creates lattice-atom interstitials (moving to the right in Fig. 35 and requiring higher energy), or it creates lattice-atom

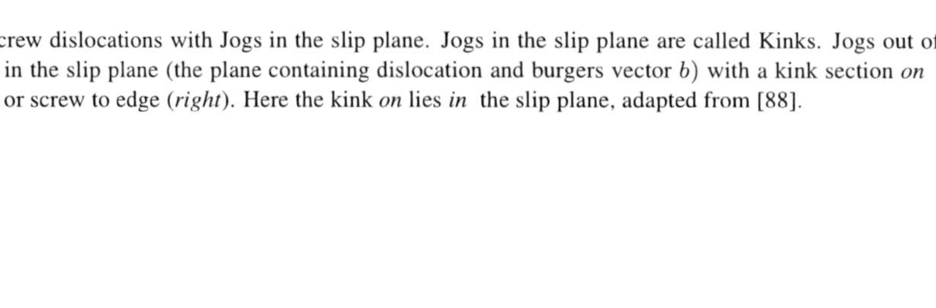

Figure 33. Edge and screw dislocations with Jogs in the slip plane. Jogs in the slip plane are called Kinks. Jogs out of the slip plane are called jogs. Dislocations *mnop* in the slip plane (the plane containing dislocation and burgers vector *b*) with a kink section *on* where its nature changes from edge to screw (*left*) or screw to edge (*right*). Here the kink *on* lies *in* the slip plane, adapted from [88].

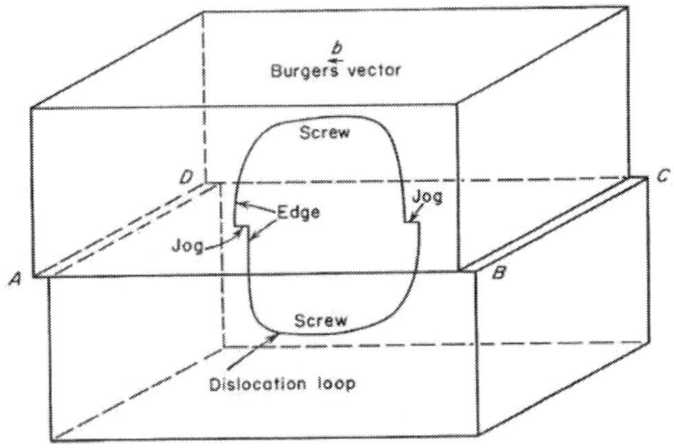

Figure 34. An *edge* dislocation parallel to BC or AD moves from left to right on slip plane ABCD, it cuts through the vertical portion (*edge*) of the dislocation loop creating jogs on the edge portion of the loop. The same would happen if a *screw* dislocation parallel to AB moved from front to back on plane ABCD: it would create the steps at BC and AD and the jogs shown, consistent with its Burgers vector b. These jogs are screw in nature (*parallel to* b), but if a dislocation cuts the screw portion of the loop, by moving on a vertical plane, it would create a jog with edge nature, adapted from [88].

vacancies (moving to the left in Fig. 35 and requiring less energy). The red area in Fig. 35 is a row of vacancies created behind the right-to-left moving screw dislocation with an edge jog. Dislocation intersection happens by millions in heavy deformation and is a source of *interior-created* vacancies (as opposed to diffusion *bringing* vacancies from surfaces or from grain boundaries).

In addition to this mechanism for creating vacancies at interior positions in the solid, electromigration, (Appendix A) drags vacancies along with migrating D^+ ions because of binding energy between vacancies and interstitial hydro-

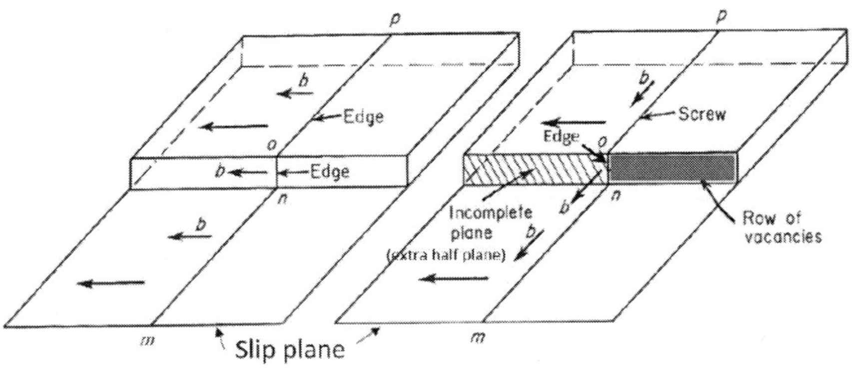

Figure 35. Dislocation *mnop* where its nature is all edge (*left*) or mostly screw (*right*) in a slip plane with a section *on* (jog). The jog *on* is ***normal*** to the main slip plane (*not a kink*) and is edge in both cases. When the jog on the right, connected to screw components, moves to the left, it must create a row of vacancies behind it (*red section*) because its nature is edge and its Burgers vector b is perpendicular to motion. This is not true of the edge dislocation and its jog on the left because of differently oriented b, adapted from [88].

gen. The volume of the nuclear active environment (NAE) is a small percent of bulk volume (Section 5), so the amount needed, by these two mechanisms, to produce a NAE is relatively small.

Dislocations intersect and tangle into dislocation cells and cell size d is controlled by dislocation density $\rho(d = K\rho^{-1/2})$ which is also related to shear flow stress $\tau(\tau = 0.57\ \mathrm{Gb}\,\rho^{1/2})$ and thus hardness, as shown theoretically by Holt [47] for all metals with high enough stacking fault energy, and then first shown experimentally for Cu [48] and subsequently most other metals [49]. Here G is shear modulus and b is Burgers vector. Dislocation cell formation occurs in unalloyed Pd and PdH as found experimentally by Deutges et al. [50]. For PdH cells form except, with hydrogen uptake there is higher dislocation density and smaller cell size ($d = K\rho^{-1/2}$ and $\tau = 0.57\ \mathrm{Gb}\rho^{1/2}$), consistent with higher hardness results (higher τ) shown above (Fig. 16). There are more vacancies from the jog mechanism when H(D) is present. High vacancy content associated with high hydrogen content found in density functional perturbation theory (DFT) supports the result of this dislocation-based vacancy producing mechanism.

Millions of such intersecting events with the associated row of vacancies stretching across the dislocation cell occur during dislocation cell formation in metals, providing a supply of vacancies large enough to support nucleation and growth of SAV phase islands within the β-phase since the volume fraction of δ is only 0.03%. These rows of vacancies are then dispersed by electromigration and very local diffusion to the eight corner positions of SAV units cells. This mechanism supplants thermal generation of vacancies and diffusion from the surface or grain boundaries which is too slow at room temperature to provide the number of vacancies needed for SAV phases. The final microstructure consists of a two-phase microstructure of predominately β with small islands of δ. Dislocation cells form under plastic deformation with dislocation density around 10^{10}–10^{12} cm per cubic centimeter [47–50, 89]. Plastic deformation is inferred to be essential for creating the nuclear active environment.

An estimate of the adequacy of this mechanism in providing vacancies can be made as follows. In a single dislocation cell of typical size of 1 μm, there will be 3636 vacancies created in the row of vacancies as jogged dislocation ABCD traverses the dislocation cell shown in Fig. 36. This is simply the ratio of the length of the row (10^{-6} m, cell size in PdH, see Fig. 5 of Ref. [50]) divided by the size of the vacancy in Pd (2.75×10^{-10} m). The number of FCC

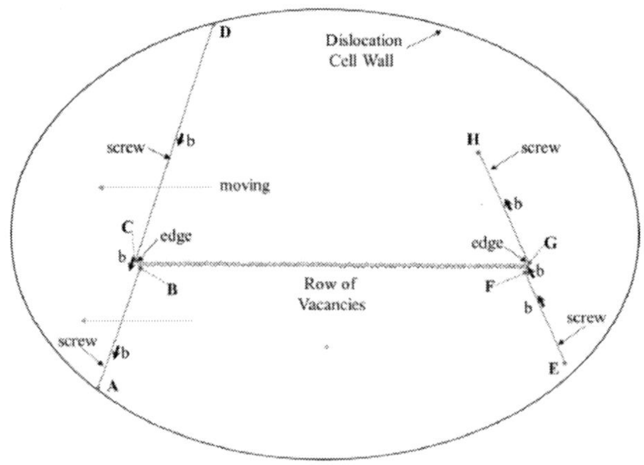

Figure 36. Dislocation ABCD, one of many jogged screw dislocations moving left across the dislocation cell dragging the edge jog component BC and creating a string of vacancies in its wake back to dislocation EFGH. The jogs BC and FG were created by the intersection of original screw dislocations AD with EH. Intersecting dislocation (EFGH) has its Burgers vector b parallel to the edge component BC, and dislocation ABCD has its Burgers vector parallel to the edge component FG, as shown. Dislocation EH was perpendicular to the glide plane of dislocation AD, assumed for the sake of illustration, to be the plane of the paper.

unit cells within this dislocation cell is 8.9×10^9 (ratio of volume of the dislocation cell, 4/3 $\pi d^3/8 = 0.52 \times 10^{-18}$ m^3, divided by the volume of the Pd FCC unit cell $= 5.89 \times 10^{-29}$ m^3, the cube of the lattice parameter): see note [90]. The lattice parameter of Pd is 3.89×10^{-10} m. Since there is one vacancy per SAV unit cell (δ-phase, Pd_3VacD_4), the number of vacancies required is the same as the number of new SAV unit cells created by this mechanism. The number of new SAV unit cells in the dislocation cell is 2.7×10^6, which is the nuclear active volume fraction of δ, ($f_v(\delta) = 0.0003$, see Section 5) times the number of FCC unit cells in the dislocation cell (8.9×10^9). Therefore the number of intersections of dislocations in the dislocation cell needed to supply the vacancies is 743. This is the ratio of vacancies need (2.7×10^6) divided by the number of vacancies per intersection (3636). For dislocation density of about 10^{11}cm per cubic centimeter [47–50,89], it would become 0.1 cm per the dislocation cell volume when proportioning the volume from a cubic cm to the volume of the dislocation cell ($d = 10^{-6}$ m, volume $= 0.52 \times 10^{-18}$ m^3). This number is expected to be about 10–100 times larger under sever plastic deformation. Using 0.1 cm, this can be thought of as 10^3 dislocations whose sum of their lengths is 0.10 cm. inside the disloction cell. If another dislocation passed through the dislocation cell, it would make 10^3 intersections with these dislocations. Since this number is greater than (or the same order as) 743, the dislocation jog mechanism is able to supply the necessary number of vacancies to form the required volume percent of δ-phase that is nuclear active. This number could be higher by a factor of 10–100 if 10–100 dislocations intersect the 10^3 dislocations of length 0.1 cm instead of just the one considered here. Therefore, diffusion of vacancies from the surface or from grain boundaries is not required to form nuclear active SAV phase (δ) at room temperature. It may not be a coincidence that the geometric row of vacancies created in this mechanism is a convenient precurser of the vacancy tubes referred to in Fig. 28.

Appendix C. X-ray Diffraction Evidence for SAV Structures

The XRD evidence for SAV relies on the ability to distinguish crystal (Bravais) lattices and lattice parameters. Simple cubic (SC) and face centered cubic (FCC) are two Bravais lattices involved in SAV structures and can be distinguished because of the structure factor [91] for each unit cell. The resultant wave scattered by all atoms of the unit cell is called the *structure factor* $\boldsymbol{F}_{\mathbf{hkl}}$ for the hkl reflection (diffraction), given by:

$$F_{\mathbf{hkl}} = \sum\nolimits_{n=1}^{N} f_n e^{2\pi i(hu_n + kv_n + lw_n)},$$

where f_n are the atomic scattering factors (ratio of the amplitude of the wave scattered by atoms to the amplitude of the wave scattered by the electron) of each atom n in the unit cell, hkl are the Miller Indices of the plane of diffraction, uvw are the fractional coordinates of atom positions within the unit cell, and N is the total number of atoms in the unit cell. $\boldsymbol{F}_{\mathbf{hkl}}$ is a complex number with both amplitude and phase, but its absolute value $|\boldsymbol{F}_{\mathbf{hkl}}|$ is a ratio of amplitudes:

$$|\boldsymbol{F}_{\mathbf{hkl}}| = \frac{\text{amplitude of wave scattered by all atoms of the unit cell}}{\text{amplitude of the wave scattered by one electron}}.$$

In addition to Bragg's law for determining when reflections at a given angle, θ, occur, the structure factor changes the intensity of a pattern of hkl reflections. Some reflections have zero intensity ($|\boldsymbol{F}_{\mathbf{hkl}}|= 0$) when adding the scattering for each atom in the unit cell. The results of $|\boldsymbol{F}_{\mathbf{hkl}}|$ for these zero intensity calculations (missing reflections) are summarized in Table 2 for selected Bravais lattices, whereas only those reflections that are *present* are listed in Table 3. The presence of the *red* (Superlattice lines) indices verses only the *black* (fundamental lines) indices in Table 3 determines if the Bravais lattice is SC or FCC. When the unit cell changes from FCC to SC Superlattice lines (red indices) appear in the XRD patterns. This was first discovered by Sykes and Evans in the gold–copper (Au–Cu) alloy with an ordered structure, $AuCu_3$ [92] and is shown in Fig. 37. XRD evidence in PdD(H) for SC instead of a FCC

Table 2. Present and missing reflections (diffracted) conditions for common Bravais lattices.

Bravais lattices type	Reflections present[a]	Reflections missing[a]
Simple	All	None
Base-centered	h and k unmixed[b]	h and k mixed[b]
Body-centered	$(h+k+l)$ even	$(h+k+l)$ odd
Face-centered[c]	h, k, and l unmixed	h, k, and l mixed
Hexagonal close packed	$(h+2k)=3$ integer, l is odd	All other cases

[a]Planes denoted by Miller Indices h, k, l defined as reciprocals of the fractional intercepts which the plane makes with the crystallographic axes.

[b]Applies to a cell with the base (face with atom at center) as the C face (plane defined by axes a and b), if base is B face (defined by a and c axes), then h and l are unmixed when reflection is present, and if base is the A face (defined by b and c), then k and l are unmixed when reflection is present.

[c]Also includes all NaCl-type structures except when $f_{\text{Na-type atom}}$ happens to equal $f_{\text{Cl-type atom}}$.

proves the unit cell is either δ-phase (Pd_3VacD_4- octahedral) or δ'-phase (Pd_3VacD_4- tetrahedral), rather than beta β-phase (PdD). But δ cannot be distinguished from δ' by XRD alone because the atomic scattering for D(H), f_D, is so small, but neutron diffraction can distinguish them. For Pd when vacancies replace corner atoms of the unit cell, it takes the same ordered arrangement as Cu_3Au. Examining this unit cell and considering corner vacancy as an atom-type, one sees that for every vacancy fully within the unit cell (1/8 of each of eight corner atoms), there are three Pd atoms fully within the unit cell (1/2 of each of the six face atoms). Clearly the vacancy-atoms, by themselves, form a simple cubic unit cell and so the Pd_3Vac ordered unit cell is SC. Its structure factor confirms SC with both fundamental and superlattice reflections. Figure 38 shows the XRD pattern for the SAV form of Pd-hydride at 500°C and H pressure of 5 GPa, after holding at 800°C for 3.5 h and another sample at 300°C and H pressure of 5 GPa, after holding at 700°C for 8.5 h from [93]. There has been separation into two phases PdH (A) and a vacancy-ordered phase Pd_3VacH_4 (B). The latter phase (B) is identified as SC from the superlattice lines. Figure 39 shows the same behavior for NiH processed at 800°C and H_2 pressure of 5 Gpa from the work of Fukai et al. [94]. Superlattice reflections indicate the crystal lattice is also simple cubic (ordered phase, Ni_3VacH_4).

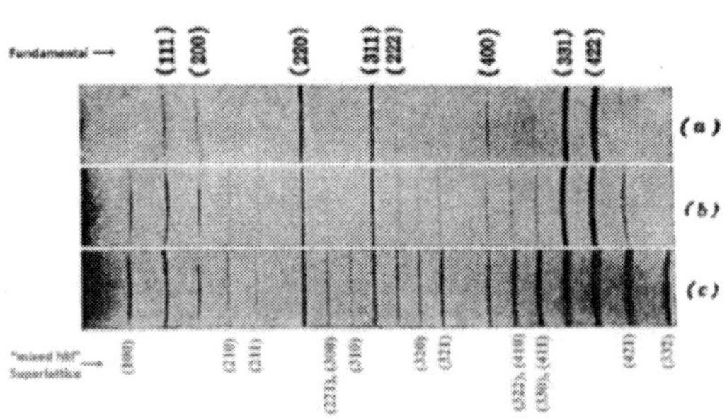

Figure 37. Powder XRD pattern (reflection line intensity recorded on film) of the development of order in an FCC Au–Cu alloy: (a) disordered (fundamental reflections), (b) partially ordered, and (c) highly ordered $AuCu_3$ with Superlattice reflections, adapted from [92].

Table 3. Reflections present for each Bravais lattices.

$(h^2 + k^2 + l^2)$ (order of reflection is increasing θ)[a]	Simple cubic (SC) Reflections present = (All *hkl*)	Face centered cubic (FCC) (including NaCl-type) Present = (h, k, and l unmixed)
1	***100*** [b]	c
2	***110***	–
3	***111***	111[c]
4	***200***	**200**
5	***210***	–
6	***211***	–
8	***220***	**220**
9	***221***	–
9	***300***	-
10	***310***	–
11	***311***	**311**
12	***222***	**222**
13	***320***	–
14	***321***	–
16	***400***	**400**
17	***322***	–
17	***410***	–
18	***330***	–
18	***411***	–
19	***331***	**331**
20	***420***	**420**
21	***421***	–
22	***322***	–
24	***422***	**422**

[a]Order: θ increases as $(h^2 + k^2 + l^2)$ since the combination of Bragg's law with d-spacing is: $\sin^2 \theta_{hkl} = \frac{\lambda^2}{4a^2}(h^2 + k^2 + l^2)$

[b]The numbers highlighted in red color *hkl* planes indicate the presence of reflections for SC structure called *Superlattice* lines.

[c] Missing reflection for FCC ($\boldsymbol{F}_{\mathbf{hkl}} = 0$), and the bold roman numbers *hkl* planes indicate FCC reflections present called *Fundamental* lines.

Appendix D. Discussion of the Construction of the Pd–D Phase Diagram (Fig. 8)

Phase diagrams dealing with heterogeneous equilibrium (equilibrium involving more than one phase, as opposed to homogeneous equilibrium, involving equilibrium within a single phase, like gas equilibrium) result from having complete equilibrium for the system: there is mechanical (uniform pressure), thermal (uniform temperature) and chemical equilibrium (same chemical potential of each component in all phases, the number of which is *P*). For heterogeneous system the chemical equilibrium condition demands that the chemical potential of each substance (each component) be identical in every phase which leads to the phase rule: $F = C - P + 2$, where F is the degrees of freedom, defined as the number of variables–temperature, pressure and composition–changes in which can be designated independently without changing the number of phases in equilibrium. Thus F is the number of intensive properties necessary to fix the state of the system. Any homogeneous portion of the system is defined as a phase. Thus different homogeneous portions at the same temperature, pressure and composition are indeed the same phase. Here the components of the system are the individual elements Pd and D. The number of components is C. In this binary system ($C = 2$) there are three primary variables: temperature, pressure and composition. The phase diagram constructed here at constant

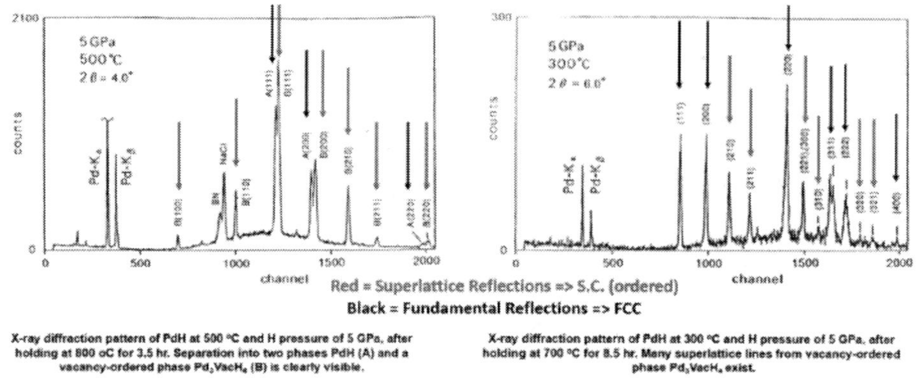

Figure 38. XRD pattern of the ordered unit cells of delta (δ), Pd_3VacD_4) and delta prime (δ'), Pd_3VacD_4) phases. The main difference is that D occupies octahedral sites in δ and tetrahedral sites in δ'. Edges of the unit cell in δ' are straight paths of open tunnels or open tubes because of vacant Pd atoms. In δ, the only atoms in tubes are D^+ ions. Superlattice reflections show the crystal lattice is SC, adapted after [93].

pressure involves only two variables, temperature and composition (thus $F = C - P + 1$). Although the complete depiction of a system involving three variables requires three dimensions, it is common practice to focus on the variables of composition and temperature which are drawn in two dimensions. Thus the equilibrium phase diagram becomes a map of the regions of thermodynamically stable phases wherein the dimensions of the map are composition and temperature. It shows the regions where the combination of temperature and composition result in phases with the lowest free energy (thermodynamic equilibrium) and are placed on the diagram for all the combinations of composition and temperature for which this is the case. The phase rule for a binary diagram ($C = 2$) specifies a one-phase region has two degrees of freedom (temperature and composition): one phase occupies an *area* on the equilibrium phase diagram, a phase field. For a two-phase equilibrium condition, there is one degree of freedom (if the temperature is fixed, then the compositions of both phases are fixed since they are in equilibrium. Two-phase equilibrium is shown on the diagram by two lines (straight or curved). These two curves in temperature-composition space representing the set of points where the two phases are in equilibrium with each other and have only one degree of freedom (either temperature or composition, but not both). And finally three-phase equilibrium can only occur at a point (phase rule

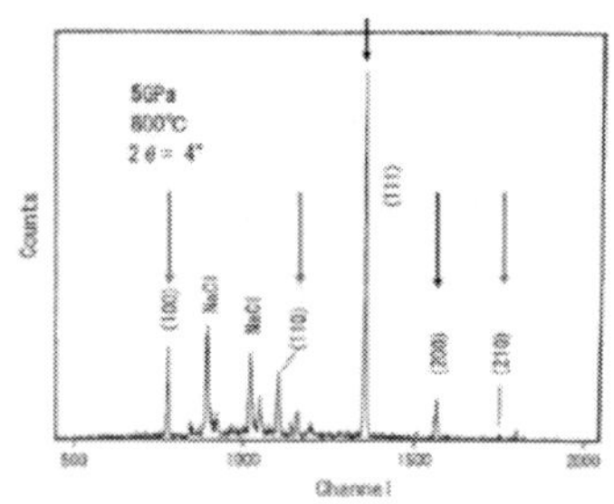

Figure 39. XRD pattern of counts (*intensity*) vs. channel (θ) of Ni_3VacH_4 at 800°C and H_2 pressure of 5 Gpa. Superlattice reflections show the crystal lattice is SC. Sample prepared via high force anvil with NaCl encapsulation to retain hydrogen. Red arrows are reflections from mixed indices (superlattice) while black arrows are from fundamental reflections, adapted after [94].

indicates zero degrees of freedom) called an invariant point. The phase sequence rule follows from these definitions and the phase rule: a single phase region must be followed by two-phase region and be subsequently followed by another single phase region, etc. when traversing the diagram along an isotherm [37].

The phase diagram in Fig. 8 can thus be constructed once the composition of a true equilibrium phase is known: that phase is placed on the map consistent with the temperature and its unique composition. There is allowance for a range of compositions over which the phase is identified as the same phase with homogeneous structure. In this case a homogeneous structure results from having the same crystal Bravais lattice. Bravais lattice is most uniquely identified from XRD patterns (satisfying Bragg's law) of the unit cell that is repeated through a region of space and therefore occupying a volume percent of the whole bulk. Phase diagrams thus also indicate the relative amount of a phase present in the bulk by the lever rule for fields of two phases. Phases occupy a volume percent of the bulk. The admission of the three new phases on the diagram is complete once their nominal (mid-range) composition has been determined from X-ray data. However, it only remains to complete an estimation of the deviation from nominal (stoichiometric) composition while still being regarded as the same phase. Future research will certainly improve on the estimates reported here as more experimental data is accrued regarding the exact positions of the lines on the phase diagram, nonetheless the qualitative layout here is firmly based in thermodynamics, phase diagram construction principles, DFT computations for free energy, and XRD for compositions, and resistivity data for crossing a phase boundary.

Starting with the previously published version of the Pd–D diagram and the newly identified phases from XRD, one proceeds as follows. These phases from the literature (the main section of the paper) γ, δ and δ' must be added to the equilibrium phase diagram according to their compositions if they are true equilibrium phases. Traversing from low to high D/Pd ratio (composition) along an isotherm, there must be a two phase equilibrium field after the β-phase (it must be $[\beta + \gamma]$), followed by a single phase field γ, followed by another two phase equilibrium field (it must be $[\gamma + \delta]$), followed by another single phase field δ, followed by another two phase equilibrium field (it must be $[\delta + \varepsilon]$). The naming of the phases follows phase diagram tradition of the order of Greek alphabet, left to right across the diagram. It was shown from the literature using DFT analysis that these phases are true equilibrium phases so they must be placed on the temperature–composition diagram according to the compositions from XRD data. These data show γ to have compositions of D/Pd ratio 0.85–1.14, based on stoichiometry of $Pd_7Vac_1D_{6-8}$, with nominal stoichiometry D/Pd $= 1.0$. The range of $D = 6$ to 8 depends on whether the central octahedral site is unfilled in both parts of the dual 4-atom unit cell (consisting of eight Pd atoms) giving a value of D_6, filled in one of the two parts D_7, or filled in both halves of the 8-atom unit cell D_8. The two phase sequence rule is derived from the phase rule; and it is used to verify the width of the $(\beta + \gamma)$-phase field, since β is in equilibrium with γ along the curves, one at 0.76 and the other at 0.85, the beginning of pure γ. The width is simultaneously determined from the data of Araki et al. [42] (where a two-phase field must exist since temperature is constant) and from the phase rule, as shown in Fig. 6. Here composition is the free variable in the phase rule: this results in the same range 0.76–0.85.

Next there must be a single-phase field γ with composition range discussed above (0.85–1.14) followed by the two-phase field $(\gamma + \delta)$. The left-hand side of this field $(\gamma + \delta)$ shares the previous border of D/Pd $= 1.14$ and extends to D/Pd ratio $= x$, where x is unsupported by experimental data at this time, but is *assumed* to be 1.24 based on the assumption that both two-phase fields on either sides of γ are symmetrical. This is quantitatively an assumption, but is not an assumption, qualitatively: the exact position of the boundary between $(\gamma + \delta)$ and δ is estimated, but its existence is recognized. The mid-range (exact stoichiometric) value of D/Pd of δ is 1.33 (from Pd_3VacD_4), but this structure exists with some deviation from perfection as to the number of D's filling the octahedral sites. It is not expected to be as wide as γ since γ is an 8-atom unit cell and δ is a 4-atom unit cell. The upper bound of δ shares the same deviation from pure stoichiometry as the lower side. The difference between δ and δ' is that δ has D in octahedral sites and δ' has D in tetrahedral sites. There are eight tetrahedral sites and only four octahedral sites per unit cell. The composition of δ' thus extends to D/Pd $= 2.66$. The boundary between δ' and δ is determined from the temperature measured during excess heat events: the resistivity of δ dropped. Since D occupies the tetrahedral sites in δ', the edges of the units cells are completely open for electron transport (low or zero resistivity). From the two-phase sequence rule, there is a two-phase region separating δ and δ' on the left along horizontal temperature traverses (isotherms), similar to the phase diagram of Au–Cu [92,95,96]: the exact width of this two-phase field is not determined here. The ε-phase is anticipated by research data [43,97–99].

Pressure isotherms confirm phase diagram boundaries in three dimensions with the variable pressure along the third dimension. The present work does not explore the variable pressure and its effects on the diagram. Research, such as that of Hagelstein [100–102], forms a more complete picture, with O-site and T-site partition functions as harmonic oscillator partition functions.

References

[1] Ian M. Robertson, P. Sofronis, A. Nagao, M.L. Martin, S. Wang, D.W. Gross and K.E. Nygren, Hydrogen embrittlement understood–2014 Edward DeMille Campbell Memorial Lecture, *ASM International, Metallurgical and Materials Transactions* B, 28 March 2015, DOI: 10.1007/s11663-015-0325-y.

[2] A.K. Eriksson, A. Liebig, Ś. Olafsson and B. Hjorvarsson, Resistivity changes in Cr/V(0 0 1) superlattices during hydrogen absorption, *J. Alloys Compd.* **446–447** (2007) 526–529.

[3] M. Khalid and P. Esquinazi, Hydrogen-induced ferromagnetism in ZnO single crystals investigated by magnetotransport, *Phys. Rev.* B **85** (2012) 134424.

[4] D. E. Azofeifa, N. Clark, W. E. Vargas, H. Solís, G. K. Pálsson and B. Hjörvarsson, Temperature- and hydrogen-induced changes in the optical properties of Pd capped V thin films, *Physica Scripta* **86** (6) (2012) 065702.

[5] S. Kala and B. R. Mehta, Hydrogen-induced electrical and optical switching in Pd capped Pr nanoparticle layers, *Bull. Mater. Sci.* (Indian Academy of Sciences) **31** (3) (2008) 225–231.

[6] H. Noh, Ted B. Flanagan, B. Cerundolo and A. Craft, H-Induced atom mobility in palladium–rhodium alloys, *Scripta Met. et Mat.* **25** (1991) 225–230.

[7] H. Noh, Ted B. Flanagan and M.H. Ransick, An Illustration of phase diagram determination using H-induced lattice mobility, *Scripta Met. et Mat.* **26** (1992) 353–358.

[8] K. Baba, Y. Niki, Y. Sakamoto, A. P. Craft and Ted B. Flanagan, The transition of the hydrogen-induced LI_2 ordered structure of Pd_3Mn to the Ag_3Mg structure, *J. Mats. Sci. Letters* **7** (11) (1988) 1160–1162.

[9] R. Balasubramaniam, Mechanism of hydrogen induced ordering in Pd_3Mn, *Scripta Met. et Mat.* **30** (7) (1994) 875–880.

[10] Scott Richmond, Joseph Anderson and Jeff Abes, Evidence for hydrogen induced vacancies in Plutonium metal, Plutonium Futures–The Science Keystone, CO, September 19–23, 2010, p. 206.

[11] M. Wen, L. Zhang, B. An, S. Fukuyama and K. Yokogawa, Hydrogen-enhanced dislocation activity and vacancy formation during nanoindentation of nickel, *Phys. Rev.* B **80** (2009) 094113.

[12] Y. Fukai and N. Okuma, Evidence of copious vacancy formation in Ni and Pd under a high hydrogen pressure, *Jpn. J. Appl. Phys.* **32** (1993) 1256.

[13] W. A. Oates and H. Wenzl, On the copious formation of vacancies in metals, *Scripta Met. et Mat.* **30** (7) (1994) 851–854.

[14] W. A. Oates and H. Wenzl, On the formation and ordering of superabundant vacancies in palladium due to hydrogen absorption, *Scripta Met. et Mat.* **33** (2) (1995) 185–193.

[15] Y. Fukai, *The Metal–Hydrogen System: Basic Bulk Properties*, 2nd Edn., Springer, Berlin, 2005, p. 216.

[16] Y. Fukai, Superabundant vacancies formed in metal–hydrogen alloys, *Physica Scripta* 2003 No. T103 (2002) 11.

[17] V.F. Degtyareva, Electronic origin of superabundant vacancies in Pd hydride under high hydrogen pressures, Presented on the Conference on Hydrogen Materials Science (ICHMS), Yalta, Ukraine, 25–31 August (2009): http://arxiv.org/pdf/1001.1525.pdf , accessed 25 November 2018.

[18] D. Tanguy and M. Mareschal, Superabundant vacancies in a metal–hydrogen system: Monte Carlo simulations, *Phys. Rev.* B **72** (17) (2005) 174116.

[19] Y. Fukai, Hydrogen-induced superabundant vacancies in metals: implication for electrodeposition, A. Ochsner, G. E. Murch and J. M. O\Q. Delgado (Eds.), *Defect and Diffusion Form* **312–315** (2011) 1106–1115.

[20] D. S. dos Santos, S. Miraglia and D. Fruchart, A high pressure investigation of Pd and the Pd–H system, *J. Alloys Compd.*

291 (1999) L1– L5.

[21] Y. Fukai and N. Okuma, Formation of superabundant vacancies in Pd hydride under high hydrogen pressures, *Phys. Rev. Lett.* **73** (12) (1994) 1640–1643.

[22] Y. Fukada, T. Hioki and T. Motohiro, Multiple phase separation of super-abundant-vacancies in Pd hydrides by all solid-state electrolysis in moderate temperatures around 300 C, *J. Alloys Compd.* **688** (2016) 404–412.

[23] M.P. Pitt and E. MacA. Gray, tetrahedral occupancy in the Pd–D system observed by *in situ* neutron powder diffraction, *Europhys. Lett.* **64** (3) (2003) 344–350.

[24] G.A. Ferguson Jr., A.I. Schindler, T. Tanaka and T. Morita, Neutron diffraction study of temperature-dependent properties of palladium containing absorbed hydrogen, *Phys. Rev.* **137** (2A) (1965) 483.

[25] N. Eliaz, D. Eliezer, D.L. Olson, Hydrogen-assisted processing of materials, *Mat. Sci. Engr.* A **289** (2000) 41.

[26] C. Zhang and Ali Alavi, First-principles study of superabundant vacancy formation in metal hydrides, *J. Am. Chem. Soc.* **127** (2005) 9808–9817.

[27] Y. Fukai, M. Mizutani, S. Yokota, M. Kanazawa, Y. Miura and T. Watanabe, Superabundant vacancy–hydrogen clusters in electrodeposited Ni and Cu, *J. Alloys Compd.* **356–357** (2003) 270.

[28] Y. Fukai, Formation of superabundant vacancies in M–H alloys and some of its consequences: a review, *J. Alloys Compd.* **356–357** (2003) 263–269.

[29] L.E. Isaeva, D.I. Bazhanov, Eyvas Isaev, S.V. Eremeev, S.E. Kulkova and Igor Abrikosov, Dynamic stability of palladium hydride: an ab initio study, *Int. J. Hydrogen Energy* **36** (1) (2011) 1254–1258.

[30] Y. Fukai and H. Sugimoto, Formation mechanism of defect metal hydrides containing superabundant vacancies, *J. Phys. Condensed Matter* **19** (2007) 436201.

[31] H. Sugimoto and Y. Fukai, Migration mechanism in defect metal hydrides containing superabundant vacancies, *Diffusion-fundamentals.org* **11** (2009) 102, pp. 1–2.

[32] L. Bukonte, T. Ahlgren and K. Heinola, Thermodynamics of impurity-enhanced vacancy formation in metals, *J. Appl. Phys.* **121** (2017) 045102-1 to -11. https://doi.org/10.1063/1.4974530.

[33] Y. Fukai and H. Sugimoto, The defect structure with superabundant vacancies to be formed from FCC binary metal hydrides: Experiments and simulations, *J. Alloys Compd.* **446,447** (2007) 474–478.

[34] R. Nazarov, T. Hickel and J. Neugebauer, Ab Initio study of H-vacancy interactions in FCC metals: implications for the formation of superabundant vacancies, *Phys. Rev.* B **89** (2014) 144108.

[35] Y. Fukai, Y. Kurokawa and H. Hiraoka, Superabundant vacancy formation and itsconsequences in metal hydrogen alloys, *J. Jpn. Inst. Met.* **61** (1997) 663–670 (in Japanese).

[36] Y. Fukai, *The Metal–Hydrogen System: Basic Bulk Properties*, 2nd Edn., Springer, Berlin, 2005, p. 10.

[37] H. Okamoto and T. Massalski, Improbable phase diagrams, *J. Phase Equilibria* **12** (2) (1991) 148–168.

[38] J.F. Shackelford, *Introduction to Material Science for Engineers*, 7th Edn., Pearson, Upper Saddle River, NJ, 2009, pp. 272,273.

[39] Tom Callahan and Fiona O'Connell, Studies in EG459**:** special topics in materials engineering–steel metallurgy, Loyola University, Maryland, Fall, 2011.

[40] *Metals Handbook*, Vol. 9, *Metallography and Microstructures*, 9th Edn., American Society for Metals, Metals Parks, OH, 1985, p. 245.

[41] M.R. Staker, The uranium–vanadium equilibrium phase diagram, *J. Alloys Compd.* **266** (1998) 167–179.

[42] H. Araki, M. Nakamura, S. Harada, T. Obata, N. Mikhin, V. Syvokon and M. Kubota, Phase diagram of hydrogen palladium, *J. Low Temperature Phys.* 134 (5/6) (2004) 1145–1151.

[43] Paolo Tripodi, Daniele Di Gioacchino and Jenny Darja Vinko, Magnetic and transport properties of PdH: intriguing superconductive observations, *Brazilian J. Phys.* **34** (3B) (2004) 1177.

[44] M. Fleischmann and M. Miles. The instrument function of isoperibolic calorimeters: excess enthalpy generation due to the parasitic reduction of oxygen, *10th Int. Conf. on Cold Fusion*, Cambridge, MA, 2003.

[45] T. Sugeno and M. Kowaka, *Mem. Inst. Scientia. Ind. Rese.*, Osaka University **11** (1954) 119.

[46] F.A. Lewis, *The Palladium Hydrogen System*, Academic Press, London, 1967, p. 45.

[47] D.L. Holt, Dislocation cell formation in metals, *J. Appl. Phys.* **4l** (1970) 3197.

[48] M.R. Staker and D L. Holt, The dislocation cell size and dislocation density in copper deformed at temperatures between 25

and 700°C, *Acta Met.* **20** (1972) 569.
[49] D. Kuhlmann-Wilsdorf, Theory of plastic deformation: properties of low energy dislocation structures, *Mat. Sci. Engr.* A **113** (1989) 1–41.
[50] Martin Deutges, Hans Peter Barth, Yuzeng Chen, Christine Borchers and Reiner Kirchheim, Hydrogen diffusivities as a measure of relative dislocation densities in palladium and increase of the density by plastic deformation in the presence of dissolved hydrogen, *Acta Mat.* **82** (2015) 266–274.
[51] J.P. Burger, D.S. MacLachlan, R. Mailfert and B. Souffaché, Electrical resistivity of PdH_x: 1 - residual resistivity, *Solid State Commun.* **17** (3) (1975) 277–280.
[52] Michael C.H. McKubre, Romeu C. Rocha-Filho, Stuart Smedley, Francis Tanzella, Jason Chao, Bindi Chexal, Tom Passell and Joseph Santucci, Calorimetry and electrochemistry in the D/Pd system, The first annual conf. on cold fusion, Univ. of Utah Research Park, Salt Lake City, Utah: National Cold Fusion Institute (1990), pp. 20–31.
[53] D. Cravens, Factors affecting the success rate of heat generation in cold fusion cells, *Proc. Fourth Int. Conf. on Cold Fusion*, Vol. 2, Calorimetry and Materials Papers, Held Lahaina, Maui, HI, 6–9 December 1993, prepared by Electric Power Research Institute, Pala Alto, CA, TR-104188-V2 (1994), p. 18-1.
[54] A.I. Schindler, R.J. Smith and E.W. Kammer, Low-temperature dependence of the electrical resistivity and thermoelectric power of palladium and palladium–nickel alloys containing absorbed hydrogen, *Proc. Int. Congress of Refrigeration*, Copenhagen, August 19–26, 1959, 10th Congress, Vol. 1, p. 74, Pergamon Press, New York, 1960.
[55] P. Tripodi, M.C.H. McKubre, F.L. Tanzella, P.A. Honnor, D. Di Giacchino, F. Celani and V. Violante, Temperature coefficient of resistivity at compositions approaching PdH, *Phys. Lett.* A **276** (2000) 122–126.
[56] S.L. Ames and A.D. McQuillan, The resistivity–temperature–concentration relationship in β-phase titanium–hydrogen alloys, *Acta Met.* **4** (1956) 609.
[57] W. Mueller, J. Blackledge and G. Libowitz (Eds.), *Metal Hydrides*, Academic Press, NY, 1968, pp. 69 and 82.
[58] F.A Lewis, *The Palladium Hydrogen System*, Academic Press, London, 1967, pp. 7, 9, 22 and 119.
[59] R.A. Oriani, The physical and metallurgical aspects of hydrogen in metals, *4th Int. Conf. on Cold Fusion (ICCF-4),* Lahaina, Maui, HI: Electric Power Research Institute, Palo Alto, CA , 1993.
[60] A. Paolone, S. Tosti, A. Santucci, O. Palumboand and F. Trequattrini, Hydrogen and deuterium solubility in commercial Pd–Ag alloys for hydrogen purification, *Chem. Engr.* **1** (2017) 14; pp.1-9 doi: 10.3390/chemengineering1020014 MDPI, Basel, Switzerland.
[61] Y. Fukai, The Metal Hydrogen System–Basic Bulk Properties, 2nd Edn., Springer, Berlin, Heidelberg, New York, 2005, p. 43.
[62] F.G. Will, K. Cedzynska, M.C. Yang, J.R. Peterson, H.E. Bergeson, S.C. Barrowes, W.J. West and D.C. Linton, Studies of electrolytic and gas phase loading of Pd with deuterium, in *Conf. Proc. of Italian Physical Society*, Vol. 33 for 'The Science of cold fusion – Proc. of Second Annual Conf. on cold fusion', T. Bressani, E. Del Giudice and G. Preparata (Eds.), Como, Italy, 29 June –4 July 1991, held at A. Volta Center for Sci. Culture, Villa Olmo, 1991, pp. 373–383.
[63] M. Srinivasan, A. Shyam, T.C. Kaushik, R.K. Rout, L.V. Kulkarni, M.S. Krishnan, S.K. Malhotra, V.G. Nagvenkar and P.K. Iyengar, Observation of tritium in gas/plasma loaded titanium samples, *AIP Conf. Proc. 228–Anomalous Nuclear Effects in Deuterium/Solid System*, 1990. Brigham Young Univ., Provo, UT: American Institute of Physics, New York, p. 514–534.
[64] David J. Nagel, Characteristics and energetics of craters in LENR experimental materials, *J. Condensed Matter Nucl. Sci.* **10** (2013) 1–1.
[65] Edmund Storms, An Explanation of low energy nuclear reactions (cold fusion), https://www.youtube.com/watch?v=SNodilc6su0, accessed May 15, 2018.
[66] L. Liu, J. Wang, S.K. Gong and S.X. Mao, Atomistic observation of a crack tip approaching coherent twin boundaries, *Scientific Reports,* Vol. 4, Article number: 4397 (2014) doi:10.1038/srep04397.
[67] P. Nordlander, J.K. Norskov, F Besenbacher and S.M. Myers, Multiple deuterium occupancy of vacancies in Pd and related metals, *Phys. Rev.* B **40** (1989) 1990–1992.
[68] R. B. McLellan and D. Zang, Kinetics of vacancy migration in hydrogenated palladium, *Scripta Mat.* **36** (1997) 1207–1210.
[69] S. Szpak, P.A. Mosier-Boss and J.J. Smith, On the behavior of Pd deposited in the presence of evolving deuterium, *J. Electronalytical Chem.* **302** (1991) 255–260.
[70] D. Letts and P.L. Hagelstein, Modified Szpak protocol for excess heat, *J. Condensed Matter Nucl. Sci.* **6** (2012) 44–54.

[71] A. Coehn, Detection (proof of the existence) of protons in metals, *J. Electrochem. (Z. Electrochem).* **35** (9) (1929) 676–680.
[72] A. Coehn and W. Specht, On the participation of protons in the electrical conduction in metals: I. Proof by potential measurements, *Zeitschrift fur Physik (J. Phys.)* **62** (1,2) (1930) 1–31.
[73] A. Coehn and H. Jurgens, On the participation of protons in the electrical conduction in metals: II. Resistance measurements, *Z. Phys.* **71** (3,4) (1931) 179–204.
[74] A. Coehn and K. Sperling, On the participation of protons in the electrical conduction in metals: III. The photographic plate as an indicator, *J. Phys. (Zeitschrift für Physik)* **83** (5,6) (1933) 291–312.
[75] B. Duhm, The diffusion of hydrogen in palladium, *J. Phys. (Zeitschrift für Physik)* **94** (1935) 434 and **95** (1935) 801.
[76] C. Wagner and G. Heller, The rate of migration of protons in the electric field in the α-phase of the palladium–hydrogen system, *J. Phys. Chem. (Z. Physik. Chem.)* **46B** (1) (1940) 242–250.
[77] M.R. Staker, private communication to Martin Fleischman, Stanley Pons, Mel Miles and Michael C.H. McKubre, at A. Volta Center for Sci. Culture, Villa Olmo, Como, Italy (29 June–4 July 1991).
[78] M.R. Staker and J.J. Antal, The use of electrostatic field to control the stoichiometry of palladium deuteride in the cold fusion process, a proposal prepared for DOE (April 1992); and a second proposal, Research on control of the cold fusion process and exploration of its nuclear origins, prepared for Dr. Thomas Passell of Materials and Chemistry Division of Electric Power Research Institute, Palo Alto, CA 94303 (18 September 1992).
[79] C. Bartolomeo, M. Fleischmann, G. Larramona, S. Pons, J. Roulette, H. Sugiura, and G. Preparata, Alfred Coehn and after: the alpha, beta, gamma of the palladium–ydrogen system, *Fourth Int. Conf. on Cold Fusion (ICCF-4)*, Lahaina, Maui, HI, USA, 1993, pp.19-1 to 19-47.
[80] F. Celani, A. Spallone, P. Tripodi, A. Petrocchi, D. di Gioacchino, P. Marini, V. di Stefano, S. Pace and A. Mancini, High power μs-pulsed electrolysis using palladium wires: evidence for a possible "phase" transition under deuterium overloaded conditions and related excess heat, *Fifth Int. Conf. on Cold Fusion (ICCF-5)*, Monte Carlo, Monaco, 9–13 April 1995, pp. 57–68.
[81] F. Celani, A. Spallone, P. Tripodi, A. Petrocchi, D. di Gioacchino, P. Marini, V. di Stefano, S. Pace and A. Mancini, Deuterium overloading of palladium wires by means of high power μs pulsed electrolysis and electromigration: suggestions of a phase transition and related excess heat, *Phys. Lett.* A **214** (1,2) (1996) 1–13.
[82] F. Celani, A. Spallone, P. Marini, V. di Stefano, M. Nakamura, A. Mancini, S. Pace, P. Tripodi, D. Di Gioacchino, C. Catena, G. D′Agostaro, R. Petraroli, P. Quercia, E. Righi and G. Trenta, High hydrogen loading into thin palladium wires through precipitate of alkaline–earth carbonate on the surface of cathode: evidence of new phases in the Pd-–H system and unexpected problems due to bacteria contamination in the heavy water, *Eighth Int. Conf. on Cold Fusion (ICCF-8)*, Lerici (La Spezia), Italy, May 21-–26, 2000, Conf. Proc., F. Scaramuzzi (Ed.), Vol. 70, pp. 181–190, 2001, Conference Proceedings of Compositori -– Società Italiana di Fisica, Bologna, Italy. ISBN 88-7794-256-8.
[83] F. Celani, O.M. Calamai, A. Spallone, A. Nuvoli, V. Andreassi, B. Ortenzi, F. Piastra, E. Righi, G. Trenta, E. Marano, P. Marini, V. di Stefano, M. Nakamura and E. Purchi, Development of a high temperature hybrid CMNS Reactor, *J. Condensed Matter Nucl. Sci.* **6** (2012) 24–33.
[84] E. Del Giudice, A. De Ninno, A. Frattolillo, G. Preparata, F. ScaramuzziA. Bulfone, M. Cola, C. Giannetti, The Fleischmann–Pons effect in a novel electrolytic configuration, *Proc. 8th Int. Conf. on Cold Fusion, ICCF-8*, Lerici, La Spezia, 21–26 May 2000, F. Scaramuzzi (Ed.), Conference Proceedings of Compositori -– Società Italiana di Fisica, Bologna, Italy. ISBN 88-7794-256-8, **70** (2001) 47–54.
[85] H. Wipf, *Electro- and Thermo-transport of Hydrogen in Metals*, G. Alefeld and J. Volkl (Eds.), Hydrogen in Metals II: Topics in Applied Physics, Vol. 29, Springer, Berlin, Heidelberg, New York, 1978, pp. 273–304.
[86] W.T. Read Jr., *Dislocations in Crystals*, McGraw-Hill New York, 1953
[87] W. Hayden, W.G. Moffatt and J. Wulff, *Structure and Properties of Materials,* Vol. III, *Mechanical Behavior,* Wiley, New York, 1965, pp. 63–64.
[88] R.E. Reed-Hill, *Physical Metallurgy Principles*, D. Van Nostrand, Princeton, NJ, 1964, pp. 139–143.
[89] M. Wen, L. Zhang, B. An, S. Fukuyama and K. Yokogawa, Hydrogen-enhanced dislocation activity and vacancy formation during nanoindentation of nickel, *Phys. Rev.* B **80** (2009) 094113.
[90] Note: The volume of the dislocation cell could also be considered as a cube of 1 μm on edge rather than a sphere, to be space filling, making its volume 1×10^{-18} m^3. Then the ratio of the volumes of 1 cm^3 to that of the dislocation cell is an even 1

$\times 10^{12}$(rather than 1.9×10^{12}) for estimating purposes.

[91] B. D. Cullity, *Elements of X-ray Diffraction*, Addison-Wesley, Reading, MA, 1956 p. 117.

[92] C. Sykes and H. Evans, The Transformations in the copper–gold alloy Cu_3Au, *J. Institute Metals* **58** (1936) 255.

[93] Y. Fukai and N. Okuma, Formation of superabundant vacancies in Pd hydride under high hydrogen pressures, *Physl Rev Lett* **73** (12 (1994) 1640–1643.

[94] Y. Fukai, Y. Shizuku and Y. Kurokawa, Superabundant vacancy formation in Ni–H alloys, *J. Alloys Compd.* **329** (2001) 195–201.

[95] P.P. Fedorov and S. N. Volkov, Au–Cu phase diagram, *Russian J. Inorganic Chem.* **61**(6) (2016) 772–775. https://doi.org/10.1134/S0036023616060061.

[96] H. Okamoto, D.J. Chakrabarti, D.E. Laughlin and T. Massalski, The Au−Cu (Gold–Copper) system, *J. Phase Equilibria (Bulletin of Alloy Phase Diagrams)* **8** (5) (1987) 454–474. DOI: 10.1007/BF02893155.

[97] H. M. Syed, T.J. Gould, C.J. Webb and E. MacA. Gray, Superconductivityin palladium hydride and deuteride at 52–61 K, Report for Queensland Micro- and Nanotechnology Centre, Griffith University, Nathan 4111, Brisbane, Australia. https://arxiv.org/ftp/arxiv/papers/1608/1608.01774.pdf , accessed 24 November 2018.

[98] I. Anderson, The Dynamics of hydrogen in metals studied by inelastic neutron scattering, in Neutron *Scattering from Hydrogen in Metals*, A. Furrer (Ed.), World Scientific, Singapore, 1994, pp. 142–167.

[99] K.P. Sinha, High temperature superconductivity in Pd–$[H (D)]_X$ system, *Natl. Academy Sci. Lett.* (India) **29** (3,4) (2006) 125–129: arXiv:cond-mat/0509596 [cond-mat.supr-con], https://arxiv.org/abs/cond-mat/0509596, accessed 25 November 2018.

[100] P.L. Hagelstein, Statistical mechanics model for the PdH and PdD phase diagram with both O-site and T-site occupation, *21st Int. Conf. Condensed Matter Nucl. Sci.* (ICCF-21), CO State University, Ft. Collins, CO, 3– 8 June 2018; *J. Condensed Matter Nucl. Sci.* (2019), in press.

[101] P.L. Hagelstein, Models for the phase diagram of palladium hydride including O-site and T-site occupation, *J. Condensed Matter Nucl. Sci.* **20** (2016) 54–80.

[102] P.L. Hagelstein, O-site and T-site occupation of α-phase $PdH_x$and PdD_x, *J. Condensed Matter Nucl. Sci.* **17** (2015) 67–90.

J. Condensed Matter Nucl. Sci. 29 (2019) 169–176

Research Article

Excess Heat is Linked to Deuterium Loss in an Aqueous Nickel LANR System

Mitchell R. Swartz* and Brian Ahern
JET Energy Inc., Wellesley Hills, MA 02481, USA

Charles Haldemann and Alan Weinberg†
Lincoln Laboratory, Massachusetts Institute of Technology, Lexington, MA, 02421, USA

Abstract

These results indicate a loss of deuterium (more precisely, "deuterons") from ordinary water when excess heat is observed in an aqueous Ni CF/LANR system (Ni/ordinaryH_2O/Pt) using a very large cathodic area. The exit gas, from the enclosed electrolytic cell, had an HD/H_2 isotopic ratio (3/2 ratio) which was significantly less than originally. When using *V*I* as electrical input power, the maximum incremental power gain was ~4 times electrical input; this occurred at lower electrical input power. The maximum excess power was circa 5 W.

Keywords: Deuterium, Deuterium loss, LANR, Palladium

1. Introduction

Many types of LANR systems exist, but few have extremely large cathodic surface areas or inline detection systems able to discern heavy hydrogen (D) from ordinary hydrogen (H). The MIT Lincoln Laboratory Electrolytic Cell system had a 3 l capacity. Electrical loading and driving were used because hard drawn nickel wire can be loaded with hydrogen to about 1×10^{-3} atomic fraction at room temperature (~20 times higher than the equilibrium level for gas phase loading). These results indicate a loss of deuterium (more precisely, "deuterons") from ordinary water when excess heat is observed in an aqueous Ni CF/LANR system (Ni/ordinaryH_2O/Pt) using a very large cathodic area. There are at least four important points. First, the loss of deuterons is linked to the excess heat (excess energy) which was repeatedly seen. Second, during these studies, the maximum excess power gains were 5–14 times electrical input when the thermoneutral correction was used. When using *V*I* as electrical input power, the maximum incremental power gain was ~4 times electrical input; this occurred at lower electrical input power. The maximum excess power

*Corresponding author. Dr. Mitchell R. Swartz ScD, MD, EE, E-mail: moac@nanortech.com.
†Presently at JET Energy Inc., Wellesley Hills, MA 02481, USA.

© 2019 ISCMNS. All rights reserved. ISSN 2227-3123

was ca. 5 W during MIT's Lincoln Lab. efforts from 1993 to 1996 [1]. The incremental excess power gains have since been increased using information and technology from JET Energy Inc. Third, the exit gas, from the enclosed electrolytic cell, had an HD/H_2 isotopic ratio (3/2 ratio) which was significantly less than originally. In addition, these results herald deuterons as the fuel in nickel-ordinary water CF/LANR systems, consistent with Swartz et al. [2,3] and the reports of helium production in Pd systems [4].

2. Experimental

2.1. Materials

The LANR electrolytic cell used, unfilled, is shown in Fig. 1. When filled, its 3 l capacity and large cathode earn its nickname: MOAC ("Mother of all Cathodes"). It has a 3 l capacity when filled. The electrolytic Ni/ordinary water/Pt system was designed to have a large electrode area. The cathode weighs 2.14 kg, and was made from No. 46 hard drawn smooth nickel wire (0.041 mm diameter, with an area of ca. 240,000 cm^2). The anode was five-folded platinized sheets of titanium with an area of $\sim$3200 cm^2, for a surface area ratio of cathode to anode of 75:1.

The electrolyte was a carbonate solution (0.6 M K_2CO_3) using laboratory distilled deionized water with no additional heavy water was added. Two internal ohmic controls were used. Figure 1 shows the LANR cell used, unfilled and not connected to the power supply or the gas venting, after its removal from Lincoln Lab. The MOAC used for this report has in the last few years evolved through a series of changes, of both design and operation, which led to consecutive increases in incremental power gain. The cell underwent incremental engineering substitutions and changes (Cells 1, 1A, 2, 2A, 3, 4) for increased incremental power gain. Some of the changes and results are listed in Table 1.

2.2. Methods

For the calorimetry, calibration was determined by ohmic controls. The putative excess heat output (if any) was determined by the delta-T at equilibrium. In these measurements, the instantaneous power gain (power amplification factor (non-dimensional)) is defined as $P_{\text{output}}/P_{\text{input}}$. Electrical input power ran to $\sim$25 W at MIT; and to >125 W at JET Energy. In addition to energy measurements, gas measurements and isotopic measurements were obtained. Gas measurement and recombiner analyses were made three ways, as shown schematically in Fig. 2.

Figure 2 shows the arrangement of the sealed MOAC cell, the collection flask(s), the extra bubbler, the mercury catch basin and mercury filled collection vessels, and the water bath with the bath thermocouple.

The three types of gas collection systems which were used. For some experiments, gas collection was done with either a combustion tube or a bubbler flame suppressor used prior to either the mass spectrometer or a platinum glow wire.

Chemical absorption was measured by a Burrell Wet Analyzer. The absorbents in the tube analyzer were KOH (sol.), $CrCl_2$ (sol.) and CuO heated to 300°C for CO_2, O_2, and H_2, respectively. Isotopic analysis was made by mass spectrometer (INFICON Quadrapole-102 V Energy) [1].

Gas flow absolute measurement was made by direct water displacement in 2000 cm^3 volumetric flasks ($\pm$0.5 cm^3), with the temperature controlled to $\pm$0.1°C, using time measurements ($\pm$0.02 s). Barometric pressure was obtained from the National Weather Service barometer and that was corrected for temperature and latitude ($\pm$0.1 mm).

The measured volume was then corrected for ambient temperature, pressure and water vapor content. Cryo-condensation was also used. The exit gas from the enclosed electrolytic cell was $\sim$100% collected by liquid nitrogen condensation. It was then sent to a recombiner.

Figure 1. The LANR electrolytic cell.

3. Results

3.1. Excess heat produced

There was excess heat (excess energy) seen, repeatedly. All calibrations had delta-T less than 0.2°C. On the other hand, for the active MOAC cell, equilibrium delta-Ts were 0.5–2.5°C above those expected from ohmic control calibration. These excess thermal outputs are shown in Fig. 3 for both DC runs and the pulsed runs for the electrolytic cell, and for the ohmic controls. These outputs are shown up to an electrical power input of about 25 W, characteristic of the (early) MIT experiments.

The maximum power gains were 5–14 times electrical input when the thermoneutral correction was used. When using $V*I$ as electrical input power, the maximum incremental power gain was ~4 times electrical input; this occurred at lower electrical input power. The maximum incremental excess power was ca. 5 W.

Improved output of active cells increased with developmental changes. In Table 1 is shown the performance of the "summary of cells assembled". It can be seen that Cells 1, 1A, 2, 2A, 3, 4 showed monotonically increasing incremental power gain. From no excess power gain to 50% increase to 14,000% (using the thermoneutral voltage correction) or 400% excess power gain (using $V*I$).

Table 1 shows information on the six versions of the MOAC were developed at MIT, and another two more at JET Energy, each with the differences in cathode, anode, and cathode to anode (C/A) ratio in columns 2,3, and 4, and other geometric arrangements. Column 5 lists the observed incremental energy gains.

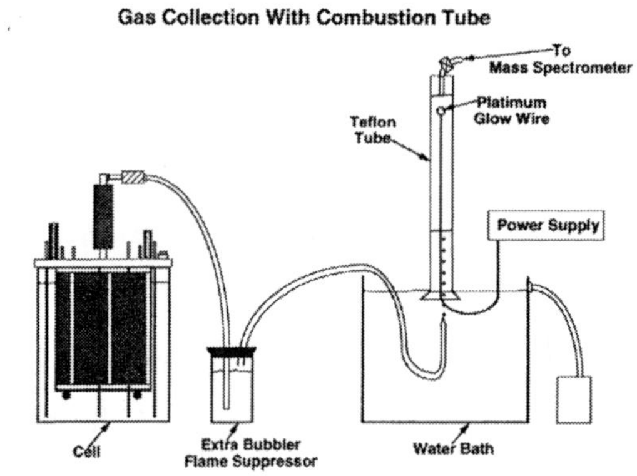

(a)

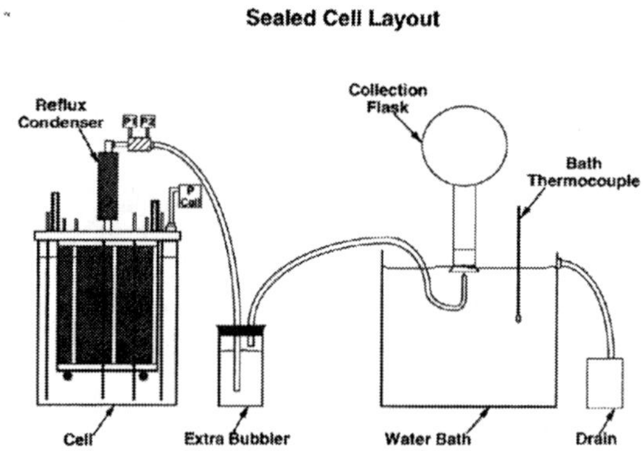

(b)

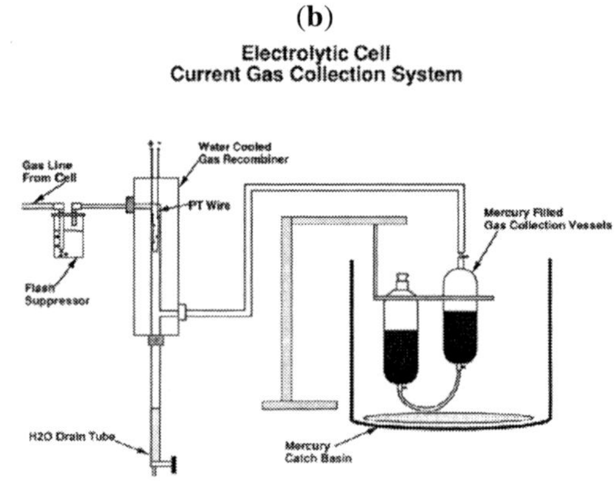

(c)

Figure 2. Several gas collection systems used (a)–(c).

Table 1. Summary of results of the LANR MOAC cell.

Cell No.	Cathode	Anode	Cathode/anode ratio	Results
1	Annealed No. 41 nickel 0.818 kg, 52,000 cm^2	Platinized titanium 100 cm^2	520:1	No excess energy
1A	Same wire Heat treated in H_2 770°C	Soft nickel sheet 3000 cm^2 Platinized titanium 100 cm^2	17:1	No excess energy
2	Hard drawn 0.5 mm nickel 16,000 cm^2	Same	5:1	5–10% excess energy
2A.	New hard drawn 0.5 cm^2 nickel 15,000 cm^2	Platinized titanium sheet 3100 cm^2	5:1	5–30% excess energy
3	Hard drawn scratched No. 44 nickel 190,000 cm^2 0.05 cm diameter	Same	61:1	20–50% excess energy
4	No. 46 hard drawn smooth nickel wire 240,000 cm^2	Same	75:1	20–1400% excess energy 4 * (V*I) input
4A,B	Same but configuration approaching Phusor®-type componen; augmented diagnostic; higher impedance solutio	Same	75:1	70 W peak excess power Demonstration of two driven electrical modes

3.2. MOAC cell gas production analysis

The exit gas from the enclosed electrolytic cell generated from 2 to 100 cm^3/min (2.8–144 l/day). The rate of exiting gas flow is shown in Fig. 4 as function of input electrical current in amperes. The 100% efficient (theoretical) line is also shown.

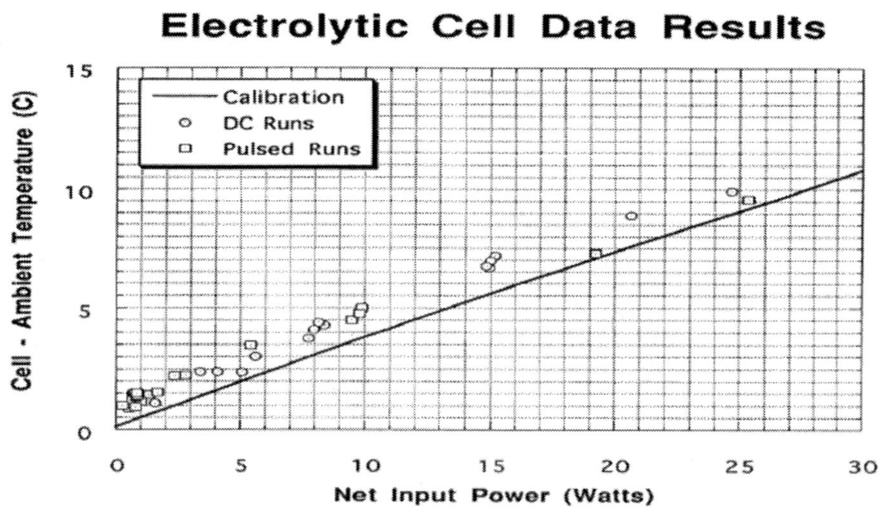

Figure 3. Heat production achieved for the LANR system and the ohmic controls electrolytic and control thermal outputs for dc and pulsed runs.

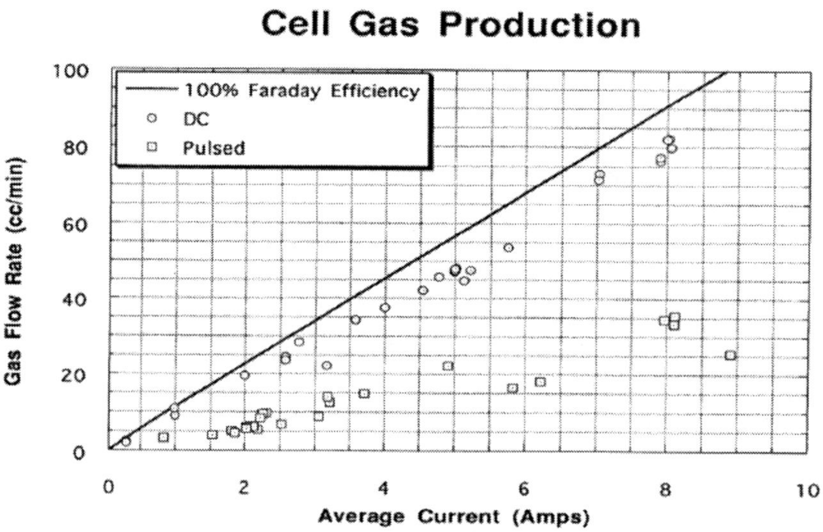

Figure 4. Electrolytic cell gas production. The rate of exiting gas flow is shown as function of input electrical current in amperes. The 100% efficient (theoretical) line is also shown.

3.3. Recombiner analysis

The recombiner liquid was analyzed (Table 2). The recombiner volume production rate was 50–100 cm^3/day, and it constituted 0.1–1.8% of the total gas flow. The residual gas from the recombiner had a content consistent with the gas measurements to $\pm 1\%$.

3.4. Isotopic ratios

Figure 5 shows the HD/H_2 isotopic ratios of the "ordinary distilled water", of the air, of the exit gas, and of the recombined water. Figure 5 also lists the test pressures, in Torr, at each location. Of special note: the HD/H_2 isotopic ratio (3/2 ratio) was more than a third less for gas leaving the cell, with some recovery in the recombiner water. This result heralds deuterons as the fuel.

4. Conclusion: Excess Heat is Linked with Deuterium Loss

These experiments produced excess power up to 5 W over considerable amounts of time. But what is the source of the energy? These experiments indicate a possible loss of deuterium from ordinary water when excess heat is observed in

Table 2. Wet gas analysis. These are the results of the analysis of air, raw cell gas, and recombination gas.

Wet gas analysis results percent				
Sample	CO_2	O_2	H_2	Residue
Air	0	21	0	79
Raw cell gas	0	32	67	1
Recombined cell gas many samples	0	18→22	0→0.2	Balance 78→82 called processed cell gas

Mass spec. analysis of processed cell gas shows N_2, A, and H_2O.

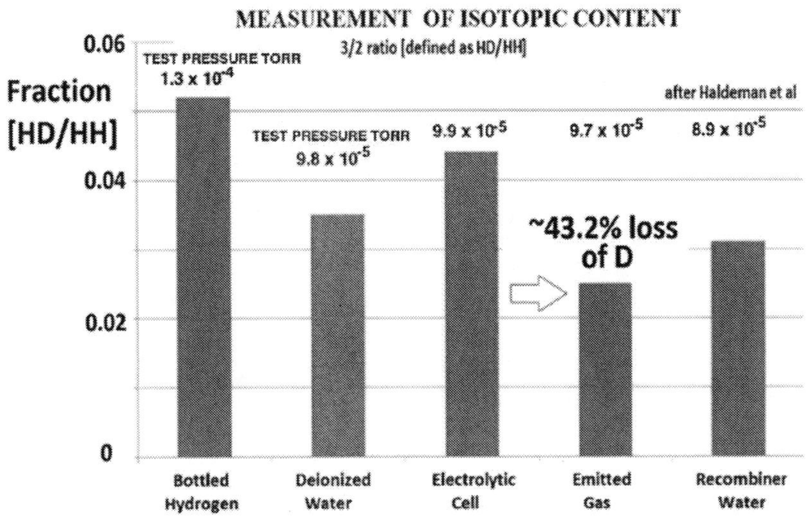

Figure 5. Results of measurement of isotopic content. The 3/2 ratios (HD/H_2 ratios) at various system locations. Also shown are the test pressure (torr) from each of the locations.

an aqueous $Ni/H_2O/Pt$ system using a very large cathodic area.

The loss of deuterium was not appreciated at the time since the investigators were looking for "hydrinos" or shrunken hydrogen (which has not been seen, and was not seen then) as the putative etiology of excess heat in LANR. The original report [1] concluded: "*We cannot prove or disprove possible explanations for (the observed) excess heat*". However, helium production, and therefore possible deuterium loss has been shown to be the actual etiology of the excess heat repeatedly observed in active, aqueous Pd LANR systems [4].

Here, the issues of electrolytic separation of H over D, and over isotopic sequestration in the nickel cathode must be ruled out. However, the change in HD/H_2 isotopic ratio, consistent with possible deuterium loss could be very significant.

These results are consistent with Swartz et al. [2,3] and probably with the reports of helium production in Pd systems [4]. Importantly, deuterons and their isotopic flow [5], including in ordinary water systems, must be considered as the fuel for active aqueous nickel CF/LANR systems.

Present studies investigate the roles, and semi-quantitative contributions to the CF/LANR activity, of electrolytic separation and metallurgic selective binding to the cathode of D versus H, the changes which occur in the nickel from electrically driving it for the desired CF/LANR reactions, and the impact of the findings of this report.

Acknowledgements

Gayle Verner and David Nagel have provided editorial assistance and significant suggestions and comments. These results, from MIT's Lincoln Lab efforts from 1993 to 1996, were conducted by Charles W. Haldeman, Alan Weinberg, C.W. E.D. Savoye, G.W. Iseler, and H.H. Clark. Some JET Energy technology were shared toward the end of that work to improve the incremental excess power output at MIT's Lincoln Lab. For the last three years supplemental work has continued at JET Energy Inc. adding further technologies.

References

[1] C.W. Haldeman, E.D. Savoye, G.W. Iseler and H.H. Clark, Excess Energy Cell Final Report, 1996.

[2] M.R. Swartz, G.M. Verner, A.H. Frank, The impact of heavy water (D_2O) on nickel-light water cold fusion systems, *Proc. ICCF9*, China, Xing Z. Li (Ed.), May 2002, pp. 335–342.

[3] M.R. Swartz, Excess power gain using high impedance and codepositional LANR devices monitored by calorimetry, heat flow, and paired stirling engines, *Proc. ICCF14*, 2008, p. 123; ISBN: 978-0-578-06694-3, p. 123, (2010); www.iscmns.org/iccf14/ProcICCF14a.pdf.

[4] M. Miles et al., Correlation of excess power and helium production during D_2O and H_2O electrolysis using palladium cathodes, *J. Electroanal. Chem.* **346** (1993) 99–117.

[5] M.R. Swartz, Quasi-one-dimensional model of electrochemical loading of isotopic fuel into a metal, *Fusion Technol.* **22** (20) (1992) 296–300.

J. Condensed Matter Nucl. Sci. 29 (2019) 177–190

Research Article

Aqueous and Nanostructured CF/LANR Systems Each has Two Electrically Driven Modes

Mitchell R. Swartz*

JET Energy Inc., Wellesley Hills, MA 02481, USA

Abstract

An important difficulty to overcome for success in this field is the recognition and control of the two different electrically driven modes for both aqueous and nanostructured CF/LANR systems. Only one state is the active, desired, excess heat (XSH)-producing state ("mode"). This is demonstrated by presenting calorimetry and other measurements of both modes during a single run, and by confirmation using CMORE spectroscopy. It is fortunate that LANR systems, when active, have distinct calorimetric and CMORE anti-Stokes-XSH linked signatures, because it explains why some CF/LANR systems fail to create "excess heat" (XSH), and reveals unwanted reactions, XSH-quenching reactions and pathways.

Keywords: Biphasic response, CMORE spectroscopy, Excess heat mode, Quenching, Two modes, Two states

1. Introduction

Lattice assisted nuclear reactions (LANR) use hydrogen-loaded alloys to create heat and other products by enabling deuterium fusion under difficult-to-achieve conditions [1]. LANR success is "rewarded" by excess heat (XSH), meaning successful generated *de novo* helium within the lattice ($\sim 10^{12}$ for every watt-second) [2]. However, there are also unwanted reactions and quenching path ways in parallel with the desired reactions of the original method which Fleischmann and Pons taught in March 1989 (aqueous, low impedance Pd/D_2O/Pt). LANR, then called "cold fusion", thus had low efficiency and poor reproducibility which created havoc for the inexperienced in metallurgy, electrochemistry, contamination avoidance, and optimal operating point (OOP) operation [3–6].

This paper reports an important difficulty to overcome for success in this field: the existence of (at least) two electrically driven modes for both aqueous and nanostructured CF/LANR systems. Two-pole calorimetry, gas measurement (O_2, H_2, pH_2O/pD_2O, barometric pressure) and Coherent Multiwavelength Optical Reflection Electric-driven (CMORE) spectroscopy have each separately revealed these two distinct states ("modes") of electrical-driven performance. Importantly, only one state is the active, desired, state, and that is the XSH-producing mode. The two states for an aqueous system during a single run are shown in Fig. 1. One mode is active and heat-producing, the other is not.

*Dr. Mitchell R. Swartz ScD, MD, EE, E-mail: moac@nanortech.com.

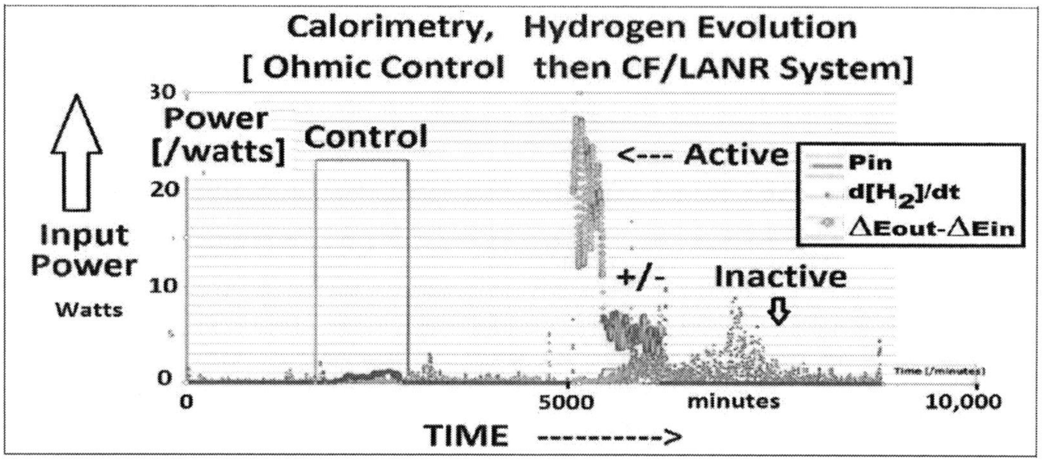

Figure 1. Two electric-drive modes revealed for aqueous NR systems by calorimetry.This figure shows the calorimetry (including incremental power gain, and the rate of XSH- and H_2-production). This aqueous Ni ordinary water system (Pt anode, PHUSOR®-type) shows the two modes in a single run. There is first a control pulse and then three levels of much lower input to the component, showing first much excess power, then an intermediate region, and then mainly hydrogen gas evolution.

2. Background

2.1. Two driven states exist in nature

Two state driven systems, with vastly different performance from each other, already exist in nature. In aviation, there are two states well known, being aerodynamic flight and stalling. A stall does not mean the engines do not work, nor does it mean that the aircraft is not moving. The implications from the performance difference are fierce. The first parallel requires one to note the similarity of shape of the OOP [7–11] (as demonstrated at ICCF-10 by papers [5,6] and open demonstrations [11]) to what is also confirmed as the shape seen for conventional aerodynamics with an airfoil (cf. Fig. 2). The second parallel might follow the fact that ONLY when vectored thrust was used to surpass the stall limit, did there finally evolve a plethora of post-stall avionic technologies. Therefore, we much pay close attention to these different modes.

2.2. Future proves past

Looking backwards, these recent discoveries have confirmed the earlier theories. The possible existence of two states was first heralded by the quasi-1-dimensional model of isotope loading [12,13,7], which thereafter yielded many contributions (codeposition [14], OOP control [7–11], HAD control [15], etc. as shown at ICCF-7 [9], ICCF-10 [5,6], and ICCF-14 [3,4]). Later, the existence of two states was clearly demonstrated for NANOR-type components when electrically driven into avalanche mode at ICCF-19 [16].

Finally, these two electrically driven states were then confirmed by dual beam coherent Raman spectroscopy of electrically driven CF/LANR components (CMORE spectroscopy), including at ICCF-20 [17,18] and ICCF-21 [19], as discussed below.

This paper augments those results with confirmation of two states and demonstrates further uniformity among both aqueous and nanostructured CF/LANR materials, and also presents aqueous experiments using the MOAC, including

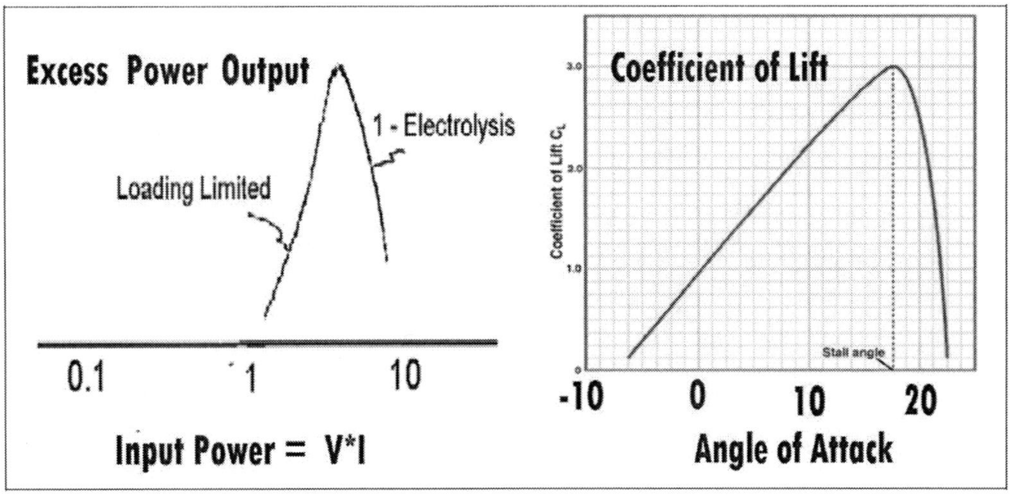

Figure 2. Similar curves describe the two states of driven systems. On the left-hand side is the OOP of aqueous Pd/heavy water vs. platinum anode, and the on the right-hand side is the lift of an aerofoil as a function of attack angle. Notice the peak performance is limited, and further input is clearly not rewarded.

both modes observed during a single run, and confirmation by CMORE spectroscopy. It is confirmed that one mode is active and heat-producing, the other is not.

2.3. CMORE scattering reveals two states

Classic Raman spectroscopy involves the inelastic scattering of light which generates new frequencies which result from the interaction of light the irradiated matter [20]. Raman anti-Stokes scattering results when the irradiated material already contains excited states, and the reflected photons are blue-shifted by the difference in energy between the excited states returning to their ground state. The major problem of the Raman effect is that it produces very weak signals because the photon conversion efficiencies are less than 10^{-18}. However, that is not the case when irradiation is made using coherent lasers which yield a much larger signal because with coherent illumination there results phase-matching conditions and quadratic dependence on the number of local oscillators. Thus, even with the same selection rules, the coherent irradiation Raman effect yields a much greater signal intensity (circa 10^6 times greater) [17–19], and it has yielded a new imaging modality for excited energy states which only appear in working active CF/LANR systems [1,3–11].

Simply put, it offers investigators of condensed matter nuclear science a new diagnostic – in addition to calorimetry and detection of classical emissions. One can easily see the distinguishing optical output in the desired correctly driven active "state" as revealed by the new diagnostic technology. Successful cold fusion systems indicate their activity by a large increase in the aS/S ratio for BOTH active aqueous and nanomaterial CF/LANR systems. There is an XSH-related anti-Stokes peak which is singular and at higher energy. This distinguishing, higher energy anti-Stokes peak (which also heralds phonon gain] is not seen in the "off" state.

As a corollary, from a materials point of view, it appears that acoustic phonons result from, or are required for, an active cold fusion process producing energy gain in its "XSH" mode in both this aqueous system and nanomaterial CF/LANR systems.

3. Experimental

3.1. Aqueous

The aqueous CF/LANR system used for this report is a Ni/H_2O/Pt system which was designed to have a very large electrode area [19,21]. The cathode weighs 4.7 pounds, made from #46 hard drawn smooth nickel wire (0.00399 cm diameter, with an area of circa 240,000 cm^2). The anode was five folded platinized sheets of titanium with an area of $\sim$3200 cm^2, for a surface area ratio of cathode to anode of 75:1. The electrolyte was a dilute carbonate solution (K_2CO_3); circa 0.1 M in distilled water of natural deuteron abundance. Two internal ohmic controls were used. This LANR cell has a 3 l capacity. The methodology of electrically driving this system was discussed at ICCF-10,14 [3–7]. It was electrically driven by a steady direct current electrical drive (2.9 V, 0.50 A; with a measured energy gain of $\sim$4.0 using $V*I$ as the electrical input) while it was irradiated by the two lasers while physically maintained in position. Using several ohmic (thermal) and other controls, the anti-Stokes to Stokes (aS/S) ratio was determined when the materials were examined in both the "off" condition including where the desired cold fusion reactions occur.

3.2. Nanomaterial

The NANOR®-type component is the two terminal component with a cylindrical shaped active CF/LANR core. The dry, preloaded NANOR®-type technology makes LANR reactions more accessible. These self-contained, two-terminal nanocomposite ZrO_2–PdNiD CF/LANR components have at their core ZrO_2–PdD nanostructured material. The NANOR® components are smaller than 2 cm length, and with 30–200 mg of active LANR material.

Their "core" contains active ZrO_2–PdD nanostructured material, loaded with additional D to loadings (ratio of D to Pd) of more than 130%, but shallow traps are not ruled out because palladium nanoparticles often have a vacancy in their center and vacancies within them. The methodology of electrically driving this system was discussed at ICCF-17 [22–24] and -18 [25–27]. Input power is defined as $V*I$. There is no thermoneutral correction in denominator.

Therefore, the observed incremental power gain is actually a lower limit. The instantaneous power gain (power amplification factor (non-dimensional)) is defined as P_{out}/P_{in}. As discussed above, the energy is calibrated by at least one electrical joule control (ohmic resistor) used frequently, and with time integration for additional energy validation. The output of the component is compared to the output of the precisely driven ohmic control.

The preloaded, stabilized components were driven by a DC voltage circuit up to 2000^+ V rail voltage. The duty cycle was split with about half going to a control portion consisting of a carefully controlled electrical DC pulse into an ohmic resistor which was used to thermally calibrate the calorimeter. We rely on three to five diagnostics to look for putative excess power gain in any sample, and time integration for putative excess power gain or excess energy gain. The presence of possible incremental power gain and possible energy gain are derived several ways to minimize the possibility of a false positive. Two of the methods use time integration for calculating total energy gain and thus command the most believability. Generally, at least three methods of verification are pooled to derive the sample activity. We use input-power-normalized delta-T (delta-T/P_{in}), input power normalized heat flow (delta-HF/P_{in}), and both single pole and double pole calorimetry, 5 and the new metric: ($\langle E_{out} \rangle$ 5 min/P_{in}), to determine the possible presence of excess energy gain.

Although small in size, this NANOR®-type preloaded LANR device is actually not *de minimus* because the LANR excess power density is more than 19,500 W/kg of nanostructured material [22–24] and the carbon footprint is zero. NANOR®-type CF/LANR quantum electronic components have enabled the way to higher instantaneous power gain, total energy gain, imaging [27], emissions [26,27], open demonstrations [23], and a better understanding of the impact of applied magnetic fields [25], electrical transconduction [16,22–27], and things that quench the desired reactions (as presented at ICCF 17–19).

Some of the results presented here were done at JET Energy and others were conducted at MIT using an independent operation with a verified system to assess the possible presence of excess energy gain. At JET Energy, a custom controlled driving system, the NANOR Explorer™ is used linked to a high voltage or current source coupled to the NANOR-type LANR system. Compared to most systems, it provides an improved method of current control, enabling an improved and better paradigm system and the ability to evolve paradigms. At MIT, the runs were driven by a programmable Keithley current source, controlled by Python. The input powers were initially designed to be below 100 mW to increase the safety at the educational institution during the month-long runs and to facilitate a rapid time constant, although some NANOR®-type components have been driven up to the two watt level.

4. Two Driven States in the Same Run

4.1. Aqueous

Figure 1 shows the calorimetry (including incremental power gain) and the rate of "XSH"- and H_2-production of an aqueous Ni ordinary water system (Pt anode, PHUSOR®-type). Two modes in a single run for this electrically driven system is revealed. One mode makes much heat, and the other makes much evolved hydrogen gas.

4.2. Aqueous – anti-Stokes with XSH

Shown in Fig. 3 are two spectra of the same LANR aqueous component resolved by dual wavelength CMORE spectroscopy. The two modes (responses) are the undriven "off"-state, and the optimal operational state, the "Desired Active Mode", where "excess" energy is being released. Note how the "XSH" mode can be distinguished in its CMORE spectra observed by a unique reflected optical backscatter along with the reflected optical beams. Note the increase of the anti-Stokes peak from the active state of an aqueous Nickel/H_2O/Pt system.

Figure 3 reveals, and confirms the existence of, the two different electrically driven states beyond "off": "on-" (not active, no XSH), and "on+" (active, with XSH) for an aqueous CF/LANR system as revealed by calorimetry (and also by hydrogen evolution and by ohmic conductivity).

4.3. Nanomaterial

Several reports demonstrated that several electrical transconduction states exist, but that only one is active, desired, and capable of producing "XSH" [1,3–11]. To determine the effectiveness of the heat source, Fig. 4 is set of curves which presents the results of the same experiment, but which plots the temperature rise (delta-T in °C) of the preloaded NANOR®-type LANR component and the ohmic control with both normalized to input electrical power. This derived value, delta-T/P_{in}, is important because it enables semiquantitative determination of the incremental power gain. Figure 4 reveals the two different electrically driven states beyond "off" : "on-" (not active, no XSH), and "on+" (active, with XSH) for a nanomaterial CF/LANR system as revealed by calorimetry (and ohmic conductivity).

The top of Fig. 4 shows the differential temperature rise normalized to input electrical power for the preloaded NANOR®-type LANR component, and for the case with no input power and for the case of input to the ohmic thermal control, located at the core. These curves show the dT/Pin ratios, which enable determination of the power gain, both in the XSH region and after the avalanche behavior, for a ZrO_2–NiD NANOR®-type component. The region of the avalanche is labeled. The x-axis represents time, and each count represents 6 s. The y-axis on the left-hand side represents electrical input power in watts. Each of the outputs is read off of the right-hand side. The y-axis on the right-hand side represents the amount of temperature rise (differential temperature increase) normalized (i.e., divided by) to the electrical input power. The units of this axis are in °C/W. A calibration pulse, used for accuracy and precision check of voltage and current measurement, are again also shown at the beginning and end (not labeled) of the run.

Because these curves plot the temperature rise normalized to input electrical power as a function of time, the ratios can be used to estimate incremental power gain. That is done here by taking the ratio of the response of the

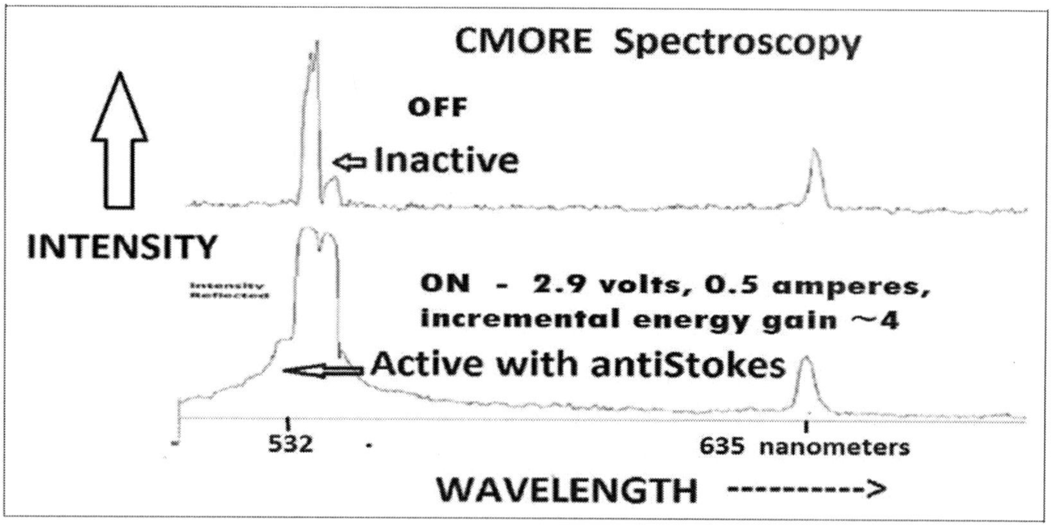

Figure 3. Two electric-driven modes revealed by CMORE spectroscopy. This figure shows CMORE spectra. The system used was an aqueous Ni ordinary water system (Pt anode). These are two CMORE spectra of a CF/LANR Aqueous Nickel/H_2O/Pt system in two electrical states; "off" top, and active XSH mode (*bottom*). Shown are the reflected optical intensities as a function of wavelength (increasing to the right side).

NANOR (green) to the ohmic control (orange). Compare the delta-T output normalized to input power for preloaded NANOR®-type LANR component to the thermal (ohmic) control, delta-T/P_{in}. By comparing that ratio, note the active preloaded LANR quantum electronic component again clearly shows significant improvement in thermal output, here input-power-normalized compared to a standard ohmic control (a carbon composition resistor). Observe that despite lower input electrical power to the NANOR®, the temperature rise normalized to input electrical power observed in the core was higher than expected, as compared to the ohmic control.

The graph therefore shows quite clearly a demonstrated active over-unity thermal output power from the NANOR®-type cold fusion (LANR) component, before the electrical avalanche.

The lower graph presents the power gain of the ZrO_2–NiD NANOR®-type component CF/LANR component as a function of time, and shows that the XSH which is generated decreases with high input power, eventually reverting to normal, ordinary, resistor-like operation after the electrical avalanche.

Note the optimal power gain of NANOR®-type cold fusion components is found far below the breakdown voltage and that the power gain decreases continuously as the electrical avalanche threshold is approached. Beyond the region of electrical avalanche, the previously active preloaded LANR quantum electronic components then give a thermal output similar to a standard ohmic control (a carbon composition resistor).

Despite driving at higher input electrical power, on other side of the electrical avalanche, these NANOR®-type components act as little more than electrical resistors which are conventional, not over-unity, and therefore are functionally "dead" with respect to producing XSH. Although this appears limiting in some ways, this phenomena does provide yet an additional control to check calorimetry beyond the measurements involving simply using a simple ohmic, thermal control. Therefore, driving a component into this region thus adds an additional verification of the actual XSH which is developed in these studies.

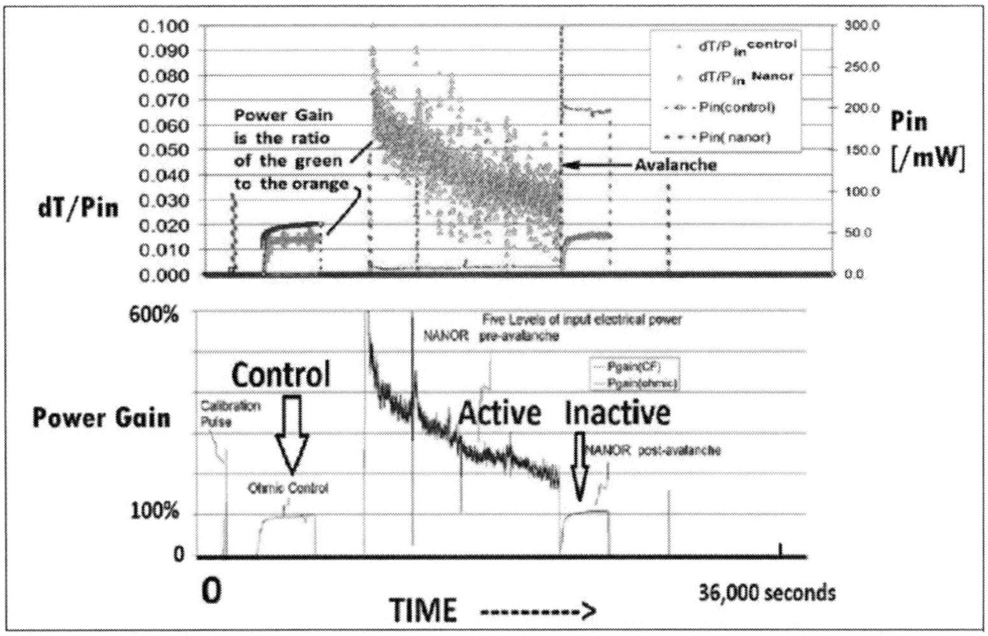

Figure 4. Two electric-drive modes revealed for a nanomaterial LANR systems nanomaterial LANR system by calorimetry. (*top*) This figure shows calorimetry (including incremental power gain, and the rate of XSH- and H_2–production). The system used was a dry preloaded ZrO_2–NiPdD NANOR®-type component (NANOR 7-7; ZrO_2NiD; results shown at ICCF-19 [16,22–24]. The delta-T/P_{in} ratios enable general estimates of the output power. The region of the avalanche is labeled. The x-axis represents time. The y–axis on the right-hand side represents electrical input power in watts. The y-axis on the left-hand side represents the amount of temperature rise (differential temperature increase) normalized (i.e., divided by) to the electrical input power. The units of this axis are in s°C/W. (*bottom*) This graph presents the power gain of the ZrO_2–NiD NANOR®-type component CF/LANR component as a function of time, and shows that the XSH which is generated decreases with high input power, eventually reverting to normal, ordinary, resistor-like operation after the electrical avalanche.

4.4. Nanomaterial – anti-Stokes with XSH

Figure 5 shows the two different electrically driven states beyond "off": "on−" (not active, no XSH), and "on+" (active, with XSH) for a nanomaterial CF/LANR system as revealed by CMORE spectroscopy, revealing the active, desired CF/LANR state by anti-Stokes emissions. Shown are the overlaid spectra of the three different electronic modes ("states"), each resolved by dual wavelength coherent electric-driven volume-enhanced reflection spectroscopy for the same preloaded ZrO_2PdD NANOR®-type CF/LANR component in three different electrical drive modes. The graph presents the output as intensity as a function of wavelength, as returned by backscatter along with the reflected optical beams from the volume-enhanced interactions. Labelled are the assignments of the anti-Stokes peaks to Zirconia and PdD.

In Fig. 5, intensity is shown as a function of wavelength, as returned by backscatter along with the reflected optical beams. One state is the unwanted electrical avalanche mode [16,22]. Another state is the desired optimal operational state (the active excess XSH mode), where excess energy is being released. The third state is the state where the electrical drive is "off". For the desired active XSH-producing state, the nanomaterial NANOR®-type CF/LANR component was properly, correctly electrically driven at 2500 V which produced an electrical current of

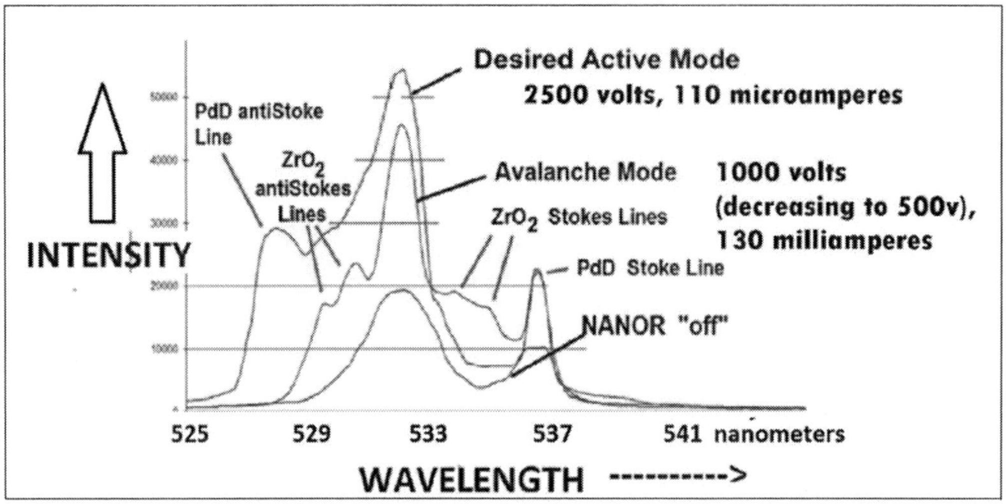

Figure 5. CMORE spectra of ZrO_2PdD NANOR®-type CF/LANR. These are the overlaid spectra of the three different electronic states optical signatures for the same preloaded ZrO_2PdD NANOR®-type CF/LANR component in three different electrical drive modes. Shown are spectra of the same NANOR®-type CF/LANR component (Nanor® 7-6) in three different electronic states, resolved by dual wavelength coherent electric-driven volume-enhanced reflection spectroscopy. Labeled are the assignments of the anti-Stokes peaks to zirconia and PdD.

about 0.11 mA. The sample maintained its high impedance (compared to that seen during avalanche mode) during the very short run, and there was no electrical avalanche quenching the desired reactions.

Note that the avalanche anti-Stokes peaks are many, and they are lower energy than the XSH mode-produced anti-Stokes peak (described below). By contrast, successful cold fusion is heralded by a large increase in the anti-Stokes to Stokes (aS/S) ratio, and the generated anti-Stokes peak for the desired and XSH-producing state is very different from the avalanche-generated many anti-Stokes peaks. That XSH-related peak is singular and at higher energy. This distinguishing, higher energy, single, anti-Stokes peak (which also heralds phonon gain) is also not seen in the off state or the avalanche (undesirable) mode. Analysis of the phonon gain heralds $\sim 7^{\pm 0.15}$ acoustic phonons assisting nuclear reactions and a core peak calculated Stokes temperature of circa 1645 K. Therefore, these findings confirm a role for PdD acoustic phonons, in the loaded lattice, during successful CF/LANR which produces XSH.

4.5. Independent confirmation

Run EF9-160410A of NANOR 8-2: ZrO_2PdD At MIT, Run EF9-160410A was the sixth run of a Series 8-2 Nanor-type LANR component. Two-pole calorimetry examined the input and output power and energy as a function of time for run EF9-160410A for an ohmic control and NANOR 8-2. Figure 6 shows the "XSH"-phase space is here represented in qualitative approximation by delta-T/P_{in}, separates out into at least three distinct groups. They are the states represented by the ohmic control, and by the NANOR-type component exhibiting pre-avalanche or post-avalanche behavior.

Figure 6 (*top*) shows the electrical impedance (resistance) of the ohmic control and then NANOR-type component at several input power levels as a function of time. The ohmic control is a nominal 1 MΩ resistor. Note that the resistor performs as approximately a more straight horizontal line. It is essentially a "constant".

Notice also that there is some avalanche breakdown in the second region of NANOR drive. There is, in fact,

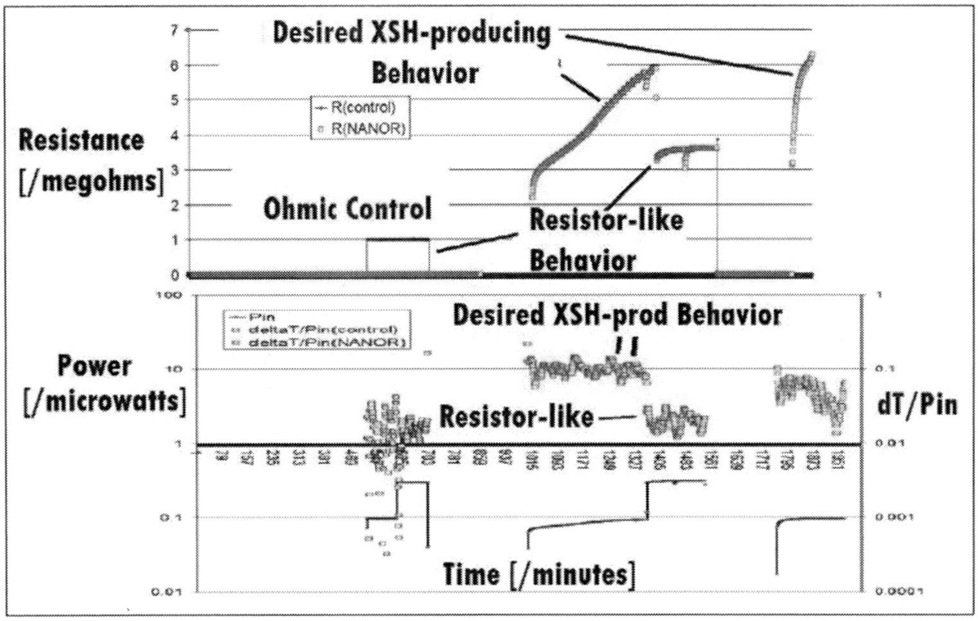

Figure 6. (*top*) The two terminal electrical impedance during run EF9-160410A for NANOR 8-2 as a function of time.(*bottom*) - $dT(*)/P_{in}$ of run EF9-160410A for NANOR 8-2 as a function of time. The star indicates that for the calorimetry, a single linear shift was used to balance dT.

synchronous loss of XSH only before the electrical avalanche, as will be seen below. After the electrical avalanche, there is a conversion to ohmic resistance behavior. These regions are clearly labeled in the figure.

Figure 6 (*bottom*) shows the result, and is the input-power-normalized incremental temperature gain ($dT(*)/P_{in}$) as a function of time for both the ohmic control and the NANOR-type component 8-2 in run EF9-160410A. There is good input power overlap and therefore Fig. 6 heralds a strong demonstration of excess power gain, and also calibrates by showing exactly what happens with electrical breakdown, and that is normal ohmic behavior.

Notice there are again two states characterizing the NANOR-type component separate out. The upper is the higher impedance XSH producing mode, and the lower is the more conductive non-XSH mode. They separate out with an incremental power gain matching the other methods of measurement.

Other NANOR type had higher incremental gains pre-avalanche, up to an estimated $100\times$ input, decreasing to $8\times$ input very close to the avalanche. Several also had confirmation by input power normalized heat flow. The two states are clearly shown in Table 1. Table 1 shows the incremental power gains pre-avalanche and post-avalanche. The incremental power gain pre-avalanche was $\sim 500\%$ over the input power. In this run, there was a very small excess energy of 13,000 μJ. These power gains for these components were evanescent, and that value decreased over the next two weeks to half that XSH by run EF9-160421A.

Table 1. Qualitative incremental power gains for NANOR 8-2.

Excess power gains for run EF9-160410A		
Method	Pre-avalanche (%)	Post-avalanche
dT/P_{in}	500	100% (normal)
$\Delta[\langle E_{\text{out}}\rangle - \langle E_{\text{in}}\rangle]/P_{\text{in}}$	250–350	100% (normal)

5. Interpretation

5.1. Origin of two different states in aqueous LANR systems

The quasi-one-dimensional (Q1D) model for hydrogen loading of an electrode describes the loading flux of hydrogen by the ratio of two energies (electric order to thermal disorder ratio). From the metallurgical point of view (POV), from the metal surface, atomic deuterons either enter the metal ("are loaded"), remain on the surface, or form diatomic deuterium gas bubbles (D_2). The gas bubbles (D_2) are undesirable producing low dielectric constant layers in front of the electrode, obstructing the electrical circuit. And so, at the metal surface, four components of deuteron flux must be considered [12–14], fundamental to the entire understanding of these phenomena. These deuteron fluxes include entry into the metal (J_{E}), movement to gas evolution (J_{G}), and an extremely tiny loss by potential fusion reactions (J_{F}). There is conservation of deuterons with the exception of a loss (J_{F}) to all putative fusion reactions, which are extremely small, even when present.

From the mathematical POV, the three components of flux are the entry of deuterons to the lattice (J_{E}), gas evolution (J_{G}), and the desired fusion reactions (J_{F}).

$$J_{\text{D}} = B_{\text{D}} * \frac{\text{d}[D(z,t)]}{\text{d}z} - \mu_{\text{D}} * [D(z,t)] * \frac{\text{d}\Phi}{\text{d}z}. \tag{1}$$

The deuteron flux, J_{D}, depends on deuteron diffusivity (B_{D}) and electrophoretic mobility (μ_{D}), and the applied electric field intensity. At any molecular site across the heavy water solution, the applied electrical energy is a tiny fraction compared to $k_{\text{B}}T$, so the deuterons migrate by drift ellipsoids of L- and D-deuteron defects in the applied electric field creating a ferroelectric inscription [28,29]. This D-defect conduction/polarization process augments other charge carriers, ionic drift, space charge polarization, and clathrates. The resultant D-defect migration produces a cathodic fall of deuterons and an E-field contraction so that most of the voltage drop is at the interface in front of the electrode surface. This concentration polarization may produce very large local electric field intensities, possibly ranging from 10^4 to 10^7 V/cm.

Dividing each flux by the local deuteron concentration yields the first order deuteron flux constants, k_{E}, k_{G}, and k_{F} (cm/s), respectively, which are the basis of the rest of the discussion, and Eq.(2).

$$k_{\text{e}} = (\mu_{\text{D}} * E) - (k_{\text{g}} + k_{\text{f}}). \tag{2}$$

Equation (2) is the deuteron loading rate equation. It relates cathodic deuteron gain from the applied electric field to the loss of deuterons from gas evolution and fusion, and teaches many things. The deuteron loading rate equation shows that the deuteron gain of the lattice (through the first order loading flux rate (k_{E})) is dependent upon the applied electric field MINUS the flux rate losses of deuterons from gas evolution (k_{G}) and fusion (k_{F}). The deuteron loading rate equation, Eq. (2), reveals that desired LANR reactions are quenched by electrolysis, which is opposite conventional "wisdom" that LANR is 'fusion by electrolysis'.

Equation (2) also heralds that LANR can be missed by insufficient loading, contamination (effecting k_{E}, by protons or salt), and by the evolution of D_2 gas, which all inhibit the desired LANR reactions [12–14,7–11], and leading to the OOP manifolds.

The modified deuteron flux equation (Eq. (3)) is Eq. (2) changed by substituting the Einstein relation.

$$k_{\mathrm{e}} = \frac{B_{\mathrm{D}} * qV}{L * (k_{\mathrm{B}} * T)} - (k_{\mathrm{g}} + k_{\mathrm{f}}). \quad (3)$$

The first term now has geometric, material factors, and the ratio of two energies (the applied electric energy organizing the deuterons divided by $k_{\mathrm{B}} * T$, thermal disorder).

The modified deuteron flux equation reveals how competitive gas evolving reactions and the applied electric field energy to thermal energy ($k_{\mathrm{B}} * T$) are both decisive in controlling the deuteron loading flux in palladium.

Successful LANR experiments are dominated by this ratio reflecting the "war" between applied electrical energy which is organizing the deuterons versus their randomization by thermal disorganization. The two terms are the first order deuteron loss rates by gas evolution and the desired fusion process(es). The Phusor®-type LANR device is a metamaterial and its physical structure enhances the metallurgic properties of loaded palladium [30]. This metamaterial change alters the electric field distribution in the high impedance solution, producing continuous deuteron flux within the loaded palladium. This is unique to this device creating a distinguishing electric field (E-field) distribution different from customary wire-wire, and other systems. This is significant and with marked more XSH.

5.2. Origin of two states in nanomaterials

Previously, we reported that the optimal power gain of NANOR®-type cold fusion components is only found below the breakdown voltage, and that the power gain decreases continuously as the electrical avalanche threshold is approached. Once the electrical avalanche has occurred, there is complete loss of the desired XSH [15,22–24]. Avalanche behavior with three regions were first observed by Swartz with ZrO–NiH NANOR®-type component. Since then, cold fusion nanomaterials, in general, and NANOR®-type LANR components (derived from them), in particular, have been repeatedly found to have distinct regions of performance on each side of the electrical avalanche. Beyond the region of electrical avalanche, the previously active preloaded LANR quantum electronic components then give a thermal output similar to a standard ohmic control (a carbon composition resistor). Therefore, this transformation of active CF/LANR components from active to inactive states has been critical to successfully controlling CF/LANR, and is critical to understanding how to engineer these systems.

For a materials POV, the nanostructured material is a composite distribution of nanostructured ferromagnetic "islands" separated among a vast dielectric zirconia "ocean". The dielectric zirconia embeds uncountable numbers of nanostructured metal ternary alloy islands. The high resistance occurs because the zirconia dielectric matrix is insulating at low voltage and it keeps the nanoscale metal islands electrically separated and prevents the aggregation of the islands. Each nanostructured island acts as a short circuit elements during electrical discharge. One hypothesis of the XSH is that these "islands" allow deuterons to form a hyperdense state in each island, where the deuterons thereafter are able to be sufficiently close together to fuse and form ^{4}He*, by some pathway not known involving paired deuterons or possibly more.

6. Conclusion

6.1. Solid evidence of two states

There is now sterling evidence of the existence of two states in electrically driven CF/LANR systems (Table 1). Two different methods (calorimetry and CMORE spectroscopy) confirm each other for both aqueous and dry preloaded LANR systems, and corroborate the existence of two electrically driven states – one inactive and the other active (XSH-producing).

As a corollary, both aqueous and nanostructured LANR systems, when active, have distinct calorimetric and CMORE anti-Stokes-XSH linked signatures. Thus, knowledge and use of these two signatures has considerable value when seeking active systems, controlling those systems, and understanding some past difficulties.

6.2. Implications of two states in CF/LANR

What is the implication of recognizing which, of two (or more) electrically driven states, is active? First, this may explain why some CF/LANR systems fail to create "XSH".

Second, it must be added to the other pitfalls involving metallurgy, electrochemistry, isotopic loading, contamination and quenching, and failure to drive at the OOP. In CF/LANR, understanding the existence of two states, and OOP Technology, opens the door to more reproducible systems. This reveals that the other reactions, most of which are unwanted and quench the desired reactions and pathways.

Third, Why is the recognition of two electrically driven states in CF/LANR very important? The most important reasons is because ONLY the more difficult-to-obtain pathway leads to the desired, sought XSH.

Table 2. Evidence of two driven states (modes)

√	Calorimetry showing XSH when active only
√	Phase change c/w XSH
√	Change in electrical conductivity c/w XSH
√	Change in anti-Stokes spectrum c/w XSH
√	OOP behavior c/w two states (suggestive)
Aqueous CF/LANR systems	
√	OOPs - indirect evidence c/w two states (ICCF-7,9,10,14)
√	Direct proof of two states (ICCF-21)
Dry Nanostructured CF/LANR systems	
√	OOPs - indirect evidence c/w two states(ICCF-17–20)
√	Direct proof of two states (ICCF-19,20)
√	Confirmed proof of two states (ICCF-21)

Acknowledgments

The author acknowledges and thanks, for their very helpful comments, editorial assistance, and support, Gayle Verner, Peter Hagelstein, Alex Frank, Florian Metzler, Brian Ahern, and Jeff Driscoll, for their help, ideas and suggestions. This effort was supported by JET Energy Inc. and the New Energy Foundation. NANOR® is a registered trademark of JET Energy, Incorporated. NANOR®-technology and other technology described here are protected by patents pending.

References

[1] M.R. Swartz, Survey of the Observed Excess Energy and Emissions In Lattice Assisted Nuclear Reactions, *J. Scientific Exploration* **23** (4) (2009) 419–436.

[2] M.H. Miles, R.A. Hollins, B.F.Bush, J.J. Lagowski and R.E. Miles, Correlation of excess power and helium production during D_2O and H_2O electrolysis using palladium cathodes, *J. Electroanal. Chem.* **346** (1993) 99–117.

[3] M.R. Swartz, Excess power gain using high impedance and codepositional LANR devices monitored by calorimetry, heat flow, and paired stirling engines, *Proc. ICCF14* **1** (2008) 123, ISBN: 978-0-578-06694-3, 123, (2010); www.iscmns.org/iccf14/ProcICCF14a.pdf.

[4] M.R. Swartz, G. Verner et al., Non-thermal near-IR emission from high impedance and codeposition LANR devices, D.J. Nagel and M.E. Melich (Eds.), *Proc. ICCF14* **1** (2008) 343, ISBN: 978-0-578-06694-3, 343, (2010); www.iscmns.org/iccf14/ProcICCF14a.pdf

[5] M.R. Swartz, Photoinduced excess heat from laser-irradiated electrically-polarized palladium cathodes in D_2O, *Condensed Matter Nuclear Science*,Peter L. Hagelstein and Scott Chubb (Eds.), *Proc. ICCF-10*, NJ, ISBN 981-256-564-6, 213-226 (2006).

[6] M.R. Swartz and G. Verner, Excess heat from low electrical conductivity heavy water spiral-wound Pd/D_2O/Pt and Pd/D_2O-PdCl2/Pt devices, *Condensed Matter Nuclear Science, Proc.ICCF-10*, World Scientific, NJ, ISBN 981-256-564-6, 29-44; 45-54 (2006).

[7] M.R. Swartz, Consistency of the biphasic nature of excess enthalpy in solid state anomalous phenomena with the quasi-1-dimensional model of isotope loading into a material, *Fusion Technol.* **31** (1997) 63–74.

[8] M.R. Swartz, Optimal operating point manifolds in active, loaded palladium linked to three distinct physical regions, D.J. Nagel and M.E. Melich (Eds.), *Proc. ICCF14* **2** (2008) 639, ISBN: 978-0-578-06694-3, 639 (2010); www.iscmns.org/iccf14/ProcICCF14b.pdf.

[9] M.R. Swartz, Optimal operating point characteristics of nickel light water experiments, *Proc. ICCF-7*, 1998.

[10] M.R. Swartz, improved electrolytic reactor performance using p-notch system operation and gold anodes, *Transactions of the American Nuclear Association*, Nashville, TN. Meeting, (ISSN:0003-018X publisher LaGrange, Ill) **78** (1998) 84–85.

[11] M.R. Swartz, Can a Pd/D_2O/Pt device be made portable to demonstrate the optimal operating point?, *Condensed Matter Nuclear Science*, Peter L. Hagelstein and Scott Chubb (Eds.), *Proc. ICCF-10*, World Scientific, NJ, ISBN 981-256-564-6, 29-44; 45-54 (2006).

[12] M.R. Swartz, Quasi-one-dimensional model of electrochemical loading of isotopic fuel into a metal, *Fusion Technol.* **22** (2) (1992) 296–300 .

[13] M.R. Swartz, Isotopic fuel loading coupled to reactions at an electrode, *Proc. ICCF4* **2** (1993) 429; *Fusion Technol.* **26** (4T) (1994) 74–77, www.lenr-canr.org/acrobat/EPRIproceedinga.pdf.

[14] M.R. Swartz, Codeposition of palladium and deuterium, *Fusion Technol.* **32** (1997) 126–130.

[15] M.R. Swartz, Investigations of heat after death , in press.

[16] M.R. Swartz, P. Hagelstein and G. Verner, Impact of electrical avalanche through a ZrO_2–NiD nanostructured CF/LANR component on its incremental excess power gain, *ICCF-19*, *J. Condensed Matter Nucl. Sci.* **19** (2016) 287.

[17] M.R. Swartz, Optical detection of phonon gain distinguishes an active cold fusion/LANR component, *J. Condensed Matter Nucl. Sci.* **20** (2016) 29–53, www.iscmns.org/CMNS/JCMNS-Vol20.pdf.

[18] M.R. Swartz and Peter L. Hagelstein, Increased PdD anti-Stokes peaks are correlated with excess heat mode, *J. Condensed Matter Nucl. Sci.* **24** (2017) 130–145.

[19] M.R. Swartz, Increase of an anti-Stokes peak at the cathode of an electrically driven, active aqueous nickel/H_2O/Pt system, *J. Condensed Matter Nucl. Sci.* **27** (2018) 1–8.

[20] C.V. Raman, *Scientific Papers of Sir C.V. Raman, The Scattering of Light*, Vol. 1, S. Ramaseshan (Ed.), India Indian Academy of Sciences, Bangalore, 1988.

[21] M.R. Swartz, Charles Haldemann, Alan Weinberg and Brian Ahern, possible deuterium loss during excess heat from ordinary water-carbonate electrolyte using nickel, **29** *J. Condensed Matter Nucl. Sci.* (2018).

[22] M.R. Swartz and P.L. Hagelstein, Energy Gain From Preloaded ZrO_2–PdNi–D Nanostructured CF/LANR Quantum Electronic Components, *J. Condensed Matter Nucl. Sci.* **13** (2014) 528, www.iscmns.org/CMNS/JCMNS-Vol13.pdf.

[23] M.R. Swartz and P.I. Hagelstein, Demonstration of energy gain from a preloaded ZrO_2–PdD nanostructured LENR quantum electronic device at MIT, *J. Condensed Matter Nucl. Sci.* **13** (2014) 516, www.iscmns.org/CMNS/JCMNS-Vol13.pdf.

[24] M.R. Swartz, G. Verner, J. Tolleson and P. Hagelstein, Dry, preloaded NANOR®-type CF/LANR components, *Current Sci.* **108** (4) (2015) 595.

[25] M.R. Swartz, G. Verner, J. Tolleson, L. Wright, R. Goldbaum, P. Mosier-Boss and P. Hagelstein, Amplification and restoration of energy gain using fractionated magnetic fields on ZrO_2–PdD nanostructured components, *J. Condensed Matter Nucl.*

Sci. **15** (2015) 66, www.iscmns.org/CMNS/JCMNS-Vol15.pdf.

[26] M.R. Swartz, Incremental high energy emission from a ZrO_2–PdD nanostructured quantum electronic component CF/LANR, *J. Condensed Matter Nucl. Sci.* **15** (2015) 92.

[27] M.R. Swartz, . Verner, J. Tolleson, L. Wright, R. Goldbaum and P. Hagelstein,, Imaging of an ative NANOR®-type LANR component using CR-39, *J. Condensed Matter Nucl. Sci.* **15** (2015) 81.

[28] M.R. Swartz, Dances with protons – ferroelectric inscriptions in water/ice relevant to cold fusion and some energy systems, *Infinite Energy* **44** (2002) 64–70.

[29] M.R. Swartz, Water is best, *Infinite Energy* **134** (2017) 16–29.

[30] M.R. Swartz and G. Verner, The Phusor®-type LANR cathode is a metamaterial creating deuteron flux for excess power gain, D.J. Nagel and M.E. Melich (Eds.), *Proc. ICCF14 Proc. ICCF14* **2** (2008) 458, ISBN: 978-0-578-06694-3, 458, (2010); www.iscmns.org/iccf14/ProcICCF14b.pdf.

J. Condensed Matter Nucl. Sci. 29 (2019) 191–201

Research Article

Light Hydrogen LENR in Copper Alloys

William H. McCarthy*
Shamrock Power, 2156 Old Middlefield Way, Mountain View, CA 94043, USA

Abstract

An example of Low Energy Nuclear Reaction (LENR) is demonstrated. A cell containing copper-boron alloy electrodes, infused with hydrogen at high temperature and moderate pressure, was tested. Calorimetry showing assumed nuclear energy well in excess of standard statistical requirements is reported. Equipment to detect far infrared emissions produced by these cells was built and is described. An electronic resonance method to reduce input is being developed to greatly improve efficiency. This should lead to a commercially viable energy source, with little or no environmental consequences. Some theoretical discussion is also given.

Keywords: Boron, Calorimetry, Capacitor, Copper, Hydrogen, Infrared, LENR, Resonance, Seebeck

1. Introduction

Low Energy Nuclear Reactions (LENR) produced in dilute copper alloys containing light hydrogen are described here. This system was selected because of the low cost of its constituents, ease of processing, and similarity to other, much more expensive, materials that have been extensively reported in the literature to produce nuclear heat. The focus here is on heat, with an excursion into thermal radiation, rather than on reaction products that are difficult and expensive to investigate.

A fundamental advantage, of LENR over other nuclear energy developments is the production of substantial power WITHOUT dangerous by-products. This research has been guided by the well-known nuclear reactions between the stable isotopes of hydrogen and the stable isotopes of boron and lithium that produce helium-four and more than 8 MeV of energy. Strong inter-particle forces in metals [as affected by solutes, dislocations, vacancies and the cloud of free electrons (in a quantum-mechanical environment)] must also be important in LENR.

2. Summary of Experimentation

The effort has three phases, each involving significantly different processes, complexity and instrumentation.

*Retired. E-mail: santander1bill@gmail.com.

(1) *Simple addition of hydrogen to a copper-based alloy*
Hydrogen was loaded into copper–lithium–boron alloys at elevated temperature (400–900°C) and pressure (about 10 atm.). The encapsulated specimens were quenched, and immediately inserted into a calorimeter to detect nuclear heat. That method had substantial experimental difficulty and showed little success. It is apparently too simple and has been discontinued.
(2) *Low–Q capacitors*
Cells containing hydrogenated copper–lithium–boron alloy electrodes separated by a liquid dielectric containing fine graphite particles have been tested. A vacuum-insulated Seebeck calorimeter was used. When the layers of copper alloy are alternately charged, electric currents are constrained to flow through the graphite particles at specific sites, resulting in microscopic variations of charge in the metal and localized electron flows. It seems that most, if not all, LENR observations involve charge dynamics. An example of anomalous heat from a low-Q capacitor is presented here. Results from another cell that had a cold worked electrode and different conductor/separator were posted at ICCF-18 [1]. That specimen showed about 9%, statistically significant, excess heat. Results from a control specimen, devoid of hydrogen, are indistinguishable from the corresponding joule calibration. Infrared images of electrodes producing anomalous heat by this method are being made.
(3) *Substantial reduction of input energy to produce useful power*
The low-Q capacitors are inherently inefficient because most of the input power is dissipated by resistance heating in the insulating layer between electrodes, and not available to stimulate nuclear reactions in the electrodes. Methods and apparatus to remedy this situation is being developed; they are described here.

3. Low-Q Capacitor

Electrodes containing less than one weight percent lithium and/or boron, as well as commercially pure copper have been tested. Hydrogenation of the electrodes is carried out in capsules pressurized to about 20 bar (300 psig) at 600–900°C. After 30 min or so, the capsule is water quenched. Dehydrogenation to prepare control electrodes is done by several vacuum (20 μm Hg) anneals at about 700°C alternated with argon pressurization (20 bar).

Test cells are prepared from these copper alloy electrodes about 0.15 mm (0.005 in.) thick, 7 mm (0.25 in.) wide and 64 mm (2.5 in.) long. The electrodes are separated by thin gauze or thread saturated with a liquid or paste dielectric containing fine graphite particles, as illustrated in Fig. 1. They are stacked alternately to resemble capacitors. Electric conductance of the cells is controlled by varying the graphite concentration (2–12 vol.%). A fine dielectric powder (e.g. diatomaceous earth) is often added to separate graphite particles. These cells are thus poor (low–Q) capacitors with an inter-electrode resistance of three to twenty ohms. Their capacitance is a few nanofarads.

4. Calorimetry

The Seebeck calorimeter used here is based on a thermopile consisting of a series string of eight to ten Type-E thermocouple junction pairs. For this spider structure, the warmer junctions are cemented at the edges and between two strips of polycarbonate that form the base of the specimen carrier. The alternate cold junctions are cemented to similar polycarbonate strips attached to the body of the calorimeter, which is maintained at ambient temperature. Thin copper channels are cemented to the carrier polycarbonate on both sides, to stiffen the assembly and disperse any temperature gradients. The specimen is clamped to one side of the carrier with small brass screws and nuts. Thin rubber pads are included to prevent short circuits, without seriously affecting the sensitivity of the calorimeter. A joule heater is attached to the opposite side of the carrier, in as symmetrical a fashion as its geometry and that of the specimen allows. That electrical resistance heater consists of nichrome wire wrapped around a strip of polycarbonate and covered with a thin layer of silicone adhesive.

The thermocouple wires have three functions in this arrangement: They mechanically support the specimen/heater carrier, they signal the voltage corresponding to the temperature difference between carrier and ambient to measure the thermal power deposited on the carrier and they provide most of the thermal hurdle that maintains that temperature difference. The sensitivity of the calorimeter can be adjusted by selecting the diameter and length of thermocouple wires. A sketch of the power sensitive part of this calorimeter is shown in Fig. 2.

The spider is enclosed in a vacuum chamber as shown schematically in Fig. 3. That chamber has a clamshell design consisting principally of two copper half-cylinders, each brazed to steel quarter-spheres on the ends. The enclosure is mechanically pumped (about 20 μm Hg) to eliminate convection and greatly minimize gas-conduction heat transfer. There is some radiation cooling of the spider assembly. At the power levels used here ($<$650 mW), however, radiation can account for no more than one-quarter of the heat passed from the spider to the enclosure shell. It was found in practice that that level of thermal radiation is insufficient to produce detectable nonlinearity in the test results.

5. Testing

Cells are activated by direct current provided by a lead-acid battery. Thus, power factor concerns are eliminated. Current is controlled by a single-transistor power supply of the author's design and construction. The same Hewlett Packard Model 3445A voltmeter is used for all current and voltage (hence power) data. This instrument provides visual digital output with least count of one millivolt. The range of power levels used here is from about 50 to 600 mW. A four-wire hookup is used to connect the power supply to the specimen or joule heater, so that voltage drops in the leads do not affect the results.. The voltmeter has very high input impedance. A device is available to reverse automatically the specimen polarity. The typical reversal period of roughly ten minutes presents no power factor concern.

Voltage is measured directly. Current is measured by the voltage drop across standardized resistors. For any test

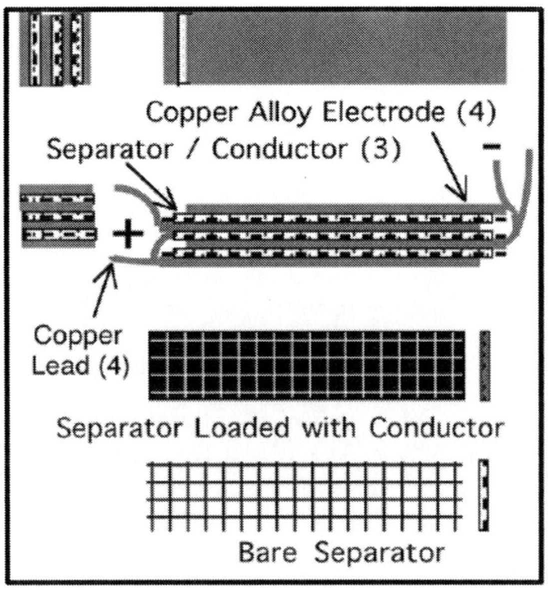

Figure 1. Cell schematic.

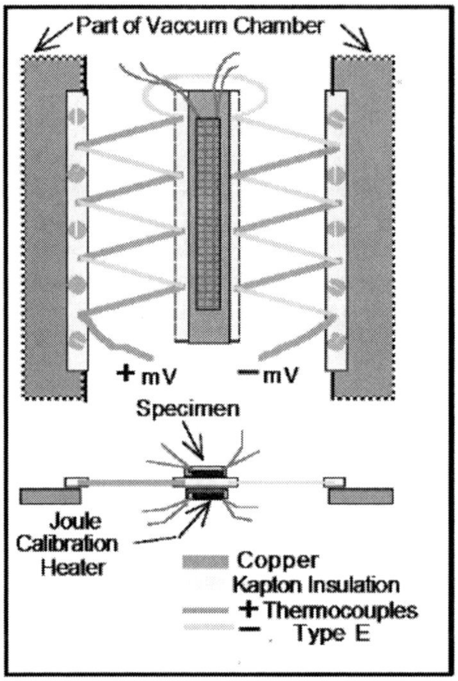

Figure 2. Seebeck calorimetry spider (heat sensor).

run, the same resistor(s) are used for both specimen and calibration heater. Joule heaters are designed to be fairly close in resistance to specimens. The standardized resistors are selected to be in the same range. Thereby, only one range of the voltmeter is normally used for joule and specimen readings (both current and voltage) over a moderate range of input power.

Thermopile output (dependent variable) is assessed using a Leeds & Northrup Model K-2 potentiometer. That 1960s era manually balanced Wheatstone bridge uses a calomel standard cell. It has a least count of 1 μV. At a typical calibration minimum input of about 50 mW, the thermopile produces about one 1 mV. Precautions and other details of the instrumentation are discussed elsewhere [1] in much more detail.

The procedure is to power the cell or heater to some fairly arbitrary power level and wait perhaps an hour for thermal equilibrium, as shown by near constancy potentiometer readings from the K-2 instrument. Then typically three readings, separated by 5–10 min, are recorded. Input power is then adjusted, and/or input is switched between specimen (or polarity) and joule heater. To get some semblance of randomness, power levels are changed arbitrarily (not in up or down series, nor in order of power level). For operational efficiency, however, specimen and calibration will usually be switched at roughly the same power level. A set of the order of 50–100 equilibrium calibration and specimen data points is thus acquired over several days.

Readings are taken while power is stabilizing and the cell is coming to equilibrium. Those data are useful to assess calorimeter performance and to estimate total anomalous energy produced during a run. Only equilibrium data are used, however. So far, only manual data recording on paper has been employed, without automated data acquisition.

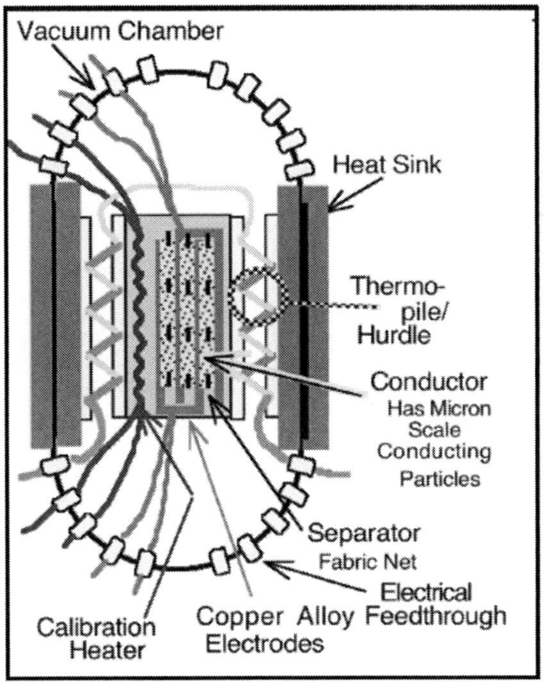

Figure 3. Calorimeter assembly.

The raw data are entered onto an Excel spreadsheet to calculate power, provide least-squares equations for calibration and specimen performance and to compile ancillary results. Data analysis and graphic displays are relied upon to define the performance of specimens and calibration heater, rather than trying to compare individual data points. Thus, the considerable trouble required for operation at precisely the same set of input levels for both heater and specimen measurements is avoided.

6. Example of Anomalous Heat

Some results from an active cell are given here. One electrode was a copper, 0.75 wt.% boron alloy hydrogenated at 820°C and 20 bar (300 psig) for 45 min, water quenched in capsule. The other electrode derived from a commercial copper tube. The separator was a layer of medical gauze about 25 μm (0.001 in.) thick. To make the conductor, it was saturated with a suspension containing nominally three micrometer (120 μin.) graphite particles in mechanical vacuum pump oil. Specimen resistance was about nine ohms to start. Figure 4 is a plot of thermopile output versus input power for that example.

Three-sigma statistical limits for the joule calibration data are defined by the band between the dashed green lines in Fig. 4. The 18 calibration data points are shown as green circles (in essentially isopower subsets of three). The green solid line is the least-squares fit of the joule calibration data. Red triangles and black squares indicate data points for the low-Q capacitor specimen. All data points, from the LENR cell, are well to the left (and above) the least-squares line of the calibration data. Above about 250 mW, the separations between the upper 3-sigma limit and the low–Q

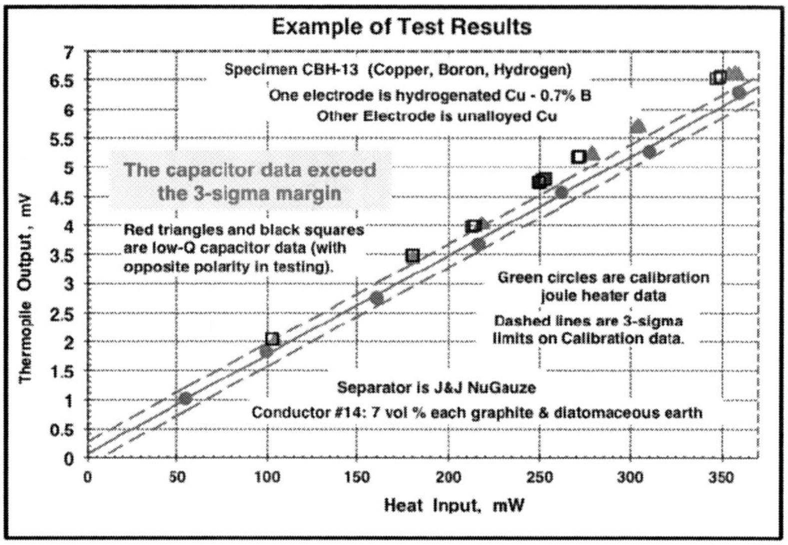

Figure 4. Test example showing anomalous heat.

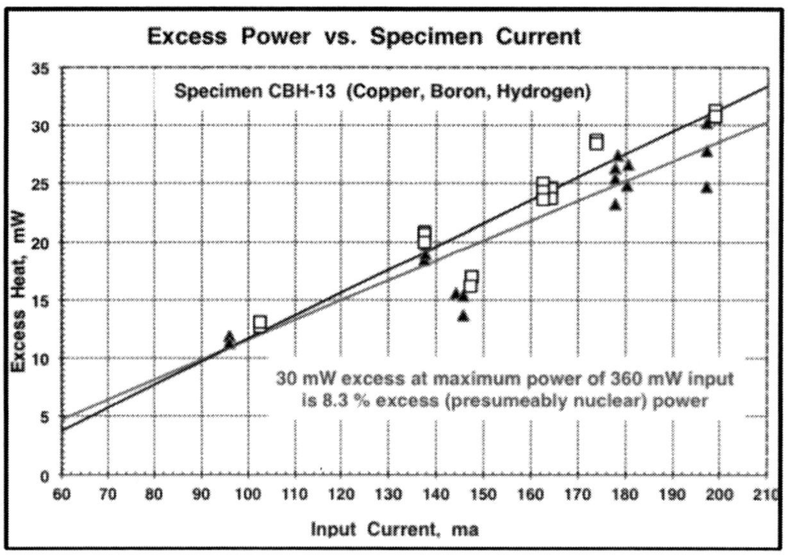

Figure 5. Anomalous power vs. input current from the data shown in Fig. 4.

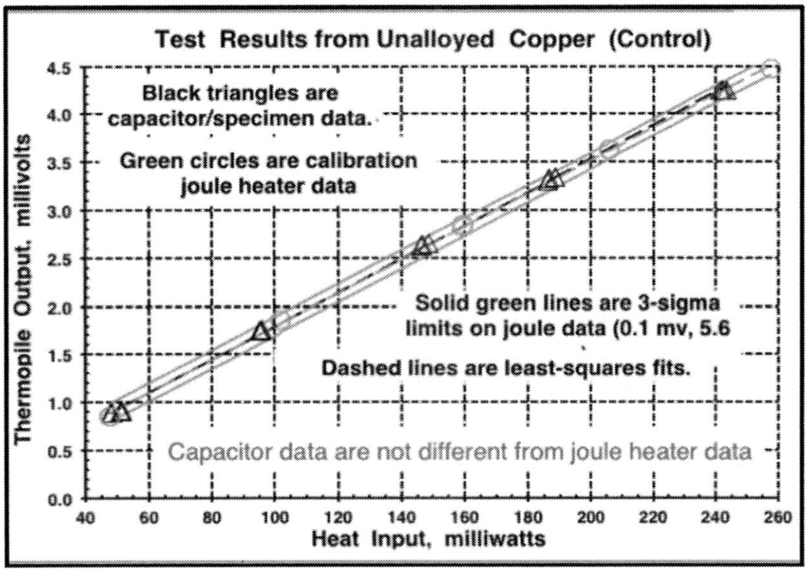

Figure 6. Test of dehydrogenated control electrodes.

data are roughly another 3-sigma amount. Therefore, the probability (assuming normal distribution and other standard statistical criteria) that the specimen and heater outputs are the same is much less than 0.3%.

Even though the low-Q capacitor results are substantially significant statistically, the magnitude of anomalous power (excess of output power over input) is small. In Fig. 5, anomalous power is plotted against input current, with red triangles for direct polarity and black squares for reverse. The assumed nuclear power reaches 30 mW at the maximum input of 360 mW. This is more than 8% of input, about one-half of which exceeds the three-sigma limit shown in the joule heater data. Nuclear power increases with input current at a roughly linear rate. Figures 4 and 5 show little difference between polarities.

7. Example of Vacuum Annealed Dehydrogenated Electrodes that Show No Anomalous Heat

A cell using commercial copper electrode materials subjected to dummy hydrogenation was prepared. Instead of high temperature pressurization with hydrogen, the electrode material was exposed at about 700°C to alternating vacuum and argon pressure. Results in the form of a plot of input power versus thermopile output are shown in Fig. 6. The joule calibration data are shown as green circles for the individual measurements (three per power level). The three-sigma band is defined by solid green lines. A dashed green line is the least-squares fit.

Specimen data are shown by black triangles. Least-squares fit for the capacitor is the dashed black line. In this case, there was no anomalous power because the dashed lines are indistinguishable, each from the other. This specimen behaved thermally exactly as did the joule heater.

The one-sided three-sigma limits for both joule heaters (Figs. 4 and 6), and for the dummy specimen are approximately equal (4.3–5.6 mW). Three-sigma limits for the active specimen are larger (6.6–15.3 mW) depending on polarity. The statistical "r" values from all five curve fits exceed 0.999.

8. A Working Hypothesis and Explanation

In researching a subject that lacks a widely accepted theory, it is helpful to create a working hypothesis of that subject's characteristics. Such a hypothesis facilitates consistency in decisions and minimizes thrashing about. One does not care if the hypothesis is precisely accurate; it can be abandoned or modified to correspond to developments. This working hypothesis is based on the well-known nuclear reactions between the stable isotopes of hydrogen and the stable isotopes of boron or lithium. These reactions produce helium-four *WITHOUT IONIZING RADIATION*. They create no dangerous emanations nor unstable nuclear residuals. The assumption is that hydrogen dissolved in some metals can react with specific light alloying elements (or tramp impurities) to produce helium and release more than 8 MeV of nuclear energy. Strong interatomic forces in metals, as affected by solutes, alloying particulates and the cloud of free electrons must also be important in LENR. This hypothesis readily explains LENR with both light and heavy hydrogen. A more complex process, involving sequential combination of four protons, is needed to explain the generation of helium from light hydrogen only.

The static forces inside metallic crystals seem insufficient trigger nuclear reactions. It is conjectured that significant motion of charges is required to cause protons (hydrogen ions: atoms stripped of their electrons) dissolved in a metal to move close enough to dissolved light element ions to fuse with the solutes which then break apart. Most, if not all, observations of LENR have taken place in dynamic systems having substantially moving electric charges. Obvious examples are acoustic bubble collapse and the plasma LENR techniques. Electron flow in the classic Pons–Fleischmann electrolytic LENR is disrupted by insulating gas bubbles on the cathode, also to cause considerable dynamism of the electronic distribution in the metal.

A proper metallurgical structure, and electric charges moving suitably therein, seems required. Edmund Storms [2] concept of a "hydroton" in a Nuclear Active Environment (NAE) seems especially cogent. The proposal there is that linear strings of oscillators are attached to micro-defects in the metal and vibrate coherently (as in a LASER). A photon is ejected simultaneously from each end of the hydroton dividing the very high energy of a nuclear reaction among a multitude of thermal oscillators to circumvent any need for hard emissions to release the energy. Dr. Storms looks to micro-cracks as the locations of these hydrotons; the current author believes dislocations are more likely sites. A suitable array of micro-cracks would be two dimensional, difficult to establish and maintain, but would be an explanation for the erratic occurrence of LENR. Dislocations, as sites for NAE, are ubiquitous, easy to control and they are curvilinear like the proposed string of oscillators.

It appears that Dr. Storms explanation for *light* hydrogen LENR on this basis is that two protons in a hydroton combine to form a deuteron which, in turn, reacts with another deuteron or a tritium ion to produce the observed result. See Table 11 in his book [2]. A direct process might be possible if a light-element ion (Li^7+ or $B^{11}+$) were *part* of the hydroton. Table 1 lists such reactions for light hydrogen. (There are similar, considerably more energetic, reactions involving deuterium and the lighter isotopes of the same metals.)

Much longer, and higher power level, tests of these LENR cells should be run to get sufficient reaction so that isotopic changes can be measured. For instance, if an electrode containing natural boron and light hydrogen were to

Table 1. Some reactions of light hydrogen that produce no ionizing radiation.

$^7Li + {}^1H$	$\longrightarrow$	$2\ {}^4He$	+17.3 MeV
$^9Be + {}^1H$	$\longrightarrow$	$2\ {}^6Li + {}^4He$	+2.1 MeV
$^{11}B + {}^1H$	$\longrightarrow$	$3\ {}^4He$	+8.7 MeV
$^{11}B + {}^1H$	$\longrightarrow$	^{12}C	+15.9 MeV

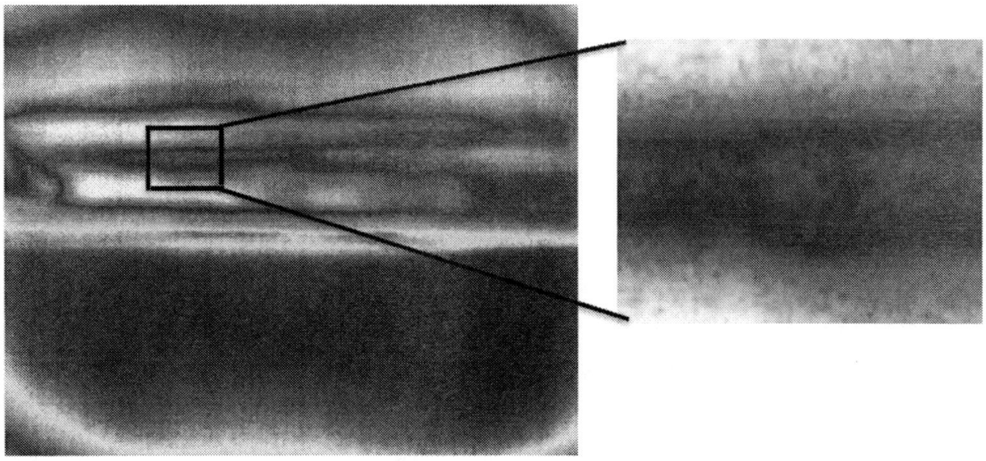

Figure 7. Preliminary infrared image.

be reacted sufficiently to show an increase in the B^{10}/B^{11} ratio, then the working hypothesis discussed above would be enhanced. It could be bolstered by an equivalent, or simultaneous, electrode containing deuterium instead, should that ratio decrease. A rough calculation shows that a cell like that discussed with Fig. 4, except using 0.25% by weight boron in copper electrodes, would have to be operated for more than 120 days at 30 mW nuclear heat to produce a change of 1% in the B^{10}/B^{11} ratio. The total nuclear energy evolved would also shine light on the nuclear chemistry.

9. Infrared Imaging of an LENR Cell

One impediment to the believability in LENR is the mystery of how a release of more than 8 MeV per nuclear event can be absorbed by heating a metal without producing any externally observable radiation. Such radiation would be in the X-ray or harder range (above about 1 keV (10^{16} Hz, wavelength less than 10 nm). If an active electrode could be imaged in the infrared through ultra violet range (10 μm–10 nm, 1 eV–1 keV) then substantial insight into the phenomenon should result. The longer the wavelength used in the imaging the better because shorter waves will degrade through phonon interaction.

A system has been set up around a digital camera with a vanadium oxide detection array that can image wavelengths from 7 to 14 μm. An evacuated Seebeck calorimeter was built along the previous design but containing a 5 mm (0.2 in.) thick zinc sulfide window that is transparent in the wavelength range of the camera. A test cell quite similar to the previous design was prepared except one somewhat transparent electrode was substituted. That electrode consists of a 2 mm (0.08 in.) thick plate of ZnS upon which a gold film has been deposited. The commercial gold film is a few μm thick and the mechanical application process resulted in many holes in the deposit. The previously used dielectrics inserted between electrodes (vacuum pump oil, vaseline) are opaque to infrared so glycerin was substituted. Ethylene glycol might have been used instead. The same fine graphite was suspended in the glycerin as previously, but in somewhat smaller concentration to minimally obstruct any infrared.

False-color images from the infrared camera focused on the active cell, as described above, are shown in Fig. 7. The larger picture on the left is approximately full size. The farther electrode of the cell is the greenish material between the red/yellow zones. The smaller image to the right is an electronic blow up of a portion of the other photo, as indicated by the black lines. It barely resolves individual picture elements (pixels) which, in the 640 × 480 array, are about 150 μm (0.006 in.) square. So far, this work has not produced evidence of soft emanations produced by LENR. Image series in video show some twinkle, but not more from the cell than from other locations. Better transparent electrodes may improve the images.

The supplier of our ZnS makes only electronically intrinsic material. If n or p semiconducting ZnS could be found, the electrical conduction difficulty might be resolved. The n or p doping would have to be carefully controlled to minimize infrared opacity. Images focused on the edge of the cell should be made. Not only would this obviate the need for a transparent electrode, but it may be that the soft emanations are directed in the plane of the electrode. If some radiation came from the edge, rather than exiting the surface, much would be explained.

10. A Resonant Cell to Substantially Reduce Input Power while Maintaining Nuclear Output

The low-Q capacitor method described above is inherently inefficient because most of the input power goes to joule heat the conductor layer between electrodes. The input could be substantially reduced, at equivalent or greater current, by using a good capacitor made of the hydrogen-infused alloy, in an electronically resonant circuit.

Such a capacitor has been made from eleven plates of hydrogenated copper 0.7% boron alloy. The plates are each about 0.12 mm (0.005 in.) thick by 5 cm (2 in.) square. Silicone saturated baking parchment is used as the dielectric between alternately connected plates, to form the capacitor. After trials of some paints and plastic films, this parchment was selected because it is thin (about 25 μm, 0.001 in.) is a good dielectric, useful above 200°C, and is fairly easy to apply. Teflon impregnated paper is available and equivalent except for its somewhat inferior dielectric constant. This capacitor, made with ten layers of parchment has a capacitance of about 12.5 nF and its resistance exceeds 20 MΩ. The total area of dielectric (about 250 cm^2, 40 in.^{2}) is about 80 times the area of the dual-leaf low-Q capacitors previously used. If the 30 mW nuclear power of the best low-Q capacitors is proportionally maintained in a scaled-up resonant system, it would take nearly 24 h to heat 500 ml of ambient water to boiling.

That moderate power is, however, sufficient to obviate any need for sophisticated calorimetry. Accordingly, a system is being developed in which heated water can be drawn off periodically and be replaced by ambient temperature water. The power is determined by the amount and temperature of the water drawn off and the heating time. This is much like many of the classic Pons/Fleischmann experiments. That method should also allow for thermoelectric generators (TEG) to be inserted thermally between reservoirs of heated and ambient water. An objective is for the TEG to generate sufficient power to actuate the input electronics, at least after equilibrium has been reached. Then all batteries and external power sources could be removed and water will continue to be heated. That should put to rest all doubt about the validity of LENR. It will also show convincingly a way toward commercialization. No clear results have yet been produced. Frequency and wave form are being optimized using an external wave generator, without connecting an inductor to produce resonance. A search is being made for a commercial wave generator/amplifier, that will be efficient and well-matched to the optimized input signal.

11. Conclusions and Summary

Low Energy Nuclear Reaction (LENR) has again been demonstrated with an example provided. Copper alloys, infused with hydrogen at high temperature and moderate pressure, were used. A process for demonstrating this effect was outlined.

An explanation of some LENR observations was provided with a working hypothesis to organize further research. The assumptions are basically that hydrogen dissolved in some metals can react with specific light alloying elements to produce helium and more than 8 MeV of nuclear energy. *These reactions produce no dangerous products.* They require suitable metallurgical structures and electric charges moving therein. Modifications to a proposal by Storms [2] involving the Nuclear Active Environment (NAE) and hydrotons were employed in this hypothesis. Equipment to image, at infrared wavelengths, cells producing anomalous energy has been assembled and was described here. An example of a photograph from that assembly was provided. The optical results have been inconclusive, so far. The effort continues. A system to radically improve LENR efficiency in copper alloys is being developed. It is expected that electronic resonance will result in devices to make LENR without any external energy source.

References

[1] William H. McCarthy, Water-free replication of Pons–Fleischmann LENR, *J. Condensed Matter Nucl. Sci.* **15** (2015) 256–267.

[2] Edmund Storms, *The Explanation of Low Energy Nuclear Reaction, An Examination of the Relationship between Observation and Explanation*, Infinite Energy Press, 2014, pp. 220–230.

J. Condensed Matter Nucl. Sci. 29 (2019) 202–210

Research Article

Nanosecond Pulse Stimulation in the Ni–H_2 System

Francis Tanzella*,†

Energy Research Center LLC, San Carlos, CA 94070, USA

Robert George and Robert Godes

Brillouin Energy Corporation, Berkeley, CA 94705, USA

Abstract

Brillouin Energy and SRI International (SRI) have been performing calorimetry measurements on the Ni(Pd–Ni/ceramic/Cu(Ni) coated tubes in a H_2 atmosphere with nanosecond pulses applied across the ceramic coating. We have been testing new materials, material fabrication techniques, and electrical stimulation methods to produce power and energy output in excess of that reported earlier. By applying fast pulses of several hundred volts and tens of nanoseconds long, the current follows the "skin-effect" principle and is concentrated at the Ni–ceramic interface but returns through the bulk of the Cu. Two stimulation methods were used – steady-state and dynamic. In the steady-state method, the pulse power is measured directly using fast oscilloscopes that record the voltage across the tube and a shunt resistor in series with the tube. The resistance of the shunt resistor is measured accurately under DC and pulse conditions. The input pulse power is determined by multiplying the calculated root-mean-square voltage and current and recorded every 10 s. Using a sophisticated model of the calorimeter with up to 15 coefficients, the power reaching the five temperature sensors is determined during simultaneous continuous ramps of both heater and pulse powers. The power emanating from the tube is determined during sequences of more frequent low voltage pulses (LVP) and compared to that found using less frequent high voltage pulses (HVP). The power determined during the more frequent LVP is set as the input power during that sequence. The power of the stimulation pulses during the less frequent HVP sequences is maintained equal to that during the more frequent LVP. Then the power calculated from the tube is divided by that calculated during the reference sequences, giving a so-called coefficient of performance (COP). We have shown an increase in both absolute LENR power produced and in COP.

Keywords: Calorimetry, Excess energy, Hydrogen gas, Nickel, Pulse stimulation

1. Introduction

For over five years, SRI International (SRI) has been performing tests on two different versions of Brillouin Energy Corp.'s low-energy nuclear reactors (LENR) [1]. We have operated these reactors independently in an attempt to

*Corresponding author. E-mail: consulting@tanzella.name.
†Formerly from SRI International, USA.

© 2019 ISCMNS. All rights reserved. ISSN 2227-3123

verify results that Brillouin has found with these reactors and others like them. We have also monitored and advised Brillouin on the results from reactors operated by Brillouin in their own laboratory. This report updates the results [2] obtained by studies in SRI's laboratory, as well as verification and validation of results obtained in Brillouin's laboratory over the past two years. Brillouin has indicated that it has designed the control systems in its reactors to drive the underlying physics of LENR, as described in its Controlled Electron Capture Reaction (CECR) hypothesis [3]. The CECR hypothesis explains how scientists at Brillouin believe their reactors generate controlled LENR reaction heat. Our study did not attempt to prove or disprove Brillouin's CECR hypothesis.

The systems tested and described in this report consist of three parts: tubes; reactors; and calorimeters. The tubes are the reactive components of the system. The reactors provide the environment and stimulation that causes the tubes to produce reaction heat. The calorimeter is used to measure the thermal efficiency and absolute heat produced by the tube-reactor system. The calorimeter was designed by both SRI and Brillouin personnel to be perfectly matched to the reactor. The results from four of these reactors are described in this report.

Brillouin's system design utilizes compensation calorimetry, in which the tube and reference temperatures are held constant by varying the input heater power while applying different types of stimulation that also input power to the reactor/calorimeter. Recently Brillouin Energy started using a thermal model based on heat input and loss identification developed by an independent commercial third party. This "dynamic" method of analysis allows us to analyze all power entering or affecting the tube as well as all power emanating from the tube based on differential equations describing temperatures and power measurements. While this requires 100 h of calibration and up to 40 h of excitation to verify a calibration, it allows testing of 12-parameter variations per hour versus one or two in the traditional steady-state method.

2. Experiment

2.1. Design

The tubes consist of a metal or ceramic substrate, which in some configurations includes a heater and thermocouple with several spray-coated layers. Generally, these coatings alternate between a hydrogen-absorbing metal and an insulating ceramic. One example is shown in Fig. 1. Other designs have used more or fewer layers. All of the layers are porous, allowing the gas(es) in the reactor chamber access to all coatings. A heater, if present, and a thermocouple are located in the center of the tube. The power to the heater is measured directly from the voltage and current supplied by the direct current (DC) power supply. A photograph, schematic, and description of the reactor/calorimeter system can be found in our earlier report [2].

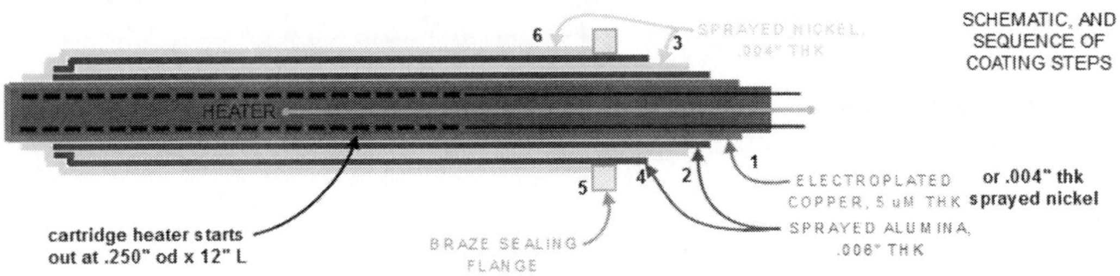

Figure 1. Example of Brillouin's recent hydrogen hot tubes.

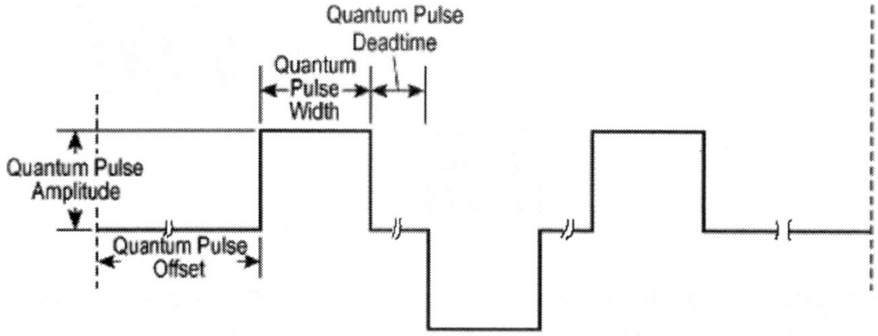

Figure 2. Example of Brillouin's "Q-Pulse".

2.2. Measurement

The outer active layer is stimulated by sending pulses through the outer layer or layers and returning electrically through the innermost layer. The nature of the pulses is such that its current travels primarily on the surface of the metal in contact with the ceramic (the "skin effect"). This effect is caused by the very fast rise time of the pulses. An example of this pulse design, which Brillouin refers to as a "Q-Pulse", is shown in Fig. 2. The pulse width is presently used is from ~30–10,000 ns with a duty cycle normally of less than 1%. More detail on the pulse trains are shown here [3].

The stimulation power imparted to the tube is measured using a circuit shown in Fig. 3. The pulse is generated by a proprietary Q-Pulse board and delivered to the tube using series and termination resistors that help match the load impedance to that of the pulse board output. Using a high-speed oscilloscope, the voltage across the end of the tube nearest the pulse board (V_1) is measured as well as the voltage across the opposite end of the tube (V_2) across the termination resistor (Z_{term}). The Z_{term} also acts as a current measuring resistor so the current is calculated as

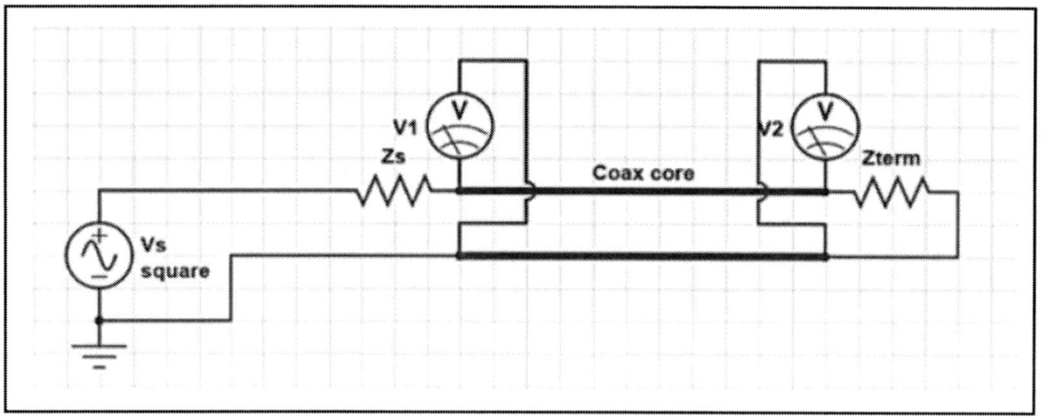

Figure 3. Pulse power measurement circuit.

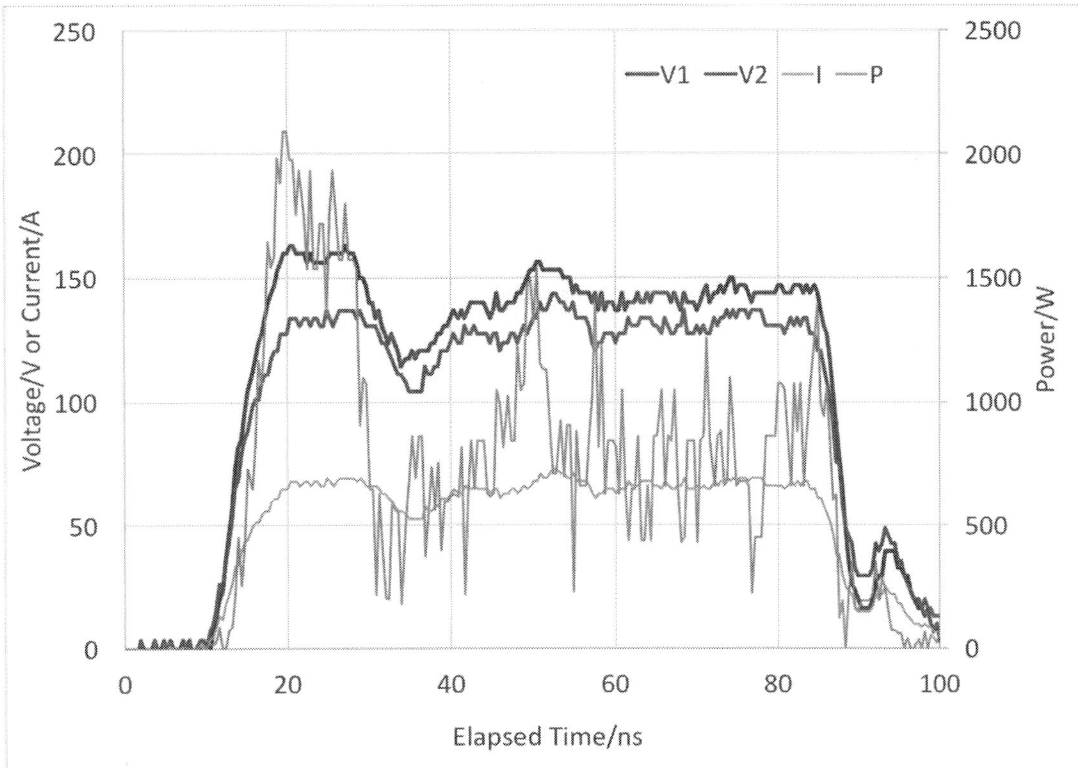

Figure 4. Measurement of the Q-pulse power across the tube.

V_2/Z_{term}. The root mean square (rms) voltage across the Z_{term} is then converted to the rms current.

The power imparted to the tube is determined using the red and blue voltage traces shown in Fig. 4. The difference between the two voltage traces is calculated after aligning them in a way that minimizes the time difference. This overestimates the power imparted to the tube by a small amount since any phase lag between voltage and current would impart less input power. The current calculated from V_2 is shown in black and the product of it with the voltage difference (power) is shown in green. It has been shown that the power calculation is essentially the same (within measurement error) whether it is calculated by multiplying the current and voltage plots point by point or by multiplying the calculated rms voltage by the rms current. This calculated power is referred to as Core Q Power.

In compensation calorimetry the heater power is varied to keep either the tube or inner block at constant temperature, which generally also keeps the other at a constant, but slightly different, temperature. The difference between the heater power with and without stimulation determines the effect of the stimulation. If this difference is greater than the stimulation that reaches the tube, then energy is being produced in the tube. Approximately 50 different parameters are collected allowing for calculation of Reaction Power (the power produced by the process induced by the pulse stimulation). Several calculation methods are possible from these parameters. In addition, two different stimulation sequences are used. In the Analysis section we describe these two sequences and the calorimetry method used for each of them.

2.3. Operation

A description of the data acquisition system with a copy of the graphical interface has been described earlier [2]. The program has several panes allowing for control of temperature, pressure, pulse voltage, pulse power, pulse width, and pulse repetition rate and gas composition. The program also collects the heater power, the pulse power at the generator (as well as at the tube), all temperatures, water flow rates, and gas pressure. The concentration of hydrogen and oxygen in the argon blanket are collected and measured. In all, 50 different parameters are collected and stored every 10 s. A sequence file can be used to automatically change any or all of the input parameters at specified intervals over a multi-day or multi-week period.

The sheath containing the tube is operated with a static fill of hydrogen, and occasionally helium or argon, gas held at constant pressure up to 15 bar. The temperature of the tube is held constant using its embedded heater and thermocouple and controlled from 200 to 600°C. The outer block temperature is held at 25°C using water flow from a Neslab® recirculating constant-temperature chiller.

The power emanating from the Q-pulse generator board, or that applied directly to the tube, is held constant as chosen by the program's front panel or the sequence file. Generally, the pulse amplitude (voltage) and pulse width are chosen. The repetition rate is adjusted automatically to maintain the chosen pulse power. Only a minor fraction of this power from the generator board reaches the tube as most of it is lost as heat in the termination resistor. This is necessary to get an accurate measure of energy actually dissipated in the tube and to match the load impedance to that of the generator thus preventing reflections that could cause measurement errors. Of that reduced power only a portion of it influences the heater power as explained in the "Measurement" subsection above. The actual pulse power is measured directly via the methodology presented above.

During this project several stimulation methods were tried to find one that can act as a blank (no excess power) using similar Core Q-Power. The DC resistive heating of the tube surface coat, was used occasionally and required changing electrical connections and using measurement hardware different from that used for the real-time Core Q-Power calculation. Ideally these methods need to be compatible with the data collection's software's calculation designed for the low duty cycle Q-pulse square waves and not require hardware changes. Some of the methods tried were: (1) straight sine waves, (2) low duty cycle square waves, and (3) large pulse widths with long rise times. Ultimately, calibration runs used Q-pulse parameters that were known not to produce LENR heat low voltage pulses (LVPs) but impart the same power to the tube as parameters expected to show LENR heat, i.e. high voltage pulses (HVPs). The hardware was modified to allow pulsing at much higher repetition rates.

Operating in power compensation mode, the computer keeps the inner tube, or outer heat spreader, temperature constant at its set point. When power is imparted from the Q-pulse, the heater power is reduced to compensate and maintain a constant temperature. Hence, the tube temperature and the inner and outer block temperatures are all held constant when using the same tube gas.

Operating at constant gas pressure, a sequence was operated from 200 to 600°C in 50°C intervals. At each temperature a given DC power was applied to the coating on the tube. This process was then repeated but applying constant power pulses varying pulse width at each temperature.

Two major methods of operation were employed, each requiring a different analysis method. The first method operated with the reactor at a steady-state temperature and input powers, which we refer to as the steady-state stimulation (SSS) method. In our second approach, the dynamic stimulation (DS) method, the heater power was ramped smoothly through a maximum and back down while smoothly ramping Q-power up and down several times. The DS method was developed to allow for many Q-pulse parameters to be tested in less time. On occasion we would interrupt the DS to allow the system to achieve a steady-state for several hours.

The SSS method was operated in power compensation mode, where the computer kept the temperature constant at either the tube or the inner block. When power was imparted from the Q-pulse, the heater power was reduced to compensate and maintain a constant temperature. Hence, when the inner and outer block temperatures are held constant, the tube temperature will respond to the stimulation. The output power (calculated from the inner minus outer block temperatures) did not change as the input power compensates for the total power emanating from the tube. The total tube power included the stimulation power and the power due to reaction heat (i.e. LENR power).

2.4. Analysis

The earlier report [2] describes two different analysis methods employed in the effort. Here we describe the stimulation and analysis methods employed most recently.

2.4.1. *SSS method*

In this method, the absolute heater power necessary to maintain constant temperature without Q-pulses present is not part of the output power calculation. We realize that only a fraction of the heater power may be imparted to the tube because the heater/thermocouple combination has measurable losses to the rest of the calorimeter and to the environment. Instead the temperature controller is instructed to keep the inner block at a constant temperature while low voltage calibration pulses are imparted to the tube and measuring the heater's response at different temperatures. The difference between the heater power with and without the LVPs voltage pulses (LVP) is called P_{drop}. At each temperature, a linear function ($P_{\text{drop}} = mP_{\text{LVP}} + b$) is determined. The b offset parameter is always insignificant and is not used in the analysis.

P_{drop}, also called heater power compensation (HPC), is determined for different amounts of LVP calibration power. This method is analogous to the traditional isoperibolic calorimeter analysis except that it substitutes heater power compensation for the temperature difference. In order to calculate Q_{reaction} as output power minus input power, we compare the heater power compensation (HPC) from LVP calibration to that from HVP stimulation. Using this LVP calibration the relationship between input power and HPC is determined so that with input pulse power the HPC can be used to back calculate the power from the pulses imparted into the tube. Q_{LVP} and Q_{HVP} are the actual Q-pulse powers measured when low voltage and high voltage HVPs are applied, respectively.

First, the linear relationship between HPC and Q_{LVP} is found by fitting a linear equation to HPC vs Q_{LVP} when Q_{LVP} is varied across the same range of powers as Q_{HVP}. These linear coefficients are then applied to the measured Q_{HVP} to calculate HPC (LVP), the amount of HPC measured at the same temperature and pulse power at low voltage, where no reaction heat is expected. Q_{reaction} is then calculated as shown in Eq. (1), where HPC (HVP) is the actual HPC measured when the HVP is applied. Equation (2) is then used to calculate COP. An alternate calculation is shown in Eq. (3). In the latter equation, the COP is calculated as the ratio of the HPC over the Q-pulse power at high and low voltage.

$$Q_{\text{reaction}} = \text{HPC(HVP)} - \text{HPC(LVP)}, \tag{1}$$

$$\text{COP} = Q_{\text{reaction}}/Q_{(\text{LVP})} = (\text{HPC(HVP)} - \text{HPC (LVP)})/Q_{\text{LVP}}, \tag{2}$$

$$\mathrm{COP} = (\mathrm{HPC(HVP)})/Q_{(\mathrm{HVP})}/(\mathrm{HPC(LVP)})/Q_{\mathrm{LVP}}. \tag{3}$$

2.4.2. *DS method*

The DS method employs a model with several components, each representing individual components of the calorimeter. Linkages between these components (and from a component to the reference room temperature) are either conductive or storage. Temperatures are measured between the heat spreader and the tube sheath (inner block) and on the outside of the cylindrical heat spreader (outer block). One differential equation (in time) models the heat imparted to the tube using a function of the difference of the tube and outer block temperature, a function of the tube and inner block temperatures, and divided by the function of the ability of the tube to store heat. Each of these three components have a coefficient that is determined fitting the temperature data to the actual power measured using LVPs as described above. A second equation does the same for the inner block. The model then yields a simple equation for power equal to a coefficient time the difference between the tube and inner block temperatures.

These functions are simple 3-coefficient binomial equations, yielding 15 possible parameters. These parameters are then used to calculate the amount of heat emanating from the tube during an attempt to produce LENR heat. A comparison between the calculated power emanating from the tube during an active run and that from the calibration run at the same temperature and with the same Core Q Power is used to determine the amount, if any, LENR heat was produced. When the DS runs were interrupted to achieve a steady-state, the amount of heater power determined to affect the tube was subtracted from both the input and output powers before calculating COP. The computer application MatLab® is used to determine the best fit parameters.

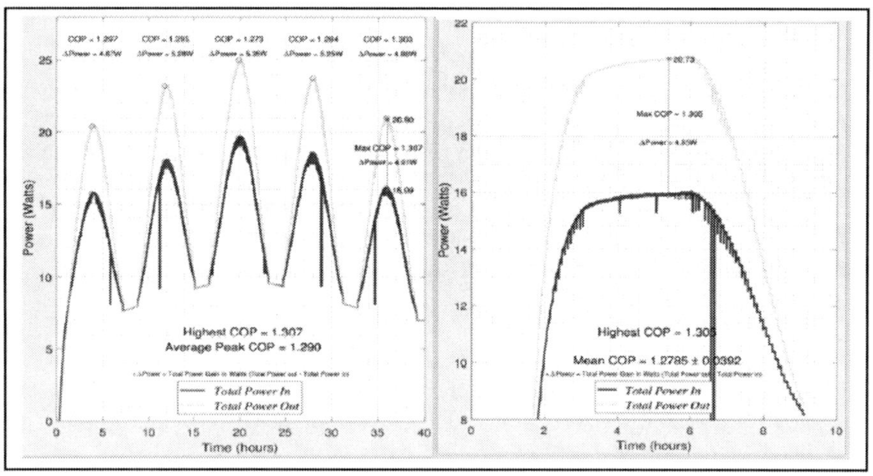

Figure 5. (a) Instantaneous COP during DS run and (b) COP during 4 h at maximum stimulation.

b]

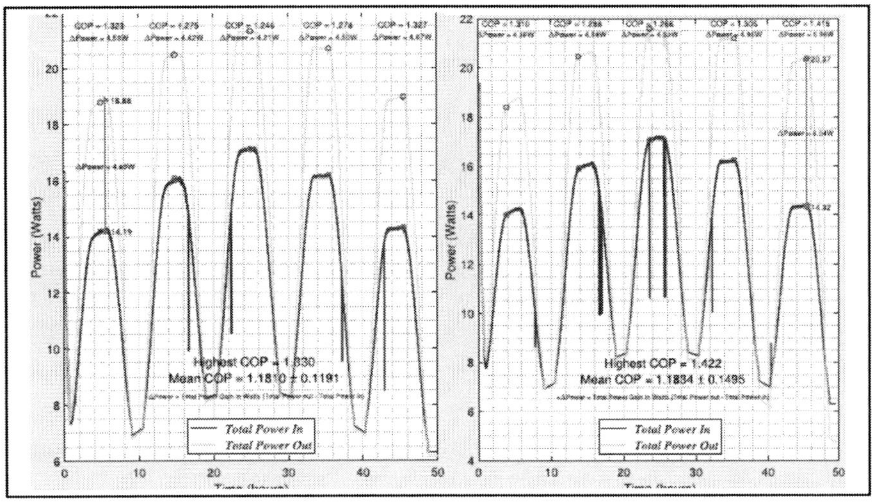

Figure 6. COP during two different DS runs with 2 h at maximum stimulation.

3. Results and Discussion

A tabulation of all experiments run to date in the IPB calorimeters is available on the Brillouin Energy web site [4]. Only a fraction of the total number of experiments will be covered in detail. The runs detailed in this analysis of SSS runs generally used a 100 ns pulse width with similar Q-power on the tube at different voltages. We attempted to keep the CoreQPow relatively constant at each temperature.

Table 1 summarizes the COP results from recent SSS runs. Using the COP as defined in Eq. (3), this shows the COP to be much greater at 250°C (1.27) than at 400°C (1.00). The power compensation/CoreQPow is very dependent on the pulse voltage at 250°C but is essentially unchanged at 400°C. Although the total pulse power from the generator is constant, the pulse power measured at the tube does vary with pulse voltage, even though we attempted to keep Core Q-Power constant by also varying the repetition rate and/or pulse width. Still, the magnitude of the power compensation is a greater percentage of the pulse power at 350 V than at 35 V. Calculations show that Q_{reaction} is greater at 350 V than at 35 V (e.g. 5.3 W versus 3.5 W in one run).

The DS method employed recently calculated two different COP results. The first was the average COP over the complete stimulation run. The other result calculated instantaneous COP, especially at the point of greatest stimulation amplitude. Two tubes showed particularly good COP's both in the average and instantaneous calculations. In general, the highest COP's were obtained using pulses of amplitude 325–350 V and $\leq$100 ns pulse width with <1% duty cycle. Minimal COP's were measured using 35–50 V and 6–10 μs wide pulses with approximately 10% duty cycle. All repetition rates (from 10 kHz–2 MHz) were adjusted to yield 4–6 W average pulse power measured on the tube. Recent runs have shown even higher COP's without the use of heater power.

Figure 5 shows an example of the instantaneous COP calculated using the DS analysis method from a recent run. To prove that this COP is stable the stimulation parameters were held constant for 4 h. This latter result is shown in Fig. 5(b). Using the method similar to that used last year is used to calculate the COP, we get significantly larger results. It is also important to note that the absolute LENR powers (Qreaction) are significantly larger than seen earlier. We also

Table 1. Summary of COP calculations from steady-state stimulation runs.

Temperature/°C	COP: IPB2-33	COP: IPB2-74	COP: IPB1-45	COP: IPB1-48
250	1.27	1.14		
275	1.4	1.15	1.11	1.13
f300	1.25	1.13	1.11	
325	1.26	1.09	1.08	1.27
350	1.05	0.94		
400	1.00	.89		

found that the best COP's were achieved between 250 and 350°C, confirming the results that were reported earlier.

The active tube, whose results are presented in Fig. 5, was removed from service at Brillouin's laboratory during December 2017, and placed in service at SRI in a different reactor during April 2018 and showed similar results. These latter plots are shown in Fig. 6.

4. Conclusion

LENR can produce thermal power when Ni and other metal-coated tubes are stimulated using fast rise-time pulses. These experiments were operated in H_2 gas from 200 to 600°C. Comparative thermal measurements were performed between heater-only power and heater and pulse power. These runs were performed in isoperibolic calorimeters operated in power compensation mode, where the heater adjusts its power to keep the inner and outer temperature-difference constant. Over 500 runs were performed on over 70 different Ni-coated tubes. Additional tubes were also tested for other experimental purposes. Recent efforts were spent optimizing tube design, stimulation protocols, calorimeter design, and calorimetric analysis methods. The accuracy and precision of these recent results have been considerably better than those reported earlier.

Reactor tubes can be transported between different laboratories and using different reactors to ta achieve very similar positive results.

Acknowledgements

I would like to acknowledge Dr. Michael McKubre (SRI Emeritus) for his work on the calorimeter design and thank Brillouin Energy scientists Jin Liu, Roger Tong and Dave Correia for their aid in the calorimetric analysis. SRI also acknowledges the Brillouin Energy Corp. for their generous support in this work.

References

[1] R. Godes, R. George, F. Tanzella and M. McKubre, Controlled electron capture and the path to commercialization, *J. Condensed Matter Nucl. Sci.* **13** (2015) 127.

[2] R. Godes, R. Herrera, C. Everleigh and F. Tanzella, Controlled electron capture: Enhanced stimulation and calorimetry methods, *J. Condensed Matter Nucl. Sci.* **24** (2017) 301.

[3] R. Godes, Drive circuit and method for semiconductor devices, US Patent 8,624,636, 2014.

[4] F. Tanzella, Isoperibolic hydrogen hot tube reactor studies, Final progress report from SRI International, July 2018, www.brillouinenergy.com/.

J. Condensed Matter Nucl. Sci. 29 (2019) 211–218

Research Article

Anomalous Isotopic Distribution of Silver in a Palladium Cathode

Jean-Paul Biberian*

Aix-Marseilles University, France

Abstract

This paper demonstrates that during cold fusion experiments by electrolysis of heavy water with palladium cathodes, there is production of silver. By Secondary Ion Mass Spectroscopy, it is shown that silver detected is not due to contamination, but probably to transmutation of palladium by absorption of a deuterium nucleus, because only the ^{107}Ag-isotope is present, and not ^{109}Ag, as in natural silver. Another important result of this study is the determination of the depth at which the reaction occurs: 1.3 μm below the cathode surface.

Keywords: Electrochemistry, Isotopes, Palladium, Silver, Transmutation

1. Introduction

Since the discovery of Cold Fusion by Fleischmann and Pons [1], an important question remains: what are the reactions responsible for excess heat? Helium-4 has been measured and correlated with excess heat [2], but also silver was detected, which was not present in the original palladium cathode [3]. Therefore, it is very likely that in addition to the fusion reaction of deuterium, there is also transmutation and production of silver by transmutation of palladium. These measurements having been made by EDX in Scanning Electron Microscopes, no isotopic analysis was available [3–5]. This is important because the best proof that silver is due to transmutation, is its possible anomalous isotopic distribution. EDX produces only chemical information. Mass spectroscopy is necessary to measure isotopic distribution.

During their stay in France at IMRA, Pons and Fleischmann did a number of very successful experiments with the ICARUS-9 calorimeter [6]. In 2001, Stanley Pons gave me a palladium cathode which according to him produced a lot of excess heat. I kept it many years. In 2018, I had access to a Secondary Ion Mass Spectroscopy (SIMS) equipment, and an analysis was made. In this paper it is shown that silver detected in this palladium electrode by SIMS has an anomalous isotopic distribution.

*Retired. E-mail: jpbiberian@yahoo.fr.

2. Experimental

The palladium cathode came from an experiment performed in an ICARUS-9 type calorimeter [6]. Figure 1 shows a schematic of the calorimeter. It is composed of a Pyrex tube 25 mm diameter, 50 cm long, inserted into a double wall aluminum casing. The two walls are separated by an insulating material. Inside the Pyrex tube the cathode is a palladium rod 10 cm long and 2 mm in diameter, surrounded by a platinum anode wire. The electrolyte is composed of 100 ml D_2O and 0.1 M LiOD. The calorimeter is placed inside a constant temperature water bath. Temperatures are measured by three pairs of thermistors located at different levels along the tube. For each pair one thermistor is located near the Pyrex tube, and the other one near the surface of the aluminum casing. Finally, one is inserted inside the Pyrex tube near the cathode to measure the temperature of the electrolyte. After several weeks of electrolysis at low current (20 mA), the current is raised, and the cell boils. There is a condenser section in the upper parts of the calorimeter, so that the water vapor condensates on the walls of the Pyrex tube and fed back to the electrolyte. The main advantage of this design is that it permits boiling for a long period of time.

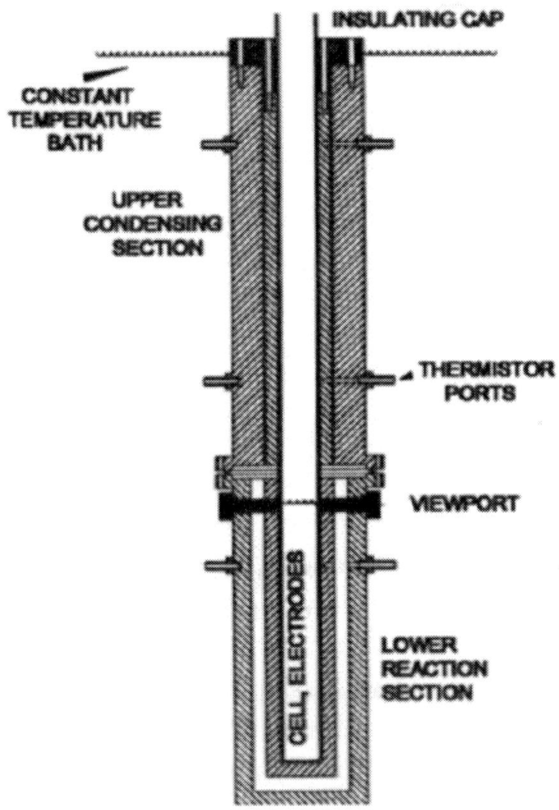

Figure 1. Schematic of the ICARUS-9 calorimeter.

3. Search for Silver

Palladium and silver are next to each other in the periodic table. The natural isotopes of palladium are: ^{102}Pd, ^{104}Pd, ^{105}Pd, ^{106}Pd, ^{108}Pd and ^{110}Pd. Those of silver are ^{107}Ag and ^{109}Ag. Therefore, there is no mass interference between Pd and Ag. Two types of mechanisms can be imagined explaining the possibility of transmutation of palladium into silver. The first one assumes the fusion of a palladium nucleus with a deuterium nucleus, and the other one, fusion of a palladium nucleus with a di-neutron. This latter mechanism supposes that a deuterium nucleus transforms itself into a di-neutron by an unknown mechanism, which can then easily fuse with a palladium nucleus. The first mechanism is difficult to imagine, due to the strong repulsive Coulomb force, without some unknown catalytic mechanism.

3.1. Assumption 1: Pd and D fuse

If the stable palladium isotopes and deuterium fuse, the following reactions would occur.

$$
\begin{array}{lclcl}
^{102}\mathrm{Pd}\ (0.96\%) + {}^{2}\mathrm{D} & \rightarrow & ^{104}\mathrm{Ag} & \rightarrow & ^{104}\mathrm{Pd}, \\
^{104}\mathrm{Pd}\ (11.0\%) + {}^{2}\mathrm{D} & \rightarrow & ^{106}\mathrm{Ag} & \rightarrow & ^{106}\mathrm{Pd}, \\
^{105}\mathrm{Pd}\ (22.2\%) + {}^{2}\mathrm{D} & \rightarrow & ^{107}\mathrm{Ag} & & \mathrm{Stable}, \\
^{106}\mathrm{Pd}\ (27.3\%) + {}^{2}\mathrm{D} & \rightarrow & ^{108}\mathrm{Ag} & \rightarrow & ^{110}\mathrm{Pd}, \\
^{108}\mathrm{Pd}\ (26.7\%) + {}^{2}\mathrm{D} & \rightarrow & ^{110}\mathrm{Ag} & \rightarrow & ^{112}\mathrm{Pd}, \\
^{110}\mathrm{Pd}\ (11.8\%) + {}^{2}\mathrm{D} & \rightarrow & ^{112}\mathrm{Ag} & \rightarrow & ^{112}\mathrm{Cd}.
\end{array}
$$

The palladium isotopes would transmute into unstable silver isotopes (except ^{107}Ag). They would quickly transmute into stable palladium isotopes, and stable ^{112}Cd. What is interesting here is the absence of ^{109}Ag, or any other mass 109 element.

3.2. Assumption 2: Pd and ^{2}n fuse

If the stable palladium isotopes and a di-neutron fuse, the following reactions would occur. Palladium isotopes would transmute into stable palladium isotopes except for ^{107}Pd which has a very long half-life, and ^{112}Pd which would quickly transmute into stable ^{112}Cd. What is interesting here also, is the absence of ^{109}Ag.

Because the masses of ^{107}Ag and ^{107}Pd are very close, it would be impossible to distinguish them with the SIMS equipment used in this study. Therefore, if palladium transmutes into silver, we should see only mass 107, which could be either silver or palladium. Therefore, with this type of analysis, it is not possible to distinguish between fusion of palladium with D or ^{2}n. However, in both cases there should be formation of ^{112}Cd. It is important to note that in both mechanisms, there is absence of mass 109.

$$^{102}\text{Pd (0,96\%)} + ^{2}\text{n} \rightarrow {}^{104}\text{Pd},$$

$$^{104}\text{Pd (11,0\%)} + ^{2}\text{n} \rightarrow {}^{106}\text{Pd},$$

$$^{105}\text{Pd (22,2\%)} + ^{2}\text{n} \rightarrow {}^{107}\text{Pd (half-life } 6.5\times 10^{6} \text{ years)},$$

$$^{106}\text{Pd (27,3\%)} + ^{2}\text{n} \rightarrow {}^{108}\text{Pd},$$

$$^{108}\text{Pd (26,7\%)} + ^{2}\text{n} \rightarrow {}^{110}\text{Pd},$$

$$^{110}\text{Pd (11,8\%)} + ^{2}\text{n} \rightarrow {}^{112}\text{Pd} \rightarrow {}^{112}\text{Ag} \rightarrow {}^{112}\text{Cd}.$$

4. SIMS Analysis

The dynamic SIMS equipment used for the analysis was a Cameca 4f. Figure 2 shows a photograph of a piece of the cathode that was analyzed, and Fig. 3 shows SEM images of the surface at various magnifications.

The Cameca 4f instrument is a dynamic SIMS, composed of an oxygen ion beam which etches the surface and a high-resolution mass spectrometer that collects the emitted ions generated by the beam. Therefore, it is possible to have mass analysis as a function of etch time. When etch time is calibrated using a calibration sample of known thickness, then it becomes possible to determine elemental composition versus depth. Masses are either atoms or molecules. It is possible that deuterium remaining in the cathode could give false identification. Indeed, a molecule composed of ^{105}Pd and D would have a mass of 107. in order to eliminate this interference, the cathode was heated at 600°C in air prior to SIMS analysis.

To calculate the depth profile, the etch rate was calibrated using a 100 nm film of palladium deposited on a silicon wafer. Not knowing the actual composition of the cathode, one piece of the cathode was hand filed in order to reach the bulk of the cathode and be analyzed by SIMS. As shown in Fig. 4, there was no silver present, only palladium. Figure 4 shows a depth profile for comparison between ^{106}Pd, ^{107}Ag and ^{109}Ag in the bulk of the cathode. We can see that there is almost no silver present.

Figure 5 shows a depth profile of the palladium cathode. Masses corresponding to ^{106} Pd are shown in black; ^{107}Ag is in red; and ^{109}Ag is in green. Natural silver has the following ratios: ^{107}Ag (51.4%) and ^{109}Ag (48.7%). The masses are almost equal. However, in Fig. 5, it can be seen that at the surface mass 107 is three times larger than mass 109.

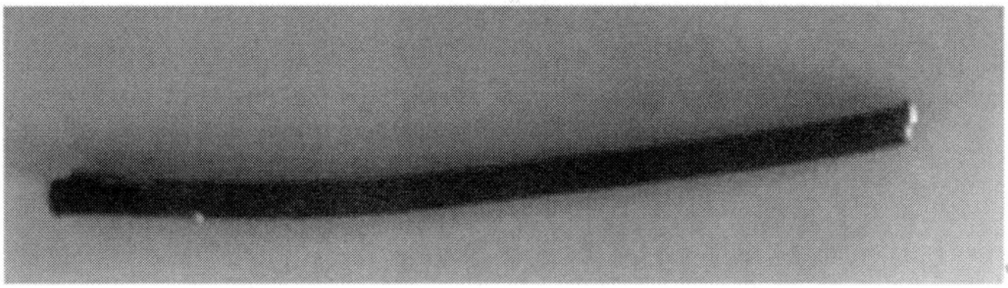

Figure 2. Photograph of a piece of the palladium cathode.

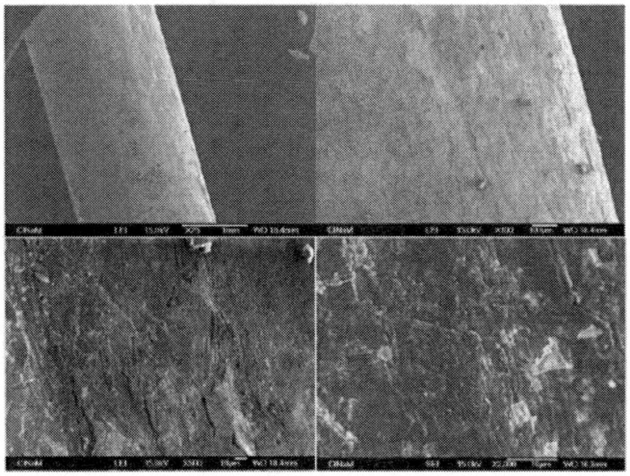

Figure 3. SEM images of the surface of the palladium cathode at various magnifications.

After about 400 s of etching, ^{107}Ag and ^{109}Ag peaks are almost equal, as expected with natural silver. This indicates that there is an excess of ^{107}Ag at the surface.

The surface of the palladium cathode is not uniform. We have discovered hot spots. In Fig. 6, the depth profile

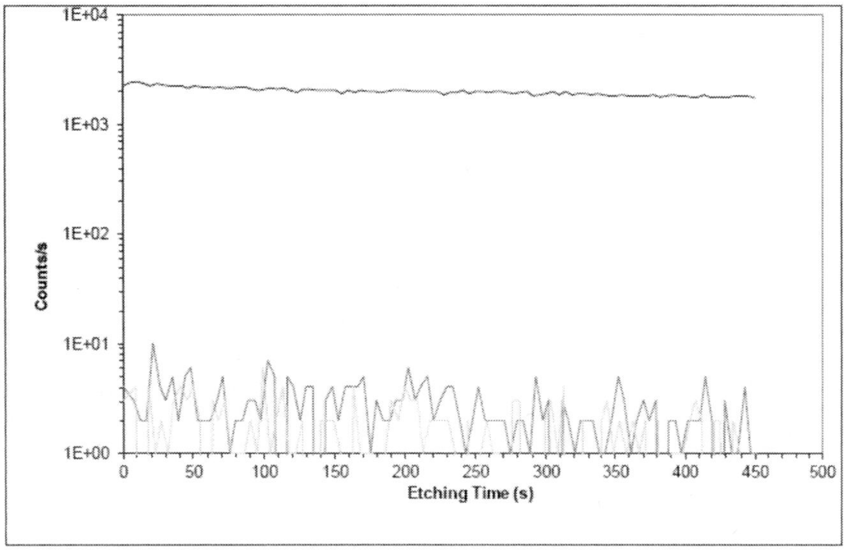

Figure 4. Depth profile of the bulk of the palladium cathode. ^{106}Pd (*black*), ^{107}Ag (*red*) and ^{109}Ag (*green*).

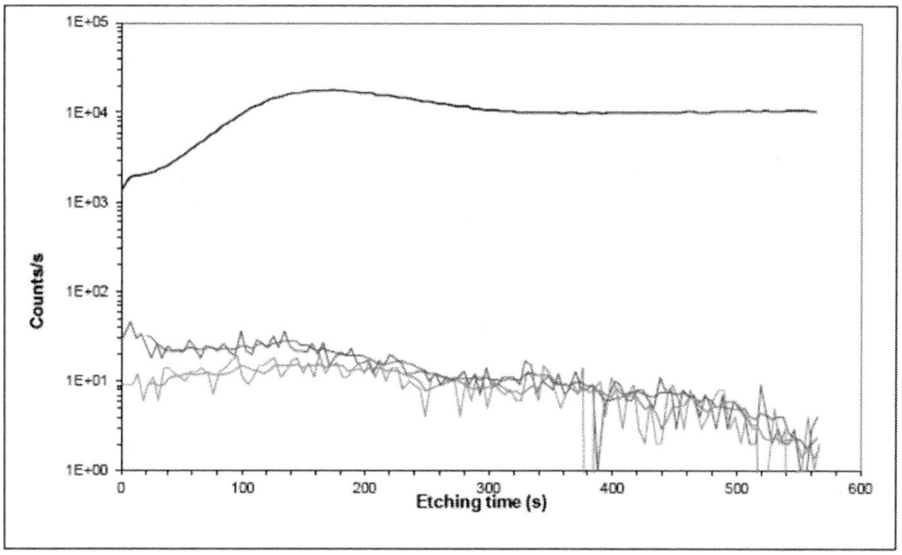

Figure 5. Depth profile of the palladium cathode. ^{106}Pd (*black*), ^{107}Ag (*red*) and ^{109}Ag (*green*).

shows that at the surface, the ratio ^{107}Ag/^{109}Ag is larger than 10. Here too, at 650 s of etching, the ^{107}Ag and ^{109}Ag peaks are equal. As both peaks should be almost equal in natural silver, we have subtracted ^{109}Ag from ^{107}Ag, so that

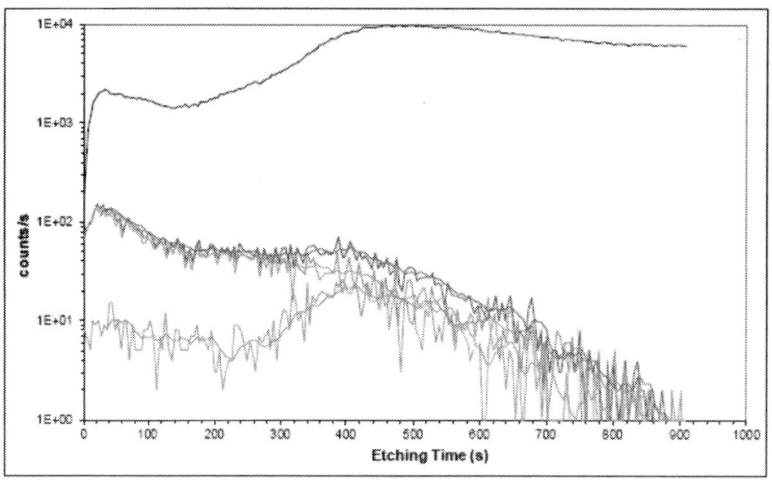

Figure 6. Depth profile of the palladium cathode. Pd-106 (*black*), ^{107}Ag (red), ^{109}Ag (*green*). The blue line shows the ^{107}Ag peak corrected for ^{109}Ag.

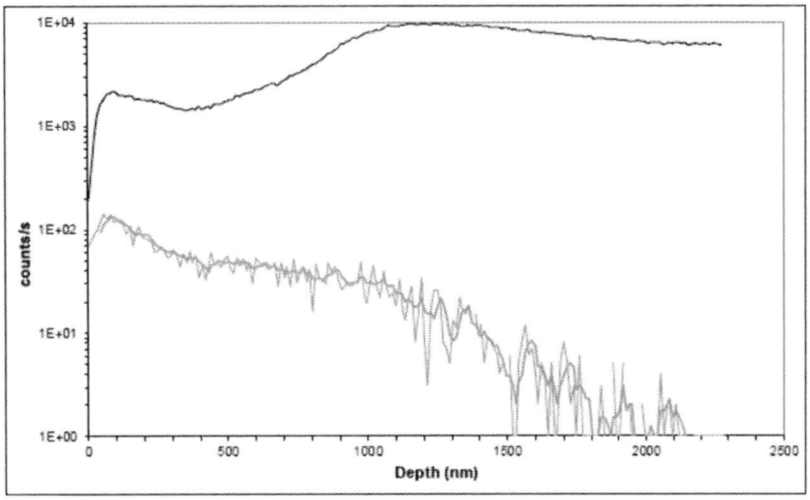

Figure 7. Depth profile of the palladium cathode. ^{106}Pd (*black*), the ^{107}Ag peak corrected for ^{109}Ag (*blue*).

the curve in blue displays only anomalous ^{107}Ag (Fig. 7). Using the etch rate calibration (2.4 nm/s), Fig. 7 shows the ^{106}Pd peak and the ^{107}Ag peak corrected of the ^{109}Ag peak by subtracting the signal of ^{109}Ag from ^{107}Ag.

It is interesting to note that after 1.3 μm, the amount of ^{107}Ag has decreased by a factor of 10. This is an indication of how far the reaction occurs under the surface of the cathode.

5. High Resolution ICP-MS Analysis

In order to check the validity of these observations, the surface of the palladium cathode was etched in aqua regia and sent for analysis to the CNRS (Centre National de la Recherche Scientifique) laboratory in Villeurbanne, France. The Inductively Coupled Plasma-Mass Spectroscopy (ICP-MS) analysis shows 3/1000 increase of ^{107}Ag from natural ^{107}Ag. Even though this increase is at the upper limit of the natural variation in ratio of the two isotopes of silver, this result confirms the SIMS analysis. The question of why we did not find the ratio ^{107}Ag/^{109}Ag $= 10$ is certainly due to the fact that it was only measured in hot spots, and not uniformly on the sample. Therefore, the ICP-MS results are compatible with the SIMS data.

6. Discussion

In this study, we have shown the presence of mass 107, but unfortunately, during the first run we did not check for mass 112. In a second run, in spite of all our attempts, we could not find another hot spot, so we could not check for the presence of mass 112.

Another question is the possible interference with ZrO_2 masses. Zirconium and ZrO_2 have the following isotopes shown below:

$$^{90}Zr\ (51.5\%) \quad ^{106}ZrO_2,$$

$$^{91}Zr\ (11.2\%) \quad ^{107}ZrO_2,$$

$$^{92}Zr\ (17.1\%) \quad ^{108}ZrO_2,$$

$$^{94}Zr\ (17.4\%) \quad ^{110}ZrO_2,$$

$$^{96}Zr\ (2.8\%) \quad ^{112}ZrO_2.$$

Therefore, mass 107, could be either ^{107}Ag or $^{107}ZrO_2$. Even though this is a potential interference, this is very unlikely, as there was no detectable Zr in the cathode.

7. Conclusion

In this paper, it is shown that during electrolysis of a palladium cathode in a heavy water electrolyte, there is production of silver. However, only ^{107}Ag is formed, which corresponds to a possible fusion of palladium atoms with deuterium atoms, with an unknown mechanism. Silver was detected only in hot spots, and not all over the electrode. Also, an important fact regarding this analysis is that silver is detected only down to 1.3 μm under the surface. This shows that in the case of this type of experiments it is a surface effect and not a bulk one that is involved.

Unfortunately, mass 112, corresponding to ^{112}Cd has not been measured in this work. This sheds some uncertainty on the above conclusion. More work should definitely be done to clarify this point.

Acknowledgement

I am grateful to Stanley Pons for having given me the palladium cathode, and to Bernard Boyer from the University of Montpellier for the SIMS analysis.

References

[1] M. Fleischmann, S. Pons and M. Hawkins, Electrochemically induced nuclear fusion of deuterium, *J. Electroanal. Chem.* **261** (1989) 301–309 and errata in Vol. 263.

[2] M.H. Miles, R.A. Hollins, B.F. Bush, J.J. Lagowski and R.E. Miles, Correlation of excess power and helium production during D_2O and H_2O electrolysis using palladium cathodes, *J. Electroanal. Chem.* **346** (1993) 99–115.

[3] W.-S. Zhang, J. Dash and Q. Wang, Seebeck envelope calorimetry with $Pd/D_2SO_4+H_2SO_4$ electrolytic cell in Pd/D_2O+H_2S, in *the 12th Int. Conf. on Condensed Matter Nucl. Sci.*, 2005. Yokohama, Japan, pp. 86–96.

[4] W.-S. Zhang and J. Dash, Excess heat reproducibility and evidence of anomalous elements after electrolysis in $Pd/D_2O+H_2SO_4$ electrolytic cells, in *the 13th Int. Conf. on Condensed Matter Nucl. Sci.*, 2007, Sochi, Russia, pp. 202–218.

[5] J. Gao, W.-S. Zhang and J.-J. Zhang, Effects of D/Pd ratio and cathode pretreatments on excess heat in closed $Pd/D_2O+D_2SO_4$ electrolytic cells, *J. Condensed Matter Nucl. Sci.* **24** (2017) 42–59.

[6] T. Roulette, J. Roulette and S. Pons, Results of ICARUS-9 experiments run at IMRA Europe, in *the Sixth Int. Conf. Cold Fusion*, Progress in New Hydrogen Energy, 1996, Lake Toya, Hokkaido, Japan, pp. 85–92.

J. Condensed Matter Nucl. Sci. 29 (2019) 219–229

Research Article

Uranium Fission Using Pd/D Co-deposition

Pamela A. Mosier-Boss* and Lawrence P. Forsley

Global Energy Corporation, San Diego, CA 92123, USA

Patrick McDaniel

University of New Mexico, Albuquerque, NM 87131, USA

Abstract

In this investigation, we explored using Pd/D generated energetic particles to fission uranium. Analysis of the CR-39 microphotographs liquid scintillation counter (LSC) alpha and beta spectra, and high purity germanium (HPGe) spectral gamma data support fissioning of U-238/U-235 implying that a hybrid fusion-fission reactor based upon Pd/D co-deposition could be feasible.

Keywords: CR-39, HPGe, Hybrid reactor, LSC, Uranium

1. Introduction

Earlier, we reported on observing the production of energetic particles during Pd/D co-deposition [1,2]. The energetic particles produced include $\geq$1.8 MeV protons (including 15 MeV protons), $\geq$7 MeV alphas, 2.2–2.5 MeV neutrons, and secondary particles from either energetic protons and/or neutrons. We have also reported on observing the production of $\geq$9.6 MeV neutrons [3]. The Java-based Nuclear Information Software (JANIS) database [4] indicates that these particles are energetic enough to fission both U-238 and U-235. In this investigation, we explored using Pd/D generated energetic particles to fission uranium. An experiment was conducted in which Pd/D co-deposition was done on a Au/U composite cathode. Both real-time high purity germanium (HPGe) and a CR-39 detector were used to monitor the reaction. A month after the reaction was completed, HPGe spectra of the U wire and the Au wire and deposit at the end of the experiment were obtained in a Compton suppressed cave. Three months after termination of the experiment, liquid scintillation measurements of the deposit as a function of time were made. The results of those measurements are discussed in this communication.

*Corresponding author. E-mail: pboss@san.rr.com.

2. Experimental

Pd/D co-deposition was done on a Au/U (natural uranium 0.71% U-235) composite cathode that was in contact with a CR-39 detector. The cell was placed inside a Pb cave close to a HPGe detector. Upon termination of the experiment, the CR-39 detector was etched as previously described [1,3] and underwent microscopic examination. A month after the experiment was terminated, HPGe spectra of the native wire as well as the Au cathode and deposit were obtained at the University of Texas, Austin. Spectra were obtained using a 30% HPGe detector in a Compton suppressed cave [5–7]. Compton suppression systems are typically used to reduce the background continuum as well as the cosmic and natural background. Three months after termination of the experiment, samples of the deposit were placed in 10 mL of scintillating (Scintiverse^{TM}) cocktail and liquid scintillation spectra were obtained periodically over a period of 550 days.

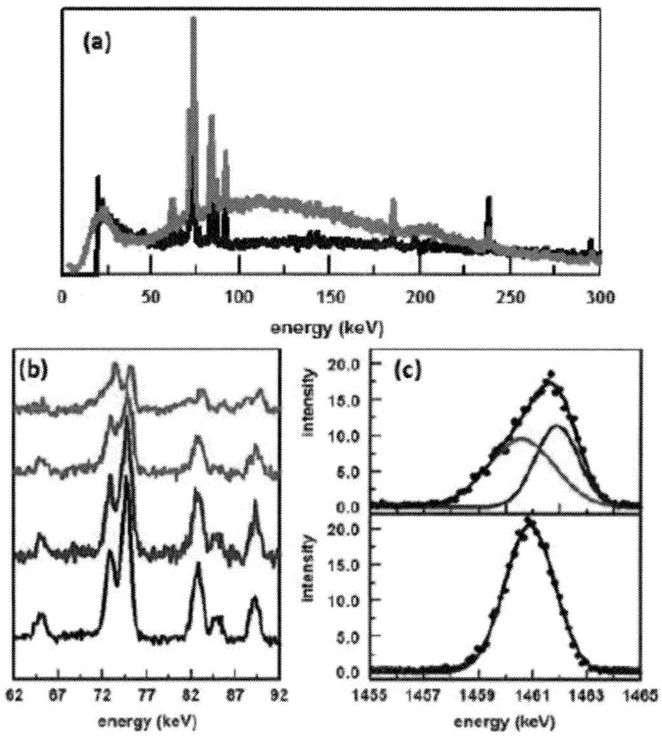

Figure 1. Gamma ray spectra obtained during the experiment where: (a) spectra of the Pb cave (*black spectrum*) and the electrochemical cell obtained on Feb. 28 at 08:30 (*gray spectrum*); (b) time normalized, baseline corrected spectra obtained at (*bottom to top*) Feb. 28 at 08:30, Feb. 28 at 13:00, Feb. 28 at 17:38, and Feb. 29 at 11:14; and (c) time normalized, baseline corrected spectra of the K-40 line obtained on Feb. 28 at 08:30 (*bottom*) and Feb. 29 at 11:14 (*top*).

3. Results and Discussion

3.1. Real-time HPGe spectra

Figure 1a shows gamma ray spectra of the Pb cave and the cell after 33.5 h of electrolysis. At this point the cell current is at –0.4 mA. The HPGe detector has an Al window that cuts off at 20 keV. This cut-off is evident in the background spectrum of the Pb cave. The spectrum of the cell shows background lines observed on top of a raised baseline. Three broad peaks at $\sim$20, 118, and 194 keV are observed on this raised baseline. With additional time, these broad peaks disappeared. The spectrum of the cell clearly shows counts below the cut-off of the Al window. These counts and the broadness of the three bands making up the baseline cannot be due to gamma rays and are caused by something interacting with the detector itself. Neutrons are the only thing that can pass through the Al window and interact with the Ge atoms of the detector.

Further evidence of neutron damage to the detector is shown in Fig. 1b,c. These spectra have been time normalized and the background has been corrected. The times the spectra were collected are indicated in the figure caption. Figure 1b shows peaks due to the Pb lines and background peaks. On Feb. 28 at 08:30 and at 13:00, the peaks are intense and cleanly resolved. By Feb. 28 at 17:28, the peaks are less intense and are broadening. On Feb. 29 at 11:14, the peaks are clearly less intense and have broadened. Figure 1c shows the K-40 line. On Feb. 28 at 08:30, the K-40 line is symmetrical and fits to a single Gaussian line. The K-40 peak observed on Feb. 29 at 11:14 has lost intensity and shows a tail on the low frequency side of the peak. Its line shape is described by the sum of two Gaussian lines, shown in Fig. 1c. The changes observed in the gamma lines are consistent with neutron damage to the HPGe detector [8].

Inelastic scattering of neutrons on Ge-74 and Ge-72 cause broad asymmetric gamma peaks at 596 and 691.3 keV, respectively, when the energy of the neutrons are $\leq$1 MeV [9]. For neutrons with energies $\geq$1 MeV, these lines broaden and flatten out. When neutron energies are greater than 4 MeV these peaks cannot be separated from the background. No peaks at 596 and 691.3 keV were observed for the electrolyzing cell. Therefore, the average energy of the neutrons generated during electrolysis is greater than 4 MeV. Another way to estimate the average neutron energy is to model the neutron elastic scattering of the Ge nuclei using the Monte–Carlo technique [10,11]. When a high energy neutron elastically scatters a germanium nucleus, it leaves an ionization trail that is proportional to the energy given to each germanium nuclei encountered. Germanium recoil spectra, based on isotropic elastic scattering, were calculated for fission neutrons (2 MeV) and 6.3–6.83 MeV neutrons. The results of these calculations are summarized in Fig. 2. The baseline measured on Feb. 28 at 08:30 more closely resembles the recoil spectrum for the 6.3–6.83 MeV neutrons than the fission neutrons. Radiation hardness refers to higher energy particles, be they gammas, protons, neutrons, etc. The degree of hardness is determined by taking the ratio of the total counts in the 100–300 keV energy range with those in the 50–100 keV energy range [12]. The higher the ratio, the harder the spectrum. As can be seen from Fig. 2, the Pd/U/D neutron energy spectrum is harder than that from fission alone. This hardness of the spectrum indicates a larger contribution of higher energy neutrons to the Pd/U/D neutron energy spectrum and that there is another source of neutrons than fissioning of uranium.

3.2. CR-39 results

The CR-39 detector in contact with the Au/U/Pd composite cathode was etched and underwent microscopic examination. Figure 3 shows some examples of elongated tracks as well as triple tracks that were observed in the detector used in the U experiment. Also shown are their corresponding DT neutron-generated tracks. More elongated and triple tracks were observed in the detector used in the U experiment than in previous Pd/D co-deposition experiments [1,3,13]. This indicates that more neutrons were generated in the Au/U/Pd/D experiment than in any Pd/D experiments. The elongated track shown in Fig. 3 is due to proton recoils. From the length of the track, it is possible to estimate that the energy of the neutron that generated it was greater than 0.114 [14]. Triple tracks are diagnostic of the carbon break-up reaction, $^{12}C(n,n')3\alpha$, which has an energy threshold for the neutron of $\geq$9.6 MeV [15]. These neutron energies are in agreement with the estimate made from the real-time HPGe measurements.

Large (40 μm long and 10 μm wide) shallow cylindrical tracks similar to the one shown in Fig. 4a were observed in the CR-39 detector used in the Au/U/Pd/D experiment. Such tracks have neither been observed in CR-39 detectors in contact with a bare U wire nor in CR-39 detectors used in Pd/D co-deposition experiments. Since these tracks have not been observed in Pd/D co-deposition experiments, uranium is the source of these tracks in the Au/U/Pd/D experiment. The large size and shallowness of the track indicates that it was caused by a high Z particle [16]. Such high Z particles would be created when uranium fissions. To verify this, a piece of native uranium wire was placed in contact with a CR-39 detector and irradiated for 4.5 h with DT neutrons. After etching, large cylindrical tracks, due to high Z fission fragments, were observed (Fig. 4b). The presence of these tracks in the detector used in the Au/U/Pd/D experiment indicates that fissioning of U has occurred and that they aren't the result of either DT neutrons alone or actinide alpha decay.

3.3. Post HPGe analysis in a Compton suppressed cave

A month after the experiment was terminated, HPGe spectra were obtained of the native U wire as well as the cathode and deposit remaining at the end of the experiment. These spectra were obtained in a Compton suppressed system at the University of Texas, Austin. Spectra are shown in Fig. 5. Identification of major lines are indicated. Clearly changes are observed in the spectrum after the U has undergone electrolysis that will be discussed *vide infra*. Gamma lines due to U-235, Th-231, and Th-234 are identified [17]. Th-231 and Th-234 arise from the decay of U-235 and U-238, respectively [18].

Figure 6 shows an expansion of the region between 60 and 120 keV. The Th and U X-ray lines are identified. It can be seen that the ratios of the Th gamma lines and the Th X-ray lines are the same for both spectra of the bare U wire and the spent cathode. Compared to the Th lines, there is clearly a decrease in the U X-ray lines in the spectrum of the spent cathode, Fig. 6. In Fig. 5, the ratio of the U-235 gamma lines and Th gamma and X-ray lines are the same for both spectra. From this it can be concluded that the decrease in the U X-ray lines cannot be due to uranium accumulating on the anode otherwise a similar decrease in the U-235 gamma lines would have been observed. Also,

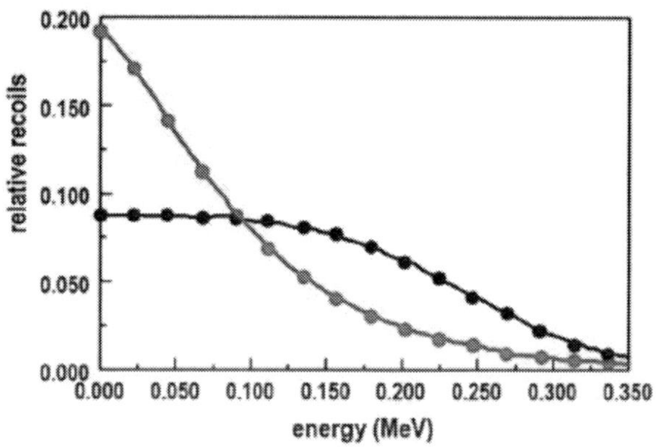

Figure 2. Germanium recoil spectra, based on isotropic elastic scattering, calculated for fission neutrons (*red spectrum*) and 6.3–6.83 MeV neutrons (*black spectrum*) using the Monte–Carlo technique.

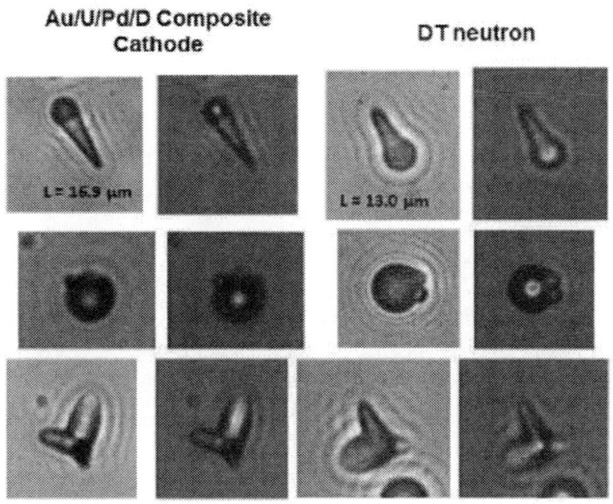

Figure 3. Photomicrographs of tracks observed in CR-39 detectors used in Pd/D co-deposition experiment conducted on an Au/U cathode and their corresponding DT neutron generated tracks. Magnification 1000× . The left-hand images were obtained by focusing the optics on the surface of the detector while the right-hand images are overlays of two images taken at two different focal lengths (*surface and bottom of the pits*). Lengths (L) of the recoil proton tracks are indicated.

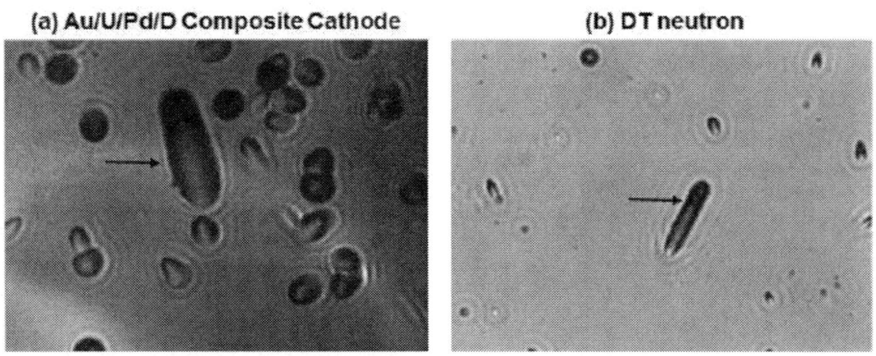

Figure 4. Photo micrographs of CR-39 detector that had been used in a Pd/D co-deposition experiment conducted on an Au/U cathode (a) and a detector that had been in contact with uranium wire exposed to DT neutrons (b). Magnification for both is 1000×. Arrows indicate tracks due to high Z fragments. The detector used in the co-deposition experiment had been etched for 6 h while the detector exposed to DT neutrons had been etched for 1 h.

no deposit was observed on the anode upon termination of the experiment. Since no changes were observed for the Th X-ray lines, the decrease in the U X-ray lines cannot be due to occlusion. Therefore, a decrease in the uranium X-ray peaks suggests that uranium has been consumed. As was discussed vide supra, the CR-39 detector next to the cathode showed large tracks attributed to large Z fission fragments (Fig. 4). These fragments came from fissioning of the uranium. Therefore, the decrease in the uranium X-ray lines provides further support that uranium has been fissioned. Furthermore, U-238 has been preferentially fissioned. Although both U-235 and U-238 will contribute to the U X-ray lines, the primary contributor is U-238 since its concentration in native U is two orders of magnitude greater than U-235. Since no change was observed in the ratio of the U-235 gamma lines compared to the Th lines, Fig. 5, it can be concluded that U-238 was fissioned.

Between 65 and 81 keV, five new lines are observed in the spectrum obtained for the spent cathode, (Fig. 6). Based on the relative intensities and positions of these lines, they are attributed to the Kα and Kβ X-ray lines of Au [17]. These are stimulated X-rays that are generated by the refilling of the K shell electron orbits ionized by the passage of charged particles or higher energy gamma rays [19].

Figure 7 shows expansions of the regions between 30–130 and 90–130 keV regions of Fig. 5. In the 90–130 keV region, three new peaks at 102.18, 105.79, and 121.37 keV are observed in the spectrum of the cathode/deposit. The peak at 121.37 keV is broader than the other two peaks and is probably comprised of overlapping peaks. Based upon the positions and relative intensities of these peaks, they have been assigned to the Kα and Kβ X-ray lines of Am [16]. Like the Au X-ray lines, these are stimulated X-ray lines for Am. As in the case of Au, these Am X-ray lines are stimulated by the alpha and gamma emissions of U-235/U-238 and their daughters. In the region between 30 and 130 keV, a small peak at $\sim$59 keV is observed. This peak is attributed to the gamma emission of Am-241, which happens to be the most intense line of Am-241 [16]. The fact that the line due to the gamma emission of Am-241 is smaller than the stimulated Am X-ray lines is probably due to absorption of this gamma emission and induced fluorescence of lower energy X-ray lines [19]. Furthermore, there should be no correlation between the magnitude of a gamma line from the nucleus and the X-ray lines from electron transitions in the shells surrounding the nucleus. The most likely reaction pathways for the production of Am-241 are [20,21]:

$$ {}^{238}_{92}\mathrm{U} + \mathrm{n} \rightarrow {}^{239}_{92}\mathrm{U} \xrightarrow[23.5\ \mathrm{min}]{\beta^-} {}^{239}_{93}\mathrm{Np} \xrightarrow[2.356\ \mathrm{days}]{\beta^-} {}^{239}_{94}\mathrm{Pu} \xrightarrow{\mathrm{n},\gamma} {}^{240}_{94}\mathrm{Pu} \xrightarrow{\mathrm{n},\gamma} {}^{241}_{94}\mathrm{Pu} \xrightarrow[14.35\ \mathrm{years}]{\beta^-} {}^{241}_{95}\mathrm{Am}. $$

The production of Am indicates that the neutron kinetic energy has to be < 1 MeV, which is the cross-section threshold favoring capture over fission in U-238, in order to create Pu-239 [4]. Production of Am indicates that there was a high flux of neutrons, which is in agreement with the CR-39 data.

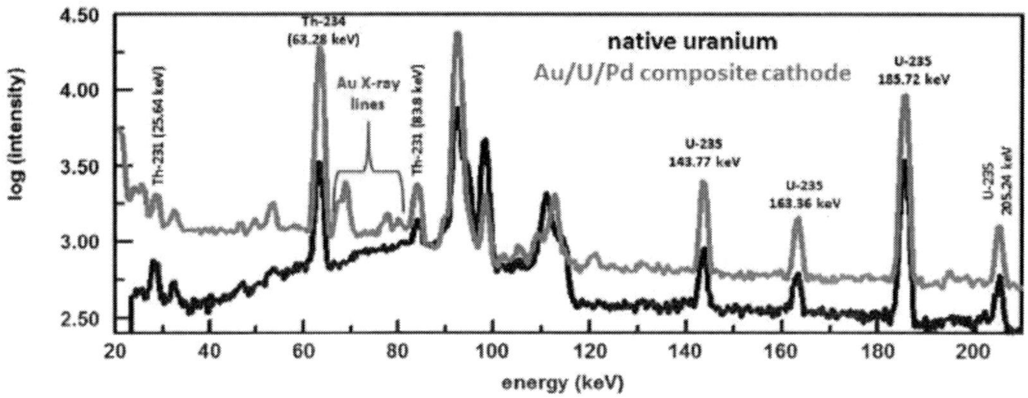

Figure 5. Gamma ray spectra of the native uranium wire (*black*) and the Au/U/Pd composite cathode (*gray*) at the end of the experiment. Peak assignments are shown [17].

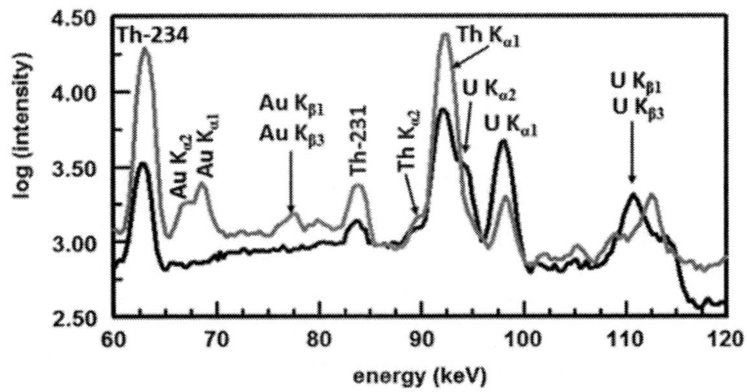

Figure 6. Gamma ray spectra, between 60 and 120 keV, of the native uranium wire (*black*) and the Au/U/Pd composite cathode (*gray*) at the end of the experiment. Peak assignments are shown [17].

3.4. Post liquid scintillation counting (LSC) analysis of the deposit

Fissioning of uranium produces several hundred radionuclides [22,23]. Most isotopes in the fission products have extra neutrons and tend to decay to more stable isotopes through β emission accompanied by γ emission. Three months after the experiment was terminated, the use of LSC was explored to measure these beta emissions. However, alpha and gammas will also stimulate photon emissions from these cocktails. The three types of radiation interact differently

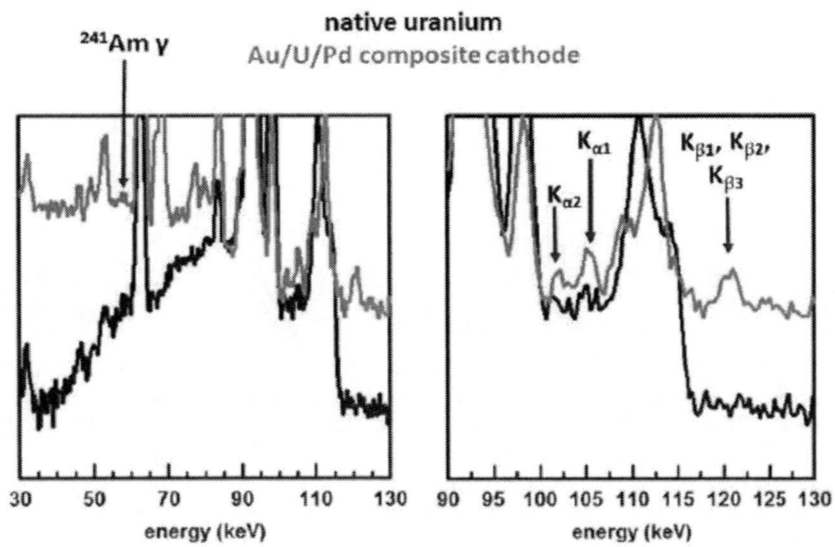

Figure 7. Expansions of the 30–130 and 90–130 keV regions of the HPGe spectra shown in Fig. 5. Peak assignments are shown.

with the liquid scintillator. For a mixture of alpha, beta, and gamma emitters, the resultant spectrum is comprised of a relatively narrow alpha peak sitting atop a beta–gamma continuum [24]. A Beckman Coulter LS6500 multipurpose scintillator counter was used to make these measurements. This counter has two diametrically opposed photomultplier tubes (PMTs). In order to be detected, the photons created by the interaction of the fluor with the radioactive nuclide need to be within the line of sight of the PMTs allowing coincidence counting.

Upon termination of the experiment, there was no evidence of the U wire in the cathode. Formation of uranium hydride/deuteride results in the complete disintegration of the structure of the original metal and the hydride/deuteride appears as a voluminous, finely divided black powder [25]. This break-up of the uranium wire during electrolysis will cause disequilibrium between U-235/U-238 and their daughters. This will cause changes in the LSC spectra as these species in the decay chains come back to equilibrium. So, besides what happened to the uranium wire during electrolysis and Pd/D co-deposition, disequilibrium will be occurring as well. To determine the effect of disequilibrium, a piece of native U wire was placed in an aqueous solution of LiCl. This caused the wire to break-up into particles which will cause disequilibrium. Some of these particles were placed in the cocktail and spectra were measured periodically over a period of 195 days. These spectra are shown in Fig. 8a. Initially, the LSC spectra of the particles show a beta–gamma continuum. Then a fairly narrow peak grows in at channel 675. This narrow peak is due to the alpha emission of U-234. It takes a while for this peak to grow in as uranium has to dissolve into solution in order for the alphas to be seen. This is because the path length for 5 MeV alpha particles in a liquid scintillator solution is approximately 50 μm [26]. Consequently, unless the particles were directly between the windows of the PMTs, no photons would be detected. As shown in Fig. 8a, no peak due to the alpha of U-238 is seen, which is due to the fact that the sample is in disequilibrium [27]. Since the half-life of U-234 ($t_{1/2} = 2.5\times 10^5$ years) is shorter than that of U-238 ($t_{1/2} = 4.5\times 10^8$ years), the U-238 peak is less intense than the U-234 peak when in disequilibrium and is obscured by the other emissions of the beta–gamma continuum [26]. With time another alpha peak grows in at channel 625. This peak is due to U-238. It keeps growing until the sample reaches equilibrium. As shown in Fig. 8a, no significant changes are observed for the rest of the LSC spectra.

Figure 8b shows LSC spectra obtained of the deposit as a function of time. It is completely different that the spectra

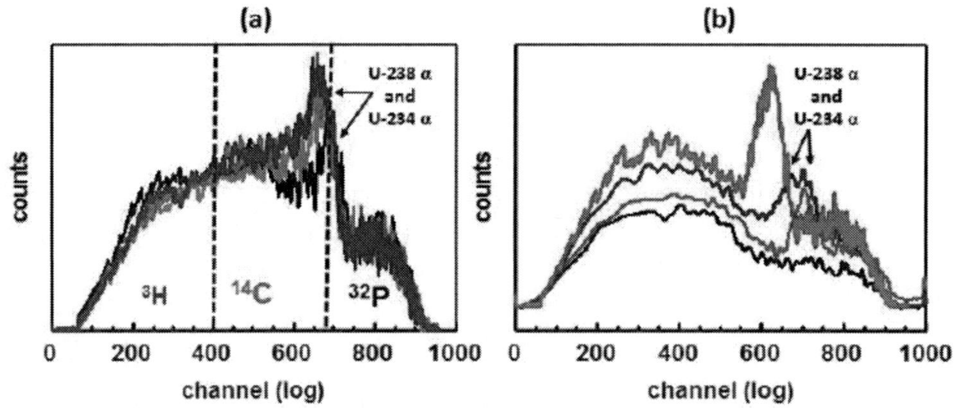

Figure 8. Liquid scintillator spectra taken as a function of time for (a) uranium particles formed by immersing a piece of native uranium wire in an aqueous LiCl solution (spectra taken after 1, 125, 182, and 195 days) and (b) the deposit as a result of the Pd/D co-deposition experiment conducted on a Au/U cathode (spectra taken after 0, 9, 85, and 356 days). Channels for ^3H, ^{14}C, and ^{32}P are indicated as are the alpha peaks for U-234 and U-238. Energies of these alphas are 4.2 and 4.7 Mev for U-238 and U-234, respectively.

obtained by the U particles shown in Fig. 8a. Initially no alpha peaks are observed. With time the alpha peaks due to U-234 and U-238 grow in as the uranium goes into solution. With additional time, another alpha peak at a lower energy than that for U-238 begins to grow in (Fig. 8b). After 355 days in the cocktail, this peak reaches maximum height. The energy of this peak is estimated to be 3.2 $\pm$ 0.3 MeV. The species responsible for this alpha peak must be long-lived. The only long-lived species that have an alpha emission near 3.2 MeV are Pt-190 ($t_{1/2}$ = 6.5 $\times$ 10^{11} years) and Gd-148 ($t_{1/2}$ = 74.6 years) [17]. Based on the intensity of the alpha peak, it is most likely due to Gd-148 as it has a much shorter half-life and it is a fission product, whereas Pt-190 is not. Besides these alphas, increases in the beta–gamma continuum are observed with time as more species leach from the deposit. These species are clearly different than what was observed for the U particles and are attributed to long-lived fission products.

4. Conclusions

The use of energetic particles produced during Pd/D co-deposition to fission uranium was explored. Real-time HPGe spectral data obtained during the early stages of the plating phase of Pd/D co-deposition on a Au/U composite cathode showed evidence of neutron damage. Using the Monte–Carlo technique, the average energy of the neutrons was estimated to be of the order of 6–7 MeV. A gamma ray spectrum obtained in a Compton suppressed Pb cave a month after the experiment showed the presence of Am-241 which is likely created by neutron capture by U-238 to form Pu-239. Pu-239 undergoes successive neutron captures to form Pu-241 which then beta decays into Am-241. The HPGe spectra also shows a decrease in the U X-ray peaks indicating that U has been consumed. Production of energetic neutrons and fissioning of U-238/U-235 were further verified by the CR-39 detector that was in contact with the Au/U/Pd composite cathode. LSC spectra showed evidence of an alpha emitter at 3.2 $\pm$ 0.3 that is likely due to Gd-148. Additional changes in the LSC spectra with time show beta–gamma emitters due to long-lived fission products. These results indicate that uranium can be fissioned by the energetic particles formed as a result of Pd/D co-deposition.

The implication of this experiment is that a hybrid fusion-fission reactor is feasible. The main advantages of such a reactor would be, (1) it does not require enrichment of U-235, (2) it does not produce greenhouse gases, and (3) it can easily be shut off by simply turning off the current to the cell. Such a reactor might also be used to dispose of nuclear wastes and long life radioactive fission products remaining in spent nuclear fuel.

Acknowledgements

This effort was funded by the Defense Threat Reduction Agency (DTRA), and JWK Corporation. The authors would like to thank Mr. Mark Morey, Dr. Jim Tinsley, and Mr. Paul Hurley of National Security Technologies, LLC, Special Technologies Laboratory in Santa Barbara, CA for exposing CR-39 detectors, and uranium in contact with CR-39 detectors, to DT neutrons. The HPGe spectra of the composite cathode and native uranium wire obtained in a Compton suppressed Pb cave were taken by Dr, Sheldon Landsberger of the University of Texas at Austin, Nuclear Engineering Teaching Laboratory. The authors would also like to thank Dr. Gary Phillips, nuclear physicist, retired from the Naval Research Laboratory, US Navy, Radiation Effects Branch for valuable discussions in interpreting the CR-39 data.

References

[1] P.A. Mosier-Boss, F.E. Gordon, L.P. Forsley and D. Zhou, Detection of high energy particles using CR-39 detectors, Part 1: results of microscopic examination, scanning, and LET analysis, *Int. J. Hydrogen Energy* **42** (2017) 416–428.

[2] A.S. Roussetski, A.G. Lipson, E.I. Saunin, F. Tanzella and M. McKubre, Detection of high energy particles using CR-39 detectors, Part 2: results of in-depth destructive etching analysis, *Int. J. Hydrogen Energ* **42** (2017) 429–436.

[3] P.A. Mosier-Boss, S. Szpak, F.E. Gordon and L.P.G. Forsley, Triple tracks in CR-39 as the result of Pd/D co-deposition: evidence of energetic neutrons, *Naturwissenschaften* **96** (2009) 135–142.

[4] JANIS software, http://www.oecd-nea.org/janis/.

[5] M. Petra, G. Swift and S. Landsberger, Design of a Ge NaI(Tl) Compton suppression spectrometer and its use in neutron activation analysis, *Nucl. Instrum. Methods* A **299** (1990) 85–87.

[6] S. Landsberger and S. Peshev, Compton suppression neutron activation analysis: past, present and future, *J. Radioanal. Nucl. Chem.* **202** (1996) 201–224.

[7] G. Nicholson, S. Landsberger, L. Welch and R. Gritzo, Characterization of a Compton suppression system and the applicability of Poisson statistics, *J. Radioanal. Nucl. Chem.* **276** (2008) 577–581.

[8] E.H. Seabury, C. Dew van Siclen, J.B. McCabe, C.J. Wharton and A.J. Caffrey, Neutron damage in mechanically cooled high-purity germanium detectors for field-portable prompt gamma neutron activation analysis (PGNAA) systems, *2013 IEEE Nuclear Science Symposium and Medical Imaging Conference*, Oct. 2013.

[9] G. Fehrehbacher, R. Meckbach, H.G. Paretzke Fast neutron detection with germanium detectors: computation of response functions for the 692 keV inelastic scattering peak, *Nucl. Instrum. Methods* A **372** (1996) 239–245.

[10] J. Ljungvall and J. Nyberg, A study of fast neutron interactions in high-purity germanium detectors, *Nucl. Instum. Methods* A **546** (2005) 553–573.

[11] M. Baginova, P. Vojtyla and P.P. Povinec Investigation of neutron interactions with Ge detectors, *Nucl. Instum. Methods* A **897** (2018) 22–31.

[12] M. Tavani, Euclidean versus non-Euclidean gamma-ray bursts, *Astrophys. J.* **497** (1998) L21–L24.

[13] P.A.Mosier-Boss, J.Y. Dea, L.P.G. Forsley, M.S. Morey, J.R.Tinsley, J.P. Hurley and F.E. Gordon, Comparison of Pd/D co-deposition and DT neutron generated triple tracks observed in CR-39 detectors, *Euro. Phys. J. Appl. Phys.* 51 (2010) 1–11.

[14] G.W. Phillips, J.E. Spann, J.S. Bogard, T. VoDinh, D. Emfietzoglou, R.T. Devine and M. Moscovitch, Neutron spectrometry using CR-39 track etch detectors, *Radiation Prot. Dosim.* **120** (2006) 457–460.

[15] A.M. Abdel-Moneim and A. Abdel-Naby, A study of fast neutron beam geometry and energy distribution using triple-α reactions, *Radiation Meas.* **37** (2003) 15–19.

[16] J.K. Pálfalvi, Y. Akatov, J. Szabó, L. Sajó-Bohus and I. Eördögh, Evaluation of solid state nuclear track detector stacks exposed on the international space station, *Radiation Prot. Dosimetry* **210** (2004) 393–397.

[17] Periodic Table Linked to ToRI data of known isotopes of each element, http://nucleardata.nuclear.lu.se/toi/perchart.htm.

[18] L.P. Forsley, P.A. Mosier-Boss, P.J. McDaniel and F.E. Gordon Charged particle detection in the Pd/D system: CR-39 SSNTD vs. real-time measurement of charged particle stimulated Pd K shell X-rays *Electrochim. Acta* **88** (2013) 373–383.

[19] Decay Chains, http://metadata.berkeley.edu/nuclear-forensics/Decay%20Chains.html

[20] J.W. Kennedy, G.T. Seaborg, E. Segrè and A.C. Wahl, Properties of element 94, *Phys. Rev.* **70** (1946) 555–556.

[21] K. Kostecka, Americium – from discovery to the smoke detector and beyond, *Bull. Hist. Chem.* **33** (2008) 89–93.

[22] G.R. Crocker, Fission product decay chains: schematics with branching fractions, half-lives, and literature references U.S. Naval Radiological Defense Laboratory USNRDL-TR-67-111, 24 Jun. 1967.

[23] B. Li, Analysis of fission products – a method for verification of a CTBT during on-site inspections, *Sci. Global Secur.* **7** (1998) 195–207.

[24] Y. Dazhu, Z. Yongjun, S. Möbius and C. Keller, Simultaneous determination of alpha and beta-emiting nuclides by liquid scintillation counting, J *Radioanal. Nucl. Chem. Lett.* **144** (1990) 63–71.

[25] T. Hashino and Y. Okajima, Mechanism of the reaction of hydrogen with uranium, *J. Phys. Chem.* **77** (1973) 2236–2241.

[26] W.J. McDowell and B.L. McDowell, *Liquid Scintillation Alpha Spectrometry*, CRC Press, Boca Raton, 1994.

[27] H.M. Pritchard and A. Cox, Liquid scintillation screening method for isotopic uranium in drinking water, in H. Ross, J.E. Noakes and J.D. Spaulding (Eds.), *Liquid Scintillation Counting And Organic Scintillators,* Lewis, Michigan, USA, 1991.

J. Condensed Matter Nucl. Sci. 29 (2019) 230–237

Research Article

Influence of Effective Microorganisms on the Activity of ^{137}Cs in the Soil Contaminated due to the Accident on the Chernobyl NPP

A.N. Nikitin*, G.Z. Gutzeva, G.A. Leferd and I.A. Cheshyk

Institute of Radiobiology of the National Academy of Sciences of Belarus, Republic of Belarus

S. Okumoto and M. Shintani

EM Research Organization Inc., Japan

T. Higa

International EM technology center, Meio University, Japan

Abstract

Microbiological soil improvers have a potential as a tool for regulation transfer of induced radioisotopes and other pollutants into crops. During the development of a method using effective microorganisms (EM) to reduce the soil-to-plant transfer of ^{137}Cs on land contaminated with radioactive cesium, an unexpected effect of EM on the reduction of the ^{137}Cs activity in soil samples was observed. Laboratory experiments were then conducted to evaluate the impact of EM and fermented organic fertilizer (EM Bokashi) on the ^{137}Cs activity in soil samples to investigate this observation. The experimental results indicate an increase in the ^{137}Cs decay rate of up to 4 times the usual decay rate corresponding to the half-life of ^{137}Cs, which is 30.17 years. Our results suggest that EM accelerates the radioactive decay of ^{137}Cs in soil.

Keywords: Cesium-137, Effective microorganisms, Radioactive decay, Radioactive isotopes, Soil

1. Introduction

Nuclear accidents are highly undesirable but potential events in the process of operating nuclear power plants. Fukushima, Chernobyl, and other accidents have caused considerable areas to become contaminated with radioactive isotopes [1,2]. One of the most critical and challenging tasks following such accidents is the decontamination of man-made radioisotopes in the soil. Current methods of decontamination are expensive and ineffective for large areas [3,4]. Therefore, it is important to find innovative technologies to clean contaminated areas.

*Corresponding author. E-mail: nikitinale@gmail.com.

© 2019 ISCMNS. All rights reserved. ISSN 2227-3123

One option to economically use farmlands contaminated with ^{137}Cs is to substantially increase the application of potassium chemical fertilizers [5,6]. However, this approach may lead to a decrease in economic efficiency compared to non-contaminated areas, as well as adverse ecological effects [7,8]. Consequently, it is critical to develop new approaches to regulate the uptake of radioactive isotopes by plants.

Soil microorganisms can affect the bioavailability of man-made radioisotopes in the soil [9,10]. Our research team (Institute of Radiobiology of the National Academy of Sciences of Belarus and EM Research Organization Inc., Japan) has investigated the possibility of using microbiological soil improver additives to decrease ^{90}Sr and ^{137}Cs transfer into crops [11–15]. Toward this goal, we tested mixtures of microorganisms known as effective microorganisms (EM). An EM is a people-friendly and environmentally safe microbial product that achieves synergistic effects by combining beneficial microorganisms that exist in nature, such as lactic acid bacteria, yeast, and phototrophic bacteria.

EM was initially developed in 1982 by Professor Teruo Higa to improve soils [16]. Following multiple studies, EM has now been in use for over 25 years in numerous fields, including sustainable agriculture, animal husbandry, and environmental conservation [17]. In addition, EM is distributed in more than 100 countries worldwide.

The results of our experiments confirm that EM and EM Bokashi reduce the transfer of ^{137}Cs and ^{90}Sr in corn, barley, lettuce, and vegetables [12,13]. We suggested two mechanisms for this effect: a decrease in the bioavailable (soluble and exchange) forms of the radioisotopes in soil under the impact of EM, and an increased migration of these radioisotopes downward in the soil profile. In our attempts to test these hypotheses, we observed an imbalance in the ^{137}Cs in the experimental system. After checking all the possible ways of losing radionuclides and all the sources of error, we suggest another hypothesis: the application of EM to soil increases the rate of ^{137}Cs radioactive decay. This study is devoted to confirming this hypothesis.

2. Materials and Methods

2.1. Soil samples

Sod-podzolic soil from the Chernobyl exclusion zone with a ^{137}Cs activity of approximately 7 kBq/kg was used in this experiment. The soil samples were dried and sieved through a 1-mm sieve. The entire volume of the soil was evenly mixed to provide a homogenous distribution of the ^{137}Cs activity and the physical and chemical properties of the soil. The soil was placed in a 100-ml container and mixed with different concentrations of EM or EM Bokashi.

2.2. Effective microorganisms (EM)

The EM (EM1®) used in the experiments was supplied by EM Research Organization Inc., Japan. For application to soil, the EM was prepared in two forms: a liquid form (an EM solution) and a solid form (EM Bokashi). The EM solution was prepared by mixing EM with sugar cane molasses and water with a ratio of 1:1:20 (v/v). The mixed ingredients were transferred to a plastic container, which was tightly closed with a plastic lid and incubated for 20–25 days at 35$\pm$ 2°C to promote fermentation. The EM solution was considered ready for use when it produced a pleasant fermentation smell and the pH was below 3.5. In Japanese, bokashi refers to fermented organic matter. EM Bokashi is an anaerobic fermentation product made from solid agricultural byproducts inoculated with EM. In EM Bokashi, organic matter serves as a growth medium for the microorganisms and provides a suitable microenvironment for EM in the soil. The EM Bokashi was prepared according to the method described in [16]. A mixture of 0.4 L of EM, 0.4 L of sugar cane molasses, and 4 L of chlorine-free tap water was added to 10 kg of wheat bran and mixed until homogeneous. The mixture was then placed in a plastic bag, which was hermetically sealed and kept under dark and warm conditions for 30 days. After the 30-day fermentation period, the EM Bokashi had a sweet-sour smell. The EM Bokashi was dried at room temperature prior to application.

2.3. Experimental layout

The different conditions of the experiment were as follows:

(1) Absolute control (dry soil).
(2) Positive control (wet soil).
(3) Positive control 2 (wet soil + organic matter (OM)).
(4) EM-1 1% + molasses.
(5) EM-1 5% + molasses.
(6) EM-110% + molasses.
(7) EM Bokashi 1%.
(8) EM Bokashi 5%.
(9) EM Bokashi 10%.
(10) EM-1 10%

There were three types of control experiments: dry soil (absolute control); wet soil with the addition of water (positive control); and wet soil with the addition of water, molasses, and wheat bran as organic matter (OM) (positive control 2). The experimental treatments were 1%, 5%, and 10% EM with molasses (Mo) and 1%, 5%, and 10% EM Bokashi. A treatment with EM but without molasses was also performed. Different exposure periods to the experimental treatments were set to 6, and 18 months. Each treatment for each exposure period was repeated 15 times.

EM Bokashi was applied to the soil only at the beginning of the experiment. The EM, water, and molasses were added to the soil at the beginning of the experiment and every six weeks. Before adding the solutions, the containers were opened to allow for natural water evaporation. For the remainder of the time, the containers were sealed.

Soil samples were kept at room temperature (20–24°C) under natural light conditions.

2.4. Measurements

The ^{137}Cs activity was measured before and after each period of the experiment. The activity of the samples was determined in the containers where the samples were kept. The samples ware positioned precisely on the same axis as the detector to obtain a more stable result.

A GX 2018 gamma-spectrometry complex CANBERRA with a coaxial germanium detector with an extended energy range was used for the measurements. The gamma-spectrometry measurement time was 600 s. The relative error in the measurements of the ^{137}Cs activity was less than 0.5%.

At the time of analysis, water was added to the samples to return the sample weight to its value at the time of the first measurement.

3. Results and Discussion

Table 1 shows the ^{137}Cs activity for each treatment at the beginning and end of the experiments during the 6, 12, and 18 months exposure periods. The ^{137}Cs activities at the beginning of the experiment were in the range of 692–929 Bq. Each sample in the experiment included precisely 100 g of evenly homogenized soil; however, we could not obtain a lower level of variability in these naturally contaminated samples.

After 6 months, the ^{137}Cs activity in the variants decreased to 686–924 Bq, and after 12 months, it decreased to 686–919 Bq. This is equivalent to 0–4.4% and 1.0–5.9% decreases with respect to the initial activity, respectively.

In the 18-month experiment, the ^{137}Cs activity in the samples decreased to 720–903 Bq. Therefore, the measured ^{137}Cs activities in the samples decreased by 0.81–4.75% after 18 months.

Table 1. The ^{137}Cs activity in the samples after different exposure periods.

Treatments	6-month exposure		12-month exposure		18-month exposure	
	May 2016	Nov. 2016	May 2016	May 2017	May 2016	Nov. 2017
Absolute control (dry soil)	924±30	924±32	929±57	919±47	890±24	883±33
Positive control (wet soil)	692±51	686±51	694±27	686±23	743±59	720±63
Positive control 2 (wet soil + molasses,and wheat bran)	845±31	824±30	842±18	811±15	836±25	812±25
EM-1 1% + molasses	912±20	893±20	916±22	861±23	917±25	873±26
EM-1 5% + molasses	901±15	879±21	915±25	886±20	904±23	880±15
EM-1 10% + molasses	898±40	860±44	901±48	869±20	916±32	874±32
EM-bokashi 1%	929±41	901±39	913±36	862±29	908±45	865±47
EM-bokashi 5%	918±29	877±26	880±34	829±32	886±25	845±21
EM-bokashi 10%	854±48	830±36	901±48	869±37	852±43	825±35
EM-1 10%	904±24	886±23	912±32	887±37	917±19	903±22

The decrease in the ^{137}Cs activity after 6 months due to radioactive decay is 1.1% because the half-life of the ^{137}Cs radioisotope is 30.17 years. The reduction rates in the dry soil and wet soil control groups were nearly the same as the usual decay rate (Fig. 1).In all the other treatment groups, the rates were higher than the usual decay rate. In particular, a significant difference was detected in the wet soil with OM, 5% and 10% EM, and 1%, 5%, and 10% EM Bokashi treatments. The difference between these treatments and the usual decay rate was 1.3–3.3% (Fig. 1).

The usual decrease in the ^{137}Cs activity for 12 months is 2.3%. The actual reduction rates were higher than the usual decay rate for all treatments except the dry soil and wet soil treatments (Fig. 2). In addition, there were significant differences in the reduction rates of the wet soil with organic matter, 1% EM, and 1% and 5% EM Bokashi treatments compared to the usual decay rate. The difference between these treatments and usual decay rate was 1.3–3.7%.

The usual decrease in the ^{137}Cs activity for 18 months is 3.4%, which is shown by the dashed line in Fig. 3. The 1% EM and 1% and 5% EM Bokashi treatments had a significantly higher reduction rate than the usual decay rate (Fig. 3). The difference between these treatments and the usual decay rate was 1.2–1.4%.

Therefore, in the 6, 12, and 18 months exposure experiments, the 1% and 5% EM Bokashi treatments consistently had significantly higher reduction rates than the usual decay rate. While not consistent, the 1% EM treatment had a significantly higher reduction rate for the 12 and 18 months exposure experiments. Therefore, the application of EM and EM Bokashi to soil appears to have an impact on the radioactive decay rate.

The physical half-life of radionuclides is known to be very stable. Therefore, this observation is counter to the current understanding of radionuclides. However, similar results have been confirmed in another experiment and in field observations.

A similar laboratory experiment was conducted in Fukushima [18]. In the 690 days of the experiment, the reduction in the ^{137}Cs activity in soil treated with EM with molasses was confirmed to be greater than the usual decay value.

Monitoring of the $^{134+137}$Cs contamination of soils in paddy fields with long-term (more than 20 years) and short-term applications of EM in the Fukushima area shows significantly more rapidly decreasing radioisotope activities compared to the physical decay value [19]. Furthermore, the rate of decreasing radioactive cesium concentration in the soil on another farm was higher in a field treated with EM-fermented cow manure compost than in an adjacent area using a chemical fertilizer [20].

According to the conventional scientific paradigm, the rate of radioactive decay is not affected by operations such as heating, the addition of water, or the addition of organic matter. It should only decrease exponentially according to the law of radioactive decay. However, according to the hypothesis of bio-transmutation [21–23], some microorganisms may alter the rate of radioactive decay.

Vysotskii and Kornilova [21] have indicated that microorganisms can accelerate the radioactive decay rate. He

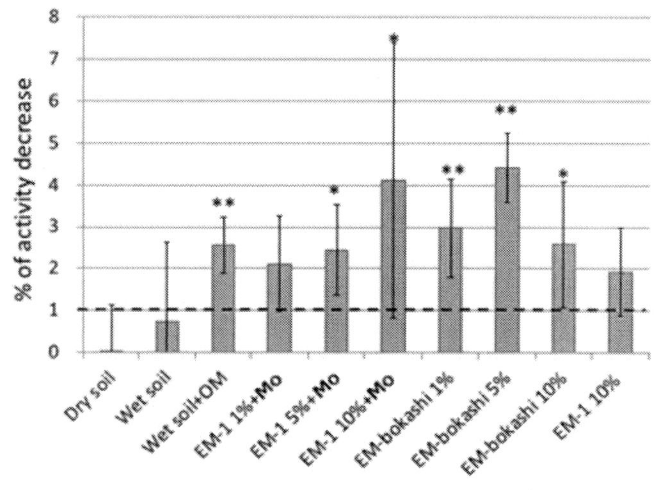

Figure 1. The reduction rate of the ^{137}Cs activity in soil samples after the 6-month exposure treatments (mean±confidence interval, $p = 0.054$). The dashed line indicates the calculated decrease in the activity due to radioactive decay with a half-life, $T_{1/2}$, of 30.17 years. * and ** indicate significant differences with respect to the calculated value at $p < 0.05$ and $p < 0.01$, respectively.

observed an increased decay rate of ^{137}Cs in experiments with an MCT (microbial catalyst transmutator) under the presence of different additional salts and propose his explanation of it. He believes that nuclear transmutations in mild

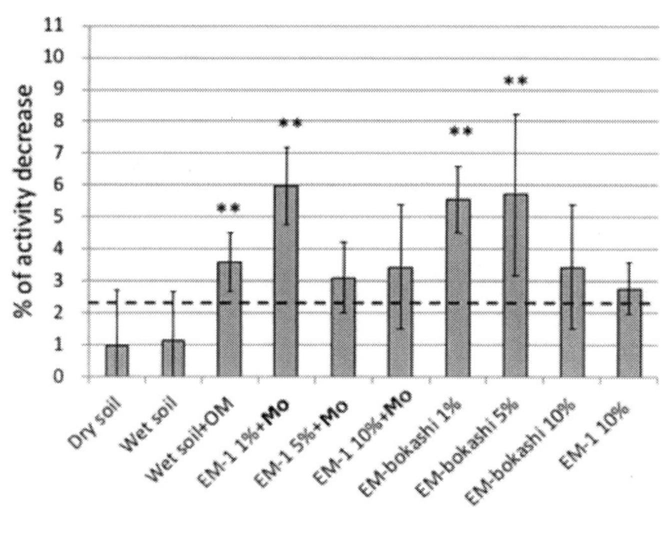

Figure 2. The reduction rate of the ^{137}Cs activity in soil samples after the 12-month exposure treatments (mean±confidence interval, $p = 0.05$). The notation is the same as in Fig. 1.

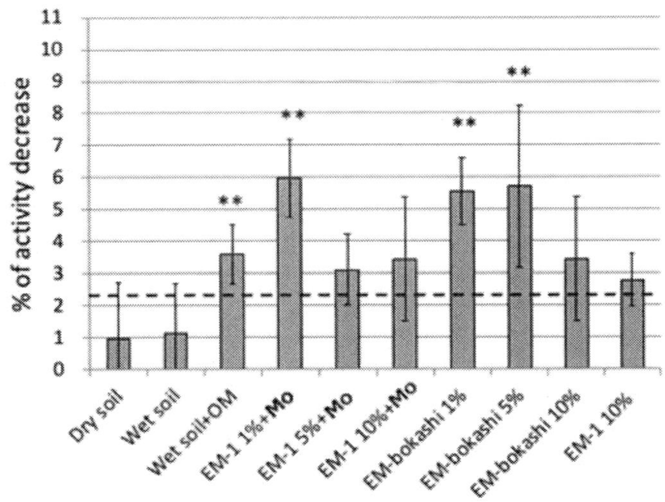

Figure 3. The reduction rate of the ^{137}Cs activity in soil samples after the 18-month exposure treatments (mean±confidence interval, $p = 0.05$). The notation is the same as in Fig. 1.

conditions are possible due to coherent correlated states of interacting particles.

Lu and Zhang [24] had shown that negative ions of hydrogen could be generated in photochemical biotic systems under the sunlight. Negative ions of hydrogen react with cesium radioactive isotope with formation of stable barium isotope same way as negative hydrogen ions can react with potassium-39 with formation of calcium-40. It is another possible explanation of the effect we seen in our experiment.

So, the obtained results suggest that EM accelerates ^{137}Cs decay in soil. Therefore, during 6 and 12 months of the experiment, soil samples with the addition of organic matter only (positive control 2) loss activity of ^{137}Cs significantly more than usual decay rate. It can indicate that not only microorganisms from EM-1 or EM-bokashi can cause the effect of increasing rate of radioactive decay of ^{137}Cs, but species of microorganism naturally inhabit the soil have a similar ability. Nothing is surprising because strains included in the EM composition were selected from the natural environment [16]. Simultaneously loss of the radioisotope activity in the soil samples with EM-1 or EM-Bokashi reached the highest level. Therefore, the impact of EM on the decay rate of ^{137}Cs is more pronounced in comparison with naturally occurring microorganisms.

The differences of ^{137}Cs activity decreasing between usual and observed levels in the soil samples with EM-1 or EM-bokashi reached the maximum level after 12 months of the exposure. The effect did not increase after 18 months of exposure. We can explain that by the death of microorganisms in the samples where EM-bokashi was added. However, new microorganisms were added repeatedly in the samples with EM-1 according to the design of the experiment.

Depletion of ^{137}Cs stock in the bio-available forms could be another explanation of the effect. Cesium firmly absorbed by clay minerals in the soil. Only a small part of the stock of this element in soil is available for plants or microorganisms intake. If we assume that the rate of the ^{137}Cs decay increases inside the cells of some species of microorganisms then we must consider only available for them part of the radionuclide. The additional analysis had shown that only 1.0–1.5% of ^{137}Cs in the soil samples was in soluble and exchangeable forms. These forms are mainly available for microorganisms. It is complicated to register the decline of soluble and exchangeable forms of ^{137}Cs in the soil samples due to intake by microorganisms on the background of 100-folds more total activity. Therefore, the decline of bio-available forms may the cause of decreasing the effect of EM on the rate of ^{137}Cs decay in the soil samples.

In view of the magnitude of these findings for mitigating radiation disasters, we need to obtain additional confirmations of this phenomenon under different conditions. If confirmed, it could be used for the remediation of areas contaminated with radioactive isotopes of cesium.

4. Conclusions

The results of the experiments suggest that EM accelerate the radioactive decay of ^{137}Cs in soil. Further studies are needed to understand the impact of EM on ^{137}Cs activity.

References

[1] Y.A. Izrael, Chernobyl radionuclide distribution and migration, *Health Phys.* **93** (5) (2007) 410–417.

[2] T.J. Yasunari, A. Stohl, R.S. Hayano, J.F. Burkhart, S. Eckhardt and T. Yasunari, Cesium-137 deposition and contamination of Japanese soils due to the Fukushima nuclear accident., *Proc. Natl. Acad. Sci.* US A **108** (49) (2011) 19530–19534.

[3] K. Wada, T. Yoshikawa and M. Murata, Decontamination work in the area surrounding Fukushima Dai-ichi nuclear power plant: another occupational health challenge of the nuclear disaster, *Arch. Environ. Occup. Health* **67** (3) (2012) 128–132.

[4] S. Fesenko, P. Jacob, A. Ulanovsky, A. Chupov, I. Bogdevich, N. Sanzharova, V. Kashparov, A. Panov and Y. Zhuchenka, Justification of remediation strategies in the long term after the Chernobyl accident, *J. Environ. Radioact.* **119** (2013) 39–47.

[5] N. Yamaguchi, I. Taniyama, T. Kimura, K. Yoshioka and M. Saito, Contamination of agricultural products and soils with radiocesium derived from the accident at TEPCO Fukushima Daiichi Nuclear Power Station: monitoring, case studies and countermeasures, *Soil Sci. Plant Nutr.* **62** (3) (2016) 303–314.

[6] N. Kato, N. Kihou, S. Fujimura, M. Ikeba, N. Miyazaki, Y. Saito, T. Eguchi and S. Itoh, Potassium fertilizer and other materials as countermeasures to reduce radiocesium levels in rice: Results of urgent experiments in 2011 responding to the Fukushima Daiichi Nuclear Power Plant accident, *Soil Sci. Plant Nutr.* **61** (2) (2015) 179–190.

[7] S. Savci, Investigation of effect of chemical fertilizers on environment, APCBEE Procedia, vol. 1, pp. 287–292, Jan. 2012.

[8] S.A. Khan, R.L. Mulvaney and T.R. Ellsworth, The potassium paradox: Implications for soil fertility, crop production and human health, *Renew. Agric. Food Syst.* **29** (1) (2014) 3–27.

[9] S. Ehlken and G. Kirchner, Environmental processes affecting plant root uptake of radioactive trace elements and variability of transfer factor data: a review, *J. Environ. Radioact.* **58** (2–3) (2002) 97–112.

[10] J.R. Lloyd and J.C. Renshaw, Microbial transformations of radionuclides: fundamental mechanisms and biogeochemical implications, *Met. Ions Biol. Syst.* **44** (2005) 205–40.

[11] G.Z. Gutzeva, G.A. Leferd and S.O. Gaponenko, Impact of mineral and bacterial preparations on lettuce productivity and accumulation of ^{137}Cs in plants, in *Proc. 5th Congr. of Belarussian Society of Soil and Agrochemistry Scientists*, Minsk, 2015, pp. 302–305.

[12] G.Z. Gutzeva, A.N. Nikitin and N.V. Telitsina, Accumulation of ^{137}Cs by biomass of barley in dependence from conditions of bacterization and mineral nutrition, in *Proc. Int. Sci. Conf. Radiobiology: Mayak*, Chernobyl, Fukushima, Gomel, 2015, pp. 72–75.

[13] N.V. Shamal, E.A. Klementieva, R.A. Korol, R K. Spirov, S.O. Gaponenko, A.N. Nikitin et al., Application of EM-technologies for growing lettuce on the contaminated soils, in *Proc. Int. Conf. Current Status and Perspectives of Innovation Development of Vegetable Growing,* Samochvalovichi, 2015, pp. 45–47.

[14] A.N. Nikitin, Prospects for the use of microbiological preparations for regulating the accumulation of man-made radionuclides by plants, in *Proc. Int. Sci. Conf. Radiobiology: Mayak,* Chernobyl, Fukushima, Gomel, 2015, pp. 157–161.

[15] N. Shamal, E. Klementjeva, R. Korol, S. Gaponenko, R. Spirov, A. Nikitin et al., Application microbiological preparation EM-1 and mineral sorbent for growing lattice on the soils contaminated by radionuclides, in *Proc. 3d Int. Conf on Radiation and Applications in Various Fields of Research*, 2015, p. 574.

[16] T. Higa, Effective microorganisms: a biotechnology for mankind, in *Proc. 1st Int. Conf. on Kyusei Nature Farming*, USDA, Washington, DC, 1991, pp. 8–14.

[17] M. Olle and I.H. Williams, Effective microorganisms and their influence on vegetable production – a review, *J. Horticultural Sci. Biotechnol.* **88** (4) (2013) 380–386.

[18] S. Okumoto, M. Shintani and T. Higa, Possibilities of effective microorganisms (EM) technology for reducing radioactive cesium contamination in soil, in *Proc. Int. Sci. Conf. Chernobyl: 30 years later,* Gomel, 2016, pp. 157–159.

[19] S. Okumoto, M. Shintani and T. Higa, Analysis of radioactive cesium in paddy fields applied with effective microorganisms (EM 1®) in Fukushima, in *Proc. Int. Sci. Conf. Radiobiology: Challenges of the XXI Century*, Gomel, 2017, pp. 20–23.

[20] S. Okumoto, M. Shintani and T. Higa, Influence on the suppression of transfer of radioactive cesium from soil to grass using cow manure compost ant its effluent fermented by effective microorganisms™, in *Proc. Int. Sci. Conf. Radiobiology*: Mayak, Chernobyl, Fukushima, Gomel, 2015, pp. 20–23.

[21] V. I. Vysotskii and A.A. Kornilova, Nuclear *Fusion and Transmutation of Isotopes in Biological Systems*, MIR, Moscow, 2003.

[22] J.-P. Biberian, Biological transmutations: historical perspective, *J. Condensed Matter Nucl. Sci.* **7** (2012) 11–25.

[23] H. Kozima, Biotransmutation as a cold fusion phenomenon, in *Proc. JCF16*, 2016, pp. 216–239.

[24] G. Lu and W. Zhang, Photocatalytic hydrogen evolution and induced transmutation of potassium to calcium via low-energy nuclear reaction (LENR) driven by visible light, *J. Mole. Catalysis* (China) **31** (5) (2017) 401–410.

J. Condensed Matter Nucl. Sci. 29 (2019) 238–248

Research Article

Comparison of NANOR®-type LANR Components to ^{238}Pu as a Heat Source for Space Flight

Mitchell R. Swartz*,†

JET Energy Inc., Wellesley Hills, MA 02481, USA

Abstract

Plutonium-238 creates heat by its natural radioactive decay to alpha particles and generates electricity using thermoelectric generators (0.57 W/g, but ∼25 mW/g when shielding and supports are considered). The technology has been proven to work on more than two dozen US space vehicles. Yet, two problems exist: (1) NASA has currently only ∼1 kg of isotope exceeding the specifications requisite for space power delivery and (2) the safety of ^{238}Pu. There is a need to consider alternative LANR systems. As a heat producing source, the Series 6 NANOR®-type components deliver up to 150 mW of excess power, and so less than a dozen of them could replace 1 g of ^{238}Pu in the power density metric. If present limitation of higher temperatures can be achieved, then the most efficient and safe deep space energy production systems in the future may include preloaded LANR components.

Keywords: Electricity production, NANOR®-type component, Plutonium, Safety, Space exploration, Space power delivery, Thermoelectricity

1. Space Exploration: Big Changes are Coming

NASA will now permit some future spacecrafts to use ^{238}Pu to heat a thermoelectric generator for electrical power, and "keep spacecraft elements warm" – which is a reversal from its previous position (Fig. 1) [1,2]. This recurrent technology will rapidly advance planetary science missions such as those planned to the far side of the moon (given its two-week lunar night [3–6]).

However, given the other dire issue associated with of ^{238}Pu, one must consider if there are other alternatives. Is it really the best or the safest? To facilitate the analysis, after reviewing both energy production sources, ^{238}Pu and lattice assisted nuclear reactions (LANR) [7–16] in general, and the NANOR®-type LANR component, in particular, Section 4.3 [15,16], several metrics will be used and addressed, including power density, energy density, maximum temperature achieved, temperature differential achieved, lifetime, and safety.

*Dr. Mitchell R. Swartz ScD, MD, EE, E-mail: moac@nanortech.com.

†Also at: Society for Planetary SETI Research.

Figure 1. A spontaneously warm pellet of pure ^{238}Pu. In the picture, the metallic pellet is red hot after its insulation (to contain the heat) was removed. As with LANR, this so-called "free energy" is actually from nuclear de-excitations.

2. Background

2.1. Plutonium is an unusual element

There are 15 different plutonium isotopes, all radioactive, emitting alphas (helium nuclei) and gamma radiation (high energy penetrating ionizing radiation) as they decay. All Pu isotopes are fissionable and two are fissile (capable of supporting a chain reaction by sustaining fission from neutrons of any energy). Each fission yields ~200+ MeV (82 TJ/kg). Plutonium-244 is the most stable isotope with the longest half-life, 82 million years, and the isotope which is usually cited in tables. It existed on Earth in trace amounts since before the atomic age only in uranium deposits, such as at Oklo, Africa. It is not found elsewhere in the Earth's crust. Things changed in the atomic age, and today many tons of plutonium drift through the Earth's biosphere because of atmospheric weapons testing which continued through the 1960s. Some of the most relevant plutonium isotopes in the atomic age are listed in Table 1.

Pure metallic plutonium has six crystalline structures, the most known for any element. The alpha phase is brittle and hard like cast iron, and ignites rapidly as it is oxidized in air to form PuO_2, which is the commercial ceramic form. The other metallic phases are each different, some only workable as alloys.

Table 1. Plutonium isotopes.

^{238}Pu	Half-life 87.5 years, alpha decay to ^{234}U, releasing 5.6 MeV
^{239}Pu	Fissile (half-life 24,000 years, alpha decay to ^{234}U)
^{240}Pu	Fertile (half-life 6,560 years, alpha decay to ^{236}U)
^{241}Pu	Fissile (half-life 14.4 years, beta decay to ^{241}Am)
^{242}Pu	Half-life 374,000 years, alpha decay to ^{238}U
^{244}Pu	Which is usually cited, and alpha decays to ^{240}U

2.2. ^{238}Pu Synthesis and production

There are two different grades of plutonium: reactor-grade and weapons-grade. They differ slightly isotopically. Both are a potential proliferation risk. Reactor-grade plutonium comes from commercial nuclear reactors. It is obtained from spent nuclear fuel, and contains many isotopes of plutonium generated after the fuel has been irradiated ("burned"') for several (3+) years.

This plutonium is important for these commercial reactors because although ^{238}Pu makes up only one or two percent of spent fuel, it still creates a significant fraction of the reactor heat. This is because of its short half-life.

The first problem is that commercial reactor-grade plutonium is not useful for producing ^{238}Pu for RTGs because of the tremendous difficultly in isotopic separation of pure ^{238}Pu. Therefore, today ^{238}Pu is obtained from uranium which is fast-irradiated during a brief 2–3 months interval within a weapons-grade plutonium production reactor.

Plutonium-238 was first synthesized by Glenn Seaborg, in 1940, using ^{238}U as a target for deuterons. In this pathway, ^{238}Np is first formed which then decays to ^{238}Pu. A second production pathway for ^{238}Pu uses irradiation of ^{237}Np, obtained from reactor fuels and target recovery projects. That irradiation forms ^{238}Np first, which then decays to ^{238}Pu. This pathway was undertaken at Savannah River. The third pathway is to have ^{238}Pu produced as a byproduct of weapons-grade ^{239}Pu production (Fig. 2). At the Savannah River weapons reactor before 1988, ^{238}Pu (contaminated with $\sim$16% ^{239}Pu) difficult-to-achieve production recovery efficiencies improved from 3 to 98% finally permitting space exploration in a plethora of space craft (Section 3.2).

Plutonium is now made in nuclear reactors and salvaged from dismantled nuclear weapons. The most common plutonium isotope formed in a typical nuclear reactor is fissile ^{239}Pu, formed by neutron capture from ^{238}U (followed by beta decay). When fissioned, it yields the same energy as the fission of ^{235}U. It takes $\sim$10 kg of nearly pure ^{239}Pu to make a weapon. This much material requires 30 MW-years of reactor time, special processing, and a three month burn of uranium.

2.3. End of ^{238}Pu production in 1988 and current prevalence

The US stopped all ^{238}Pu production at the Savannah River Site reactors in 1988. In June 2000, the US and Russia agreed to each dispose of 34 tons of weapons-grade plutonium by 2014. To make up the difference in availability and

Figure 2. Irradiated neptunium at the ORNL fuel core. The blue color is Cherenkov radiation which emanates from spent fuel before it is removed from the High Flux Isotope Reactor (HFIR).

that required, since 1992, the USA bought ∼16.5 kg ^{238}Pu from Russia for American spacecraft, until production and sales ceased in 2009. The US neptunium inventory was transferred to Idaho National Laboratory (INL) in 2008. In October 2016, Russia suspended its Agreement upon "a (Russian) presidential decree" [17].

Today, the total world generation of new reactor-grade plutonium is ∼70 tons per year. It is extracted from 1300 tons of used, burned fuel, but as discussed above, it is contaminated and not useful for space and similar applications. In the US, ^{238}Pu inventory includes NASA (civil space) and other national security applications. As of March 2015, a total of 35 kg (77 pounds) of ^{238}Pu was available for civilian space uses. However, only 1 kg has had ^{239}Pu and the other contaminants removed resulting in an isotopic distribution meeting NASA specifications requisite for space power delivery.

2.4. Future ^{238}Pu plans for production

In February 2013, a small amount of ^{238}Pu was produced by the Oak Ridge's High Flux Isotope Reactor, and by December 2015, Oak Ridge National Laboratory reported that its researchers had successfully produced 50 g (1.8 ounces). To recover from the paucity of useful ^{238}Pu, the US production rate target is 0.4 kg now and 1.5 kg ^{238}Pu per year by 2025 [18]. However, generating larger quantities is quite complicated and involves moving a lot of radioactive material around (Fig. 3).

In March 2017, Ontario Power Generation (OPG) and its Venture Arm, Canadian Nuclear Partners, announced plans to produce ^{238}Pu as a second source for NASA [19]. The plan is to use PNNL's ^{237}Np targets, shipped to CNL's Chalk River Laboratories in Ontario, and there to be assembled into reactor bundles, then to have them sent to OPG's Darlington High Flux Isotope Reactor, there to be irradiated for ∼72 days to generate ^{238}Pu.

Final removal requires the bundles returned to CNL for disassembly and chemical processing. The DOE's alternate complicated pathway uses a plutonium pipeline involving waypoints of ORNL, INL and Los Alamos National Laboratory. There is also a public–private partnership led by a company called Technical Solutions Management (TSM).

3. The Use of Plutonium in Space

3.1. The use of plutonium in space as a heat source

Obviously, space exploration needs heat and electricity, but solar irradiant power is too weak and just not feasible at the solar system rim. Therefore radioisotopes are one logical alternative to be used to heat instrumentation and to produce electricity. There are many radioisotopes known, but only 22 are safe enough, plentiful enough, and produce enough heat for the useful purpose of heating a deep-space probe (US National Academy of Sciences, 2009). Plutonium is a major contender, and a third of the heat generated by most nuclear power plants comes from plutonium.

Plutonium-238 creates heat by its natural radioactive decay to alpha particles. They collide with the nearby materials heating them up through collisions. Each gram of ^{238}Pu has 1/238 mol, and therefore 2.53×10^{21} plutonium atoms. The activity (decay rate of this sample) is 634 GBq with a decay rate of 634 billion per second. Each emitted alpha starts with 5.59 MeV and delivers 8.96×10^{-13} J by its deceleration in the surrounding material which creates *de novo* heat. As it does, therefore each gram of ^{238}Pu spontaneously generates 0.568 W of thermal power ("excess" thermal power). This continues decreasing through the half-life of 87.7 years.

In summary, the specific decay thermal power of ^{238}Pu is 0.57 W/g which enables its use as a heat source, and as an electricity source in Radioisotope Thermoelectric Generators (RTGs) [20,21]. However, in the real world, RTGs necessarily have other materials including shielding which yields a net useful specific power of about 25W/kg plutonium. This makes the real world, effective decay thermal power of ^{238}Pu only ∼25 mW/g. As discussed below, that thermal power is less than most Series 6 NANOR®-type components' excess power outputs (Section 5).

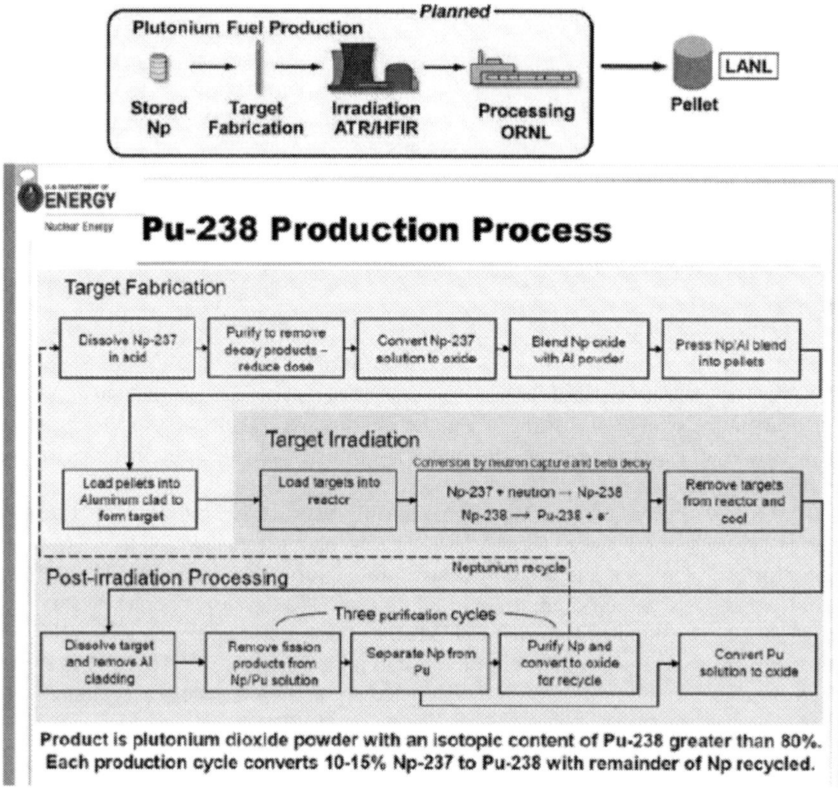

Figure 3. Complex pathway of Planned ^{238}Pu Production. *Top*: overview bottom: DOE Process for the planned production of ^{238}Pu.

3.2. ^{238}Pu Heat production in space flight

Today, from Earth to the rim of the solar system, radioisotope thermoelectric generators (RTGs) are used to heat and provide electricity to spacecraft through many decades. Plutonium is the material of choice for heat and electrical power for space missions. More than two dozen US space vehicles, and many planetary and galactic probes, have depended on ^{238}Pu heated RTGs.

These include the twin Voyager 1 and Voyager 2 probes still moving beyond the rim of the solar system, the Cassini Saturn orbiter, the Viking 1 and Viking 2 Mars landers, the Curiosity Mars rover, Pioneer 10 and 11, the SNAP-27 RTG on the moon for science experiments by the Apollo astronauts, and Transit Navy navigation satellites.

The Viking landers had 14 kg of ^{238}Pu for 30 W output. NiCd batteries (28 V, 8 Ahr) were used to store the energy, and to enable higher power outputs when needed. The Cassini spacecraft carried three generators with 33 kg of plutonium oxide providing 870 W in orbit around Saturn. In the Pathfinder, Sojourner, and Polar Landers, the RTGs were used solely for heating, with electricity provided by solar cells.

The future will include New Horizons missions and the Mars Science Laboratory for long-term nuclear power generation.

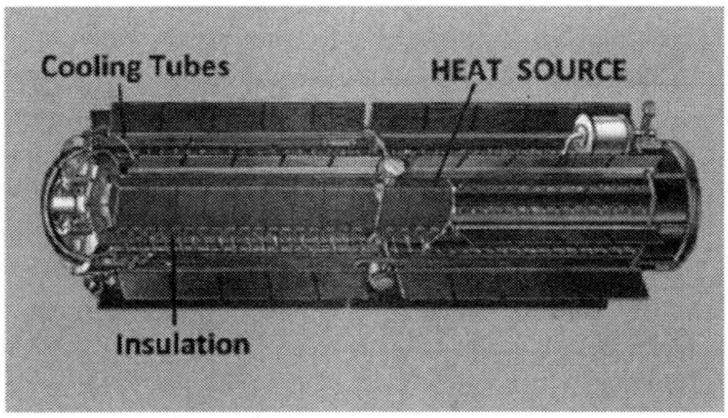

Figure 4. Actual RTG used in a US space probe (after GE GPHS-RTG).

3.3. ^{238}Pu Heat production and RTGs

Plutonium-238 is therefore used as a heat source in RTGs. RTGs are reversed thermocouples, which convert the heat into its most important fungible product: electricity (Fig. 4). The first RTG used ^{210}Po as the heat source. It was invented in 1954 by Ken Jordan and John Birden. On January 1, 1957, they entered into an Army Signal Corps contract (R-65-8- 998 11-SC-03-91) and began research using ^{210}Po.

3.4. ^{238}Pu in future NASA missions

The US 2019 Planetary science budget of $2.2 billion offers new opportunities. Within it, there is funding for NASA's current planned RTG, which is called the Multi-Mission Radioisotope Thermoelectric Generator (MMRTG). It will contain 4.8 kg of ^{238}Pu as the ceramic oxide, with the radioisotope divided into eight 290 W RTG subunits. The total of 4.8 kg ^{238}Pu oxide will produce 2 kW thermal power to generate 110 W electric power (2.7 kWh/day). The first MMRTGs are scheduled to go to Curiosity and the Mars 2020 rover. Another may go to Dragonfly, which is a finalist for a "New Frontiers" future mission to the moon.

4. Competing Heat-Production Systems

4.1. LANR systems and components

There are few choices because of plutonium has availability and experience, albeit the safety issues (*vide infra*). The other choices are conventional, unobtainable, and CF/LANR systems. Alkaline fuel cells were used in Apollo and on the Space Shuttle. But they are neither an efficient energy production or energy storage system. Conventional nuclear energy production systems are of two types, fission and fusion. In fission, fissile elements (e.g. plutonium or uranium) have their heavy nuclei divide. They have been discussed elsewhere and have issues of radioactivity (direct and induced), safety, and lack of energy source (e.g. photovoltaics) or being energy storage and not production. Therefore, there is a need/opportunity to consider CF/LANR systems.

4.2. Lattice assisted nuclear reaction (LANR) systems

Lattice assisted nuclear reactions (LANR, also known as cold fusion and LENR, low energy nuclear reactions) are an alternative to plutonium and other radionuclides for space travel [7–11]. LANR uses hydrogen-loaded alloys to create heat and other products by enabling deuterium fusion to form an excited *de novo* helium nucleus at near room temperature [14]. Under difficult-to-achieve conditions, aqueous LANR systems produce several watts to fractions of a kilowatt of excess power. Nanomaterial LANR systems and specifically NANOR®-type portable components provide excess power in the milliwatt range. All of these systems are clean, without the harmful penetrating radiation emissions and radioactive waste that characterize other (hot) fusion sources [15,16]. The "excess heat" observed is derived from de-excitation of $^4He^*$ to its ground state via the lattice phonons with direct energy transfer to the lattice. LANR excess power density is more than 19,500 W/kg, with no radioactivity and zero carbon footprint (cf. Fig. 6) [12–16].

Conceptually, LANR systems are often described by their energy production in terms of the "microscopic" amounts of helium produced, as a reaction product [12]. In actual working aqueous LANR systems, the most significant other reaction product by mass is anodic-generated molecular oxygen [10], and it increases with time. This is very important for manned spacecraft because later, when the LANR activity decreases to a few percentage of the initial activity over time, for a 3 l system, this changes to $\sim 3.1 \times 10^{13}$ J of heat, 732,000 l of molecular oxygen (needed for life and fuel cells) and only $\sim$1 l of *de novo* helium-4 per year.

Prof. George Miley (University of Illinois, USA) has been pivotal in suggesting and promoting cold fusion cells for use in outer space [7]. His devices are multi-layered thin films of palladium and nickel in a heavy water solution. Superconducting quantum interference devices (SQUID) have confirmed ultra-dense states of deuterons within palladium crystal defects after repeated loading by deuterons from the heavy water [22].

4.3. The NANOR®-type LANR component

Dr. Mitchell Swartz and associates have crafted both aqueous [12,13] and dry nanostructure [15,16] LANR components. Some have been constructed into preloaded, dry LANR devices capable of functioning as actual portable heaters. They have been shown in open demonstrations at MIT since 2012, and currently are fabricated to perform at the multiwatt level. For the lower power, nanostructured materials (Series 6 materials) the excess energy gain compared to driving input energy is up to 20–80 times input electrical power; with reasonable reproducibility and controllability. These would be very useful for space flight and distributed energy sources, especially because they are preloaded and driven by a DC electrical current. These devices deliver heat (or electricity) when wanted and not at a natural unchanging decay rate.

5. Analysis and Comparison by Metrics

With that background, this section continues with the important metrics to semiquantitatively determine whether, and if so –how, the NANOR®-type component might be important for space exploration, and how does it compare to ^{238}Pu for that and other applications.

5.1. Metric 1 – Power density

As a heat producing source, the Series 6 NANOR®-type components delivered up to 150+ mW of excess power with circa 60–200 mg of active material was at the core (Fig. 5). As discussed above, each gram of ^{238}Pu generates 0.568 W unshielded, and $\sim$25 mW shielded, thermal power per gram. Therefore, with respect to power density, the ^{238}Pu– and NANOR®-type components are comparable in output power.

5.2. Metric 2 – Energy density

With respect to energy density, it is important to first consider lifetime, and then apply the power density. In this model, to the first order, the degradation of materials and thermoelectric converters (from gamma and alpha damage from the

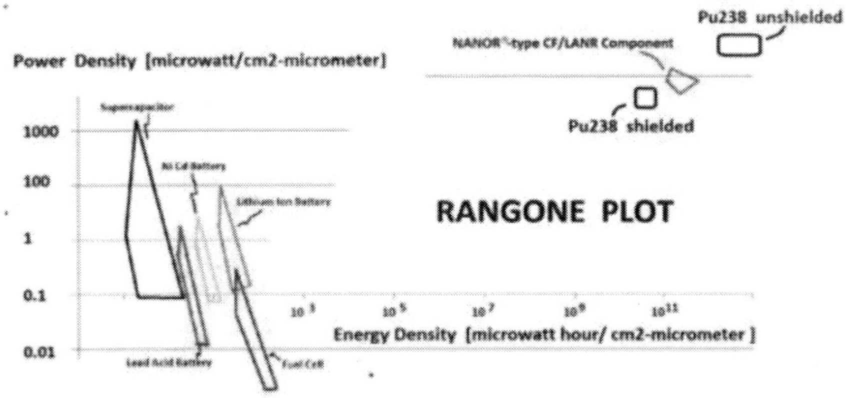

Figure 5. This graph is a Rangone-type plot of both the power density and the energy density of typical energy production and energy conversion devices/materials. These ranges from supercapacitor, to the lead acid battery, to the fuel cell, to the NANOR®-type cold fusion (lattice assisted nuclear reaction) component, to ^{238}Pu. It can be seen that the latter two are one of the most efficient energy density devices known. Only the NANOR®-type component using LANR has zero carbon footprint.

^{238}Pu) will be ignored, but are an issue. It is also important to recognize that the analysis of the complete energy density available in the cold fusion systems is approximate. It is extremely unlikely that all of the deuterium could be used, so the numbers below, although derived on the quantity of deuterium in each of the components, are absolutely upper limits to what is available.

5.3. Metric 3 – Lifetime of energy production

Although the specific excess power is small, and the fact that the longevity of ^{238}Pu clearly exceeds the lifetime of NANOR®-type technology at this time, a comparative analysis of the two for potential lifetime "excess" energy produced (and thus energy density) is still needed. The energy density of ^{238}Pu is derived by the product of its half-life (87.5 years) and its (exponentially decreasing) power density. This is considerable, and as a result the Voyager spacecraft remain in communication after 41 years.

If we consider the dry preloaded NANOR®-type component, then with a typical Series 6 component delivering 150 mW excess power, the ~300 micromoles of D in this unit could theoretically deliver approximately 100 MJ over 21 years. Although this is shorter than the ^{238}Pu half-life, the aqueous LANR systems show that this can be greatly increased.

For example, if we consider the aqueous MOAC #3 nickel system [23,24], which is filled with 3000 ml of ordinary water. It can deliver circa 40 W excess power, and there is enough deuterium (55 mmol) to run for 13.3 years delivering 16.7 GJ excess energy. If we consider the future anticipated MOAC #4, with palladium cathode and very high impedance heavy water, and only the use of the lower limit of 40 W excess power (since there are approximately 300 mol of deuterium) this would make available 90.3 TJ over 71,600 years.

5.4. Metric 4 – Maximum temperature achieved

It is critically important to consider the peak temperature and temperature differential achieved. Both are relevant because, ultimately, it is the temperature differential that determines the thermoelectric conversion efficiency to make electricity, the most fungible, therefore important, type of energy.

Plutonium in space using PuO_2 pellets achieves ~1000°C peak local temperature. It is not clear if the NANOR® type components can achieve that at this time; mainly because so far the main metric in our experimental runs have been to maximize the incremental power gain factor, followed by reproducibility and the total power output achievable (in that order).

With respect to the maximum temperature, the LANR components, both aqueous and dry preloaded, have never been driven to the peak temperatures achievable for a number of reasons (safety unknown, protection of the reagents, etc.). However, the NANOR type component, a dry preloaded LANR material, is from a line of systems/materials that do have the possible characteristics heralding that the local temperature might, under suitable engineered conditions, be comparable to that achieved for ^{238}Pu in this metric, too. First, successful aqueous PHUSOR type systems have achieved 155°C, outside of 50 ml aqueous surrounding; with the system electively clamped because of loss of heavy water [13]. Second, near infrared examinations of emissions from the core of the active material when excess heat was observed indicate a color temperature between ~500 and 1000 K [25]. Third, coherent Raman anti-Stokes measurements have suggest that the effective temperature during excess heat, at core, is ~1645 K (~1372°C) [26,27].

5.5. Metric 5 – Maximum temperature differential achieved (ΔT)

The high temperature is needed to maximize the delta-T (ΔT) which determines the electrical conversion efficiency. This efficiency is a key important and limiting issue. Worse, the temperature differential in space is even more difficult to achieve. The reasons include that any additional mass for adds launch challenges and also because outer space being a vacuum necessarily makes it thermally insulating for conduction and convection. As a result, typical spacecraft cold junction temperatures are reported circa 300°C. This makes achieving a large magnitude ΔT as difficult to achieve in space as it is here on Earth.

The ΔT is the present limiting factor of many technologies including the important Advanced Stirling Engines that NASA has been developing for several decades requiring a ΔT of 500°C; which produce only 25% thermoelectric conversion efficiency and are so far unflown. Similarly, plutonium in space using PuO_2 pellets achieve only ~7.5% thermoelectric conversion efficiency at their ~1000°C the peak local temperature.

5.6. Non-ΔT systems

In this metric, it is important to consider the possibility of direct current conversion which does not requiring large ΔT. Some NANOR-type components have been crafted to elicit direct current conversion which does not require large ΔT. They use preloaded nanomaterials but only generate extremely low levels of electric power, directly [28]. Other suggestions range from MHD for de-excited alphas to electron capture directly.

5.7. Metric 6 – ^{238}Pu - Safety issues

The metric of safety involving ^{238}Pu is a very serious problem. It has the highest hazard number (#152) of all 256 known radionuclides [3]. Environmental issues, including potential contamination and human toxicity, limit the use [29]. The main threat to humans comes from its inhalation, especially when the particulate size is less than 10 μm. Additionally, as a consequence of the shorter half-life is that ^{238}Pu is about 275 times more radioactive than ^{239}Pu (i.e. 17.3 curies (640 GBq/g) compared to 0.063 curies (2.3 GBq)/g).

When ^{238}Pu was made during WW2 and the cold war, the weapons component project's Special Metallurgical Building was nicknamed "Snake Mountain" because of the secondary environmental problems that developed [30]. As a result, work stopped there in 1968.

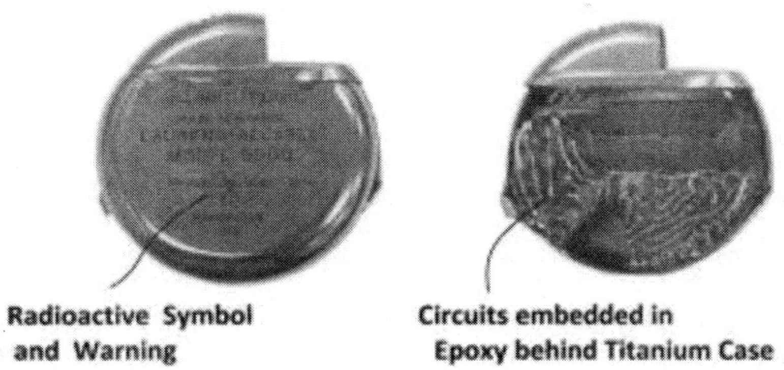

Figure 6. ^{238}Pu Cardiac pacemaker showing identification used (2.75 in. diameter, plutonium removed).

Despite these issues, with engineering ^{238}Pu has been used in several hundred cardiac pacemakers since 1966 (Fig. 6). Dose rates at the surface of the pacemaker was 15 mrem per hour, from gamma rays and neutrons. The patient's whole body exposure was circa 0.1 rem per year. This use stopped after recognition of post-cremation package leakage failures. Thus, ^{238}Pu was judged to be just too dangerous for this biomedical application.

6. Conclusion

There are substantial heat and electricity requirements to fully explore space and other planets, including for colonization. Future missions are being planned as passengers continue to patiently and eagerly wait their turn to fly. Lattice assisted (or enabled) nuclear reactions (LANR) [7–16] may have significant advantages over present systems, since they provide ecologically clean energy and potential oxygen production sources. If the maximum core temperature can be raised sufficiently, or direct electrical conversion achieved at higher efficiency, then one of the most efficient and safe deep space energy production systems in the future may be preloaded LANR components.

Acknowledgements

The author acknowledges and thanks Gayle Verner and David Nagel for their editorial assistance, and significant suggestions and comments. This effort was supported by JET Energy Inc. and Nanortech, Inc. NANOR® and PHUSOR® are registered trademarks. Technologies described here are protected by patents issued and pending.

References

[1] NASA to Allow Nuclear Power Systems for Next Discovery Mission, Jeff Foust, SpaceNews Writer, March 22, 2018.
[2] Mitch Ambrose, Oak Ridge scientists produce first ^{238}Pu in 28 years, *Phys. Today*, 4 February 2016. https://physicstoday.scitation.org/do/10.1063/PT.5.1064/abs/ (2016). https://www.space.com/40037-nasa-to-allow-nuclear-power-systems-for-next-discovery-mission.html.
[3] The Toxicity of Plutonium, Medical Research Council, HMSO London, 1975 (ISBN: 0114500304).
[4] Management of separated plutonium, *The Royal Society*, February 1998 (ISBN: 9780854035144).
[5] Plutonium Fuel: An Assessment, OECD/NEA Paris, 1989 (ISBN: 9789264132658).
[6] Radioisotope Power Systems: An Imperative for Maintaining US Leadership in Space Exploration, US National Academy of Sciences, 2009.

[7] G.H. Miley, Xiaoling Yang and Eric Rice, Distributed power sources for mars, *Mars*, 2009, pp. 213–239, ISBN 978-3-642-03629-3.

[8] M.R. Swartz, Energy sources for propulsion to, and for distributed use on mars, *J. Space Exploration* **4** (2015) 2.

[9] M.R. Swartz, T. Schuster, G. Verner, J. Gyllinsky and J. Tolleson, Paraterraforming Mars – I. Heat, electricity and oxygen, *Infinite Energy* **131** (2017) 14.

[10] M.R. Swartz, Paraterraforming Mars – II. Fueling colonies on mars (and ceres) by LANR and subsurface ordinary water-ice, *Infinite Energy* **131** (2017) 19.

[11] M.R. Swartz, NanorSat Spacecraft, *Infinite Energy* (2018).

[12] M.R. Swartz, Survey of the observed excess energy and emissions in lattice assisted nuclear reactions, *J. Scientific Exploration* **23** (4) (2009) 419–436.

[13] M.R. Swartz, Excess power gain using high impedance and codepositional LANR devicesmonitored by calorimetry, heat flow, and paired stirling engines, D.J. Nagel and M.E.Melich (Eds.), *Proc. ICCF14,* 1, 2008, p. 123, ISBN: 978-0-578-06694-3, 123, (2010), www.iscmns.org/iccf14/ProcICCF14a.pdf.

[14] M. Miles et al., Correlation of excess power and helium production during D_2O and H_2O electrolysis using palladium cathodes, *J. Electroanal. Chem.* **346** (1993) 99–117.

[15] M.R. Swartz, G. Verner, J. Tolleson and P. Hagelstein, Dry, preloaded NANOR® -type CF/LANR components, *Current Sci.* **108** (4) (2015) 595, http://www.currentscience.ac.in/Volumes/108/04/0595.pdf.

[16] M.R. Swartz and P.I. Hagelstein, Demonstration of energy gain from a preloaded ZrO_2–PdD nanostructured LENR quantum electronic device at MIT, *J. Condensed Matter Nucl. Sci.* **13** (2014) 516, www.iscmns.org/CMNS/JCMNS-Vol13.pdf.

[17] World Nuclear News,
http://www.world-nuclear-news.org/NP-Russia-suspends-plutonium-agreement-with-USA-04101601.htm.

[18] Tracey l. Bishop, Deputy Assistant Secretary for Nuclear Infrastructure Prog Office of Nuclear Energy, DoE, before the House Committee on Science, Space and Technology Subcommittee on Space US House of Representatives, on October 4, 2017; https://fas.org/nuke/space/powering.pdf, p. 21.

[19] Assessment Plutonium production alternatives,
https://www.energy.gov/sites/prod/files/NEGTN0NEAC_PU-238_042108.pdf.

[20] R.R. Furlong and E.J. Wahlquist, US space missions using radioisotope power systems, *ANS, Nucl. News* **4** (2) (1999) 26–34, http://www3.ans.org/pubs/magazines/nn/pdfs/1999-4-2.pdf.

[21] https://en.wikipedia.org/wiki/MHW-RTG and https://en.wikipedia.org/wiki/Radioisotope_thermoelectric_generator.

[22] George, H. Miley, Xiaoling Yang, Hugo Leon, Prajakti Joshi Shrestha and Heinz Hora, Small power cells based on LENR, *NIST Nanotech* **3** (2009) 115–117, ISBN 978-1-14398-1784-1, https://briefs.techconnect.org/wp-content/volumes/Nanotech2009v3/pdf/10255.pdf.

[23] M.R. Swartz, Increase of an anti-Stokes peak at the cathode of an electrically-driven, active aqueous nickel/H_2O/Pt system, *J. Condensed Matter Nucl. Sci.,* to appear in Vol. 29.

[24] M.R. Swartz, Charles Haldemann, Alan Weinberg and Brian Ahern, Possible deuterium loss during excess heat from ordinary water-carbonate electrolyte using nickel, *J. Condensed Matter Nucl. Sci.*, to appear in Vol. 29.

[25] M.R. Swartz, Gayle Verner and Alan Weinberg, Non-thermal near-IR emission linked with excess power gain in high impedance and codeposition Phusor-LANR devices, in *Proc. the 14th Int. Conf. on Condensed Matter Nucl. Sci.,* D.J. Nagel and M.E. Melich (Eds.), p. 343, 2010.

[26] M.R. Swartz and Peter L. Hagelstein, Increased PdD anti-Stokes peaks are correlated with excess heat mode, *J. Condensed Matter Nucl. Sci.* **24** (2017) 130–145.

[27] M.R. Swartz, Optical detection of phonon gain distinguishes an active cold fusion/LANR component, *J. Condensed Matter Nucl. Sci.* **20** (2016) 29–53, www.iscmns.org/CMNS/JCMNS-Vol20.pdf.

[28] M.R. Swartz and C. Entenmann, Direct electricity from LANR nanostrucured NANOR devices, in preparation.

[29] Some of the significant criticism has been discussed on Roger Launius' Site,
https://launiusr.wordpress.com/2015/10/30/protesting-cassinis-launch/.

[30] cf. Snake Mountain; https://en.wikipedia.org/wiki/Plutonium-238.

J. Condensed Matter Nucl. Sci. 29 (2019) 249–259

Research Article

A Simple Calculation of the Inter-nucleon Up-to-down Quark Bond and its Implications for Nuclear Binding

N.L. Bowen*

Colorado Mountain College, Glenwood Springs, Colorado, CO, USA

Abstract

This paper describes an interesting and potentially significant phenomenon regarding the properties of up and down quarks within the nucleus, and how the possible inter-nucleon bonding of these quarks may affect the bonding energy of the nuclear force. A very simple calculation is used, which involves a bond between two inter-nucleon up and down quarks. This simple calculation does not depend on the type or mechanism for the bond. Furthermore, this simple calculation does not specify the shape or structure for the nucleus. This calculation only examines the energy of all possible up-to-down inter-nucleon bonds that may be formed within a quantum nucleus. A comparison of this total energy is made to the experimental binding energy with excellent results. The potential significance of this finding is discussed.

Keywords: Inter-nucleon bond, Nuclear binding energy, Up and down Quarks

1. Introduction

The nuclear force is defined as the force which binds the protons and neutrons together within a nucleus. One of the currently accepted models of the nuclear force is the liquid-drop model [1]. This model of the nuclear force uses the Weizsäcker formula [2] to predict the binding energies of nuclides. The Weizsäcker formula is a curve-fitting formula that uses five parameters, plus one conditional logic statement, in order to achieve its results. These parameters are selected to empirically curve-fit the equation to the experimental data. The liquid drop model is considered to be a "semi-classical" model of the nuclear force [3], rather than a quantum model.

Another currently accepted model of the nuclear force is the shell model, which uses magic numbers to explain certain nuclear behavior. The nuclear shell model is similar to the electronic shell model, which describes the electrons orbiting around an atom. The nuclear shell model, however, cannot predict the. nuclear binding energy, and it makes no attempt to do so in any description of this theory.

A third currently accepted model of the nuclear force is the residual chromo-dynamic force model (abbreviated as the RCDF model in this paper). Before describing this residual chromo-dynamic force, it is useful to mention a few

*E-mail: nbowen@coloradomtn.edu.

specifics about quantum chromo-dynamics (QCD). Quantum chromo-dynamics hypothesizes that the three valance quarks of protons and neutrons possess an attribute called color charge–either red, green, or blue. Historically, a contradiction of the quantum mechanical basis of nucleon properties with the Pauli exclusion principle led to the concept of the color charge for quarks [4]. (It should be noted that the words red, green, and blue are simply the names of the color charges, and do not imply any type of physically visual hue for the quarks. Also, the term "charge", when referring specifically to the color charge, is not related to electrical charge.) Quantum chromo-dynamics states that a strong bond is formed among the three color charges of the quarks *inside* the nucleon [5].

The residual chromo-dynamic force model assumes that the chromo-dynamic force also has a weaker residual force *outside* of the nucleon. The RCDF model states that this residual force forms an inter-nucleon bond, bonding the two different nucleons together. The inter-nucleon bond is formed by the residual color charges of the quarks.

While the RCDF model is considered to be the *mechanism* for nuclear bonding, the model is unable to duplicate the experimental bonding energy curve. This inability of the RCDF model to reproduce any salient nuclear behaviors currently is attributed to the extreme difficulty of modeling the multi-body interactions of the three color charges [6,7].

The problem with the derivation of nuclear forces from the residual QCD force is two-fold. First, each nucleon consists of three quarks, which means that a system of two nucleons is already a six-body problem. Second, because the chromodynamics force between quarks has the feature of being very strong compared to the lower energy scale of the residual chromo-dynamic force, this extraordinary strength makes it difficult to find converging mathematical solutions. The six-quark problem can be solved with brute computing power, by putting the six-quark system on a four dimensional lattice of discrete points: three dimensions of space and one of time. This method has become known as lattice quantum chromo-dynamics, or lattice QCD. However, such calculations are computationally very expensive and are not normally used as a standard nuclear physics tool [7]. Only the liquid drop model, with the five empirical-fit parameters, can duplicate the experimental binding curve.

The brute-force method for the computer calculations in lattice QCD puts each quark in a lattice by assigning to it an x, y, z, and t position, and then attempts to determine the binding energy. This is done through extremely complex mathematical models and often using Monte-Carlo simulations [8]. Because of the computational difficulties, binding energies of only the smallest nuclides, $A < 14$, have been attempted. Thus, the RCDF model remains largely unverified when testing its binding energy predictions against experimental data.

2. Properties of Up and Down Quarks

From QCD theory, we know there are six different flavors of quarks: up, down, strange, charm, top, and bottom. Of these six different flavors, only two flavors are found in the stable matter of neutrons and protons: the up and down quarks [9]. (The terms of up and down do not imply any specific orientation with regard to spatial direction, and are simply the names of these types of quarks.)

An up quark has an electric charge that is +2/3 the charge of a proton, and it also contains a positive magnetic moment, which is parallel to of the spin of the nuclide. The up quark has a spin of 1/2 and a mass of about 0.3% of the proton. The color charge of an up quark can be either red, green, or blue.

A down quark has an electric charge that is –1/3 the charge of a proton, and it contains a negative magnetic moment, which is anti-parallel to of the spin of the nuclide. The down quark has a spin of 1/2, and a mass of about 0.6% of the proton. The color charge of a down quark can be either red, green, or blue.

The magnetic moments of the up and down quarks are estimated to be +1.85 and –0.97, respectively, in units of nuclear magnetons. The electrical charges of the proton and neutron are completely contained within the quarks. The proton is comprised of two up quarks and one down quark, giving it a net charge of +1. The neutron is comprised of one up quark and two down quarks, giving it a net charge of 0.

The quarks inside of a proton and neutron have both attributes of flavor and color. Each quark inside of a proton or neutron is one of six types: up and red, up and green, up and blue, down and red, down and green, or down and blue [5]. Both the neutron and the proton contain one each of the three different color charges: red, green, and blue. Thus in terms of the color charges, there is no difference between the proton and the neutron; the only difference in the quark characteristics between a proton or neutron resides in the number of up and down quarks. Any bond between the different color charges is also a bond between some combination of the up and down quarks. Thus, the quantum assumptions that are made in the RCDF model about the possibility of an inter-nucleon bond between the residual color charges of the quarks are also applicable to the formation of an inter-nucleon bond between up and down quarks.

3. A Simple Calculation Involving Inter-nucleon Up-to-down Quarks

Using the concept of the inter-nucleon quark-to-quark bond, and applying this concept to the up and down quarks, an interesting and potentially significant set of data emerges.

Table 1 shows a spread sheet, with a representative sample of stable nuclides. The following information is listed in this spread sheet for every nuclide:

- The nuclide name.
- The number of nucleons, A, which is the sum of neutrons plus protons.
- The number of protons, Z.
- The number of neutrons, N.
- The experimental binding energy in units of MeV, as obtained from the nuclear tables [10].
- The experimental binding energy per nucleon.
- Other columns, which are described below.

A plot of the Experimental Binding Energy per nucleon is shown in Fig. 1; this plot is similar to the usual diagrams for the nuclear binding found in textbooks.

4. An Additional Constraint for Inter-nucleon Quark-to-quark Binding

Given the consideration that the color charges contained within a nucleon does not inherently distinguish between a neutron or proton, an examination of an inter-nucleon bond being formed only between an up and a down quark is an appropriate possibility to explore. Specifically, this additional constraint is that the inter-nucleon quark-to-quark bond must between an up and a down quark; it cannot be between two up quarks or two down quarks. If this additional constraint is made to the RCDF model, then an interesting pattern emerges with regard to the binding energy.

For any given nuclide, the number of inter-nucleon up-to-down quark pairs can be determined, based on how many up and down quarks each nuclide has. For each nuclide in the table, this calculation is made, as shown in Eq. (1).

$$\begin{gathered}\text{Number of up quarks} = (Z \times 2) + (N \times 1),\\ \text{Number of down quarks} = (Z \times 1) + (N \times 2),\\ \text{Number of possible pairs} = \text{the smaller of these two numbers.}\end{gathered} \tag{1}$$

This information has been incorporated into three additional columns in Table 1, showing the number of up quarks, the number of down quarks, and the number of possible up-to-down quark pairs, for each of the nuclides.

For simplicity of this very quick and easy calculation, it is assumed that every bonded pair of up-to-down quarks has the same bonding energy. Thus, *just for this simple calculation*, the equation for the binding energy of a nuclide is the number of inter-nucleon up-to-down quark pairs times the binding energy per pair, as shown in Eq. (2).

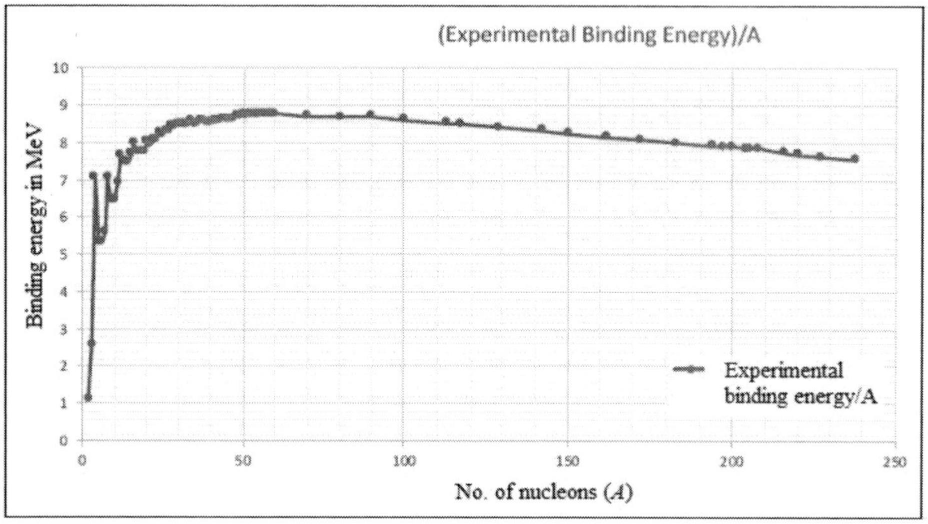

Figure 1. Plots of the experimental nuclear binding energy per nucleon (blue) and the simple calculated binding energy (orange), taking into consideration the quantum rules of hard-core repulsion and zero electric dipole moment. This is based on the number of possible quantum-allowed up-to-down quark pairs and a fixed binding energy per bonded pair.

$$\text{Calculated binding energy} = (\text{number of pairs}) \times (\text{binding energy per bonded pair})\,. \tag{2}$$

The binding energy per bonded pair is the same value for all of the nuclides, and this parameter is selected to match the empirical data. Shown is Fig. 2 is a plot of this binding energy per nucleon for a classical (non-quantum) object. Note that for Fig. 2, neither the *type* of bond nor the *structure* of these bonds comes into consideration; this is simply the number of possible pairs times a fixed binding energy per bonded pair.

5. Quantum Considerations

A nucleus is a quantum object, and being so, certain quantum rules must apply. A known phenomenological feature of the nuclear force is the QCD hard-core repulsion. The hard-core repulsion states that nucleons, such as a proton or neutron, cannot overlap in their spatial location [11,12]. The application of this phenomenon to this simple calculation reduces the number of bonds for only the two smallest nuclides. To prevent this overlap, hydrogen ^{2}H can have only one bond instead of three, and helium ^{3}He can have only three bonds instead of four. Three other stable nuclides are affected by this, the ones in which Z is odd and $N = Z$. These are ^{6}Li, ^{10}B, and ^{14}N. Other nuclides are not affected by the application of this rule, for this simple calculation.

Table 1. A representative sample of nuclides, with the nuclide name, A, Z, N, experimental binding energy of the nuclide, and the experimental binding energy per nucleon. Also shown is the number of up quarks, down quarks, and the number of possible inter-nucleon up-to-down quark bonded pairs.

Nuclide	A	Z	N	Exp. binding energy (MeV)	Exp. binding energy/A	No. of up quarks	No. of down quarks	No. of up-down bonds,classical	No. of up-down bonds, quantum
^{2}H	2	1	1	2.225	1.11	3	3	3	1
^{3}He	3	2	1	7.718	2.57	5	4	4	3
^{4}He	4	2	2	28.296	7.07	6	6	6	6
^{5}He	5	2	3	26.626	5.33	7	8	7	4
^{6}Li	6	3	3	31.995	5.33	9	9	9	6
^{7}Li	7	3	4	39.245	5.61	10	11	10	7
^{8}Be	8	4	4	56.5	7.06	12	12	12	9
^{9}Be	9	4	5	58.165	6.46	13	14	13	10
^{10}B	10	5	5	64.751	6.48	15	15	15	12
^{11}B	11	5	6	76.205	6.93	16	17	16	13
^{12}C	12	6	6	92.162	7.68	18	18	18	15
^{13}C	13	6	7	97.108	7.47	19	20	19	16
^{14}N	14	7	7	104.659	7.48	21	21	21	18
^{15}N	15	7	8	115.492	7.7	22	23	22	19
^{16}O	16	8	8	127.619	7.98	24	24	24	21
^{17}O	17	8	9	131.762	7.75	25	26	25	22
^{18}O	18	8	10	139.808	7.77	26	28	26	23
^{19}F	19	9	10	147.801	7.78	28	29	28	25
^{20}Ne	20	10	10	160.65	8.03	30	30	30	27
^{21}Ne	21	10	11	167.406	7.97	31	32	31	28
^{22}Ne	22	10	12	177.77	8.08	32	34	32	29
^{23}Na	23	11	12	186.564	8.11	34	35	34	31
^{24}Mg	24	12	12	198.257	8.26	36	36	36	33
^{25}Mg	25	12	13	205.587	8.22	37	38	37	34
^{26}Mg	26	12	14	216.681	8.33	38	40	38	35
^{27}Al	27	13	14	224.952	8.33	40	41	40	37
^{28}Si	28	14	14	236.537	8.45	42	42	42	39
^{29}Si	29	14	15	245.01	8.45	43	44	43	40
^{30}Si	30	14	16	255.62	8.52	44	46	44	41
^{31}P	31	15	16	262.917	8.48	46	47	46	43
^{32}S	32	16	16	271.78	8.49	48	48	48	45
^{33}S	33	16	17	280.422	8.5	49	50	49	46
^{34}S	34	16	18	291.839	8.58	50	52	50	47
^{35}Cl	35	17	18	298.21	8.52	52	53	52	49
^{36}S	36	16	20	308.71	8.58	52	56	52	49
^{36}Ar	36	18	18	306.716	8.52	54	54	54	51
^{37}Cl	37	17	20	318.784	8.62	54	57	54	51
^{38}Ar	38	18	20	327.343	8.61	56	58	56	53
^{39}K	39	19	20	333.724	8.56	58	59	58	55
^{40}Ar	40	18	22	343.81	8.6	58	62	58	55

^{40}Ca	40	20	20	342.053	8.6	60	60	60	57
^{41}K	41	19	22	351.619	8.6	60	63	60	57
^{42}Ca	42	20	22	361.895	8.6	62	64	62	59
^{43}Ca	43	20	23	369.828	8.6	63	66	63	60
^{44}Ca	44	20	24	380.96	8.7	64	68	64	61
^{45}Sc	45	21	24	387.849	8.6	66	69	66	63
^{46}Ca	46	20	26	398.772	8.7	66	72	66	63
46Ti	46	22	24	398.194	8.7	68	70	68	65
^{47}Ti	47	22	25	407.072	8.7	69	72	69	66
^{48}Ca	48	20	28	415.992	8.7	68	76	68	65
^{48}Ti	48	22	26	418.699	8.7	70	74	70	67
^{49}Ti	49	22	27	426.841	8.7	71	76	71	68
^{50}Ti	50	22	28	437.78	8.8	72	78	72	69
^{50}Cr	50	24	26	435.047	8.7	74	76	74	71
^{51}V	51	23	28	445.842	8.7	74	79	74	71
^{52}Cr	52	24	28	456.345	8.8	76	80	76	73
^{53}Cr	53	24	29	464.287	8.8	77	82	77	74
^{54}Cr	54	24	30	474.009	8.8	78	84	78	75
^{54}Fe	54	26	28	471.765	8.7	80	82	80	77
^{55}Mn	55	25	30	482.075	8.8	80	85	80	77
^{56}Fe	56	26	30	492.257	8.8	82	86	82	79
^{57}Fe	57	26	31	499.905	8.8	83	88	83	80
^{58}Fe	58	26	32	509.945	8.8	84	90	84	81
^{58}Ni	58	28	30	506.456	8.7	86	88	86	83
^{59}Co	59	27	32	517.314	8.8	86	91	86	83
^{60}Ni	60	28	32	526.842	8.8	88	92	88	85
^{70}Zn	70	30	40	611.08	8.7	100	110	100	97
^{70}Ge	70	32	38	610.519	8.7	102	108	102	99
^{80}Se	80	34	46	696.867	8.7	114	126	114	111
^{80}Kr	80	36	44	695.438	8.7	116	124	116	113
^{90}Zr	90	40	50	783.895	8.7	130	140	130	127
^{100}Ru	100	44	56	861.929	8.6	144	156	144	141
^{113}Cd	113	48	65	963.557	8.5	161	178	161	158
^{113}In	113	49	64	963.091	8.5	162	177	162	159
^{117}Sn	117	50	67	995.623	8.5	167	184	167	164
^{129}Xe	129	54	75	1087.65	8.4	183	204	183	180
^{142}Ce	142	58	84	1185.28	8.4	200	226	200	197
^{142}Nd	142	60	82	1185.15	8.4	202	224	202	199
^{150}Sm	150	62	88	1239.25	8.3	212	238	212	209
^{150}Gd	150	64	86	1236.39	8.2	214	236	214	211
^{162}Dy	162	66	96	1323.88	8.2	228	258	228	225
^{172}Yb	172	70	102	1392.76	8.1	242	274	242	239
^{183}W	183	74	109	1465.53	8	257	292	257	254
^{194}Pt	194	78	116	1539.58	7.9	272	310	272	269
^{197}Au	197	79	118	1559.4	7.9	276	315	276	273

^{200}Hg	200	80	120	1581.2	7.9	280	320	280	277
^{204}Hg	204	80	124	1608.7	7.9	284	328	284	281
^{204}Pb	204	82	122	1605.3	7.9	286	326	286	283

Quantum mechanics also states there can be no net electric dipole moment for the nuclide [13,14]. For this second quantum rule, three more bonds must be subtracted from the number of bonds available, in order to remove the electric dipole moment. Without stating any specific configuration for the nuclide in this very simplified calculation, this reduction of bonds can be best understood from the fact that the electric charge distribution of the nuclide must not have a net difference in electrical charge for any of the three spatial dimensions, x, y, or z. To prevent an electric dipole moment, a bond is broken in each of these three dimensions, so that the net charge is symmetric about the x, y, and z axes. This quantum requirement removes three of the classically allowed bonds. This rule applies to all nuclides, except for the very smallest nuclides, ^{2}H, ^{3}He, and ^{4}He. The inclusion of these two quantum rules is shown in Table 2. As before for this simple calculation, the calculated binding energy is the number of bonds times a fixed energy per bond. The energy per bond is the one selected parameter; for this simple calculation, it is 6.000 MeV per bond. These data are plotted in Figs. 3 and 4. In Fig. 3 all of the stable nuclides are shown, out to lead ^{204}Pb. In Fig. 4, only the first 50 nuclides are shown, to show the detail there. When there is more than one stable nuclide for a given A, these are shown as well in Figs. 3 and 4. To re-iterate, this is a very quick and easy calculation, involving only the counting of bonded inter-nucleon up-to-down quark pairs. This calculation does not specify the arrangement of the nucleons or the mechanism of the bond. It is just a simple counting of the quantum-allowed bonds.

6. Discussion

The excellent reproduction of the experimental data for these simple calculated results is impressive, especially considering that there is only one variable that must be selected, instead of five variables as in the Weizsäcker formula. The excellent reproduction of the experimental data is especially impressive considering that other currently accepted nuclear theories cannot easily duplicate this curve.

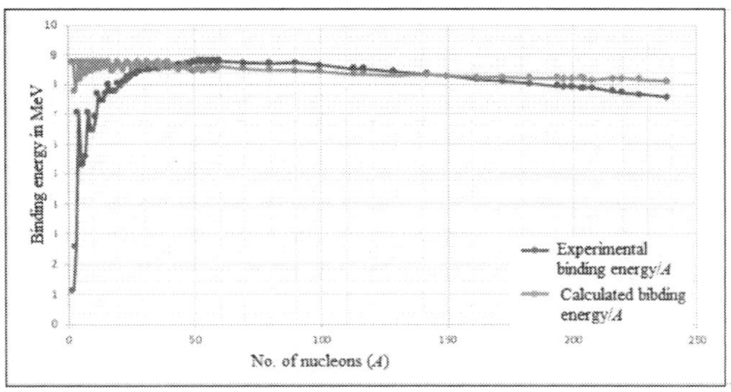

Figure 2. Plots of the experimental nuclear binding energy per nucleon (*blue*) and the simple calculated binding energy (*orange*), taking into consideration the quantum rules of hard-core repulsion and zero electric dipole moment. This is based on the number of possible quantum-allowed up-to-down quark pairs and a fixed binding energy per bonded pair.

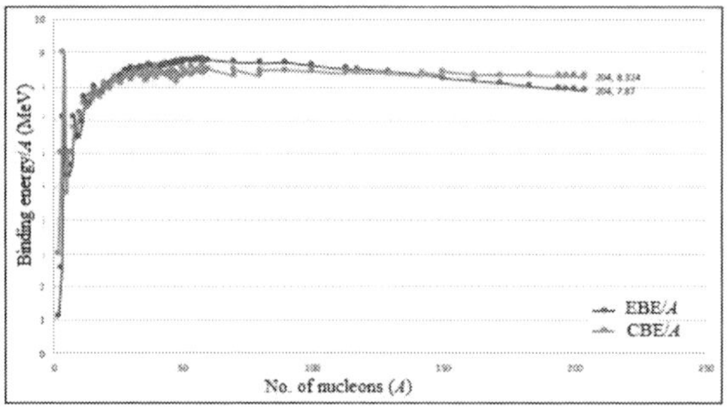

Figure 3. Plots of the experimental nuclear binding energy per nucleon (*blue*) and the simplistic calculated binding energy (*orange*), taking into consideration the quantum rules of hard-core repulsion and zero electric dipole moment, showing only the first 50 nuclides.

The residual chromo-dynamic force for quark-to-quark bonding is one possibility for the mechanism of this bond. Another possibility for this bond becomes apparent when it is recalled that the up quark has a charge +2/3 charge of a proton, the down quark has a –1/3 charge of a proton, and they both carry a magnetic moment. These electromagnetic properties of the up and down quarks create an attractive electromagnetic force between the up and the down quarks. The strength of this electromagnetic force is dependent only on the minimum proximity between the up and down quarks engaged in a bond. (Historically, it was believed that the strength of the electromagnetic force had an upper limit. However, this misconceived notion is now known to be invalid.)

The inter-nuclear bond is most likely some type of combination of both the electric charge and the color charge of the quarks, but the relative percentages of these two contributions is not postulated here. However, regardless of the relative percentages, the electromagnetic component of this bond must be taken into consideration. A more detailed analysis of the electromagnetic part of this inter-nucleon up-to-down quark bonding can easily be made.

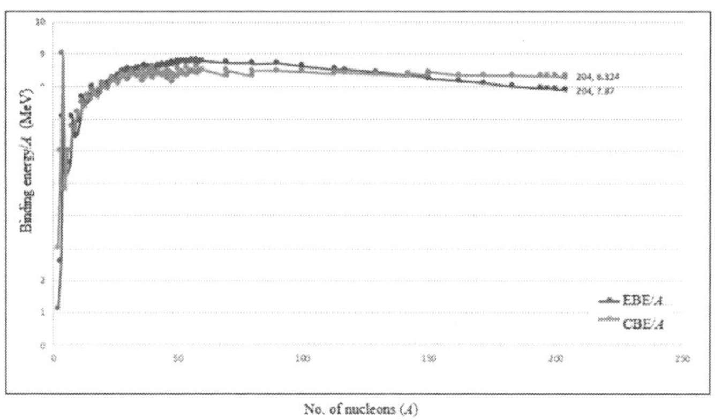

Figure 4. A plot of the experimental nuclear binding energy per nucleon, for a representative sample of nuclides.

Nuclide	A	Z	N	Experimental binding energy (MeV)	Experimental binding energy/A	No. of up quarks	No. of down quarks	No. of up-down bonds, classical	No. of up-down bonds, quantum	Calculated binding energy	Calculated binding energy/A	%Error
^{2}H	2	1	1	2.225	1.11	3	3	3	1	6	3	
^{3}He	3	2	1	7.718	2.57	5	4	4	3	18	6	-133.2
^{4}He	4	2	2	28.296	7.07	6	6	6	6	36	9	-27.23
^{5}He	5	2	3	26.626	5.33	7	8	7	4	24	4.8	9.86
^{6}Li	6	3	3	31.995	5.33	9	9	9	6	36	6	-12.52
^{7}Li	7	3	4	39.245	5.61	10	11	10	7	42	6	-7.02
^{8}Be	8	4	4	56.5	7.06	12	12	12	9	54	6.75	4.42
^{9}Be	9	4	5	58.165	6.46	13	14	13	10	60	6.667	-3.15
^{10}B	10	5	5	64.751	6.48	15	15	15	12	72	7.2	-11.2
^{11}B	11	5	6	76.205	6.93	16	17	16	13	78	7.091	-2.36
^{12}C	12	6	6	92.162	7.68	18	18	18	15	90	7.5	2.35
^{13}C	13	6	7	97.108	7.47	19	20	19	16	96	7.385	1.14
^{14}N	14	7	7	104.659	7.48	21	21	21	18	108	7.714	-3.19
^{15}N	15	7	8	115.492	7.7	22	23	22	19	114	7.6	1.29
^{16}O	16	8	8	127.619	7.98	24	24	24	21	126	7.875	1.27
^{17}O	17	8	9	131.762	7.75	25	26	25	22	132	7.765	-0.18
^{18}O	18	8	10	139.808	7.77	26	28	26	23	138	7.667	1.29
^{19}F	19	9	10	147.801	7.78	28	29	28	25	150	7.895	-1.49
^{20}Ne	20	10	10	160.65	8.03	30	30	30	27	162	8.1	-0.84
^{21}Ne	21	10	11	167.406	7.97	31	32	31	28	168	8	-0.35
^{22}Ne	22	10	12	177.77	8.08	32	34	32	29	174	7.909	2.12
^{23}Na	23	11	12	186.564	8.11	34	35	34	31	186	8.087	0.3
^{24}Mg	24	12	12	198.257	8.26	36	36	36	33	198	8.25	0.13
^{25}Mg	25	12	13	205.587	8.22	37	38	37	34	204	8.16	0.77
^{26}Mg	26	12	14	216.681	8.33	38	40	38	35	210	8.077	3.08
^{27}Al	27	13	14	224.952	8.33	40	41	40	37	222	8.222	1.31
^{28}Si	28	14	14	236.537	8.45	42	42	42	39	234	8.357	1.07
^{29}Si	29	14	15	245.01	8.45	43	44	43	40	240	8.276	2.04
^{30}Si	30	14	16	255.62	8.52	44	46	44	41	246	8.2	3.76
^{31}P	31	15	16	262.917	8.48	46	47	46	43	258	8.323	1.87
^{32}S	32	16	16	271.78	8.49	48	48	48	45	270	8.438	0.65
^{33}S	33	16	17	280.422	8.5	49	50	49	46	276	8.364	1.58
^{34}S	34	16	18	291.839	8.58	50	52	50	47	282	8.294	3.37
^{35}Cl	35	17	18	298.21	8.52	52	53	52	49	294	8.4	1.41
^{36}S	36	16	20	308.71	8.58	52	56	52	49	294	8.167	4.76
^{36}Ar	36	18	18	306.716	8.52	54	54	54	51	306	8.5	0.23
^{37}Cl	37	17	20	318.784	8.62	54	57	54	51	306	8.27	4.01
^{38}Ar	38	18	20	327.343	8.61	56	58	56	53	318	8.368	2.85
^{39}K	39	19	20	333.724	8.56	58	59	58	55	330	8.462	1.12
^{40}Ar	40	18	22	343.81	8.6	58	62	58	55	330	8.25	4.02
^{40}Ca	40	20	20	342.053	8.55	60	60	60	57	342	8.55	0.02
^{41}K	41	19	22	351.619	8.58	60	63	60	57	342	8.341	2.74
^{42}Ca	42	20	22	361.895	8.62	62	64	62	59	354	8.429	2.18
^{43}Ca	43	20	23	369.828	8.6	63	66	63	60	360	8.372	2.66
^{44}Ca	44	20	24	380.96	8.66	64	68	64	61	366	8.318	3.93
^{45}Sc	45	21	24	387.849	8.62	66	69	66	63	378	8.4	2.54
^{46}Ca	46	20	26	398.772	8.67	66	72	66	63	378	8.217	5.21
^{46}Ti	46	22	24	398.194	8.66	68	70	68	65	390	8.478	2.06
^{47}Ti	47	22	25	407.072	8.66	69	72	69	66	396	8.426	2.72
^{48}Ca	48	20	28	415.992	8.67	68	76	68	65	390	8.125	6.25
^{48}Ti	48	22	26	418.699	8.72	70	74	70	67	402	8.375	3.99
^{49}Ti	49	22	27	426.841	8.71	71	76	71	68	408	8.327	4.41
^{50}Ti	50	22	28	437.78	8.76	72	78	72	69	414	8.28	5.43
^{50}Cr	50	24	26	435.047	8.7	74	76	74	71	426	8.52	2.08
^{51}V	51	23	28	445.842	8.74	74	79	74	71	426	8.353	4.45
^{52}Cr	52	24	28	456.345	8.78	76	80	76	73	438	8.423	4.02
^{53}Cr	53	24	29	464.287	8.76	77	82	77	74	444	8.377	4.37

^{54}Cr	54	24	30	474.01	8.78	78	84	78	75	450	8.333	5.07
^{54}Fe	54	26	28	471.77	8.74	80	82	80	77	462	8.556	2.07
^{55}Mn	55	25	30	482.08	8.77	80	85	80	77	462	8.4	4.16
^{56}Fe	56	26	30	492.26	8.79	82	86	82	79	474	8.464	3.71
^{57}Fe	57	26	31	499.91	8.77	83	88	83	80	480	8.421	3.98
^{58}Fe	58	26	32	509.95	8.79	84	90	84	81	486	8.379	4.7
^{58}Ni	58	28	30	506.46	8.73	86	88	86	83	498	8.586	1.67
^{59}Co	59	27	32	517.31	8.77	86	91	86	83	498	8.441	3.73
^{60}Ni	60	28	32	526.84	8.78	88	92	88	85	510	8.5	3.2
^{70}Zn	70	30	40	611.08	8.73	100	110	100	97	582	8.314	4.76
^{70}Ge	70	32	38	610.52	8.72	102	108	102	99	594	8.486	2.71
^{80}Se	80	34	46	696.87	8.71	114	126	114	111	666	8.325	4.43
^{80}Kr	80	36	44	695.44	8.69	116	124	116	113	678	8.475	2.51
^{90}Zr	90	40	50	783.9	8.71	130	140	130	127	762	8.467	2.79
^{100}Ru	100	44	56	861.93	8.62	144	156	144	141	846	8.46	1.85
^{113}Cd	113	48	65	963.56	8.53	161	178	161	158	948	8.389	1.61
^{113}In	113	49	64	963.09	8.52	162	177	162	159	954	8.442	0.94
^{117}Sn	117	50	67	995.62	8.51	167	184	167	164	984	8.41	1.17
^{129}Xe	129	54	75	1087.6	8.43	183	204	183	180	1080	8.372	0.7
^{142}Ce	142	58	84	1185.3	8.35	200	226	200	197	1182	8.324	0.28
^{142}Nd	142	60	82	1185.1	8.35	202	224	202	199	1194	8.408	-0.75
^{150}Sm	150	62	88	1239.3	8.26	212	238	212	209	1254	8.36	-1.19
^{150}Gd	150	64	86	1236.4	8.24	214	236	214	211	1266	8.44	-2.39
^{162}Dy	162	66	96	1323.9	8.17	228	258	228	225	1350	8.333	-1.97
^{172}Yb	172	70	102	1392.8	8.1	242	274	242	239	1434	8.337	-2.96
^{183}W	183	74	109	1465.5	8.01	257	292	257	254	1524	8.328	-3.99
^{194}Pt	194	78	116	1539.6	7.94	272	310	272	269	1614	8.32	-4.83
^{197}Au	197	79	118	1559.4	7.92	276	315	276	273	1638	8.315	-5.04
^{200}Hg	200	80	120	1581.2	7.91	280	320	280	277	1662	8.31	-5.11
^{204}Hg	204	80	124	1608.7	7.89	284	328	284	281	1686	8.265	-4.81
^{204}Pb	204	82	122	1605.3	7.87	286	326	286	283	1698	8.324	-5.77

Table 2. A representative sample of nuclides, showing quantum-allowed bonded up-to-down quark pairs for each nuclide, taking into consideration the QCD hard-core repulsion and the elimination of a nuclear dipole moment.

A detailed analysis would include the addition of the energy due to all electric charges interacting with each other, which would be a double summation of the interaction of each electric charge with every other electric charge [15]. Similarly, this more detailed analysis would also include the variation of the electromagnetic bond due to the vector orientation of the magnetic moments. Additionally, the energy of the magnetic moments interacting with each other, should be included, which is a double summation over all magnetic moment vectors [16]. And finally, the kinetic energy of the quantum spin of the nuclide should also be included in this overall binding energy calculation [17,18]. For more detailed and accurate calculations to be done, however, the lowest energy configuration of the nuclide must be specified before the interaction energies can be accurately calculated.

7. Conclusion

An extremely simple calculation of the inter-nuclear up-to-down quark bonding has been made, giving excellent results in duplicating the nuclear bonding energy curve, using only one parameter rather than five. The resulting errors for nuclides going up to lead ^{204}Pb are on the order of a few percent. The average error from $A = 12$ to $A = 50$ is less than 2%. Also, due to the similarities of this concept to the residual chromo-dynamic force model, the existence of inter-nucleon up-to-down quark bonding cannot be relegated as inconceivable or implausible. An obvious implication of these results is that some significant part of the nucleon-to-nucleon force is electromagnetic. The excellent reproduction of experimental data strongly suggests that the inter-nucleon up-to-down quark bonding is a concept that should be seriously considered and more thoroughly examined by mainstream nuclear physics.

References

[1] G. Gamow, Gamow's description of the liquid drop model, *Roy. Soc.* **123** (1929) 373–390.
[2] C.F. von Weizsäcker, Zur theorie der kernmassen. *Zeitschrift für Physik* (in German) **96** (7–8) (1935) 431–458.
[3] J.L. Basdevant, J. Rich and M. Spiro, *Fundamentals in Nuclear Physics*, Springer Science and Business Media, LLC., New York, 2005. p. 67.
[4] O.W. Greenberg, Color charge degree of freedom in particle physics, *Compendium of Quantum Physics*, Springer, Berlin, Heidelberg, 2009, pp. 109–111. doi: 10.1007/978-3-540-70626-7_32.
[5] R. Eisberg and R Resnick, *Quantum Physics of Atoms, Molecules, Solids, Nuclei, and Particles*, Wiley, New York, 1998,. p. 683.
[6] U.G. Meisner, Modern theory of nuclear forces A.D. 2006, *Euro. Phys. J.* A **31** (2007 397–402, doi: 10.1140/epja/i2006-10170-1.
[7] R. Machleidt, Nuclear Forces. *Scholarpedia* **9(??)** (2014) 30710, doi: 10:4249/scholarpedia.30710.
[8] S. Duane, A. Kennedy, B. Pendleton and D. Roweth, *Phys. Lett.* B 195 (1987) 216.
[9] J. Griffiths, *Introduction to Elementary Particles* Wiley, New York, 1987, p. 47
[10] National Nuclear Data Center, information extracted from the NuDat 2 database, http://www.nndc.bnl.gov/nudat2/.
[11] F. Wilczek, Particle physics: hard-core revelations, *Nature* **445** (2007) 156–157,. doi: 10.1038/445156a.
[12] N. Ishii, S. Aoki and T. Hatsuda, Nuclear force from lattice QCD, *Phys. Rev. Lett.* **99** (**??**) (2007) 022001, doi: 10.1103/PhysRevLett.99.022001.
[13] J.D. McGervey, *Introduction to Modern Physics.* Academic Press, New York, , 1971, p. 479.
[14] C.A. Bertulani, *Nuclear Physics in a Nutshell*, Princeton University Press, Princeton New Jersey, 2007, p 105.
[15] G.E. Owen, *Electromagnetic Theory*, Allen and Bacon, Boston, 1963, p. 23.
[16] K.Yosida, *Theory of Magnetism*, Springer, New York, 1996, p. 13.
[17] K.S. Krane, *Introductory Nuclear Physics.* Wiley, New York, 1988, p. 144.
[18] J.D. McGervey, *Introduction to Modern Physics*, Academic Press, New York, 1971, p. 498.

J. Condensed Matter Nucl. Sci. 29 (2019) 260–274

Research Article

Atomic Nuclei Binding Energy

Philippe Hatt*

Route Gouvernementale, 154/B-1950 Kraainem, Belgium

Abstract

In 1936 Bethe and Bacher and in 1938 Hafstad and Teller predicted that α-particle structures could be present in atomic nuclei. In the course of developing a theory of nuclear structure based on the assumption of closest packing of clusters of nucleons, Linus Pauling found that the magic numbers have a very simple structural significance. He assumed that in nuclei the nucleons may, as a first approximation, be described as occupying localized 1s orbitals to form small clusters. These small clusters, called spherons, are usually helions (i.e. α-particles), tritons and dineutrons. In nuclei containing an odd number of neutrons, an ^{3}He cluster or a deuteron may serve as a spheron. The close-packed-spheron model differs from the conventional liquid-drop model of the nucleus in having spherons rather than nucleons as the units. This is a simplification: ^{154}Gd, for example, is described in terms of 45 spherons, rather than 154 nucleons. This enables to determine the systematic of binding energy in a much simpler way than the approach based on individual nucleons. The authordeveloped that idea, i.e. having clusters as basic bricks within the nucleus instead of nucleons. So, the authorconsidered the binding energy of α-particle and of Deuterium, Tritium, ^{3}He and the way these spherons are bonded instead of the bonding between individual nucleons. According to that hypothesis the nuclei of the various elements are constituted out of α-particles and other nucleons grouped in order to form sub nuclei bound together by four types of bonds called NN, NP, NNP, and NPP. Nevertheless, my purpose is not about looking for a new 3D model of atomic nucleus structure. It is the reason why the author favored an approach trying to breakdown the binding energy value of each element and its isotopes in several sub values indicated above. So, this process is considering only unidimensional binding energy values. This binding energy distribution approach in the nuclei is essential to the comprehension of LENR process.

Keywords: Alpha particle, Deuterium, Dineutron, ^{3}He, Tritium

1. Introduction

In line with Pauling's [1] view on the nuclear structure, allowing him to determine some clusters within the nucleus he called spherons, the author tried to organize the binding energy of the nucleus in a similar way. The sub-nuclei The author took into considerations are these spherons particles which are linked together with four types of bonds determined in the following way.

*Independent Researcher. E-mail: pcf.hatt@gmail.com.

© 2019 ISCMNS. All rights reserved. ISSN 2227-3123

- Deuterium like bond, called NP with value 2.2246 MeV, linking a neutron of one α-particle with a proton of a second α-particle, or a neutron or proton outside an α-particle to that α-particle.
- Tritium like bond, called NNP with value 8.4818 MeV, linking three nucleons of three different α-particles, or one or two nucleons outside an α-particle to one or two α-particles.
- ^{3}He like bond, called NPP with value 7.718 MeV, having a similar function as NNP.
- A dineutron bond the author called NN, with value 4.9365 MeV and linking two neutrons not being located within the same α-particle. This bond value is deduced from the α-particle binding energy, as it will be shown in the following.
- The α-particle bond with value 28.325 MeV [2].

2. Light Nuclei Binding Energy

With these five bonds: α (actually binding energy of the α-particle) NN, NP, NNP, and NPP the author could determine the binding energy (quoted E_{B}) of the n α nuclei and later of all stable nuclei. When these bonds occur any other time, they are quoted NN/2, NP/2, NNP/2, and NPP/2. Also, as NN and NP are constantly oscillating, the author determined the mean of these two bonds as an "A" (A for average) bond = NN/2 + NP/2.

2.1. Examples of n α nuclei binding energy (E_{B})

E_{B} ^{16}O $= 4$ E_{B} $\alpha+$ 4A, i.e. 4 times the binding energy of α-particle + 4 (NN/2 +NP/2) = 127.622 MeV,

E_{B} ^{20}Ne $= 5$ E_{B} $\alpha+$ A + 2NPP = 160.6416 MeV,

E_{B} ^{24}Mg $= 6$ E_{B} $\alpha+$ 2A + NN + NNP + NPP = 198.2474 MeV,

E_{B} ^{28}S i= 7 E_{B} $\alpha+$10A + NN/2 = 236.5488 MeV,

E_{B} ^{32}S $= 8$ E_{B} $\alpha+$ 4A + 4NPP = 271.7942 MeV,

E_{B} ^{36}Ar $= 9$ E_{B} $\alpha+$ 8A + 3NPP = 306.7234 MeV,

E_{B} ^{40}Ca =10 E_{B} $\alpha+$ 6A + NN + 2NNP + 2NPP = 342.0694 MeV.

Moreover, one can see the kinship between these nuclei, for example:

E_{B} ^{16}O versus E_{B} ^{32}S, E_{B} ^{32}S = 2 E_{B} ^{16}O - 4A + 4 NPP,

E_{B} ^{40}Ca versus E_{B} ^{16}O and E_{B} ^{24}Mg, E_{B} ^{40}Ca = E_{B} ^{16}O + E_{B} ^{24}Mg+ NNP + NPP.

2.2. Examples of n α nuclei isotopes binding energy

E_{B} ^{12}C $= 3$ E_{B} $\alpha+$ NN + NP = 92.136 MeV,

E_{B} ^{13}C $= 3$ E_{B} $\alpha+$ NN + NP + NP/2 + NPP/2 = 97.107 MeV,

E_{B} ^{14}C $= 3$ E_{B} $\alpha+$ 1.5 NP + 2NNP = 105.274 MeV,

E_{B} ^{15}C $= 3$ E_{B} $\alpha+$ NP + 2.5 NPP = 106.495 MeV,

E_{B} ^{16}C $= 3$ E_{B} $\alpha+$ NP + 2.5 NPP + NNP/2 = 110.735 MeV.

E_{B} ^{14}N $= 3$ E_{B} $\alpha+$ NNP/2 + 2 NPP = 104.652 MeV,

E_{B} ^{15}N $= 3$ E_{B} $\alpha+$ 2NN + 2NP + NNP + NPP = 115.497 MeV,

E_B ^{16}N $= 3$ E_B $\alpha+$ 1.5NN + 2.5NP + NNP +1.5NPP = 118.000 MeV.

E_B ^{16}O $= 4$ E_B $\alpha+$ 2NN + 2NP = 127.622 MeV

E_B ^{17}O $= 4$ E_B $\alpha+$ 1.5NN + 1.5NP + NPP = 131.760 MeV

E_B ^{18}O $= 4$ E_B $\alpha+$ 2NN + 4NP + NPP = 139.789 MeV

Remark: for all these results the differences between experimental and calculated values are less than 0.030 MeV (Basis: "The Ame 2012 atomic mass evaluation").

3. Theoretical Basis for These Calculations

3.1. Definition of lines

On my website (www.philippehatt.com) the author propose a structure for the neutron based on the following:

- A core constituted with 1800 electron masses divided into 18 × 100 masses, i.e a mass of 1800 × 0.5109989461 MeV = 919.798103 MeV.
- Some adjacent structures having a total mass of 38.637343 electron masses which are constituted with the following "lines":

18 + 1800 + 18,
18 + 1800 + 17.
So, there are 71 lines (18 + 18 + 18 +17). These lines are alternating two by two so that there are 18 + 1800 + 17.5, i.e 35.5 lines at each time. This number of 35.5 lines (71/2) corresponds to 38.637343 electron masses. Each line has a mass of 1.088375859 electron masses (38.637343 electron masses / 35.5) or 0.5561589 MeV. This is the standard unit of bond mass in MeV. So, 38.637343 electron masses × 0.5109989461 MeV = 19.7436415 MeV. The average mass per line in MeV is
19.7436415 /35.5 = 0.5561589 MeV.

The value of the binding energy of Deuterium, Tritium and He3 is determined on basis of these lines:

- E_B deuterium = NP = 4 lines: 4 × 0.5561589 = 2.2246 MeV,
- E_B tritium = NNP = 15.25 lines: 15.25 × 0.5561589 = 8.4814 MeV,
- E_B ^{3}He = NPP = 13.875 lines: 13.875 × 0.5561589 = 7.7167 MeV,
- 2NN = 17.75 lines = 9.871821 MeV, i.e. 19.7436415/2 MeV.

The value of α-particle binding energy is 28.29566 MeV according to "The Ame 2012 atomic mass evaluation". In adding the four values determined above one obtains:

E_B deuterium	2.2246 MeV,
E_B tritium	8.4814 MeV,
E_B ^{3}He	7.7167 MeV,
2NN	9.8718 MeV,
E_B α	28.2945 MeV.

The difference between the two values (Ame value versus calculated value according to my theory) is about 1 keV. So, the addition of these four bonds values determine the value of the α-particle binding energy. These bonds values are the following converted into lines values:

2NN = 17.75 lines,
NP = 4 lines,
NNP = 15.25 lines,
NPP = 13.875 lines.

These values are correlated:
1.25 NP + NN = 13.875 = NPP,
5 NN = 44.375 = 2 NNP + NPP,
2 NN - 1.25 /2 NP = 15.25 = NNP.

NP = 4 × 0.5561589 = 2.224636 MeV, difference with CODATA value is 0.00007 MeV.
NNP = 15.25 × 0.5561589 = 8.48142347, difference with CODATA value is 0.0004 MeV.
NPP = 13.875 × 0.5561589 = 7.716705, difference with CODATA value is 0.00134 MeV.

The last difference is more significant than the two others, nevertheless acceptable.
So, the binding energy of α-particle according to my theory is calculated as follows:

50.875 (=17.75 + 4 +15.25 + 13.875) × 0.5561589 MeV (value of 1 line) = 28.29458 MeV, the difference between CODATA value being 0.00108 MeV, mainly due to the NPP difference. These values ($E_{\mathrm{B}}\ \alpha$, NP, NNP, and NPP) are constituting the basic values for calculating the binding energy of all the nuclei elements.

3.2. Hypothesis for the nucleosynthesis

In order to determine the binding energy of the different light nuclei it was relied on the following hypothesis: each nucleus (starting from $^4_2$He) has a substructure made up of α-particls. The N and P supplementary to α-particls can create $^2_1$H, $^3_1$H and $^3_2$He substructures or clusters. Thus, once the nucleus consists of 2N and 2P, its structure is that of the α-particle one, and when it consists of $x\alpha$ + 2N + 2P, its structures become x + 1) α. According to that hypothesis the nucleus can be conformed to in the following ways:

$x\alpha$+ 1N	$x\alpha$+ 1P				In case of one nucleon supplementary to $x\alpha$
$x\alpha$+ 2N	$x\alpha$+ 1N + 1P	$x\alpha$+ 2P			In case of two nucleons supplementary to $x\alpha$
$x\alpha$+ 3N	$x\alpha$+ 2N + 1P	$x\alpha$+ 1N + 2P	$x\alpha$+ 3P		In case of three nucleons supplementary to $x\alpha$
$x\alpha$+ 4N	$x\alpha$+ 3N + 1P	$x\alpha$+ 2N + 2P	$x\alpha$+ 1N + 3P	$x\alpha$+ 4P	In case of four nucleons supplementary to $x\alpha$

This conforms to the Ikeda diagram [3] showing how the structure of light-α conjugate nuclei can be considered as comprised of α-clusters.

The stability of the atomic nucleus depends on its composition. The stability is maximum if it is entirely composed with α-particles, insofar as these particles have stable bonds among them.

When the structure of the nucleus is formed by x particles α to which nucleons are added, this structure is stable as long as the parity between N and P remains stable. Anyway, there is a tendency to respect this symmetry by transformation of N in P or P in N, usually with the emission of particles $\beta-$ and $\beta+$. If the nucleons supplementary to the α-particles have an odd number as a value, N prevails over P and the substructure or cluster at issue is N or 2N + P in case of a stable structure. When the number of nucleons supplementary to $x\alpha$ is equal to or more than 4 (e.g. 3N + P), the nucleus has a tendency of creating a supplementary α-particle through the transformation of N into P or vice versa. The stable light nuclei are therefore formed mainly by α-particles.

This hypothesis could be confirmed for the lightest nuclei for which one can verify a parity or nearly parity between N and P. For the heaviest nuclei, one should consider another hypothesis.

3.3. Binding energy for each nucleus

According to the hypothesis developed in the former points the binding energy of every nucleus is the sum of the binding energy of its different substructures and the binding energy among these substructures.

3.4. The proton binding energy

On my website the author propose a structure of proton which is similar to that one of neutron. This structure is based on a core constituted with 1800 electron masses as for the neutron and some adjacent structures having a total mass of 36.11216 electron masses (called PP bond for proton–proton bond).

The mass values of neutron and proton which are correlated between themselves (see my website www.philippehatt.com), are the following:

N = 1838.637343 electron masses,
P = 1836.11216 electron masses.

So, the binding energy of α-particle according to that hypothesis is:

$$(38.637343/2 + 36.11216) \times 0.5109989461 = 28.3250964 \text{ MeV} = 2\text{NN} + \text{PP}.$$

The difference between 28.3250964 MeV ($E_{\mathrm{B}}\ \alpha = 2$ NN + PP) and 28.29458 MeV (value of α-particle binding energy based on the lines, i.e 50.875 lines, see Section 3.1) is equal to 0.03052 MeV. So, my hypothesis is that the NP, NNP, NPP bonds of α-particle rearrange themselves to produce the PP bond of value 36.11216 electron masses within a given α-particle in a nucleus including more than one α-particle. In other terms the "free" α-particle binding energy has a value of 2NN + NP + NNP + NPP bonds and the "bound" α-particle binding energy has a value of 2NN + PP bonds.

3.5. Values used for the present calculations

This is the reason why the author used the following data in his theory:

28.325 MeV for binding energy of α-particle,

4.9365 MeV for NN binding energy (in fact the half of 2 NN),

2.2246 MeV for NP (deuterium like) binding energy,

8.4818 MeV for NNP (tritium like) binding energy,

7.7180 MeV for NPP (^{3}He like) binding energy.

The three last ones are the values of Ame 2012 atomic mass evaluation. They are very close to the author's own values (see Section 3.1). This choice avoids discussions on "values". On choosing the value of 28.325 MeV as binding energy of α-particle the author gave my reasons (see Section 3.4), as well as for NN bond which is part of α-particle bond. All these values are good approximations of the "real" values and are enough precise to determine the binding energy of every nucleus to be compared with the values given in the Ame 2012 atomic mass evaluation. The author am able to determine each nucleus binding energy with less than 30 keV difference only. Moreover, and far more important as the precision of the data, is the demonstration of the kinship between the nuclei and between the isotopes of each element.

One can notice that kinship in the following, showing some examples of nuclei binding energy values.

Remark: the NN value (4.9365 MeV) is comparable to the mass of quark down (4.8 MeV) and the NP value (2.2246 Mev) is comparable to the mass of quark up (2.4 MeV).

4. Examples of Binding Energy Distribution

4.1. Oxygen 16 ($^{16}_{8}$O)

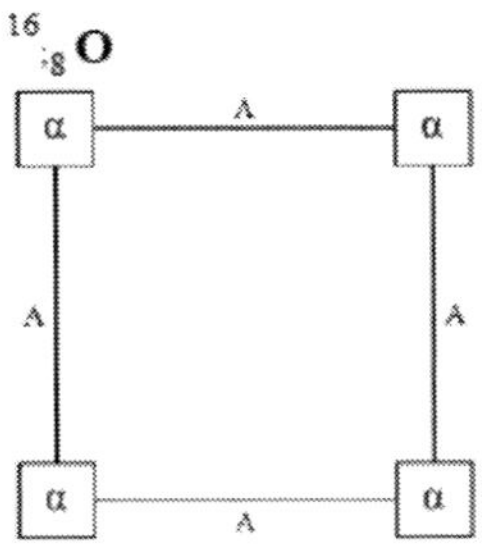

$E_B = 4\ E_B\ \alpha + 4A =$ 127.6222 MeV+ 0.003

A = NN/2 + NP/2

Four α-particles are linked with four A bonds. An A bond is equal to (NN/2+NP/2) bonds. Within an A bond the NN bond and the NP bond are constantly alternating, oscillating.
This alternative movement creates the most solid bonds in a nucleus. So, having only A bonds is the most favorable condition for nucleus stability

4.2. Figure 2 – Oxygen 17 ($^{17}_{8}$O)

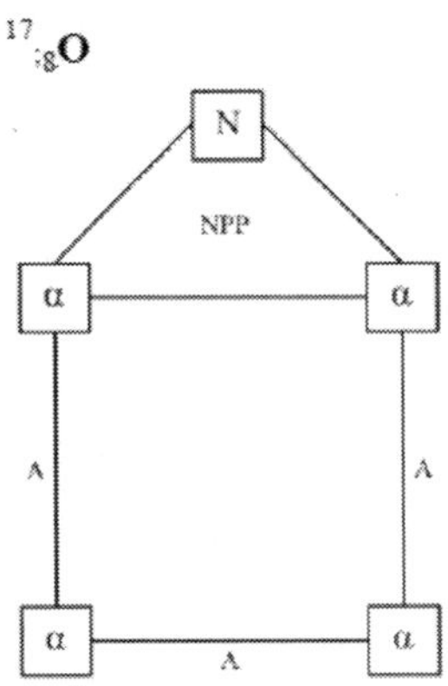

$E_B = 4\ E_B\ \alpha + 3A +$ NPP = 131.7596 MeV - 0.003

$^{17}_{8}$O = $^{16}_{8}$O− A + NPP

^{17}O has a supplementary neutron in its structure. This induces the replacement of one A bond by a NPP bond linking the neutron to two protons located in two different α-particles.
$^{17}_{8}$O = $^{16}_{8}$O− A + NPP

4.3. Figure 3 – Oxygen 18 ($^{18}_{8}$O)

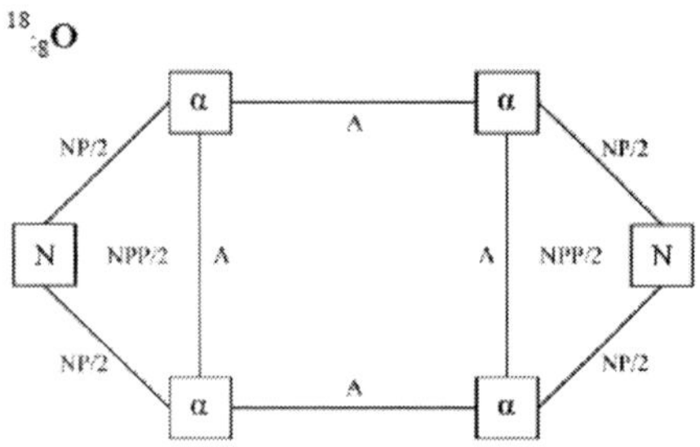

$E_B = 4\ E_B\ \alpha + 4A + 2NP + NPP = 139.7894$ MeV

$^{18}_{8}O = ^{17}_{8}O + A + 2NP = ^{16}_{8}O + NPP + 2NP$

^{18}O has two supplementary neutrons attached to the ^{16}O structure by one NPP bond and two NP bonds.

$^{18}O = ^{17}O + A + 2NP = ^{16}O + NPP + 2NP$

4.4. Figure 4 – Fluorine 19 ($^{19}_{9}$F)

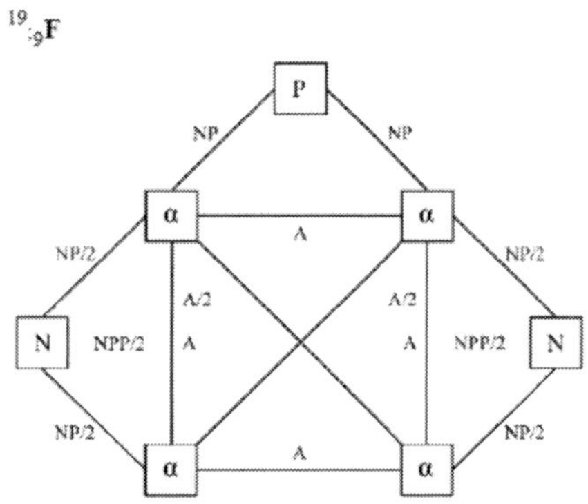

$E_B = 4\ E_B\ \alpha + 5A + 4NP + NPP = 147.8192$ MeV $+ 0.018$

$^{19}_{9}F = ^{18}_{8}O + A + 2\ NP$

^{19}F has a proton in excess of ^{18}O. One A and 2NP bonds are added to ^{18}O structure. $^{18}O + A + 2NP$. $^{19}_{9}F = ^{16}O + A + 4NP + NPP$. Compared with ^{16}O one A bond is added to the core structure, the other bonds are linking the two N (NP + NPP/2) and the P (2NP) with that core structure.

5. Examples of Other Nuclei Binding Energies

One can now start the study of all other nuclei. It will be seen that the bonds of α type, as well as $^{2}_{1}$H, $^{3}_{1}$H, $^{3}_{2}$He and NN are sufficient to explain the binding energy of all nuclei, which pleads for predominance of α-particles in the nuclei. The atomic nuclei are therefore constituted of sub-nuclei, essentially $^{4}_{2}$He but also $^{2}_{1}$H, $^{3}_{2}$He and $^{3}_{1}$H or simply of N or P, somewhat like molecules, being understood that the bonds can be more or less strong.

More examples are to be found in the book on Atomic Nuclei Binding Energy [5].

5.1. ^{32}S binding energy distribution

^{32}S is composed with two ^{16}O

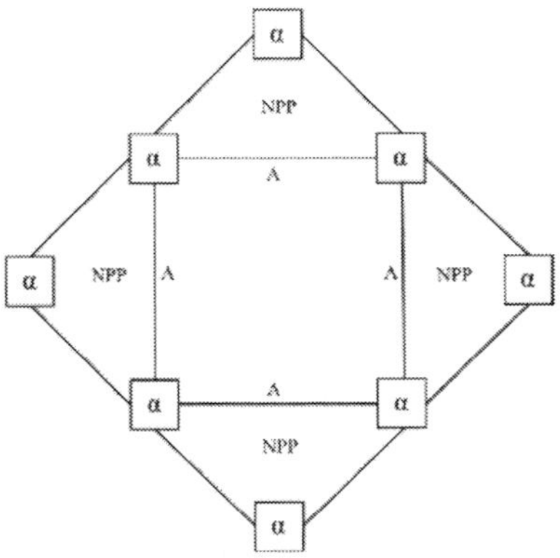

E_B =8 E_B α+ 4A + 4NPP = 271.7942 MeV+ 0.014

^{32}S = 2×016 − 4A + 4NPP

The two ^{16}O structures combine in the following way: one ^{16}O structure stays unchanged; the second ^{16}O structure bonds the first one like shown in the figure. As a consequence, three α-particles are at stake and the A bonds are changed in NPP bonds.

Remark: Up to ^{40}Ca the author uses the notation from nuclide to nuclide as seen above, i.e. boxes to represent the α-particles. As the number of α-particle is growing, it is better to use a different system in the diagrams (see below for Cu diagrams) for a practical reason of saving space.

5.2. ^{63}Cu binding energy distribution among ^{63}Cu

$^{63}_{29}$Cu	14α, 6N, 1P supplementary			E_B in MeV = 551.3847
Stable				
Nat. abundance: 69.2 %				
	14	x	28.325	396.5500 MeV
	7	x	4.9365	34.5555
	7	x	2.2246	15.5722
	0	x	8.4818	0
	7	x	7.7180	54.0260
	6	x	4.9365	29.6190
	6	x	2.2246	13.3476
	0	x	8.4818	0
	1	x	7.7180	7.7180
				551.3883 MeV
				+ 0.004

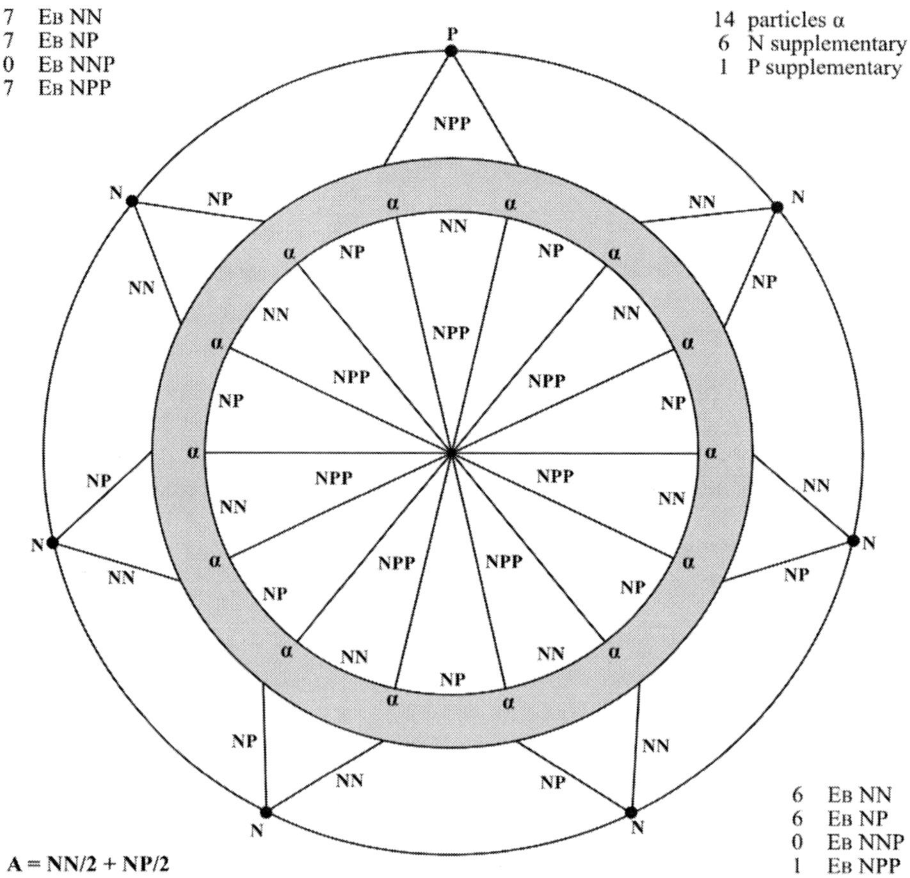

Figure 1. Binding energy distribution among ^{63}Cu.

The α-particles are directly bonded with NN and NP bonds and cross bonded with NPP bonds. The 6 N supplementary are linked to α-particles with NN and NP bonds.

NN and NP are oscillating so that each bond equals NN/2 + NP/2 = A.

The P supplementary is linked with one NPP bond to α-particles.

5.3. ^{64}Cu binding energy distribution

The ^{64}Cu bonds differ from ^{63}Cu ones by one NNP cross bond replacing one NPP cross bond.

The 7 N supplementary are linked to α-particles with 7 (NN+NP) bonds.

$^{64}_{29}$Cu	14α, 7N, 1P supplementary	$E_{\rm B}$ in MeV = 559.3008
	14×28.325	
	7×4.9365	
Lifetime: 12.9 h	7×2.2246	396.5500 MeV
	1×8.4818	34.5555
	6×7.7180	15.5722
		8.4818
		46.3080
	7×4.9365	
	7×2.2246	34.5555
	0×8.4818	15.5722
	1×7.7180	0
		7.7180
		559.3132 MeV+0.012

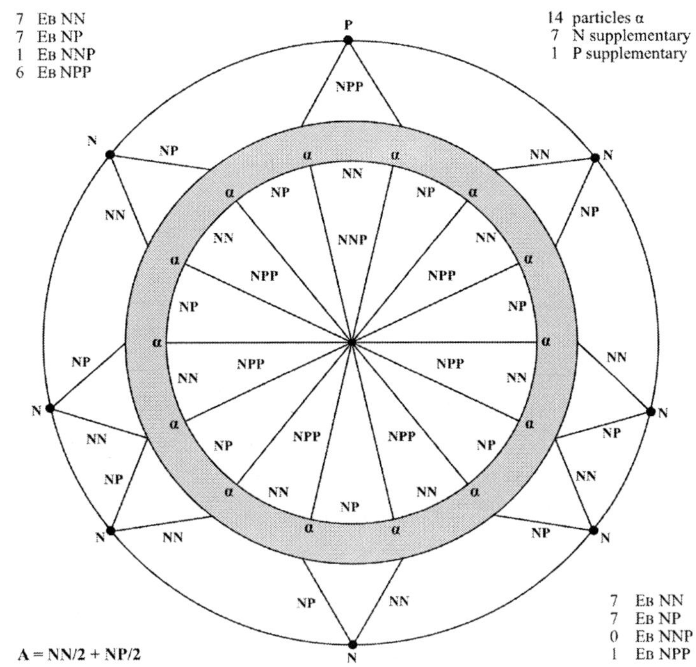

Figure 2. Binding energy distribution among ^{64}Cu.

5.4. ^{65}Cu binding energy distribution

^{65}Cu bonds structure is similar to ^{63}Cu one.

Nevertheless 7 NNP cross bonds are replacing the 7 NPP cross bonds of ^{63}Cu.

Also, the P is linked on one time to two α-particles by one NPP bond as normally (first case), and on another time to two N, forming as such one NNP bond (second case). Therefore, the two NP bonds linking the two N to the two α-particles are not "allowed", hence two NP/2 bonds for the normal binding (first case).

$^{65}_{29}$Cu	14α, 8N, 1P supplementary	E_B in MeV = 569.2112
	14 × 28.325	
	7 × 4.9365	
Stable	7 × 2.2246	396.5500 MeV
Nat. abundance: 30.8 %	7 × 8.4818	34.5555
	0 × 7.7180	15.5722
	8 × 4.9365	8.4818
	7 × 2.2246	59.3726
	0.5 × 8.4818	0 39.4920
	0.5 × 7.7180	15.5722
		4.2409
		3.8590
		569.2144 MeV + 0.003

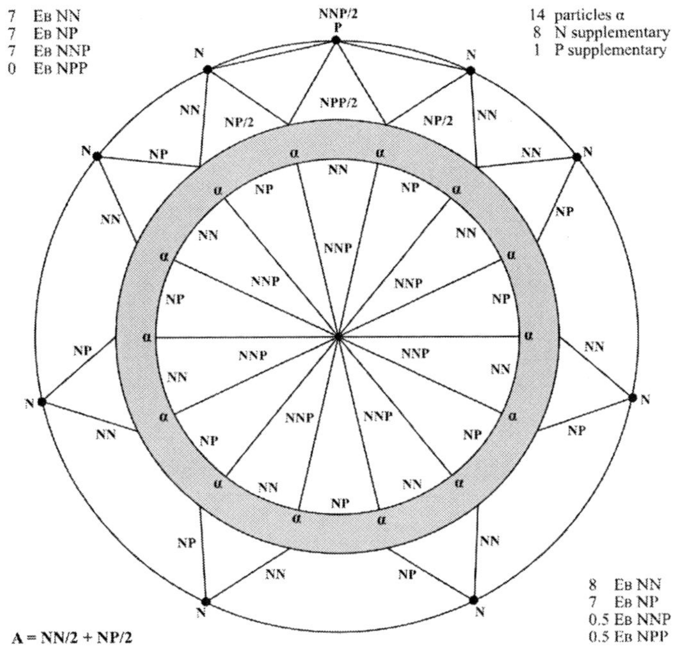

Figure 3. Binding energy distribution among ^{65}Cu.

6. Binding Energy Distribution Among ^{55}Mn to ^{68}Zn Stable Nuclei

See above a summary of binding energy values of ^{55}Mn to ^{68}Zn stable nuclei. ^{63}Cu and ^{65}Cu values correspond to that ones of Figs. 1–3. Table 1 is to read according to these examples. Table 2 shows the corresponding values of binding energies.

7. Summary

(1) These results are obtained by comparing binding energy values of several nuclei, especially isotopes of the same element and by breaking down these values in NP, NNP, NPP and α-particle binding energy values. The α-particle binding energy value is also broken down in 2 NN and PP values. Actually, NN is active inside and outside α-particle, PP being active only within α-particle. One single process is used, i.e. looking step by step, isotope element after isotope element, for binding energy differences between the various isotopes. So, one can determine the binding energy value of every element or its isotopes.

(2) My geometrical schemas are not designed to build a structure of nuclei but are destined to be a visual support for my research, especially to see the kinship between the binding energy distribution within the various nuclei. For instance, in case of ^{16}O (see Sections 4.1–4.4) the figure is based on four α-particles bound by four equal bonds the author calls "A", actually a simplification for NN/2 + NP/2. If a neutron is added it becomes ^{17}O. So, the author looks for a bond linking the new neutron and two nucleons located within two α-particles. This is the state the closest to ^{16}O. The author has the choice between NNP and NPP. It is NPP which fits, so the

Table 1. Binding energy distribution among ^{55}Mn to ^{68}Zn stable nuclei.

		E_B Core				E_B supplementary N, P			
Nucleus	Structure	NN	NP	NNP	NPP	NN	NP	NNP	NPP
$^{55}_{25}$Mn	12 α	6	6						
		2	2		4				
	6 N, 1 P								7
$^{54}_{26}$Fe	13 α	6.5	6.5						
		5.5	3.5		1				
	2 N					2	2		
$^{56}_{26}$Fe	13 α	6.5	6.5						
		6	5	0.5	0.5				
	4 N					4	4		
$^{57}_{26}$Fe	13 α	6.5	6.5						
		6.5	6.5						
	5 N								5
$^{59}_{27}$Co	13 α	6.5	6.5						
			1		6				
	6 N, 1 P								7
$^{58}_{28}$Ni	14 α	7	7						
		5.5	6.5		0.5				
	2 N					2	2		
$^{60}_{28}$Ni	14 α	7	7						
		4.5	4.5		2.5				
	4 N					4	4		
$^{61}_{28}$Ni	14 α	7	7						
		4	4	0.5	2.5				
	5 N					5	5		
$^{62}_{28}$Ni	14 α	7	7						
		4.5	4.5	1	1.5				
	6 N								6
$^{63}_{29}$Cu	14 α	7	7						
					7				
	6 N, 1 P					6	6		1
$^{65}_{29}$Cu	14 α	7	7						
				7					
	8 N, 1 P					8	7	0.5	0.5
$^{64}_{30}$Zn	15 α	7.5	7.5						
		6	9						
	4 N								4
$^{66}_{30}$Zn	15 α	7.5	7.5						
		3	3	0.5	4				
	6 N					6	6		
$^{68}_{30}$Zn	15 α	7.5	7.5						
		8	7						
	8 N								8

Table 2. Binding energy values for ^{55}Mn to ^{68}Zn in MeV.

Binding Energy	Nucleus	E_B value (MeV)	Difference with exp. Value (MeV) [4]
E_B	^{55}Mn	482.087	+0.011
E_B	^{54}Fe	471.749	– 0.015
E_B	^{56}Fe	492.258	– 0.001
E_B	^{57}Fe	499.909	+0.004
E_B	^{59}Co	517.331	+0.017
E_B	^{58}Ni	506.470	+0.011
E_B	^{60}Ni	526.842	– 0.004
E_B	^{61}Ni	534.663	– 0.003
E_B	^{62}Ni	545.270	+0.008
E_B	^{63}Cu	551.388	+0.004
E_B	^{65}Cu	569.214	+0.003
E_B	^{64}Zn	559.096	– 0.002
E_B	^{66}Zn	578.146	+0.010
E_B	^{68}Zn	595.392	+0.006

author takes that one arbitrarily. The author is aware of that "theoretical failure", my purpose being not to build a theory on strong nuclear force but rather to find simplicity in the "jungle" of hundreds of nuclear bonds in order to explain better the LENR process. Actually, the author uses three bonds which pre-exist to the α-bond, i.e. NP, NNP, NPP, and a fourth one deduced from α-bond, i.e. NN.

(3) On my method: it is not based on a theory. Instead, the author made mind experiments. As said above the author has the choice to use a few bonds each time a new neutron or proton is entering a nucleus. The author chooses that one which "fits". This unconventional way is comparable to the work of a chemist looking for several solutions in his experiments and validating that one which fits best. Moreover, the author is looking at the compliance of the solution for one nucleus with the solution for another nucleus in order to avoid discrepancies, especially between isotopes. The author is also taking care of symmetry within a given nucleus and between nuclei. Indeed, my work is not addressing the three-dimensional model of nuclei in the sense that the author is not looking for a structure of these nuclei but rather for the distribution of binding energy within them. Nevertheless, my work could be complementary to those dealing with this topic. A comparison with 3D nuclei models could be relevant.

8. Conclusion

The distribution of binding energy in each nucleus and each isotope as shown above is fundamental for understanding the LENR process, and especially the transmutation process. It allows to determine how the binding energy evolves, nucleus after nucleus, isotope after isotope. This is essential for LENR process comprehension as the difference in the distribution of binding energy between the elements present at the beginning of the reaction and at the final stage is directly related to the energy released. Many more examples are displayed in my book on Atomic Nuclei Binding Energy [5].

References

[1] L. Pauling, *Science* **150** (1965) d 297.
[2] Philippe Hatt, *J. Condensed Matter Nucl. Sci.***26** (2018) 45–53.
[3] K. Ikeda, N. Takigawa and H. Horiuchi, *Prog. Theor. Phys*. Suppl. (1968) E 464.
[4] The Ame 2012 atomic mass evaluation.
[5] See www.philippehatt.com.

J. Condensed Matter Nucl. Sci. 29 (2019) 275–285

Research Article

The Enthalpy of Formation of PdH as a Function of H/Pd Atom Ratio

Edmund Storms*

Kiva Labs, Santa Fe, NM 87501, USA

Abstract

A Seebeck calorimeter is used to measure the bond energy between H and the lattice in the Pd–H system as a function of H/Pd atom ratio during electrolysis. The bond energy was found to become increasingly endothermic in excess of about $PdH_{0.75}$. This energy is found to be sensitive to repeated gain and loss of hydrogen and to the bond disruption caused by reduction in thickness. The study used Pd having various purities and in one case a single crystal. In addition, a new method is described to measure the H/Pd atom ratio during electrolysis, which is used to show how rapidly H reacts with Pd.

Keywords: Bond energy, Enthalpy of formation, Palladium hydride, Reaction rate, Single crystal

1. Introduction

The bond energy between the PdH structure and the contained H atoms is directly measured in real time as a function of H/Pd ratio from zero to the maximum H content using an accurate calorimeter and the electrolytic method. This method is applied to several types of Pd including commercial Pd, extra pure Pd, and a zone-refine single-crystal of Pd after each is subjected to several treatments. These treatments include repeated loading–deloading cycles, annealing at 900°C, and reduction in thickness. The bond energy is found to be sensitive to purity, treatment, and H/Pd ratio. Good agreement is achieved with published measurements after certain treatments are used. This study supports the previous conclusion [1–4] that formation of beta-PdH(D) becomes endothermic at an unknown hydrogen content, as shown in Fig. 1. The partial enthalpy of formation as a function of H/Pd atom ratio shows that bonding between H and the lattice becomes repulsive at H/Pd atom ratios above about 0.75. This conclusion is important to correctly understand how PdH functions in many situations including when the low-energy nuclear reaction (LENR) occurs.

In order for the enthalpy of formation to be properly measured and understood, additional behaviors need to be measured at the same time. These include the effects produced when hydrogen reacts with Pd (loading) and as the H_2 is removed (deloading), which results in the formation of excess volume. Residual strain also has an effect. Each of these variables can affect the enthalpy of formation and is studied as part of this work.

*E-mail: storms2@ix.netcom.com.

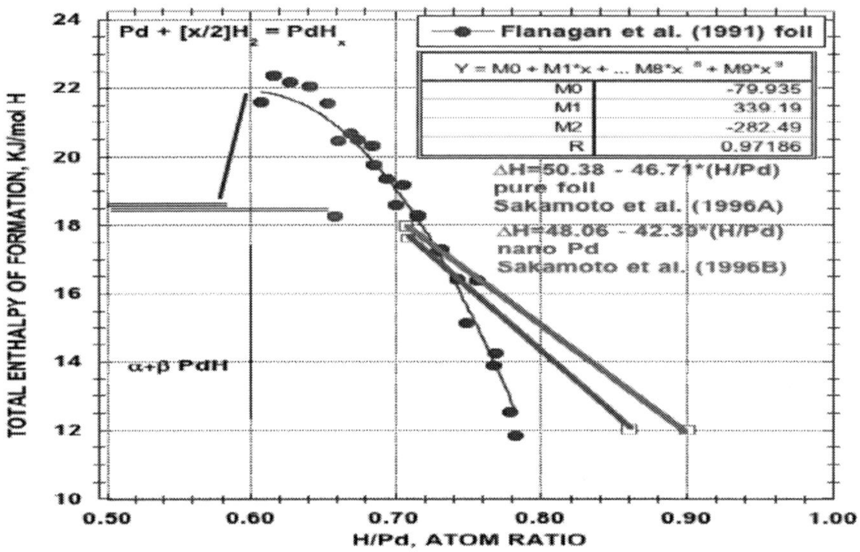

Figure 1. Total enthalpy of formation for PdH as a function H/Pd ratio. A positive sign is used to indicate that energy is given off when PdH forms, as is conventional when energy is measured using a calorimeter. In each study the Pd was loaded and deloaded 10 times in order to increase its reactivity to H_2 gas before the measurements were made using a calorimeter and H_2 gas. The negative slopes within the beta-phase region show that while creation of the beta-phase releases energy, further addition of H absorbs energy. The tabulated values provided by Flanagan et al. [1] were fit by a quadratic equation in order to better define the behavior of the partial enthalpy of formation. The studies by Sakamoto et al. [2] fit their own values to straight lines, which are plotted here.

2. Experimental Method and Results

Three methods are used to measure the average H/Pd ratio during loading. These methods use weight gain (WG), orphaned oxygen (OO), and a new method using the recombiner temperature (RT). The weight of H in the PdH_x is measured every minute for seven minutes after current is stopped and the sample is placed on a 5-place balance. The weight is extrapolated back to zero time using the square root of time. The orphaned oxygen method measures the extra oxygen left behind after the H released from H_2O reacts with Pd. The volume of oxygen is measured by displacing oil on to a balance, from which is calculated the moles of H reacted with Pd. The temperature of the recombiner is used to determine how much of the decomposed H_2O is present as unreacted H_2 and O_2, which when compared to the amount of H_2O that is decomposed by applied current, results in a value for how much H reacted with Pd. A typical comparison between the three methods is shown in Fig. 2.

Many samples experience an initial time during which every H provided by electrolysis reacts with and enters the Pd, an example of which is shown in Fig. 3 for the high purity Pd sheet. The duration of this time is brief compared to the total loading process and is sample dependent. The rate of reaction slowly decreases, finally reaching zero when no further net addition of H to the sample is possible. This limit occurs when the rate of addition equals the rate of loss. Hydrogen continues to react and enters the structure at this limit but it leaves as H_2 gas just as rapidly from certain well-defined sites [5]. Many of these sites are cracks and other physical flaws in the surface.

Each sample was studied several times after repeated loading and deloading. This treatment causes the physical shape of the sample to change so that a plate increases in thickness while its length and width shrink after each

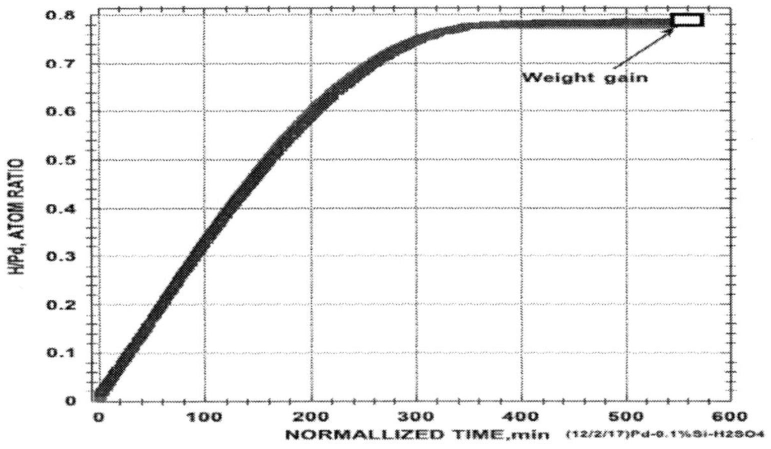

Figure 2. Comparison between the average H/Pd atom ratio obtained using the RT, OO, and WG methods with values measured every minute. The electrolytic current is 0.101 A. The delay caused by the calorimeter reaching equilibrium is avoided by applying power to a heater matching the electrolytic power. After the temperature in the calorimeter has reached constant value, heater current is turned off and electrolytic current is applied. This method allows the system to reach equilibrium within 5 min rather than after 60 min. Good agreement between the methods gives confidence in the value. Occasional lack of agreement reveals important information about the nature of the electrolytic process.

loading–deloading cycle. This process is known to result in significant changes in the physical and chemical behaviour [7–11]. An example of the increase in thickness is shown in Fig. 5. As shown below, this treatment affects the

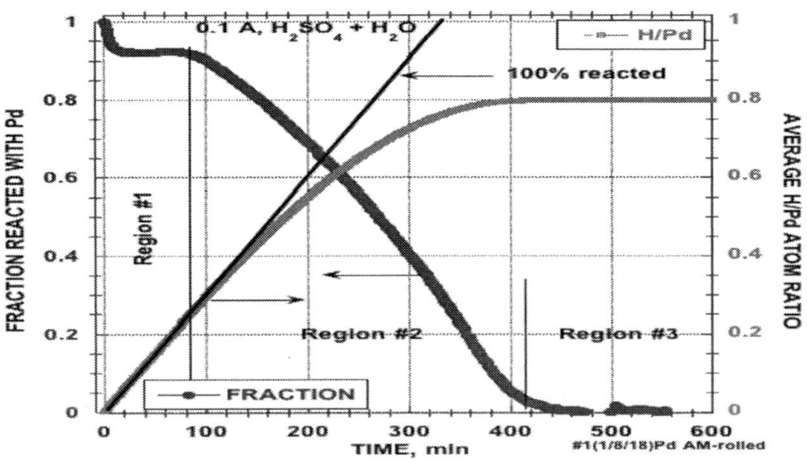

Figure 3. Example of fraction of H reacting as a function of time and the resulting average H/Pd ratio. Initially, nearly every H released from H_2O reacts with Pd (*Region* #1). Upon reaching the H/Pd ratio limit, the net amount of H reacting with the PdH is zero (*Region* #3) because H_2 leaves the sample as fast as it is applied by electrolytic action. Because every sample shows different behavior, the H/Pd ratio can not be obtained using the theoretical loading behavior as has been done by some researchers in the past.

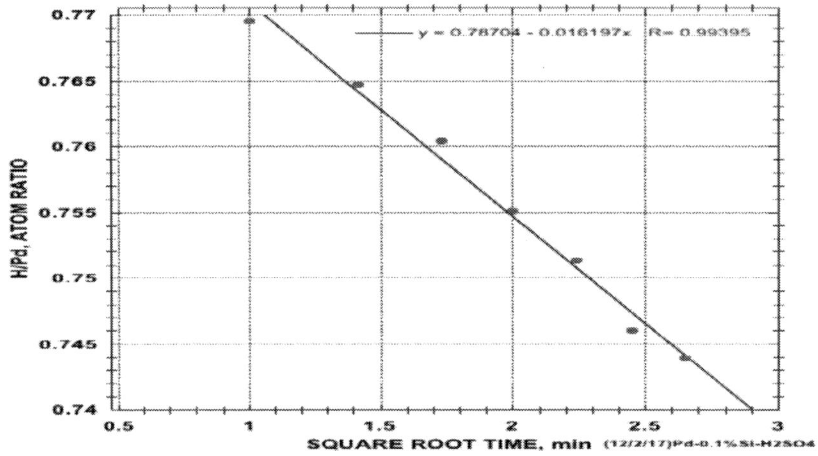

Figure 4. Example of loss of H after an electrolytic current of 0.101 A is stopped and the sample is allowed to lose H_2 in air. Weight is measured every minute using a 5-place balance with conditions of 0.75 atm and 20°C (lab pressure and temperature). The slope of this line is called the loss rate. This loss rate occurs in air, in acetone, and in the electrolyte when electrolytic current is stopped. Extrapolation to zero time gives the H/Pd ratio before electrolysis was stopped. This linear behavior continues for many hours and changes slope when enough H has been lost to cause the alpha phase to form on the surface.

bond energy in the material, which complicates interpretation of measurements made after repeated loading–deloading cycles are applied.

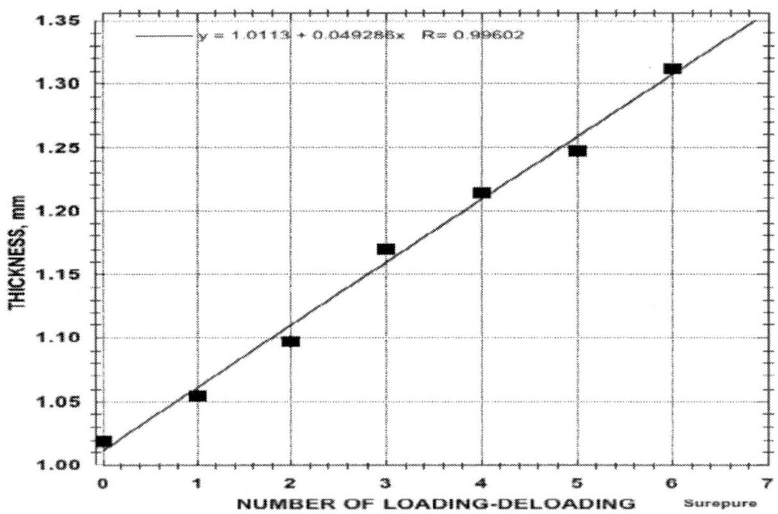

Figure 5. Effect of loading–deloading cycles on the thickness of palladium sheet. Although the other dimensions are reduced in value, the average volume increases. Each sample shows a different amount of expansion.

Figure 6. Picture of the opened calorimeter with the orphaned oxygen apparatus for measuring the H/Pd atom ratio of the cathode. The calorimeter contains a sealed Pyrex electrolytic cell and a fan. All insides of the box are covered by 54 thermoelectric convertors that are maintained at a constant reference temperature of 15 $\pm$ 0.01°C by flowing water. Measurements are made using NI hardware and LabView. Water enters and leaves the top though the hinges.

Once the H/Pd ratio reaches the saturation limit, most H_2 is lost from cracks or flaws that penetrate the surface [6]. This loss rate was measured by weighting the sample in air as a function of time immediately after electrolytic current is stopped. An example is shown in Fig. 4. This loss rate provides another way to evaluate the variables that limit the amount hydrogen the beta phase can acquire.

A Seebeck-type calorimeter [12], shown in Fig. 6, is used to determine the amount of net heating power involved in the electrolytic process. This calorimeter has a measured uncertainty over the full range of applied power (0–25 W) of $\pm$0.010 W. At the same time, the rate of reaction between H_2 and Pd is obtained by measuring the temperature of a catalyst used to convert excess H_2 and O_2 to water within the electrolytic cell. This novel method allows the moles of H_2 that have not reacted with Pd to be determined, hence are available to react with O_2. The number of moles of hydrogen that have reacted with Pd is calculated by subtracting this number from the number of moles made available by the electrolytic current.

Figure 7 shows a comparison between the power measured by the calorimeter and the rate of reaction between hydrogen and the Pd cathode based on using the temperature of the recombining catalyst.

Values for the total enthalpy of formation measured after a series of loading–deloading cycles are compared in Fig. 8. Values for the alpha phase show a steady and dramatic reduction in value as result of the treatment.

In contrast, the values measured at higher H contents show a tendency to become only slightly smaller as result of loading–deloading. The effect of loading–deloading cycles on the behavior of the beta phase is not obvious when the data are viewed this way because the behavior of the alpha phase affects subsequent values. Nevertheless, addition of H to beta-PdH shows the expected downturn (Fig. 1) in values as H/Pd is increased from 0.6 to 0.8.

The disruption of bonds produced by change in shaped produced by rolling to reduced thickness also causes significant reduction in enthalpy of formation, as shown in Fig. 9. This effect is eliminated by deloading at 200°C in air after which the sample is again reacted with hydrogen. Deloading and annealing at 900° C caused no significant

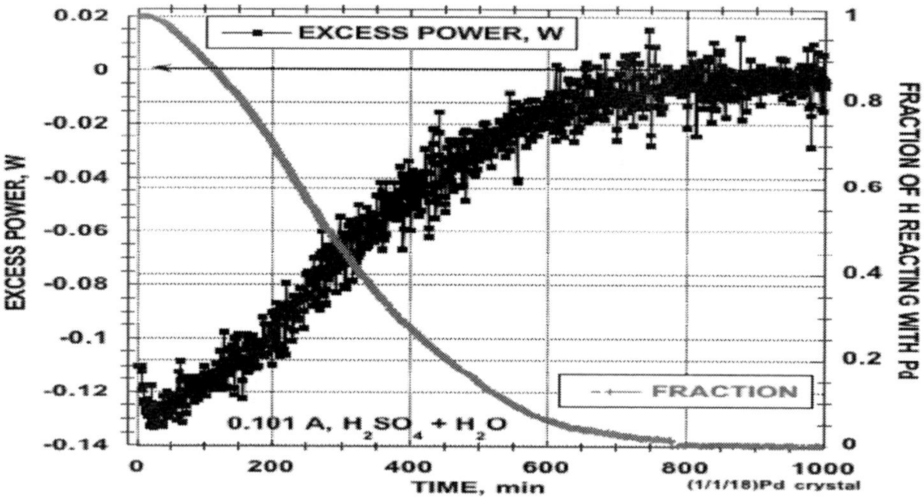

Figure 7. Variation of measured power and the fraction of H reacting with Pd based on the recombiner temperature using a single-crystal rod of Pd. The points taken during the initial 5 min are ignored because the calorimeter is close to but not at steady state. Values are measured every minute.

change in behavior. The apparent phase boundaries may not be real because the samples are expected to suffer from non-uniform hydrogen content.

Initially, each of the samples showed a slightly different behavior but remained in basic agreement with the behavior reported by Flanagan et al. [1] as shown in Fig. 10. This basic agreement

indicates that the composition gradient expected to be present in the samples studied here is not having a serious effect on the measured values for the enthalpy of formation at the composition limits. Nevertheless, the behavior between these limits is expected to show an unknown effect caused by this gradient.

The partial enthalpy of formation can be calculated for each H/Pd ratio by using the values measured during each 1 min interval. These values are equivalent to the slope of the enthalpy of formation, such as potted in Figs. 1, 8–10. This value can also be considered a measurement of the bond energy between the lattice and each H atom. The bonding behavior is more easily seen in Fig. 11 where the energy involved when H is added at each H/Pd ratio is plotted as a function of average H/Pd ratio for the initial and final loading–deloading cycles of the sample described in Fig. 8. This sample has a lower limit to the beta phase boundary that is apparently much smaller than the accepted value ($PdH_{0.6}$), perhaps because the H content is not uniform. Nevertheless, the transition to the endothermic behavior near H/Pd = 0.75 is consistent with the behavior all the samples studied here.

3. Discussion

Loading followed by deloading has very little effect on the bonding within the beta phase, with the alpha phase experiencing the greater effect. In addition, the behavior shows no evidence for formation of the proposed stable phase described as Pd_3vacH_4 by Fukai [13,14] or the phases proposed by Long et al. [15] based on complex calculations. Nevertheless, the apparent increased rejection of H by the lattice would suggest eventual formation of a less stable structure as the H/Pd ratio is increased beyond the range studied here.

All samples, regardless of their nature or treatment, exhibit endothermic behavior when the H/Pd ratio exceeds

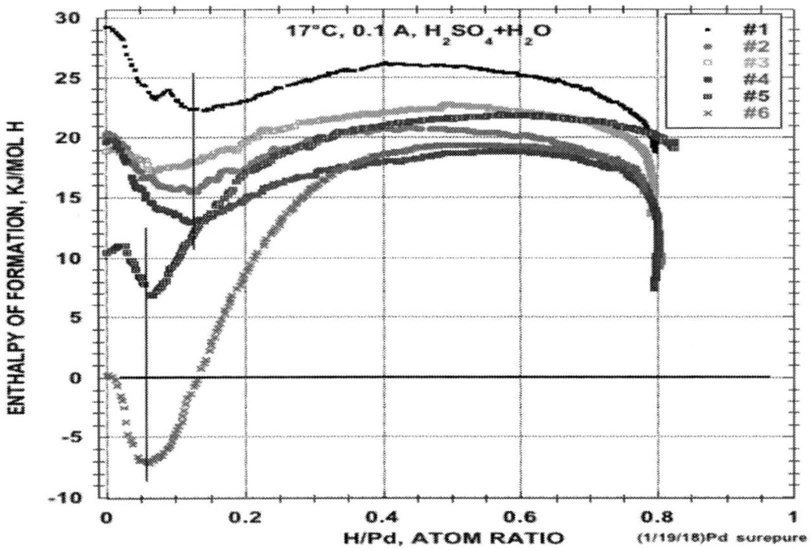

Figure 8. Total enthalpy of formation as a function of H/Pd for the various loading–deloading cycles that resulted in the thickness change shown in Fig. 5. The values apply to the reaction Pd + $(x/2)H_2 = PdH_x$.

approximately 0.75. This behavior can be compared the resistance ratio (Fig. 12), which starts to decrease near this composition. These two behaviors may have a relationship as proposed below.

Conversion of the alpha to the beta phase requires all the atoms of Pd to move further apart and 60% of the resulting

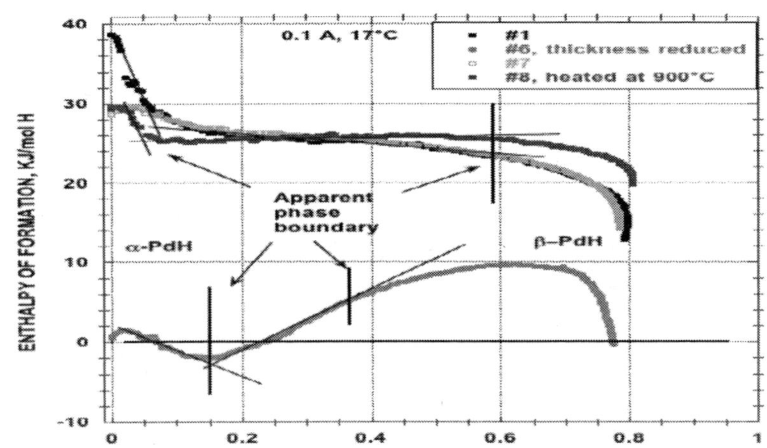

Figure 9. Effects of various treatments on the energy released by the total amount of H reacted at the average H/Pd atom ratio. The distortion in bond angles and distance caused by the rolling process caused an especially large reduction in bond energy between the lattice and the H ion.

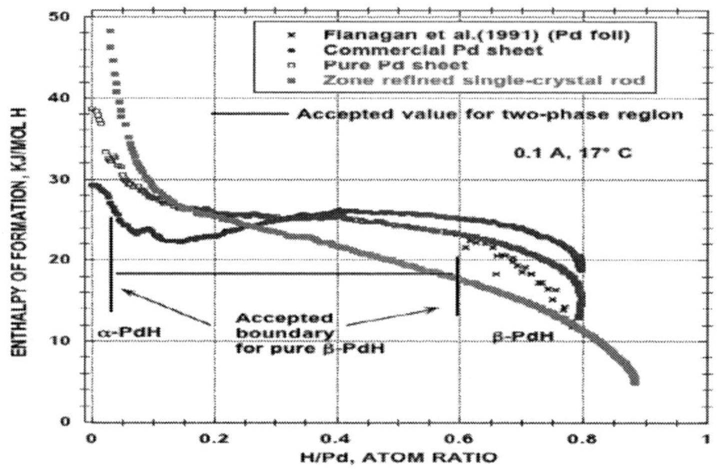

Figure 10. Enthalpy of formation measured before loading and deloading treatment was applied compared to values obtained by Flanagan et al. [1]. when a composition gradient was not present.

octahedral sites to be occupied in a random way by H. This process causers the resistivity to increase. This change in resistivity is proposed to result from electrons being moved from the conduction band to the bonding energy state. As result, the bond energy initially increases between the H and the lattice.

This behavior changes when the ratio exceeds 0.75. Thereafter, added electrons are proposed to enter the conduction band as free electrons because the energy states associated with the Pd atoms would have filled. Consequently,

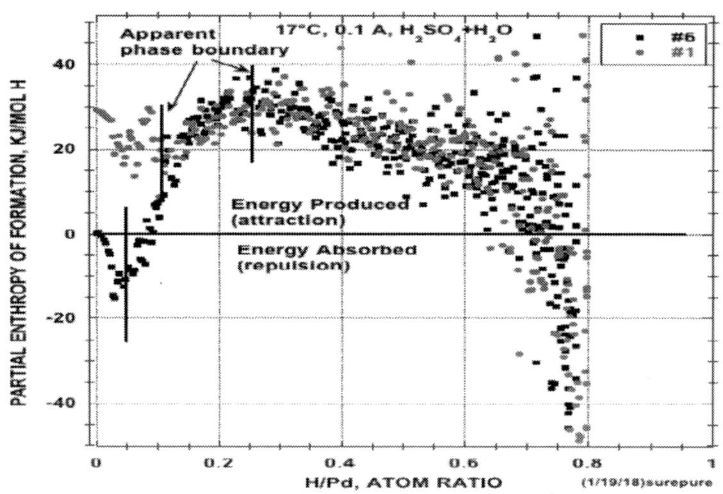

Figure 11. Partial enthalpy of formation for the first loading compared to the behavoir after the final loading–deloading cycle.

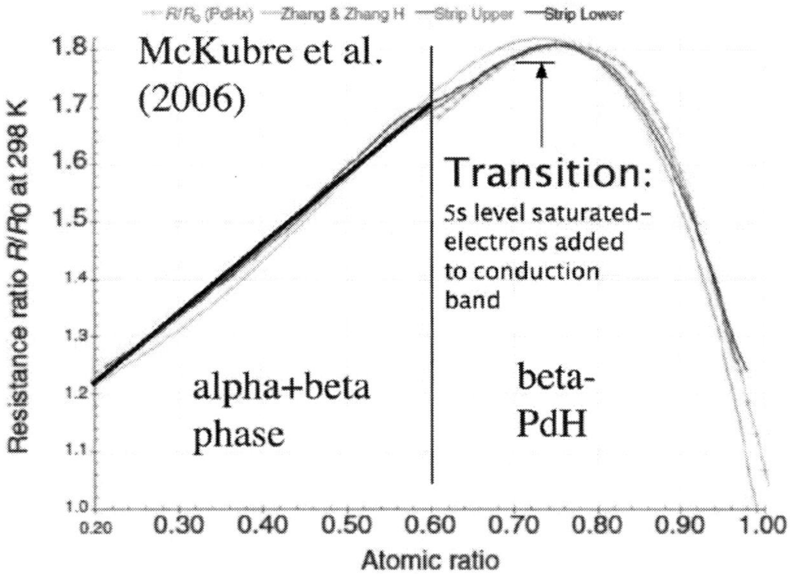

Figure 12. Resistance ratio ($R_{(PdH)}/R_{(Pd)}$ for PdH as a finction of H/Pd atiom ratio [18]. The region in the $\alpha + \beta$ two-phase region needs to be described by a straight line, as is added here, because the behavior results from the combined average of two phases having two different but constant resistances.

further addition of H and its electrons would require energy in order to promote them to the Fermi Energy, thereby making the process endothermic as is observed. The added electrons would then enter the conduction band where they would reduce the resistivity, as is observed. In fact, these electrons are found to be superconducting at low temperatures [16].)

It is important to note that the loading and deloading process affects both the bond energy as well as the resistance [17]. This correlation in behavior suggests the resistance is influenced by the same condition and mechanism that affects bond energy. Consequently, these two behavious appear to be related as result of the proposed electron transfer.

This need to apply energy to achieve high loading hinders the ability to achieve greater H contents and helps to explain why high pressure H_2 is required to reach greater H contents. This goal is further hindered because crack formation, which results from stress produced when H reacts with Pd, creates avenues from which H_2 gas can easily escape without being hindered by the electrolytic process. The number of such cracks is apparently related to the amount of excess volume a particular sample will experience when it reacts with hydrogen, as demonstrated by Storms [19–21]. His work shows that the more excess volume produced by the loading and deloading process, the less hydrogen a sample would retain at the upper limit. This excess is related to the amount of thickness expansion experienced by a sheet of Pd, an example of which is shown in Fig. 5. Consequently, this measurement can be used to test whether a piece of Pd can be expected to achieve high loading.

Storms [9,19,22] showed how the LENR process is favored when only a small amount of excess volume is produced when β-PdD forms. In other words, LENR is less likely to occur when many large gaps are present in the material as revealed by the excess volume measurement. How is this observation relate to the Storms theory [23,24] of nano-cracks being the site of LENR? Samples having a large concentration of large cracks or gaps cannot form the required small gaps because the large cracks would dissipate all the stress that is required to cause additional crack formation having the required small gap. Consequently, absence of excess volume means that large cracks are not present, which reduces the loss rate of H_2, thereby allowing a greater H/Pd atom ratio to be achieved. Consequently, the increased probability to cause LENR is actually related to the absence of large cracks, not to the ability to achieve a high H/Pd ratio. In other words, it is easy to reverse the cause and effect relationship between these two variables that apparently affect LENR, thereby placing emphasis on the wrong variable for improving the ability to cause LENR. Unfortunately, the conditions that affect excess volume formation are not known and, consequently, cannot yet be controlled.

4. Conclusions

This work reveals how the bond energy between H and the PdH structure changes and becomes endothermic when the hydrogen content is increased above a critical value. This bond energy is sensitive to loading–deloading cycles and to distortions caused by reduction in thickness using the rolling method. The work shows the relationship between bond energy and electrical resistance, both of which might be affected by the same variables.

A new and very effective way to measure the H/Pd ratio during electrolysis is described using the temperature of the recombiner catalyst. This method can be used with other methods to measure the H/Pd ratio in order to achieve better accuracy.

Many attempts have been made to extract additional details about bonding and atom arrangement in PdH by using various theoretical models, but without considering how the partial enthalpy of formation changes with H/Pd ratio or the effect of the loading and deloading process on the behavior of the alpha phase. This work shows the importance of these variables.

Various issues about how to cause LENR are explored. Achieving a high D/Pd ratio is shown not to be as important as preventing formation of large cracks and flaws in the structure.

References

[1] T.B. Flanagan, W. Luo and J.D. Clewley, Calorimetric enthalpies of absorption and desorption of protium and deuterium by palladium, *J. Less-Common Met.* **172–174** (42) (1991).

[2] Y. Sakamoto, M. Imoto, K. Takai and T. Yanaru, Calorimetric enthalpies in the b-phase regions of Pd black-H(D) systems, in *Sixth Int. Conf. on Cold Fusion, Progress in New Hydrogen Energy*, M. Okamoto (Ed.), Vol. 1, New Energy and Industrial Technology Development Organization, Tokyo Institute of Technology, Tokyo, Japan, Lake Toya, Hokkaido, Japan, 1996, pp.162–170.

[3] Y. Sakamoto, M. Imoto, K. Takai, T. Yanaru and K. Ohshima, Calorimetric enthalpies for palladium–hydrogen (deuterium) systems at H(D) contents up to about [H]([D])/[Pd] = 0.86, *J. Phys.: Condens. Mater.* **8** (1996) 3229.

[4] T. Kuji, W.A. Oates, G. Bowerman and T.B. Flanagan, The partial excess thermodynamic properties of hydrogen in palladium, *J. Phys.* F. **13** (1983) 1785–1800.

[5] J.A.S. Green and F.A. Lewis, Overvoltage component at palladized cathodes of palladium and palladium alloys prior to and during bubble evolution, *Trans. Faraday Soc.* **60** (1964) 2234.

[6] E.K. Storms, *The Science of Low Energy Nuclear Reaction*, World Scientific, Singapore, 2007, 312 p.

[7] Y.-G. Jung and Y. Sakat, Damage mechanism and deterioration of hydrogen storage ability on Pd at hydrogen absorption–desorption multi-cycles, *J. Soc. Mat. Sci. Japan* **49** (2000) 1242–1248.

[8] M.V. Goltsova, Reverse hydride transformations in the Pd–H system, *Int. J. Hydrogen Energy* **31** (2006) 223–229.

[9] E.K. Storms, My life with cold fusion as a reluctant mistress. *Infinite Energy* **4**, 42 (1999).

[10] R. Feenstra, R. Griessen and D.G. de Groot, Hydrogen induced lattice expansion and effective H–H interaction in single phase PdH, *J. Phys. F., Met. Phys.* **16** (1986) 1933.

[11] J.F. Lynch, J.D. Clewley and T.B. Flanagan, The formation of voids in palladium metal by the introduction and removal of interstital hydrogen, *Phil. Mag.* **28** (1973) 1415.

[12] E. Storms, Anomalous energy produced by PdD, *J. Condensed Matter Nucl. Sci.* **20** (2016) 81–99.

[13] Y. Fukai, Formation of superabundant vacancies in M–H alloys and some of its consequences: a review, *J. Alloys Compounds* **356-357** (2003) 263–269.

[14] Y. Fukai and N. Okuma, Evidence of copious vacancy formation in Ni and Pd under a high hydrogen pressure, *Jpn. J. Appl. Phys.* **32** (1993) L1256–1259.

[15] D. Long, M. Li, D. Meng, Y. He, I.T. Yoon, R. Ahuja and W. Luo, Accounting for the thermo-stability of PdH_x ($x = 1 - 3$) by density functional theory, *J. Hydrogen Energy* (2018).

[16] P. Tripodi, D. Di Gioacchino and J.D. Vinko, A review of high temperature superrconducting property of PdH system, *Int. J. Modern Phys.* B **21** (2007) 3343–3347.

[17] P. Tripodi, N. Armanet, V. Asarisi, A. Avveduto, A. Marmigi, J.D. Vinko and J.-P. Biberian, The effect of hydrogenation/dehydrogenation cycles on palladium physical properties, *Phys. Lett.* A **373** (2009) 3101–3108.

[18] M.C. McKubre and F. Tanzella, Using resistivity to measure H/Pd and D/Pd loading: method and significance, in *Condensed Matter Nucl Sci., ICCF-12*, A. Takahashi, K. Ota, Y. Iwamura (Eds.), World Scientific, Yokohama, Japan, 2005, p. 392.

[19] E.K. Storms, A study of those properties of palladium that influence excess energy production by the Pons–Fleischmann effect, *Infinite Energy* **2** (1996) 50.

[20] E.K. Storms, Some thoughts on the nature of the nuclear-active regions in palladium, in *Sixth Int. Conf. Cold Fusion*, Progress in New Hydrogen Energy, M. Okamoto (Ed.), Vol. 1, New Energy and Industrial Technology Development Organization, Tokyo Institute of Technology, Tokyo, Japan, Lake Toya, Hokkaido, Japan, 1996, p. 105.

[21] E.K. Storms and C. Talcott-Storms, The effect of hydriding on the physical structure of palladium and on the release of contained tritium, *Fusion Technol.* **20** (1991) 246.

[22] E.K. Storms, Measurements of excess heat from a Pons–Fleischmann-type electrolytic cell using palladium sheet, *Fusion Technol.* **23** (1993) 230.

[23] E. Storms, How basic behavior of LENR can guide a search for an explanation, *J. Condensed Matter Nucl. Sci.* **20** (2016) 100–138.

[24] E.K. Storms, *The Explanation of Low Energy Nuclear Reaction*, Infinite Energy Press, Concord, NH, 2014, 365 p., updated e-version available at Amazon.com.

J. Condensed Matter Nucl. Sci. 29 (2019) 286–297

Research Article

Reaction of the Hydrogen with Air During the Desorption of Palladium Hydride

Jacques Ruer*, David J. French† and Douglas Yuill
SFSNMC, France

Abstract

Palladium cathodes can be loaded with hydrogen or deuterium by electrolysis. Sometimes, when a piece of loaded palladium is removed from the cell, it turns red hot. A simplified numerical model confirms that high temperatures can be reached. The process requires a catalytic metal surface. The combination of a detonation in the gas phase and of palladium self-heating may offer an alternative scenario to explain the meltdown of the Fleischmann and Pons cell operated with a 1 cm cube palladium cathode.

Keywords: Diffusion, Explosion, Hydride dissociation, Ignition, Meltdown, Palladium, Self-heating, 1 cm cube

1. Introduction

When a piece of palladium that has been loaded by electrolysis with hydrogen (or deuterium) is left in the air, it sometimes heats up on its own. On rare occasions, the piece of Pd becomes red hot. Because such experiments are most frequently conducted during the search for non-conventional energy, it is tempting to think that the heating is caused by such a reaction.

However, another explanation is possible: The oxygen from the surrounding air oxidizes the hydrogen seeping out of the metal. The chemical reaction between hydrogen and oxygen can be catalyzed by the metal surface.

Self-heating events have been observed by several researchers. A few examples are given here:

- F.A. Lewis wrote [1]: *"One related experimental observation is that the hydrogen contained in palladium wire electrodes which had been coated with palladium black could be so rapidly oxidized after the specimen had been dried out in the atmosphere that they were momentarily raised to red heat (J.A.S. Creen, unpublished). The atmospheric oxidation of the hydrogen in palladium had been noted, however, in his initial studies by Graham (1868) and also in other early work (see, e.g., Troost and Hautefeuille, 1874)".*

*Corresponding author. E-mail: jsr.ruer@orange.fr.
†Deceased.

- Edmund Storms reports [2]: *"The behavior of PdD when exposed to air is very unpredictable. Some samples lose D_2 or H_2 rapidly and some lose none at all. Only a few get very hot when the D_2 or H_2 reacts with O_2 on the surface. Although some impurities on the surface are especially effective in catalyzing the reaction, such as sulfur, other samples react for no apparent reason. During my studies involving hundreds of samples, I have had no more than six that self-heated when exposed to air".*
- Melvin Miles was experimenting in 1995 with a 2 cm long bar palladium cathode [3]. The electrode was clearly outgassing hydrogen in the cell because there were bubbles forming. The palladium electrode was lifted out of the cell and placed on the lab counter top overlying its electrical lead. Then, for an unclear reason, it started to heat up. It actually became cherry red. The electrical connecting wire, made of nickel, was welded to the palladium electrode and did not detach as it would if the connection had been made by solder. The electrode remained cherry hot for some considerable number of minutes. Miles imagined that, at that elevated temperature, the out-gassing must have been accelerated contributing to more vigorous combustion.
- Jean-Paul Bibérian relates similar experiences with Pd ribbons [4].

Based on these experiences, the heating of palladium samples is a rare event. However, the authors think that when the process starts the temperature reached can be high. The purpose of this paper is to show the results obtained by a simple model that simulates such effects. The numerical model uses a simple set of hypotheses in order to determine the temperature range that can be reached via the degassing/oxidation process. It does not predict the temperature evolution with great accuracy because too many parameters would have to be taken into consideration for that.

2. Hypotheses

2.1. General concept of the model

It is supposed in the following that the Pd sample has been loaded with H or D via electrolysis in H_2O or D_2O based electrolyte. At the start of the calculation, the piece of PdH_x contains x mole of H (or D) per mole of Pd. The piece is supposed homogenous, with a uniform distribution of H through the solid sample. At the beginning of the process considered here, the electrolysis is interrupted and the Pd cathode is removed from the bath, dried and exposed to ambient air at room temperature.

It is assumed that:

- The oxygen contained in the air reacts readily with the hydrogen that comes to the surface by diffusion.
- All the H (D) atoms at the surface are consumed so that the concentration of H (D) right on the surface is zero.
- The presence of oxygen on the surface is maintained by diffusion so that the oxygen concentration is not a limiting factor.
- The water vapor leaves the surface, so that the reaction rate is not hindered by an accumulation of the reaction product.

The heat of reaction between the hydrogen in the oxygen is dissipated via several mechanisms:

- Radiation from the solid surface into the environment considered as a black body at 300 K.
- Convection in the surrounding air.
- The balance contributes to the reheating of the metal.

The heat diffuses by conduction within the solid. The temperature has an influence on the H diffusion rate, which is taken into account.

2.2. Heat of reaction

The hydrogen dissolved (absorbed) in the lattice is in atomic form. The heat of reaction of the gas that leaves the metal with oxygen can be calculated according to the first principle of thermodynamics [5], as illustrated in Fig. 1.

The initial state is: H absorbed (dissolved) in the Pd lattice.
The final state is: H_2O in the atmosphere surrounding the metal.

The overall reaction is sketched in Fig. 1 by the path 1. The path between these two states can be detailed along the paths 2 and 3:
Path 2:

- Adsorption of atomic H on the surface: $H_{abs} \rightarrow H_{ads}$.
- Recombination of molecules at the surface: $2H \rightarrow H_{2\,ads}$ (adsorbed on Pd surface).
- Desorption of the molecule into the air $H_{2\,ads} \rightarrow H_2$ (gas released in the atmosphere).

Path 3:

- Reaction with O_2: $H_2 + \frac{1}{2}O_2 \rightarrow H_2O$ (vapor in air).

Each step is accompanied by an energy change. From the above reactions, the last one (path 3) is well known, because this is the combustion of hydrogen in air. The corresponding heat release is –242 kJ/mol H_2 or –249 kJ/mol D_2. In order to quantify the heat involved by the other steps, we can consider the reverse process that is following path 2 in reverse. When Pd is exposed to H_2, the following phenomena occur:

- adsorption of H_2 on the Pd surface,
- dissociation of the H_2 molecule into two atoms of H,
- dissolution of the atomic H into the Pd.

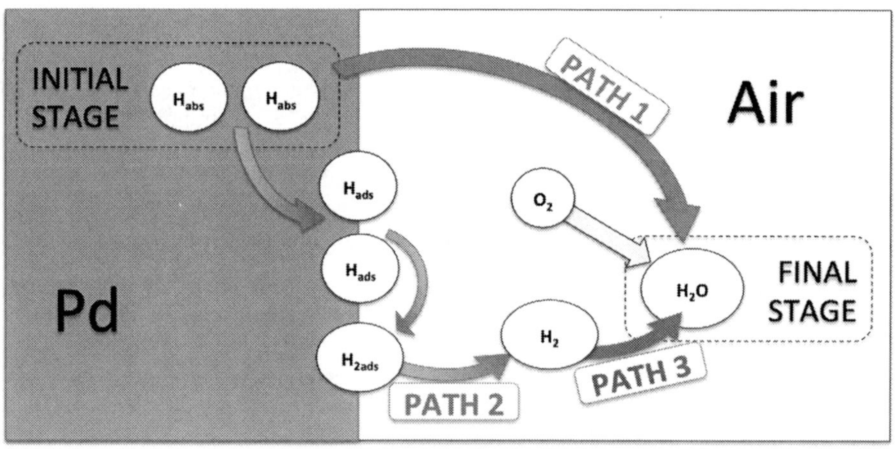

Figure 1. Determination of the reaction energy involved in the formation of H_2O from H atoms dissolved in the metal according to the first principle of thermodynamics – The direct path 1 can be dissociated into the virtual paths 2 and 3.

The dissociation of the H_2 molecule is highly endothermic (436 kJ/mol H_2). However, the dissolution is accompanied by a large energy release that more than compensates the energy required for the dissociation. The literature gives the enthalpy of the overall reaction: –40 kJ/mol H_2 (–35 kJ/mol D_2) [5].

The phenomena that take place during the dissolution of H_2 in Pd are exactly opposite to steps 1–3 discussed in the above (path 2). Therefore, the enthalpy associated to path 2 is the opposite of the enthalpy released during the dissolution. As a conclusion, the heat of reaction when the gas dissolved in the metal escapes and burns in air is:
Hydrogen:

$$2\mathrm{H_{abs}} + \frac{1}{2}\mathrm{O_2} \to \mathrm{H_2O}, \qquad \Delta H = -242 + 40 = -202\ \mathrm{kJ/mol\ H_2}. \tag{1}$$

Deuterium:

$$2\mathrm{D_{abs}} + \frac{1}{2}\mathrm{O_2} \to \mathrm{D_2O}, \qquad \Delta H = -249 + 35 = -214\ \ \mathrm{kJ/mol\ D_2}. \tag{2}$$

2.3. Diffusion of hydrogen in palladium

The coefficient of diffusion of hydrogen in palladium depends on the temperature but also on the hydrogen loading ratio. In this simplified model, the diffusion law is represented by the following relationships. According to [6] we have:

$$\mathrm{D_H} \quad = 2.5 \times 10^{-7}\,\exp(-21\,800/RT)(\mathrm{m^2 s^{-1}}), \tag{3}$$

$$\mathrm{D_D} \quad = 1.7 \times 10^{-7}\,\exp(-19\,900/RT)(\mathrm{m^2 s^{-1}}). \tag{4}$$

These values are for solid pieces of palladium.

2.4. Properties of palladium

The model supposes that the piece of metal behaves as if it were pure palladium. The presence of hydrogen in the lattice is not taken into account on the properties listed below [7]. The role of temperature is also disregarded.

Specific mass: 12 030 kg $\mathrm{m^{-3}}$,
Heat capacity: $c = 0.24$ J $\mathrm{g^{-1}\ K^{-1}}$ or 2952$\times$ 10^6 J $\mathrm{m^{-3}\ K^{-1}}$,
Thermal conductivity: $\lambda_{\mathrm{Pd}} = 75\ \mathrm{Wm^{-1}K^{-1}}$,
Normal spectral emissivity: $\varepsilon_{\mathrm{Pd}} = 0.49$.

These values valid for Pd are also considered for PdH_x to simplify the problem. Hydrogen concentration: For a hydride corresponding to the formula PdH_x, the equivalent standard volume of molecular di-hydrogen contained in the metal is:

$$C_0 = x\,1294\ \mathrm{Nm^3H_2/m^3Pd}. \tag{5}$$

For $x = 0.7$ we find $C_0 = 906$ volumes of gas per volume of metal, a value frequently noted in the literature.

3. Model of the Cylindrical Configuration (2D)

The numerical model uses a finite difference method and works as follows:

The cylindrical metal rod is subdivided into concentric layers (see Fig. 2). Initial temperature is $T_a = T$ ambient in the whole piece of metal. The initial H concentration is uniform along the radius and corresponds to the loading ratio PdH_x selected as input. H concentration at metal/air interface is considered zero at all time. There is therefore a concentration gradient that drives the H out of the metal. H_2O concentration on the surface is zero at all time. The model calculates by time steps, sufficiently small to guarantee the numerical stability.

All the hydrogen flowing out of the surface reacts with oxygen. Oxygen availability for the reaction is supposed sufficient without further analysis. The time constant of the reaction of the hydrogen adsorbed at the surface with oxygen is supposed to be negligible compared to the rate of diffusion of hydrogen towards the surface, so that the process is governed by the diffusion rate of hydrogen in the lattice. The heat of reaction Q_0 is split into several fluxes.

- Q_1: radiation loss to the environment,
- Q_2: convection in the surrounding air,
- Q_3: the balance modifies the temperature of the outermost metal layer.

The heat flow is computed in each layer to obtain the new temperature distribution. The coefficient of diffusion is recalculated at each step in each layer according to the new temperature.

The heat conductivities of palladium λ_{Pd} and air λ_{air} are constant, as well as the Pd heat capacity c. A more accurate model should revise these hypotheses, taking into account the influence of the temperature on the heat capacity and the heat conductivity. Another improvement would be to take into account the influence of the temperature and the hydrogen concentration on the heat of dissociation of the hydride, as shown for example in [8]. This sophistication is not introduced here.

This set of equations is introduced in the numerical model. The model gives the distribution of the temperature and hydrogen content in the rod at each time step. The evolution of the temperature along the time is also obtained.

The model is a finite difference explicit method written in Visual Basic. It is known that numerical errors introduced by this method are proportional to the time step and the square of the space step [9]. On the other hand, if the space step is diminished in an attempt to improve the accuracy, the time step must also be reduced, otherwise the numerical model is unstable. In our case a space step $\Delta x = 5 \times 10^{-4}$ m is sufficient to obtain a valuable result. The corresponding time step is $\Delta t = 10^{-3}$ s. Smaller space and time steps slightly increase the peak temperature calculated due to the chemical reaction but considerably increase the computing time. For example, for a 10 mm diameter rod with $\Delta x = 5 \times 10^{-4}$ m and $\Delta t = 10^{-3}$ s the peak temperature is 879°C. When Δx is divided by 2 and Δt by 4 the peak reaches 912°C. The difference is not considered essential in the present study because even the set of physical data used is an approximation. In any case, the exact behavior must depend on many parameters that are out of reach in such a model, like effective activity of the metal surface. The surface characteristics can be very different from one sample to another one. Assuming that all hydrogen atoms are immediately oxidized is a strong hypothesis. It may approximately correspond to the rare events when self-heating was observed. However, in most cases there is no heating, probably because the surface does not readily catalyze the hydrogen oxidation. This model is a simplified one that can only shed some light on the major phenomena that may occur. It cannot exactly predict the temperature evolution.

4. Simulation Results

Figures 3 and 4 show the temperature of cylindrical rods for different parameters. The temperature rises slowly during the first minutes, then very rapidly until a peak is reached. The temperature progressively returns to the ambient. The calculation shows that the temperature at each moment is practically uniform between the axis and the surface. Figure 3

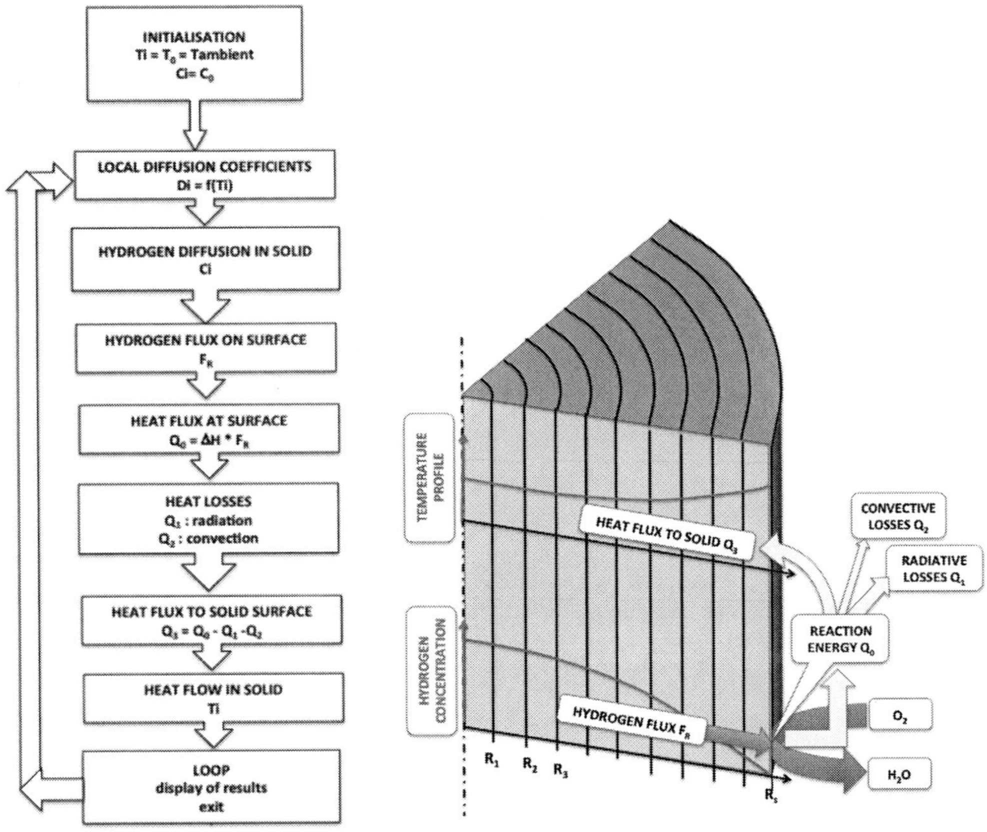

Figure 2. Flow chart of the model and schematic representation (figure for a 2D configuration).

shows that the surface peak temperature is higher in the case of a 5 mm diameter rod than a 10 mm rod, but the heat burst lasts longer in the larger sample. Figure 4 shows that the process is faster and the peak temperature higher when the initial loading increases. This reflects the major influence of the temperature on the hydrogen diffusion coefficient.

Figure 5 illustrates the evolution in a 10 mm diameter rod of the local deuterium content for an initial loading of 0.5. The initial diffusion flow is largest near the surface because of the sharp concentration slope. The reaction progressively increases the temperature. The high temperature accelerates the diffusion rate, so that most of the hydrogen flows out within a few minutes. Once most of the hydrogen is gone, the temperature drops, and so does the diffusion rate. The end of the de-loading is slow.

Figure 6 shows the variations of the deuterium inventory in the metal with time. When the loading ratio is high the large peak temperature accelerates the diffusion. As a result, the final content after cooling down is lower when the initial loading increases. The same figure displays the gas content when the metal surface is supposed to remain cold. This can be the case if the surface is non-reactive so that the hydrogen does not react with air. The desorption is slow.

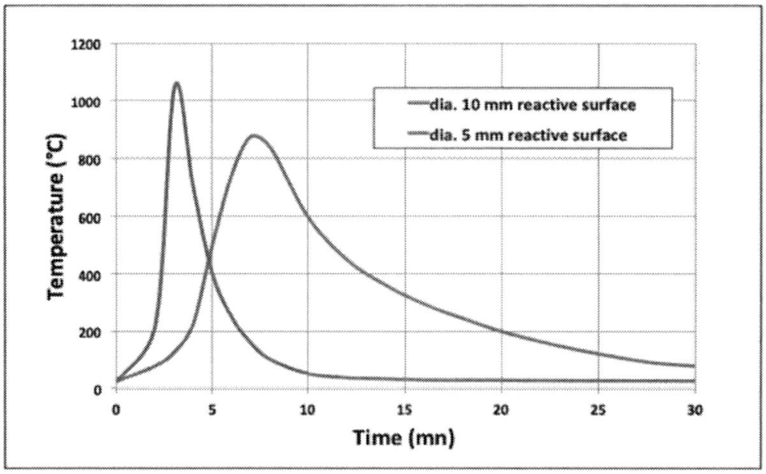

Figure 3. Temperature evolution at the surface of different diameter rods – Initial deuterium loading ratio: $x = 0.9$.

The model was adapted for a 3D spherical configuration. In the case of a 10 mm diameter sphere with a deuterium loading ratio $x = 0.9$ the calculated peak temperature is 982°C instead of 875°C for a 10 mm diameter rod, and the peak is reached after 3 min instead of 7 min. Although the calculated values cannot be considered precise, the relative magnitude is probably meaningful.

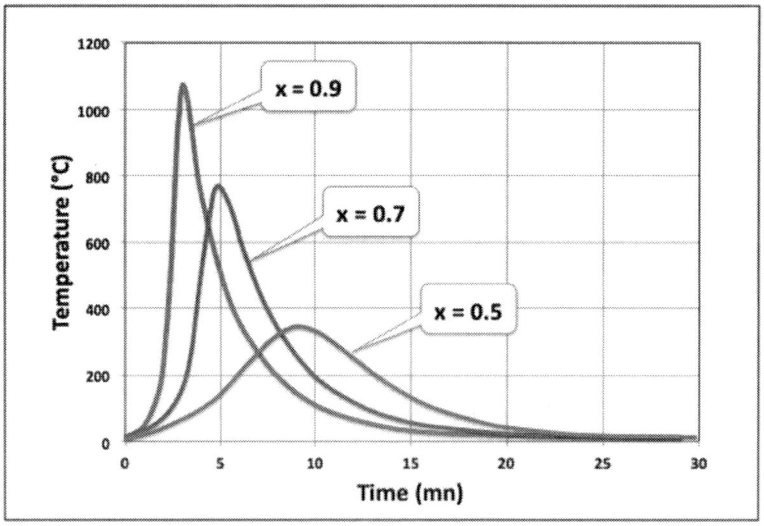

Figure 4. Influence of the initial deuterium loading ratio – Rod diameter: 5 mm.

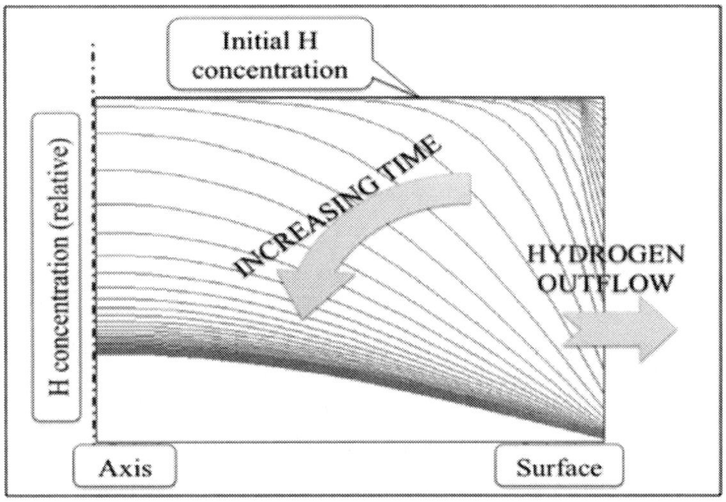

Figure 5. Example of evolution of the local deuterium content with time – Diameter: 10 mm. D distribution plotted every 30 s.

5. Conclusions of the Simulation

The model confirms what has been observed: A palladium sample loaded with hydrogen (or deuterium) can heat up when it is exposed to the air. The gas diffuses out of the metal and reacts at the surface with oxygen. The reaction increases the temperature of the metal. The higher temperature enhances diffusion, so that the process is

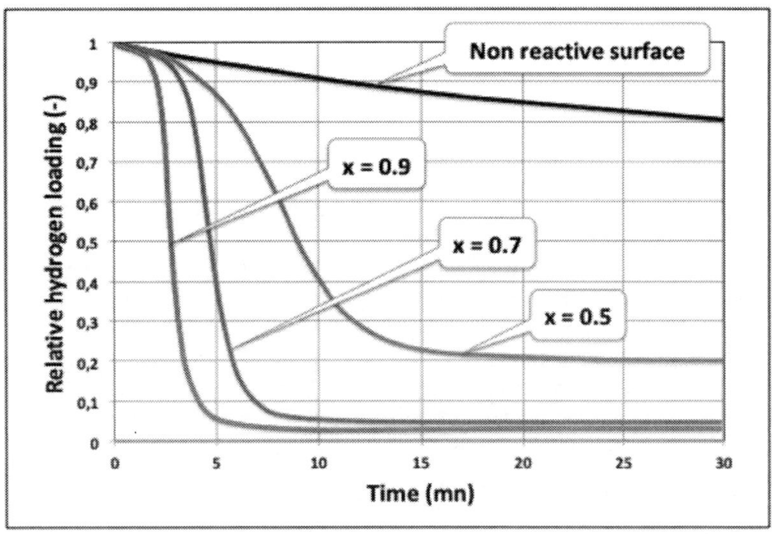

Figure 6. Ratio of the inventory of deuterium to the initial one. 10 mm diameter rod.

auto-accelerated. The reaction slows down once the hydrogen is depleted.

It remains to be clarified if the catalytic nature of the Pd is able to trigger the reaction of all the gas atoms coming to the surface or only a fraction of them, and if other conditions are necessary. Small samples heat up faster than large ones. The maximum temperature increases with the initial loading and the final hydrogen content left after cooling is lower. Cylinders and spheres react in a similar fashion, but the desorption is faster in 3D configuration (spheres) than in a 2D one (cylinders). Cubes being also a 3D configuration probably have the same behavior as spheres. The model is a simple one, so that the results presented here should only be considered as indicative.

6. Speculative Scenario to Explain the Meltdown that Occurred at Fleischmann and Pons Laboratory in 1985

In March 1985, Fleischmann and Pons were conducting an electrolysis experiment with a cathode made of a 1 cm cube piece of palladium. The cell exploded. There was no direct witness of the event that occurred at night during a weekend. This is acknowledged in the original Fleischmann and Pons paper [10]:

We have to report here that under the conditions of the last experiment, even using D_2O alone, a substantial portion of the cathode fused (melting point 1544°C), part of it vaporized, and the cell and contents and a part of the fume cupboard housing the experiment were destroyed.

Charles Beaudette describes the scene in his book [11]: *An early experiment consisted of a one-centimeter cube of palladium suspended in a flask of heavy water containing dissolved lithium metal*

Kevin Ashley was a graduate student of Pons in the chemistry department. He witnessed the scene the morning after the meltdown. This one morning I walk in, the door is open and Pons and Fleischmann are in the room with Joey. The lab is a mess and there is particulate dust in the air. On this lab bench are the remnants of an experiment. The bench was one of those black top benches that was made of very, very hard material. There were cabinets under one end of the bench, but the experiment was near the middle where there was nothing underneath. I was astonished that there was a hole through the thing. The hole was about a foot in diameter. Under the hole was a pretty good sized pit in the concrete floor. It may have been as much as four inches deep.

Because such damage requires a large amount of energy, it is often discussed within the CMNS community that this event was a strong evidence of nuclear reactions in this type of experiment. See for instance [12,13].

Another scenario is proposed here. This is admittedly a speculative one because there is no evidence left after so many years to prove it is correct. The experiment involved electrolysis "*in a flask of heavy water*" (see Fig. 7A). Electrolysis produces a stoichiometric mix of hydrogen and oxygen. This gas is highly explosive. In a previous paper, the author analyzed an explosion that occurred during a similar experiment [14]. The explosion can take the form of a rather benign deflagration. Under some circumstances, a phenomenon called Shock Wave Amplification by Coherent Energy Release (SWACER) transforms the explosion in a violent detonation able to develop shock wave pressures in excess of 40 bar. It is here supposed that the quantity of explosive gas contained in the cell was sufficient to detonate (Fig. 7B). The shock wave is stronger along its travelling path than on the other directions. If the shock wave was directed downwards it may have contributed significantly to the damage of the bench.

Once the cell was destroyed, the cathode fell on the bench, covered by debris from the explosion. It is further speculated that the palladium cube started to heat up, following the process explained above. The piece of metal kept hot for several minutes, sufficient to burn a hole in the bench, if it was not already broken by the explosion (Fig. 7C). All debris fell and accumulated on the floor, some covering the glowing piece of palladium. The concrete floor was subjected to a thermal shock during several minutes. In this scenario, the concrete may have exhibited some blistering (Fig. 7D). Further investigation could determine the relationship between the heat input and the extent of blistering. It is interesting to note that the large bas-relief at Stone Mountain near Atlanta was carved in granite via the use of oxy-fuel burners [15]. In this case the oxy-fuel burners produced sustained heating. It might be instructive to do some

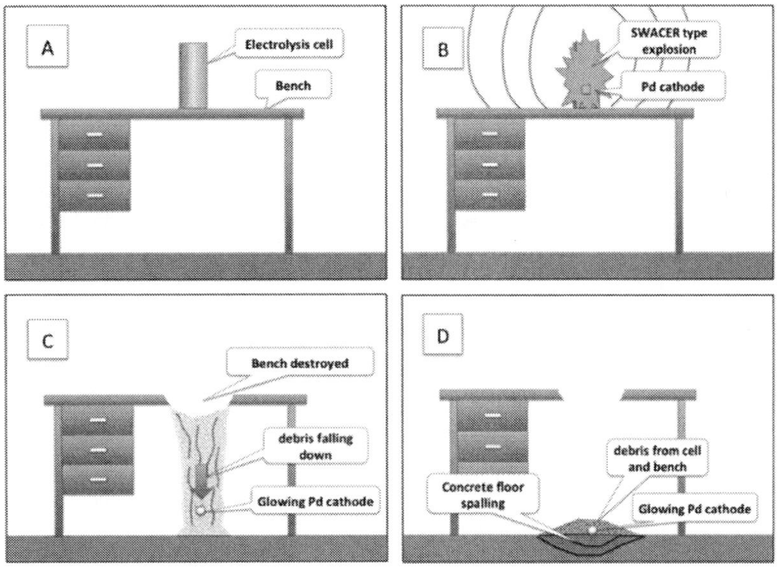

Figure 7. Potential scenario of the Fleischmann and Pons explosion event (A) Experimental setup, (B) SWACER type explosion in the gas phase of the cell, (C) debris on the floor and Pd self-heating and (D) spalling of the concrete floor.

tests to check if a piece of glowing metal can have similar effects.

7. General Conclusions

Palladium cathodes loaded with hydrogen or deuterium can heat up once exposed to air. Such events are rare but have been observed by several researchers. The simplified numerical model shows that temperature levels in accordance with the experiences can be attained. The process requires that the hydrogen present at the metal surface react readily with the oxygen from the air. The metal catalytic activity is the key that triggers the reaction (or fails to trigger it).

The combination of a detonation in the gas phase of the cell and of the self-heating of the palladium piece of metal may offer an alternative explanation of the event that occurred at Utah University in 1985 when Fleischmann and Pons were conducting an electrolysis experiment with a 1 cm cube palladium cathode.

Acknowledgements

The authors thank Jean-Paul Bibérian, Melvin Miles and Edmund Storms, who shared their experiences with us, as well as Nicolas Armanet for his input with valuable information regarding palladium hydrides.

Post-scriptum

The authors acknowledge they were tempted by David French to do this study. Unfortunately, David passed away on December 2, 2018. His co-authors want to dedicate this work to his memory.

References

[1] F.A. Lewis, *The Palladium Hydrogen System*, Academic Press, London, 1967, pp. 64–65.
[2] E. Storms, private communication.
[3] M. Miles, private communication.
[4] J.P. Bibérian, private communication.
[5] Andreas Andreasen, predicting formation enthalpies of metal hydrides, Risø report R-1484, October 2004.
[6] T.B. Flanagan, Hydrogen in palladium and its alloys, in *Proc. Symposium on Electrochemistry and Materials Science of Cathodic Hydrogen Absorption and Adsorption*, 1995, Brian E.Conway and Gregory Jerkiewicz (Eds.), The Electrochemical Society, ISBN-10 1566770785.
[7] C. Cagran and G. Pottlacher, Thermophysical properties of palladium, *Platinum Metals Review* **50** (3) (2006) 144–149.
[8] J.M. Joubert and S. Thiébaut, Thermodynamic assessment of the Pd–H-T system, *J. Nucl. Materials* **395** (2009)79–88.
[9] Wikipedia, Finite difference method https://en.wikipedia.org/wiki/Finite_difference_method.
[10] M. Fleischmann, S. Pons and M. Hawkins, Electrochemically induced nuclear fusion of deuterium, *J. Electroanal. Chem.* **261** (1989) 301.
[11] C.G. Beaudette, *Excess Heat: Why Cold Fusion Research Prevailed* , 2nd Edn., Oak Grove Press, 2002.
[12] E-Cat World – A remarkable account of the Pons–Fleischmann explosion – https://e-catworld.com/2014/12/03/a-remarkable-account-of-the-pons-fleischmann-explosion-russ-george/.
[13] D. Nagel and A. Moser, High energy density and power density events in lattice-enabled nuclear reaction experiments and generators, *J. Condensed Matter Nucl. Sci.* **19** (2016) 219–229.
[14] J. Ruer and J. P. Bibérian, Reanalysis of an explosion in a LENR experiment, *J. Condensed Matter Nucl. Sci.* **26** (2018) 76–97.
[15] Wikipedia, "Stone Mountain".

Comment by a Reviewer

This paper, like others from Ruer, is a valuable contribution to the literature on LENR. However, there is one aspect of the current paper that deserves additional consideration.

In Section 7, the author wrote "In March 1985, Fleischmann and Pons were conducting an electrolyze experiment with a cathode made of a 1 cm cube piece of palladium. The cell exploded." He also stated "Another scenario is proposed here. This is admittedly a speculative one because there is no evidence left after so many years to prove it is correct." It is unclear to this reviewer why the author speculates on the existence of an explosion, given the reports on damage caused by the event by eye witnesses and others, who have talked to eye witnesses of the damage. The damage to the bench under the experiment would require a tightly directed explosion. While explosions can be directed by confinement, as in a gun, or by using shaped charges, there were neither of those conditions in the Fleischmann–Pons experiment. An explosion in the head space above the electrolyte in the glass cell would most likely produce more or less isotropic damage. In the unlikely case that a propagating shock front did develop, what the author calls a Shock Wave Amplification by Coherent Energy Release (SWACER), it could have been directed either upward or downward, as he assumes. If directed downward toward the bench, the shock wave would have encountered the electrolyte, cubic cathode, anode and glass cell, which would have deflected some of the shock energy into a larger solid angle. That would have resulted in a lesser shock over a larger area on the surface of the bench. Hence, even if there was a downward directed explosion, its interaction with the experimental cell and its contents would have caused it to be much less focused. A broad hole in the bench was reported. But, that is at odds with what Fleischmann told Forsley, namely that the hole in the bench was about the diameter of a thumb.

The damage to the floor is better documented. It is noted in the autobiography by Chase Peterson, the President of the University of Utah at the time of the incident. This reviewer talked to Kevin Ashely about the event. He was a student in that building, who saw the damage to the Fleischmann–Pons experiment on the morning it was discovered.

He validated the reported diameter of about eight inches (20 cm) for the damage in the floor. That damage implies either of two scenarios. The first is a narrowly directed blast that made it through the bench, and still had enough energy to produce the reported damage in the floor in a very short time. The second possibility is much more likely. It is a hot piece of Pd burning a small hole through the bench, and then sitting on the concrete while continuing to produce nuclear energy. This second scenario is remarkable, but much more plausible than an explosion that, most likely, would have gone out in all directions.

In summary, this reviewer believes that it is wrong to invoke an explosion to explain the damage to the experiment, bench top and concrete floor. A runaway nuclear event in the cubic palladium cathode could have destroyed the cell, burned through the bench and damaged the concrete, if it continued to produce substantial heat for a significant time.

J. Condensed Matter Nucl. Sci. 29 (2019) 298–308

Research Article

Development of a Sensitive Detection System for the Measurement of Trace Amounts of ^{4}He in Deuterium, Hydrogen, and Other Gasses

Malcolm M. Fowler*

McFarland Instrumentation Services Inc., Espanola, NM 87532, USA

Thomas N. Claytor†

High Mesa Technology, Los Alamos, NM 87544, USA

Abstract

Many LENR systems have been reported to generate small amounts of ^{4}He from experiments that employ deuterium as part of the experiment. There are few analytical methods that can detect trace amounts of ^{4}He in the presence of D_2. Most commercial magnetic sector mass spectrometers are large and expensive and, while they have excellent $\Delta m/m$ resolution, they still require some additional separation of He, such as a getter system to achieve ppb sensitivity. We have developed a lower cost, compact system that allows us to measure ^{4}He down to sub ppm levels in D_2 and other gasses. This system utilizes a column of activated carbon cooled with liquid nitrogen (LN_2) that effectively absorbs everything but helium. Post absorption, the system uses an MKS Microvision residual gas analyzer (RGA) and a Stanford Research Systems (SRS) RS 100 RGA to check for helium purity. The helium eluted from the column is quantified by a small magnetic sector mass spectrometer tuned to mass 4. A typical sample size required to achieve ppb sensitivity to ^{4}He is 50 cm^3 at 50 Torr. Calibrations have been done with air, ^{4}He in D_2 at various concentrations and show a 3% variation from standard sample to standard sample. We will discuss the typical operation of this instrument and show results from various calibrations using different carrier gases.

Keywords: Chromatographic gas separation, Helium analysis, LENR product analysis, Mass spectrometry, Trace gas analysis

1. Introduction

We have been involved with many aspects of the “cold fusion” phenomena (LENR) since its announcement in 1989 and, for some time, we have gotten inquiries about analyzing for small amounts of helium in deuterium and/or hydrogen. About 2 years ago we started to look at what it would take to do these measurements and we present some of

*Corresponding author. E-mail: Malcolm-Fowler@zianet.com.
†E-mail: Claytor@att.net.

© 2019 ISCMNS. All rights reserved. ISSN 2227-3123

Table 1. Masses of light ion.

	Neutral mass	Positive ion mass
Atomic hydrogen	1.0078	1.0073
Molecular hydrogen	2.0157	2.0151
Deuterium	2.0141	2.0136
Tritium	3.0160	3.0155
HD	3.0219	3.0214
^{3}He	3.0160	3.0155
Molecular deuterium	4.0282	4.0277
^{4}He	4.0026	4.0021

the results of that development effort here. The separation and quantification of helium in the presence of hydrogen isotopes is quite difficult because the masses of the various species of interest are quite close in mass as is shown in Table 1. For instance, ^{4}He and D_2 differ in mass by only 0.025 mass units

We considered a variety of methods that would allow the measurements. Included were chemical removal of hydrogen isotopes. This approach included reacting the sample with excess oxygen, possibly with a catalyst. One can also pass the sample over hot copper oxide to produce water that is then trapped out or otherwise absorbed. Also considered was the use of a getter to absorb hydrogen and perhaps other constituents in the sample. Experience with the removal of tritium from He^3 has shown that the getter approach is not attractive as it is slow and requires circulation of the sample over the getter for some time to achieve quantitative separations. These methods all require consumable materials that make the methods somewhat unattractive. The sample often requires additional processing before final determination can be made. Several optical spectroscopic methods were considered, but it seemed that they did not have the required sensitivity and the methods were often compromised by other trace constituents in the sample.

We finally chose to look at a procedure based on gas chromatography. In this method, there is one or more columns, cooled with liquid nitrogen, that absorb essentially all components of the gas sample except for helium. Initially, we started with two columns, supplied by Hewlett-Packard [1], a molecular sieve column filled with MS-5A sieve followed by a second column filled with activated charcoal. The MS-5A column contained 100 g of 45/60 mesh molecular sieve 5A, while the activated charcoal column contained about 20 g of activated charcoal. Both columns were about 0.5 in. in diameter by 20 in. long. The normal procedure is to pump out the traps while being heated to about 200°C. At this temperature both water and carbon dioxide can be removed from the MS-5A molecular sieve trap. Once the pressure in the traps is below 0.001 Torr, the traps are cooled with liquid nitrogen (LN) and the sample is bled into the traps where most gasses are absorbed. Helium from the sample is not absorbed and is pushed through the traps by the incoming sample gas and ultimately is fed directly into the Residual Gas Analyzer (RGA) system. After some time, we found that a single activated charcoal column was adequate for the separations we wanted.

2. Experimental

Initially, data were collected from the SRS RGA operating in the mode where the partial pressures of several trace constituents were recorded as a function of time. The integral of the ^{4}He signal, over the time that the helium was eluting from the cooled columns, gives a result that is proportional to the ^{4}He concentration in the sample. In this mode, the minimum detectable signal is about 5×10^{-10} Torr. After operating in this mode for a while, it became clear that the RGA did not have the sensitivity to make the low-level measurements that were desired. We then turned to the Alcatel ASM-110 leak detector which had much lower background and higher sensitivity. The shortcoming of this instrument is that it only measures mass 4 with poor resolution. The low-level analog output from the ASM-110 leak detector was recorded as a function of time using a Tektronix TDS-744 digital scope. Typically, measurements were made at a rate of 5 per second with a total record length of 5000 s, resulting in 25 000 points per measurement run. The scope was also run in the "high resolution" mode. In this mode the signal is averaged for the time between measurement samples resulting in about 12-bit resolution as opposed to 8-bit resolution in the normal mode. To improve the data acquisition, we switched from the TDS-744 scope to a Tektronix DPO-7254. The DPO scope has four input channels, much better I/O capabilities, and a fully capable windows computer on board. Another RGA, made by MKS, was added to the system that has a range of only 0–6 amu, but with about 50 times the mass resolution of the SRS RGA. This RGA has good resolution and can distinguish between D_2 and ^{4}He. Short comings for this RGA are relatively slow mass scan and mass tailing to the low side of a mass peak. In the case of D_2 and ^{4}He, the deuterium is the heavier mass and that peak tails to lower mass and overlaps the helium peak. If the deuterium is at a greater concentration relative to the helium, it quickly becomes very difficult to measure the helium. Notwithstanding these shortcomings, the MKS RGA is quite useful in looking for contamination in ^{4}He samples to be analyzed.

Once we started using both the RGA and the leak detector at the same time it became clear that the background of the leak detector was lower than the RGA, resulting in greater sensitivity. One difficulty with the leak detector was that the analog signal was small (25–50 mV) and the output followed range changes on the leak detector. This meant that careful records had to be kept so that after the run the recorded data could be properly scaled to give a smooth curve. In addition, the small signal made the data more susceptible to noise and the records often had to be noise filtered prior to analysis.

As a further development, it was discovered that there was a high-level output (0–100 V) available from the ASM-110 leak detector. In addition, this output does not change if the range of the ASM-110 is changed. To make use of this output, we paralleled all four inputs of the DPO7254 scope and operated the four inputs at 0.1, 0.5, 2.0, and 10 V/division. In this way I could record signals as large as 100 V yet retain good precision for signals as small as a few 10's of mV. The result is that the recording system has more than three orders of magnitude dynamic range without changing ranges. A final addition was to replace the DPO7254 scope with a Keysight 34465A 6.5-digit voltmeter. This voltmeter logs the data and can directly output the data to an Excel workbook.

3. Runs with Large Nitrogen and Air Samples

We have looked at pure nitrogen to verify that there is no helium signal from the nitrogen or from background in the RGA. Initially, 10 cm^3 samples of nitrogen that were at local atmospheric pressure (635 Torr) were analyzed to verify that there was no leakage through the traps. We injected three successive 10-cm^3 samples with no evidence of leakage through the traps. The traps were regenerated by heating the traps to 200–250°C while pumping on them. It was surprising that, even with heating, it took several hours to remove most of the absorbed gas. In cases of air samples, H_2O and CO_2 were the last to come off and this was presumably from the molecular sieve trap. In an effort to speed up the elution of the ^{4}He from the columns, we removed the MS-5A column, leaving only the activated charcoal column. We found that the absorption of sample gasses was still quantitative. In addition, the time for elution was decreased, improving background uncertainty.

Next, we investigated the capacity of the traps and for this, we used 1725 cm^3 of nitrogen at 1001 Torr for a total of 2.27 l-atm. of nitrogen. This amount of sample is far larger than anticipated for most samples, but we wanted to verify that the traps had adequate capacity to trap and retain the bulk of the sample.

Again, the absorption of the nitrogen was quite rapid and there was no leakage through the traps. In Fig. 1 is shown the absorption of the 2.27 l-atm. of nitrogen, starting at a pressure of 1001 Torr. It is interesting to note that more than 90% of the nitrogen sample was absorbed in less than ten minutes and 99.9% was absorbed in about 20 min.

Next, we analyzed air samples and tried to detect helium in air. The sample size was 1725 cm^3, but the pressure was 636 Torr, so that sample (1.44 l-atm.) was only about 63% of the size of the pure nitrogen sample. This air sample contained 2.03×10^{17} atoms of helium (0.007593 atm.cm^3 of He). When this sample was bled into the cooled traps, I observed a good helium signal in the RGA less than 5 min after start of introduction of the sample into the traps.

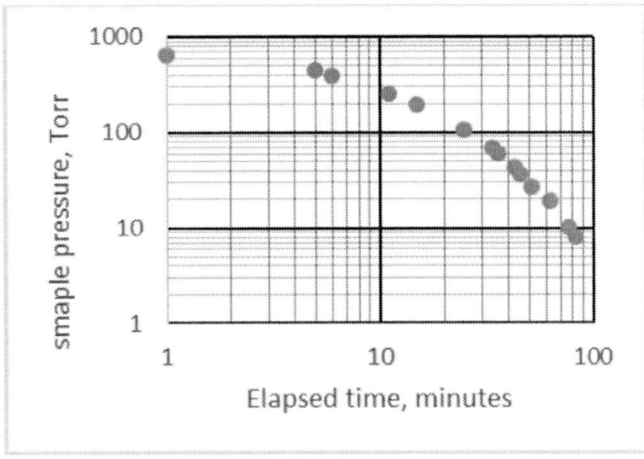

Figure 1. Absorption of pure nitrogen in cold traps.

Figure 2 shows the absorption of the air sample in the cooled traps. It is of interest here that the absorption of air is quite a bit slower than for the absorption of pure nitrogen. In this case, it took about 100 min to absorb 90% of the sample while it took only about 10 min for the pure nitrogen sample. The precise reason for this is not known at this time.

Next, we ran two air samples that were each 350 cm^3 @ 200 Torr. Each of these samples were much smaller than the earlier samples. These samples each contained 0.000485 atm. cm^3 of helium. In Fig. 3 is shown the time history of the partial pressure of helium and several other gasses that make up the background in the RGA. The introduction of the air sample was started at about 8 min and the helium elution was essentially complete by about 25 min after the

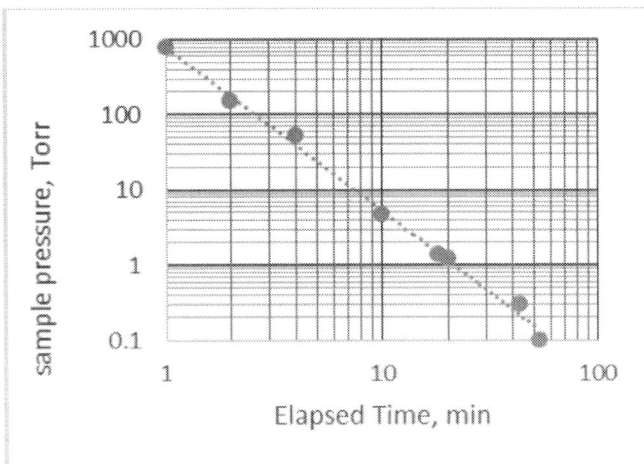

Figure 2. Absorption of air in cold traps.

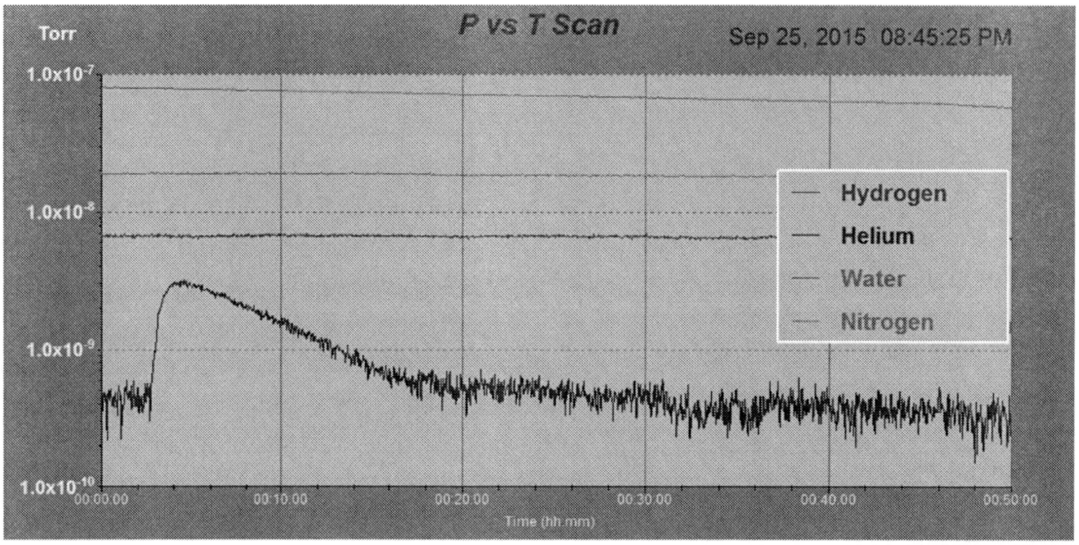

Figure 3. RGA analysis of effluent from cooled traps injected with air.

start if the introduction of the air. Figure 4 shows the net helium signal for the air sample after subtracting the averaged background. The integral of this curve is related to the helium concentration in the air sample. For this sample, the partial pressure was sampled every 2 s and the first 250 samples (500 s) were averaged to get the "before" background of 2.697×10^{-10} Torr. The "after" background was obtained by averaging 250 samples starting at 2000 s and gave a value of 2.75×10^{-10} Torr. If one sums up the data before background subtraction to get the uncorrected gross

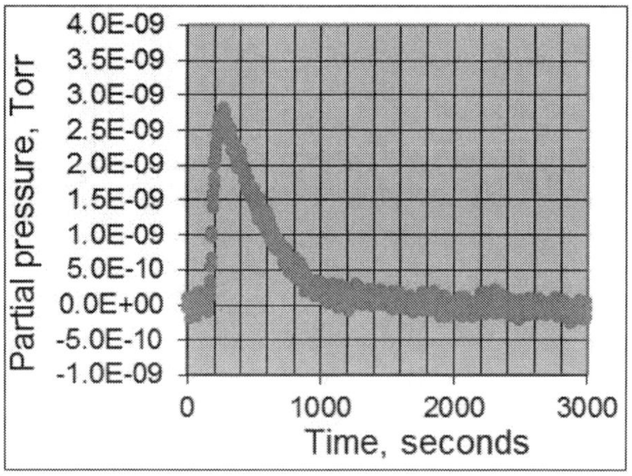

Figure 4. Net ^{4}He signal from 350 cm^3 of air at 200 Torr measured with an RGA.

signal, one obtains 6.71×10^{-7} Torr sample. The unweighted average of these two values (2.723×10^{-10} Torr) was then subtracted from each of the data points. The net data points were then summed to get the net helium signal, and this gave a value of 4.68×10^{-7} Torr sample and the difference between this and the net signal is the background contribution of 2.03×10^{-7} Torr sample. Thus, the background represents about 30% of the total signal for this run. The uncertainty associated with the background is about 28% so this contributes an uncertainty in the net signal of 8.4%.

4. Run with a 10 cm^3 air sample at 2585 Torr

This sample was one of the smaller air samples run to date and is similar in size to those expected from LENR experiments. For this run the Alcatel AMS-110 leak detector was used it to provide the backing pump for the RGA turbo. In this way, relatively quantitative helium measurements as the leak detector could be calibrated with a known helium leak and we could monitor the signals from both the RGA as well as the leak detector.

Figure 5 shows the net helium signal from the RGA after subtracting the averaged background. For this sample, the partial pressure was sampled every 2 s. We averaged 100 s of data just before the sample was injected to get a "before" background of 4.79×10^{-10} Torr. The "after" background was obtained by averaging 100 s of data samples starting at 1900 s and gave a value of 5.35×10^{-10} Torr. The unweighted average of these two values (5.066×10^{-10} Torr) was then subtracted from each of the data points. The net data points were then summed from 500 to 1500 s to get the net helium signal and this gave a value of 9.63×10^{-8} Torr sample. Each sample was 2 s so to convert "Torr Samples" to Torr s one need to multiply by 2, and this gives a net He signal of 1.93×10^{-7} Torr s. If one sums up the data shown in Fig. 5 to get the uncorrected gross signal, one obtains 3.50×10^{-7} Torr sample and the difference between this and

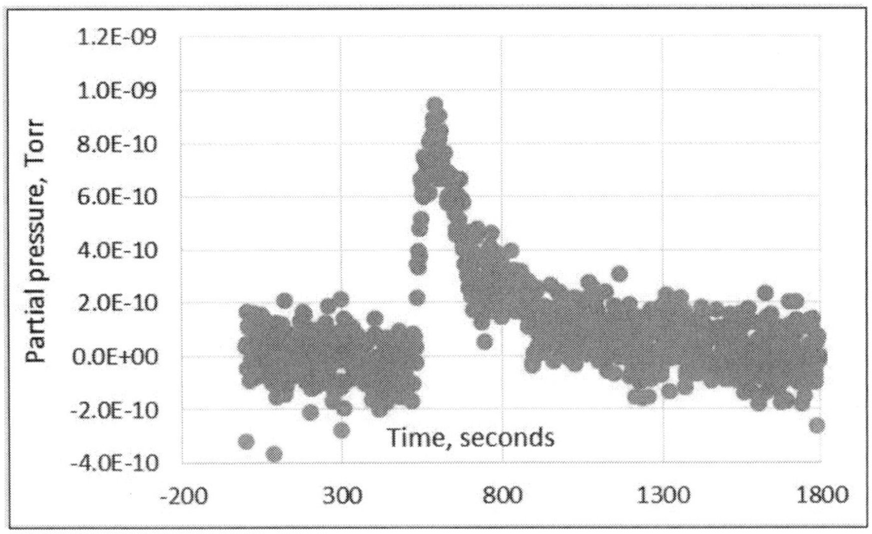

Figure 5. Net He-4 RGA signal from 10 cm^3 of air at 2585 Torr.

the net signal is the background contribution of 2.53×10^{-7} Torr sample. The background represents about 72% of the total signal for this run. These data are quite noisy and that limits the accuracy as well as the detection limit for helium. We have estimated that the uncertainty in the net signal is on the order of 50%. It seems that for this size of sample the detection limit is on the order of 6×10^{-8} Torr s corresponding to about 2 ppm of helium.

We looked at the data from the leak detector for the same sample and found it to be much less noisy as well as having a very small background. Figure 6 shows the data from the leak detector. The background for the leak detector when connected to the RGA and the traps is about 1×10^{-10} atm cm^3/s. If the signal is integrated from about 450 to 2000 s the result is 1.5×10^{-7} atm. cm^3. When the helium signal is integrated over the same period, the result is 3.73 $\times 10^{-5}$ atm. cm^3. In this case, the background is only about 0.4% of the total signal. From these data, we estimate we could measure a much lower concentration of helium with the leak detector than with the RGA. We estimate that we could easily detect 20 ppb of helium in a 10 cm^3 sample at 2585 Torr pressure.

5. Calibration

We can verify that the calibration of the helium leak detector is correct using a known calibrated leak. Figure 7 shows the response of the leak detector to a calibrated leak that has a leak rate of 4.0×10^{-8} atm. cm^3/s. The integral of the measured sample "leak rate" over time gives the number of atm. cm^3 of He that have "leaked" into the system. The ratio of this to the total number of atm. cm^3 in the sample gives the concentration of helium in the sample.

We can also prepare known standards of helium that are similar in concentration to that expected in the samples and then analyze the sample and standard using the same volumes and pressures for both. Figure 8 shows the measured peak area for several different pressures of the 12.8 ppm standard in the 50 cm^3 sample volume. The concentration of helium in the sample is given by the ratio of the integrated signal, over time for the standard, to that for the sample. Several standards were prepared that have equal amounts of hydrogen and helium. The hydrogen allows a check to be made that no hydrogen isotopes are leaking through the chromatographic procedure.

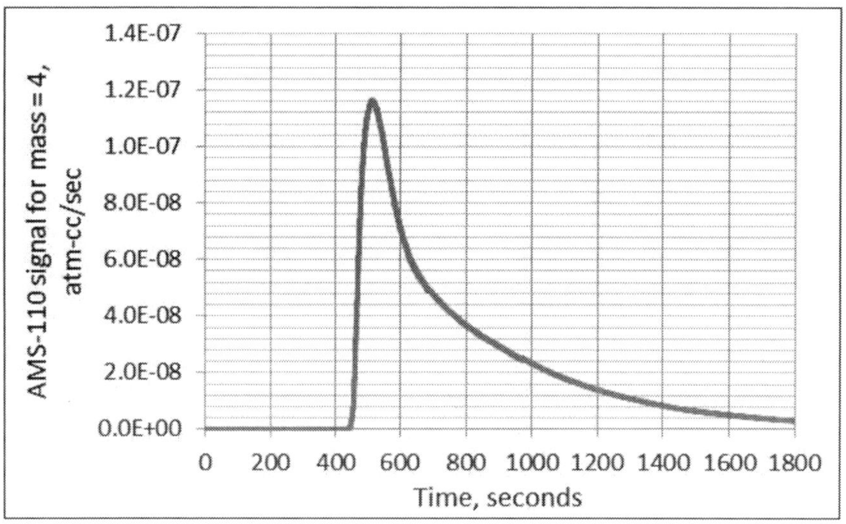

Figure 6. AMS-110 signal for mass=4 from 10 ^{3} of air at 2585 Torr.

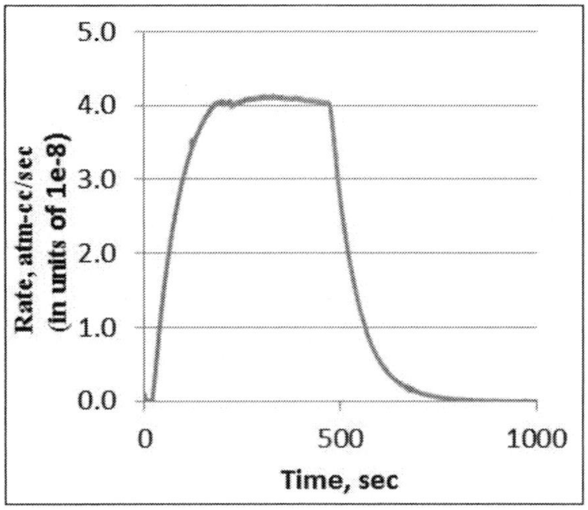

Figure 7. Measured He signal from standard leak vs. time.

Another approach would be to use He^3 as a spike and then use isotopic dilution to get at final sample concentration, but we have not implemented this yet.

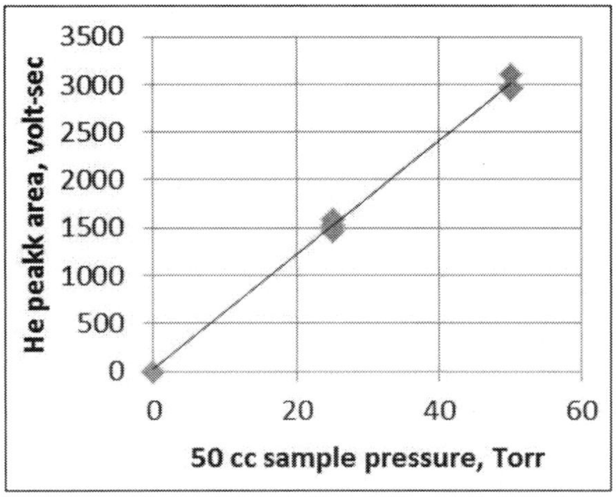

Figure 8. Calibration curve for 12.8 ppm He standard.

6. Current Performance

Figure 9 shows the current system response to a sample that was 1250 Torr cm^3 with a ^{4}He concentration of 12.8 ppm. This sample drives the output to nearly 80% of full scale, so if we wanted to measure more concentrated samples, we would need to use a smaller sample (lower pressure). The background of the ASM-110 leak detector corresponds to about 1×10^{-10} atm. cm^3/s and that gives a signal of about 10 mV, so with a full-scale range of 100 V, we have a dynamic range of four orders of magnitude. With a background "leak rate" of 1×10^{-10} atm. cm^3/s and an ^{4}He elution time of about 100 s, we can estimate that the minimum amount of ^{4}He we can detect is about 1×10^{-8} atm. cm^3, or 2.5×10^{11} atoms of ^{4}He in a sample that is 1250 Torr cm^3. This would correspond to about 6 ppb in the sample. This is a lower estimate compared the earlier estimate of 20 ppb, due mainly to the shorter elution time.

Certainty more modern mass spectrometers could lower this detection limit by quite a bit, but it does not seem necessary for LENR investigations especially when one considers that ambient air is about 5 ppm ^{4}He. Many, if not most, experiments are not well isolated from air. This is also true for the components and chemicals that are used, so it is not surprising that small amounts (ppm) of ^{4}He are found in most samples. As of late 2018, we have run nearly 200 samples, most of which were standards, pure gasses, and backgrounds. We have recently run several samples of pure argon and pure nitrogen. The argon samples gave an average value for ^{4}He of 0.202 ± 0.006 ppm while the hydrogen samples gave an average value of 0.033 ± 0.007. The quoted uncertainties are estimates of the statistical uncertainties. The total uncertainty is probable larger due to the uncertainty in the background for measurements close to the detection limit.

7. Summary

A system has been developed that can measure small amounts of helium in air samples. Currently, we can measure the amount of helium in air at ambient levels (5.26 ppm) with an uncertainty of about 10% with a sample of about 100 cm^3 at 1 atm. pressure to make the measurement. The minimum detectable amount of helium is estimated to be

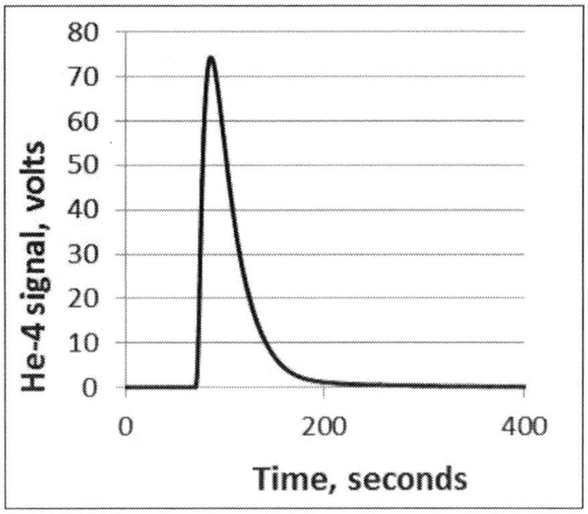

Figure 9. ASM-110 signal from 50 cm^3 of 12.8 ppm ^{4}He at 25 Torr.

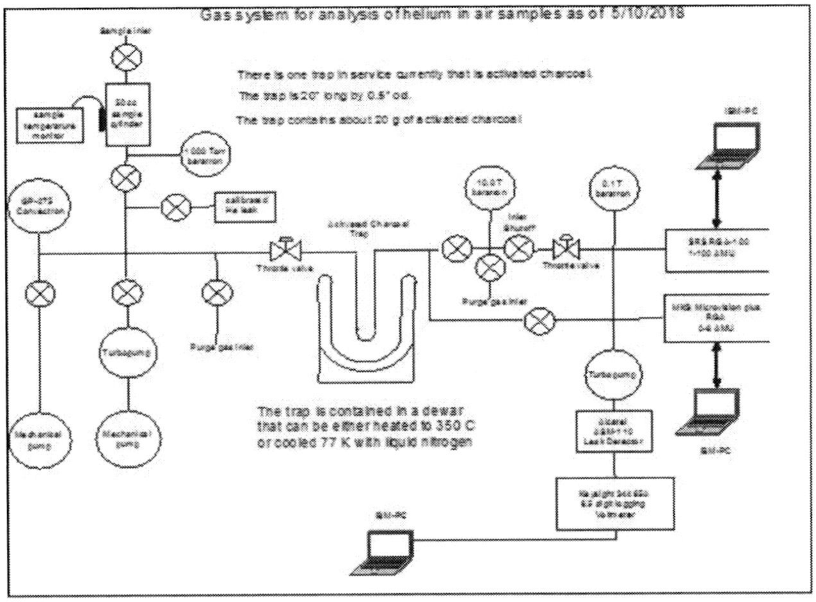

Figure 10. Diagram of the system for trace analysis of helium in gas samples as of 1/1/2019.

about 0.3 ppm for the same size sample. The measurement requires about 1 h to complete, but then requires several hours to warm and regenerate the traps to prepare for the next sample. So far, we have demonstrated that we can detect and measure helium in nitrogen and other gasses with sensitivities quite a bit better than the goal of 1 ppm. Uncertainties are driven mainly by the relatively long time that one must integrate the signal. If there is a relatively stable background, the relative contribution of this background to the total signal increase with time and eventually dominates the overall uncertainty. Efforts to improve uncertainty are centered around two areas; first, one could, in theory, decrease the elution time of the He and thereby reduce the time required for signal integration. At this time, this does not seem too feasible, but changing the geometry of the columns should help here. Next, we can be more rigorous in preconditioning the columns to reduce the overall background. If this can be done in a reproducible manner, then this could lead to considerable improvement. Another approach would be to model the shape of the elution curve. If this could be done, then one might be able to extract a few shape parameters from the recorded data that would allow estimation of the area of the elution curve. I have investigated this a bit and, so far, this does not seem too promising.

8. Future Efforts

We have fabricated a tube furnace that is connected to the analysis system with the goal of being able to heat foil samples to desorb and measure contained helium. What we have found so far is that above about 300°C the quartz tube becomes sufficiently permeable that we see significant signal from atmospheric helium. This is also true for 304 stainless steel at higher temperatures. 316 stainless steel might perform better but we have not tried this yet. We are now investigating methods to protect the heated tube from atmospheric air with coaxial tubes that are either evacuated or filled with some inert gas such as argon.

Acknowledgements

We would like to acknowledge the continued support and encouragement for this work by: Mr. Thomas Francis Darden II, J.D, Mr. J.T. Vaughn, and Mr. Dewey Weaver of Industrial Heat, LLC, 111 East Hargett Street, Suite 300, Raleigh, NC 27601 USA.

References

[1] Hewlett-Packard, MS-5A part number 5060-9084, Activated Charcoal part number 5060-9094.

J. Condensed Matter Nucl. Sci. 29 (2019) 309–333

Research Article

Modeling and Simulation of a Gas Discharge LENR Prototype

Bob Higgins* and Dennis G. Letts

Industrial Heat, 310 S. West Street, Raleigh, NC 27603, USA

Abstract

Modeling of a thermal system is described using circuit analogues and the highly developed SPICE simulator operating in nonlinear transient mode. Thermal circuit modeling is shown to provide a good fit in a Seebeck calorimeter, even for complex experimental thermal behavior. Component values are either derived or extracted for the calorimeter. The resulting model is used to correct for experimental measurement deficiencies, and to deconvolve the excess heat (XH) waveform (residual) from the output response. Model computation time is less than 5 s for typical circuits on an average contemporary personal computer. Also, during the course of modeling the calorimeter, experimental evidence was found for the obscure phenomenon of nonFourier heat transfer.

Keywords: Deconvolution, Non-Fourier heat propagation, SPICE, Thermal modeling

1. Introduction

There are several options to model thermal systems – from the most fundamental formulations as finite element discretizations to simplified analytic parameter fits. Finite element analysis is the most complex to setup, and for the Letts calorimeter, modeling could take years of development and simulation to achieve a satisfactory result. On the other end of the spectrum, analytic parameter fits provide little correspondence to the physical attributes of the system, provide little understanding of the cause for behavior, and limitations to the applicability of the fit are not always apparent.

Between these two extremes lies thermal circuit modeling. This technique uses simple electrical circuit analogues for fundamental heat sources, temperature sinks, thermal flow resistance, and thermal storage. System models are constructed with connections of these components having direct correspondence to the physical system. The resulting thermal "circuit model" can represent the complete dynamic thermal response of the system to an experimental heat waveform input. Agreement between experiment and simulated data can be remarkable. The response can be computed in less than 1 s using a modern personal computer.

The preferred simulator for such modeling is SPICE[a]. The SPICE simulator is a nonlinear circuit simulator available for free from several vendors with an excellent graphical user interface. The SPICE based simulator used for

*Corresponding author. E-mail: bhiggins@industrialheat.co.

[a] SPICE – Simulation Program with Integrated Circuit Emphasis. A common core electrical engineering simulation tool based upon numerical integration, and capable of handling nonlinear circuits in transient simulation mode.

this project was SIMetrix/SIMPLIS™ Elements (hereafter referred to simply as SIMetrix). It is a free version of their commercial simulation environment having full functionality but limited number of simulation nodes. This limitation in maximum nodes is unlikely to impact most thermal modeling problems. The SIMetrix modeling suite [1] has many desirable features including support for variables (for computed component values), non-linear elements, and the capability to drive the model sources from experimental data.

First, the basics of thermal circuit modeling are described. This is followed by application of thermal circuit simulation (using SIMetrix) to model the Letts gas discharge reactor system and its surrounding calorimeter.

2. Thermal Modeling with Electrical Circuit Analogues

Only a small subset of the SPICE supported model elements are required for ordinary thermal modeling. The fundamental components are as follows.

- *Thermal resistance* – resistance to heat flow across a temperature difference.
- *Thermal capacity* – heat storage; fundamental in dynamic thermal modeling.
- *Heat source* – this could be any source of heat, including electrical dissipation.
- *Temperature sink* – a place in the circuit where the temperature is known or fixed.

The following sections describe each of these and how they correspond to their electrical analogues.

2.1. Thermal resistance

The thermal resistance is the simplest and most fundamental thermal modeling component. It represents the heat flow as a function of temperature difference. Figure 1 shows the correspondence between the electrical quantities and the thermal quantities. Understanding this correspondence directly leads one to understand how circuit simulation is used with thermal quantities.

In Fig. 1, the normal resistor's electrical quantities are shown in blue. The definition of the electrical resistance, R_e, via Ohm's law is

$$R_e = \frac{\Delta \text{ voltage across element}}{\text{current through element}} \text{ (V/A)}.$$

A voltage difference across the resistor causes a flow of charge through the resistor and the rate of flow of charge is a current.

In the thermal resistance, heat flows through the resistance depending linearly (normally) upon the temperature difference across the resistive flow path.

$$R_{th} = \frac{\Delta \text{ temperature across element}}{\text{heat in watts}} \text{ (°C/W)}.$$

The correspondence is that heat flow in watts is equivalent to current and temperature corresponds to voltage in the electrical domain. And, with those two analogues ($I \leftrightarrow W$, and $V \leftrightarrow T$), the thermal resistance in °C/W corresponds directly to electrical resistance in ohms.

Thermal resistance is used to model any heat flow. The thermal resistance may be low if the conducting path is through a metal, or it may be a high resistance if the conducting path is through gas.

Thermal resistance may not be a simple constant. In fact, when modeling heat flow to ambient over large temperature ranges, the best model will be a nonlinear resistance (versus temperature) to account for heat flow via convection

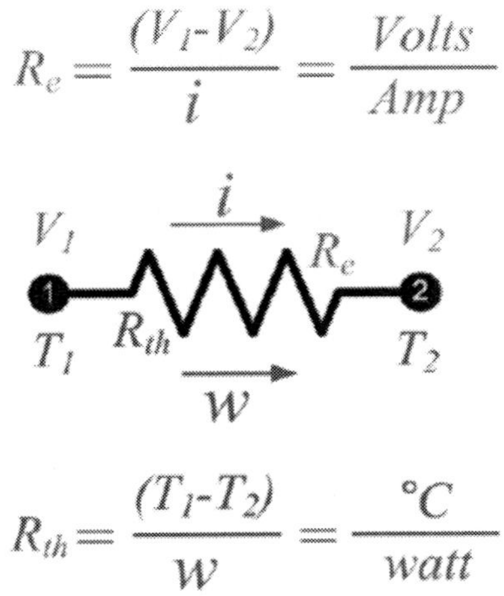

Figure 1. Thermal resistance correspondence to electrical resistance.

and radiation. By utilizing test experimental data over a wide temperature range, a good fit can be obtained for the nonlinear resistance using a simple low-order polynomial representation of the resistance – even when the temperature range being modeled spans more than 1000°C.

The electrical analogue equivalence described above suggest that introducing heat into a circuit is done by introducing a current in the electrical circuit – e.g. using a current source. A temperature sink (a place/node where the temperature is known or constant) may be implemented simply as a voltage source (a source that has a forced voltage corresponds to a thermal source that has a forced temperature).

These fundamental correspondences ($I \leftrightarrow W$ and $V \leftrightarrow T$) set the stage for the whole of circuit based thermal simulation.

2.2. Thermal capacity

Thermal capacity is the ability of an object (mass) to store heat. The copper block in the Letts calorimeter has a high thermal capacity – its mass is high. Heat (Q) is commonly measured in joules. Electrical current is the rate of flow of the charge. Heat flow, in joules/second, is equivalent to watts. A mass that stores heat accumulates heat like a capacitor accumulates charge. For an object with a given thermal capacity, the more heat one adds, the higher the temperature of the object.

Interestingly, electrical capacitors are flexibly used in electrical circuits with both terminals free for interconnection in the circuit. However, capacitors in thermal circuit modeling are always connected with one of the leads connected to the reference (ground) node. Thermal capacity is the fundamental element needed to model the dynamics in a thermal system.

Thermal capacity can readily be estimated based upon the material and it's mass. For example, consider the copper block in the calorimeter. Pure copper has a heat capacity of 24.44 J/(mol °C). The copper molar mass is also listed as 63.546 g/mol. Thus, the heat capacity is 0.385 J/g °C. Knowing the weight of copper in the central calorimeter block, the thermal capacity of the block can be calculated as $C_{\mathrm{th}} = 0.385$ (block weight in g) in J/°C. How this capacity is used in thermal modeling will be described in subsequent sections.

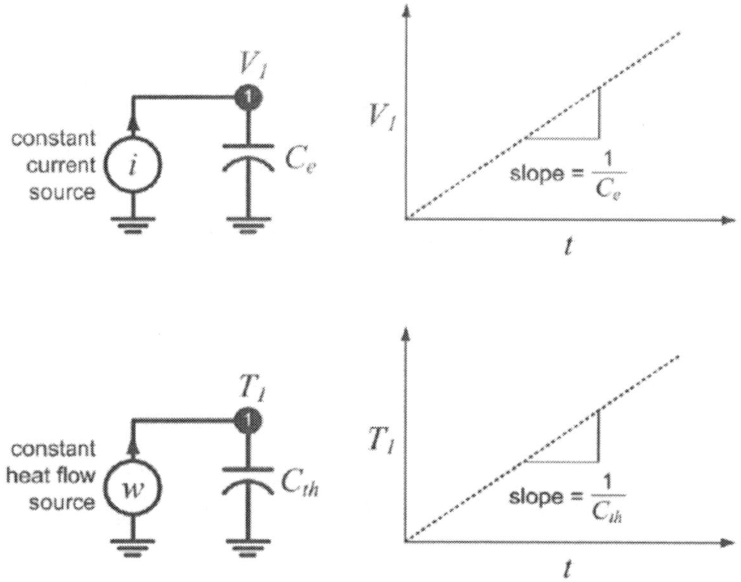

Figure 2. Thermal capacity correspondence to electrical capacity.

2.3. Heat sources

The thermal model is used to compute the thermal response of the system (at various measurement points) to thermal inputs, primarily heat sources. Heat may come from many places in the system, and regardless of type, it can be modeled as an electrical current source. As Fig. 2 suggests an electrical constant current source models a constant source of heat in watts. However, a current source can also be a time dependent waveform. If an ideal waveform is to be tested, the current source may be configured in the simulator as a current waveform generator.

The SIMetrix SPICE simulator provides a very useful general current waveform generator – one that can be driven from a file. This source is called a PWL source for "Piece-Wise Linear". A file name is supplied for the source – within this text file, samples of (time (in seconds) current (in amps)) are supplied, one sample in each record (see Fig. 3). The simulator interpolates linearly between the supplied samples as it computes the response. Interpolation is required because SPICE does not simulate the response with uniform time steps (and uniform time steps are not required for the PWL data file). The SPICE transient solver varies the time steps automatically as needed to maintain the accuracy of the numerical integration. The utility of the PWL source is that heat input into the experiment is typically recorded as part of the sampled data. This exact experimental input can be extracted as a 2-column text file and used to drive the model with the actual heat input waveform that was used in the experiment. Parameter values (thermal resistance and capacity values, and/or circuit topology) are adjusted such that the simulated output response agrees with measured output response obtained when being driven with the actual experimental input heat waveform.

How heat sources are employed will be described in application to the Letts calorimeter modeling is shown in Table 1.

Table 1. Text data file for driving a PWL source.

0.0	27.95
60.0	27.97
120.0	27.98
180.0	27.97
240.0	28.02
300.0	27.96

2.4. Temperature sinks

In almost all thermal systems there will be certain points in the model where the temperature is either held constant, or it is known. This temperature "sink" may be modeled using an electrical voltage source to ground (the thermal analogue of voltage is temperature). Thus, a voltage source in the model represents a node at known temperature. For example, suppose there was a place in the system that was water cooled and regulated to 28°C. This could be modeled in the electrical circuit as a voltage source to ground with a voltage of 28 V.

For the common case where the temperature of a thermal sink is not constant, but varies with time across the experiment, and presuming its temperature is recorded, a PWL voltage source may be used to force exactly this behavior within the model. The experimentally measured data for the temperature at that modeled location is extracted into a 2-column text file and used to drive the PWL voltage/temperature source during simulation. This useful feature will be frequently incorporated into thermal models.

2.5. Thermoelectric (Seebeck) generator modules

In the Letts calorimeter, heat flow is measured by proxy in the generated output voltage from a set of thermoelectric generator (TEG) modules. Figure 3 shows a sample cross-section of these modules in the Letts calorimeter between the copper block of the calorimeter and the finned aluminum heat sinks.

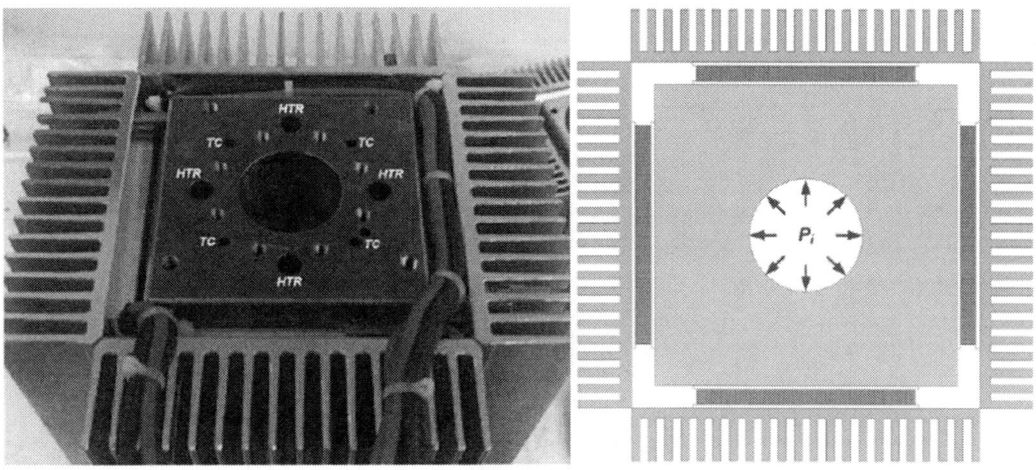

Figure 3. Letts calorimeter cross-section showing TEG module location.

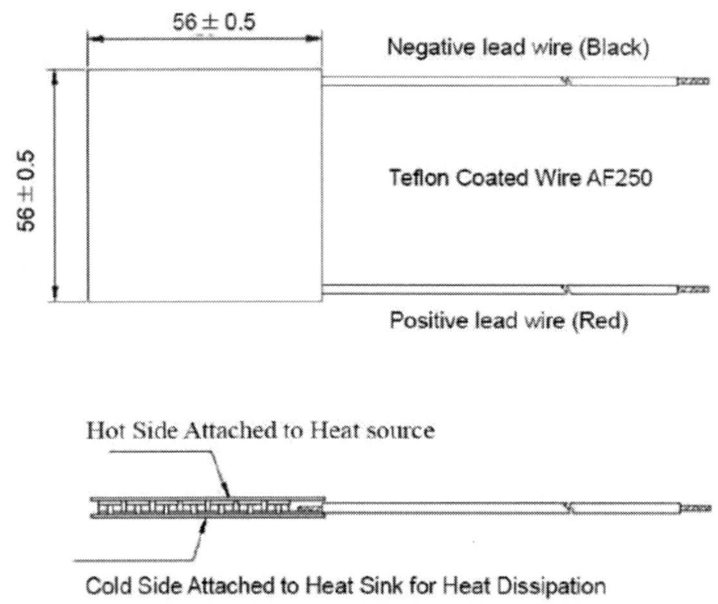

Figure 4. TECTEG TEG1-12611-6.0 TEG module drawing.

Figure 4 shows the module dimensions used in the Letts calorimeter. The drawing also shows the modules having thin ceramic (typically alumina) electrically insulating but thermally conductive plates with semiconductor "chunks" between the plates. The modules are designed to have heat flow from the large area of the hot side ceramic plate to the cold side ceramic plate. Thus, from the hot side to the cold side in a circuit model for this element, there

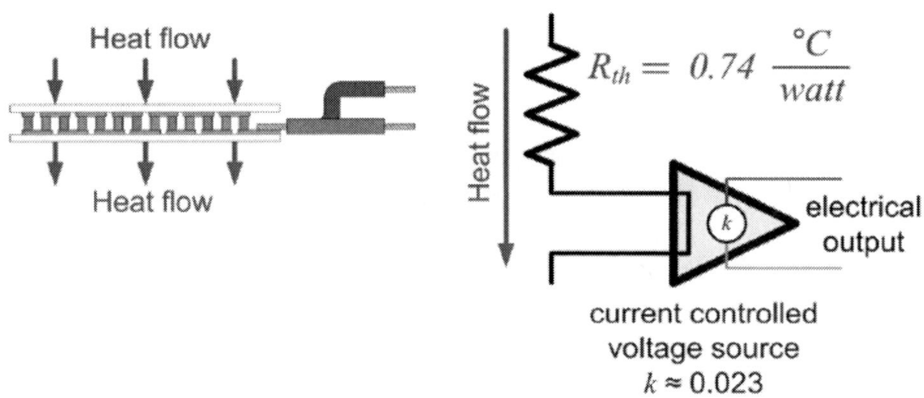

Figure 5. Simplified model for TEG module.

will be a thermal resistance. From the manufacturer's specification, the thermal resistance is ~0.74°C/W. The open circuit voltage generated by the module is ~23.0 mV/W passing through the module, but will be a somewhat nonlinear function of the temperature difference across the module.

The model for this element is a mixed electrical and thermal circuit because it has a thermal input/output and an electrical output. The mixed thermal and electrical representation is no problem for the simulator – SPICE does not keep track of what is thermal and what is electrical (the user must keep track). A simple model for the TEG module is used for the Letts calorimeter because only its open circuit output voltage is measured (modulation of the thermal resistance by the electrical load is ignored). For this case, the module is represented as a thermal resistance and a current controlled voltage source as in Fig. 5.

3. Overview of the Letts Discharge Reactor and Calorimeter

The Letts gas discharge reactor (LENR Tube, LT) and its surrounding calorimeter are described in another paper [2], and is only briefly described here. The core reactor is a 28 cm long by 2.5 cm diameter coaxial stainless steel gas discharge vessel having a central molybdenum wire anode and an interior tube surface (cathode) electroplated with the active LENR metal (typically Pd, see Fig. 6). The tube is evacuated and then filled with ~10 torr of D_2 or H_2 (for comparison). The power source for the discharge is a 200 mA constant current source which typically operates at about 330 V DC (66 W input) when the discharge has been struck. It was found that the desired pressure (10 torr) must be in place before the discharge is struck. Heat is taken from the system radially from the outside of the tube envelope.

Around the LT is a large copper block (photo in Fig. 3) with a large central bore for inserting the LT assembly; and eight small holes on a common radius for four inserted resistive heater cartridges ("HTR"), and four thermocouples ("TC"). The heater cartridges are used to raise the quiescent temperature of the copper block and LT. Calibrations of the system are normally taken with the resistive cartridge heaters set at a fixed power, or changed to keep the total input power constant (discharge power + resistive heater cartridge power).

The purpose of the large copper block is to thermally sum the heats coming from the LT and the resistive block heater cartridges, and distribute the heat to flow through the TEG modules into the finned aluminum air cooled heat sinks, with the heat flow registered as an output voltage on the TEG modules. The copper block, TEG modules, and aluminum heat sink comprise the core of the calorimeter. To help maintain repeatability, the calorimeter is housed in constant temperature commercial refrigerator with a set point temperature of 28°C.

Typical experimental protocol involves filling the LT with D_2, establishment of LT discharge, looking for excess heat (XH), and if XH is found, evacuate the LT and repeat with H_2 to show quenching of the XH – presuming and proving isotopic dependence of the XH effect.

Note that modeling the temperature of the aluminum heat sinks inside the temperature controlled chamber is unnecessary for some experiments when thermocouples are present to measure their temperatures. In the model, this can be represented by forcing the temperature of the heat sinks to be the experimentally recorded value, minimizing unknowns in the model.

4. Thermal Model Circuit Development

Thermal model circuits are generally much less complex than most electrical circuits because the model is comprised of such few element types: thermal sources, thermal resistances, and thermal capacitances (which are always to ground). The TEG modules are somewhat unique to the application, but are simple fundamental sub-circuits.

In development of a thermal circuit model it is desirable to recognize which heat flows can be combined into one element (or path). For example, consider the copper block at the heart of the Letts Calorimeter as shown in Fig. 7. The block and its coupling of heat is symmetric around the center to its four sides; I.E. heat is taken out of all four

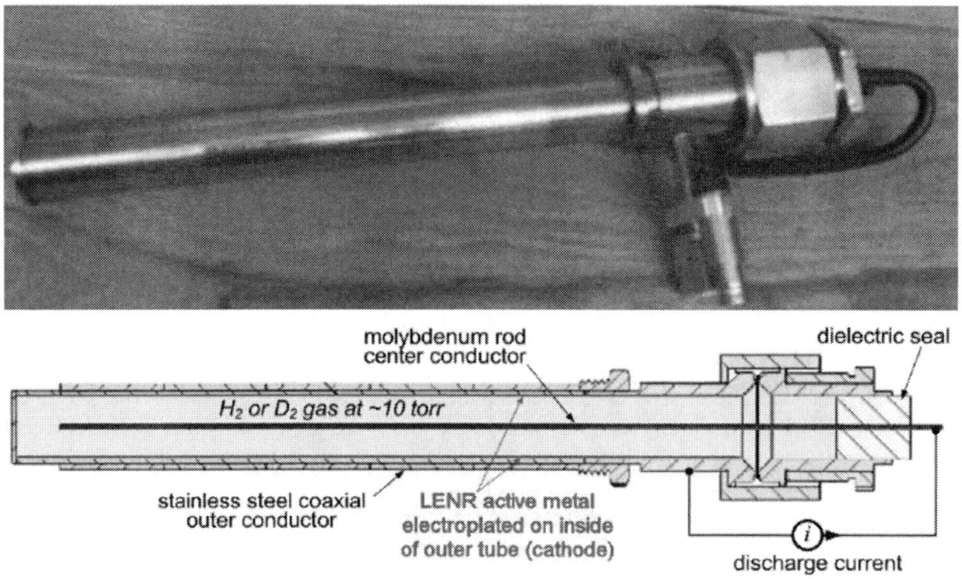

Figure 6. Gas discharge "LENR Tube" (LT).

sides in the same way. Heat enters radially distributed around the block center from the outer shell of the LENR tube (LT). Provided the remainder of the system remains symmetric, all of the heat conduction path resistors are essentially parallel and may be modeled by just one equivalent combined thermal resistance. The intention is not to model the heat flow distribution on each side, since there are no measurements to support that modeling. The resistor flow paths on a single side can be condensed into a single flow path for the circuit model. All of the LT heat will come in at P_{i} and pass through a single combined/effective thermal resistance that will model the combination path to all four sides. This is a tremendous simplification of the model circuit – it means that only one value (R_{equiv}) will have to be found that provides a good match to the measured response.

Also, consider that each side of the copper block will have five TEG modules arranged along its length (see Fig. 8), through which the heat will flow. The module electrical outputs are wired in series. There is no need to put five modules in the model – put a single TEG model and presume all of the heat from the five TEG modules flow through the one module – the heat flow will be five times that going through the single module and the voltage output of the single effective model TEG module will be five times that of a single module. The same is true for each of the other module sides – these can again be combined into a single module representation with all of the heat from the 20 TEG modules flowing through a single equivalent module – a substantial simplification!

Only when it is found that this contraction of the model produces a circuit whose response cannot match that of the experimental device should one consider expanding the circuit into more of a 2-dimensional model. Try a simple circuit first – it is usually found that even a simple circuit model can represent a great deal of complexity in the output response.

Problems with matching the experimental and simulated responses are not always caused by the simplification into a single path model – sometimes other phenomena are present. High temperature difference and air flow convection and radiation mechanisms can combine to make the resistance non-linear – meaning that the resistance is smaller for

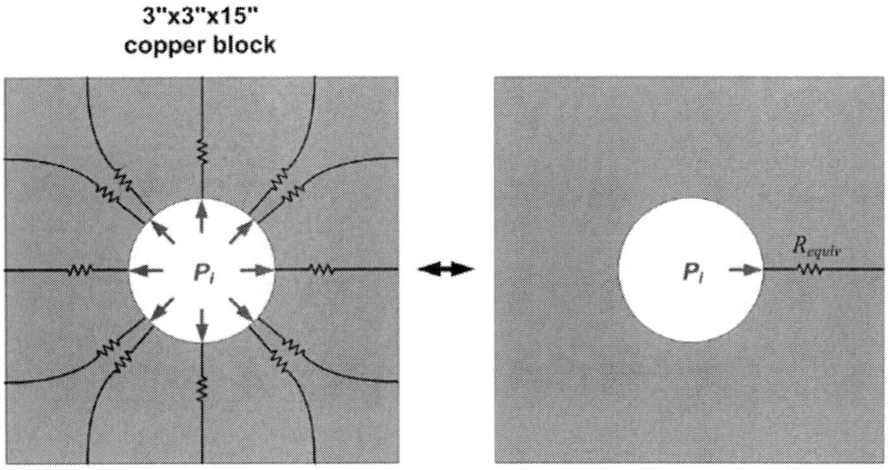

Figure 7. Thermal model for copper block.

a large temperature difference than it is for a small temperature difference. The fix for this problem is to use a single non-linear resistance rather than trying to find topology changes in the circuit. Note that the thermal resistance of a path within a metal is seldom non-linear over the temperature ranges likely to be modeled.

Try to maintain a direct 1:1 correspondence between your circuit and the physical structure. Another rule of thumb is to start simple and add additional heat paths as needed to achieve a match between the experimental response and the simulated response.

Figure 8. TEG modules along length of copper block.

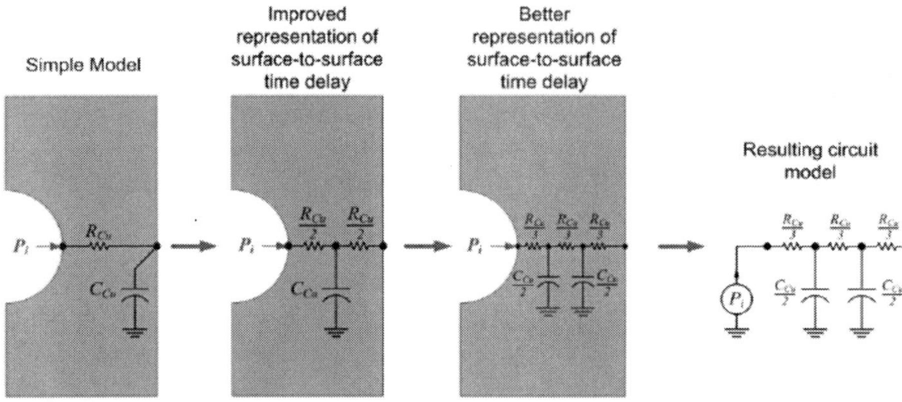

Figure 9. Dynamic thermal model for copper block.

4.1. Thermal model dynamics

The dynamic response of the model is primarily due to heat storage in thermal capacitances. The model of Fig. 7 of the copper block did not include the thermal capacitances that would have provided the dynamics of the heat flow. Generally, the dynamic response is sensitive to the $R-C$ product time constant, and much less sensitive to the absolute accuracy of either R or C.

The simple model for the copper block would be an $R-C$ as shown in Fig. 9 (circuit at the far left). However, the nature of the flow from the inner surface where the heat is applied from the LT to the outer surface of the block, where the heat is extracted through the TEG modules, is of a more distributed nature. Thus, an improved model comprises breaking the $R-C$ into multiple sub-sections – a more distributed $R-C$ delay wherein the total resistance and capacitance has not changed. Seldom is there an advantage to breaking the circuit into more than two capacitances and three thermal resistances.

Consider that the heat source is represented by a constant current source (current being the analogue of heat flow in watts).

4.2. Modeling the core copper block

Holes for the heater elements in the block and for the thermocouples are ignored – the absolute value of the thermal capacity is of less importance than the time constant of the system, which will be extracted from experimental data. Figure 10 shows the dimensions of the core copper block. Its volume is computed as 128.3 in.^{3} = 2102.9 cm^3. Copper's density is listed as 8.96 g/cm^3 – the total weight of the copper block is $\sim$18.8 kg. In Section 2.2, the thermal capacity of copper was found to be 0.385 J/(g °C); thus, the thermal capacity for this copper block is 7254 J/°C. In the model for the copper block, this will translate to a circuit capacitance of 7254 F (but it is divided in the distributed $R-C$ modeling).

4.3. Copper block with TEG modules and finned heat sinks

Heat from the Cu block is expected to largely pass through the TEG modules to the finned Al heat sinks, and subsequently into the forced air flowing past the heat sinks (primary path, see Fig. 3). As heat flows through the TEG

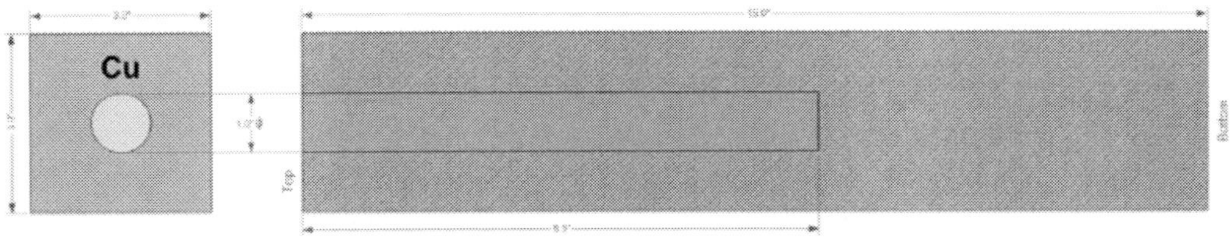

Figure 10. Copper block dimensions for mass calculation.

modules, the TEG voltage is recorded as a proxy of the heat flowing through the module. There will bea small heat leakage path from the Cu block directly to the ambient air that by passes measurement by the TEG.

Figure 11 shows the simple model for the Cu block assembly with its Al heat sinks. C_1 is the thermal capacitance calculated for the Cu block – it is partitioned into 2. Resistor $R1$ is partitioned into three pieces as part of the same distributed $R - C$ delay network. The primary heat conduction path is through thermal resistance, $R4$, for parallel combination

Of the thermal resistances of all 20 TEG modules. The per-module resistance was stated in Section 2.5, Fig. 4, to be 0.74°C/W. For 20 modules the thermal resistance is reduced by a factor of 20, so R4 in the model would be (0.74/20) $=$ 0.037 Ω (a very small value!) The unknown for the TEG modules is the k_s thermoelectric conversion constant that will be determined by extraction so as to match test experimental data.

Thermal resistances R3 and R5 represent the heat leakage flows from the Cu block to ambient and to the Al heat sink, respectively. R3 and R5 are unknown and will be difficult to separately extract from the test data. Initially R5 will be ignored and all of the heat leakage will be lumped into R3.

An important simplification is shown in Fig. 12. Since the data acquisition records the heat sink temperature, the

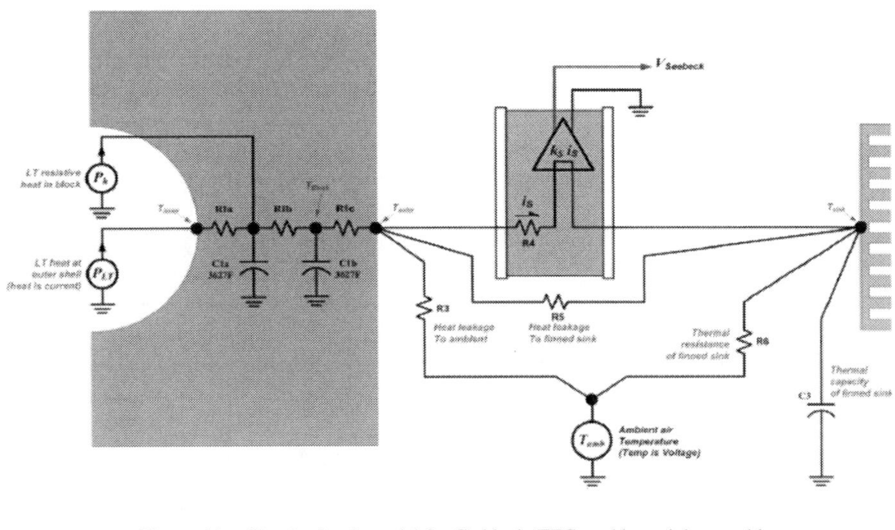

Figure 11. Simple circuit model for Cu block, TEG, and heat sink assembly.

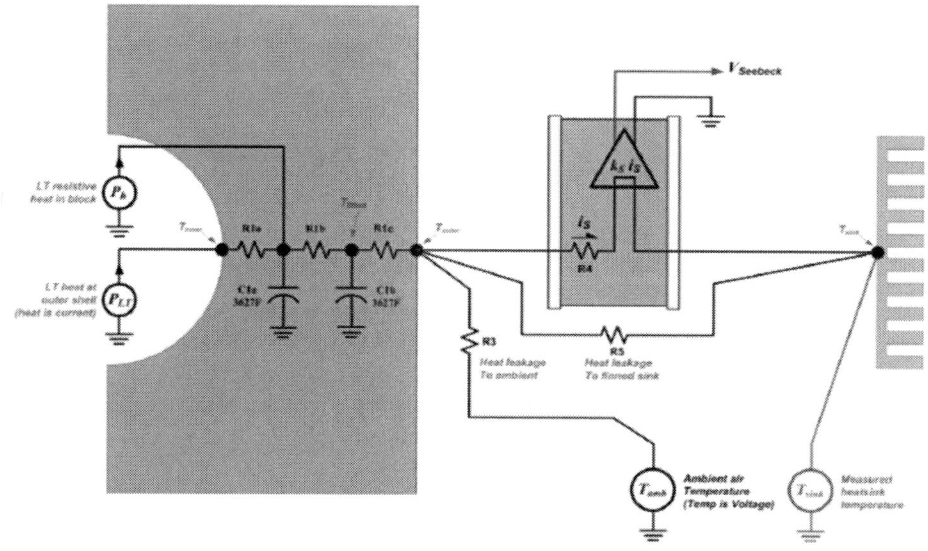

Figure 12. Reduced circuit model.

model can be driven with a voltage source (temperature) to constrain the circuit to have the experimental temperature for the heat sink. This will remove heat sink thermal behavior from the model – its state variable (its temperature) is known/forced as a function of time. Once the R1, R3 (ignore R5 for now), and k_s are determined, then R6 and C3 may be extracted by choosing values for them that cause the measured heat sink temperature to reproduce the experimental value of T_{sink} vs. time without a source driving the heat sink node.

"Parameter extraction" is determination of the model circuit values by iterating the unknown model circuit values to cause the simulated output(s) to match the measured experimental output(s). A first estimate is made for the model element values and the simulator is run with the known/measured values for the input waveform(s), and provides simulated output waveform(s). The simulated output waveform(s) are compared to the measured/experimental output waveforms, and the model element values are adjusted to bring the simulated output(s) to closer to the measured waveform(s). This process is repeated until sufficient agreement is reached (Fig. 13). When agreement is reached, the model parameter values are said to have been "extracted" – this is the objective.

Once the parameters have been extracted, the model is ready for use. Referring to Fig. 14, if the simulator is driven with known input waveforms (and 0 for XH), then the simulated outputs will match the experimentally measured null experiment output waveform(s). I.E. the model will represent the known/expected response from the known inputs.

Suppose a new experiment is run, this time with an unknown LENR XH being generated sometime during the experiment. For this experiment, the model can take the known/measured power inputs and provide the expected output corresponding to those inputs if there is no XH. The modeled output may be used to extract the waveform for XH – but it is not a simple problem. In general, this problem is a class of deconvolution, which is well known to be an "ill-conditioned problem". It is ill-conditioned because of noise in the measured data. If the system comprises a delay (typical), it will exhibit a low pass response – the XH high frequencies will be attenuated in the measured output and the signal-to-noise ratio (SNR) for the high frequencies of the waveform will decline continuously as the frequency

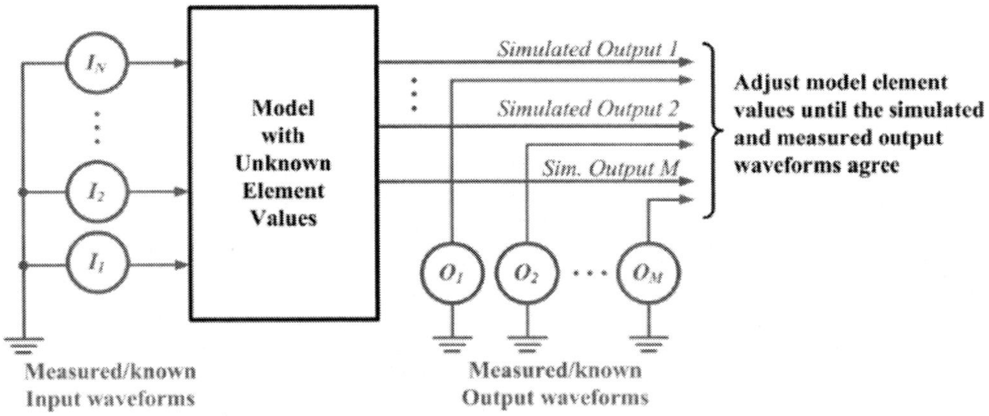

Figure 13. Model parameter extraction.

increases. At some frequency within the unknown XH waveform, the response in the output data will just be noise. Reconstruction of the high frequency components of the XH from the noisy output waveform will, by virtue of the noise, become arbitrary. Thus, one must limit the reconstruction of the XH (in extraction) to a component frequency range having real supporting data (SNR) in the output waveform. The better the output can be measured (lowest possible noise), the better the XH waveform can be extracted/reconstructed.

Because the system is likely to be non-linear, deconvolution cannot be done by Fourier techniques. Generally the solution requires iteration – take a guess at the XH waveform, find the output (experimental- simulated) difference, and construct a better guess at the XH utilizing the difference. Repeat this process until the simulation using the XH

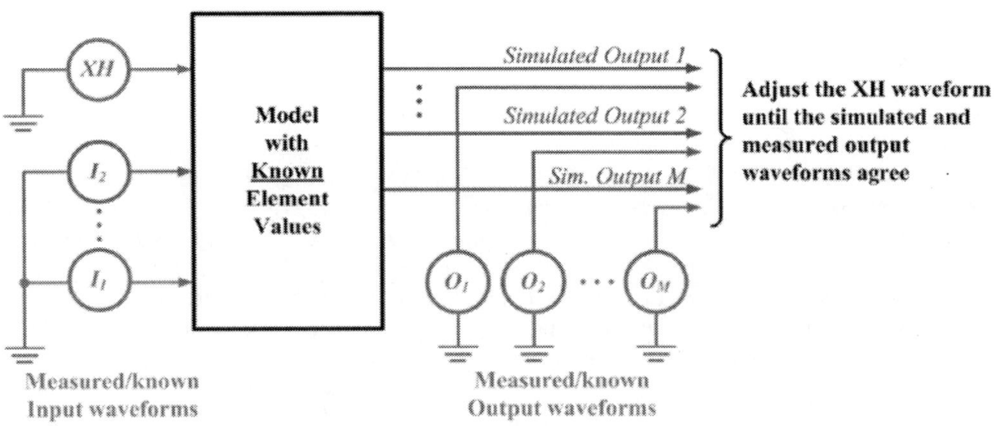

Figure 14. XH extraction.

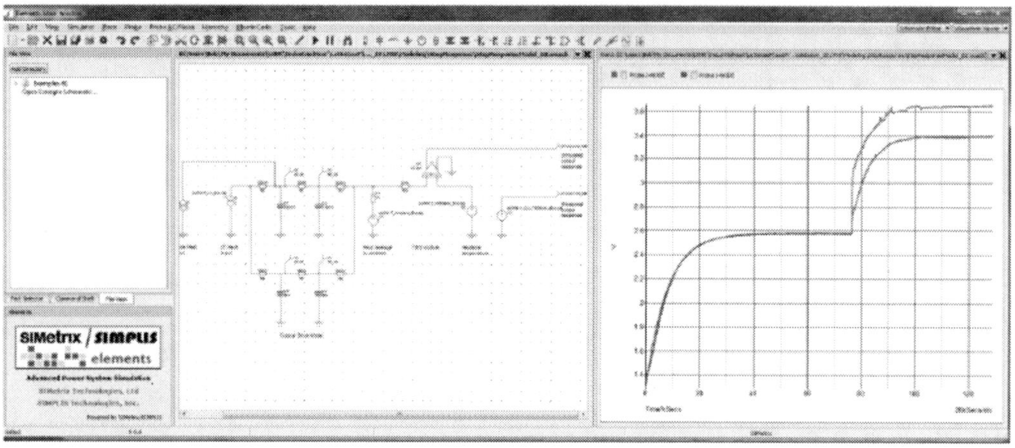

Figure 15. Step response of calorimeter.

waveform creates a simulated vs. experimental output difference of essentially zero. With this process, the XH versus time waveform may be extracted.

Fortunately, an easier way to perform this extraction using the power of the simulator exists, and will be described below.

5. Modeling With The SIMetrix SPICE Simulator

A description of how to use the SIMetrix SPICE simulator is deferred to its well developed manuals. Figure 15 shows a model entered into the SIMetrix SPICE schematic editor. The circuit is easily entered with the graphical interface, and graphical outputs are managed in another window (windows are un-dock-able). Any of the currents/powers and voltages/temperatures may be probed after the simulation has completed by selecting a probe type, and clicking on the desired branch/node in the schematic.

Early on, an experimental run was created with a canonical step heat input waveform. The step input excites the natural response of the system. A curious effect was noted in the experimental data (see Fig. 16). In this figure, the blue curve is the actual step heat input into the calorimeter (non-ideal). The red curve is the heat proxy measured by the calorimeter via the TEG voltage output (approximately calibrated to translate to power measured). What is seen is that more than one time constant is present. The output initially rises quickly, with approximately 40% of the heat registered in a very short time. Following the initial rise, the remainder of the heat is registered in the output with a longer time constant as expected ($\sim$119 min). Close examination (see Fig. 17 zoom) showed the faster time constant to be $\sim$2.7 min – 44 $\times$ faster! The initial rise is so fast that it is poorly sampled by the 1 min sample period of the data. While the initial rise is fast compared to the main propagation of the heat through the block, it is still slow compared to a "speed of sound" effect. As the model development progressed, it became clear that this second propagation mode was real, repeatable, and present in the response at all excitation levels.

The plots were sent to the manufacturer of the TEG modules who suggested that the response could be due to non-linearity in the power flow to voltage output conversion. A university professor suggested that the response appeared to be a "non-Fourier" heat transfer mode – poorly documented in the literature [3]. The existence of the second mode, as opposed to a non-linear response, was confirmed by excitation with a stair-step power waveform in place of the

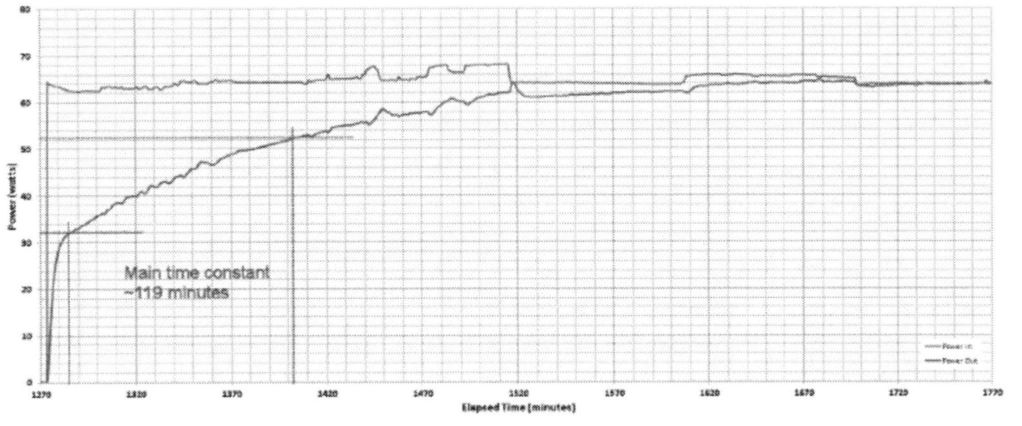

Figure 16. Step response of calorimeter at TEG output.

single large step (see Fig. 18). Each of the individual small steps showed the dual time constant which would not have occurred if the response shape were due to non-linearity in the TEG. While this anomalous, non-Fourier heat flow mode is interesting, it is simply noted here that there is more than one heat propagation mode in the system. The second propagation mode just adds additional complexity to the model.

Data from a calibration run having stepped electrical power applied to the resistive heaters in the copper block (LT still present, but no discharge current supplied) was used in model development. The resistive heaters were driven in a stair-step (DC) in power steps of 20 W from 0 to 200 W (Fig. 18).

Figure 19 shows an early single propagation mode model for the calorimeter (after the benefit of iterative parameter adjustment). On the left, I2 is the heat input versus time as a PWL current source playing out the measured data for the

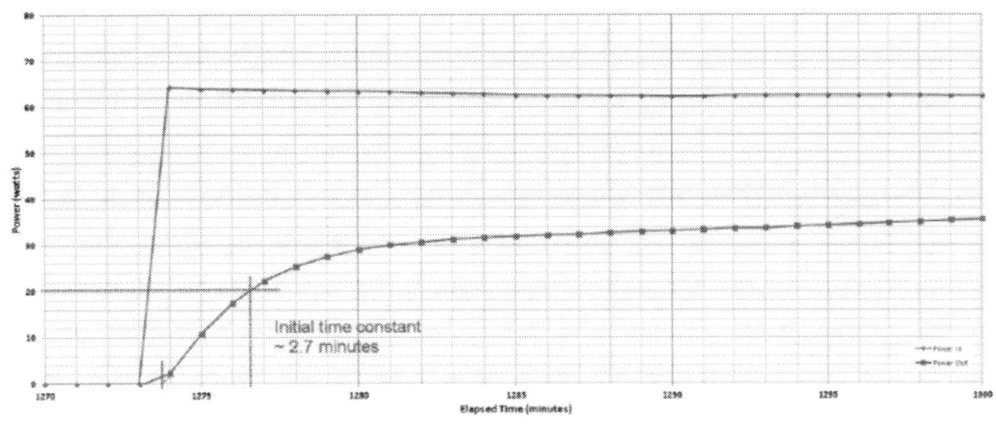

Figure 17. Anomalous fast rise.

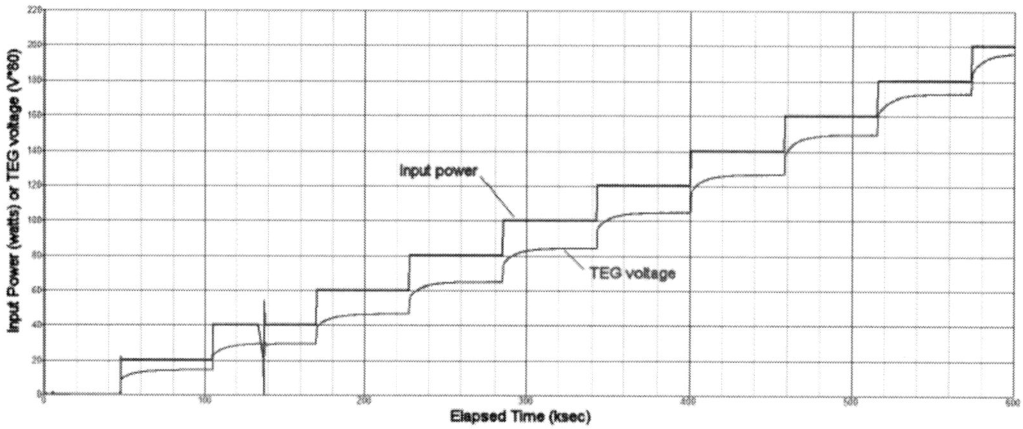

Figure 18. Stair-step excitation and experimental response.

input heater power. V1 is the ambient temperature versus time; and V2 is the block's heat sink temperature versus time (being driven by the measured data). V3 is simply playing out the measured TEG output voltage for graphing. Around these sources is the model for the block and its TEGs. The output voltage from the TEGs was found to be non-linear, and a non-linear transfer function block (ARB1) was included in the TEG model. The small symbols labeled "IC" stand for "Initial Condition". These provide the simulator with starting conditions ($t = 0$) for the state variable – in this case the temperature of the node. The IC is placed as a component in the schematic, but it is actually a simulator directive.

Figure 20 shows the simulated TEG output voltage in response to the experimental input power at I2 and compares it with the experimental TEG voltage supplied at V3 (for convenient graphing only). The response shows a good match at the settled points for the 0–200 W stepped input power.

However, one can see the difference in the step response dynamics between the single mode simple model and the

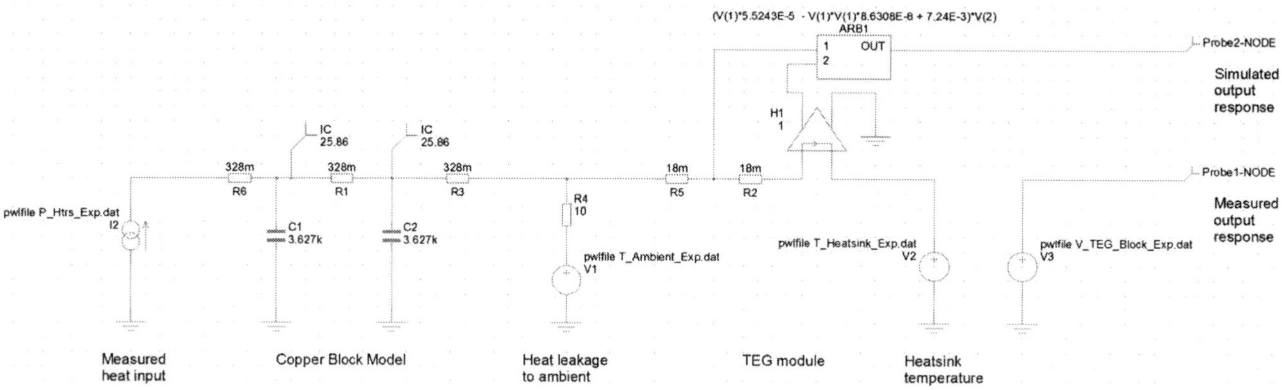

Figure 19. First pass single heat flow mode model.

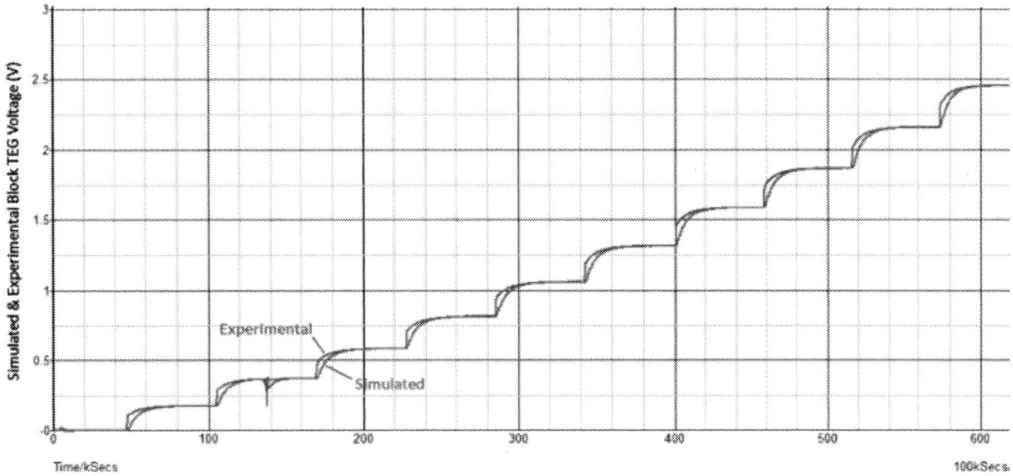

Figure 20. Response of first pass single mode model.

actual behavior of the real experimental copper block – the experimental data rises much faster initially at each step due to the previously described second mode.

This fast early rise can be clearly seen in the zoom of one of the steps in Fig. 21. To achieve a good match between simulation and experiment, a second delay/propagation mode must be included in the model.

Including a second propagation mode in the model began with a parallel $R-C$ track in the model for the copper

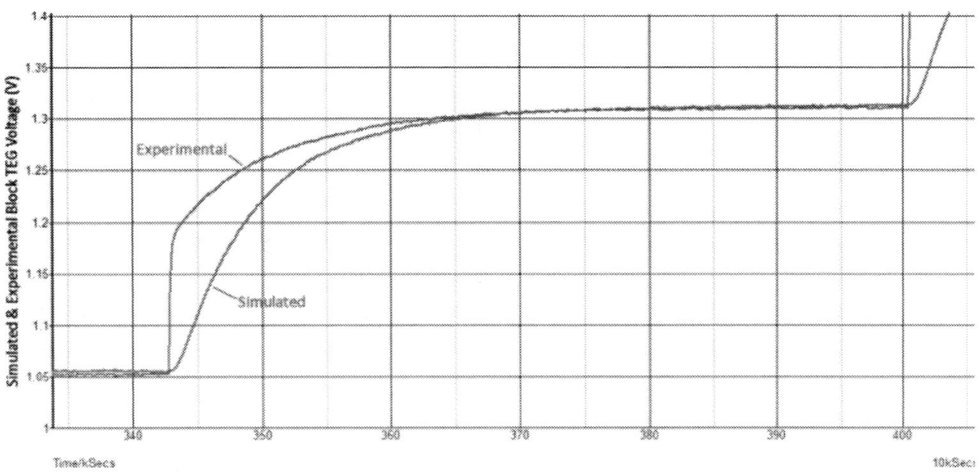

Figure 21. Response of first pass single mode model (zoom).

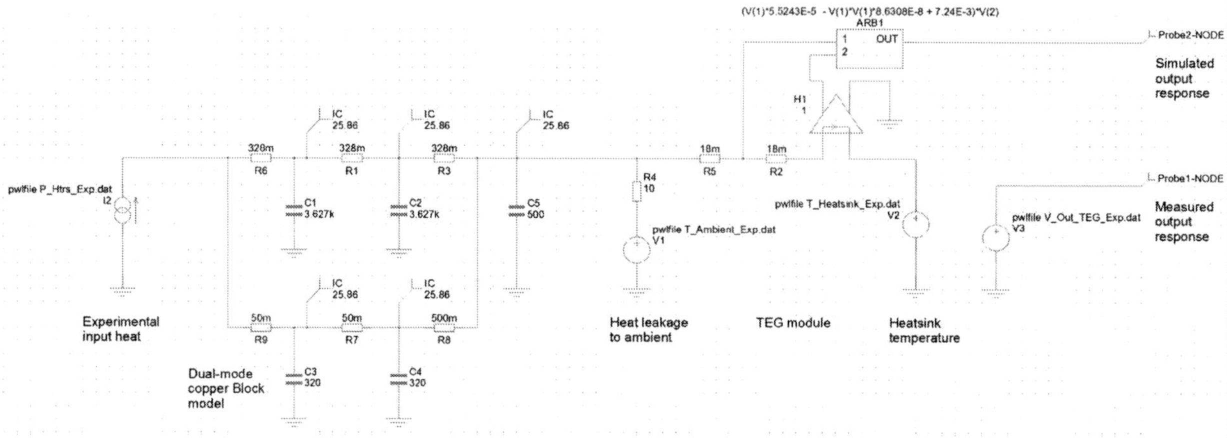

Figure 22. Second pass dual mode model.

block. Figure 22 shows a second schematic having the parallel $R - C$ tracks. When Fig. 22 was simulated, it clearly showed both the fast and the ordinary modes. The amount of the input delivered into each mode was determined by the size of R9/R7/R8 compared to R6/R1/R3. Adjustment of the values proceeded to provide a better match of the output dynamics near the beginning of the steps, but it was found there were not enough element values to change to provide a smooth transition from one mode to the other as was actually seen in the experimental response.

While the dual $R - C$ delay path model shown in Fig. 22 provided a visualization of the transition from a single-mode model to a dual-mode model, the topology did not provide a good match for the dynamics of the steps. Figure 22 circuit model was abandoned in favor of a new heuristically derived topology (Fig. 23) that would provide a better match for the response at the expense of having lesser visualization of the presence of the two modes. Figure 24 shows the good match between the experimental and the modeled response to the stair-step power input. Note that the glitch in the stair-step input at ~139 ks was a power failure that was recovered. Using the actual measured input data to drive the model (using a PWL source) insures that this event is included in the input. Figure 25 is a zoom at a step in the

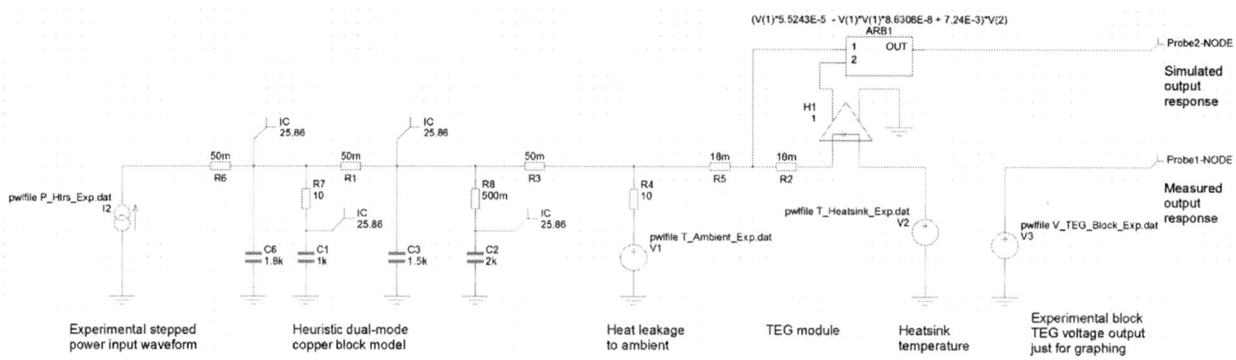

Figure 23. Third pass heuristically derived model.

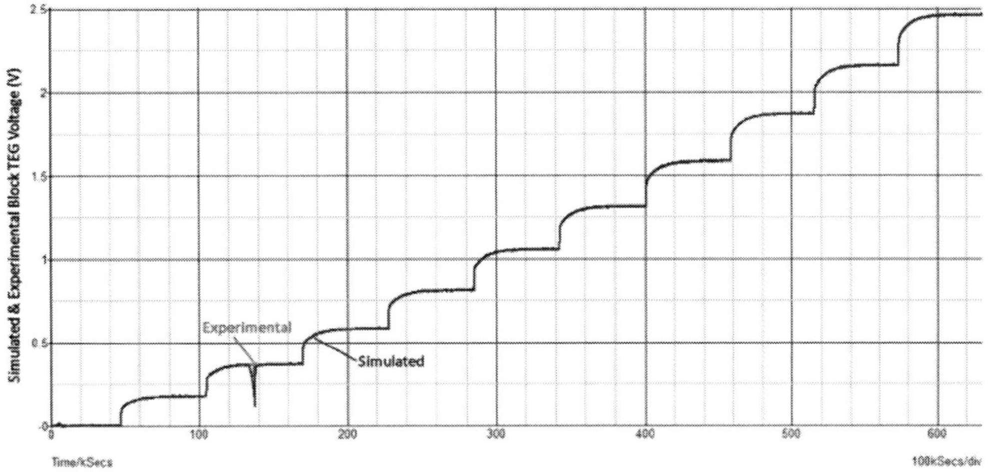

Figure 24. Stair-step response of third pass model.

power input, showing the match to the dual mode dynamics.

Up to this point, it was presumed that all of the input heat was going into the copper block for simplicity. Only ~5% of the heat exits the calorimeter through the "Lid" – the top cover of the calorimeter having its own TEG modules. The fourth pass of the model addresses the addition of the Lid.

To include the Lid in the model, circuitry was added to divide the incoming heat between the block section and the

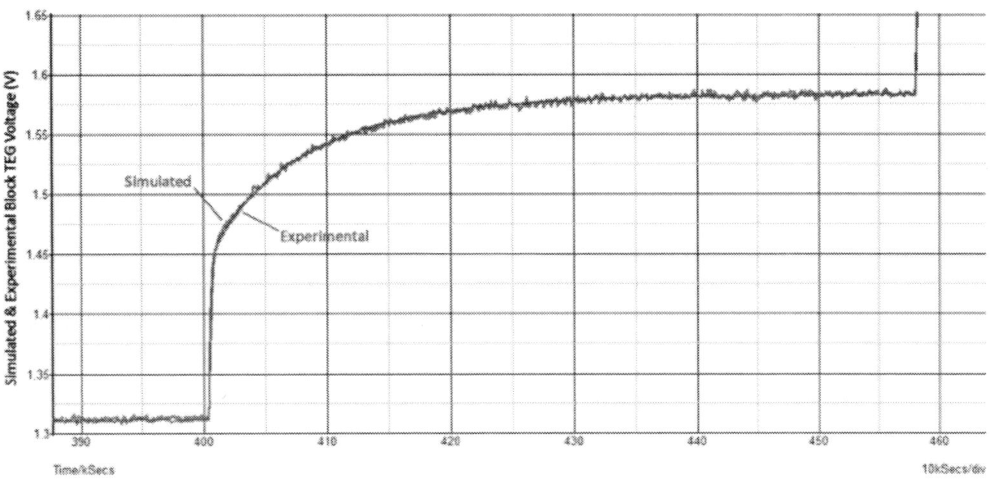

Figure 25. Response of third pass model to stair step (zoom).

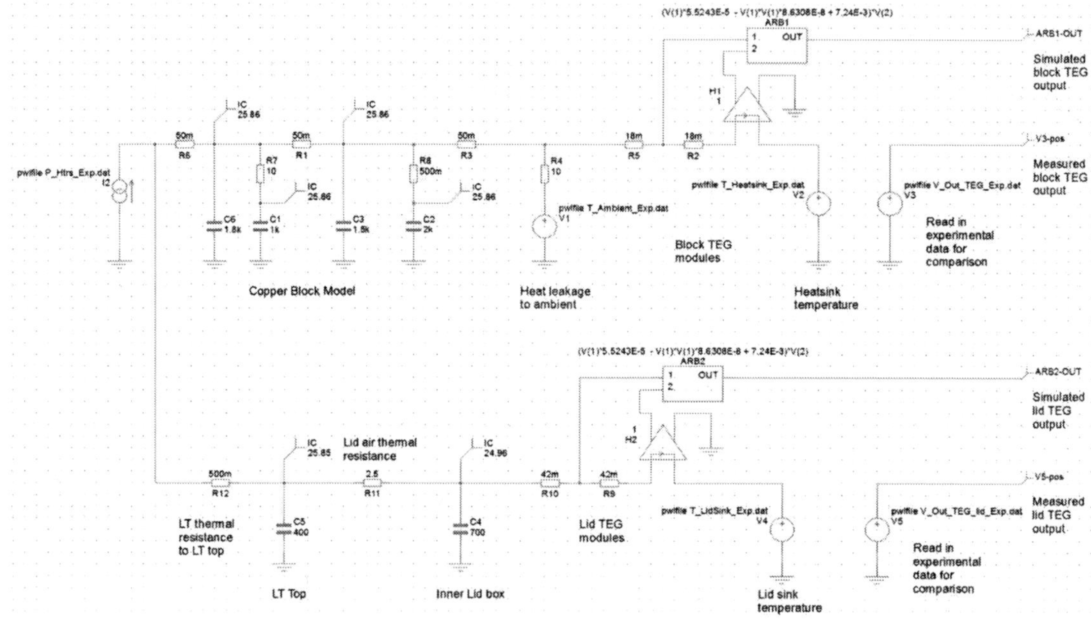

Figure 26. Schematic of the fourth pass model including the Lid.

Lid sub-model. A second output was created from the Lid TEG module(s). Heat is delivered to the Lid by conduction from the bottom portion, radiation from the LT top, and convection through the air in the Lid. Figure 26 shows the fourth pass model having elements to model both the block and Lid. The model for the block section has been adjusted slightly to account for the heat coupled into the Lid. The TEG modules were modeled the same for the Lid, but the thermal resistance was changed due to fewer modules. Non-linear resistances were used to model the heat flow in the air in the Lid to better fit the experimental data.

Figure 27 shows the excellent fit of the simulation to the block TEG voltages in the fourth pass model over the entire 200 W stepped range.

5.1. Extracting the XH wave form

The goal of this modeling effort is to provide a means to extract, in a calibrated way, the XH occurring within the Letts calorimeter. Once the model has been extracted, determination of the XH waveform is a problem in deconvolution, as mentioned in Section 4.3. Fortunately a solution has been found to use the power of the simulator to perform this deconvolution with no manual iteration. The technique also has promise for single parameter extraction in model development.

Consider the hierarchical schematic in Fig. 28. The extracted model for the LT and calorimeter are packed into the "Seebeck SubModel" block with its inputs provided at the left and the outputs (block and Lid TEG voltages) on the right. The ARB1 block in the figure simply combines (sums) the block and Lid TEG voltages and subtracts the sum of the experimental block and Lid voltages – producing a difference at each instant in simulation time of the simulated and experimental responses.

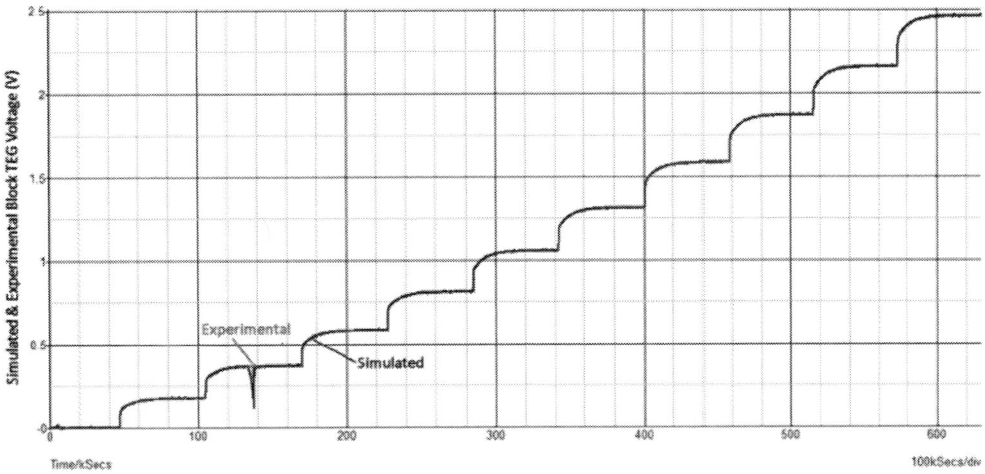

Figure 27. Fourth pass model block TEG response.

The high gain amplifier G1 (voltage in, current out) provides a negative feedback to the heat input node (left side of the SubModel block), attempting to force the difference between the simulated and experimental results to zero. If there is no difference between the modeled and experimental heats (a null experiment), then G1 will supply no feedback correction. If there is an XH , the experiment will have more output power than the model, and G1 will supply (via negative feedback) the heat necessary to balance the model output with the measured output. The XH waveform would be the feedback current.

To test whether this circuit can extract an XH waveform, source I1 (an ideal waveform generator) is placed in the model as an injected XH waveform in an otherwise null experiment. If I1 is zero, no XH is detected in the feedback current. Figure 29 shows the feedback current XH extraction when the I1 waveform is a triangle wave of mock XH

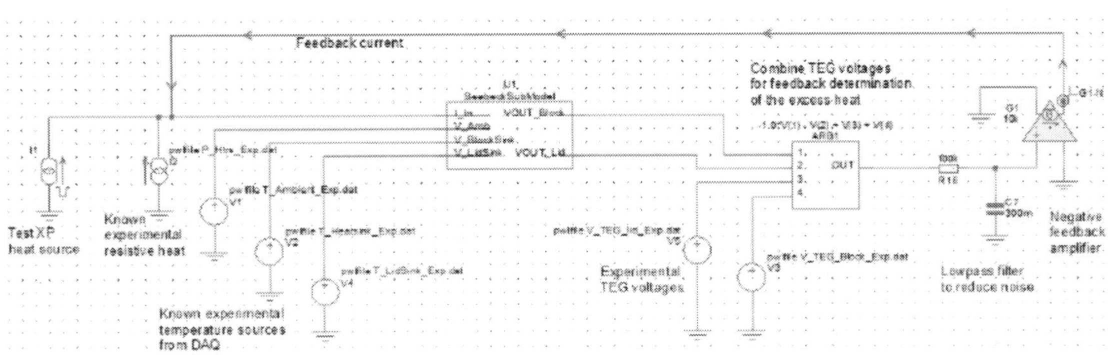

Figure 28. Feedback solution for XH.

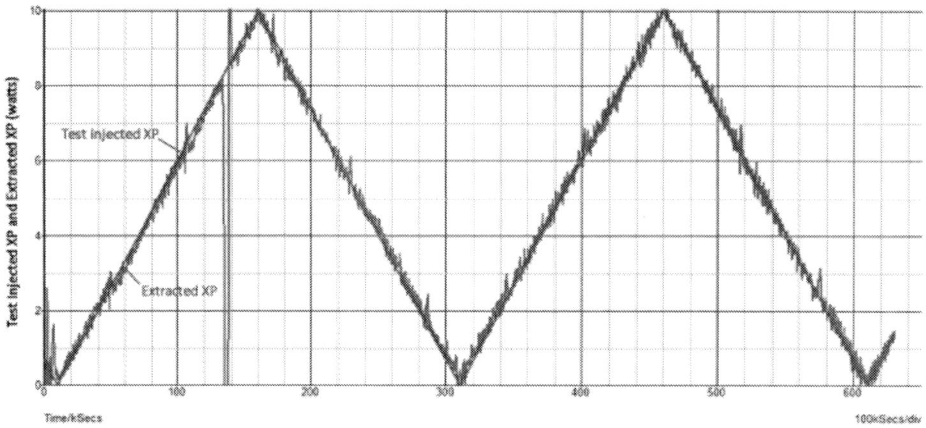

Figure 29. Feedback extraction of mock XH waveform.

power. While there is noise in the feedback waveform (expected), the feedback does a good job of reconstructing the mock XH that was injected.

Note the low pass filter between the ARB1 block and the G1 amplifier. This limits the bandwidth of the feedback signal and hence the representation of the XH waveform. A lower bandwidth of this low pass section will reduce the noise, while increasing the feedback gain at G1 increases the noise. For each system, a trade-off will have to be made between feedback gain at G1 which reduces the error in the waveform, and the low pass bandwidth that reduces the noise and the high frequency components of the extracted XH waveform. Since this extraction used driving data from a null experiment, some of the noise is quantization noise from the coarsely sampled measured data that is linearly interpolated by the simulator. Sampling at a smaller period would reduce this noise.

As further verification, a null LT was created having a resistive heater in place of the discharge in the tube. As a simulation experiment, the heat to this null LT heater was driven with a stepped waveform and the model was tested in its ability to recover the heat that had been applied – the results are shown in Fig. 30. As can be seen, the input waveform comprised very high bandwidth steps, and aside from the ringing produced by feedback around the low pass response (common in deconvolution), the technique faithfully recovers the resistive LT heater waveform.

6. Additional Modeling Considerations

As modeling of actual experiments was attempted, a number of other factors affecting the response were noted:

- It was discovered that the wires leading to the heater cartridges were not copper for the full length up to the inserted cartridge – there was in one case up to 1 m of nickel wire. Heat generated in the resistive loss of this wire was not deposited into the copper block, but it was measured as part of the input heat. This effect can be included in the model, but it would be better to make a measurement of the heater voltage with 0-current wires tapped onto the heater cartridge as it enters the copper block (a 4-wire connection).
- Un-modeled heater lead wire resistance will lead to calibration scale error.
- A heat storage effect was found inside the discharge tube (LT) itself. While there may only be 10 torr of gas inside the tube, it is heated to an effective temperature of $\sim$2500°C when the discharge current is turned ON.

This heating not only has a rise time, but the heat is released when the discharge is turned OFF. This can be modeled.

- There is a small heat removed when the gas is evacuated from the LT. If evacuation occurs even minutes after the discharge is stopped, there is little heat lost to the system in evacuation. However, when the evacuated tube is re-loaded with gas, heat may be removed due to injection of cool gas, or Joule–Thomson gas expansion may add or removing heat from the system (J–T expansion is generally a cooling effect except for H_2, He, or Ne). This effect could be added to the model.
- Discharge current flowing when the gas pressure is less than about 2 torr in the Letts LT geometry causes two problems. The glow forms closer to the top of the LT and disproportionally heats top and Lid of the calorimeter. Second, the thermal conductivity of the gas declines and the center conductor of the LT can get hot causing thermal conduction to the top of the LT. Thermal imaging during discharge with the tube outside the calorimeter is a good diagnostic to determine the safe operating region where heat will be deposited along the length of the LT.

7. Experimental Versus Model-extracted XH

Letts experiment 780 is provided as an example of the difference between XH ("Directly extracted XP" in Fig. 31) calculated using the simple differential calibration, and the XH extracted using the model ("residual" in Fig. 31).In the figure, the simulation extracted XH is labeled "residual" because it is that which remains in the feedback power (current) when all known heat sources have been accounted – it is presumed to be LENR heat. The experimental protocol is complex due to attempts to find the triggers of the XH , and to quench the excess by substitution of light hydrogen for deuterium.

With the aid of the model, simulation extracts the true behavior of the XH flow in this experiment. The extracted XH proves to be lower in magnitude than that found from simple differential calibration. The excess power waveform (the residual) shows fewer endothermic regions and of smaller magnitude. Many of these endothermic excursions are caused by evacuation and gas change in the LT during the experiment; the Joule–Thomson thermodynamics of which

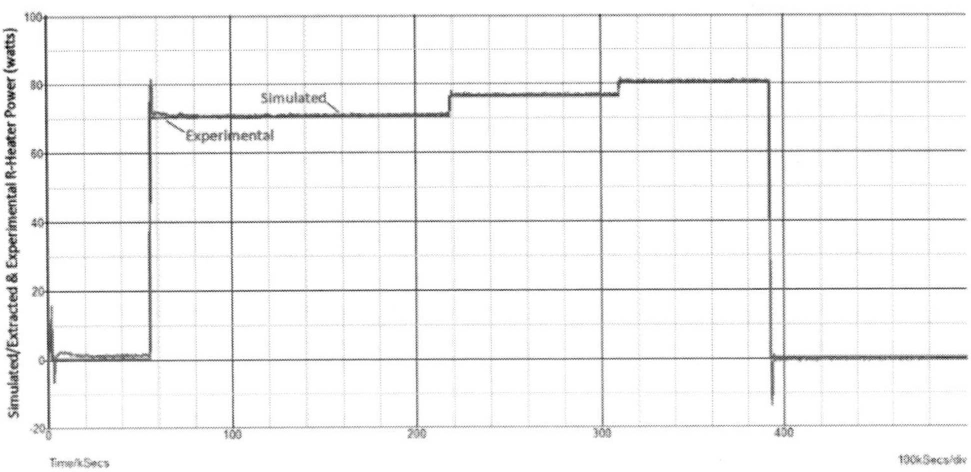

Figure 30. Simulation extracted null LT heater waveform.

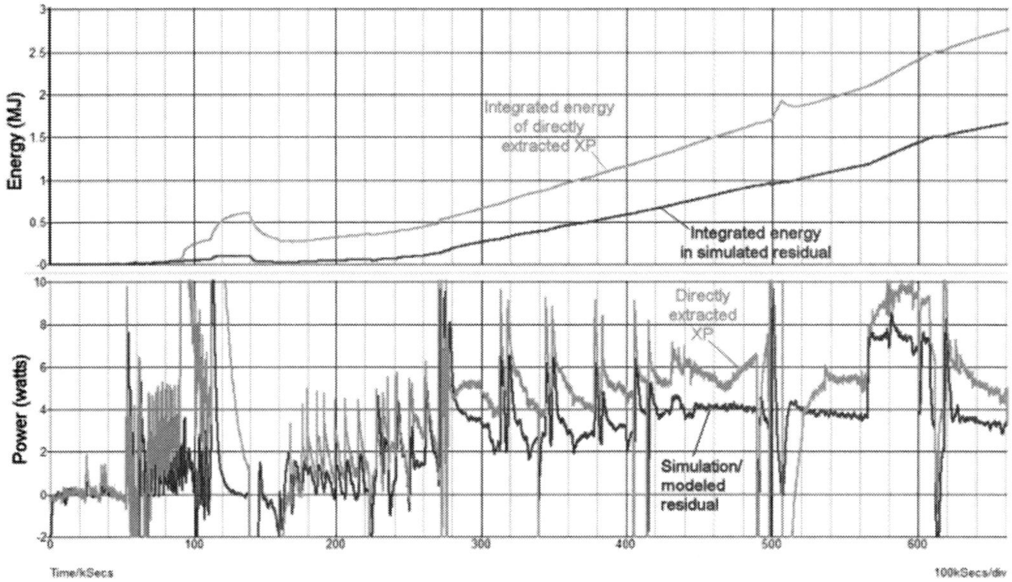

Figure 31. Experiment 780 XH .

are not entirely captured in the present model. Notice that when excess power is present in the simulation extracted residual, the excess power turns ON quickly, and is much flatter than what is found from the excess power taken directly from the differential calibration. This is because the simulation de-embeds the long thermal dynamics of the copper block in the calorimeter from the response waveform.

8. Conclusion

A technique is presented showing how electrical circuit analogues may be used to model a complicated thermal system. The SPICE simulator may be used in transient simulation mode to evaluate the modeled response. Use of the SIMetrix SPICE simulation environment is convenient, easy to use, and free for simulation of small size problems. The electrical circuit analogue technique for thermal modeling requires some effort to extract a topology and component values to provide a good fit to the response of the apparatus, but the resulting simulation is highly efficient – almost always requiring less than 5 s of computation on an ordinary 2018 personal computer. During the course of model development, substantial understanding of the measurement system and its limitations will be developed that may be missed if only differential calibration is utilized and new sources of possible error and inaccuracy may be overturned. Additionally, opportunities for system improvement become obvious during the modeling.

Model extraction using purpose run null experiments improves the quality and accuracy of the extracted excess power (residual) compared to use of a single calibration run and differential calibration for residual extraction. Essentially the modeling incorporates information from the calibration runs which is all ultimately brought to bear in computing the actual XH .

An interesting thermal propagation mode was found in the copper block – possibly a rarely documented non-Fourier heat transfer mode. The TEG modules, as thermal flow sensors, have a very fast response and are capable of detecting such fast thermal propagation modes. It is noted that F. Piantelli implicates such a non-Fourier thermal propagation mode in the stimulation of LENR in his nickel rod experiments [4,5].

Acknowledgements

The apparatus being described in this paper is the work of Dennis Letts from Dennis Letts Lab. in Austin, TX. The author is grateful to Dennis Letts for the data supplied and for null experiments explicitly run to support the modeling. The author would like to thank Industrial Heat for their generous support of this modeling effort. Also, the author thanks the reviewer for his thorough reading and thoughtful comments.

Note: The Industrial Heat standard for official confirmation of any XH claims requires verification by a major independent lab with recognized, credible, and skeptical researchers.

References

[1] SIMetrix/SIMPLIS Elements Simulator, Available for free from SIMetrix Technologies at: www.simetrix.co.uk/downloads/download-elements.html.

[2] D.G Letts and D.J. Cravens, Building and testing a high temperature Seebeck calorimeter, *J. Condensed Matter Nucl. Sci.* ICCF-21 (2018).

[3] N.I. Kobasko and SH. E. Guseynov, An explanation of the nature of thermal waves "a poker effect" on the basis of hyperbolic heat conductivity equation analysis and existence of free electrons in metals, *Recent Researches in Circuits and Systems* (2014), ISBN 978-1-61804-108-1, WSEAS Press, pp. 167–172.

[4] S. Focardi, V. Gabbani, V. Montalbano, F. Piantelli and S. Veronesi, Large excess heat production in Ni–H systems, *IL Nuvo Cimento* **A 111** (11) (1998) 1233–1242.

[5] F. Piantelli, Private conversation with Piantelli, 2016.

[6] M. Marz and P. Nance, Thermal modeling of power-electronic systems, *Infineon Technologies Application note*, available at: https://pdfs.semanticscholar.org/438f/69381bbdaf56d853f082e17c8d3c51a0ef6e.pdf.

[7] F. Christiaens and E. Beyne, Transient thermal modeling and characterization of a hybrid component, *Proc. Elec. Components and Tech. Conf.* IEEE 46 (1996), DOI 10.1109/ECTC.1996.517388.

J. Condensed Matter Nucl. Sci. 29 (2019) 334–347

Research Article

Building and Testing a High Temperature Seebeck Calorimeter

Dennis G. Letts* and Dennis J. Cravens

Industrial Heat/Lettslab, Austin, TX 78727, USA

Abstract

A high temperature Seebeck calorimeter capable of operating at 200–300°C was built and tested. The testing used a glow discharge tube containing plated palladium on the side walls and a molybdenum central anode. Running with deuterium gas and high voltages demonstrated excess thermal power at levels of 5–10 W but no excess power was observed when natural abundance hydrogen was used. In addition to the normal Seebeck measurements, excess power was also observed using back-off power measurements of the enclosure heater containing the Seebeck. A resistive control heater was used to verify the system.

Keywords: Anomalous energy, Gas discharge reactor, High temperature Seebeck fabrication, Seebeck calibration, Seebeck calorimetry

1. Introduction

The design intent of the LENR prototype is to provide quality Seebeck-style calorimetry at reactor temperatures of 200–300°C. The reactor is a stainless-steel cylinder, 1-inch OD, 7/8-inch ID, one foot long, which is inserted into a copper rectangular block during operation. hydrogen-absorbing metals can be plated onto the interior walls of the cylindrical reactor or tubular metal/alloy inserts can be placed inside the cylinder for testing. Thermoelectric devices are mechanically attached to the sides, top, and bottom of the copper block. Aluminum heat fins are attached to the outside surface of the thermoelectric devices to provide a constant temperature interface. A commercial freezer surrounds the reactor and Seebeck assembly, capable of providing a constant temperature environment. A typical thermal enclosure set point is 28°C. The system was designed by Dennis Letts with additional design and prototyping by Mike Guerrina and Carlos Jobe of Apparent Technologies Inc., Austin, Texas.

1.1. Experimental setup (Fig. 1)

The system consists of three main components. (1) The instrument rack containing power supplies to drive the reactor, to heat the copper block, to power the thermal enclosure heaters, and to actuate electronic valves. (2) The thermal

*Corresponding author. E-mail: DLetts@industrialheat.co.

Figure 1. The instrument rack on the far left holds power supplies for system heat and fans plus the KEPCO 1000–0.2 MG power supply for the discharge. The rack also holds the lab computer that controls all instruments using LabView.

enclosure houses the high temperature Seebeck and the reactor; the enclosure holds ambient temperature at 28 $\pm 0.1^\circ$C. (3) A vacuum pump capable of reaching 9×10^{-8} Torr with a pressure control and measurement manifold attached to the side of the thermal enclosure. All vacuum connections are Swagelok VCR quality, ranging from 1/4 in. to 1 in. in diameter. Nickel gaskets with retainer clips are used on all vacuum connections. Various instruments are placed on top of the thermal enclosure for convenience; instruments shown are a Fluke 105B oscilloscope to monitor the discharge voltage and current, Keithley multimeters to provide redundant voltage and current readings, a magnetic ballast sits on top of the Keithley multimeters to provide current and voltage stability when the discharge is running. This serves to eliminate current spiking and makes the input power readings stable and reliable. The final instrument is an HP RF generator used with 4–10 W amplification to occasionally attempt RF triggering of exothermic reactions in the reactor [1]. The amplified RF is coupled to the high voltage DC discharge power by using a bias T built by Barth Electronics of Boulder City, Nevada.

1.2. Design and fabrication details

The LENR tube (LT) is shown in Fig. 2. The tube is made from 316L stainless steel, about 11 in. long, 1-inch OD and 7/8-inch ID. The end is orbital welded. The tube is sealed with a 1-inch VCR nut and gas is introduced to the reactor through a 1/4 in. stainless tube fitted with a 1/4 in. VCR connector. A 5 kV high voltage connector conducts high voltage DC to the reactor. Positive DC goes to the anode and the case is grounded negative, serving as the cathode. The reactor is plated in a separate bath using normal methods of electrochemical plating, discussed in more detail below. The reactor can reach high vacuum at 9×10^{-8} Torr. Motor magnets are fitted around the reactor body and provide a 230 G magnetic field at the reactor inner surface (not shown). The magnetic field at the inner wall was measured using an Alpha Lab model GM-1-ST gaussmeter. Magnets were placed around an open tube of 316L stainless steel with

Figure 2. Figure shows the reactor used for the experiments – a 316L stainless steel tube approximately one-foot in length with an outside diameter of one-inch. Swagelok VCR connectors are used to seal the unit capable of holding vacuum to 10^{-9} Torr. High voltage DC is supplied to the reactor via a 5-kV connector obtained from MPF Products. Clamshell motor magnets (not shown) are placed around the reactor during operation and provide 230 G on the inner wall of the reactor.

identical OD and wall thickness as the reactor to facilitate the measurement. A small field probe was inserted into the stainless-steel tube and the measurement was made at the inner wall with the magnets in place.

The reactor and its attached magnets are inserted into a central bore hole drilled into the $3 \times 3 \times 15$ in. copper rectangular block. The block is equipped with four Watlow heaters and four Omega thermocouples. The Watlow heaters can dissipate up to 800 W of power but the maximum power dissipated in practice is 300 W.

The thermoelectric devices are from Tecteg Manufacturing, Ontario, Canada; the devices in use are rated to 350°C and other devices can be obtained that can handle 850°C. The lower temperature devices were chosen because they had a shorter lead time. The thermoelectric devices are attached to the copper block using machined clips as shown in Fig. 4. Not all of the surface area are covered but the conversion from heat flow to voltage is very consistent and repeatable from experiment-to-experiment. The aluminum heat fins are attached to the copper block using Bellville washers to provide a constant force and consistent thermal contact. The washers are shown in Fig. 4 (inset).

Thermocouple and block heater placements are shown in Fig. 5. Thermal enclosure temperature is controlled by running the commercial freezer compressor in opposition to the heat produced by the Seebeck and a 450 W Watlow heater positioned above the fans against the back wall of the enclosure. Additional air stirring is provided by a large freezer fan in the ceiling of the enclosure. Enclosure temperature variation is less than ± 0.1°C (Fig. 6).

The 1/4 in. Watlow heater shown in Fig. 5 right balances against the freezer compressor to provide stable enclosure temperature. Interestingly, when excess power appears in the reactor, it flows out of the reactor and into the enclosure to assist the Watlow heater. Power provided to the Watlow heater declines (i.e. the power "backs off") in quantitative agreement with the excess power reported by the TEC devices. This feature provides a dual-method calorimeter using independent measurement methods.

Figure 7 shows the Seebeck calorimeter in its thermal enclosure, with its bolted down lid equipped with TEC devices and heat fins. A Helmholtz coil surrounds the Seebeck and allows the imposition of up to a 40 G field onto the reactor inside the Seebeck. The coil can be used to stimulate LENR reactions or to cancel the earth's magnetic field.

Figure 3. The left portion shows the copper block core of the Seebeck calorimeter. The reactor and attached magnets are placed into the central borehole. The larger 1/4 in. holes in the block hold the Watlow resistance heaters that provide minimum heating desired for the reactor. The 1/8th-inch holes are for Omega thermocouples. The right portion shows the reactor in place. A machined lid fits over the reactor and is bolted in place during operation.

Figure 4. Figure shows how the thermoelectric devices are attached to the copper block using Bellville washers, shown in the inset. The Bellville washers provide a constant force to the heat fins as they press against the thermoelectric devices.

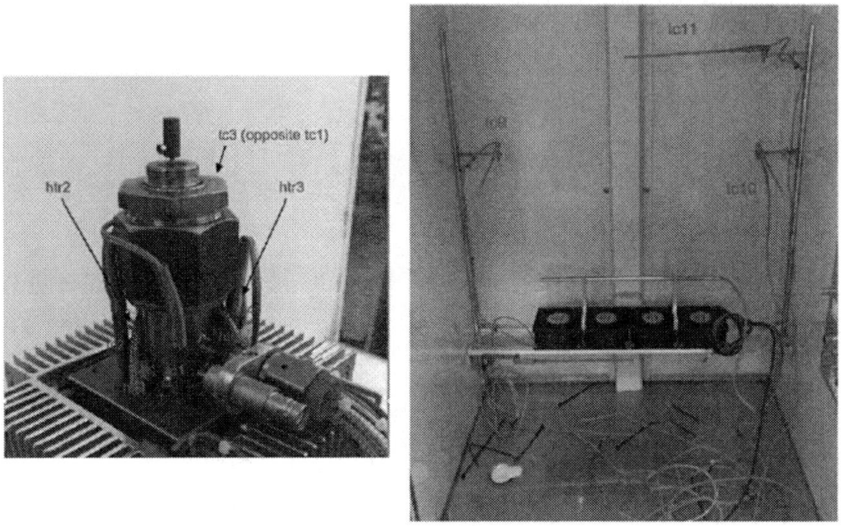

Figure 5. Left-hand side shows the positioning of four thermocouples (tc1–tc4) and four Watlow resistance heaters (htr1–ht4). Right-hand side shows the placement of four thermocouples (tc9–tc12) to monitor and control thermal enclosure temperature. Note the Watlow resistance heater above the four stirring fans that is used to keep the thermal enclosure at a constant temperature, as shown in Fig. 6.

The coil is controlled by Labview. The TEC devices are wired in series and voltage is sent to a datalogger through the enclosure panel, as shown by the red and black cabling. There are 12 thermocouples that are also read by a datalogger via the yellow type K connectors shown in Fig. 7. Electrical power to drive the enclosure heater, fans, and the reactor at high voltage is also passed through the panel. A single 1/4 in. vacuum/gas line connects to the reactor through

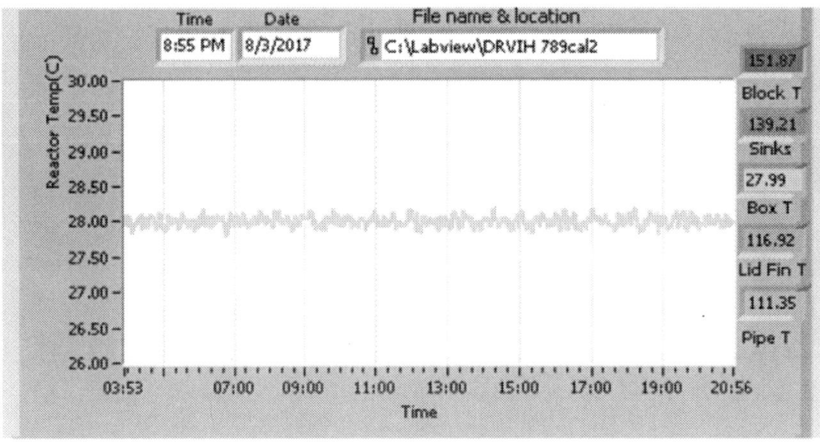

Figure 6. Figure shows the average temperature of the thermal enclosure over an 18-hour period.

Figure 7. Figure shows the Seebeck calorimeter with the bolted lid in place, circulating fans, and data cables connected at the enclosure interface. The Helmholtz coil can be used as a listening coil for RF/acoustic signals emanating from the reactor or to apply a 40-gauss magnetic field to the reactor. Of note: the reactor produces a 3.5 kHz signal when operating and has been detected using the Helmholtz coil and verified with an Agilent spectrum analyzer.

the side panel; three electronic valves can be actuated to selectively pump the reactor down in pressure, or to add deuterium/hydrogen to a desired pressure. In operation, a layer of high temperature insulation surrounds the Seebeck to boost internal temperature.

2. Calibration and Experimental Results

The calibration produced a calibration equation of $P(W) = 0.0651\times$ mV −167, where mV is the total voltage output of the Seebeck in millivolts less a constant term of 167. This equation was applied to the Seebeck output voltage and the result is shown in Fig. 10.

We also completed a calibration that combined data points using discharge power and resistive heater power with data points using resistive power only. This result is shown in Fig. 11.

Reactor #780 was plated with 0.48 g of palladium, on the inner wall of the cylindrical reactor. This was done using normal plating methods at 1 A, 50°C. 16 g of $PdCl_2$ 5% solution (Alfa-Aesar) was plated, which contained approximately 0.48 g of palladium. Plating depth was estimated at 5–10 μm over about 6 in. of the 11 in. reactor. Two-piece Clam-shell motor magnets were place around the exterior of the reactor during plating and operation. The magnetic field at the interior wall was measured at 230 G. From previous work, it is thought that an external magnetic

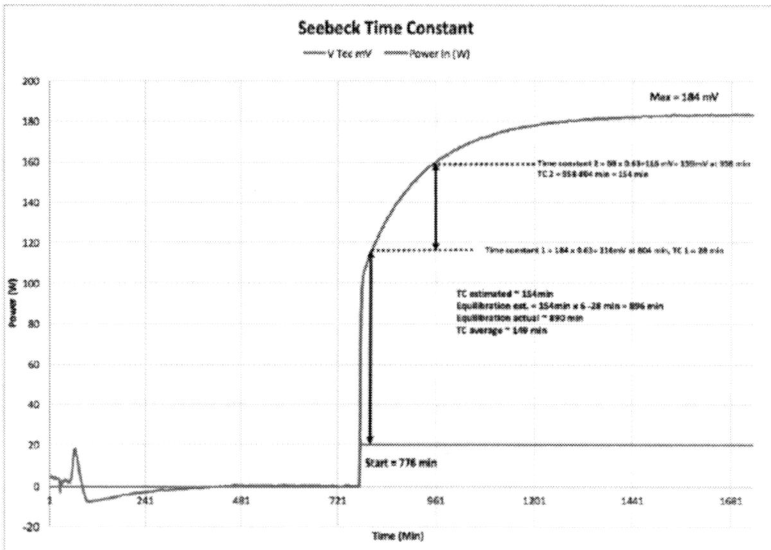

Figure 8. The Seebeck has dual time constants, the reason for which is not clear. The TEG manufacturer suggests it may be due to the fact that the thermoelectric devices are a composite of BiTe and PdTe. This plot is NOT for calibration, only to show the time constants of the Seebeck.

field influences excess power production via the Larmor precession frequency [2].

As Fig. 10 shows, the experiment began with light hydrogen in the reactor and a discharge power applied to the reactor of about 72 W (350 V, 200 mA). Power was in balance with zero excess power reported for about 15 h, demonstrating the calibration was sound. Near minute marker 1000, light hydrogen was pumped out of the reactor and replaced with deuterium. Excess power remained near zero as the high voltage was cycled off and on. Near minute marker 3600 deuterium was added to the reactor and power-out began to increase over the input power supplied to the discharge. As the experiment proceeded, discharge voltage was cycled off and on and deuterium was added to the reactor. Excess power reached a typical value of 5–6 W near minute markers 6000–7000, reaching a brief maximum of 10 W near minute marker 9600.

Experiment DRVIH 780 was significant because it clearly showed that a high temperature Seebeck could report relatively small power gains at higher than electrochemical temperatures, and that deuterium produces excess power but ordinary hydrogen does not.

2.1. Independent confirmation of excess power

During this experimental campaign, a single 450 W resistance heater was used to keep the thermal enclosure at a constant temperature (Fig. 6). It was also observed that when excess power/energy was present, electrical power/energy delivered to the heater declined over the same time period. This indicated that the excess power/energy produced by the reactor was assisting the resistance heater. Figure 12 shows the plot where this quantitative agreement can be observed.

The confirmation of excess power/energy is indicated by the gray shaded areas at the far end of the plot shown in

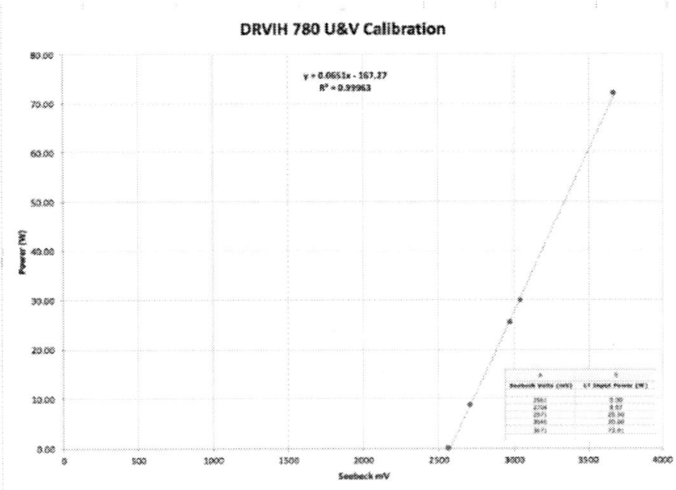

Figure 9. A calibration was performed for experiment 780 by adding power to resistive block heaters on top of a steady 200 W that provided baseline operating temperature of about 160°C. The result was a linear calibration plot shown in the figure. The Seebeck reports 2561 mV at zero discharge power because resistance heater power is supplied to provide a minimum operating temperature. Discharge power is applied in addition to baseline heater power.

Fig. 12. Excess power averaged 6 W over the minute markers 8600–10,974. Maximum excess power of 10 W was observed in this time period as well. The integrated excess power (gray area) under the red trace was 899 kJ; the inte-

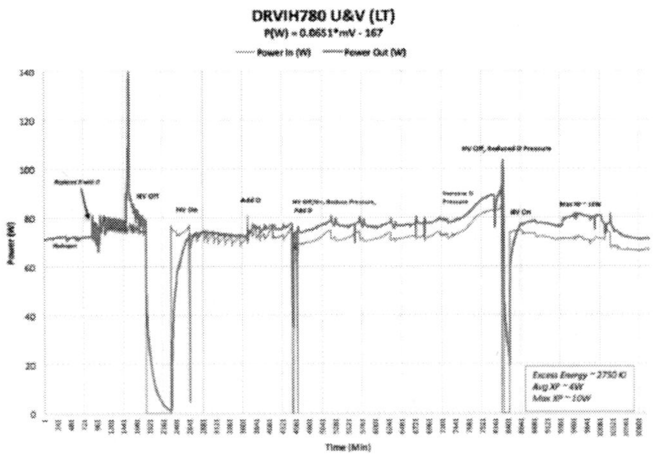

Figure 10. Figure shows a long run that produced significant excess power and energy. Of note is the beginning of the experiment where light hydrogen was present in the reactor. After the light hydrogen was replaced with deuterium, excess power appeared and slowly developed over the duration of the experiment that lasted for 7.6 days (11,000 min).

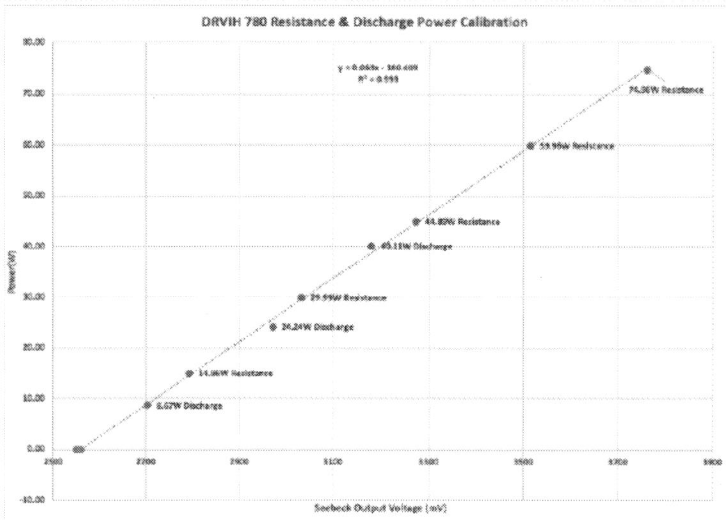

Figure 11. Figure shows a calibration made using resistive heating power alone and resistive heating power plus discharge power. The resistive heater only method shown in Fig. 9 produced a higher power output than the method shown in figure by about 1% over the length of the experiment. We chose to use the calibration based on resistive heating power but both calibration methods demonstrated significant excess power and energy.

grated decline in heater power (gray area) above the blue trace was 893 kJ. The difference in these two measurements is 0.7%. This confirms the excess power reported by the Seebeck using an independent measurement.

2.2. DRVIH 780 U&V confirmation via box heater power

To check the validity of the excess power confirmation method using the enclosure heater shown in Fig. 5 (right), Dr. Dennis Cravens of Cloudcroft, New Mexico suggested making a larger resistance heater to be placed inside the reactor tube. The eleven-inch reactor and six-inch resistance heater are shown in Fig. 13. The OD of the heater is approximately 0.86 in., which permits insertion of the heater into the reactor with a close sliding fit. The idea was to provide a simulation of heat generation in the reactor without using high voltage discharge methods. The heater, shown in Fig. 13, is rated at 450 W but only needed to provide 70–80 W of heat to simulate discharge power used to drive experiment 780 U&V.

DRVIH 790 was conducted in August 2017 to demonstrate that the amount of power/energy produced in the reactor could be observed as a decline in power/energy provided to the thermal enclosure heater shown in Fig. 5 (right).

2.3. Resistive power calibration runs

To help verify the measurements and to vet the system, a series of resistive heater calibration experiments were run. Although the calorimeter appears to be reproducible and correctly reports power over a wide range, a series of resistive runs were designed to make sure that there are no systematic errors unique to the use of high voltage or the LENR heat observed.

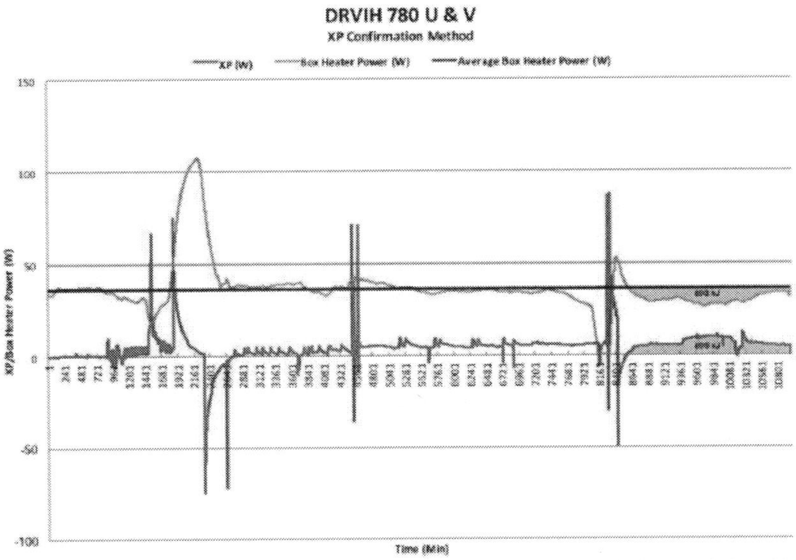

Figure 12. At the end of the plot shown in figure, excess power was produced at 5–7 W. This power was sufficient to assist the thermal enclosure heater, allowing Labview to "back-off" or reduce electrical power supplied to the resistance heater. In this case, the reduction of heater power was quantitatively equal to the excess power produced by the reactor, within experimental error. Note that gray areas in the plot show that the energy gain in the reactor was the same as the reduction of energy supplied to the heater.

A 15.2-ohm commercial heater (Tempco[a]) was machined to snuggly fit within the bore of the LENR tube (LT). This provided the ability to closely match the location and power levels of heat deposited via the LT during the 780

[a]Tempco Electric Heater Corp., http://www.tempco.com/Default.htm.

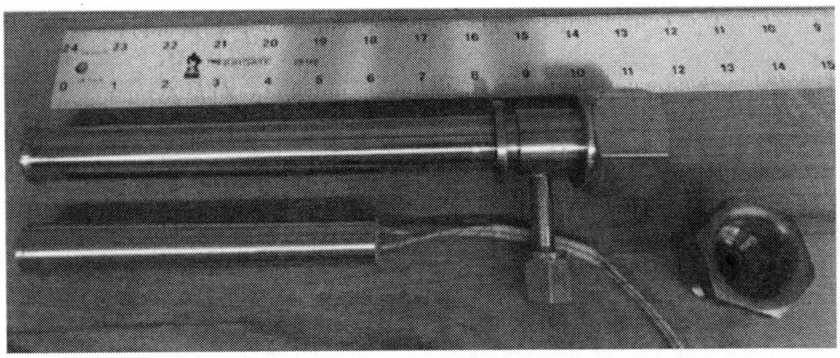

Figure 13. Figure shows the reactor and a custom fitted resistance heater used to verify that energy released in the reactor will assist in heating the thermal enclosure. In short, this is a method to simulate the release of excess power in the reactor and observe its effect on the thermal enclosure resistance heater.

U&V runs.

2.4. 200-Watt block heater calibration

Recall that 200 W of power applied to the large copper block holding the LT was common to the 780 U&V runs and most other experiments. The 200 W of power provides a higher operating temperature, which is conjectured to enhance power output from LENR reactors. The LT used for experiment DRVIH 780 U&V was replaced with an LT holding the resistive heater cartridge, and the baseline voltages from the TEG's were observed. The average input to the sum of the four block heaters was 200.45 W over the run with a standard deviation of 0.06 W. Using the same calibration constants for this resistive run as the 780U&V run, the Seebeck reported 201.79 W with a standard deviation of 0.06 W. This is a difference of 1.34 W.

Recall that there were multiple (10) experiments conducted between the 780U&V run and the DRVIH 790 resistive heater run with a time span of about 9 months. This shows that even with different tubes, configurations, and multiple opening and closing of the calorimeter, the power conversion from TEG voltage to output heat held to within 1.34 W on the 200 W heating of the block. This means the drift/repeatability of the calorimeter is within 0.5%, even with the slight differences between tube geometry. It should be noted that although the resistive heater element was matched closely to the LT, the heat parsing between the lid and copper block body are expected to be slightly different. The results of the block heating alone were then used as a recalibration point for the remaining resistive trials.

2.5. 200-Watt block heater with 70 W applied to the resistive LT heater cartridge

Once the recalibration was done for the resistive block heater input, it was possible to evaluate the claimed excess heat from the active 780 U&V run; 200 W of power was applied to the block heaters, and 70 W was applied to the LT resistive heater cartridge to match the electrical input powers applied to the 780 U&V run, which was 200 W to the block heaters and approximately 70 W to the LT in high voltage gas discharge. The average input to the LT resistive heater cartridge was 70.11 W with a standard deviation of 0.033 W and the calorimeter reported 70.23 W with a standard deviation of 0.053 W. That is to say, with the resistive block heater calibration in place from the previous block heater-only run, there was 0.12 W variation – well within an assumed conservative 0.5 W error bar. This showed that the system reported correctly over the reported range of experiment 780 U&V values.

2.6. A secondary check using the box heater data

In addition to the Seebeck readings, the box heater back-off power method can be used as a confirming measurement. The temperature control heater back-off measurements are thought to be less reliable than the Seebeck calorimeter measurements. The calorimeter is placed inside a constant temperature enclosure to allow for a stabilized outer reference temperature. The power required to keep the enclosure at a constant temperature can be used as a reality check for the measured power output from the calorimeter. The power applied to the heater cartridge used to control the chamber temperature is physically and temporally removed from the reactor, and hence potentially not as accurate as the calorimeter measurement in assessing the calorimeters total heat output. The amount of heat generated by the reactor and block heaters is supplied to the thermal enclosure. Thus, as the calorimeter emits more heat, less power input to the enclosure heater is required to maintain the enclosure temperature set point. A record of the power used by the enclosure heater can thus provide a proxy estimate of the power produced by the LENR reactor.

In the case of the 70 W of power applied to the LT resistive heater cartridge, the box heater back-off method reported 70.27 W with a standard deviation of 1.2 W over the run. When 76 W was applied to the resistive heater, the back-off method reported 75.91 W with a standard deviation of 0.46 W over the run (see Fig. 14). Again, this is

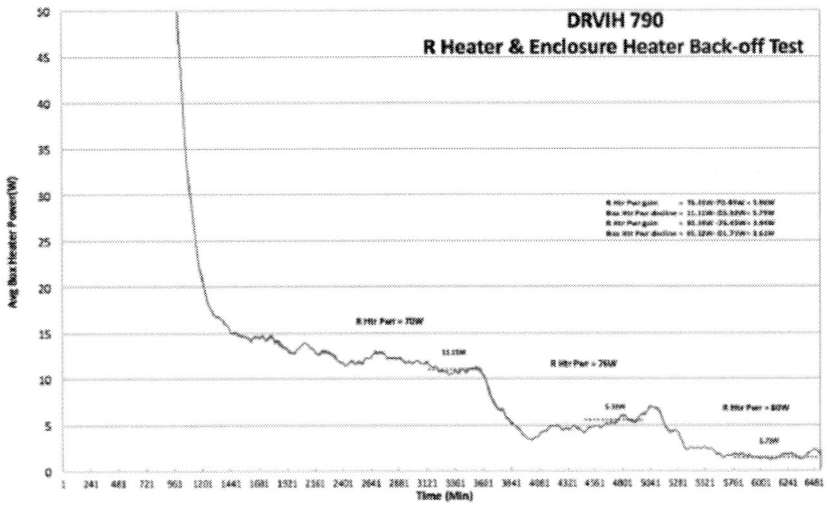

Figure 14. The plot shown in the figure shows how the thermal enclosure heater declines in power as a known electrical power is released inside the reactor using the resistance heater shown in Fig. 13.

a secondary measurement but convincingly establishes the ability of declining enclosure heater power as a means to estimate excess power produced by experiment DRVIH 780 U&V.

The plot shown in Fig. 14 demonstrates that power/energy supplied to the thermal enclosure heater declines in close quantitative agreement with the power/energy dissipated in the resistance heater located inside the reactor.

The tests revealed that when power to the resistance heater was increased by 5.96 W, the power supplied to the box temperature control heater declined by 5.79 W. The box heater power was averaged over a 500-minute interval. Energy added to the resistance heater was 5.9 6 W $\times$ 30,000 s/1000 $=$ 178.8 kJ; energy reduction to the enclosure heater $=$ 5.79 W $\times$ 30,000 s/1000 $=$ 173.7 kJ. Agreement is within 3%. Accuracy might be improved with longer hold times at each step.

3. Evaluation of the Signal-to-Noise + Error Ratio

To evaluate the signal-to-noise ratio (SNR), the best data choice is DRVIH 790, the run with block heat and a resistive cartridge heater inserted in an empty LT tube. In this experiment, the output excess power (the residual) should be identically 0. Any deviation from 0 W of excess power (XP) is either error or noise. Figure 15 shows the XP from experiment 790. In this experiment, there are large dips and peaks that are associated with transients in the response of the system. These transient peaks and dips are excluded from the calculation of the standard deviation (σ) in this experiment with the caveat that one must only use the resulting statistic associated with settled regions in other experiments when comparing signal-to-noise.

The regions highlighted in yellow in Fig. 15 were those used to compute the standard deviation (σ) for the DRVIH 790 null experiment. The value calculated in Excel was $\sigma = 0.736$ W.

If this experimental value for the error + noise is applied to the XP from experiment DRVIH 780 U&V, the SNR can be calculated as XP/σ. This curve of SNR is plotted in Fig. 16.

As shown in Fig. 16, more than half of experiment DRVIH 780 U&V has excess power better than 6σ and in the

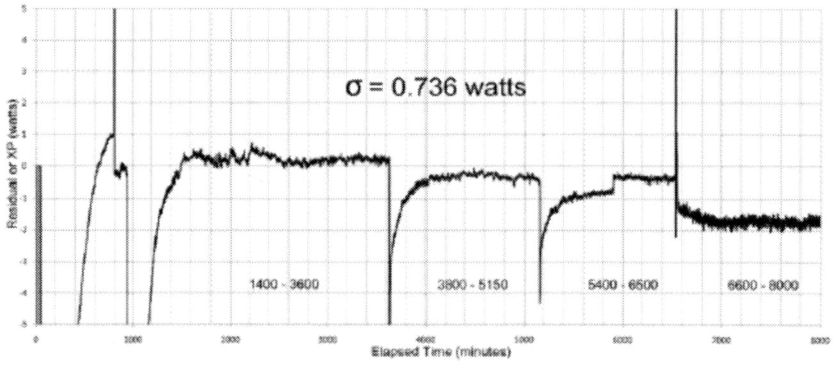

Figure 15. Figure shows the standard deviation for power produced in the reactor using the resistance heater shown in Fig. 13. We view the standard deviation as a reasonable measure of calorimeter stability.

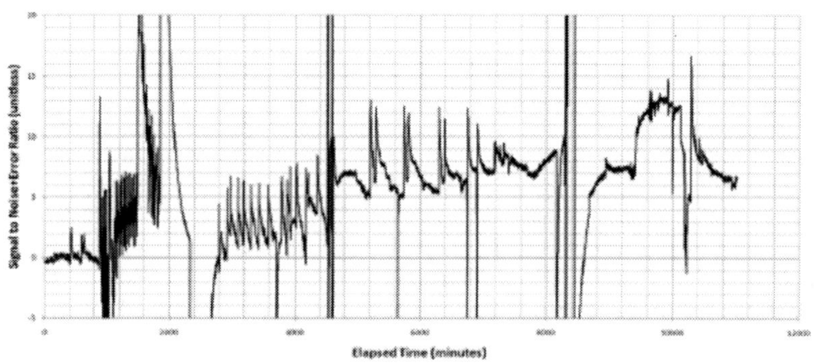

Figure 16. Figure shows the computed SNR over the entire experiment, which ranged from 6 to 12 σ, suggesting a very credible signal.

11-hour period from 9450 to 10,120 min has a SNR better than 12σ. The SNR for the excess heat in DRVIH 780 U&V provides good confidence that the measured excess heat is not a consequence of noise or error.

4. Conclusions

A high temperature Seebeck calorimeter can be successfully designed and constructed to make power measurements for gas-based LENR devices at or above 200°C. Glow discharge reactors are effective in loading and triggering exothermic reactions in deuterated metals, although the authors do not yet claim mastery over the methods.

The tests conducted by the authors support the idea that palladium deuteride produces excess power but palladium hydride does not. This is consistent with our previous work and results reported by the LENR community.

The work discussed in this paper also demonstrated that an acceptable redundant power measurement method is available by measuring and recording electrical power provided to the resistance heater used to maintain a constant ambient temperature within the system's thermal enclosure. It was demonstrated that power developed or applied within the reactor was quantitatively reflected in the reduction of electrical power consumed by the enclosure's resistance heater. This observation provides a dual-method calorimeter for LENR heat measurements.

Acknowledgements

The authors gratefully acknowledge the support of Industrial Heat (the IH standard for official confirmation of any excess heat claims requires verification by a major independent lab with recognized, credible, and skeptical researchers) of Raleigh, North Carolina. Special thanks to Mr. Tom Darden, Mr. J.T. Vaughn, and Mr. Dewey Weaver.

References

[1] J. O'M Bockris, R. Sundaresan, Z. Minevski and D. Letts, Triggering of heat and sub-surface changes in Pd–D systems, *Fourth Int. Conf. on Cold Fusion*, Lahaina, Maui, Vol. 2, 1994, p. 1.1.

[2] J. Cobine, *Gaseous Conductors*, 2nd edn., Dover, New York, 1958, pp. 1–289.

[3] J. Linhart, *Plasma Physics*, 2nd revised edition, North-Holland, Amsterdam, 1961, pp. 1–257.

[4] D. Letts, A method to calculate excess power, *Infinite Energy* **19** (112) (2013) 63–66.

J. Condensed Matter Nucl. Sci. 29 (2019) 348–357

Research Article

Effective LENR in Weakly Ionized Gas Under the Action of Optimal Pulsed Magnetic Fields and Lightning (Theory and Experiments)

Vladimir Vysotskii*and Mykhaylo Vysotskyy

Kiev National Shevchenko University, Ukraine

Abstract

The features and mechanism of LENR production in a system of free particles under the action of a weak impulse (e.g., the action of a pulsed magnetic field) are considered. It is shown that in such a system effective formation of coherent correlated states of particles with the accompanying very sharp increase in the energy fluctuations take place. The amplitude of these fluctuations exceeds by many orders the average thermal energy of the particles and can reach 10–50 keV and more. This mechanism fully explains the nuclear reaction both in the atmosphere during a lightning discharge, and laboratory experiments conducted using electric discharges.

Keywords: Coherent correlated states, LENR in gas, Lightning, Nuclear transmutation of isotopes

1. Application of Coherent Correlated States in LENR Processes

Numerous successful LENR experiments, some of which have confidently emerged from the "child" age of laboratory experiments and have manifested themselves at the industrial level, up to now have not been based on a reliable theoretical model that adequately explains non-trivial results that are not consistent with the traditional models of nuclear physics.

This problem has been actively discussed over the past 20 years. There are many different theoretical models that should explain the results of successful experiments. It should be noted that the majority of these models are of a highly specialized nature. Each of them is aimed at explaining a specific experiment in a specific environment involving specific nuclei. As a rule, the theoretical model proposed for one particular experiment (with one set of conditions and parameters) is not applicable to another experiment (with another set). We believe that such approach (the existence of a set of uncoordinated theoretical models) is not correct and has very limited area of application.

Moreover, each of these models is aimed at explaining one unique property of LENR: the anomalously high probability of a tunneling effect at low particle energy. In this case, other unique properties of such reactions (ban on the formation of radioactive daughter isotopes and very strong suppression of concomitant gamma radiation) are not

*Corresponding author. E-mail: vivysotskii@gmail.com.

actually considered. In our works [1–9] a completely different approach has been proposed and developed: a unified theoretical model (method of formation of coherent correlated states (CCS) of interacting particles), which allows us to explain in details all successful LENR experiments, without exception. The physical basis of this method is related to the Schrödinger–Robertson uncertainty relations [10,11]:

$$\sigma_A\sigma_B \geq |\langle[\widehat{A}\widehat{B}]\rangle|^2/4(1-r^2) \equiv G^2|\langle[\widehat{A}\widehat{B}]\rangle|^2/4, \tag{1}$$

$$r = \sigma_{AB}/\sqrt{\sigma_A\sigma_B}, \quad \sigma_C = \langle(\widehat{C} - \langle C\rangle)^2,$$
$$\sigma_{AB} = (\langle\widehat{A}\widehat{B} + \widehat{B}\widehat{A}\rangle)/2 - \langle A\rangle\langle B\rangle,$$

where r is a correlation coefficient determining the degree of mutual coupling of the quantities A and B in a particular superposition state, described by the wave function $\Psi(q)$, and $0 \leq |r| \leq 1$, G is the coefficient of correlation efficiency. More clearly, the influence of the correlation at $|r| \to 1$ characterizes a parameter $G = 1/\sqrt{1-r^2}$ that varies in the interval $1 \leq G < \infty$ and which can be called the coefficient of correlation efficiency [6,9,12–14].

In the case $A = q$, $B = p$, $\langle q\rangle = 0$, $\langle p\rangle = 0$ relations (1) take the form

$$\delta q\delta p \geq \hbar/2\sqrt{1-r^2} \equiv G\hbar/2,$$

$$r(t) = \langle q\widehat{p} + \widehat{p}q\rangle/2\delta q\delta p, \quad \delta q = \sqrt{\sigma_q} \equiv \sqrt{\langle q^2\rangle}, \quad \delta p = \sqrt{\sigma_p} \equiv \sqrt{\langle p^2\rangle}. \tag{2}$$

From a formal point of view, the presence of a correlation in the uncertainty relation can be taken into account by replacing $\hbar \to \hbar^* \equiv \hbar/\sqrt{1-r^2} \equiv G\hbar$. In totally uncorrelated state (at $r \to 0$) we have $\hbar^* \to \hbar$ and formula (2) takes the form of standard Heisenberg uncertainty relations $\delta q\delta p \geq \hbar/2$.

This result directly corresponds to the approximate formula for the transparency coefficient of the barrier in the presence of CCS

$$D_{r\neq 0} \approx \exp\left\{-\frac{2\sqrt{1-r^2}}{\hbar}\int\limits_R^{R+L(E)}\sqrt{2M\{V(q)-E\}}\,\mathrm{d}q\right\} = (D_{r=0})^{\sqrt{1-r^2}} \equiv \sqrt[G]{D_{r=0}}. \tag{3}$$

An estimation of the efficiency of such approximation was carried out in [8–10]. In these works it was shown that it is possible to increase the transparency of the barrier at low energy (e.g., at room temperature) at a real deformation of the potential well (which leads to CCS formation) from a negligible values $D_{r=0} \leq 10^{-100} - 10^{-300}$ to an acceptable values $D_{r\neq 0} \approx 10^{-1} - 10^{-10}$, which are sufficient for the efficient nuclear fusion of both light and heavy isotopes and elements. A non-trivial physical reason for the sharp increase of the transparency of the barrier for a particle in a CCS is the interference of mutually correlated particle momentum fluctuations corresponding to different components of the non-stationary superposition state in which the particle is located. The result of this interference is the formation of giant fluctuations in momentum and kinetic energy, which leads to a reasonable increase in the transparency of the barrier. This effect clearly follows from a simple analysis.

The total current fluctuation of the momentum of a particle in a potential well with N levels in the superposition state is described by the expression:

$$\Delta\vec{p}(t) = \sum_n^N \Delta\vec{p}_n(t). \tag{4}$$

The corresponding dispersion of this momentum is determined by the formula

$$\sigma_p = \left\langle \left\{ \sum_n^N \Delta\vec{p}_n(t) \right\}^2 \right\rangle = N \left\langle (\Delta\vec{p}_n)^2 \right\rangle + N^2 \left\langle \Delta\vec{p}_n \Delta\vec{p}_m \right\rangle \tag{5}$$

The average total kinetic energy of a particle in such well is characterized by formula

$$\langle \Delta T \rangle = \langle \Delta\vec{p}(t)^2 \rangle / 2M \; = N^2 \left\langle \Delta\vec{p}_n \Delta\vec{p}_m \right\rangle / 2M + N \left\langle (\Delta\vec{p}_n)^2 \right\rangle / 2M. \tag{6}$$

From the last formulas follows that at the absence of correlation of quantum states of a particle at different levels of quantized motion (for the case $\langle \Delta\vec{p}_n \Delta\vec{p}_m \rangle = \langle \Delta\vec{p}_n^{\,2} \rangle \delta_{nm}$), we have a "standard" (usual) result: the total average energy of a particle in a system of energy levels is equal to the sum of energies at all levels where a particle can be located

$$\langle \Delta T \rangle = \sum_{n=1}^N \left\langle (\Delta\vec{p}_n)^2 \right\rangle / 2M = N \left\langle (\Delta\vec{p}_n)^2 \right\rangle / 2M \equiv N \langle \Delta T_n \rangle \sim N. \tag{7}$$

However, at the presence of such correlation (at $r \neq 0$), we have a fundamentally different result

$$\langle \Delta T \rangle = N \left\langle (\Delta\vec{p}_n)^2 \right\rangle / 2M + N^2 \left\langle \Delta\vec{p}_n \Delta\vec{p}_m \right\rangle / 2\tilde{M} \sim N^2 \tag{8}$$

corresponding to a very significant increase of the mean kinetic energy at $N \gg 1$.

The most optimal method of CCS formation is connected with optimal modulation of non-stationary parabolic potential $V(q,t) = M\omega^2(t)q^2/2$ for discussed particle (in fact, modulation of effective frequency $\omega(t)$). On the basis of solution of non-stationary Schrödinger equation an explicit form of the correlation coefficient for this particle can be found [15]

$$r = \mathrm{Re}\left\{ \varepsilon^* \frac{\mathrm{d}\varepsilon}{\mathrm{d}t} \right\} / \left| \varepsilon^* \frac{\mathrm{d}\varepsilon}{\mathrm{d}t} \right|, \qquad r^2 = 1 - 1/\left| \varepsilon^* \frac{\mathrm{d}\varepsilon}{\mathrm{d}t} \right|^2. \tag{9}$$

In a similar way it is possible to find the compression coefficient k determining the ratio of the dispersions (variances) of the complex dimensionless coordinates ε and the momentum $\mathrm{d}\varepsilon/dt$ of the particle

$$k = \sigma_q/\sigma_p = |\varepsilon/(\mathrm{d}\varepsilon/\mathrm{d}t)|^2 \tag{10}$$

and also the values of these dispersions

$$\sigma_q \geq \; (\hbar/2)\sqrt{k/(1-r^2)}, \qquad \sigma_p \geq (\hbar/2)\sqrt{1/k(1-r^2)}. \tag{11}$$

In these formulas $\varepsilon(t)$ is a dimensionless (normalized to $q_0 = \sqrt{\hbar/M\omega_0}$) complex coordinate of the particle, which is a solution of motion equation of a classical oscillator with a variable frequency

$$\frac{\mathrm{d}^2\varepsilon}{\mathrm{d}t^2} + \omega^2(t)\varepsilon = 0, \qquad \varepsilon(0) = 1, \qquad \left. \frac{\mathrm{d}\varepsilon}{\mathrm{d}t} \right|_0 = i, \qquad \omega(0) = 1, \tag{12}$$

$\omega(t)$ is a dimensionless frequency normalized to a characteristic frequency ω_0; t-dimensionless (normalized to ω_0^{-1}) time.

The simplest practical method of CCS formation is associated with different regimes of deformation (modulation) of a non-stationary harmonic oscillator, in the parabolic field (parabolic well) of which the particle under consideration is located. This modulation leads to an optimal mutual phasing of the different eigen states of the particle in this well. Such a formation can be obtained with a monotonic asymptotic decrease or increase of the oscillator frequency [1,2,6,9], with a change of this frequency in a limited interval [3–9] or with its periodic modulation at the absence [6–9, 12–14,17] or at the presence of random force and frequency fluctuations [12], at external pulse action [18,19].

These mechanisms can be produced in different systems: with external irradiation of crystals with a spatial alternation of heavy and light nuclei [4,20], in the process of growth or squeezing of nanocracks on a surface of metal hydrides [5–9], at the orientational motion of particles in crystals and molecular gas [21,22], in biophysical processes [21–26].

2. Formation of Coherent Correlated States and Effective Low Energy Reactions with the Participation of Charged Particles at Action of Pulsed Magnetic Field

In this paper we consider one additional practical method of LENR production based on the formation of CCS in a gaseous or plasma environment under the action of a pulsed magnetic field on free ions. This system is fundamentally different from the usually considered LENR models, which are based on a specific type of interaction in a solid target. The universal method of LENR production using of coherent correlated states, which have been considered for a long time (e.g. [1–20]), makes it possible to successfully solve also such a problem.

It is well known that the state of a charged particle in a magnetic field $H(t)$ is analogous to the state of a particle in a harmonic oscillator with the appropriate energy spectrum and characteristic frequency.

$$E_n = \hbar\omega(t)(n + 1/2), \qquad \omega(t) = |q|H(t)/Mc \tag{13}$$

and the wave functions $\Psi_n = C_n H_n(\varepsilon)\,\mathrm{e}^{-\varepsilon^2/2}$, where $H_n(\varepsilon)$ is Hermite polynomial. Under the action of a pulsed magnetic field

$$H(t) = H_0(1 + f(t)), \qquad f(t) = g\,\mathrm{e}^{-(t-t_0)^2/2\tau^2}, \quad t_0 \gg \tau \tag{14}$$

on a charged particle, the characteristic cyclotron frequency has the similar form

$$\omega(t) = \omega_0(1 + f(t)). \tag{15}$$

For the convenience it is assumed that the particle is situated in a weak permanent magnetic field H_0 (e.g., the Earth's magnetic field) before the impulse action starts and after its completion. The results of calculation of correlation coefficient $r(t)$ and formation of a CCS using a pulsed magnetic field (14) and the system of equations (9)–(12), (14), and (15) are shown in Figs. 1 and 2.

Figure 1 shows the dependence of the current $r(t)$ value of the correlation coefficient versus time t and maximum $|r(t)|_{\max}$ value versus duration of pulse τ for relative amplitude $g = 5$ of the magnetic field pulse. From these data it can be seen that under such influence a large values of the current $|r(t)|_{\max} = 0.9972$ and averaged over time $\langle|r(t)|\rangle = 0.952$ correlation coefficient are achieved at a specific pulse duration $\tau_{\text{opt}} = 0.112/\omega_0$. If the pulse duration is longer or shorter, both values ($|r(t)|_{\max}$ and $\langle|r(t)|\rangle$) very rapidly decreases.

From these data it can be seen that under such influence great values of the current $|r(t)|_{\max} = 0.9972$ and averaged over time $\langle|r(t)|\rangle = 0.952$ correlation coefficients are achieved at a specific pulse duration $\tau_{opt} = 0.112/\omega_0$.

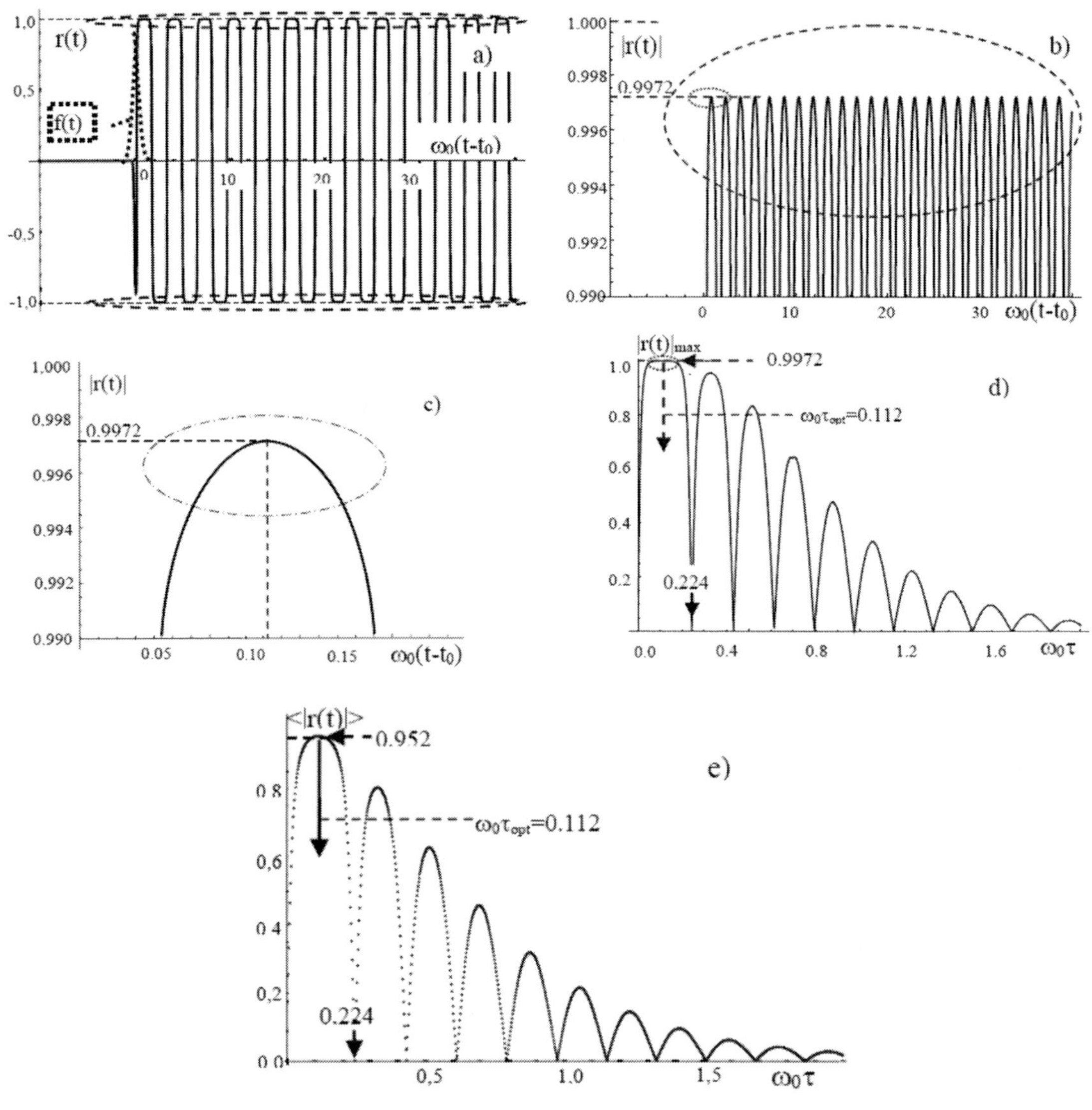

Figure 1. Dependence of the correlation coefficient $r(t)$ versus time t and duration τ of the frequency modulation pulse $f(t)$ (18) at the relative amplitude of magnetic field pulse $g = 5$: (a) general view of $r(t)$ together with $f(t)$; (b) and (c) fragments of $|r(t)|$ in the area $|r(t)| \approx 1$; (d) and (e) dependence of $|r(t)|_{\max}$ and $\langle |r(t)| \rangle$ versus τ.

If the pulse duration is longer or shorter, the both values ($|r(t)|_{\max}$ and $\langle |r(t)| \rangle$) rapidly decreases. To determine the general trend of such effect, the process of formation of a CCS with a different relative amplitudes of the magnetic

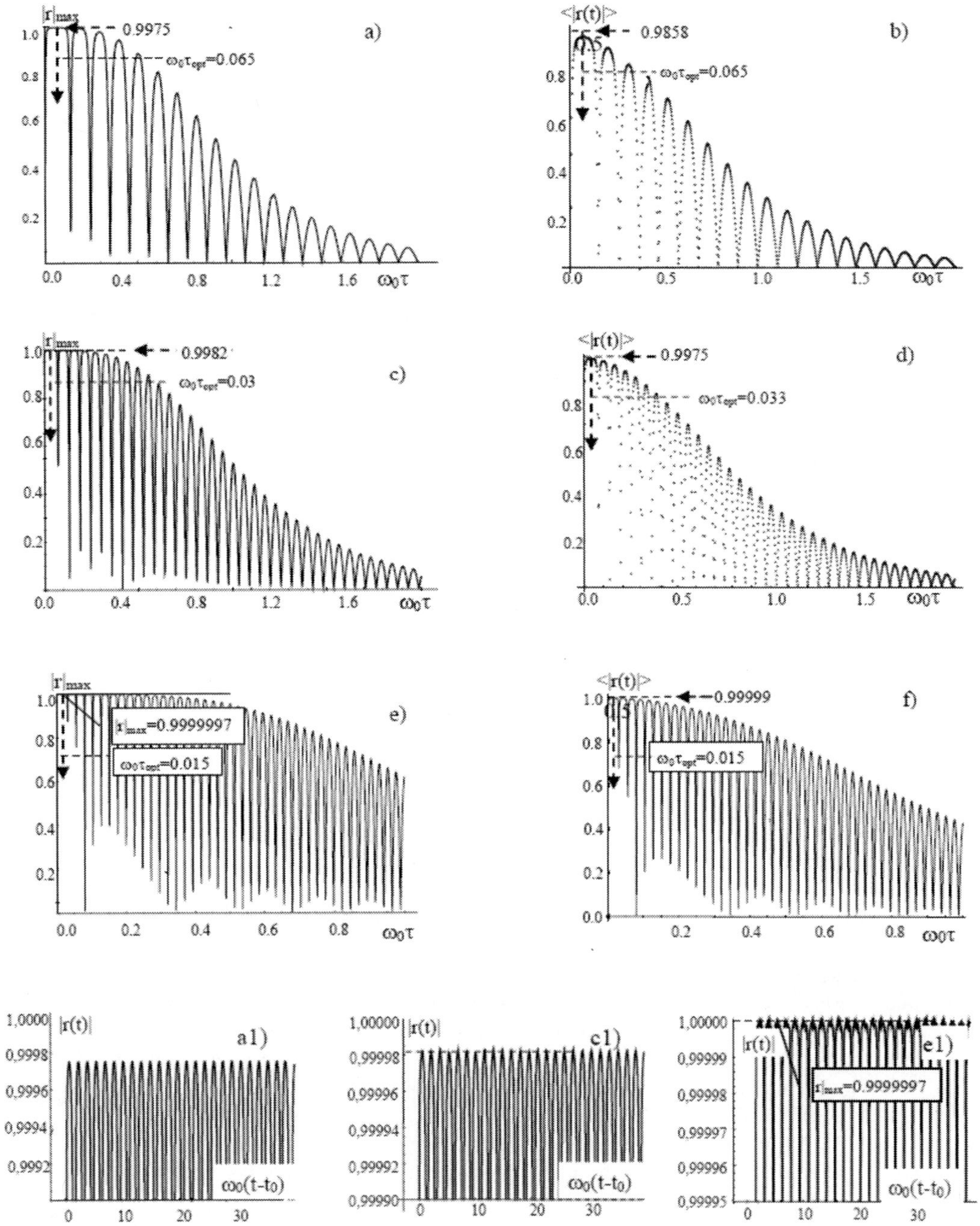

Figure 2. Dependence of the maximum $|r(t)|_{\max}$ (a, c, e) and time-averaged $\langle|r(t)|\rangle$ (b, d, f) correlation coefficient versus duration of the pulse of the frequency modulation (18) for different amplitude of this pulse: $g = 10$ (a), (b); 20 (c), (d); 50 (e), (f). The lower row is the time-dependent structure of the correlation coefficient $|r(t)|$ that correspond to the main (first) maximum of values $|r(t)|_{\max}$ and $\langle|r(t)|\rangle$ at different g and τ: $g = 10$ and $\tau = 0.065/\omega_0$(a1); $g = 20$, $\tau = 0.033/\omega_0$(c1); $g = 50, \tau = 0.015/\omega_0$(e1).

field pulse ($g = 10, 20, 50$) were considered. These results are presented in Fig. 2.

From the data $\omega_0\tau_{opt} \approx 0.112; 0.065; 0.03; 0.015$ presented in Figs. 1 and 2 and corresponding to different relative amplitudes of the magnetic field pulse g =5;10;20;50 for realization of the maximum values of $|r|_{\max}$ and $\langle|r(t)|\rangle$ for a given initial frequency ω_0, it can be seen that the functional dependence of the optimal duration τ of magnetic pulse and the maximum frequency $\omega_{\max} = \omega_0(1+g)$ can be written in the form of a universal relation

$$g\omega_0\tau \approx \omega_{\max}\tau \equiv \tau|q|H_{\max}/Mc \approx 0.6 - 0.7, \tag{16}$$

which is convenient for the analysis of real experiments. It is easy to understand the reason for this condition.

It was shown in [6,9,13] that the maximum efficiency of CCS formation in a non-stationary harmonic oscillator and the possibility of achieving large values of the magnitudes $|r|_{\max}$, $\langle|r(t)|\rangle$ and $G_{\max}$ corresponds to the situation when the frequency Ω_M of the controlled periodic change of the parameters of this oscillator (modulation frequency) in equal to double to the natural frequency of this oscillator during the formation of the CCS. Under this condition, the requirement of optimal synchronization of the phases of the eigenfunction of the particle in the potential well is satisfied. If we apply this result to the problem under consideration, is corresponds to the condition $\Omega_{\max} = 2\omega_{\max}$.

On the other hand, the Fourier spectrum of the pulse (14), (15) has the form

$$f(\Omega) = \tau g \exp(-\Omega^2\tau^2/2), \tag{17}$$

it remains almost constant at $\Omega \leq \sqrt{2}/\tau$ and sharply decreases at $\Omega > \sqrt{2}/\tau$.

Comparison of the last two formulas leads to a relation $\omega_{\max}\tau \approx 1/\sqrt{2}$, which completely coincides with [16]. In other words, the condition for optimal CCS formation coincides with the requirement of the maximum spectral density of the function of modulation (14) at the optimum modulation frequency, which leads to synchronization of all fluctuations of momentum and energy. It can be seen from these data that with an increase of the pulse amplitude and, accordingly, with a decrease of its duration, the correlation coefficient and the transparency of the potential barrier increase very rapidly.

If we use, for example, the actual duration of electric discharge pulses $\tau \approx 10^{-7}$s, then, in order to form the CCS of free hydrogen ions (protons) with mass M_p , the optimum value of the maximum strength of the pulsed magnetic field produced by the electric discharge, according to (17), should be equal to $H_{\max}^{(p)} \approx 600$ Oe. For the case of shorter magnetic field pulses, a corresponding increase of the amplitude of these pulses is needed. It should be noted that the use of shorter pulses is the most optimal option, since the effect of dephasing of the CCS due to the accidental collisions in the gas is substantially weakened. This question is considered in detail in the papers [12,18]. From the calculation follows that with such action of magnetic field with $g = 50$ on free protons or deuterons a short-term CCS formation with a very high correlation coefficient $|r|_{\max} = 0.99999997$ and correlation efficiency coefficient $G_{g=50} = 1290$ takes place. Even more impressive results can be obtained with the additional magnification $g \gg 50$ taking into account a corresponding decrease of the duration and increase of the amplitude of the magnetic field pulse. Such increase of $|r(t)|$ and G leads to a very substantial increase in the transparency of any Coulomb barrier and the possibility of nuclear fusion at low energy during the interaction of considering particle (proton) with closely spaced nuclei. This, in particular, can be similar nuclei, which are moving in the same magnetic field, or nuclei of different atoms and molecules, which are situated near the considered moving nucleus. With such a value of the coefficient of correlation efficiency, there is an increase of the transparency of the barrier from very small value $D_{r=0} \leq 10^{-100} - 10^{-300}$ to an acceptable values $D_{r\neq 0} \approx 10^{-1} - 10^{-10}$ at which the probability of nuclear reactions is very large.

It should be noted that the daughter products of such reactions (including fast neutrons and alpha particles) were recorded many times during the experiments with a gas discharge. In particular, the results of neutron detection experiments at a level of 2200 neutrons/pulse in a nanosecond electric discharge in gaseous deuterium at low pressure are presented in [27]. Such regime corresponds to the reaction d(d,He^3)n in the volume of the deuterium. The structure of the discharge pulse in this experiment is very close to the Gaussian function (18) with parameters $\tau \approx$ 1ns, $J_{\max} \approx$ 11 kA . In this work, the accelerator mechanism was considered under assumption that neutron generation occurs due to the acceleration of deuterons and their interaction with deuterated cathode. However, in this work it is noted that at a deuterium pressure of about 1 Torr, neutron generation is observed when using not only deuterated cathodes, but also cathodes that do not contain deuterium. In this case, the authors assumed the possibility of a reaction due to strong heating of the gas medium by the means of a shock wave. Such an argument is rather questionable, since a shock wave can not heat the gas to a thermonuclear temperature.

This result is explained very well on the basis of the formalism of coherent correlated states, which are formed under optimum impulse action. If we compare the parameters of this experiment with the optimal condition (25), which ensures large values of the quantities, $|r|_{\max}$, $\langle|r(t)|\rangle$ and $G_{\max}$ accordingly, the greater probability of tunneling and the production of the nuclear reaction, it is clear that this condition will be satisfied near the tip of an acute electrode at a small distance from its axis or in region $R \leq 1$ mm of local electronic explosive emission. The small size of these regions leads to a relatively small total neutron yield. We note that when the amplitude or duration of the discharge current pulse is increased by about 10 times, this condition will be satisfied even near the outer surface of the cylindrical tubular electrode used in this work, with a radius , that should lead to a sharp increase of the yields of the reaction. An even greater number of neutrons (at a level of 6000 neutrons/pulse) was recorded in experiments [28] with an electric discharge in ordinary air. The authors did not propose a valid model of the observed effect. This effect is well explained if it is assumed that it is caused by the formation of coherent correlated states of deuterons and protons in air. We note that deuterium enters the air in the form of molecules $\mathrm{D_2}$, and, in a larger amount, in the composition of water vapor with a total concentration $n_{\mathrm{D}} \geq 10^{12} - 10^{13}\,\mathrm{cm}^{-3}$. Such particles, which are in a correlated state, can stimulate in the air many reactions of the type

$$\begin{gathered} \mathrm{d}+\mathrm{d}=\mathrm{T}+\mathrm{p}, \qquad \mathrm{d}+\mathrm{d}=\mathrm{He}^3+\mathrm{n}, \qquad \mathrm{C}^{12}+\mathrm{n}=3\mathrm{He}^4+\mathrm{n}', \\ \mathrm{C}^{13}+\mathrm{p}=\mathrm{N}^{14}, \qquad \mathrm{C}^{12}+\mathrm{d}=\mathrm{N}^{14}, \qquad \mathrm{N}^{15}+\mathrm{p}=\mathrm{O}^{16}, \\ \mathrm{N}^{14}+\mathrm{d}=\mathrm{O}^{16}, \qquad \mathrm{O}^{18}+\mathrm{p}=\mathrm{F}^{19}, \end{gathered} \tag{18}$$

part of the products of which (in particular, He^4) were registered in this experiment.

Based on the parameters of the whole pulse ($J_{\max} \approx 10 - 15\,kA, \tau \approx 100\,\mathrm{ns}$), the criterion (16) was satisfied at a distance $R \approx 1.5\,\mathrm{cm}$ from the current axis. At the same time, in this experiment, a short pulse of neutron generation with a duration 20–30 ns was synchronized with the concomitant X-ray radiation and was observed at the leading edge of the current pre-pulse at $\tau \approx 220\,\mathrm{ns}$ and $J_{\max} \approx 550\,A$, which agrees with (16) at $R \approx 0.2$ cm and predicts the possibility of the reactions (18) proceeding directly near the outer surface of the discharge cord. The similar effects of the self-similar formation of the CCS can also explain the mechanism of neutron generation in air during a lightning discharge [29,30] due to the synthesis with the participation of deuterium, which is part of water vapor.

References

[1] V.I. Vysotskii and S.V. Adamenko, Correlated states of interacting particles and problems of the Coulomb barrier transparency at low energies in nonstationary systems, *J. Tech. Phys.* **55** (5) (2010) 613.

[2] V.I. Vysotskii, M.V. Vysotskyy and S.V. Adamenko, Formation and application of correlated states in non-stationary systems at low energy of interacting particles, *J. Exp. Theoret. Phys.* **114** (2) (2012) 243–252.

[3] V.I. Vysotskii, S.V. Adamenko and M.V. Vysotskyy, The formation of correlated states and the increase in barrier transparency at a low particle energy in nonstationary systems with damping and fluctuations, *J. Exp. Theoret. Phys.* **115** (4) (2012) 551–566.

[4] V.I. Vysotskii, S.V. Adamenko, M.V. Vysotskyy, Subbarier interaction of channeling particles under the self-similar excitation correlated states in periodically deformed crystal, *J. Surface Investigations* **6** (2) (2012) 369–374.

[5] V.I. Vysotskii and S.V. Adamenko, Low energy subbarrier correlated nuclear fusion in dynamical systems, *J. Condensed Matter Nucl. Sci.* **8** (2012) 91–104.

[6] V.I. Vysotskii and M.V. Vysotskyy, Coherent correlated states and low-energy nuclear reactions in non stationary systems, *Euro. Phys. J.* A **49** (8) : 99 (2013) 1–12.

[7] V.I. Vysotskii, The problem of creation a universal theory of LENR, *Infinite Energy* **18** (108) (2013) 30–35.

[8] V.I. Vysotskii and M.V. Vysotskyy, Application of coherent correlated states of interacting particles in non-stationary controlled LENR, *Infinite Energy* **112** (2013) 71–76.

[9] V.I. Vysotskii, S.V. Adamenko and M.V. Vysotskyy, Acceleration of low energy nuclear reactions by formation of correlated states of interacting particles in dynamical systems, *Ann. Nucl. Energy* **62** (2013) 618–625.

[10] E. Schrodinger, Ber. Kgl. Akad. Wiss., Berlin, S24, 1930, p. 296.

[11] H.P. Robertson, *Phys. Rev.* A **35** (1930) 667.

[12] V.I. Vysotskii and M.V. Vysotskyy, Correlated states and transparency of a barrier for low-energy particles at monotonic deformation of a potential well with dissipation and a stochastic force, *J. Exp. Theoret. Phys.* **118** (4) (2014) 534–549.

[13] V.I.Vysotskii, M.V.Vysotskyy, S.V.Adamenko, Application of correlated states of interacting particles in nonstationary and periodical modulated LENR systems, *J. Condensed Matter Nucl. Sci.* **13** (2014) 624–636.

[14] V.I.Vysotskii, M.V.Vysotskyy, Formation of correlated states and optimization of nuclear reactions for low-energy particles at nonresonant low-frequency modulation of a potential well, *J. Exp. Theoret. Phys.* **120** (2) (2015) 246–256.

[15] V.V. Dodonov, A.B.Klimov and V.I. Man'ko, Low energy wave packet tunneling from a parabolic potential well through a high potential barrier, *Phys. Lett.* A **220** (1996) 41–48.

[16] V.I. Vysotskii and M.V. Vysotskyy, The formation of correlated states and optimization of the tunnel effect for low-energy particles under nonmonochromatic and pulsed action on a potential barrier, *J. Exp. Theoret. Phys. J. Exp. Theoret. Phys.* **121**(4) (2015) 559–571.

[17] V.I. Vysotskii and M.V. Vysotskyy, Coherent correlated states of interacting particles – the possible key to paradoxes and features of LENR, *Current Science* **108** (40) (2015) 524–530.

[18] V.I. Vysotskii and M.V. Vysotskyy, The formation of correlated states and tunneling at low energy at controlled pulse action on particles, *J. Exp. Theoret. Phys.* **125** (2) (2017) 195–209.

[19] V.I. Vysotskii and M.V. Vysotskyy, Universal mechanism of realization of nuclear reactions at low energy, *RENSIT* **9** (1) (2017) 21–36.

[20] V.I. Vysotskii, M.V. Vysotskyy and S. Bartalucci, Features of the formation of coherent correlated states and nuclear synthesis in the interaction of slow particles with crystals and free molecules, *J. Exp. Theoret. Phys.* **127** (3) (2018) 479–490.

[21] V.I. Vysotskii and A.A. Kornilova, Nuclear fusion and transmutation of isotopes in biological systems, Monograph, MIR, Moscow, 2003, 302 p.

[22] V.I. Vysotskii and A.A. Kornilova. *Nuclear Transmutation of Stable and Radioactive Isotopes in Biological Systems,* Pentagon Press, India, 2010, 187 p.

[23] V.I. Vysotskii A.A.Kornilova and I.I. Samoylenko, A method of producing stable isotopes due to nuclear transmutation type low-temperature nuclear fusion of elements in microbiological cultures, Patent RU 2052223C1, 1995-10-18; https://patents.google.com/patent/RU2052223C1/ru.

[24] A.A. Kornilova and V.I. Vysotskii, Method for purifying water of radionuclides, Int. patent WO 2015156698 A1, PCT/RU2014/000273, Date of publication 15 Oct. 2015.

[25] V.I. Vysotskii and A.A. Kornilova, Microbial transmutation of Cs-137 and LENR in growing biological systems, *Current Science* **108** (4) (2015) 636–640.

[26] V.I. Vysotskii and A.A. Kornilova, Transmutation of stable isotopes and deactivation of radioactive waste in growing biological systems, *Ann. Nucl. Energy* **62** (2013) 626–633.

[27] M.I. Lomaev, B.A. Nechaev, V.N.Padalko etc., *J. Tech. Phys.* **57** (2012) 124–132.

[28] A.V. Agafonov, A.V. Bagulya, O.D. Dalkarov, M.A. Negodaev, A.V. Oginov, A.S. Rusetskiy, V.A. Ryabov and K.V. Shpakov, Observation of neutron bursts produced by laboratory high-voltage atmospheric discharge, *Phys. Rew. Lett.* **111** (2013) 115003.

[29] A.V. Gurevich, V.P. Antonova, A.P. Chubenko, A.N. Karashtin, G.G. Mitko, M.O. Ptitsyn, V. A. Ryabov, A. L. Shepetov, Yu. V. Shlyugaev, V.I.ldanova and K.P. Zybin, Strong flux of low-energy neutrons produced by thunderstorms, *Phys. Rev. Lett.* **108** (2012) 125001.

[30] B.Zh. Zalikhanov, From an electron avalanche to the lightning discharge, *Phys. Part. Nucl.* **47** (2016) 108–133.

J. Condensed Matter Nucl. Sci. 29 (2019) 358–367

Research Article

Using the Method of Coherent Correlated States for Production of Nuclear Interaction of Slow Particles with Crystals and Molecules

Vladimir Vysotskii* and Mykhaylo Vysotskyy
Kiev National Shevchenko University, Ukraine

Sergio Bartalucci
INFN – Laboratori Nazionali di Frascati, Frascati, Italy

Abstract

In this paper, the possibility and characteristics of effective nuclear fusion based on the interaction of low energy proton beams with the nuclei on a crystal surface or gas of free molecules are discussed. It is shown that this effect can be explained by the process of formation of coherent correlated states, which take place during the interaction of moving protons with lithium molecules.

Keywords: Coherent correlated states, Fusion on accelerated particles, LENR channels automatic sorting, LENR in gas and crystals

1. Introduction

Traditionally, it has been believed that LENR ("Low Energy Nuclear Reactions") can occur only in condensed medium, such as a crystal or liquid, and can be produced only with a special form of the static local (interatomic) environment of interacting particles. This is reflected, in particular, in the "official" term for this area of nuclear physics: "Condensed Matter Nuclear Science". In our opinion the more correct and more universal term is LENR.

It is a very important fact that LENR effects not only take place in "traditional" crystals and solids, but also in low-temperature plasmas and in other media, which has no relation to condensed matter. One of the very interesting variants of such interaction is described in [1]. It is a low-energy variant of accelerator fusion, where nuclear fusion is induced by the proton beam with the energy no higher than several hundred electron volts, which is much lower than the energies of 30–100 keV, typical of the standard accelerator fusion. In these experiments the reaction of fusion

*Corresponding author. E-mail: vivysotskii@gmail.com.

$$Li^7 + p = 2He^4 + 17.3 \text{ MeV} \tag{1}$$

was observed with a count rate $J \approx 157$ pulse/s of fast alpha particles He^4 with an energy of about 8.6 MeV in the case of a lithium target (made of natural lithium) and the beam of protons with energy 448 eV (current $i_p \approx 50\,\mu A$). This result cannot be explained within standard models of a nuclear reaction involving accelerated particles because at such energy the coefficient of barrier transparency and cross-section of reaction are very small ($D(E) \approx 10^{-50}$ and $\sigma(E) \approx 10^{-74}\text{cm}^2$).

If we consider that mean free path of protons with such low energy in the lithium crystal even in the planar channeling regime, it is shorter than $\langle L \rangle \approx 10{-}20$ nm. In such case, the differential $\mathrm{d}W_f/\mathrm{d}z$ and total W_f probabilities of the fusion reaction per proton, as well as the total intensity of the reactions J for the beam with the current $i_p \approx 50\,\mu A$, are also very small:

$$\mathrm{d}W_f/\mathrm{d}z = \sigma\langle n_{Li}\rangle \approx 3\times 10^{-50}\text{cm}^{-1}, \quad W_f \approx \langle L\rangle\sigma\langle n_{Li}\rangle \approx 10^{-56}, \quad J \approx i_p W_f \approx 10^{-41}\ \text{s}^{-1}. \tag{2}$$

These values are incommensurably insignificant in comparison with the results of the experiments.

In the same work [1] authors performed another variant of this experiment, when a proton beam with the energy varied from 50 eV to 5 keV and the current $i_p \approx 100\,\mu A$ passes through lithium vapor. This experiment has no relation to the condensed matter.

The results of this experiment are very close to the results of experiments with lithium foil and are shown in Fig. 1. It can be seen that the nuclear interaction of moving protons with lithium vapor is characterized by a sharp peak with proton energy in a narrow range near 500 eV, where the maximum count rate of alpha particles reached $J \approx 4\times 10^4\ \text{s}^{-1}$. A typical spectrum of alpha particles in these experiments is presented in Fig. 2.

The most amazing feature of this spectrum is the complete absence of alpha particles with the energies of 1.7 and 2.3 MeV, which correspond to a fusion reaction

$$Li^6 + p = He^3 + He^4 + 4 \text{ MeV} \tag{3}$$

with the participation of Li^6 isotope! On the other hand, the "standard" calculation showed that the probabilities of both reactions ((1) and (3)) at a given energy should be approximately the same, which completely contradicts the experimental data.

We assume (and this will be proved below) that all features of the course of the reaction of accelerating fusion at such low energy are fully justified, and can be explained by the use of the method of coherent correlated states (CCS), which are automatically formed in the systems under consideration, both with lithium foil and lithium vapor.

2. Features of Formation of Coherent Correlated States of the Particles During its Passage Through the Crystal and Molecules of Gas

The method of CCS well explains the above-mentioned features and anomalies of nuclear reactions with the participation of low energy protons and lithium isotopes. This method provides a high probability of LENR and can be applied with the same efficiency to different experiments (including processes in solids, liquids, gas, plasma, at action of non-stationary magnetic fields, in geological and biological systems etc). It should be noted that the CCS method makes it possible to explain different LENR paradoxes on the basis of standard quantum mechanics and modern nuclear physics without involving fantastic heuristic models.

The physical basis of this method is related to the Schrödinger–Robertson uncertainty relations [2,3]

$$\delta p\delta q \geq \hbar/2\sqrt{1-r_{\rm pq}^2} \equiv G_{\rm pq}\hbar/2, \quad \delta E\,\delta t \geq \hbar/2\sqrt{1-r_{\rm Et}^2} \equiv G_{\rm Et}\hbar/2, \tag{4}$$

where $|r| \leq 1$ is a *correlation coefficient* and $G = 1/\sqrt{1-r^2}$ is a *coefficient of correlation efficiency* [6–10]. In works [4–19], it was shown that the mechanism of the formation of CCS with the sharp increase of up to $G \geq 10^3 - 10^4$ can be produced when the particle is localized in the field of a nonstationary harmonic oscillator and for different regimes of modulation of parameters of harmonic oscillator.

In this case if the particle is localized in a potential well with a width L, for which $\delta q \approx L/2$, then the kinetic energy fluctuations

$$\delta T^{(\min)} = (\delta p)^2/2m = G_{\rm pq}^2\hbar^2/2mL^2 \tag{5}$$

are very large.

The methods for CCS calculations in different systems have been considered in detail in [4–19]. The simplest and the most adequate method involves analyzing the state of the considered particle in nonstationary parabolic field. In this case, the correlation coefficient can be calculated on the basis of the solution of equations

$$\begin{array}{l} \frac{\mathrm{d}^2\varepsilon}{\mathrm{d}t^2} + \omega^2(t)\varepsilon = 0, \quad \varepsilon(0) = 1, \qquad \frac{\mathrm{d}\varepsilon}{\mathrm{d}t}\Big|_0 = i, \quad \omega(0) = 1, \\ r = \mathrm{Re}\left\{\varepsilon^*\frac{\mathrm{d}\varepsilon}{\mathrm{d}t}\right\} / \left|\varepsilon^*\frac{\mathrm{d}\varepsilon}{\mathrm{d}t}\right|, \quad r^2 = 1 - 1/\left|\varepsilon^*\frac{\mathrm{d}\varepsilon}{\mathrm{d}t}\right|^2, \quad G = 1/\sqrt{1-r^2} \end{array} \tag{6}$$

Here, $\omega(t)$ is the dimensionless frequency in units of the characteristic frequency ω_0 and t is the dimensionless time in units of ω_0^{-1}. In particular, in works [9–12] it was shown that the maximum rate of G increase in the case of harmonic modulation $\omega(t) = 1 + g\,\cos\,\Omega t,\ |g| \ll 1$ (in dimensional form $\omega(t) = \omega_0(1 + g\,\cos\,\Omega t)$) corresponds to the condition, when the normalized oscillator modulation frequency Ω is two times greater than the average frequency ω_0 of the oscillator (see Fig. 3).

This mechanism and the results of analysis can be successfully used for proton moving with the velocity ν through a channel formed by a sequence of periodically arranged N pairs of atoms (fragment of the crystal lattice with longitu-

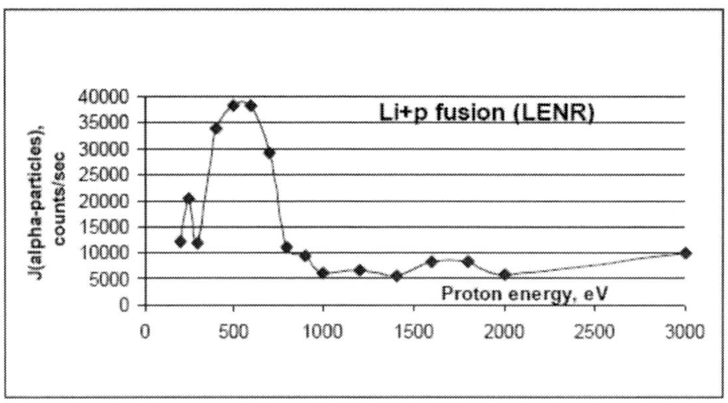

Figure 1. Intensity of the generation and detection of fast alpha particles versus the energy of protons passing through lithium.

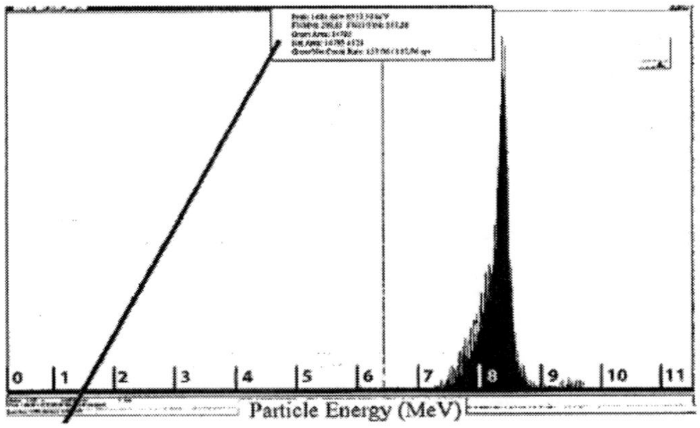

Figure 2. Experimental spectrum of detected alpha particles generated during Li+p fusion with participation of 500 eV protons passing through vapor of natural lithium.

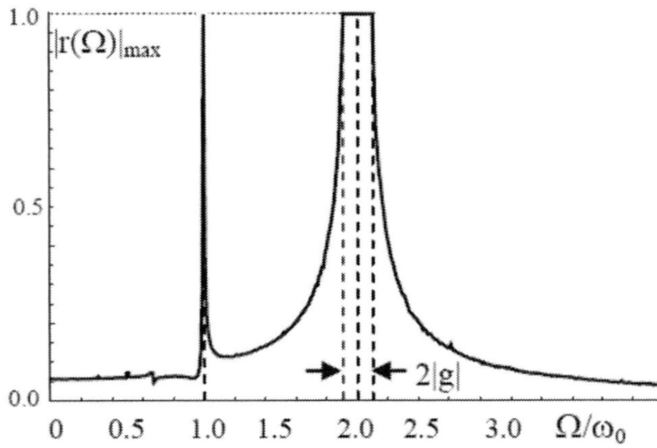

Figure 3. Resonance structure in the dependence of the maximum correlation coefficient on the modulation frequency $\omega(t) = \omega_0(1 + g \cos \Omega t)$ of the parameters of nonstationary potential well $V(x, t) = m_p x^2 \omega^2(t)/2$.

dinal a_z and transverse a_x lattice periods), which are located at the points $z_n = (n - 1/2)a_z$, $n = 1, 2, \ldots$ (see Fig. 4, *top*) and pairwise oriented perpendicularly to the direction of moon of the particle.

By analogy with the physics of channeling of positively charged particles, the electric field inside this channel can be approximately described by a parabolic potential (in the transverse direction). In contrast to the analysis of "traditional" channeling, which is valid only for sufficiently fast particles and involves the longitudinally averaged atomic potential of crystal axes and planes as the basic model, this analysis involves the real (not averaged) periodic potential $V(x, z) = V(x, z \pm a_z)$ with the period a_z in the direction of motion of the particle. In this model the character of proton motion in the periodic crystal field $V(x, z)$ corresponds to the inhomogeneous (along the longitudinal coordinate z)

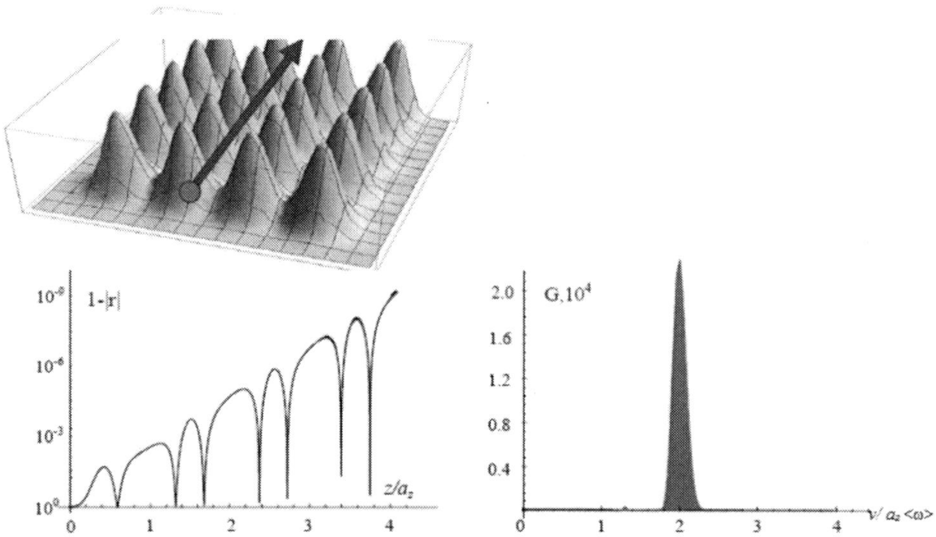

Figure 4. *Top*: the scheme of proton motion through a crystal of lithium; bottom: dependence of the coefficient of correlation $|r|$ at optimal velocity $\nu = 2\langle\omega\rangle a_z$ versus distance z/a_z and the coefficient of correlation efficiency versus proton velocity in a lithium crystal at $z = 4a_z$.

harmonic oscillator

$$V(x,z) = m_{\rm p}x^2\omega^2(z)/2 = \frac{m_{\rm p}x^2\omega_{\max}^2}{2}\sum_{n=1}^{N}\exp\{-[z-(n-1/2)a]^2/u^2\}, \quad |x| \le a_x, \ z \ge 0. \tag{7}$$

In the rest system of the proton such motion corresponds to a nonstationary harmonic oscillator with nonstationary potential energy

$$V(x,z) = \frac{m_{\rm p}x^2\omega^2(t)}{2} \tag{8}$$

and nonstationary characteristic frequency

$$\omega(t) = \omega_{\max}\left\{\sum_{n=1}^{N}\exp\left[-((\nu/a_z)t + 1/2 - n)^2K^2\right]\right\}^{1/2}, \quad K = a_z/u, \quad t \ge 0. \tag{9}$$

Here

$$\omega_{\max} = \sqrt{8\langle V_{\max}\rangle K/a_x^2 m_{\rm p}\sqrt{\pi}}, \quad \langle V_{\max}\rangle \equiv \langle V(x = \pm a_x/2, z)\rangle,$$

$$\omega(t=0) = \omega_{\max}{\rm e}^{-K^2/8} = \langle\omega\rangle\sqrt{K/\sqrt{\pi}}{\rm e}^{-K^2/8},$$

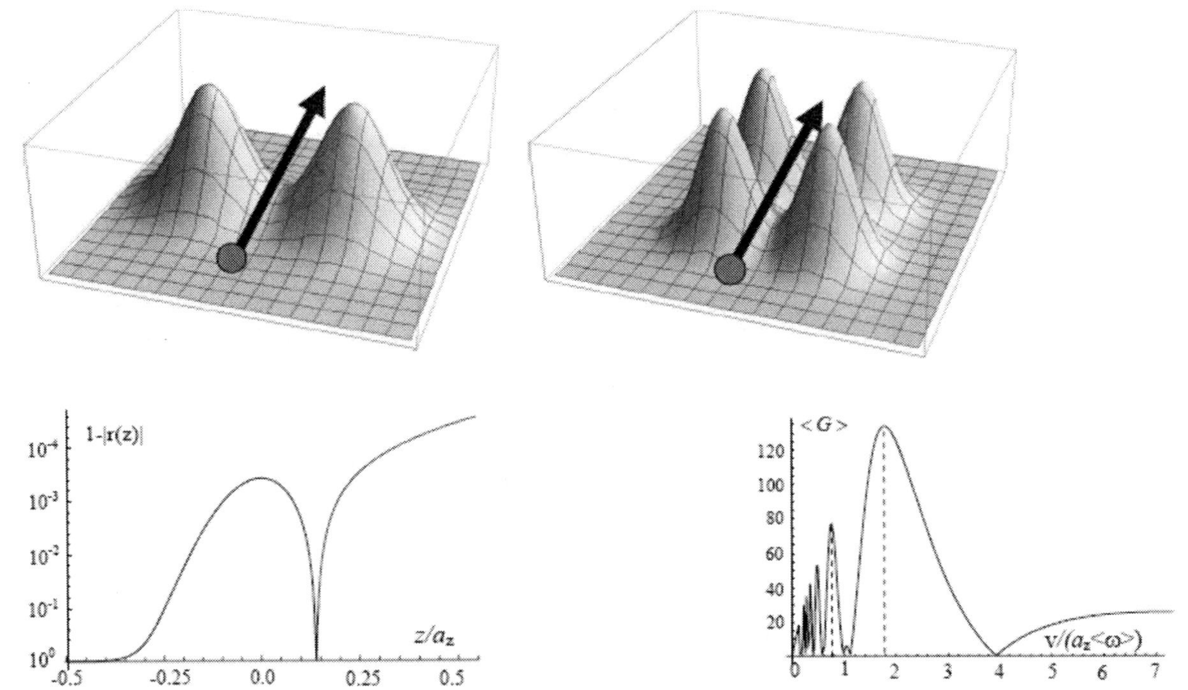

Figure 5. *Top*: the scheme of proton motion through free Li_2 and $(Li_2)_2$ molecules; *bottom*: (left) dependence of the coefficient of correlation $|r|$ at optimal velocity $\nu = 2\langle\omega\rangle a_z$ versus distance z/a_z from center of molecule and (right) the coefficient of correlation efficiency versus proton velocity (both graphs correspond to D_2 molecule at the condition $K = a_z/u = 5$).

$$\langle\omega\rangle = \omega_{\max}\sqrt{u\sqrt{\pi}/a_z} = \omega_{\max}\sqrt{\sqrt{\pi}/K}.$$

The structure of metallic lithium corresponds to a bcc lattice with a period of 3.5 Å, a distance of 3.03 Å between the nearest nuclei and averaged potential of the crystal axis $\langle V\rangle \approx 20$ eV. The average frequency (equal to the channeling frequency in the averaged potential of the crystal channel) calculated with these parameters (a_z =3–3.5 Å, $u \approx$ 0.4–0.5 Å is equal to $\langle\omega\rangle = \omega_{\text{chan}} \approx (5-6)\times 10^{14}\ \text{s}^{-1}$.

The coefficient of correlation efficiency for the motion of the particle inside the crystal between two axes with periodically arranged atoms has been calculated by solving of Eqs. (6) with the use of explicit time dependence (9) of the variable frequency $\omega(t)$. The calculation has been performed for the initial segment of the trajectory of the particle including four pairs of atoms and with the parameters $K = a_z/u = 4, 5, 6,$ which are close to the parameters of the lithium crystal. The results of this calculation for the case $K = a_z/u = 5$ are shown in Fig. 4 (*bottom*).

In this case the optimal condition for CCS formation corresponds to the optimal proton velocity $\nu_{\text{opt}} \approx 2a_z\langle\omega\rangle$ and optimal longitudinal kinetic energy of protons

$$T_{\text{opt}} = \frac{m_{\text{p}}\nu_{\text{opt}}^2}{2} = 2m_{\text{p}}a_z^2\langle\omega\rangle^2 \approx 400-600\ \text{eV} \quad \text{at } K\text{=4-6}. \tag{10}$$

These results are in very good agreement with the experimental data [1] discussed above. At such energy of the proton, the value of the correlation efficiency coefficient is equal to $G_{\max} \approx 23\,000$, which corresponds to giant fluctuations of the kinetic energy $\delta T^{(\min)} \approx 50 - 80$ keV in the transverse direction. At such energy of the proton, the transparency of the Coulomb barrier increases from the initial very small value $D(E) \approx 10^{-50}$ (which corresponds to the energy of 500 eV) to a large value $D(E) \approx 0.01$. In this case the nuclear fusion parameters correspond to the values,

$$\mathrm{d}W_{\mathrm{f}}/\mathrm{d}z = \sigma \langle n_{\mathrm{Li}} \rangle \approx 3 \times 10^{-6}\ \mathrm{cm}^{-1}, \quad W_{\mathrm{f}} \approx \langle L \rangle \sigma \langle n_{\mathrm{Li}} \rangle \approx 10^{-12}, \quad J \approx i_{\mathrm{p}} W_{\mathrm{f}} \approx 10^{3}\ \mathrm{s}^{-1}, \tag{11}$$

which are in good agreement with the experimental data.

It is very important that since CCS are formed very rapidly (within several crystal lattice periods), the deceleration of protons and their collisions with atoms of the crystal lattice can be disregarded.

Let us consider the features of the formation of CCS during the interaction of protons with the molecules of vapor lithium. In the composition of this vapor, about half or more of the lithium atoms are in the form of molecules Li_2 and $(Li_2)_2$ [19]. The distance between the nuclei in the lithium molecule is 2.67 Å. The potential energy in the volume of each of these molecules can be treated as the time-dependent harmonic oscillator for the charged particle moving through a molecule. Consequently, coherent correlated states can be formed in this motion.

This process can be calculated using the same Eqs. (6) and the dependence (7) of the variable frequency $\omega(t)$ in the rest system of the particle under the formal condition $N = 1$. For the comparability of the results for an individual molecule and a crystal, the calculation has been performed for an interval of the space localization of a moving particle in the field of the molecule that corresponds to one period a_z in the crystal. The results of the numerical calculation for $K = a_z/u = 5$ are shown in Fig. 5 (*bottom*).

It follows from these results that the optimal velocity and energy of the particle is slightly different from the same parameters in the case of the crystal and have different values for different ratios of a_z and u. In particular, with the same parameters of atoms as in the case of the lithium molecules, we obtain

$$K = 6, \quad T_{\mathrm{opt}} \approx 400 - 450\ \mathrm{eV}; \quad K = 5, \quad T_{\mathrm{opt}} \approx 500 - 550\ \mathrm{eV}; \quad K = 4, \quad T_{\mathrm{opt}} \approx 600 - 650\ \mathrm{eV}, \tag{12}$$

The physical reason for such difference is associated with the above-mentioned features of the formation of CCS at the modulation $\omega(t) = \omega_0(1 + g\cos\,\Omega t)$ of the parameters of the time-dependent oscillator by monochromatic action. As mentioned above and presented in Fig. 2, the probability of this process is maximal when the modulation frequency Ω is two times greater the average frequency of oscillations of the particle in this potential well in the absence of modulation.

In the case of single molecules, the spectrum of modulation in the coordinate system co-moving with the particle corresponds to the continuous spectrum of a normalized single pulse (for the Gaussian distribution of the potential energy, the spectrum is also Gaussian). In the case of the infinite crystal, the spectrum of modulation consists of a set of discrete lines at frequencies determined by the lattice period and velocity of the particle. All these lines are located inside the spectrum of a single momentum. In the case of a finite fragment of the crystal lattice, each of the discrete spectral lines is broadened and becomes a band with width inverse to the length of the fragment. This spectral analysis obviously indicates that the optimal velocity is that at which the necessary spectral component of the Fourier spectrum of the potential energy is maximal. It is obvious that this condition for different fragments of the potential field at different u values can be satisfied at slightly different velocities and energies of the particle; these values are identical only for a very long lattice.

According to the results presented in Fig. 5, at the ratio $K = 5$ of the considered parameters, which are close to the parameters of the real Li_2 molecule, the coefficient of correlation efficiency for the proton moving at this energy through the lithium molecule is $G_{\max} \approx 140$. At $K = 6$ the value $G_{\max}$ is even greater. At these $G_{\max}$ values, the cross section and probability of the $(\mathrm{Li}^7, \mathrm{p})$ reaction are very large. This is the result of the fact that in this case it is necessary to take into account that the effective quantization of the moving proton in the time-dependent potential well existing inside the molecule occurs only for the transverse component of the momentum $p_x = p \sin\theta$, which depends on the angle θ of entry of the particle to the space between atoms, and, correspondingly, for the transverse energy associated with this component. We recall that the formation of CCSs concerns these transverse components of the momentum and kinetic energy. If the initial transverse component of the total kinetic energy $T_{\mathrm{opt}} \approx 400 - 600$ eV is, e.g., $T_x = p_x^2/2m_{\mathrm{p}} = 1 - 10$ eV (for this, the proton should be incident at the angle $\theta \approx 2^\circ - 10^\circ$ to the normal to the axis of the molecule), the effective fluctuation of this energy appearing in the process of CCS formation is $\delta T \approx G^2 T_x \approx 10 - 100$ keV. This ensures a high efficiency of the fusion reaction even on individual molecules and is in good agreement with the experimental data for lithium vapor presented in Fig. 1. The real efficiency of fusion can be obtained by averaging over the mutual orientation of lithium molecules and the proton beam.

For a model cluster of two Li_2 molecules, the correlation coefficient in the case of the proton moving with the energy $T_{\mathrm{opt}} \approx 400-600$ eV (Fig. 2) at the same real value $K = 6$ is as high as $|r|_{\max} \approx 0.999999$, which corresponds to the correlation efficiency $G_{\max} \approx 700$ and also very large efficiency of fusion.

It is easy to see that other features of LENR (first of all the absence of daughter radioactive isotopes and very strong selection of channels of nuclear reactions) also follow from the peculiarities of CCS. Different aspects of these features were considered in detail in our earlier works [11,12,15–18].

In particular, from the basic relations (4) it follows that the possibility of producing LENR with virtual kinetic energy $\delta E \equiv \delta T_{|r|\neq 0}$ is limited by the conservation laws for the entire system. Very important is the fact that this virtual energy "exists" in the given system (that is, it can have a certain influence on different processes) a finite time δt. As a consequence, any process using δE can be produced only if during the reaction carried out by this virtual energy, the reaction energy ΔE that is not less than δE is released and the time of "return" to the system under consideration of this virtual energy (in fact, the duration of the reaction with the release of energy) does not exceed the value of δt.

This result, with respect to the nuclear reaction, corresponds to the fact that the total time of the reaction T_{total} (including the time of approach of the particle to the barrier t_1, the duration of the passage through the barrier t_2 and the time of the reaction itself with the release of energy T_{reaction}) should not exceed δt. This requirement, taking into account the very short duration δt of a large fluctuation amplitude $\delta T_{|r|\to 1}$, imposes very stringent conditions on such processes and automatically excludes the possibility of nonoptimal reactions. These conditions fully explain the complete prohibition on reaction (3) and "permission" for reaction (1).

Let us take as an estimate that for the rapid production of these reactions it is necessary for proton to have the energy $\delta E \approx 10$ keV. In the case of using the Heisenberg uncertainty relation $\delta E\, \delta t_{r=0} \geq \hbar/2$ such fluctuation can exist for a time $\delta t_{r=0} \approx \hbar/2\delta E \approx 5 \times 10^{-21}$ s. At this energy, the minimum total reaction time is equal to $T_{\mathrm{total}} = T_{\mathrm{reac}} + t_1 + t_2 \approx T_{\mathrm{reac}} + L(\delta E)/\nu(\delta E) \approx 10^{-18}$ s.

It can be seen from the reaction schemes (1) and (2) that for such value of T_{total} the necessary condition $T_{\mathrm{total}} < \delta t_{r=0}$ is not satisfied for both reactions and they are totally impossible in "usual" (noncorrelated) state.

In the correlated state, with an achievable value $|r| \approx 0.99999$, the same energy fluctuation $\delta E \approx 10$ keV can exist during $\delta t_{r=0.99999} \approx \hbar/2\delta E\sqrt{1-r^2} \approx 2.5 \times 10^{-18}$ s. Comparing this value with the total duration $T_{\mathrm{total}} \approx 10^{-18}$ s of reaction (1) $\mathrm{Li}^7 + \mathrm{p} = 2\mathrm{He}^4$, we come to the conclusion that $T_{\mathrm{total}} \approx 10^{-18}$ s $< \delta t$ (i.e. the total reaction time is less than the fluctuation time that stimulates this reaction), and the flow of such a reaction is consistent with the law of conservation of energy and the corresponding uncertainty relation.

In contrast, for reaction (2), the opposite condition $T_{\mathrm{total}} \approx 10^{-13}$ s $\gg \delta t$ takes place and such reaction is absolutely impossible due to the formation of CCS.

These results fully coincide with the data of both the above-mentioned paper [1] and very detailed experiments [20] conducted for 32 days in Lugano for the examination of the A. Rossi installation, in which very efficient processing of the Li^7 isotope was observed, which is characterized by a short reaction time, and complete absence of reactions involving the Li^6 isotope.

Similarly, it is easy to show that the same selection rule prohibits the production of LENR in reaction channels involving any other isotopes and elements that have a longer reaction time exceeding δt. It is obvious that, taking into account the smallness of δt even in systems with a large correlation coefficient, the reactions passing through the stage of formation of long-lived radioactive isotopes fully fall under this prohibition.

In conclusion, it should be noted that in the framework of a general methodological approach the use of CCS allows us to uniquely describe, calculate and predict the main features of nuclear reactions at low energy of interacting particles. Such analysis is fully applicable to nuclear processes occurring both in "standard" environments for such experiments (gas, metal hydrides, low-pressure plasma, liquid, crystals) and in more exotic objects and systems (in biological macromolecules and their nonstationary ensembles [21,22], in a fluid in the presence of cavitation phenomena [23], in active Mossbauer spectroscopy [24], in astrophysics in the process of combined gravitational-Coulomb collapse [25] and even under the action of a pulsed magnetic field of thunderous times series with the participation of deuterium in the natural isotopic abundance of the atmosphere and subsequent generation and detection of neutrons [17,26–28].

References

[1] S. Lipinski and H. Lipinski, Hydrogen–ithium fusion device, Int. Patent WO2014/ 189799 A9 (2013).

[2] E. Schrodinger, *Ber. Kgl. Akad. Wiss.*, Berlin, **S24** (1930) 296.

[3] H.P. Robertson, *Phys. Rev.* A **35** (1930) 667.

[4] V.V. Dodonov, E.V. Kurmishev and V.I. Man'ko, Generalized uncertainty relation and correlated coherent states, *Phys. Lett.* A **79** (1980) 150–152.

[5] VV. Dodonov and V.I. Man'ko, Invariants and correlated states of nonstationary quantum systems, *Trans. FIAN* (Russia) **183** (1987) 71.

[6] V.V. Dodonov, A.B. Klimov and V.I. Man'ko, Physical effects in correlated quantum states, *Trans. FIAN* (Russia) **200** (1991) 1991.

[7] V.I. Vysotskii and S.V. Adamenko, Correlated states of interacting particles and problems of the Coulomb barrier transparency at low energies in nonstationary systems, *J. Tech. Phys.* **55** (5) (2010) 613.

[8] V.I. Vysotskii, M.V. Vysotskyy and S.V. Adamenko, Formation and application of correlated states in non-stationary systems at low energy of interacting particles, *J. Exp. Theoret. Phys. (JETP)* **114** (2) (2012) 243–252.

[9] V.I. Vysotskii, S.V. Adamenko and M.V. Vysotskyy, The formation of correlated states and the increase in barrier transparency at a low particle energy in nonstationary systems with damping and fluctuations, *J. Exp. Theoret. Phys. (JETP)* **115** (4) (2012) 551–566.

[10] V.I. Vysotskii, S.V. Adamenko and M V. Vysotskyy, Subbarier interaction of channeling particles under the self-similar excitation correlated states in periodically deformed crystal, *J. Surface Invest.* **6** (2) (2012) 369–374.

[11] V.I. Vysotskii and M.V. Vysotskyy, Coherent correlated states and low-energy nuclear reactions in non-stationary systems, *Euro. Phys. J.* A **49** (8) (2013) 1–12.

[12] V.I. Vysotskii, S.V. Adamenko and M.V. Vysotskyy, Acceleration of low energy nuclear reactions by formation of correlated states of interacting particles in dynamical systems, *Ann. Nucl. Energy* **62** (2013) 618–625.

[13] V.I. Vysotskii an M.V. Vysotskyy, Correlated states and transparency of a barrier for low-energy particles at monotonic deformation of a potential well with dissipation and a stochastic force, *J. Exp. Theoret. Phys. (JETP)* **118** (4) (2014) 534–549.

[14] V.I. Vysotskii and M.V. Vysotskyy, Formation of correlated states and optimization of nuclear reactions for low-energy particles at nonresonant low-frequency modulation of a potential well, *J. Exp. Theoret. Phys. (JETP)* **120** (2) (2015) 246–256.

[15] V.I. Vysotskii and M.V. Vysotskyy, The formation of correlated states and optimization of the tunnel effect for low-energy particles under nonmonochromatic and pulsed action on a potential barrier, *J. Exp. Theoret. Phys. (JETP)* **121**(4) (2015) 559–571.
[16] V.I. Vysotskii and M.V. Vysotskyy, Coherent correlated states of interacting particles – the possible key to paradoxes and features of LENR, *Current Sci.* **108** (4) (2015) 524–530.
[17] V.I. Vysotskii and M.V. Vysotskyy, The formation of correlated states and tunneling at low energy at controlled pulse action on particles, *J. Exp. Theoret. Phys. (JETP)* **125** (2) (2017) 195–209.
[18] V.I. Vysotskii and M.V. Vysotskyy, Universal mechanism of realization of nuclear reactions at low energy, *RENSIT (Radioelectronics Nanosystems Information Technologies)* **9** (1) (2017) 21–36.
[19] V.I. Vysotskii, M.V. Vysotskyy and S. Bartalucci, Features of the formation of coherent correlated states and nuclear synthesis in the interaction of slow particles with crystals and free molecules, *J. Exp. Theoret. Phys. (JETP)* **127** (3) (2018) 479–490.
[20] Expertise in Lugano, http://www.sifferkoll.se/ sifferkoll/wp-content/uploads/2014/10/LuganoRe portSubmit.pdf.
[21] V.I. Vysotskii and A.A. Kornilova, Transmutation of stable isotopes and deactivation of radioactive waste in growing biological systems, *Ann. Nucl. Energy* **62** (2013) 626–633.
[22] V.I. Vysotskii and A.A. Kornilova, Microbial transmutation of Cs-137 and LENR in growing biological systems, *Current Sci.* **108** (4) (2015) 636–640.
[23] A.A. Kornilova, V.I. Vysotskii, N.N. Sysoev and A.V. Desyatov, Generation of intense X-rays during ejection of a fast water jet from a metal channel to atmosphere, *J. Surface Invest. X-ray, Synchrotron and Neutron Techniques* **4** (6) (2010) 1008–1017.
[24] V.I. Vysotskii, V.P. Bugrov, A.A. Kornilova, R.N. Kuzmin and S.I. Reyman, The problem of gamma-laser and controlling of Mössbauer nuclei decay (theory and practice), *Hyperfine Interaction* **107** (1997) 277–282.
[25] S.V. Adamenko and V.I. Vysotskii, Neutronization and protonization of nuclei – two possible ways of the evolution of astrophysical objects and the laboratory electron-nucleus collapse, *Found. Phys. Lett.* **19** (1) (2006) 21–36.
[26] A.V. Agafonov, A.V. Bagulya, O.D. Dalkarov, M.A. Negodaev, A.V. Oginov, A.S. Rusetskiy, V.A. Ryabov and K.V. Shpakov, Observation of neutron bursts produced by laboratory high-voltage atmospheric discharge, *Phys. Rew. Lett.* **111** (2013) 115003.
[27] A.V. Gurevich, V.P. Antonova, A.P. Chubenko, A.N. Karashtin, G.G. Mitko, M.O. Ptitsyn, V.A. Ryabov, A.L. Shepetov, Yu. V. Shlyugaev, V.I. Idanova and K.P. Zybin, Strong flux of low-energy neutrons produced by thunderstorms, *Phys. Rev. Lett.* **108** (2012) 125001.
[28] B. Zh. Zalikhanov, From an electron avalanche to the lightning discharge, *Phys. Part. Nucl.* **47** (2016) 108–133.

J. Condensed Matter Nucl. Sci. 29 (2019) 368–375

Research Article

Generation and Detection of Undamped Temperature Waves at Large Distance in LENR Related Experiments

Vladimir Vysotskii*

Kiev National Shevchenko University, Ukraine

Alla Kornilova, Timothy Krit and Sergey Gaydamaka

Moscow State University, Russia

Abstract

In this paper we describe the process of detecting undamped (self-channeled) high-frequency heat waves generated and propagated in air from the cavitation of a water jet in a closed chamber, and the results of action of these waves on low energy nuclear fusion in a remote deuterated polycrystalline titanium sample with grain sizes of not more than 50 microns. These waves are formed on the reverse side of the metal target, which is affected by the jet of water in a state of cavitation, and are characterized by strictly defined frequencies (in air under normal conditions and different humidity, the minimum frequency of such a wave is equal to MHz) [1–7] Such waves can propagate in air for a long distance (in the laboratory - more than 2 meters and this distance was limited only by the size of our laboratory). Under the influence of such waves to the remote target, effective quasicontinuous nuclear dd-fusion with a concomitant generation of alpha-particles process takes place.

Keywords: Cavitation, Distant dd–fusion, LENR, Temperature waves, Thermal relaxation

1. Generation and Detection Of Undamped Temperature Waves At Large Distance

During water cavitation experiments simultaneously with the detection of X-rays we have detected previously unknown undamped thermal waves in air at a great distance.

The photograph of the experimental setup, its schematic and the view of signals registered in air by broadband acoustic piezoceramic detector at different distances L from the outer surface of the target made of tungsten [8] are presented in Fig. 1.

These unknown waves in the air beyond the metallic target were recorded by a wide-band acoustic receiver – a TsTS-19 (VA-500) piezoelectric 20 mm in diameter with a resonance frequency of $\omega_{\text{res}} = 1$ MHz. The receiver was moved along the facility axis in the distance range from 5 mm up to 198 cm from the outer surface of the metallic

*Corresponding author. E-mail: vivysotskii@gmail.com.

© 2019 ISCMNS. All rights reserved. ISSN 2227-3123

target. Measurements were conducted with steps of 1 cm. The receiver during reference measurement was turned relative to the facility axis at angles of 0°C, 20°C, 90°C, and 180°C. The experiments were carried out with the use of a tungsten target (or in some experiments, molybdenum).

In these experiments, we detected with confidence at all distances a frequency-ordered system of quasimonochromatic signals in the frequency interval $\omega \approx 75 - 900$ MHz and more! It is very important to note that the amplitude of these waves is relatively independent of their frequency over a very wide range.

On the other hand, the detection of an acoustic signal even with minimal frequency $\omega \approx 75$ MHZ at distances $L = 1 - 200$ cm from the back side of the target is the paradox that cannot be explained by standard acoustics.

It is easy to see that these are not standard acoustic waves of high frequency. The standard formula for absorption (dissipation) of ultrasound

$$\delta(T) = \frac{\omega^2}{2\rho\{c(T)\}^3}\left[\left(\frac{4}{3}\eta + \xi\right) + \lambda(T)\left(\frac{1}{C_V} - \frac{1}{C_p}\right)\right] \quad (1)$$

has shown that at such high frequency the coefficient of ultrasound absorption in the air is very large ($\delta \geq 10^4 \text{cm}^{-1}$) and the mean free path $\bar{l} \leq 1$–$2\ \mu$m is very small! Here $c(T) \approx (331.3 + 1.21T^0) \times 10^2$ cm/s is the velocity of acoustic waves, ρ the density of air, $c_p = 1000$ (J/kg K) and $c_v = 717$ (J/kg K) is the specific heat of air at room temperature and normal pressure at constant pressure and volume, $\eta = 1.9 \times 10^{-5}$ (Pa s) and $\xi = 17.2 \times 10^{-6}$ (Pa s) – coefficients of shear and bulk viscosity of air, temperature in °C.

This result becomes still more surprising when we take into account the relative low resonance frequency of the used broadband acoustic detector $\omega_{res} \approx 1$ MHz, which is 75 times less than the frequency of the recorded signal! This condition is evidence of a much lower recording efficiency, which is proportional to the Q factor of acoustic resonance in the detector. From this it is evident that the real amplitude of the signal in the distant acoustic detector was very great.

This paradox can be resolved if we assume that it is not usual acoustic (ultrasound) waves but the special type of thermal (temperature) waves - undamped thermal waves [1–7]

The existence of undamped temperature waves that can propagate without dissipation in environments with small time τ of local temperature relaxation was theoretically predicted for the first time in [1–3]. It was shown that such waves have a minimal characteristic eigen frequency $\omega_{opt} \approx \pi/2\tau$ and can be excited in an environment under the influence of short heat pulses with duration ($\Delta t < \tau$). This relaxation time (the time of thermalization) differs significantly for different media.

In a plasma the relaxation time to the equilibrium (Maxwellian) distribution within the electronic subsystem in a small space region is equal to

$$\tau^{ee} \approx \sqrt{m_e}(k_B T_e)^{3/2}/4\pi\Lambda n_e e^4, \quad (2)$$

where n_e, m_e, t_e, respectively, the electron concentration, electron mass and electron temperatures, $\Lambda \approx 15$ is the Coulomb logarithm. For the ionic subsystem in a plasma, the relaxation time is equal to

$$\tau^{ii} \approx \sqrt{m_i/m_e}\tau^{ee}. \quad (3)$$

A typical relaxation time for metals and semiconductors is determined by relaxation of degenerate electron gas and corresponds to $\tau \approx 10^{-14} - 10^{-12}$ s [1–3].

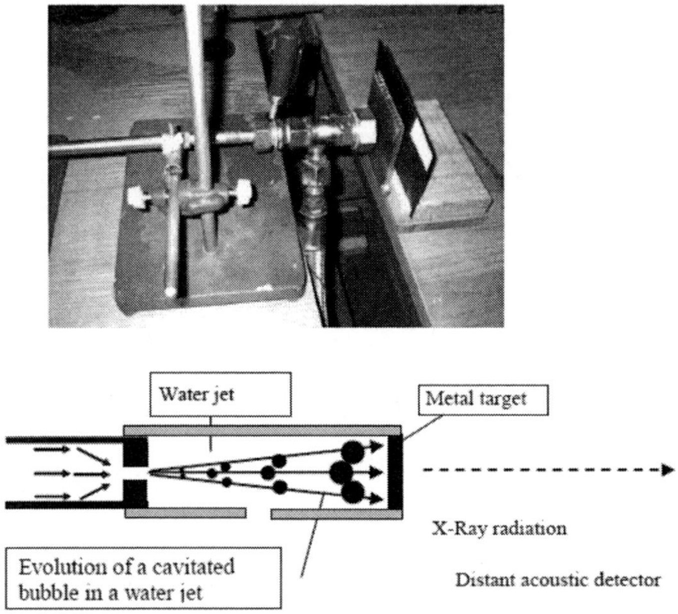

(a)

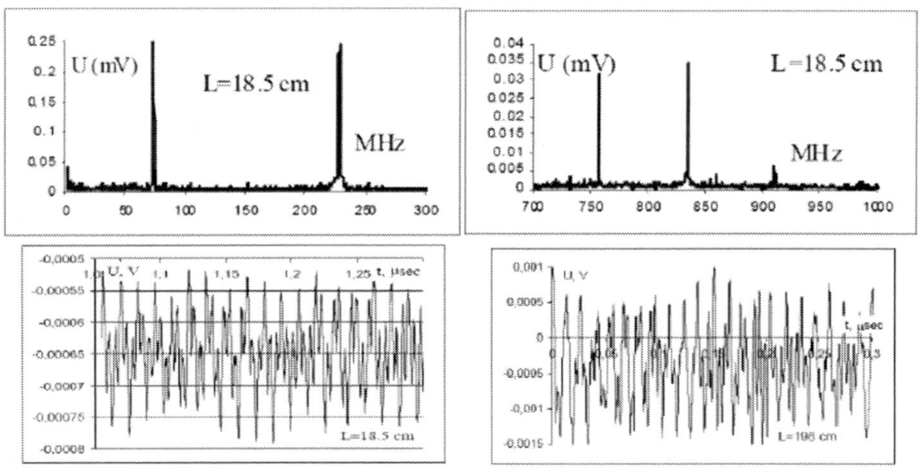

(b)

Figure 1. General view and scheme of experimental setup and the view of signals (frequency spectrum and structure of time dependence) registered in air by broadband acoustic piezoceramic detector at distances $L = 18.5$ cm and $L = 198$ cm from the outer surface of the tungsten target.

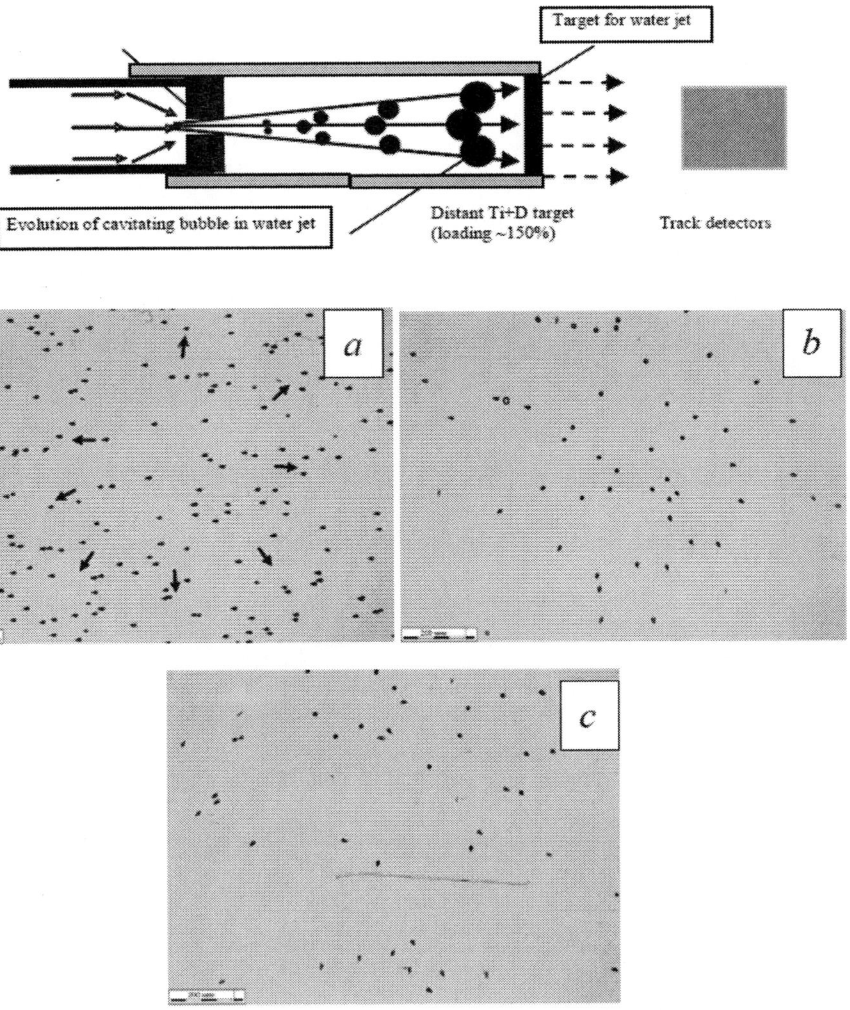

Figure 2. Experimental setup and microscopic track analysis of the spatial distribution of tracks and the direction of motion of alpha particles in samples subjected to the action of a thermal wave for 40 min when the track detector is located near the end and lateral (cylindrical) surfaces of the deuterated sample (Fig. 2a,b). (c) Corresponds to a control measurement using an alpha emitter based on a combination of three radionuclides (U^{233}, Pu^{238} and Pu^{239}).

In air and liquid, the duration of the maxivelization process $\tau \approx 10/n\langle\sigma(\nu)\nu\rangle$ is determined by the cross section $\sigma(\nu)$ of elastic scattering, the current velocity ν of the medium particles, taking into account their heating by the thermal wave and the concentration n of these particles. When the temperature and, especially, the density and composition of the air change (e.g., in the presence of water vapor), the value τ can vary over a wide range ($\tau = 10^{-7} - 10^{-8}$ s).

In the framework of the density matrix formalism, this process corresponds to the relaxation of diagonal and nondiagonal elements of the density matrix.

If we take into account the finite (non-zero) time of local thermodynamic relaxation, then the heat transfer equation with delay takes the form [1–3]

$$\frac{\partial T(x,t+\tau)}{\partial t} = G\frac{\partial^2 T(x,t)}{\partial x^2}, \tag{4}$$

where G is the coefficient of thermal diffusivity. The solution of this equation with small time τ of local temperature relaxation is superposition of colliding thermal waves [1–3]

$$T(\omega,x,t) = A_\omega \exp\left(-\kappa\frac{\cos\,\omega\tau}{\sqrt{1+\sin\,\omega\tau}}x\right) e^{\{i(\omega\tau-\kappa\sqrt{1+\sin\,\omega\tau})\,x\}} + B_\omega \exp\left(\kappa\frac{\cos\,\omega\tau}{\sqrt{1+\sin\,\omega\tau}}x\right) e^{\{i(\omega\tau+\kappa\sqrt{1+\sin\,\omega\tau})\,x\}}, \tag{5}$$

$$\cos\omega\tau \geq 0, \tag{6}$$

which fundamentally differ from solution of standard' heat equation without this relaxation (delay)

$$T(\omega,x,t) = A_\omega e^{-\kappa x} e^{i(\omega t-\kappa x)} + B_\omega e^{\kappa x} e^{i(\omega t+\kappa x)} \tag{7}$$

which determines the thermal waves with very strong damping coefficient $\delta \equiv \kappa = \sqrt{\omega/2G}$. For typical air parameters ($G \approx 0.15$ cm^2/s), the damping coefficient of such a wave at $\omega \approx 1$ MHz is equal $\delta \approx 1700\,\text{cm}^{-1}$. When the frequency is increased to the experimental value $\omega \approx 75$ MHz, this coefficient increases to a very large value $\delta \approx 2 \times 10^4\ \text{cm}^{-1}$, for which the mean free path does not exceed $\bar{l} = 1/\delta \approx 0.5$ m km!

For a system with relaxation the damping coefficient and phase velocities of colliding waves

$$\delta = \kappa\cos\,\omega\tau/\sqrt{1+\sin\,\omega\tau}, \quad \nu_{\text{p}} = \pm\sqrt{2G\omega/1+\sin\,\omega\tau} \tag{8}$$

depend on the thermal diffusivity G, time delay τ and frequency ω of the wave [1–3]. Waves with the frequencies $\omega_{\text{opt}(n)} = (n+1/2)\pi/\tau$ corresponding to the conditions $\cos\omega_{\text{optr}}\tau = 0$, $|\sin\omega_n\tau| = 1$ are characterized by the real wavenumbers $k \equiv \kappa\sqrt{\omega/2G}$ and a complete lack of damping, which corresponds to $\delta \equiv 0$.

2. Stimulation of Effective Fusion Under the Action of Undamped Heat Waves on a Remote Target

We have conducted many experiments studying the action of undamped thermal waves on a distant target made of deuterated polycrystalline titanium (loading about 150%). The diameter and length of the cylindrical target were about 1 cm. It was shown for the first time that the action of such waves on the remote target leads to the generation of alpha particles in LENR reaction of dd-fusion. To conduct the alpha-track analysis, a plastic detector of polycarbonate (polyallyl diglycol) type CR-39 with a density of 1.3 g/cm^3 was used. The thickness of the TASTRAK® detector (Track Analysis Systems Ltd., Bristol, UK) was 1 mm. During the experiments, the detector was placed at a distance of 5 mm behind the back surface of the sample (relative to the source of the thermal wave).

Experimental setup and photos of the fragments of track detectors after its exposure for 40 min near the nuclear target are shown in Fig. 2 (a,b) – LENR experiments with the action of thermal waves, 2(c) – control experiment with similar track detector and laboratory alpha-source on the base U^{233}, Pu^{238} and Pu^{239} of and alpha-active isotopes [8]). The experiments were carried out when the track detectors were located near (~5 mm) the end (2a) and lateral (2b) surfaces of the cylindrical target. The thermal wave acted on the front (opposite to the position of the detector) target surface.

It can be seen from Fig. 2a that the trajectory of motion of the detected particles was characterized by a central symmetry, which agrees well with the assumption of an axially symmetric expansion of the products of the nuclear reaction (the axial direction of emission of these particles corresponds to the geometry and orientation of the target).

3. Conclusion

In our opinion the possible mechanism for low energy reaction optimizing and the course of this reaction is associated with the formation of coherent correlated states [9–23] of deuterons in nonstationary microcracks (formed during the loading and migration of deuterium in the matrix of titanium or existing between individual grains of a polycrystalline titanium target) that change (e.g., when compressed and then decompressed) under the action of shock waves generated by action of thermal wave. In the process of such deformation, the coherent correlated states of the protons present in the volume of such nanocracks are formed.

When the deuteron is localized in an interatomic space typical for condensed media with a period $a = 1.5$ Å, the energy fluctuation in the coherent correlated state exceeds the value [11–23]

$$\delta E^{(\min)} \approx G^2\hbar^2/2Ma^2 \approx 50 - 100 \text{ keV} \tag{9}$$

that even at this lower threshold is much more that the temperature which is planned for future TOKAMAKS. Here $G = 1/\sqrt{1-r^2} \approx 10^4$ is the realistic coefficient of correlation efficiency, r – coefficient of correlation [11–23]. It should be noted that the realistic amplitude of this energy fluctuation can significantly exceed this minimal value.

Another method, which makes it possible to determine the increase in the probability of a nuclear reaction in the presence of such correlation, consists in modifying

$$D_{G\gg 1}(E) \approx \{D_{G=1}(e)\}^{1/G}, \tag{10}$$

the standard expression $D_{G=1}(E)$ for the Coulomb barrier transparency coefficient. This substitution was confirmed by comparing it with the result of an exact quantum-mechanical calculation [11,15,20] In this case, the effectiveness of LENR in the presence of correlation follows from a simple estimation. If the barrier transparency at low energy in usual noncorrelated state is very small (e.g. $D_{G\gg 1}(E)) \approx 10^{-100}$ then in the realistic case of a correlated state with $G = (2-10) \times 10^3$ it increases to a large value $D_{G=2500}(E) \approx 0.01 - 0.1$, which provides very effective nuclear fusion. What type of nuclear reaction has been observed in these experiments is a very important question, as well as what type of particles have been detected by track detectors.

It is well known that for a high deuteron energy a large and approximately the same probabilities has two reactions

$$\mathrm{d+d = p + t} + 4.03 \text{ MeV}, \tag{11a}$$

$$\mathrm{d+d = n + He^3} + 3.27 \text{ MeV}. \tag{11b}$$

The cross section for these reactions at high energy is about 0.09 b. The third possible reaction

$$\mathrm{d} + \mathrm{d} = \mathrm{He}^4 + 23.8\ \mathrm{MeV}. \tag{11c}$$

at a high energy of interacting particles has a very small probability (its cross section is 10^{-26}b). A fundamentally different situation can occur in the case of formation of a correlated state of interacting particles.

In such a correlated state, the probability of reaction (11c) due to the specific use [15–18] of virtual energy (9) may exceed the probability of standard reactions (11a) and (11b). This is mainly due to the fact that in the case of the coherent correlated state, a very specific situation takes place – very large fluctuations of energy (9) can exist for a long time $\delta t \geq G\hbar/2\delta E^{(\min)}$, which opens the way to realization LENR using large virtual energy (9). This is the direct results of the modified uncertainty relation – the relation of the Schrödinger–Robertson [9–22]

$$\delta E \delta t \geq G\hbar/2. \tag{12}$$

We note that for the case of uncorrelated states with $G = 1$, such a process is impossible because of the very short lifetime of this fluctuation. For this reason, fast reactions whose duration satisfies the condition become quite real. Among such reactions, there may be a reaction (11c).

It is also important to note that the parameters and main characteristics of the alpha tracks in the experimental (2a,b) and control (2c) detectors were almost identical, which is a good confirmation of the course of the reaction with the generation of alpha particles.

These investigations will be continued, but even at this stage it is obvious that the method of such remote stimulation of nuclear fusion opens up new opportunities and prospects for controlled nuclear fusion at low energy.

References

[1] A.O. Vasilenko, V.I. Vysotskii and V.B. Vassilenko, Heat transfer equation with delay for media with thermal memory, *Int. J. Sci.: Basic and Appl. Res. (IJSBAR)* **12** (1) (2013) 160–166.

[2] V.I. Vysotskii, A.O. Vasilenko and V.B. Vassilenko, Generation and propagation of undamped temperature waves under pulse action on a target surface, *J. Surface Investigation* **8** (2) (2014) 367–373.

[3] A.O. Vasilenko, V.I. Vysotskii, V.B. Vassilenko and M.V. Vysotskyy, Nonequilibrium thermal effects during pulsed action on conducting medium, *Inorganic Materials: Appl. Res.* **6** (3) (2015) 199–204.

[4] V. I. Vysotskii, A. A. Kornilova, A. O. Vasilenko and V.I.Tomak, Detection and investigation of undamped temperature waves excitation under water jet cavitation, *J. Surface Investigation* **8** (6) (2014) 1186–1192.

[5] V.I. Vysotskii, A. A. Kornilova and A.O. Vasilenko, Observation and investigation of X-ray and thermal effects at cavitation, *Current Sci.* **108** (4) (2015) 114–119.

[6] V.I. Vysotskii, A.A. Kornilova, A.O. Vasilenko, T.B. Krit and M.V. Vysotskyy, The prediction, observation and study of long-distant undamped thermal waves generated in pulse radiative processes, *Nucl. Instr. Method. Phys. Res.* B **402** (2017) 251–255.

[7] V.I. Vysotskii, A.A. Kornilova, T.B. Krit and M.V. Vysotskyy, On the long-range detection and study of undamped directed temperature waves generated during the interaction between a cavitating water jet and targets, *Surface* **11** (4) (2017) 749–755.

[8] A.A. Kornilova, V.I. Vysotskii, Y.A. Sapozhnikov, I.E. Vlasova, S.N. Gaidamaka, A.A. Novakova, V.M. Avdyukhina, I.S. Levin, M.V. Vysotskyy, E.I. Hait and N.H. Volkova, The problem and realization of stable generation of alpha particles by deuterated titanium, located in the thermal wave field, *Eng. Phys.* **5** (2018) 13–22 (in Russian).

[9] E. Schrodinger, *Ber. Kgl. Akad. Wiss.*, Berlin **S24** (1930) 296.

[10] H.P. Robertson, *Phys. Rev.* A **35** (1930) 667.

[11] V.I. Vysotskii and S.V. Adamenko, Correlated states of interacting particles and problems of the Coulomb barrier transparency at low energies in nonstationary systems, *J. Tech. Phys.* **55** (5) (2010) 613–621.

[12] V.I. Vysotskii, M.V. Vysotskyy and S.V. Adamenko, Formation and application of correlated states in non-stationary systems at low energy of interacting particles, *J. Exp. Theoret. Phys. (JETP)* **114** (2) (2012) 243–252.

[13] V.I. Vysotskii, S.V. Adamenko and M.V. Vysotskyy, The formation of correlated states and the increase in barrier transparency at a low particle energy in nonstationary systems with damping and fluctuations, *J. Exp. Theoret. Phys. (JETP)* **115** (4) (2012) 551–566.

[14] V.I. Vysotskii, S.V. Adamenko and M.V. Vysotskyy, Subbarier interaction of channeling particles under the self-similar excitation correlated states in periodically deformed crystal, *J. Surface Investigations* **6** (2) (2012) 369–374.

[15] V.I. Vysotskii and M.V. Vysotskyy, Coherent correlated states and low-energy nuclear reactions in non stationary systems, *Euro. Phys. J.* A **49** (8) (2013) 1–12.

[16] V.I. Vysotskii, S.V. Adamenko and M.V. Vysotskyy, Acceleration of low energy nuclear reactions by formation of correlated states of interacting particles in dynamical systems, *Ann. Nucl. Energy* **62** (2013) 618–625.

[17] V.V.I. Vysotskii and M.V. Vysotskyy, Correlated states and transparency of a barrier for low-energy particles at monotonic deformation of a potential well with dissipation and a stochastic force, *J. Exp. Theoret. Phys. (JETP)* **118** (4) (2014) 534–549.

[18] V.I. Vysotskii and M.V. Vysotskyy, Formation of correlated states and optimization of nuclear reactions for low-energy particles at nonresonant low-frequency modulation of a potential well, *J. Exp. Theoret. Phys. (JETP)* **120** (2) (2015) 246–256.

[19] V. I. Vysotskii, M.V. Vysotskyy, The formation of correlated states and optimization of the tunnel effect for low-energy particles under nonmonochromatic and pulsed action on a potential barrier, *J. Exp. Theoret. Phys. (JETP)* **121** (4) (2015) 559–571.

[20] V.I. Vysotskii and M.V. Vysotskyy, Coherent correlated states of interacting particles – the possible key to paradoxes and features of LENR, *Current Sci.* **108** (4) (2015) 524–530.

[21] V.I. Vysotskii and M.V. Vysotskyy, The formation of correlated states and tunneling at low energy at controlled pulse action on particles, *J. Exp. Theoret. Phys. (JETP)* **125** (2) (2017) 195–209.

[22] V.I. Vysotskii, M.V. Vysotskyy and S. Bartalucci, Features of the formation of coherent correlated states and nuclear synthesis in the interaction of slow particles with crystals and free molecules, *J. Exp. Theoret. Phys. (JETP)* **127** (3) (2018) 479–490.

[23] V.I. Vysotskii, The problem of creation a universal theory of LENR, *Infinite Energy* **18** (108) (2013) 30–35.

J. Condensed Matter Nucl. Sci. 29 (2019) 376–391

Research Article

Electron Quasi-particle Catalysis of Nuclear Reactions

Anthony Zuppero*

Tionesta Applied Research Corporation, Sequim, Washington, USA

Thomas J. Dolan†

Nuclear, Plasma, and Radiological Engineering Department, University of Illinois, Urbana, IL 61801, USA

Abstract

Our model applies solid state, nuclear, and quantum mechanics principles to the molecular chemistry process. We show how most of our predicted transmutation products are consistent with experimental data from a wide variety of LENR experiments, and how they can be triggered. Chemical physics recently discovered a new type of chemical reaction that concentrated most of the energy of reactants into electrons that were originally trapped between reactants. The reaction leaves the reaction product molecule relatively cool. Considering the rules of solid state physics, we apply the pattern of these chemical reactions to nuclear reactions, referred to as "Lattice Enabled Nuclear Reactions" (LENR, also called "Low Energy Nuclear Reactions"). The predicted nuclear energy release also concentrates the energy in electrons, energized inside the nuclear product. The nuclear products are predicted to be "cold", implying non-radioactive, ground state. In some cases the excitation energy of the compound nucleus is sufficient to fracture it into more isotopes that were not originally present. The total attraction energy (coulomb plus nuclear) can overcome the quantum kinetic energy repulsion of the squeezed electron quasi-particles (Heisenberg Uncertainty Principle) when the effective electron mass exceeds a threshold value. Our model is consistent with cold nuclear fusion reactions catalyzed by muons in isotopes of hydrogen (Alvarez, UC Berkeley, 1956). We identify mechanisms to create transient, sufficiently elevated effective mass electron quasi-particles. According to our model some of these reactions should produce highly energetic neutral helium or helium-3, but they are difficult to measure. Our model predicts no "cold fusion" of deuterium plus deuterium into helium, even though it predicts copious, energetic helium emission. We will describe the model principles and compare its predictions with data from various reactions. Our model also predicts that certain LENR reactions should transmute radioactive fission products into normal elements, neutralizing the radioactivity. Other possible applications include process heat, hydrogen production, direct generation of electricity, and space propulsion.

Keywords: Electron catalysis, Electron scattering, Kinetic energy of confinement, Transmutation, Vibrationally promoted electron emission

*Corresponding author. E-mail: azuppero@gmail.com.

†E-mail: dolantj@illinois.edu.

© 2019 ISCMNS. All rights reserved. ISSN 2227-3123

1. Chemical Physics Basis

Our reaction model is based on chemical physics research performed at University California –Berkeley and Santa Barbara. The energy of a fuel (f) and a reactant (R) binding together by molecular attraction is transferred directly to an electron (e^-) trapped between them (Fig. 1).

The reaction partitions the RF binding energy into electron kinetic energy, internal vibration of the Rf molecule (if any remains), and a small recoil energy of the molecule. This "Vibrationally Promoted Electron Emission" is a direct conversion of binding energy to electron kinetic energy [1,2].

Figure 2 shows only the binding energy, not the coulomb attraction energy. At the moment of ejection, the electron is born with both reactant binding and electron coulomb bonding energy. The electron expends the coulomb bonding energy while escaping the coulomb potential, and it carries away the reactant binding energy, minus the small product molecule recoil (and internal vibration energy, if any remains). Appendix A discusses the chemical physics reactions.

The electron satisfies the Heisenberg Uncertainty Principle (HUP) as it is confined in the potential well of width σ_x between the reactants R and f. Its repulsive momentum (also called "kinetic energy of confinement") counteracts the attractive potentials of molecular binding and coulomb attraction (discussed below). When the attractive potential is insufficient to overcome momentum repulsion, there is no transition, and a vibrationally excited molecule decays in familiar ways, such as by thermalizing with its environment or by vibrational de-excitation radiation.

When applying the model to the *nuclear* reactants, such as hydrogen and nickel, the coulomb attraction is insufficient to bring the reactants close enough for the nuclear potential to act, and the transition is forbidden. However, we will show that the momentum repulsion can be reduced if the effective electron quasi-particle mass can be temporarily increased. Then, when the reduced momentum repulsion is barely exceeded by the attractive potentials, R and f bind, and the excitation energy of the compound nucleus de-excites. The resultant nucleus would be in the ground state (or have a very long half-life).

After discussing the system Hamiltonian and kinetic energy of confinement we will derive the effective mass threshold for the electron quasi-particle to trigger catalysis. Then we will discuss how to transiently elevate some of the electron quasi-particle masses.

2. Hamiltonian and Quantum Kinetic Energy

When an electron is confined to a small volume, its momentum increases in accordance with the Heisenberg uncertainty principle, creating a back-pressure against the confining walls. This momentum repulsion is also called "kinetic energy of confinement (KEC)."

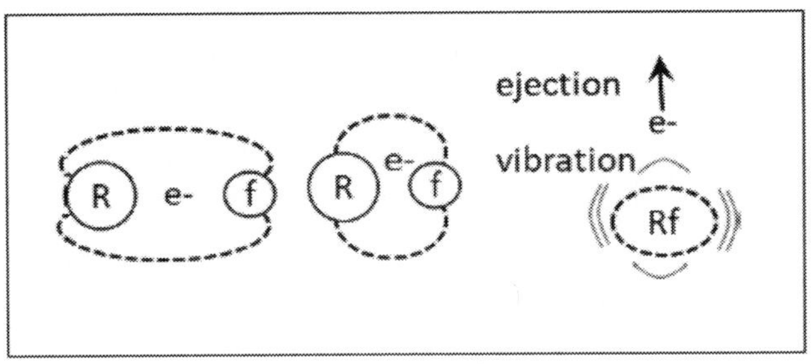

Figure 1. Molecular binding of f and R ejects an electron, leaving the Rf molecule in a low vibrational state.

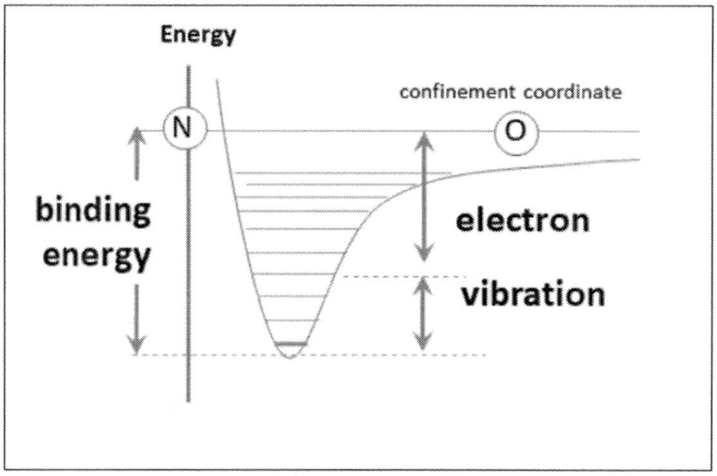

Figure 2. Potential energy of nitrogen and oxygen molecular system vs. "confinement coordinate" (representing molecular separation distance).

The one-dimensional Hamiltonian may be written

$$H = T_\mathrm{i} + T_\mathrm{e} + V_\mathrm{nuc} + V_\mathrm{e}, \tag{1}$$

where V_e is the three-body Coulomb potential, V_nuc an independent nuclear binding energy, T_i the kinetic energy of the ions, and T_e is the kinetic energy of the electron. T_e becomes very high when the electron is squeezed between narrow boundaries. V_e is relatively weak ($\sim$ eV) with a chemical range of order 100 pm. The nuclear binding potential V_nuc is relatively strong (MeV), with a very short range (several fm). The ion chemical turning points occur where the attractive term $V_\mathrm{e} + V_\mathrm{nuc}$ is balanced by the repulsive momentum from kinetic energy of confinement, bringing $T_\mathrm{i} = 0$. A nuclear reaction is possible only if the attractive potential energy is greater than the repulsive KEC:

$$|V_\mathrm{e} + V_\mathrm{nuc}| > T_\mathrm{e} \qquad \text{at } x = x_\mathrm{n}, \tag{2}$$

where x_n is the nuclear force radius.

We can estimate the kinetic energy of the electron from the Robertson–Schrödinger relation (modern form of HUP), which relates electron quasi-particle momentum p to its spatial confinement dimension. The confinement dimension σ_x is approximately the ion separation distance x:

$$\sigma_x^2 \sigma_\mathrm{p}^2 = (\hbar/2)^2 K(n), \tag{3}$$

where the variances are

$$\sigma_x^2 = \langle x^2 \rangle - \langle x \rangle^2, \qquad \sigma_\mathrm{p}^2 = \langle p^2 \rangle - \langle p \rangle^2$$

,

and $\hbar$ is the reduced Planck constant. And for ground states K (ground state) ~ 1, for a quantum harmonic oscillator stationary state $K(n) = (2n+1)^2,\ n = 0, 1, \ldots$; for a Gaussian initial condition or for a coherent state $K = 1$; for a particle in a box $K(n) = (n^2\pi^2/3 - 2),\ n = 1, 2, \ldots$

We assume a Born–Oppenheimer approximation where the electron motion accommodates fast compared to ion motions. This implies an oscillatory motion with $\langle p \rangle = 0$, and $\sigma_{\mathrm{p}}^2 = \langle p^2 \rangle$. The expected value of the electron KEC is therefore

$$\langle T_{\mathrm{e}} \rangle = \langle p^2/2m \rangle = \hbar^2 K(n)/8m\sigma_x^2. \tag{4}$$

If m is heavier, then repulsive $\langle T_{\mathrm{e}} \rangle$ is reduced, and the ion turning point occurs at a smaller value of x. Combining Eq. (4) with Eqs. (1) and (2), and solving for the threshold electron quasi-particle mass m_{th}, we find

$$m_{\mathrm{th}} = \hbar K(n)/8\sigma_x^2 |V_{\mathrm{e}} + V_{\mathrm{nuc}}| \qquad \text{at } x = x_{\mathrm{n}}. \tag{5}$$

A trapped electron needs to acquire m_{th} in the chemical molecule formed by (f e^- R) or by (p muon d). We ask: "What is the probability density of the chemical electron quasi-particle at a dimension just inside the nuclear force region?" This is the procedure used in K-capture physics, where the normal electron has a density inside a normal nucleus. When energy can be lowered by a proton in the nucleus absorbing the electron to form a neutron, a new nucleus appears, and neutrinos are emitted. The absorbed electron is a K shell electron.

In our LENR case we use the same procedure as for K-capture analysis, but the fractional electron penetration is from the chemical molecule instead of from a K shell electron. The tri-body Ve term is analyzed elsewhere. [3].

When an electron's mass $m < m_{\mathrm{th}}$, its probability density in the nuclear region is only imaginary and no reaction is possible. When $m > m_{\mathrm{th}}$ a finite real component of the probability density is possible by tunneling. We estimate the tunneling probability for several example LENR cases. When we plot the total energy as a function of confinement parameter for different values of effective mass we can visualize how $m > m_{\mathrm{th}}$ can trigger a reaction (Fig. 3).

When the effective mass exceeds the threshold (bottom curve Fig. 3) the relative position of the ground state of the nuclear system lowers to below the zero of energy and matches the energy level of the chemical state. Figure 4 shows the resonant electron tunneling can then occur to the nuclear force region between reactants. Upon tunneling, the reactants transfer their kinetic energy to the electron when they collide with the electron. The electron scatters, leaving the reactants together with only the energy of the ground state. This means the reactants bind in the ground state and the energy is transferred to the electron. The electron catalyzed the nuclear reaction.

Note that the electron need not be confined at nuclear dimension. The total energy allows the electron to appear as if it collided with a moving wall that gave it the entire binding and bonding energy, the nuclear plus coulomb potential energy in the region of the nucleus. The result is a prompt transition to a bound nucleus in the ground state and an unbound electron. (The model is discussed further in Appendix B.)

We will discuss how to achieve m_{th}, calculate it for many reactions, and compare the model predictions with experimental data.

3. Increase of Electron Quasi-particle Mass

Instead of solving the Bloch equations for an electron quasi-particle in a crystallite, and then computing an effective mass, we use the quasi-particle shortcut: use solid state effective mass. The effective mass of a ballistic electron quasi-particle in a periodic crystal conductor is proportional to the inverse of curvature of its Band Structure diagram. The effective mass of an electron quasi-particle is

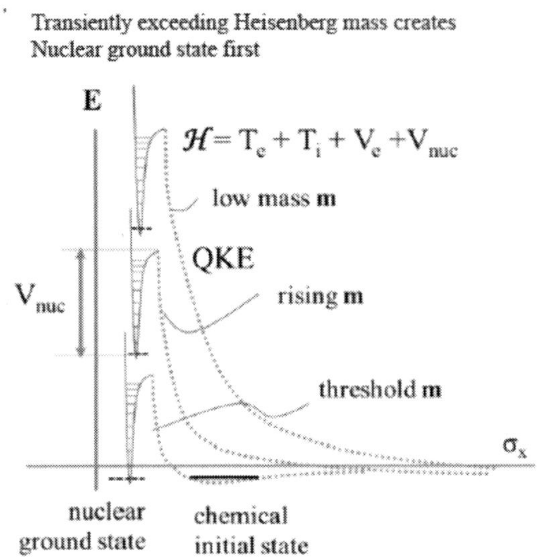

Figure 3. Total energy vs. confinement parameter for three different quasi-particle masses.

$$m = \hbar^2/(\partial^2 E/\partial k^2) = \hbar/\text{curvature}, \quad (6)$$

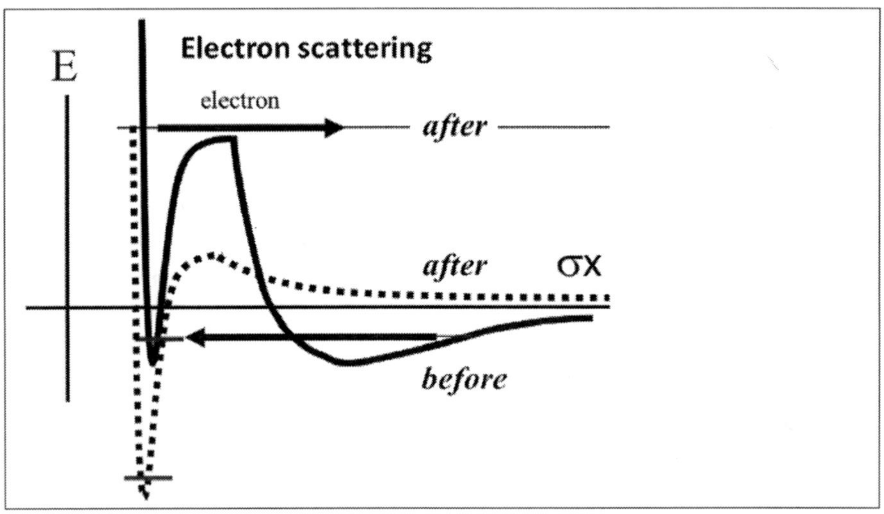

Figure 4. Potential energy transfer from binding and bonding potentials to electron quantum kinetic energy is stimulated the instant an electron tunnels to nuclear region.

where $\hbar$ is the reduced Planck constant [4].

The quasi-particle can only have energy and crystal momentum values on the diagram. Electrons can be placed in the region of an inflection point by simultaneous injection of particular values of crystal momentum (phonons) and electron energy into a crystallite region. One of many ways to inject a spread of crystal momentum values and energy values into a crystal's band structure is adsorbing or desorbing a nucleus, which may produce a spectrum of energies and momenta in the crystal, Fig. 5.

More inflection points are created in a crystal with facets. Therefore, crystals in the process of disintegrating or integrating create a splatter of inflection point targets. This means adsorbing or desorbing nuclei and disintegrating and integrating crystallites may place some fraction of the electrons sufficiently near an inflection point to usefully raise their effective masses, Fig. 6.

When we increase the effective mass, we also increase the apparent density of electrons in the chemical region.

4. Example Reactions

Whenever we see the same potential energy schematic or diagram, we suspect the solutions are similar. We notice that the proton–muon–deuteron picture is like that of the chemical research (Fig. 7). The negative muon brings p and d close enough to create the transition state where the nuclear force can bind them. The muon has a mass well above the threshold. It should immediately eject the muon, which is a transient heavy electron. The muon would carry away the binding energy leaving the merged product (pd), which is helium-3. This was observed [5–9].

Table 1 lists some other reactions of interest. Some nuclear reactions have the same f e R pattern, but with several fuel ions and electrons, such as

$$4f4e^{-}R \rightarrow [4fR] + \text{energy}$$

.

If electron quasi-particles could acquire sufficient mass, then most LENR transmutations could be understood.

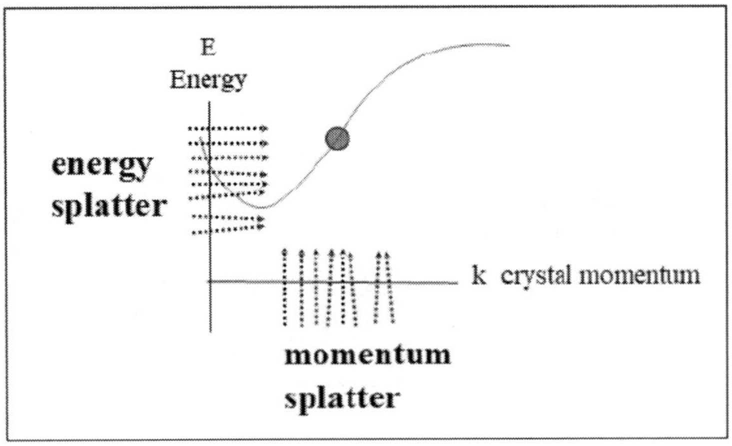

Figure 5. Splatter of energy and momentum can place some electrons near the inflection point.

Table 1. Selected chemical and nuclear tri-body reactions.

f	e^-	R	$\longrightarrow$	fR
H	e^-	Metal	$\longrightarrow$	HM
OH	e^-	Ag	$\longrightarrow$	OH-Ag
N	e^-	O	$\longrightarrow$	NO
O	e^-	CO	$\longrightarrow$	CO_2
p	μ^-	d	$\longrightarrow$	^3He
p	e^-	^{62}Ni	$\longrightarrow$	^{63}Cu
2p	$2e^-$	^{62}Ni	$\longrightarrow$	^{64}Zn
d	e^-	^{105}Pd	$\longrightarrow$	^{107}Ag
4d	$4e^-$	^{133}Cs	$\longrightarrow$	^{141}Pr

A revealing example uses only protons with a lattice reactant to produce helium and isotopes. The example suggests the process is not a fusion of protons or deuterons together, but instead a binding of protons or deuterons with a lattice reactant. For example, the model predicts that two protons will bind with ^{62}Ni or ^{64}Ni to generate Fe, Cu, Zn, and Co. Reference [8] quantified the relative amounts of the isotopes to the original Ni. Brillouin Energy (2015) reported the observation of helium. The 2H + ^{64}Ni + 2 heavy electrons with effective mass $\sim$ 35 electron rest masses would create excited state Zn with 13 MeV within the ^{64}Zn product nucleus and decay by one of several branches. Helium with 11.8 MeV and ^{62}Ni can be ejected, or two helium sharing 4.8 MeV and ^{58}Fe can be ejected. Other branches may emit Cu, Cr, Ti, and C and have been variously reported. Similarly, 2H + ^{62}Ni can produce ^{60}Ni + helium or ^{56}Fe + 2 He.

Helium production from proton–metal reactions as predicted by our 1D model appears to confirm the process is not fusion, cold or hot. It is instead a form of nuclear vibrationally promoted electron emission with the energy re-used inside the reacting nucleus.

During electrolysis of H and D with Pd Alexandrov observed both ^4He and ^3He [9]. This is consistent with our prediction of a catalytic reaction where H and D merge with Pd to make an intermediate Cd compound nucleus and containing two electrons with about 20 MeV total. This energy is enough that it could fracture the Cd into stable parts: Pd and ^3He. Effective mass threshold for H–D–Pd reactions is of order 12 – 14. In this reaction palladium is a

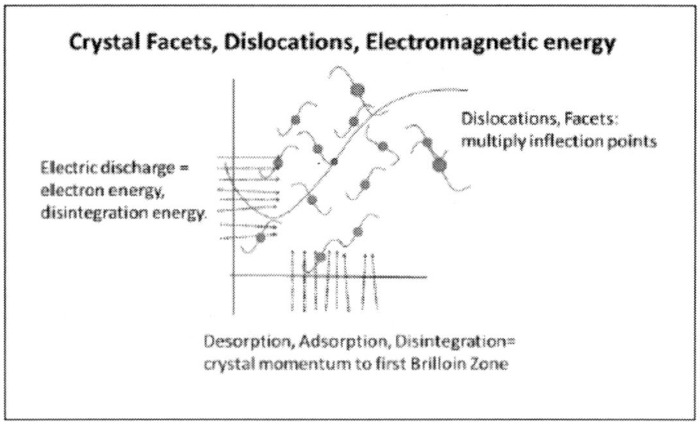

Figure 6. Multiple crystal facets introduce more and different band curves, with new inflection points.

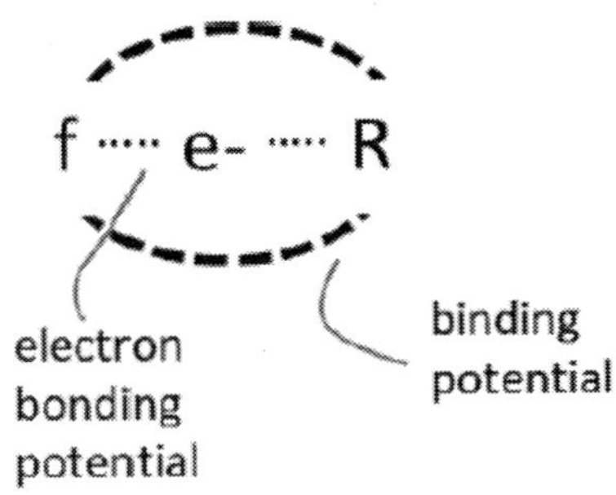

Figure 7. The molecular chemistry model and the muon catalysis model.

catalyst that is regenerated. The mixture also produced the expected $D + D + 2e^{-} + Pd \rightarrow (Cd^{*} + 2e^{-}) \rightarrow Pd + {}^{4}He$. Effective mass threshold for deuteron–Pd reactions are of order 10–13. Another example is $H + H + 2e^{-} + {}^{104}Pd \rightarrow (Cd^{*} + 2e^{-}) \rightarrow {}^{102}Pd + {}^{4}He$. Effective mass threshold for proton–Pd reactions are of order 22–29. Here, the Pd is a reactant and a different Pd isotope product.

5. Rules for Producing LENR in Crystallites

Rule 1: *The Mass–energy difference* between reactants and product must be exothermic. (The mass of the starting reactants must be greater than the mass of the products.)
Rule 2: *Inject a splatter of crystal momentum and energy* to raise the electron quasi-particle mass m above the threshold mass m_{th}. Expect either energetic electron emission or heat resulting from their thermalization.
Rule 3: *Ignore the normal electrons*, including the core electrons, around nuclei. Muon–nucleus particle physics observed that the heavy electrons may approximately ignore the normal and low mass electrons.
Rule 4: *Highest atomic number attracts heavy electrons* Muon–nucleus particle physics definitively observed that heavy electrons are strongly attracted to the nuclei with the highest number of protons. The isotopes with the highest proton number should either react with the heavy electrons or thermalize them without reaction.
Rule 5: *Evaluate possible emissions using "excitation energy"* that is available in the compound nucleus. The binding energy is transferred to multi-MeV electron quasi-particles inside the nucleus. Stable subunits of the nucleus, such as alpha particles, can be emitted if the excitation energy is above the alpha binding energy.
Rule 6: *Consider electron capture* of the injected heavy electrons when electrons have > 0.8 MeV. Expect this energy to escape the "excess heat" measurements, because the lowest mass absorbs the highest energy, neutrinos are almost zero mass and they rarely interact, so that neutrino emission is nearly undetectable.

We have considered many different LENR reactions, including:

- When deuterium (D) was reacted by electrolysis with natural palladium-105 (${}^{105}Pd$), silver-107 (${}^{107}Ag$) appeared [10].
- When cesium, barium, strontium, calcium and tungsten (Cs, Sr, Ba, Ca, and W) were separately reacted with deuterium using deuterium adsorption and desorption, the isotope products looked like nuclei with 2, 4 or 6 more deuterons had been added (${}^{141}Pr$, ${}^{96}Mo$, Sm, Ti, Os, and Pt) [11–13].
- Praseodymium (${}^{141}Pr$) appeared, as if cesium (${}^{133}Cs$) merged with four deuterium [12]. Iron (${}^{56}Fe$) appeared as if calcium (${}^{44}Ca$) absorbed six deuterium to make iron (${}^{56}Fe$), with intermediates titanium (${}^{44}Ti$) and

chrome (^{52}Cr) [12].
- When a titanium fuse was exploded in light water, Ni, Zn, Cu, and Fe appeared [14].
- In electrolysis of a mixture of heavy and light water, both ^{3}He and ^{4}He were observed [9,15].
- When dislocations, crystallite boundaries and imperfections were included, LENR reactions increased [16].
- When rubidium salt was included in a light water electrolysis, radioactive emissions consistent with the three-day half life of ^{87}Y were observed [17].
- When $LiAlH_4$ was included in a nickel reaction, all the nickel isotopes reacted with lithium and hydrogen fuels in combination, resulting in the expected Fe, Cu, Co and other isotopes. The Al also reacted with Li and H, generating Cl and Si. [8,18–20].

Our model predictions are consistent with the transmutation data from each of these experiments.

Two types of alpha emitting reactions should occur with about 1/10 of the isotopes in the periodic table. The hydrogen reaction is ^{A}X + 2H $\rightarrow ^{A-2}$X +^{4}He. The deuterium reaction is ^{A}X +2n*D $\rightarrow ^A$X +n*^{4}He, where n is an integer. These reactions can be almost invisible.

Many of the emitted alpha particles (He^{++}) may pick up electrons from the surrounding dense cloud of electrons (both heavy and light) and emerge as energetic ($\sim$10 MeV) neutral atoms. These energetic neutral atoms' (ENA) ranges should be much longer than the range of charged alpha particles, making them difficult to measure.

A few radioactive isotopes should be produced, even though the theory predicts that most the products are "born in the ground state." The hydrogen-nickel reactions (2H + Ni $\rightarrow$ isotopes + energy) seem to react with all the Ni isotopes. Some of the intermediate nickel isotope products should be "stable" on nanosecond time scales, but could be beta emitting and electron capture isotopes. Such products look like a "ground state" because of their stability (seconds to mega years) on the time scale of nuclear reactions (10^{-20} s).

These intermediates should be beta emitters or electron capture isotopes. One would not see the K-alpha radiation of the usual K-capture process because the externally supplied electron is already inside the nucleus and does not come from the K shell. One would not see much beta emission because electrons have such a short range in real materials. Many ejected electrons are thermalized in the electron cloud around the emitting nucleus during tens of fs, resulting in heat. However, Karabut and Saavatimova have reported isotopic evidence of the radioactive beta and electron capture isotopes in conducting targets of proton glow discharges with energies under about 3 keV [21–25]. Lipoglavsek apparently observed evidence for the expected 5.3 MeV electron from proton–electron–deuteron transmutations [26–28].

A third branch might explain a small fraction of observed isotopes: the production of neutrons from energetic electrons inside the nucleus colliding with protons, which would be a weak interaction inside the nucleus.

6. Reaction Branches

Three branches have been identified:

(1) transmutations to other stable isotopes emitting electron quasi-particles as energy,
(2) prompt fracturing to another branch, emitting energetic stable products (such as He), possibly charge neutral,
(3) prompt electron capture to stable products emitting neutrino energy.

7. Conclusions

Our tri-body catalysis model assumes that a particular crystal momentum and electron energy addition can transiently elevate some electron *quasi-particle masses* in a crystalline lattice containing reactants. The size of chemical entities is inversely proportional to the quasi-particle mass. When the effective mass exceeds a threshold value $m_{\rm th}$, the size of chemistry shrinks and the chemical separation distance can become small enough that electron tunneling into the region of the forces binding the reactants can apparently catalyze reactions.

The recently discovered *chemical physics* of Vibrationally Promoted Electron Emission (VPEE) reported by LaRue [1] provides the physics for a model to use nuclear binding potentials in place of chemical potentials to bind reactants together. Instead of chemical reactant binding, nuclear reactant bindings (transmutations) can be stimulated when the shrinking of chemical size exceeds a threshold. VPEE with elevated effective mass quasi-particles provide a model to characterize and predict the rates and products of hydrogen/reactant interactions in the solid state.

Alvarez [5] provided the first example of nuclear reactions stimulated by a shrinking of the size of chemical entities when he used a *mu meson* as the bonding electron in a hydrogen–heavy electron–reactant molecule. The result was similar to its chemical counterpart. A new nucleus can be formed from its reactants and nuclear binding energy is delivered to electrons inside the region of nuclear forces. The electron quasi-particles may eject and thermalize to heat within the lifetime of a ballistic electron in the lattice, with the transmuted product remaining in its ground state. Or the electron may deliver the energy to the product itself, causing partitioning and ejection of stable subnuclei, also transmutation products.

Reactions of *protons* binding with *nickel* isotopes should result in a form of fission of the zinc product into helium and transmutation nuclei. The helium and the isotopes have been observed. No deuterons need be involved, and no fusion, hot or cold, need be invoked.

Solid state *methods* to transiently *elevate* the electron effective mass are identified. In each case of a claimed transmutation stimulated by chemical means, one or more of the identified methods to transiently elevate electron quasi-particle mass appear to be activated.

Helium-3 and Helium-4 productions were predicted for mixed light and heavy hydrogen *electrolysis* with *palladium*. The intermediate product was a cadmium, energized with $\sim$20 MeV electrons inside the nuclear force region. The final products include the original palladium and helium-3 from H + D + Pd and helium-4 from D + D + Pd reactions, which Alexandrov reported.

When *multiple pairs* of deuterium were reacted with various isotopes including Cs, Sr, Ba, Ca, and W, the observed ground state isotope products looked like nuclei with 2, 4 or 6 more deuterons had been added (^{141}Pr, Mo, Sm, Ti, Os, and Pt). The reaction fits the heavy electron model for "*shrinking* of the size of chemistry" and of the VPEE process to bind the reactants into ground state product nuclei.

Reaction rates are based on an elevated effective mass, *evanescent* electron *wavefunction density* when a chemical electron with scatters from the nuclear region between the reactants. The elevated residence time of the heavy electron between the attracting reactants (which must include the electron to be attractive) allows the reactants to converge to the smaller size of the product nucleus. The electron is scattered when it collides with the converging reactant nuclei, and appears either as an ejected electron or as electron energy within the nuclear product, typically inducing fission-like emissions.

Appendix A. Relevant Chemical Physics Experiments

Chemical physics discovered a fundamentally new type of reaction during the 2000s. Figure 8 shows the energy of a fuel (f) and a reactant (R) binding together by molecular attraction is transferred directly to an electron (e^-) passing between them.

Key experiments at UC Berkeley and UC Santa Barbara demonstrated direct conversion of the chemical binding energy into the kinetic energy of a single electron. The observed reaction seemed to violate the Born–Oppenheimer Approximation. It is now accepted that vibrational energy can transfer directly into a single electron's kinetic energy.

Hellberg observed both 1–2 eV electrons and photons when "f" comprised chlorine molecules, Cl_2, which adsorbed on "R", a potassium K surface [29]. The merged "molecule," "fR," was Cl_2–K or Cl–K. The 1–2 eV electron received almost all the energy available. The Cl_2 on the K surface was only expected to lose energy by vibrational decay or photon emission, one quantum at a time, not all at once. Instead, the reaction ejected a single energetic electron taken from the K surface when the two came in contact. The two potentials attracting the fuel and reactant together include the electron coulomb bonding and the Cl_2-K attraction's binding potential [29].

Figure 9 shows how Nienhauss provided hydrogen radicals, H, to a 6 nm thin Ag silver surface [30]. The H was supposed to form a bond with the silver and slowly lose energy to thermal electrons one vibrational quantum at a time. Instead, an electron from the silver was ejected with energy above the Schottky barrier ($\sim$ 0.6 eV). The silver formed a Schottky barrier diode with the silicon underneath it. The electron energy must have been $\geq$ 0.7 eV, because Nienhauss measured a short circuit current across the diode. The expected energy was much lower, in the 1 meV range for phonons and thermal range (0.026 eV), for electrons [30].

As shown in Fig. 10, Huang reported energizing an NO molecule using a laser to stretch the molecule to nearly its breaking point [31]. A free N radical was placed just close enough to a free O radical for them to bind and become a highly vibrationally excited NO molecule. When the molecule contacted the gold vacuum chamber walls, the NO molecule suddenly lost most of its vibration energy to a single electron that jumped on the NO molecule from the gold, and then immediately ejected with "too much energy." The prevailing theory had expected that the energy went into the thermal bath of electrons [31].

Ji, Zuppero, Somorjai and Gidwani reacted carbon monoxide CO with an oxygen radical O on a 2 nm thin Pd catalyst on an n-TiO_2 semiconductor substrate (Fig. 11). Prof. Somorjai declared that they observed an "impossible"

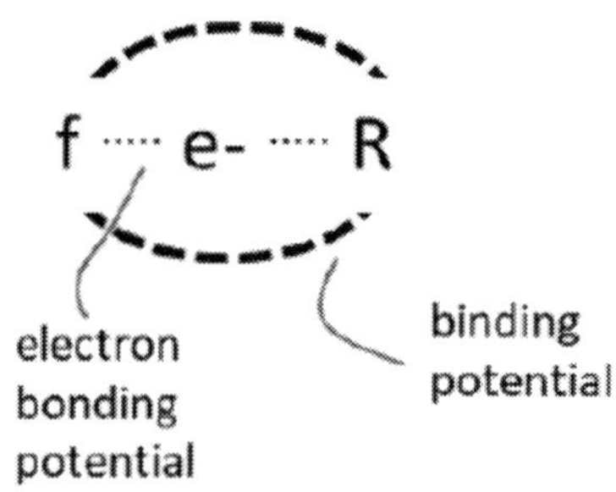

Figure 8. At the moment a sufficiently heavy electron forms a three-body with attracting reactants f and R that can bind, prompt binding may occur and the electron may be ejected with most of the energy.

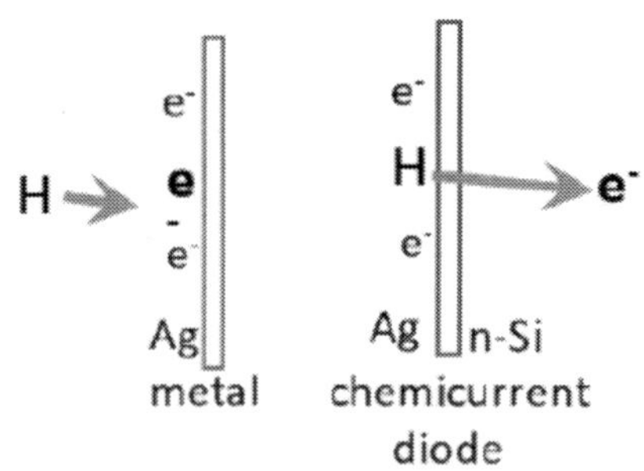

Figure 9. A fuel radical H adsorbs on a reactant "molecule", a conducting Ag metal substrate, energizing and ejecting an electron with the H-metal binding energy.

forward voltage across the diode of 0.68 eV, about half of the available $\sim$ 1.5 eV from the CO and O adsorbed on the catalyst. This confirmed that a single electron, not a thermal bath, received the energy [32].

LaRue and White reported direct observation of the electron energy (Fig. 12), confirming a partition of binding energy between an NO vibration and the electron energy. They confirmed that the energy was deposited into *a single electron*. The key observation was electrons with the entire binding energy, upsetting the prevailing theories. This was a new effect, which they called *"Vibrationally Promoted Electron Emission" (VPEE)* [1].

Appendix B. Effective Electron Quasi-particle Mass at Nuclear Dimensions

We apply our simple chemical binding model to the nuclear case for binding nuclei such as p and d together into helium-3. The trapped electron repulsive momentum (derived from "kinetic energy of confinement") increases to very high values when the confinement parameter σ_x (representing ion separation distance) is small, according to the HUR, so the ion separation remains at many picometers, while the nuclear force range x_n is only a few femtometers. If we could make some electron quasi-particles heavier, then their repulsive momentum would be reduced, and the ion

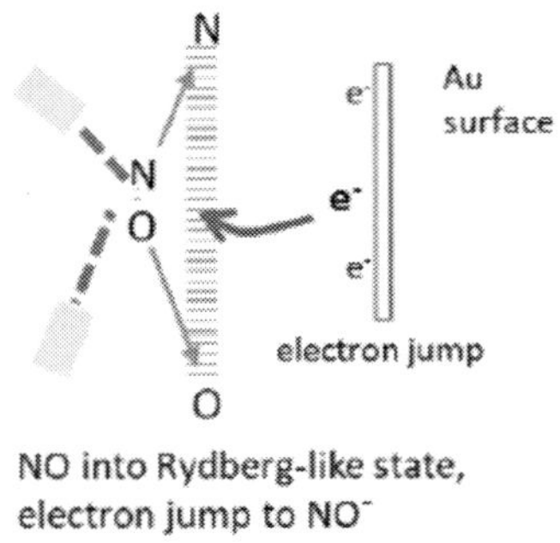

Figure 10. A laser excited an NO molecule to a high vibration state (nu 12) as it collided with an Au conducting chamber surface where an electron attached to the NO, was energized by the NO vibration energy, and the electron was promptly ejected with a major fraction of the vibration energy.

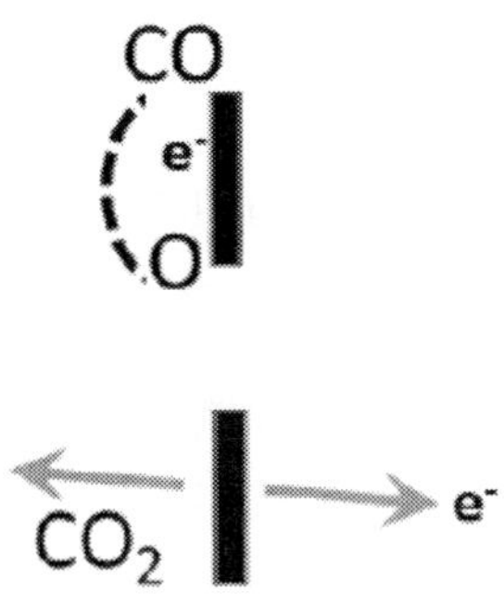

Figure 11. Fuels CO and O adsorbed on a thin catalyst surface formed CO2 gas, energizing an electron between them with sufficient energy to surmount the Schottky barrier between the catalyst and a semiconductor under the catalyst.

separation would decrease. The separation distance would still be far greater than x_n, but it might be possible for part of the quasi-particle wave function to tunnel into the x_n region, if the total attractive potential exceeded the repulsive momentum there. This condition defines the required heavy quasi-particle mass threshold $m_{\rm th}$(Eq. (5)).

The key issue is whether the effective electron mass can actually be increased and effective at nuclear dimensions. We hypothesize that the crystallite lattice might cause the electron wavefunction to have oscillations with relatively high slope and curvature, even if it does not form a bond between f and R. The transition state does not need a bond. We would need the slope and curvature of a wavefunction segment to be locally above the global variance limits of the Robertson–Schrődinger equation (Eq. (3)). The slope of the wavefunction is the momentum operator, $p = (\hbar/i)\mathrm{d}\psi/\mathrm{d}x$ and the quantum kinetic energy is the curvature, $E = (\hbar/i)^2(\mathrm{d}^2\psi/\mathrm{d}x^2)/2m$. Our concept would produce a wave function with elevated slopes and curvature like quantum well states in conducting layers like those studied by the groups of Chiang and of Matsuda [33,34]. Such electron quantum well states might impart momentum and energy as if electron effective mass in the region between the f and R were either reduced or elevated. Researchers studying electrons in few-atom quantum well monolayers theoretically predicted and experimentally observed a Fabry–Perot interferometer-like modulation of the electron wavefunction near boundaries of crystallites, before it damped out deeper into the crystal [33,34]. Each peak in the modulation could locally affect electron momentum.

Chiang notes that the crystal acts like a multiple-reflection periodic array, similar to the multiple reflections of light in a Fabry–Perot etalon. Chiang experimentally explored a one-dimensional equivalent of a 3D Fabry–Perot etalon formed by crystallite conducting atom layers. The electron wave function replaces light in the equations.

Multiple reflections of the delocalized electron quasi-particle wavefunction from the crystal ions are one way to look at the Bloch theorem. The Bloch equation for a crystallite is:

$$\Psi_m(kr) = \frac{1}{\sqrt{N}} \sum_n a_m(R_n, r)\, \mathrm{e}^{ik\cdot R_n}. \tag{B.1}$$

The R_n is the vector from electron to the nth ion, r is the vector electron position, k is the crystal momentum, the $\Psi_{\rm m}$ is the electron wavefunction. The form of the wave function is like that of a 3D Fabry–Perot etalon, with periodic spacing terms. Limiting the range of "n" to include only a region within a radius equal to the collision mean free path of the electron imposes a limit on the wavefunction, resulting in an envelope. The electrons outside the electron ballistic mean free path are assumed to be not interacting.

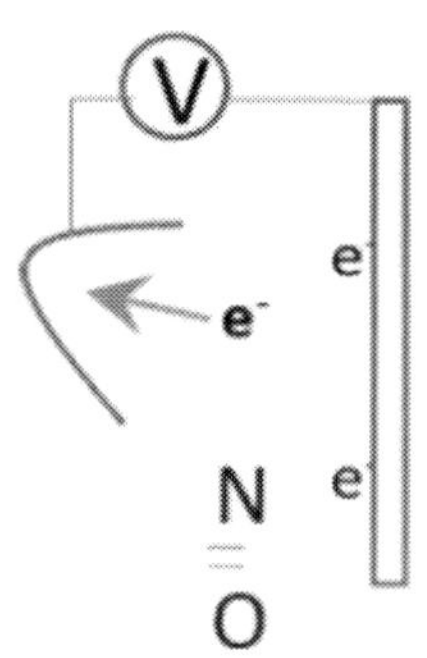

Figure 12. After a highly vibrationally excited NO molecule contacted a conducting surface and energized an electron, it was ejected into an electron detector with as much as the entire binding energy.

One purpose of the electron quantum well work in conducting films by Chiang, by Matsuda, and by Paggel [35] is to precisely define the band structure (energy vs. crystal momentum). The band structure diagram, which can constructed using the Bloch conditions, represents a solution of the wave equation, given the "non-interacting" condition. Thus, we believe that using the "effective mass" given by Eq. (6) is a valid shortcut to evaluating the momentum and energy transfer properties of a transient heavy electron in a crystallite. The $a_{\rm m}()$ terms in Eq. (A1) cause multiple peaks to develop within the envelope. The slope at each peak is steeper than for a normal electron, changing the momentum operator $p = \hbar/i{\rm d}\psi/{\rm d}x$. Each peak represents an electron quasi-particle density, but has a slope somewhat proportional to the number of reflections from periodic crystalline nuclei. The slope and curvature relate to the effective mass.

Chiang provides a guiding approximation for each peak: $\psi(z) \propto \sin(n\pi z/d)$, where n is an integer quantum number, z distance into the quantum well film and d is the film thickness. The associated momentum is $p \propto {\rm d}\psi/{\rm d}z \propto (n\pi/{\rm d})\cos(n\pi z/d)$ could be large in very thin films at high n. Chiang provides an example, 24 monolayer envelope with multiple peaks. The slope of Ψ at the peaks is roughly *n* times that of a normal electron (Chiang Eq. (3)).

The elevated effective mass electron is valid at a chemical dimension σ_x . The envelope determines the confinement dimension σ_x, which is always at chemical dimension. The electron quasi-particle cloud centroid is at chemical dimensions, but a small, evanescent portion may tunnel to nuclear dimensions x_n if the effective mass exceeds the threshold value $m_{\rm th}$. We have estimated $m_{\rm th}$ and the tunneling probabilities for several LENR cases [3]. When the electron effective mass exceeds $m_{\rm th}$, the system can undergo prompt collapse down to the size of the compound nucleus, ejecting one or more electron quasi-particles with high energies (~5 MeV per nucleon reacted).

Thus, we believe that the chemical physics model of vibrationally promoted electron emission may be adapted to heavy electron catalysis of LENR transmutations.

References

[1] J.L. LaRue, Tim Schaefer, Daniel Matsiev, Luis Velarde, N. Hendrik Nahler, Daniel J. Auerbach and Alec M. Wodtke, Electron kinetic energies from vibrationally promoted surface exoemission: evidence for a vibrational autodetachment mechanism, *J. Phys. Chem.* A **115** (2011) 14306–14314, dx.doi.org/10.1021/jp205868g.

[2] J.D. White, J. Chen, D. Matsiev, D.J. Auerbach and A.M. Wodtke, Vibrationally promoted electron emission from low work-function metal surfaces, *J. Chem. Phys.* **124** (6) (2006) 064702.

[3] A.C. Zuppero and T.J. Dolan, Heavy electron catalysis of nuclear reactions, *J. Condensed Matter Nucl. Sci.* (2018), submitted.

[4] Charles Kittel, *Introduction to Solid State Physics*, 8th Edn., Wiley, Hoboken, NJ, 2005, ISBN-0: 047141526X.

[5] L.W. Alvarez, H. Bradner, F.S. Crawford Jr., J.A. Crawford, T.P. Falk-Vairant, M.L. Good, J.D. Gow, A.H. Rosenfeld, F. Solmitz, M.L. Stevenson, H.K. Tichoy and R.D. Tripp, Catalysis of nuclear reactions by mu mesons, American Physical Society Meeting, Monterey, CA, Dec. 28, 1956, *Phys. Rev.* **105** (1957) 1127.

[6] J.D. Jackson, Catalysis of nuclear reactions between hydrogen isotopes by mu_minus mesons, *Phys. Rev.* **106** (1957) 330.

[7] E.J. Bleser, E.W. Anderson, L.M. Lederman, S.L. Meyer, J.L. Rosen, J.E. Rothberg and I.-T. Wang, Muonic molecules in liquid hydrogen, *Phy. Rev.* **132** (6) (1963) 2679–2691.

[8] Yu.N.J. Bazhutov, *J. Condensed Matter Nucl. Sci.* **13** (2014) 29–37.

[9] Dimiter Alexandrov, Nuclear fusion in solids – experiments and theory, Abstract, *ICCF-21*, Fort Collins, CO, June 3–8, 2018.

[10] Jean-Paul Biberian, Anomaloous isotopic composition of silver in a palladium electrode, Abstract, *ICCF-21*, Fort Collins, Colorado, June 3–8, 2018.

[11] Y. Iwamura, T. Itoh, M. Sakano, S. Sakai and S. Kuribayashi, Low energy nuclear transmutation in condensed matter induced by D_2 gas permeation through Pd complexes: correlation between deuterium flux and nuclear products, *Proc. 10th Int. Conf. on Cold Fusion*, 2003, Cambridge, MA, USA, Accessed at http://www.lenr-canr.org/acrobat/IwamuraYlowenergyn.pdf.

[12] Yasuhiro Iwamura, Takehiko Itoh, Y. Terada and T. Ishikawa, Transmutation reactions induced by deuterium permeation through nano-structured pd multilayer thin film, *Transactions of the American Nuclear Society*, Vol. 107, San Diego, CA, November 11–15, 2012.

[13] Yasuhiro Iwamura, Mitsuru Sakano and Takehiko Itoh, Elemental analysis of Pd complexes: effects of D_2 gas permeation, *Jpn. J. Appl. Phys.* **41** (2002) 4642–4650, Part 1, No. 7A, July 2002.

[14] L.I. Urutskoev, Review of experimental results on low-energy transformation of nucleus RECOM Inst. Kurchatov, Moscow, *Annales de la Fondation Louis de Broglie* **29** (3) (2004) 1149.

[15] Michael McKubre, Francis Tanzella, Paolo Tripodi and Peter Hagelstein, The emergence of a coherent explanation for anomalies observed in D/Pd and H/Pd system: evidence for 4He and ^{3}He production, in *8th Int. Conf. on Cold Fusion*, 2000, Lerici (La Spezia), Italy: Italian Physical Society, Bologna, Italy.

[16] Czerski Konrad, Influence of crystal lattice defects and the threshold resonance on the deuteron–deuteron reaction rates at room temperature, Abstract, *ICCF-21*, Fort Collins, Colorado, June 3–8, 2018.

[17] R. Bush and R. Eagleton, Evidence for electrolytically induced transmutation and radioactivity correlated with excess heat in electrolytic cells, with light water rubidium salt electrolytes, EPRI, *Proc. Fourth Int. Conf. on Cold Fusion*, 1994.

[18] Giuseppe Levi, Evelyn Foschi, Bo Höistad, Roland Pettersson, Lars Tegnér and Hanno Essén, Observation of abundant heat production from a reactor device and of isotopic changes in the fuel, Accessed at URL: http://lenr-canr.org/acrobat/LeviGobservatio.pdf. www.sifferkoll.se/sifferkoll/wp-content/uploads/2014/10/LuganoReportSubmit.pdf.

[19] K.P. Rajeev and D.J. Gaur, Evidence for nuclear transmutations in Ni–H electrolysis, *J. Condensed Matter Nucl. Sci.* **24** (2017) 278–283.

[20] G.H. Miley and J.A. Patterson, Nuclear transmutations in thin-film nickel coatings undergoing electrolysis, *J. New Energy*, **1**(3) (1996) 5, http://www.lenr-canr.org/acrobat/MileyGHnucleartra.pdf.

[21] I.B. Savvatimova, Transmutation of elements in low-energy glow discharge and the associated processes, *J. Condensed Matter Nucl. Sci.* **6** (2012) 181–198.

[22] I. Savvatimova, Y. Kucherov and A.B. Karabut, Cathode material change after deuterium glow discharge experiments, in *fourth Int. Conf. on cold fusion*, 1993, Lahaina, Maui: Electric Power Research Institute 3412 Hillview Ave., Palo Alto, CA 94304.

[23] I. Savvatimova, Decay of tungsten under low-energy deuterium discharge and creation of more light isotopes 1, creation of more light elements in tungsten irradiated by low energy deuterium ions, in *8th Int. Workshop on Anomalies in Hydrogen/Deuterium Loaded metals*, Catania, 13–18 October, 2007.

[24] I. Savvatimova, Gennady Savvatimov and Alla Kornilova, Decay of tungsten under low-energy deuterium discharge and creation of more light isotopes 2, gamma/x ray emission of tungsten cathodes before, during and after deuterium discharge, *8th Int. Workshop on Anomalies in Hydrogen/Deuterium Loaded Metals*, 13–18 October 2007, Sheraton Catania, Hotel and Conf. Center; Via Antonello da Messina 45, 95020 Cannizzaro(CT), Sicily, Italy.

[25] A.B. Karabut, *8th Int. Workshop on Anomalies in Hydrogen/Deuterium Loaded Metals*, 13–18 October 2007, Sheraton Catania, Hotel and Conf. Center, Via Antonello da Messina 45, 95020 Cannizzaro(CT), Sicily, Italy.

[26] M. Lipoglavšek, S. Markelj, M. Mihovilovîc, T. Petrovîc, S. Štajner, M. Vencelj and J. Vesîc, Observation of electron emission in the nuclear reaction between protons and deuterons, *Phys. Lett.* B 773 (2017) 553–556.

[27] A. Cvetinovic, M. Lipoglavsek, S. Markelj and J. Vesic, Molecular screening in nuclear reactions, *Phy. Rev.* C **92** (2015) 065801. .

[28] S.A. Rakityansky, S.A. Sofianos, L.L. Howell, M. Braun and V.B. Belyaev, Nonradiative proton-deuteron fusion in stellar plasma, *Nucl. Phys.* A **613** (1997) 132–146.

[29] L. Hellberg, J. Stromquist, B. Kasemo and B.I. Lundqvist, *Phys. Rev. Lett.* **74** (1995) 4742.

[30] H. Nienhaus, H.S. Bergh, B. Gergen, A. Majumdar, W.H. Weinberg and E.W. McFarland, Hole pair creation at Ag and Cu surfaces by adsorption of atomic hydrogen and deuterium, *Phy. Rev. Lett.* **82** (2)(1999) 446.

[31] Yuhui Huang, Charles T. Rettner, Daniel J. Auerbach and Alec M. Wodtke, Vibrational promotion of electron transfer, *Science* **290** (6) (2000) 211.

[32] Ji, Xiao Z. and Gabor A. Somorjai, Continuous hot electron generation in Pt/TiO_2, Pd/TiO_2, and Pt/GaN catalytic nanodiodes from oxidation of carbon monoxide, *J. Phys. Chem.* B **109** (2005) 22530–22535.

[33] T.-C. Chiang, *Surface Sci. Reports* **39** (2000) 181–235.

[34] Iwao Matsuda, Takehiro Tanikawa, Shuji Hasegawa, Han Woong Yeom, Kensuke Tono and Toshiaki Ohta, *Euro. J. Surface Sci. Nanotech.* **2** (2004) 169.
[35] J.J. Paggel, T. Miller and T.-C. Chiang, *Science* **283** (1999) 1709–1711.

J. Condensed Matter Nucl. Sci. 29 (2018) 392–400

Research Article

Calculation of the Boosted Spin–orbit Contribution to the Phonon–Nuclear Coupling Matrix Element for ^{181}Ta

Peter L. Hagelstein*

Massachusetts Institute of Technology, Cambridge, MA, USA

Abstract

Modern deformed shell model single–proton wave functions are used to evaluate the boosted spin–orbit contribution to the phonon–nuclear interaction for the 6.237 keV transition in ^{181}Ta. This is the lowest energy E1 transition from the ground state among the stable nuclei, and is important in connection with proposed excitation transfer and up-conversion experiments. The value resulting for the magnitude $|\langle 9/2^-|\hat{a}|7/2^+\rangle|$ is 1.3×10^{-6}. The wave functions are also used to evaluate the radiative decay rate which is found to be about 2×10^5 higher than the experimental value, which is strongly hindered. The large interference effect found in the Nilsson model for this transition does not occur in the deformed nuclear model we used. One possibility is that the Nilsson model provides a better physical description, and another possibility is that the problem is more fundamental. We speculate about different approaches that might lead to model wave functions relevant to low-energy transition energies, and also for radiative decay rates.

Keywords: Deformed nucleus, E1 radiative decay rate, Phonon–nuclear coupling, Phonon–nuclear matrix element, ^{181}Ta

1. Introduction

After nearly 30 years following the announcement of the Fleischmann–Pons experiment [1,2] there has emerged no consensus as to what microscopic reaction or what physical mechanisms are involved, in spite of the hundreds of theoretical papers that have been put forth over the years. Many in our field are convinced that the origin of the effect is nuclear; however, the absence of energetic nuclear radiation commensurate with the energy produced means that conventional nuclear detectors cannot be used for the determination of a reaction mechanism.

Our efforts in recent years have focused on phonon–nuclear coupling [3], in which the raising or lowering of a nuclear state is coupled to the creation or destruction of a phonon, due to the relativistic boost correction of the nucleon-nucleon interaction in an oscillating nucleus [4]. The boost correction is known in the literature, it is stronger than the coupling derived from external electric and magnetic fields at the nucleus due to the motion of surrounding nuclei, and it has not been considered previously in the literature in connection with the slow oscillations of nuclei in condensed matter. We have interpreted observations of unexpected non-exponential decay of X-ray and gamma lines

*E-mail: plh@mit.edu.

in excitation transfer experiments with a radioactive ^{57}Co source in our lab in terms of excitation transfer mediated by high-frequency phonons as a result of this relativistic boost correction interaction. Models that we have studied for the down-conversion of large energy quanta originating with the D_2/^{4}He transition (followed by subdivision) accomplish the conversion through a great many sequential non-resonant excitation transfer steps which transfer energy from nuclei to phonons while quantum coherence is maintained.

At this point we have a fundamental Hamiltonian, we can make use of perturbation theory to evaluate the indirect coupling associated with excitation transfer [5], and we have some results from models for up-conversion and down-conversion. The models and the results are expressed in terms of the phonon–nuclear coupling matrix elements, which will need to be evaluated explicitly in order to develop quantitative predictions from the theory. We have reported previously some early efforts to evaluate matrix elements of the phonon–nuclear interaction [6,7], which were done prior to the analysis of Ref. [4] and which will need to be done again with the improved version of the interaction.

The phonon–nuclear interaction has selection rules for nuclear transitions consistent with that of an electric dipole, which focused our attention initially on low-energy nuclear electric dipole (E1) transitions. For up-conversion low-energy transitions are favored since fewer quanta are required to produce excitation. For excitation transfer the indirect interaction favors high frequency phonons and low energy nuclear transitions. The lowest energy nuclear E1 transition from a stable ground state of all the nuclei occurs in ^{181}Ta and is listed in the BNL Nudat2 database as involving an excited state at 6.237 keV. We put in some effort to develop an excitation transfer experiment with a ^{181}W source evaporated on a ^{181}Ta plate, but ran into difficulty obtaining radioactive ^{181}W. Nevertheless, this experiment in our view remains a potentially important one (even if we are not able to implement it at this time). In support of this experiment we chose some time ago to focus on a calculation of the phonon–nuclear coupling matrix element. In this report we summarize briefly the associated model, calculation, and result.

2. Nonspherical Potential Model for Proton Orbitals

The ^{181}Ta nucleus is known to be strongly deformed in the ground state and first excited state, based on the large nuclear quadrupole moments [8]. Consequently, the simplest relevant model that we might use is a Nilsson model [9] (see also [10] for earlier work on deformed nuclei).

Prior to the work of Nilsson and collaborators, much progress had been made on shell models for nuclear orbitals in an independent particle model based on spherical potential models. In such models a nonrelativistic Hamiltonian containing the kinetic energy, potential energy, and spin–orbit interaction was used to model proton and neutron orbitals [11,12], which was successful in understanding the shell structure effects in nuclei, and the ordering of the different subshells.

2.1. Deformed models from the Dudek group

Once the model is extended to deformed nuclei, then new issues arise. For example, a deformed nucleus can rotate, which allows for a quantitative explanation of the energy levels of observed rotational levels. From a purely practical perspective, the question of what deformed potential to use must be addressed, along with the parameterization of the spin–orbit potential. Fortunately, there has been much subsequent work on the problem, where models have been optimized to match well with known nuclear levels and quadrupole moments [13–15]. These models have been implemented in a Fortran program that has been made available to the scientific community [16].

It is useful to discuss briefly the difference between the older potential models used by the nuclear physicists of the 1950s, and the more modern potential models of the Dudek group used in since the 1970s. It was not appreciated at the outset of our work the importance of this issue. In the 1950s the focus was largely on models based on deformed harmonic oscillator potentials [9], due in part to the more accessible associated mathematics, and in part from the reasonably good observed agreement with experiment for the resulting single–particle models. The more modern deformed potentials of the Dudek group are essentially deformed Woods–Saxon potentials, which are flat in the interior of the nucleus, and which taper off to zero rapidly at the nuclear boundary. This kind of model is much closer to models long used for nuclear scattering calculations, and would be expected to provide a more realistic description of the interaction of an outer nucleon with the core.

2.2. Deformed potential Hamiltonian

The Hamiltonian for a proton in a deformed potential according to these models is made up of terms accounting (in order) for the proton kinetic energy, strong force potential, Coulomb potential, nuclear spin–orbit interaction, and Coulomb spin–orbit interaction according to

$$\hat{H} = -\frac{\hbar^2\nabla^2}{2M} + U(\rho,z) + V_{\mathrm{C}}(\rho,z) - \lambda\left(\frac{\hbar}{2Mc}\right)^2 \nabla U \cdot \hat{\boldsymbol{\sigma}} \times \hat{\mathbf{p}} + \frac{\hbar}{2M^2c^2}(\nabla V_{\mathrm{C}} \times \hat{\mathbf{p}}) \cdot \hat{\mathbf{s}}, \tag{1}$$

where M is the proton mass; $\hat{\boldsymbol{\sigma}}$ a vector operator made up of Pauli matrices; $\hat{\mathbf{p}}$ the proton momentum operator; $\hat{\mathbf{s}}$ the proton spin operator; V_{C} is the Coulomb potential associated with a uniform charge distribution within the boundary of the nucleus; and c is the speed of light. The scaling parameter λ for the nuclear spin–orbit interaction has been determined empirically by optimizing predicted orbital energies with observed orbital energies. The model strong force potential U is a taken to be a deformed Woods–Saxon potential

$$U = \frac{V_0}{1 + \exp\left[\mathrm{dist}/a\right]} \tag{2}$$

with fitting parameters V_0 and a; where "dist" in this formula is the distance between the position vector and the deformed nuclear surface given by

$$R(\theta) = c' R_0\left[1 + \beta_2 Y_{20}(\theta,\phi) + \beta_4 Y_{40}(\theta,\phi)\right] \tag{3}$$

with Y_{lm} a spherical harmonic function. The deformation parameters R_0, β_2 and β_4 define the shape of the deformed nuclear potential, with c' serving as a normalization constant to preserve the nuclear volume.

2.3. Coupled-channel model

Early in our study we were not aware that any codes were available, so we developed our own numerical models. We began with simple spherical models based on the Woods-Saxon potential and spin–orbit interaction, which very quickly gave us orbital energies in agreement with literature values. The extension of the model to the deformed case involved a great deal of work. After implementing and discarding different numerical implementations (we had high hopes for an elegant iterative coupled channel scheme), we ended up making use of a coupled channel approach that makes use of a brute force sparse matrix eigenvalue solution for the radial equations.

The model assumes a finite basis approximation of the form

$$\Psi = \sum_{m_s}\sum_{l,m} |s, m_s\rangle Y_{lm}(\theta,\phi)\frac{P_{lmsm_s}(r)}{r}, \tag{4}$$

which leads to coupled channel equations that we can write as

$$\begin{aligned} EP_{lmsm_s}(r) =& \frac{\hbar^2}{2M}\left[-\frac{d^2}{dr^2}+\frac{l(l+1)}{r^2}\right]P_{lmsm_s}(r) \\ &+\sum_{l'}\langle lmsm_s|U+V_C|l'msm_s\rangle P_{l'msm_s}(r) \\ &+\sum_{l'm'}\sum_{m'_s}\langle lmsm_s|U_{\rm so}+V_{\rm so}|l'm'sm'_s\rangle P_{l'm'sm'_s}(r), \end{aligned} \tag{5}$$

where $U_{\rm so}$ and $V_{\rm so}$ are the nuclear and Coulomb spin–orbit interactions. The implementation of this model involved a substantial technical effort. In our implementation we approximated the spin–orbit interaction through terms involving radial derivatives and omitted the angular derivatives. This saved much work as well as making the code run faster, for what appears to be a minor loss in accuracy.

3. Phonon–nuclear Coupling Matrix Element

In order to evaluate the phonon–nuclear coupling matrix element, we require a determination of the deformation parameters, as well as a specification of the boost correction.

3.1. Determination of the deformation parameters

The choice of the β_2 parameter can be made by looking for a level crossing (6.237 keV is small compared to the MeV shifts that occur when β_2 is varied between 0.15 and 0.25) between the 7/2$^+$ ground state and the 9/2$^-$ excited state (where the state notation here is I^p, with I the nuclear spin and p the parity). A value of $\beta_4 = -0.038$ was reported for ^{181}Re in Ref. [17], which we adopted for ^{181}Ta. The associated curve crossing occur close to $\beta_2 = 0.250$.

3.2. Quadrupole moment

According to Ref. [18] we can write for the intrinsic quadrupole moment

$$Q_0 = \frac{3}{\sqrt{5\pi}} eZR^2\beta_2\left(1+\pi^2\left(\frac{a}{R}\right)^2+\frac{2}{7}\sqrt{\frac{5}{\pi}}\beta_2\right) \tag{6}$$

with $a = 0.54$ as relevant to the Pb region, and

$$R = 1.2A^{1/3} = 6.788 \text{ fm}. \tag{7}$$

The spectroscopic quadrupole moment $Q_{\rm s}$ is related to the intrinsic quadrupole moment through

$$Q_{\rm s} = \frac{3K^2 - I(I+1)}{(I+1)(2I+3)} Q_0. \tag{8}$$

For the 7/2+ ground state $K = I = 7/2$ and

$$Q_{\rm s} = \frac{7}{15} Q_0. \tag{9}$$

A deformation parameter of $\beta_2 = 0.247$ is consistent with an intrinsic quadrupole moment of 7.37 eb, which is close to the value used in [19], and a bit higher than the spectroscopic values listed in [8].

Note that according to [20] the ratio of the quadrupole moment of the first excited state to the ground state is

$$\frac{Q\left[\frac{9}{2}\right]}{Q\left[\frac{7}{2}\right]} = 1.133 \pm 0.010. \tag{10}$$

3.3. Code comparison

After we had developed preliminary versions of the coupled channel code, we were interested in arranging for tests in order to debug the code. At this point we located the Dudek group code, downloaded it, and got it running on our system. We might have made use of it for matrix element calculations, except that it was not developed for this purpose, and we did not put in sufficient effort to extract the wave functions. Instead we decided to carry out matrix element calculations with our code, which had been developed specifically for this purpose.

Results from the two codes for the $7/2^+$ and $9/2^-$ states are shown in Table 1, including deformation, and with matched underlying spin–orbit models. The agreement is reasonable, with the differences due primarily to the use of an incomplete spin–orbit interaction in the coupled channel code.

3.4. Boosted spin–orbit interaction

The phonon–nuclear interaction in the models we have studied comes about primarily as a boost correction of the nucleon–nucleon interaction. Ideally we would prefer a derivation of the Nilsson model from first principles, and then apply a boost correction systematically to all of the different interactions. However, the Nilsson model is an empirical model, with potential parameters fitted to experimental data. Consequently at present a systematic boosting of all contributing interactions is not possible.

Instead, what is possible is to work with a boosted version of the spin–orbit interaction. This should provide a lower limit within an independent particle picture (the independent particle picture itself will be the focus of some discussion below). The nuclear contribution to the spin–orbit interaction can be written as

$$U_{\text{so}} = -\lambda \left(\frac{\hbar}{2Mc}\right)^2 \nabla U \cdot \boldsymbol{\sigma} \times \hat{\mathbf{p}}. \tag{11}$$

Table 1. Orbital energies from the Dudek code compared with results from the coupled channel code with 100 grid points for a deformed version of the problem ($\beta_2 = 0.25$ and $\beta_4 = -0.038$) and matched spin–orbit models.

Index	E_{Dudek} (MeV)	E_{code} (MeV) $l_{\text{max}} = 10$
7/2+ (1)	–13.3163	–13.455392
7/2+ (2)	–6.3483	–6.289249
7/2+ (3)	–1.9020	–2.067574
9/2− (1)	–6.3458	–6.425660
9/2− (2)	2.9235	2.941132

We can arrange for a boosted momentum according to

$$\hat{\mathbf{p}} \rightarrow \hat{\pi} + \frac{\hat{\mathbf{P}}}{A}, \tag{12}$$

where $\hat{\mathbf{P}}$ is the center of mass momentum and where $\hat{\pi}$ is the relative momentum. For the correction we can write

$$\Delta U_{\text{so}} = -\lambda \left(\frac{\hbar}{2Mc}\right)^2 \nabla U \cdot \boldsymbol{\sigma} \times \frac{\hat{\mathbf{P}}}{A} = \lambda \left(\frac{\hbar}{2Mc}\right)^2 \boldsymbol{\sigma} \times \nabla U \cdot \frac{\hat{\mathbf{P}}}{A} = \hat{\mathbf{a}} \cdot c\hat{\mathbf{P}} \tag{13}$$

with the $\hat{\mathbf{a}}$ operator given by

$$\hat{\mathbf{a}} = \lambda \frac{\hbar^2}{4AM^2c^3} \boldsymbol{\sigma} \times \nabla U. \tag{14}$$

3.5. Matrix element of the $\hat{\mathbf{a}}$ operator

Making use of the proton orbitals discussed above, we calculated the magnitude of the matrix element for the $\hat{\mathbf{a}}$ operator for a one proton transition between the first excited state and the ground state with the result

$$\left| \left\langle \frac{9}{2}^- \middle| \hat{\mathbf{a}} \middle| \frac{7}{2}^+ \right\rangle \right| = 1.3 \times 10^{-6}. \tag{15}$$

4. Radiative Decay

A check on the model can be provided by making use of the deformed model wave functions for the evaluation of the radiative decay rate. While conceptually simple, we will encounter a number of complicated issues as a result.

4.1. Calculation of the single-particle decay rate

For the radiative decay rate we have used the length form according to

$$\gamma = \frac{4}{3}\alpha \left(\frac{N}{A}\right)^2 \frac{\omega^3}{c^2} \left| \left\langle \frac{9}{2}^- \middle| \mathbf{r} \middle| \frac{7}{2}^+ \right\rangle \right|^2, \tag{16}$$

where the electric dipole matrix element was evaluated using the orbitals discussed above. In this formula N is the number of neutrons in the nucleus, and Z is the number of protons. The decay rate that results is

$$\gamma = 3.9 \times 10^8 \text{ s}^{-1}. \tag{17}$$

4.2. Decay rate from experiment

The half-life of the ^{181}Ta 6.237 keV excited state is 6.05 μs according to the Nudat2 database, and the internal conversion coefficient is listed as 70.5. From this we can estimate the radiative decay rate from experiment to be

$$\gamma_{\rm expt} = \frac{\ln 2}{6.05\ \mu{\rm s}} \times \frac{1}{1+70.5} = 1.6 \times 10^3\ {\rm s}^{-1}. \tag{18}$$

We see that the deformed shell model estimate for the radiative decay rate is high by more that 2×10^5.

4.3. Discussion

This disagreement between the single–particle theory in general and experiment is of course well known in the literature [21], and occurs at one level or another with all low-energy electric dipole transitions (and also with magnetic dipole transitions, and transitions of other multipolarity). The question is what produces the hindrance.

There is probably agreement that if a sufficiently large set of basis states could be assembled, the it should be possible to make use of (massive) multi-configurational calculations to calculate radiative decay rates accurately (assuming a suitably accurate nuclear potential model). In atomic physics calculations one can do well with generally with a modest number of configurations for simple atoms and ions, largely due to the presence of a strong nuclear potential, goodness of an average potential due to the other electrons, and relatively weak correlation effects. In the nuclear problem there is no dominant central potential, and the nucleon–nucleon interaction is strong, which maximizes correlation effects.

An astonishing advance on the problem of modeling radiative decay rates was reported in the 1950s when Nilsson and coworkers made use of deformed nuclear potential models to compute radiative decay rates for single proton and neutron transitions, where interference effects led to a reduction in the associated matrix elements and correspondingly large reductions in the predicted rates. However, there remained substantial differences between the rates predicted with Nilsson orbitals and experiment to account for. Motivated by the recently developed BCS theory for superconductivity, a similar approach was proposed to account for the residual correlation effects in nuclei. It became possible to account systematically for the decay rates of low-energy nuclear transitions then taking into account deformation and correlation effects [22].

At the beginning of this study we were aware of some of this progress, and the presumption was that by taking advantage of more modern and sophisticated deformed potential models we would have better wave functions and obtain even closer agreement. But as discussed above, the deformation parameters which give the best results for ^{181}Ta do not lead to the same interference associated with the Nilsson model. It is possible to find regions in the associated deformation parameter space where the requisite interference occurs, but this is not particularly close to where our optimization has landed us.

So, the relevant question at this point is how might we interpret the resulting situation?

One possibility is that the Nilsson model is the better physical model, that interference effects occur just as described by the model, and that we are not going to be able to get reliable answers for radiative decay rates from the deformed potential model of Dudek and coworkers. In this case we should not pursue further work with the Dudek model, and instead turn our attention to the Nilsson model with pairing corrections. This could be done, and probably should be done since it would not involve a particularly large investment of effort beyond what has been put in already.

Another possibility is that the problem is more fundamental. As has been pointed out by Cook [23] and others it is hard to see how an independent particle model should apply under conditions where the nucleus is dense, and where the mean free path of an incident nucleon is on the order of a fermi. While a lattice model might be a good approximation for some calculations, the nuclear density is too low for the nucleus to be a solid [24]. Instead, we expect the nucleus to be a Fermi liquid. There has been much development of Fermi liquid theory for nuclei in the literature [25], but this kind of model does not appear to have been used for low-energy E1 transitions in recent years. Other approaches have been successful in modeling E1 transitions over a wide energy range including the low-energy region generally [26,27].

It may be that nucleons do not travel freely throughout the interior of the nucleus, but instead their movement is hindered due to the dense packing of neighboring nuclei. In such a picture, a nucleons is localized some of the time, but tunnels to exchange with a neighboring nuclei. It may be possible to develop an empirical dispersion relation to model this effect, which would have an impact on the energy level spacing for low-lying states. It would be relatively easy to develop and test this kind of single-particle model.

Were we to think of the nucleus from the point of view of a lattice with the frequent coherent exchange of positions, then it would follow that a large reduction of the matrix element for the current should be expected relative to an independent particle picture due to the associated correlation between the nucleons. The question is whether it is possible to develop model many-particle wave functions of this kind, and evaluate matrix elements with them.

5. Summary and Conclusions

Proton wave functions were computed for a proton in a modern deformed nuclear potential to model the boosted spin–orbit contribution to the phonon–nuclear interaction between the ground state and first excited state of ^{181}Ta. The numerical implementation was based on a coupled channel formulation, and we made use of an incomplete spin–orbit interaction to reduce the effort level and run time. The energy eigenvalues are in reasonably good agreement with results from the Dudek group code for this problem and others.

The magnitude of the â matrix element that results is 1.3×10^{-6}, which is on the low end of the range of what we had expected. We might expect boosted versions of the central and tensor interactions to lead to a larger phonon–nuclear interaction; however, at present there is no derivation of the Nilsson or other deformed potential models that we might use to evaluate the associated contributions.

As a check on the model we calculated the radiative decay rate, and found that it was not hindered as in the Nilsson model. That low-energy electric dipole transitions are strongly hindered relative to spherical harmonic oscillator model estimates in nuclei is of course well known. This is modeled in the Nilsson theory through interference effects (which are not present in the Dudek deformed potential model), and also with corrections due to correlation effects. The concern of course is that there may be corresponding reductions in the â-matrix element due to interference and correlation effects. It would be worthwhile to make use of an appropriate Nilsson model for the â-matrix element calculation, a project of interest to us in the future.

The accurate calculation of phonon-nuclear matrix elements for low-energy transitions is made difficult due to the absence of relevant simple models that a predictive. We are considering a number of possibilities, including Fermi liquid models, an empirical dispersion relation, and the possibility of developing new many-particle models based on a lattice picture with coherent exchange.

References

[1] M. Fleischmann, S. Pons and M. Hawkins, *J. Electroanal Chem.* **201** (1989) 301; errata **263** (1990) 187.

[2] M. Fleischmann, S. Pons, M.W. Anderson, L.J. Li and M. Hawkins, *J. Electroanal. Chem.* **287** (1990) 293.

[3] P.L. Hagelstein, Current status of the theory and modeling effort based on fractionation, *J. Condensed Matter Nucl. Sci.* **19** (2016) 98–109.

[4] P.L. Hagelstein, Quantum composites: a review and new results for condensed matter nuclear science, *J. Condensed Matter Nucl. Sci.* **20** (2016) 139–225.

[5] P.L. Hagelstein, Phonon mediated nuclear excitation transfer, *J. Condensed Matter Nucl. Sci.*, in press.

[6] P.L. Hagelstein and I.U. Chaudhary, Coupling between a deuteron and the lattice, *J. Condensed Matter Nucl. Sci.* **9** (2012) 50–63.

[7] P.L. Hagelstein and I.U. Chaudhary, Central and tensor contributions to the phonon-exchange matrix element for the $D_2/^4$He transition, *J. Condensed Matter Nucl. Sci.* **11** (2013) 15–58.

[8] N.J. Stone, Table of nuclear magnetic dipole and electric quadrupole moments, *IAEA INDC International Nuclear Data Committee*, Report INDC(NDS)-0658 (2014).

[9] S.G. Nilsson, Binding states of individual nucleons in strongly deformed nuclei, *Dan. Mat. Fys. Medd.* **29** (1955) 1–69.

[10] A. Bohr The coupling of nuclear surface oscillations to the motion of individual nucleons, *Dan. Mat. Fys. Medd.* **26** (1952) 1–40.

[11] O. Haxel, J. Hans, D. Jensen and H.E. Suess, On the "magic numbers" in nuclear structure, *Phys. Rev.* **75** (1949) 1766.

[12] M.G. Mayer, Nuclear configurations in the spin–orbit coupling model. I. Empirical evidence, *Phys. Rev.* **78** (1950) 16–21.

[13] J. Dudek and T. Werner, New parameters of the deformed Woods–Saxon potential for A = 110–210 nuclei, *J. Phys. G: Nucl. Phys.* **4** (1978) 1543–1561.

[14] J. Dudek, A. Majhofer, J. Skalski, T. Werner, S. Cwiok and W. Nazarewiczs, Parameters of the deformed Woods–Saxon potential outside A =110–210 nuclei, *J. Phys. G: Nucl. Phys.* **5** (1979) 1359–1381.

[15] J. Dudek, W. Nazarewiczs and T. Werner, Discussion of the improved parametrisation of the Woods–Saxon potential for deformed nuclei, *Nucl. Phys.* A **341** (1980) 253–268.

[16] S. Cwiok, J. Dudek, W. Nazarewiczs, J. Skalski and T. Werner, Single–particle energies, wave functions, quadrupole moments and g-factors in an axially deformed Woods–Saxon potential with applications to the two-centre-type nuclear problems, *Computer Phys. Commun.* **46** (1987) 379–399.

[17] R. Bengtsson, J. Dudek, W. Nazarewicz and P. Olanders, A systematic comparison between the Nilsson and Woods–Saxon deformed shell model potentials, *Physica Scripta* **39** (1989) 196–220.

[18] G. Neyens, Nuclear magnetic and quadrupole moments for nuclear structure research on exotic nuclei, *Rep. Progr. Phys.* **66** (2003) 633–689.

[19] D. McLoughlin, S. Raboy, E. Deci, D. Adler, R. Sutton and A. Thompson, Distribution of electrical charge in the nucleus of ^{181}Ta, *Phys. Rev. C* **13** (1976) 1644–1663.

[20] G. Kaindl, D. Salomon and G. Wortmann, Quadrupole splitting of the 6.2-keV γ rays of ^{181}Ta in rhenium metal, *Phys. Rev. Lett.* **28** (1972) 952–955.

[21] C.F. Perdrisat, Survey of some systematic properties of the nuclear E1 transition probability, *Rev. Modern Phys.* **38** (1966) 41–94.

[22] V. Feifrlik and J. Rizek, The single particle E1 transitions with $\Delta K = 1$ in deformed nuclei, *Nucl. Phys.* A **121** (1968) 153–160.

[23] N.D. Cook, Nuclear and atomic models, *Int. J. Theoret. Phys.* **17** (1978) 21–32.

[24] V. Canuto and S.M. Chitre, Is nuclear matter a quantum crystal?, *NASA Report* (1973).

[25] J.W. Holt, G.E. Brown, J.D. Holt and T.T.S. Kuo, Nuclear matter with Brown–Rho–scaled Fermi liquid interactions, *Nucl. Phys.* A **785** (2007) 322–338.

[26] E. Litvinova and N. Belov, Low-energy limit of the radiative dipole strength in nuclei, *Phys. Rev.* C **88** (2013) 031302.

[27] K. Sieja, Electric and magnetic dipole strength at low energy, *Phys. Rev. Lett.* **119** (2017) 052502.

J. Condensed Matter Nucl. Sci. 29 (2018) 401–415

Research Article

Statistical Mechanics Models for PdD_x and PdH_x Phase Diagrams with both O-site and T-site Occupation

Peter L. Hagelstein*

Massachusetts Institute of Technology, Cambridge, MA, USA

Abstract

The phase diagram of PdH_x was first understood with the development of statistical mechanics models in the 1930s, where hydrogen atoms were modeled as occupying octahedral sites with an O-site energy that depended on the loading. In an earlier study we made use of a generalization of this kind of model to include both O-site and T-site occupation to model loading in the alpha phase, loading in the beta phase, and to develop a mean field model for the phase diagram of PdH_x. Here we extend the modeling to develop a model phase diagram for PdD_x. Since the PdD_x phase diagram is less studied that the PdH_x phase diagram, the selection of isotherms is an issue, and we encountered minor technical issues in the digitization of published isotherm data. It was possible to develop good extrapolations to low loading (required for our phase diagram optimization), but we found that a reliable extrapolation of isotherms at high temperature to high loading was not possible. Consequently, after some exploration we found that "reasonable" global phase diagram models could be obtained with the introduction of a "guiding" point, where we introduce an estimate for a single pressure at high temperature and high loading to constrain the O-site energies and resulting phase diagram. This underscores the need for experimental measurements in this regime in order to understand the phase diagram. We were successful in optimizing a mean field phase diagram model for PdD_x using this approach. We made use of the isotherms from our earlier modeling for PdH_x to develop a new phase diagram model for PdH_x making use of the same basic approach. The resulting O-site energy curves are qualitatively similar, with a somewhat deeper binding for H at high loading (attributed to the increased lattice expansion), and larger spread in O-site energies as a function of temperature (attributed to the reduced number of configurations accessible at low energy due to the larger excitation energy of the lower mass hydrogen atom).

Keywords: PdD phase diagram, PdH phase diagram, mean-field statistical mechanics model, O-site and T-site occupation of H in Pd, empirical alpha-phase model for PdD_x

1. Introduction

We are interested in the development of a statistical mechanics model for the phase diagram of PdD_x in connection with simulating loading and excess heat in the Fleischmann–Pons experiment [1,2]. Such a model would likely have wider applications; for example, to allow for the estimation of the deuterium chemical potential in PdD_x; to help in the

*E-mail: plh@mit.edu.

© 2018 ISCMNS. All rights reserved. ISSN 2227-3123

development of kinetics models for the deuterium evolution reaction; and to aid in the interpretation of experiments generally involving PdD_x.

What seemed to be a straightforward project at first glance turns quickly into a complicated venture. Where PdH_x has long been studied, PdD_x has received less attention. Where the phase diagram of PdH_x is in reasonable shape, the phase diagram for PdD_x has been less studied. Considerable theoretical work has been devoted to statistical mechanics models for PdH_x, but much less effort has been devoted to the problem of PdD_x.

Early experimental work established that hydrogen isotopes occupy octahedral sites near room temperature. However, it is not possible to model loading in the α-phase of PdH_x or PdD_x accurately over a wide range of temperature based on a statistical mechanics model with only O-site occupation. Instead, models with both O-site and T-site occupation can fit the experimental data [3] very well [4]. Some neutron diffraction experiments show evidence for T-site occupation at higher loading; however, there are also neutron diffraction experiments which do not (see the discussion and references in [5]).

What happens at high loading near room temperature is very much unclear. There are gas loading experiments and electrochemical experiments in which the D/Pd loading ratio is thought to exceed unity [6–8]. Yet there is no agreement within the hydrogen in metals community that palladium is capable of being loaded with H or D above unity under experimentally accessible conditions. And if it is possible for the loading to exceed unity, there is no agreement as to where additional interstitials go, as there are only as many octahedral sites as palladium atoms. One possibility is that double occupation of O-sites occurs. Another is that a phase change occurs with H or D going into T-sites. We have explored models in which T-site occupation occurs when most of the O-sites are filled [9].

A gaping hole in the currently available experimental data sets occurs for high loading and high temperature (it is much easier to measure isotherms near room temperature and at low temperature for high loading). Continuing isotherm measurements at high loading involves high pressure, which becomes problematic at the very high pressures required. As a result, isotherms for PdH_x at elevated temperature simply are not available at high pressure. The isotherm data sets available for PdD_x are less complete than for PdH_x.

In earlier modeling work with PdH_x we made use of empirical extrapolations in an attempt to fill in missing parts of isotherms [10]. At low loading, the extrapolations appear to have worked well. At high loading this was perhaps ill-advised, as there is no reason to believe that the extrapolations used should be accurate.

Another headache is that Pd vacancies are stabilized at high loading, so that phase changes to superabundant vacancy phases are expected. Fukai and Okhuma showed that highly loaded NiH and PdH undergo a phase change where one out of four host metal lattice atoms are missing (after having diffused elsewhere) at elevated temperature [11,12]. A further brief discussion relating to vacancy phases appears in Appendix A. In the models considered here and in our earlier work we assume that the palladium atoms fully occupy their positions in an FCC lattice, and that the conversion to superabundant vacancies has not yet occurred. This assumption impacts only the part of the phase diagram at very high loading; and is relevant to the majority of experiments in which either the temperature is too low for Pd self-diffusion, or else the loading is too low to stabilize vacancies at elevated temperature.

For PdH_x we developed statistical mechanics models based on a generalization of Lacher's model [13] to include O-site and T-site occupation. Such models might be considered to be (simple) mean-field models, in contrast to more sophisticated models that have been published in recent years. Instead of making use of isotherm data to optimize a physical underlying model Hamiltonian, we were interested in the development of a simpler empirical mean-field model that might provide for a statistical mechanics engine that we could use for estimating the hydrogen chemical potential at different temperatures and loadings in numerical simulations.

We have been interested in extending our statistical mechanics models for PdH_x to PdD_x for some time. In this brief report we summarize some of our recent efforts to this end. The first issue to deal with is that the published isotherms are generally not particularly consistent, so that assembling a suitable phase diagram for PdD_x from experimental data is problematic. In response we selected a subset of the available data minimally inconsistent to work with. We made use once again of an extrapolation to low loading as in the case of the PdH_x phase diagram. For PdD_x this seems to work well enough. At high loading we initially attempted to extrapolate as before; however, this time the data sets do not extend as far and we were not able to develop reliable extrapolations. We tried developing phase diagram models based on high pressure data only for the lowest temperature used; however, the optimization of the fitting coefficients tended to produce natural extrapolations to high temperature which did not seem particularly physical. In the end what seemed most plausible was to provide a single "guiding" point at high pressure and high loading which resulted in O-site energy curves free of "pinching" or large dispersal near a loading of unity. As a result, the parameterized isotherms that result are "reasonable" at high loading and high temperature within the framework of the model, and not informed by experimental data. In essence the phase diagram model by itself provides for an extrapolation to high loading at high temperature. This underscores the need for new experimental measurements of isotherms in this regime.

Given that extrapolations to high loading are problematic, especially at high temperature, we were motivated to revisit the phase diagram for PdH_x and develop an analogous phase diagram model free of such extrapolations. Once again a single "guiding" point at high pressure and high loading is used to constrain the phase diagram and O-site energies at high loading and at high temperature to be "reasonable".

2. Isotherms

Studies of isotherms with experimental data for PdD_x appear in Refs. [3,14–27]. We made use of the online WebPlot-Digitizer of Ankit Rohatgi to develop numerical versions of isotherms from published plots. (Note that in some cases the available online versions of the experimental papers contain plots that are neither precisely vertical nor rectangular, so that some effort was needed to assemble numerical isotherm data.) It became clear from digitizing and comparing isotherms from different references that there were issues with consistency. In some cases isotherms taken at different temperatures would cross, and in other cases isotherms taken at the same temperature would be significantly offset in pressure. The construction of a phase diagram directly from data sets from many groups as was done for PdH_x [28] does not work as well for PdD_x.

2.1. Choice of isotherms

After numerous tries with different isotherm data sets, in the end we decided to work with: the 298 and 323 K isotherms of Sakamoto et al. (1996) [25]; the 348 K isotherm of Kibria et al. (1998) [26]; and the set of isotherms at 393, 433, 473, 533, 573, and 613 K of Wicke and Blaurock (1987) [22]. We note that the 348 K isotherm of Kibria et al. does not seem to be particularly consistent with the data sets of Sakamoto et al. or of Wicke and Blaurock. Even so, this choice of isotherms seemed better than other sets that we worked with earlier in the study.

2.2. Extrapolation to low loading

In order to construct a model phase diagram along the lines discussed in previous publications, we need to extrapolate the isotherms down to lower pressure and loading (which allows us to develop appropriate initial conditions for the fitted isotherms). To do this we made use of extrapolated isotherms based on

$$\ln P(\text{atm}) = \left(\frac{a_{-1}}{T} + a_0 + a_1 T\right) + \left(\frac{b_{-1}}{T} + b_0 + b_1 T\right)\theta + 2\ln\frac{\theta}{1-\theta}. \tag{1}$$

We made use of α-phase isotherm data published by Clewley et al. (1977) [3], Kleppa and Phutela (1982) [18], and by Lasser (1984) [20]. The fitting parameters were optimized to give

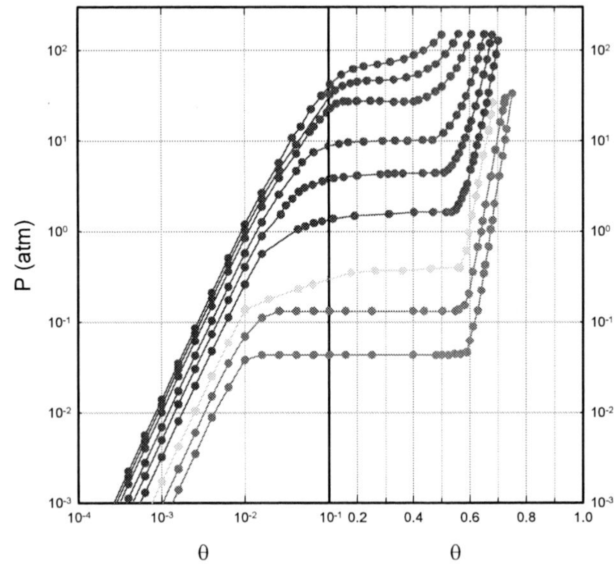

Figure 1. Desorption isotherms extrapolated to lower pressure and loading. The circles indicate the loading and pressure data points digitized from [25] in red, [26] in light green, and from [22] in blue; along with extrapolations of the data into the α-phase.

$$a_{-1} = -2375.64\,\mathrm{K}, \quad a_0 = 15.4305, \quad a_1 = -0.00324156\,\mathrm{K}^{-1},$$

$$b_{-1} = 2034.1\,\mathrm{K}, \quad b_0 = -46.2616, \quad b_1 = 0.0372747\,\mathrm{K}^{-1}, \tag{2}$$

The results are shown in Fig. 1.

2.3. Data points at high loading

In previous work we extrapolated isotherms to higher loading making use of simple extrapolation formulas [10]. As mentioned in the Introduction, we attempted a similar approach with the isotherms under discussion in this section. Unfortunately, in this case there were problems developing a suitable extrapolation (since the PdD$_x$ isotherm data does not go to as high pressure). Instead, it seemed best to make use only of the model results at high loading at 300 K from [5] to serve as an informed extrapolation at room temperature.

3. PdD$_x$ Phase Diagram Model

Our efforts toward developing a model phase diagram for PdH$_x$ included working with numerous different types of models, as discussed in [10]. In the end the results deemed to be best were obtained from a mean field model based on an O-site to T-site energy difference which depends on loading (and not on temperature), and on an O-site energy that depends on both loading and temperature. For the PdD$_x$ phase diagram we decided to optimize a similar model, expecting qualitatively similar results since PdH$_x$ and PdD$_x$ are closely related theoretically.

In the course of our study we developed more than 50 optimizations of a variety of different models (and also with different isotherm data sets). The results obtained for PdD$_x$ were found not to be particularly close to the result obtained previously for PdH$_x$. This motivated us to develop new optimizations for the PdH$_x$ phase diagram using a similar approach, hoping to resolve the differences. In the end good results were obtained as discussed in Section 3.1.

3.1. Model for the O-site energy

The optimization of a phase diagram model was done following the approach discussed in [10]. For a generalized Lacher type of mean-field model the O-site energy $E_{\rm O}$ is optimized to minimize the difference between the model chemical potential and the chemical potential that corresponds to the experimental isotherm data. For the study reported here we made use of an O-site energy that depends on loading and temperature according to

$$E_{\rm O}(\theta, T) + \frac{E_{\rm D}}{2} = \left[a_0 + a_1\theta + a_2\theta^2 + a_3\theta^3 + a_4\theta^4 + a_5\theta^5 + a_6\theta^6\right] + T\left[b_0 + b_1\theta + b_2\theta^2 + b_3\theta^3 + b_4\theta^4 + b_5\theta^5 + b_6\theta^6\right], \tag{3}$$

where $E_{\rm D}$ is the D_2 dissociation energy. This corrects the definition of the parameterization of the O-site energy from [10] (where the $E_{\rm D}/2$ factor was omitted).

3.2. Results

As mentioned in Section 1, what seemed to be the best results were obtained where we allowed the model to fit the experimental and α-phase extrapolated data. However, left to itself the optimization would in some cases produce a pinching or a deviation of the O-site energies at high loading. This provided motivation to introduce a single "guiding" point at high temperature and high pressure to minimize these effects. The fitted phase diagram that results in shown in Fig. 2. One can see two added points near room temperature at high loading consistent with the models discussed in [5], along with a high temperature "guiding" point at high pressure.

The a_j fitting parameters used for the O-site energy are (in meV) are as follows:

$$a_0 = -72.435, \quad a_1 = -249.999, \quad a_2 = 87.897, \quad a_3 = -39.679,$$

$$a_4 = 103.705, \quad a_5 = 230.615, \quad a_6 = -214.613, \tag{4}$$

and the b_j fitting parameters are (in meV/K)

$$b_0 = 0.02810, \quad b_1 = 0.03778, \quad b_2 = 0.09514, \quad b_3 = -0.15121,$$

$$b_4 = -0.01107, \quad b_5 = 0.00714, \quad b_6 = 0.03747. \tag{5}$$

The resulting O-site energies are shown in Fig. 3.

4. PdH_x Phase Diagram Model

As discussed above we carried out new optimizations for the PdH_x phase diagram using the same approach hoping for qualitatively similar results in terms of the resulting mean field model. The resulting phase diagram is shown in Fig. 4.

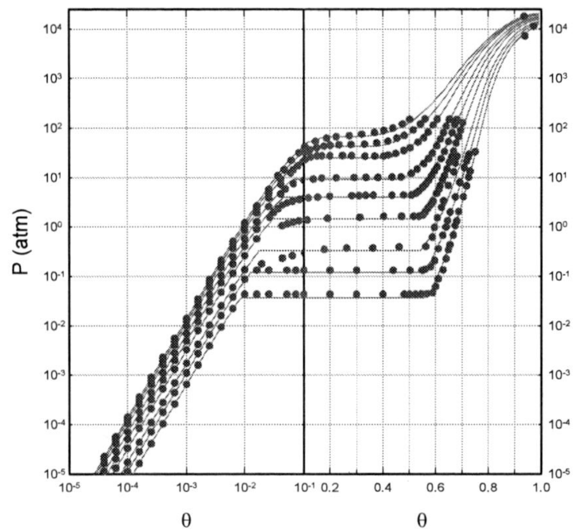

Figure 2. Model isotherms (*red lines*) and input isotherm points (*blue circles*) for PdD_x.

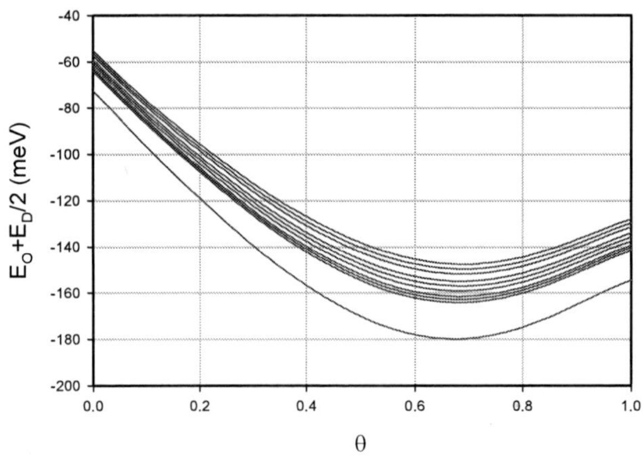

Figure 3. Model O-site energies in meV as a function of loading for (from lowest to highest curves) T = 0,298, 323, 348, 393, 433, 473, 533, 573, and 613 K.

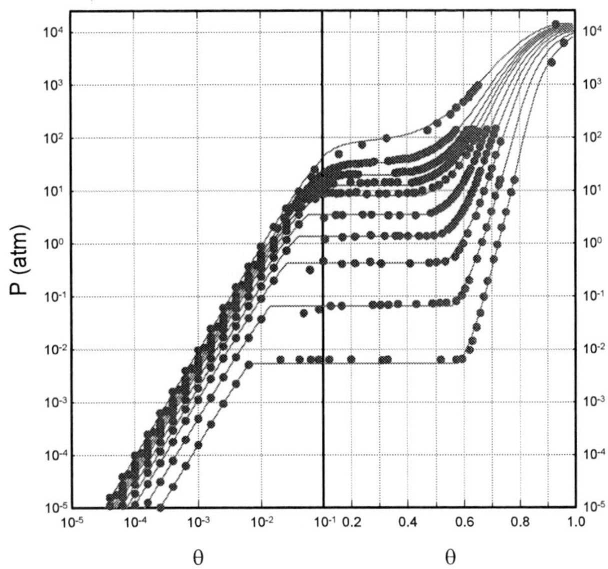

Figure 4. Model isotherms (*red lines*) and input isotherm points (*blue circles*) for PdH_x. The isotherms from lowest to highest correspond to temperatures of for (from lowest to highest curves) $T = 293, 343, 393, 433, 473, 516, 543, 573, 613$, and 710 K.

The a_j fitting parameters obtained in this case (in meV) are

$$a_0 = -74.835, \quad a_1 = -260.600, \quad a_2 = 56.644, \quad a_3 = 15.128,$$

$$a_4 = 78.713, \quad a_5 = 211.893, \quad a_6 = -195.118, \tag{6}$$

and the b_j fitting parameters in (meV/K) are

$$b_0 = 0.02791, \quad b_1 = 0.08690, \quad b_2 = 0.01451, \quad b_3 = -0.07286,$$

$$b_4 = -0.00988, \quad b_5 = 0.00620, \quad b_6 = 0.00635. \tag{7}$$

O-site energies for this model are shown in Fig. 5.

5. Discussion

The general approach discussed in our previous efforts to construct phase diagram models for PdH_x in [10] seems sound (with the exception of making use of extrapolations to fill in where experimental data is lacking at high loading and at high temperature). Consequently, we thought that the extension of the approach to PdD_x should result in a reasonable phase diagram model. The biggest issues we encountered in this project were the selection of isotherm data, and technical issues associated with developing a faithful digitization.

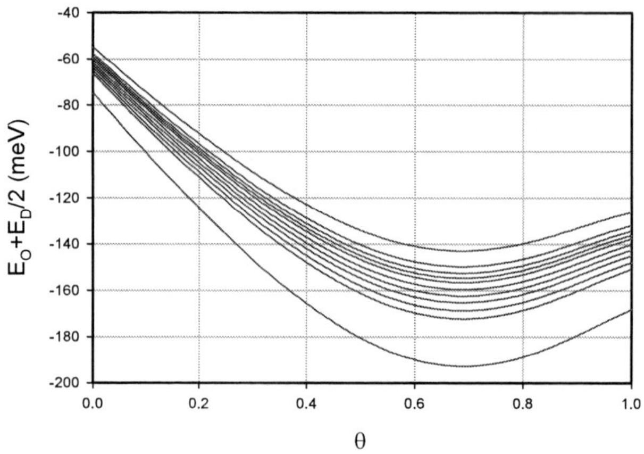

Figure 5. Model O-site energies in meV as a function of loading for (from lowest to highest curves) $T = 0$, 293, 343, 393, 433, 473, 516, 543, 573, 613, and 710 K.

We made use of our earlier work on isotherms at high loading near room temperature [5] to provide estimates for pressure at two loading points for both PdD_x and PdH_x. We tried a variety of "guiding" points at high loading and at high temperature, seeking a "reasonable" solution in both cases. What was found was that if the guiding point was too low, then the E_O curves would droop near $\theta = 1$; and if the guiding point was too high then the E_O curves began to diverge. Once again, we would prefer to work with experimental data for this region, but as yet none is available.

The phase diagram models that result in both cases match the isotherm data quite well, with unweighted mean square errors I (see [10]) on the order of 2.2×10^{-5} for PdD_x and 1.4×10^{-5} for PdH_x. In our earlier work we appeared to get a spread at low loading that could be interpreted as associated with the shift in the Fermi level, but this effect does not show up in the models and optimization reported on here.

We had expected the O-site energy curves should be similar for the two cases extrapolated to $T = 0$. The agreement is good at low loading, but the PdH_x O-site energy shows a minor deviation, ending up more than 10 meV lower at a loading of unity. This is likely a result of the larger lattice constant for PdH_x as the loading increases. The biggest difference between the two models obvious from the E_O vs. θ plots is that there is a larger spread in the E_O curves at different temperatures for PDH_x than for PdD_x, which is likely due to the larger number of low-energy configurations available to PdD_x as the loading increases.

Appendix A. Role of Superabundant Vacancy Phases

A reviewer has asked about the role of the superabundant vacancy phase in PdD_x, which is relevant since vacancy phases are preferred thermodynamically at high loading.

People had long modeled solid state systems based on (simple) atom-atom potentials, until the early 1980s when it was observed that embedded atom models that took into account in addition the background electron density could connect with experiment [31]. Following this, embedded atom models were developed for helium and hydrogen in metals and applied systematically to the analysis of light interstitials, as well as for hydrogen and helium in vacancies. Estimates that resulted for the binding energy of H and D in vacancies [32,33] indicated that vacancies are traps for hydrogen isotopes in Ni and Pd. Two measurements of the monovacancy enthalpy for bulk Pd has been measured giving 1.85 eV [34] and 1.5 eV [35]. If there is an interstitial hydrogen or deuterium in an O-site next to a Pd atom, the vacancy formation energy is reduced by more than 200 meV (and further reduced with more interstitial neighbors). This suggests that at a D/Pd loading near unity the vacancy formation energy will be much reduced, and that vacancies should form spontaneously.

However, the spontaneous formation of superabundant vacancies has not been reported in electrochemical experiments near room temperature. For the lattice to rearrange itself into one of the superabundant vacancy phases, the host Pd atoms need to diffuse to other locations. The atomic self diffusion coefficient for Pd in bulk Pd is very low [36,37], so that a conversion to a vacancy phase would be expected to take a prohibitively large number of years (unless the phase change to a vacancy phase were to proceed like a more conventional rearrangement type of phase change not involving vacancies, but as yet there is not evidence for this). We would expect faster self-diffusion in PdD_x [38–40], but we would not expect the self-diffusion to be sufficiently fast to change the conclusion here. To arrange for a faster conversion, Fukai and Okhuma used an elevated temperature between 700 and 800 °C [11] where the conversion to the vacancy phase took several hours.

Consequently, even though the superabundant vacancy phases are expected to be preferred thermodynamically at high loading, a PdD_x sample would not be expected to convert to the vacancy phase in a relevant time scale (although a focused experimental clarification of this point would be of great interest). Under these conditions the use of a phase diagram based on α-phase, β-phase and miscibility gap region for FCC Pd as presented in this work is relevant and useful.

We have noted previously that special arrangements need to be made in order to develop superabundant vacancies in an electrochemical experiment near room temperature. Since the Pd atoms in the cathode would have a negligible self-diffusion coefficient (and would not be expected to move much), we might consider the addition of new Pd through codeposition as a way to make a vacancy phase layer if the D/Pd loading is sufficiently high when the codeposition occurs. This was offered as an interpretation [41] for the rapid development of excess heat in the Szpak codeposition experiment [42]. Letts demonstrated a prompt excess heat heat following Pd codeposition at high current density [43]. The idea in this experiment is that co-deposition at high current density allows for a high loading near the surface, but the Pd co-deposition rate is kept low by limiting the number of Pd^{2+} ions in the electrolyte in order to allow time for the Pd atoms at the surface to move to locations that minimize energy. Whether superabudant vacancies are produced in the Letts co-deposition experiment has not been clarified experimentally yet.

Appendix B. Partial Enthalpy of Formation

A reviewer wonders about partial enthalpy in the comment:

The data of Flanagan et al. as well as Sakamoto et al. both show that the partial enthalpy of formation of hydrogen becomes endothermic at some composition within the beta phase. This behavior would be expected to cause the log P(H_2) vs, 1/T relationship to change sign at the H/Pd ratio where the exothermic–endothermic transition occurred. This behavior is not included in the evaluation shown here. Why not?

This is an interesting question worth considering. We can write

$$\Delta H = \frac{R}{2}\left(\frac{\partial \ln p}{\partial (1/T)}\right)_\theta \tag{B.1}$$

adapted from Eq. (3.21) in Speiser's review [44]. This in a sense provides a measure of how much change in pressure

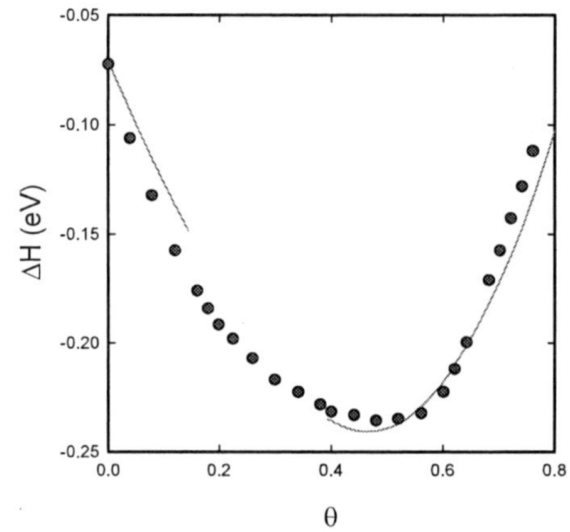

Figure 6. Model partial enthalpy of formation at 590 K (*red line*) as a function of H/Pd outside of the model miscibility gap; partial enthalpy from Flanagan and Luo (*blue points*).

occurs at a specific loading going from an isotherm at one temperature to a nearby isotherm at a slightly different temperature.

In an experiment that focuses on a single isotherm at a single temperature, a determination of ΔH would then allow for an extrapolation to temperatures not measured. It would extend the utility of a single temperature measurement, as well as providing data that can be used to exercise statistical mechanics models.

The same relation can be used to develop systematic estimates for ΔH when a more complete set of data is available; for example as is the case when dealing with the isotherms of a phase diagram. In the model described in this paper, the fitting parameters are optimized so that the model isotherms match the phase diagram data as closely as possible. If the fit is accurate, then we would expect the partial enthalpy to be equivalently accurate.

In response to the reviewer's comment we used the model to calculate ΔH above the critical point for Fig. 1 from Flanagan and Luo [45]. There are technical issues in this comparison, as the model critical point (around 635 K) lies above the critical point experimental from experiment (around 566 K). We compare with a model isotherm calculated at 590 K. The model result outside of the model miscibility gap is in reasonable agreement with the published curve as shown in Fig. 6.

The change in sign of ΔH in this case occurs near $\theta = 0.90$, so the effect of interest in the reviewer's comment is included in the model, although it is a weak effect. One can see a bunching of the isotherms at high temperature and at high loading; which is due primarily to the deviation of the fugacity from the pressure, and in small part from the change in sign of ΔH at high temperature.

Appendix C. Octahedral and Tetrahedral Sites

The reviewer has commented about octahedral and tetrahedral sites:

The H, D atoms in the alpha phase are proposed to occupy random sites located in the interstitial spaces between the Pd atoms. These sites have tetrahedral characteristics. Octahedral sites form only after the Pd atoms shift to form the beta phase. Therefore, the statement that octahedral sites are involved in the evaluation of the alpha phase is

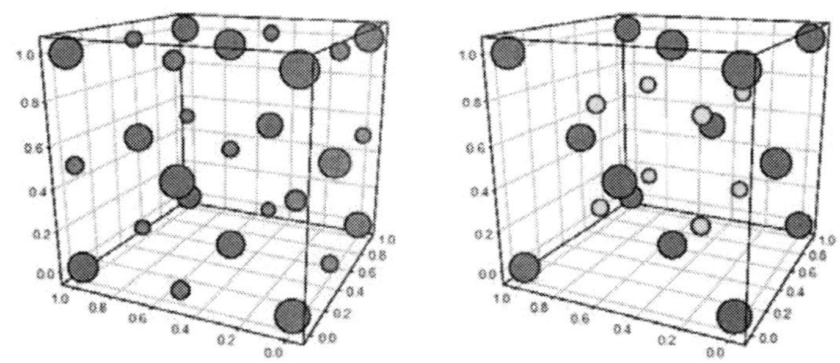

Figure 7. Octahedral (*left*) and tetrahedral (*right*) site locations for interstitial hydrogen isotopes in an FCC metal lattice. The locations of the metal atoms are indicated by the larger gray circles. The O-site positions are indicated by the smaller pink circles on the left; and the T-site positions are indicated by the smaller light green circles on the right.

confusing.

Given the controversy associated with octahedral and tetrahedral site occupation in PdH_x and PdD_x, this issue is one well worth additional discussion.

Appendix C.1. Site locations

Stoichiometric PdH at low temperature is thought to occur in an FCC NaCl type of structure, with the Pd atoms and Pd atoms at alternating positions on a cubic lattice as shown in the left schematic of Fig. 7. The tetrahedral sites occur at interior locations, as indicated in the schematic on the right.

The metal lattice positions in these figures are at

$$\left(0,0,0\right), \left(\frac{1}{2}a, \frac{1}{2}a, 0\right) \left(a, 0, 0\right), \left(0, a, 0\right), \cdots$$

and equivalent lattice sites (we have included redundant locations here), where a is the lattice constant. The octahedral sites are located at

$$\left(\frac{1}{2}a, 0, 0\right), \left(0, \frac{1}{2}a, 0\right), \left(0, 0, \frac{1}{2}a\right), \left(\frac{1}{2}a, \frac{1}{2}a, \frac{1}{2}a\right), \cdots$$

and equivalent lattice sites (again with redundant locations included). The tetrahedral sites are at

$$\left(\frac{1}{4}a, \frac{1}{4}a, \frac{1}{4}a\right), \left(\frac{3}{4}a, \frac{1}{4}a, \frac{1}{4}a\right), \left(\frac{1}{4}a, \frac{3}{4}a, \frac{1}{4}a\right), \left(\frac{1}{4}a, \frac{1}{4}a, \frac{3}{4}a\right), \left(\frac{3}{4}a, \frac{3}{4}a, \frac{1}{4}a\right), \cdots$$

and equivalent lattice sites (also with redundant locations).

What distinguishes an octahedral site or a tetrahedral site is the position relative to the sites occupied by the metal atoms. There is a single octahedral site in the bulk for each metal atom site, and there are two tetrahedral sites for each metal atom.

Appendix C.2. Site occupation from experiment

The reviewer has suggested that hydrogen and deuterium occupy tetrahedral sites in palladium at low loading in the α-phase, while only at higher loading in the β-phase do octahedral sites form, and that hydrogen and deuterium occupy octahedral sites in the β-phase. In order to make progress on the associated issues we must turn to experiment.

Worsham [46] carried out neutron diffraction experiments at room temperature, and reported that from these experiments octahedral site occupation was dominant in the beta phase for both powder PdH_x and PdD_x. Nelin [47] focused on powder PdH_x and PdD_x at elevated temperature, concluding that hydrogen and deuterium are in octahedral sites, even in the α-phase well above room temperature.

Neutron diffraction works best for this problem when the interstitial occupation is large, which hinders α-phase measurements near room temperature or below. In this case a different approach is required, and ion channeling experiments have proven to be effective. Carstanjen [48] made use of deuterium ion channeling and fusion reactions to determine that deuterium atoms occupy octahedral sites in $PdD_{0.007}$ at low temperature.

Appendix C.3. O-site and T-site occupation in the α-phase

The reviewer has written in addition:

The relationship between O-site and T-site is complex in the alpha phase. Here the O-site is initially much smaller than in the beta phase and, consequently, less likely to be occupied compared to the T-site. Nevertheless, occupancy of the O-site increases with temperature and H/Pd atom ratio, until the alpha and beta phases eventually merge as the temperature is increased above a critical value. Nevertheless, at lower temperatures, this conversion of alpha to beta is a step function with only a very small amount of H in the alpha phase being required to cause the conversion. Once conversion takes place, essentially all of the H in the alpha phase moves to the O sites, thereby causing the lattice to abruptly expand.

Given the prominence of interpretations involving different sites in some of the earlier literature, it seems appropriate to consider this issue here. A comparable picture is described in the case of titanium hydride put forth by E. J. Goon (see the brief discussion in [49]). An early neutron diffraction study in PdH reported different structures at low temperature in β-phase PdH, which was interpreted as due to O-site occupation above 55 K, and a migration to T-site occupation below 55 K [50].

This argument figures prominently in connection with resistance measurements in [51]. Subsequent neutron observations did not confirm the presence of tetrahedral occupation, as discussed in [52,53].

In principle the channeling experiments of Carstanjen should be sufficient to settle the issue. However, attention should be drawn to the measurements of Bugeat and Ligeon, who also observed O-site occupation in Pd loaded in the α-phase by implantation [54]. In this case O-site occupation was observed at low temperature, but when the sample was warmed T-site occupation was observed. The interpretation presented by the authors for this observation was that when the temperature was increased the hydrogen diffused from O-sites to vacancy sites to occupy T-sites. In a monovacancy in Ni (which is similar to the monovacancy in Pd) the O-site lies 3 meV below the T-site [33]. With nearly degenerate O-sites and T-sites in the Pd monovacancy, a more modern interpretation of the Bugeat and Ligeon experiment is that both O-site and T-site occupation of the monovacancies gives rise to some T-site signals.

Note also that while the O-site may be smaller in the α-phase than in the β-phase, the same is true of the T-site. Even so, the associated differences in the zero-point energy much less than the difference that we found for the empirical O-site to T-site excitation energy [5].

So, in this case we are not in agreement with the picture suggested by the reviewer.

References

[1] M. Fleischmann, S. Pons and M. Hawkins, *J. Electroanal. Chem.* **201** (1989) 301; errata **263** (1990) 187.

[2] M. Fleischmann, S. Pons, M.W. Anderson, L.J. Li and M. Hawkins, *J. Electroanal. Chem.* **287** (1990) 293.

[3] J.D. Clewley, T. Curran, T.D. Flanagan and W.A. Oates, Thermodynamic properties of hydrogen and deuterium dissolved in palladium at low concentrations over a wide temperature range, *J. Chem. Soc., Faraday Trans. 1: Phy. Chem. Condensed Phases* **69** (1973) 449–458.

[4] P.L. Hagelstein, O-site and T-site occupation of α–phase PdH_x and PdD_x, *J. Condensed Matter Nucl. Sci.* **17** (2015) 67–90.

[5] P.L. Hagelstein, Empirical models for octahedral and tetrahedral occupation in PdH and in PdD at high loading, *J. Condensed Matter Nucl. Sci.* **17** (2015) 35–66.

[6] B. Baranowski, S.M. Filipek, M. Szustakowski, J. Farny and W. Woryna, Search for cold fusion in some Me–D systems at high pressure of gaseous deuterium, *J. Less Common Metals* **158** (1990) 347–357.

[7] I.F. Silvera and F. Moshary, Deuterated palladium at temperatures from 4.3 to 400 K and pressures to 105 kbar: Search for cold fusion, *Phys. Rev. B* **42** (1990) 9143–9146.

[8] M.C.H. McKubre, The importance of replication, *Proc. ICCF14* (2008) 673–688.

[9] P.L. Hagelstein, Statistical mechanics models for PdH_x and PdD_x, *J. Condensed Matter Nucl. Sci.* **24** (2017) 87–97.

[10] P.L. Hagelstein, Models for the phase diagram of palladium hydride including O-site and T-site occupation, *J. Condensed Matter Nucl. Sci.* **20** (2016) 54–80.

[11] Y. Fukai and N. Okhuma, Evidence of copious vacancy formation in Ni and Pd under a high hydrogen pressure, *Jpn. J. Appl. Phys.* **32** (1993) L1256–L1259.

[12] Y. Fukai and N. Okhuma, Formation of superabundant vacancies in Pd hydride under high hydrogen pressures, *Phys. Rev. Lett.* **73** (1994) 1640–1643.

[13] J.R. Lacher, A theoretical formula for the solubility of hydrogen in palladium, *Proc. Roy. Soc. London, Series A, Math. Phys. Sci.* **161** (1937) 525–545.

[14] L.J. Gillespie and W.R. Downs, The palladium–deuterium equilibrium, *J. Am. Chem. Soc.* **61** (1939) 2496–2502.

[15] T.B. Flanagan, Absorption of deuterium by palladium, *J. Phy. Chem.* **65** (1961) 280–284.

[16] M.J.B. Evans and D.H. Everett, Thermodynamics of the solution of hydrogen and deuterium in palladium, *J. Less Common Metals* **49** (1976) 123–145.

[17] R. Lässer, Palladium–tritium system, *Phys. Rev. B* **26** (1982) 3517–3519.

[18] O.J. Kleppa and R.C. Phutela, A calorimetric-equilibrium study of dilute solutions of hydrogen and deuterium in palladium at 555–909 K, *J. Chem. Phys.* **76** (1982) 1106–1110.

[19] R. Lässer and K.-H. Klatt, Solubility of hydrogen isotopes in palladium, *Phys. Rev. B* **28** (1983) 748–758.

[20] R. Lässer, Solubility of protium, deuterium, and tritium in the α phase of palladium, *Phys. Rev. B* **29** (1984) 4765–4768.

[21] W.A. Oates, R. Lässer, T. Kuji and T.B. Flanagan, The effect of isotopic substitution on the thermodynamic properties of palladium–hydrogen alloys, *J. Phys. Chem. Solids* **47** (1986) 429–434.

[22] E. Wicke and J. Blaurock, New experiments on and interpretations of hysteresis effects of $Pd–D_2$ and $Pd–H_2$, *J. Less Common Metals* **130** (1978) 351–363.

[23] D.H.W. Carstens and W.R. David, Equilibrium pressure measurements in the beta region of palladium protide and palladium deuteride, *Los Alamos Report* LA-11456-MS, 1989.

[24] T.B. Flanagan, W. Luo and J.D. Clewley, Calorimetric enthalpies of absorption and desorption of protium and deuterium by palladium, *J. Less Common Metals* **172–174** (1991) 42–45.

[25] Y. Sakamoto, K. Takai, I. Takashima and M. Imada, Electrical resistance measurements as a function of composition of palladium–hydrogen(deuterium) systems by a gas phase method, *J. Phys.: Condensed Matter* **8** (1996) 3399–3411.

[26] A.K.M.F. Kibria, T. Tanaka and Y. Sakamoto, Pressure-composition and electrical resistance-composition isotherms of palladium–deuterium system, *Int. J. Hydrogen Energy* **23** (1998) 891–897.

[27] S.V. Dyomina, M.V. Glagolev and A.I. Vedeneev, Equilibrium pressure over palladium and its alloys, *Int. J. Alternative Energy and Ecology* **S2** (2003) 83–84.

[28] F.D. Manchester, A. San–Martin and J.M. Pitre, The H–Pd (hydrogen–palladium) system, *J. Phase Equilibria* **15** (1994) 62–83.

[29] R. Lässer, Isotope dependence of the phase boundaries in the PdH, PdD and PdT systems, *J. Phys. Chem. Solids* **46** (1985) 33–37.

[30] J.-M. Joubert and S. Thiébaut, Thermodynamic assessment of the Pd–H–D–T system, *J. Nucl. Materials* **395** (2009) 79–88.

[31] N. Esbjerg and J.K. Norskov, Dependence of the He-scattering potential at surfaces on the surface electron-density profile, *Phys. Rev. Lett.* **45** (1980) 807–810.

[32] S. M. Myers, P. Nordlander, F. Besenbacher and J.K. Norskov, Theoretical examination of the trapping of ion-implanted hydrogen in metals, *Phys. Rev. B* **33** (1986) 854–863.

[33] F. Besenbacher, B. Bech Nielsen, J.K. Norskov, S.M. Myers and P. Nordlander, Interaction of hydrogen isotopes with metals: Deuterium trapped at lattice defects in palladium, *J. Fusion Energy* **9** (1990) 257–261.

[34] K. Maier, G. Rein, B. Saile, P. Valenta and H.E. Schaefer, Positron annihilation measurements in Cu, Ni, Pd, and Pt in thermal equilibrium, *Proc. Fifth Int. Conf. on Positron Annihilation*, Sendai, R.R. Hasiguti and K. Fujiwara (Eds.), 1979.

[35] V.E. Zinovev, R.P. Krentsis and P.V. Geld, Thermal diffusivity and thermal conductivity of palladium at high temperatures, *Sov. Phys. Solid State* **11** (1969) 685–687.

[36] N. L. Peterson, Isotope effect in self-diffusion in palladium, *Phys. Rev.* **136** (1964) A568-A574.

[37] J.B. Adams, S.M. Foiles and W.G. Wolfer, Self-diffusion and impurity diffusion of fee metals using the five-frequency model and the embedded atom method, *J. Materials Res.* **4** (1989) 102–112.

[38] R.B. MacLellan, The kinetic and thermodynamic effects of vacancy interstitial interactions in Pd–H solutions, *Acta Materialia* **45** (1997) 1995–2000.

[39] Y. Fukai, The structure and phase diagram of M–H systems at high chemical potentials – High pressure and electrochemical synthesis, *J. Alloys and Compounds* **404–406** (1995) 7–15.

[40] Y. Fukai, Hydrogen induced enhancement of atomic diffusion in metals, *Defect Diffusion Forum* **297–301** (2010) 132–141.

[41] P.L. Hagelstein, Bird's eye view of phonon models for excess heat in the Fleischmann–Pons experiment, *J. Condensed Matter Nucl. Sci.* **6** (2012) 169–180.

[42] S. Szpak, P.A. Mosier-Boss and J.J. Smith, On the behavior of Pd deposited in the presence of evolving deuterium, *J. Electroanal. Chem.* **302** (1991) 255–260.

[43] D. Letts and P.L. Hagelstein *J. Condensed Matter Nucl. Sci.* **6** (2012) 44–54.

[44] R. Speiser, The thermodynamics of hydrogen in metals, *Metal Hydrides*, W. M. Mueller, J. P. Blackledge, and G. G. Libowitz (Eds.), Academic Press, New York, 1968.

[45] T.B. Flanagan and S. Luo, Thermodynamics of hydrogen solution and hydride formation in binary Pd alloys, *J. Phase Equilibria and Diffusion* **28** (2007) 49–57.

[46] J.E. Worsham Jr., M.K. Wilkinson and C.G. Shull, Neutron-diffraction observations on the palladium–hydrogen and palladium–deuterium systems, *J. Phys. Chem. Solids* **3** (1957) 303–310.

[47] G. Nelin, A neutron diffraction study of palladium hydride, *Phys. Stat. Sol. (b)* **45** (1971) 527–536.

[48] H.D. Carstanjen, J. Dünstl, G. Löbl and R. Sizmann, Lattice location and determination of thermal amplitudes of deuterium in α-$PdD_{0.007}$ by channeling, *Phys. Stat. Sol. (a)* **45** (1978) 529–536.

[49] G.G. Libowitz, The nature and properties of transition metal hydrides, *J. Nuclear Materials* **2** (1960) 1–22.

[50] G.A. Ferguson, Jr., A.I. Schindler, T. Tanaka and T. Morita, *Phys. Rev.* **137** (1965) A483–A487.

[51] G. Bambakidis, R.J. Smith and D.A. Otterson, Electrical conductivity versus deuterium concentration in palladium, *Phys. Rev.* **177** (1969) 1044–1048.

[52] J.K. Jacobs and F.D. Manchester, Thermal and motional aspects of the 50 K transition in PdH and PdD, *J. Less Common Metals* **49** (1976) 67–73.

[53] H.W. King and F.D. Manchester, A low-temperature X-ray diffraction study of Pd and some PdH alloys, *J. Phys. F: Metal Phys.* **8** (1978) 15–26.

[54] J.P. Bugeat and E. Ligeon, Lattice location and trapping of hydrogen implanted in FCC metals, *Phys. Lett.* **71A** (1979) 93–96.

J. Condensed Matter Nucl. Sci. 29 (2019) 416–439

Research Article

Investigation of Electron Mediated Nuclear Reactions

Andras Kovacs* and Dawei Wang

Fiskarsin Voima Oy, Nordanvik 142, 02580 Sjundeå, Finland

Pavel N. Ivanov

Moscow Engineering Physics Institute, Kashirskoe shosse 31, Moscow, 115409 Russia

Abstract

We investigated the concept of electron mediated nuclear reactions. The analysis of experimental reaction data indicates burst-like reactions and continuous energy production phases, which appear to be two distinct processes. We discuss the signatures of electron states which are highly localized around the nucleus. A theory of electron transitions into such highly localized states is proposed, with good correspondence to the observed reaction dynamics. Understanding the underlying nuclear reaction requires further investigation. With respect to applications, the discovered process implies the possibility of sustainable energy production from fuels comprising hydrogen, nickel, and lithium.

Keywords: Electron zitterbewegung, Highly localized electron states, Nuclear fuel, Nuclear energy production

1. Introduction

Up to now, an elementary particle intermediated nuclear reaction process has been conceptualized and investigated only in the context of neutron-mediated reaction of heavy nuclei. Here, we investigate the concept of electron mediated nuclear reactions. This class of nuclear reactions appears to involve an electron–nucleus configuration where the electron and the nucleus are in close proximity, and the reaction generates energetic electrons.

In this paper, we discuss Ni containing materials as possible fuels for electron mediated nuclear reactions. Observation of nuclear energy production from nickel–hydrogen and nickel–deuterium systems has been well established in multiple experiments [1–4]. In a previous publication [4], we have described experiments with Ni and Li containing mixtures, that showed experimental evidence of an exothermic nuclear reaction, which appears to be a continuous and well controllable reaction process. However, the theory of such reaction process is not yet settled. By focusing on the role of excited electron states in these experiments, we expect to progress towards uncovering the underlying reaction process.

*Corresponding author. E-mail: andras.kovacs@broadbit.com.

© 2019 ISCMNS. All rights reserved. ISSN 2227-3123

This paper is organized as follows. In Section 2, we discuss the experimental signatures of electron-mediated nuclear reaction processes with nickel containing fuels. In Section 3, we discuss in detail the possibility of highly localized electron states where an electron orbits a proton or deuteron at 0.383 pm proximity. Conclusions are drawn in Section 4.

2. Analysis of Nickel Fueled Reactor Experiments

Previous nickel fuel related observations of exothermic nuclear energy production also involved the presence of hydrogen [1,2,5], deuterium [3,8], or aqueous electrolysis [19]; therefore, the general assumption was that some type of hydrogen or deuterium involving fusion reaction plays a role in the underlying nuclear process. We have investigated lithium–nickel–copper containing fuel mixtures as possible fuels for electron mediated nuclear reactions [4], and we showed experimental evidence for an exothermic nuclear reaction. In our setup, hydrogen was introduced in the form of LiOH, produced over lithium during the assembly and sealing of fuel containers in ambient air. In the near future we plan to perform a LiOH-free counter-experiment in order to be certain that hydrogen plays a role in the initiation of this nuclear reaction.

As documented in [4], the observed reaction is initially a sequence of sudden heat and radiation producing bursts, which we are able to periodically re-initiate by temperature cycling. Figure 6 highlights some sudden temperature jumps. We note that the bursts appear to have originated from the lithium-rich molten phase. Such bursts imply chain reaction-like reaction dynamics. The observed bursts had varying strength and duration, and the measured exothermic heat production has been several hundred Watts. During a strong burst in our experiment, we observed radio-frequency signal generation with uniform power spreading in the 1–10 MHz frequency range; such flat radio-frequency power spectrum is an expected signature of decelerating charged particles. At the same time, a Geiger counter placed 0.5 m from the fuel container indicated a radiation level 40 times that of the background.

In [6], the gamma spectrum was monitored during the heating of Ni and $LiAlH_4$ fuel mixture, at similar temperature conditions as in our experiment. In that experiment, at a certain point, a strong burst event occurred whose gamma radiation spectrum is shown in Fig. 1, with the red dots indicating the signal of interest. The overall signal-to-background count ratio is 5.44. Because of the similar experimental conditions and the transient (i.e. burst-like) radiation, we anticipate that the gamma spectrum shown in [6] has the same origin as in our experiment.

Such bursty reaction dynamics was also observed in Ni–H and Ni–D systems, in the initial experimental phase. Figure 7 of Ref. [8] shows intense heat-producing bursts during the heat treatment of both Ni–H and Ni–D fuels. Based on continuous monitoring of H and D gas pressure, the authors conclude that the energy production during these bursts is too high for any possible chemical process. Figure 6 of Ref. [16] shows heat-producing bursts during the gas treatment of Ni–H fuel, as well as the coincident burst-like pressure increases. These burst type processes may be signatures of the same underlying nuclear reaction.

Several experiments achieved continuous energy production from Ni–H [1,2] and Ni–D [3] fuels. The contrast between the bursty and continuous reaction dynamics is shown in Fig. 2. A careful measurement of the radiation spectrum during continuous Ni–H energy production process [1] shows the absence of any measurable radiation signal above 100 keV. In a nickel electrolysis experiment, the observation of low-level X-ray emission in the 20 keV energy range and lasting several months after the experiment has been reported 19. Therefore, the bursty and continuous nuclear processes can be distinguished both through the different reaction dynamics and different gamma spectrum. However, occasional spontaneous overheating of copper–nickel alloy wires during Ni–H energy production runs has also been reported in [2]. Such sudden overheating indicates possible transitions between a continuous energy production process and the initial bursty reaction process.

It is useful to analyze in detail the transition from a bursty process into a continuous process, which has also been observed in detail in our experiments. Figure 21 of Ref. [4] shows the overall temperature evolution during

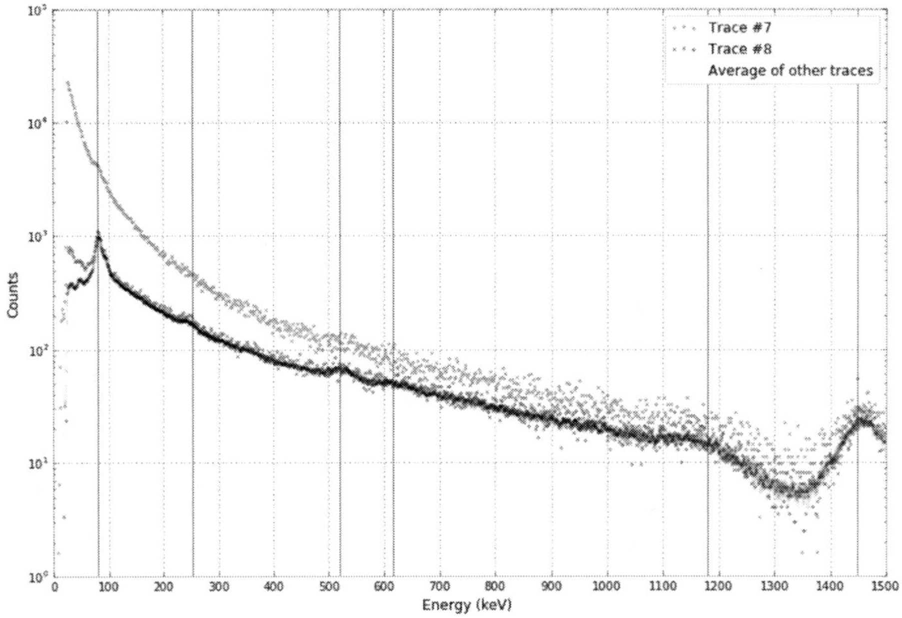

Figure 1. The gamma spectrum measurement of a burst from Ni+$LiAlH_4$ fuel during heating, produced by the authors of [6].

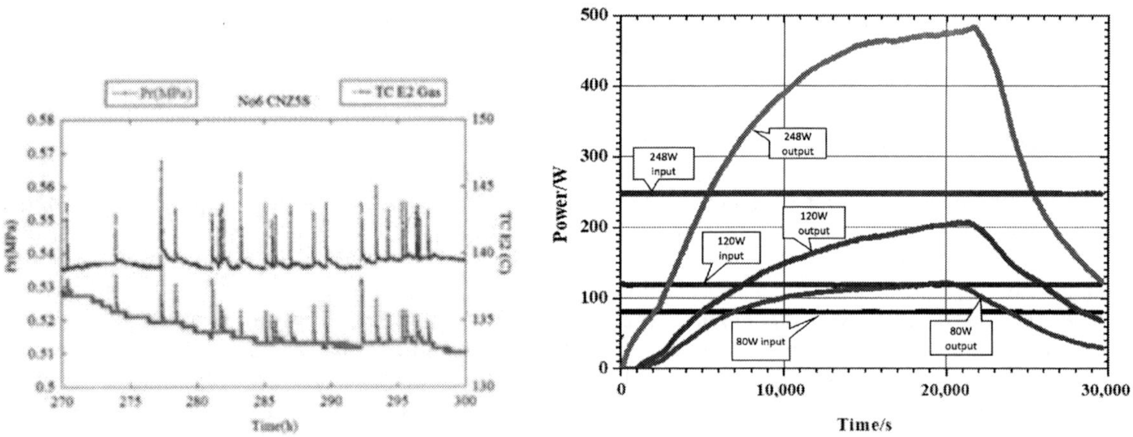

(a) (b)

Figure 2. The contrast between bursty (*left*), reproduced from [16]) and continuous (*right*), reproduced from [3]) reaction dynamics with nickel based fuel.

this process. After about half an hour at the operating temperature regime, the continuous reaction starts up and is maintained for approximately three hours. Figure 3 shows the temperature evolution of the heating and cooling phases

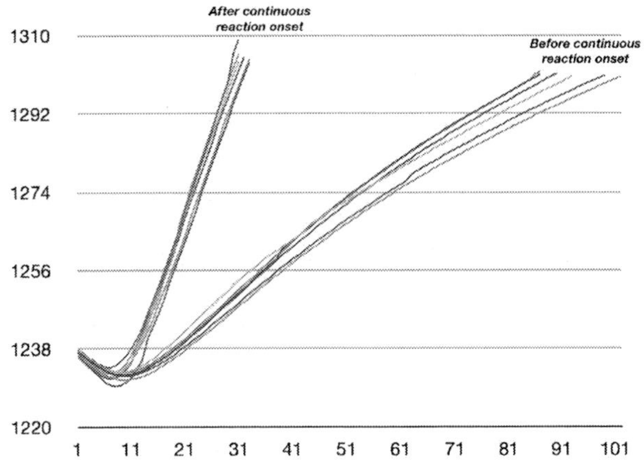

(a)

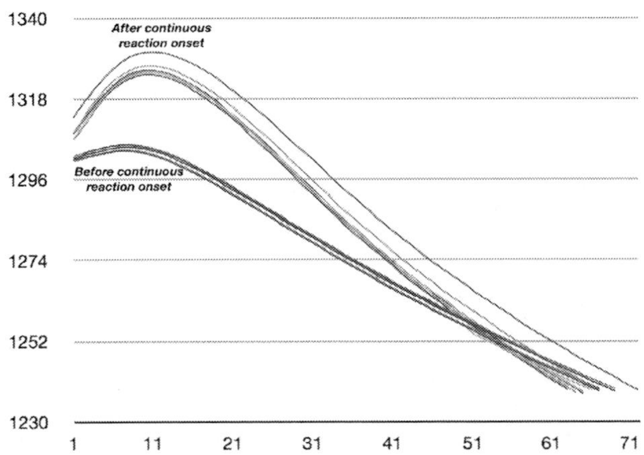

(b)

(c)

Figure 3. Temperature (°C) vs. time (s) charts during transition from bursty to continuous process with Ni–Cu–Li fuel mixture. (a) heating phases (constant heating power). (b) Cooling phases (zero heating power). The labels (c) indicate cycle no. relative to the starting of continuous reaction.

in the six cycles before and after the continuous reaction start-up. The negative (vs. positive) numbers of labels indicate cycle numbers before (vs. after) the starting of continuous reaction.

The continuous reaction starts up right away at the beginning of a certain heating cycle (i.e. "cycle 1"), as can be seen from the steeper slope already at the beginning. The fairly constant slope during the heating cycles, as well as the very similar slope among the positive numbered cycles show that the reaction produces constant exothermic power during the heating cycles. Signatures of the burst-like reaction (i.e. sudden temperature jumps) can be seen in some negative numbered preceding cycles, and also simultaneously with the continuous reaction in some positive numbered cycles. Cycles 1, 2, and 5 show very visible temperature jumps. In the cooling phase, the first observable feature is a temperature overshoot in the positive numbered cycles, relative to the temperature evolution of the negative numbered cycles. This indicates that the continuous reaction stays active for some seconds after the heating shut-off. Subsequently, the cooling rate is somewhat faster in the positive numbered cycles than in the negative numbered cycles. This is the consequence of local heat generation during the continuous reaction, which was picked up by the nearby thermocouple. It means that the average reactor temperature at the end of heating is somewhat lower in the positive numbered cycles than in the negative numbered cycles, because the electric heating was turned on for a much shorter time. The dynamics of the temperature falling phase indicates little or no exothermic reaction occurring during the temperature fall. Therefore, the reaction dynamics is actually semi-continuous, i.e. the continuous exothermic power is present mainly during the heating phases.

The longer-term dynamics of this continuous reaction was analyzed by constructing the frequency spectrum of the temperature signal's autocorrelation function. Figure 4 shows this spectrum for the initial part of the operating temperature regime, over the 20 minutes segment prior to the semi-continuous reaction onset, and then the spectrum for the duration of a semi-continuous exothermic reaction. The top part of Fig. 4 has less sharp auto-correlation peaks, even though the transients at the start of operating temperature regime were excluded. This may be partly caused by remaining transients of the initial operating temperature regime, and partly caused by the bursty temperature jumps in the initial phase. The continuous reaction part shows very sharp auto-correlation peaks and the absence of high-frequency noise. Since the heating program is controlled via temperature feedback, this data proves that the reaction dynamics are highly ordered, and remain nearly constant from cycle to cycle. There must be some reaction control parameter, which regulates the reaction rate to such a constant value. Perhaps the rate of nickel influx into the molten phase is such a control parameter. In summary, we have shown that the exothermic nuclear reaction of nickel-fueled reactors appears to be electron-mediated. In the initial phases it is a burst-like process, and may subsequently transition into a continuous reaction process, which is highly controllable. It is clearly seen from the data that the initial reaction bursts are a distinct process from the subsequent semi-continuous reaction process.

We also investigated a lithium-nickel-aluminum fuel composition, enclosed in a welded stainless steel container. The approximate Li:Ni:Al atomic composition was $1:10:1$. Figure 5 shows the temperature evolution at the fuel container (green curve) and at the edge of reactor (yellow curve), as well as the heating power evolution (blue curve). The horizontal axis shows the elapsed experiment time (seconds); after a slow temperature ramp-up, the left edge of the figure corresponds to the start of a constant 1350°C temperature program. The thermocouple at the fuel container is used for the temperature feedback control. Initially, the heating power is gradually reduced as the reactor transitions from being heated up to maintaining the target temperature. The first vertical dashed line indicates the first reaction signature, which slightly raises the reactor temperature while the heating power is reduced. The next vertical dashed line indicates the second reaction signature, which further raises the reactor temperature while the heating power is even faster reduced. The falling heating power is corroborated by the falling temperature at the edge of the reactor. A run-away reaction has occurred shortly after the red dashed line, which has melted down a large segment of the stainless steel container and destroyed the heating wires. The overall dynamics of this run-away reaction is very similar to the bursty reactions documented in [4], and the presence of electromagnetic emissions has also been detected. In the context of the previously published Li–Ni–Cu fueled experiments, this Li–Ni–Al experiment indicates that the active

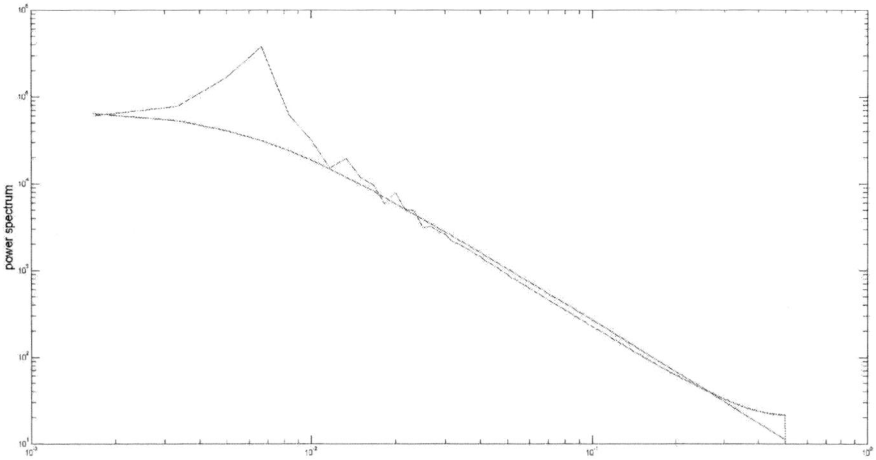

(a)

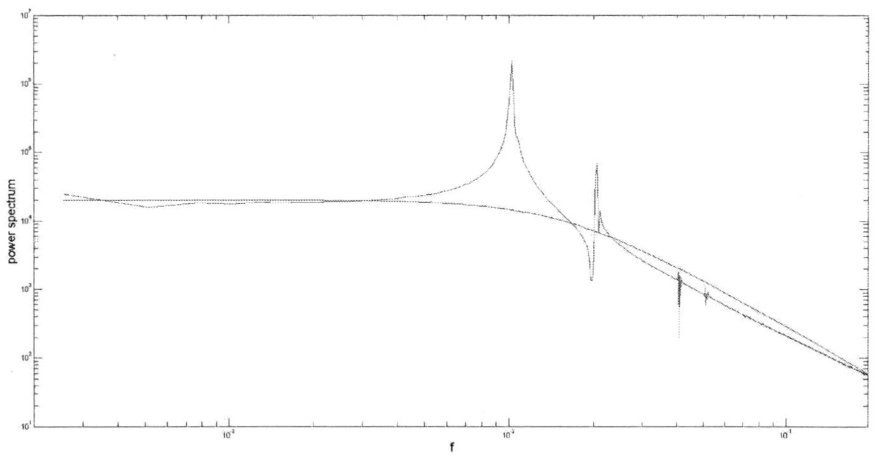

(b)

Figure 4. Autocorrelation frequency spectrum during the continuous reaction process with Ni–Cu–Li fuel mixture. (a) The spectrum prior to the continuous reaction and (b) the spectrum during the continuous reaction.

fuel component is indeed nickel and/or lithium.

In the Li–Ni–Cu experiments, the reaction is observed starting from the 1200°C limit (as shown in Fig. 6), which corresponds to the lowest melting temperature of the employed Constantan alloy. In the Li–Ni–Al experiments, the reaction is observed starting from the 1350°C temperature limit (as shown in Fig. 5); this corresponds to the melting temperature of the $AlNi_3$ phase, which is the lowest melting point phase within the nickel-rich nickel–aluminum alloys.

We noted in our experiments, that heating the fuel above approximately 1200°C is accompanied by electromagnetic noise while the temperature is rising. Many thermocouple-sensed electromagnetic disturbances reported in [4] are

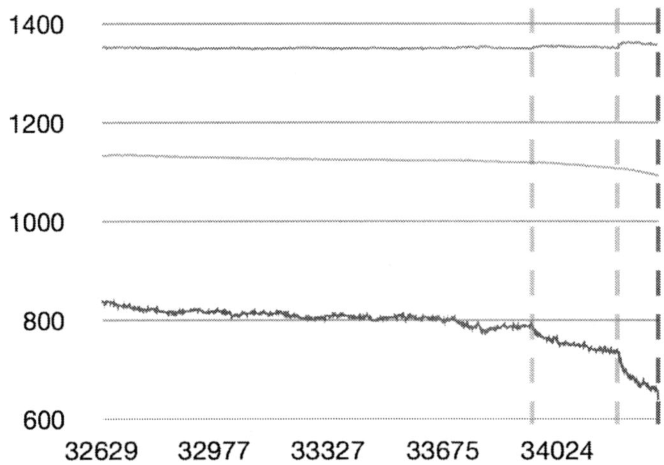

Figure 5. Temperature and heating power evolution of a reactor containing Li–Ni–Al fuel composition. The green curve shows the temperature (°C) at the fuel container, the yellow curve shows the temperature (°C) at the edge of reactor, and the blue curve shows the applied heating power (W). The horizontal axis indicates the elapsed time (s).

therefore not yet attributes of the actual reaction signal, but presumably signatures of such noise generating events. Figure 6, which is adapted from [4], shows the distinction between such non-exothermic precursor events and actual exothermic reaction signatures. The non-exothermic nature of these noise generating events can be seen from the constant slope of the temperature rise. We emphasize that such noise has been absent from calibration runs in the same temperature regime, and is therefore a genuine electromagnetic emission signal of the fuel sample.

As reported in [4], the reaction onset has been very predictable throughout multiple experimental runs, without any excessive waiting times. It requires further studies to determine whether the noise generating events, the sudden reaction bursts, and the continuous reaction processes are precursor events to each other or independent phenomena.

3. Discussion

For the possibility of electron-mediated nuclear reactions, the challenge is to identify the electron configuration which allows much stronger electron–nucleus interaction than the ordinary interaction between the inner shell electrons and the nucleus. It is indeed known that the electron environment can impact, e.g., the rate of nuclear electron capture; this effect has been most extensively studied for ^{7}Be. For instance, Table I. of Ref. [9] shows the half-life ^{7}Be in different environments, where the electron capture rate of ^{7}Be can indeed be changed by its environment. An electron configuration involving close electron–nucleus proximity would enhance the probability of electron capture. Most importantly, a close proximity electron–proton or electron–deuteron configuration would allow small inter-nuclei distance between such quasi-neutron and some other nucleus, thereby enabling catalyzed fusion reactions.

A mechanism for high probability nucleus-to-electron energy transfer is a pre-condition for the production of energetic electrons, which produce the observed braking signatures. In muon-catalyzed p–D fusion experiments, it has been observed that the muon carries away the fusion energy in most reactions, suggesting that a similar effect might arise under close electron–nucleus proximity. Until recently, no similar effect has been observed with electrons. However, it was recently observed that certain environments cause a shift in the de-excitation pathway of a fused

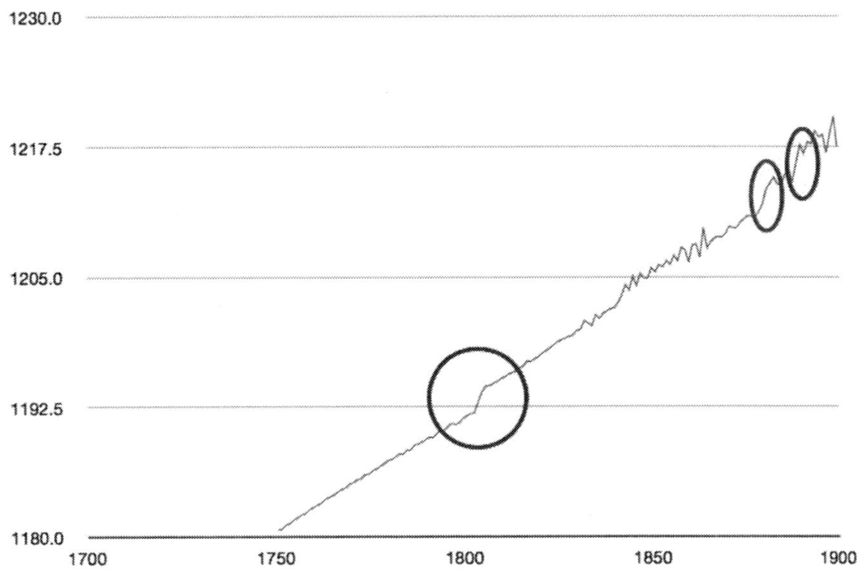

Figure 6. Temperature (°C) vs. time (s) measurement, reproduced from [4]. The rapid exothermic events are indicated by red circles. Electromagnetic emissions from non-exothermic processes, which appear as noise, are observable in-between.

^{3}He nucleus from gamma photon emission towards electron acceleration [7]. Specifically, all of the nuclear reaction energy of some p-D fusion events has been carried away by energetic electrons. The results presented in [7] therefore point to the existence of a very special configuration with close electron-nucleus proximity, resulting in strong enough electron–nucleus interaction for electrons to carry away the nuclear excitation energy.

Based on the above motivation, we introduce the idea of excited electron states, where electrons are in very close proximity around the nucleus. Usually, the excitation of electrons results in more delocalization, as electrons are pushed into higher energy orbitals. For the electron to be in a highly excited and highly localized state, the nature of excitation must be different from ordinary orbital excitations. We present below a discussion of such electron transition process.

3.1. Concept of close proximity electron–nucleus configurations

Our investigation of highly localized electron configuration is based on the relativistic quantum physics of the electron. The suggestion of a ring-shaped circulating electron structure has been originally made in 1915 by A. Parson. Such toroidal current structure is characterized by the electron's anapole moment (also referred to as toroidal moment) and charge radius parameters. The relativistic quantum mechanics-based calculation of the electron's toroidal current loop radius and charge radius is generally referred to as the electron's "zitterbewegung". Since Schrödinger's first zitterbewegung calculations in 1930, this structure was studied by many researchers of stochastic electrodynamics; a recent summary of the electron's relativistic dynamics is published in [10]. A quantum field theory based approach for calculating a gauge invariant expression for the electron's anapole moment and charge radius has been published in [11]; this study shows that the electron's zitterbewegung can also be derived through quantum field theory. The toroidal electron structure is sketched in Fig. 7; the electron current is circulating in both toroidal and poloidal directions, and the electron is locally moving at the speed of light. This light-speed movement of electromagnetic fields around the circulation axes is the electron's wave-like aspect; it is described by the electromagnetic wave equation. The displacement of electromagnetic fields into the orthogonal direction to this toroidal plane is the electron's particle-like aspect; it is described by the equations of relativistic particle dynamic, i.e. the Dirac equation. Reference [10] discusses the connection between these two aspects; i.e. it shows how relativistic particle dynamics arises from the wave-like electromagnetic current loop. We recommend that the reader becomes familiar with zitterbewegung related concepts described in [10], and validates that the presently discussed toroidal wave-like electron aspect is complementing the electron's particle-like dynamics, which is described by the Dirac equation. To avoid misunderstandings in the following paragraphs, we advise the reader to consider this wave-particle duality of the electron structure; i.e. the complementary roles of the electromagnetic wave equation describing electron oscillations within the toroidal zitterbewegung plane, and of the Dirac equation which describes many orders of magnitude slower quantum mechanical oscillations in the perpendicular direction to the zitterbewegung plane. In this study, we work with the virial theorem, which remains valid for the expectation values even in the quantum regime. When we refer to "orbit radius R", we mean its expectation value.

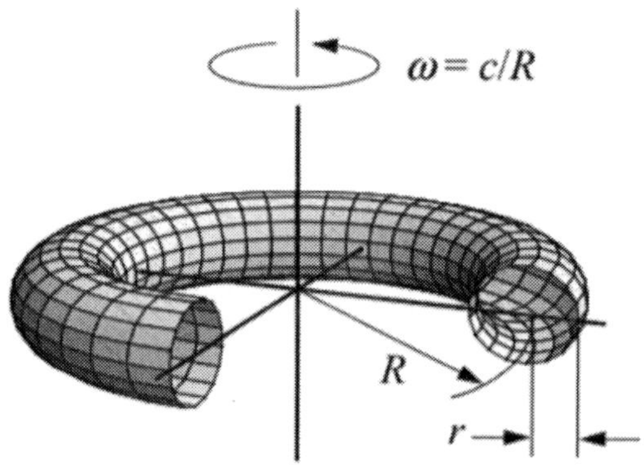

Figure 7. Visualization of the toroidal electron structure. R is the reduced Compton wavelength, and r is the charge radius.

At the thermal energy scale, the presence of electron zitterbewegung is revealed only by magnetic fields which cause a precession of its wave-like current loop circulation; this effect is detected as the electron spin [10], and in spectroscopy it is known as the Zeeman effect. Through the Zeeman effect, a signature of the toroidal electron structure has been experimentally observed for over 100 years, but for most of these years its origin was simply treated as an "inherent property" of the electron. As the electron energy level increases, it's dynamics must eventually be described relativistically, considering both particle-like mechanical motion and wave-like current loop motion. In the following paragraphs we investigate the dynamics of the electron at the reduced Compton wavelength scale, and consider whether it is possible for the electron to have a stable state at such distance around a proton. We refer to this hypothetical state as the "zitterbewegung orbit", and in this context the word "orbit" refers to both particle-like and wave-like motions of the electron charge. When an electron ring is located around a proton at the reduced Compton wavelength radius of $R_0 = 0.38616$ pm, its electrostatic potential is $U_{p0} = -3.728$ keV. Before falling into the proton's electrostatic potential well, the total electron energy is denoted as E_{total}. The initial energy of the electromagnetic field corresponding to the electron's wave-like motion is $W_{\mathrm{em}0} = 510.999$ keV.

At an orbit radius R around the proton, the potential energy is $U_{\mathrm{p}} = U_{\mathrm{p}0}(R_0/R)$ and the electromagnetic field energy of purely wave-like motion is $W_{\mathrm{em}} = W_{\mathrm{em}0}(R_0/R)$. For the magnetic field, this relation directly follows from the formula derived in [10]; $W_{\mathrm{magnetic}} = (\hbar v_{\mathrm{wave}}/2R)$. Since the magnetic and electric field energies of the electron's wave-like motion are equal, the same relation holds for the electric field energy. It is seen from this equation that it requires large energy to compress the electron into smaller orbit than R_0. When the electron is in the state of purely wave-like circulation at an orbit radius R, the following energy balance equation holds:

$$W_{\mathrm{em}} - W_{\mathrm{em}\,0} = E_{\mathrm{total}} - U_{\mathrm{p}}. \tag{1}$$

Based on the principle of particle-wave duality, in an equilibrium state we require an equivalence between a purely wave-like zitterbewegung motion, and a simultaneous wave-like zitterbewegung plus relativistic particle-like motion. This equivalence means that in both cases the electron has the same orbit and its zitterbewegung stays centered around the nucleus. In the first case, the wave-like motion has an instantaneous speed vector $\vec{c}_{\mathrm{em}}$, while in the latter case there are two orthogonal instantaneous speed vectors: the wave-like $\vec{v}_{\mathrm{wave}}$ and the particle-like $\vec{v}_{\mathrm{kinetic}}$, with $c^2 = v_{\mathrm{wave}}^2 + v_{\mathrm{kinetic}}^2$. The two descriptions yield the same trajectory if $\vec{c}_{\mathrm{em}} = \vec{v}_{\mathrm{wave}} + \vec{v}_{\mathrm{kinetic}}$ always holds true. An other way to express this equivalence is to require that upon reaching orbit radius R, the electron's particle-like rotation plus zitterbewegung rotation must be equal to a purely wave-like zitterbewegung rotation; i.e. the two descriptions become indistinguishable as the electron establishes an equilibrium state. Figure 8 illustrates the equivalence between these two descriptions of the electron.

Since the electron is at a steady distance R from the proton, and moves at

$$\gamma = \frac{1}{\sqrt{1 - \frac{v^2}{c^2}}},$$

the relativistic formulation of the virial theorem applies to its particle-like motion:

$$E_{\mathrm{kinetic}} = -U_{\mathrm{p}}\frac{\gamma}{\gamma + 1}. \tag{2}$$

The above formula is derived from the relativistic expression of the virial term: $\frac{1}{2}pv = \frac{1}{2}\beta\gamma m_0 c \cdot \beta c$. Considering that

$$E_{\mathrm{kinetic}} = (\gamma - 1)m_0 c^2,$$

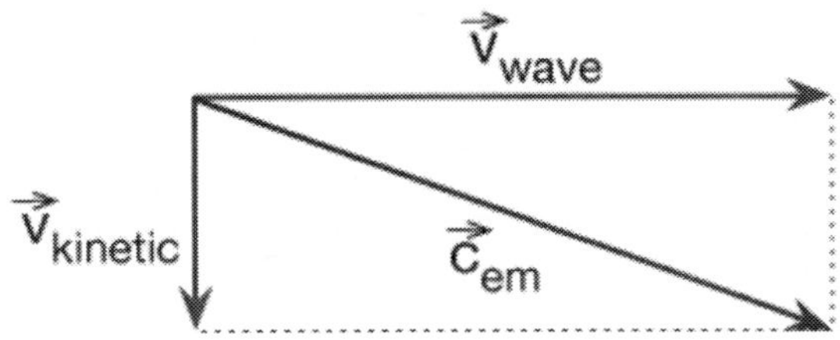

Figure 8. Illustration of the equivalence between instantaneous electron speed vectors in the case of purely wave-like zitterbewegung rotation (*blue*) and the particle-like case of composition between mechanical plus wave-like zitterbewegung rotations (*red*).

we get

$$\frac{1}{2}pv = \frac{\gamma+1}{2\gamma}E_{\text{kinetic}},$$

from which the above formula is derived. Formula (2) is valid for a motion along a straight line. The electromagnetic field energy of the complementing wave-like zitterbewegung current loop is

$$W_{\text{wave}} = W_{\text{em}}\frac{v_{\text{wave}}}{c}. \tag{3}$$

The total energy difference between a purely wave-like state and a wave-like plus relativistic particle-like state is

$$\Delta E = W_{\text{wave}} + E_{\text{kinetic}} - W_{\text{em}}. \tag{4}$$

Equation (2) cannot yet be exact, because it is applicable only to movements along straight line, while the particle-like electron motion is along a circular orbit. We therefore refine Eq. (2) by taking into account also the Thomas precession effect, which causes the circular orbit's angular speed to change as $\omega \rightarrow \gamma\omega$ in the frame of the electron. If the proton could be suddenly removed, the electron would continue its path along a straight line, without any instantaneous change of its momentum or kinetic energy with respect to the lab frame. However, its lab frame speed would instantaneously change because of the removal of Thomas precession. Let γ and β describe the electron's Lorentz factor and light speed fraction obtained according to Eq. (2). As discussed above, the Thomas precession effect does not change the electron's momentum or kinetic energy, therefore $p = \beta\gamma m_0 c$ and $E_{\text{kinetic}} = (\gamma - 1)m_0c^2$ However, since the electron precesses γ times faster in its own frame than in the lab frame, its lab frame speed becomes $v = \frac{\beta c}{\gamma}$. Using these formulas for p, v, and E_{kinetic}, the following refined kinetic energy formula is obtained from the relativistic formulation of the virial theorem:

$$E_{\text{kinetic}} = -U_{\text{p}}\frac{\gamma^2}{\gamma+1}. \tag{5}$$

Figure 9 shows E_{total} and ΔE as a function of R, calculated from Eqs. (3)–(5). Based on the above stated wave-particle equivalence, we require $\Delta E = 0$. Why is $\Delta E = 0$ in an equilibrium state? The meaning of an equilibrium state is that small perturbations around the equilibrium do not change the energy of the system, e.g. like gravitational energy equilibrium at the top of a hill or in the bottom of a valley. The electron's particle-like aspect is its movement in perpendicular direction to the zitterbewegung plane. The $\Delta E = 0$ condition means that small perturbations of the zitterbewegung orbit state do not change the energy of the system.

This $\Delta E = 0$ condition is met at a negative binding energy, i.e. $E_{\text{total}} > 0$. We note the interesting coincidence that the zitterbewegung radius has shrinked from its natural reduced Compton wavelength value by exactly one electron charge radius; i.e. by 2.82 fm. So far in this analysis, we have neglected the magnetic electron-nucleus interactions. The following refined calculation considers the also Lorentz force experienced by the electron due to the proton's magnetic field.

To minimize the magnetic potential, the electron's and proton's magnetic moments align their directions. Consequently, the proton's magnetic moment is perpendicular to the zitterbewegung plane. The proton-originating magnetic field experienced by the electron is therefore also perpendicular to the zitterbewegung plane, and has the following magnitude:

$$B = \frac{\mu_0}{4\pi}\frac{\mu_{\text{p}}}{R^3} = 2.5 \times 10^4\ \text{T}.$$

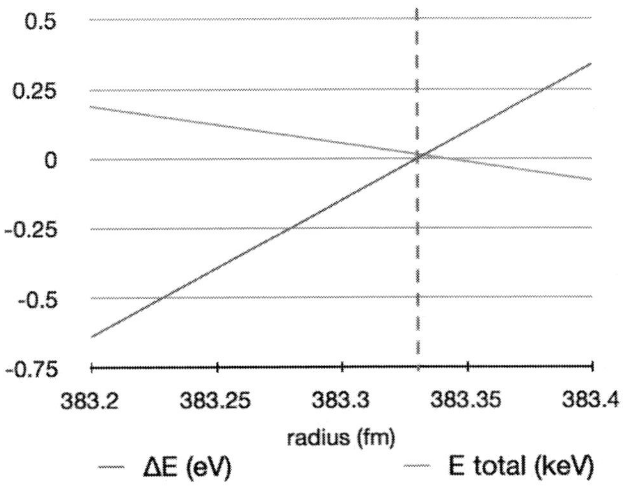

Figure 9. Relativistic estimation of ΔE and E_{total} as a function of the orbit radius. The dashed line indicates the equilibrium condition of $\Delta E = 0$.

The above discussed electrostatic estimation of the equilibrium state gives $\beta = 0.08558$. The Lorentz force experienced by the electron is radial, and has a magnitude of $F_{\text{L}} = ec\beta B = 1.028 \times 10^{-7}$ N. The radial Coulomb force experienced by the electron is

$$F_{\text{C}} = \frac{-U_{\text{p}}}{R} = 1.57 \times 10^{-3}\ \text{N},$$

which is four orders of magnitude larger than the magnetic force. Although the virial theorem is not applicable to a magnetic potential, since the magnetic force is so much smaller than the electrostatic force, and since the two forces are parallel, the magnetic effect can be treated as a linear perturbation of the electric potential. The effective force felt by the electron is $F = F_{\text{C}} + F_{\text{L}} = (1 + 6.55 \times 10^{-5}) \times F_{\text{C}}$. Equating the radial force with the radial derivative of potentials, we get

$$F_{\text{C}} = \frac{-1}{R} U_{\text{p}} \quad \text{and} \quad F_{\text{L}} = \frac{-2}{R} U_{\text{M}}.$$

Therefore at a given radius $F_{\text{L}}/F_{\text{C}} = 2U_{\text{M}}/U_{\text{p}}$. In other words, there needs to be twice as much Coulomb potential as magnetic potential in order to have the same force effect. Using linear perturbation, this additional force can be incorporated into Eq. (5) by making $U_{\text{p}} \to (1 + 2 \times 6.55 \times 10^{-5}) \times U_{\text{p}}$ substitution. Considering that in the above estimation $U_{\text{p}} = -3.756$ keV, this additional force effect corresponds to $\Delta U_{\text{p}} = 0.49$ eV.

Figure 10 shows the E_{total} and ΔE values obtained after also taking into account the Lorentz force effect, calculated again from Eqs. (3) and (4), and the $U_{\text{p}} \to U_{\text{p}} + 0.49$ eV adjusted Eq. (5). The $\Delta E = 0$ condition is met at $E_{\text{total}} = 81$ eV. This energy value is our final theoretical estimation for the required transition-initiating electron kinetic energy in case of a proton nucleus.

The obtained result shows that the required transition-initiating electron kinetic energy depends on the nuclear magnetic moment. The deuteron's magnetic moment is 0.857 nuclear magnetons, which is significantly weaker than

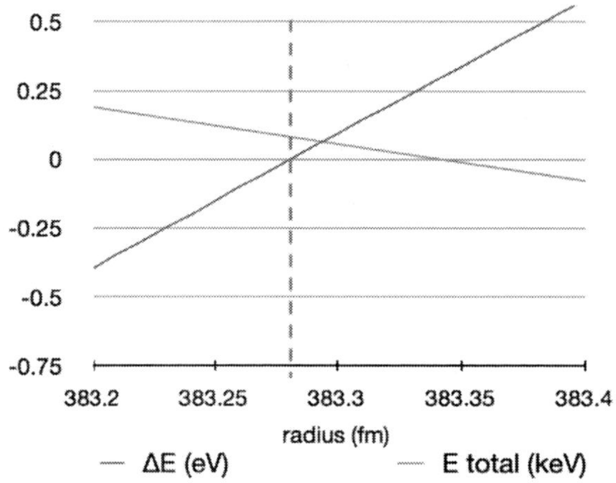

Figure 10. Final estimation of ΔE and E_{total} as a function of the electron orbit radius around a proton, taking into account both electric and magnetic interactions. The broken line indicates the equilibrium condition of $\Delta E = 0$.

the proton's magnetic field. Using this magnetic moment value for the equilibrium state calculation, we obtain $\Delta U_{\text{p}} = 0.15$ eV and $E_{\text{total}} = 35$ eV total energy for a zitterbewegung orbit around a deuteron.

In the preceding paragraphs we refined the total energy estimation by also taking into account the effect of the proton's magnetic field on the electron orbit. In this paragraph we consider how the electron's magnetic field effects the proton in the center. The strong induced magnetic field at the center of the electron orbit interacts with the proton's zitterbewegung motion, causing it to precess around the magnetic field lines. This type of nuclear precession is equivalent to the electron precession described in Section 3.2 of [10], and is routinely exploited in nuclear magnetic resonance imaging devices; this precession frequency is half of the photon frequency causing nuclear magnetic resonance at a given magnetic field strength. The induced precession of the proton's zitterbewegung motion is causing a Zeeman split in the proton's energy levels, and the proton assumes the lower energy level. This lowered proton energy level creates a restoring force for maintaining the equilibrium state; i.e. the electron's zitterbewegung orbit is now a magnetically stabilized meta-stable state. Figure [11] illustrates the closely bound electron–proton system in such zitterbewegung orbit state. This analysis shows that there is a meta-stable equilibrium electron orbit around a proton at the reduced Compton wavelength distance scale, but it has a positive total energy. We emphasize that the positive total energy of this zitterbewegung orbit state implies that at ordinary temperatures electrons occupy the lower energy Bohr orbit state around a proton, where $E_{\text{total}} = -13.6$ eV.

The electron's meta-stable zitterbewegung orbit around a proton, or other light nucleus, may be understood as a fundamental relativistic state. Furthermore, [10] discusses the compatibility between the photon-like zitterbewegung motion and the Heisenberg uncertainty principle. We therefore propose that at a certain electron kinetic energy level the electron's zitterbewegung motion may localize itself as a relativistic orbit around a light nucleus. In case of a proton, our theoretical estimation gives $\approx$80 eV for the transition-initiating electron kinetic energy level, while for the deuteron we obtain $\approx$35 eV energy level. This proposition allows us to correctly predict the reaction dynamics of nickel-fueled reactors. Using relevant experimental data, in the following sections we will precisely identify this required electron kinetic energy level.

Regarding the probability of energetic electron output vs. γ photon output upon nuclear de-excitation, the impact of zitterbewegung orbit presence may be estimated through the methodology shown in [27], specifically through the

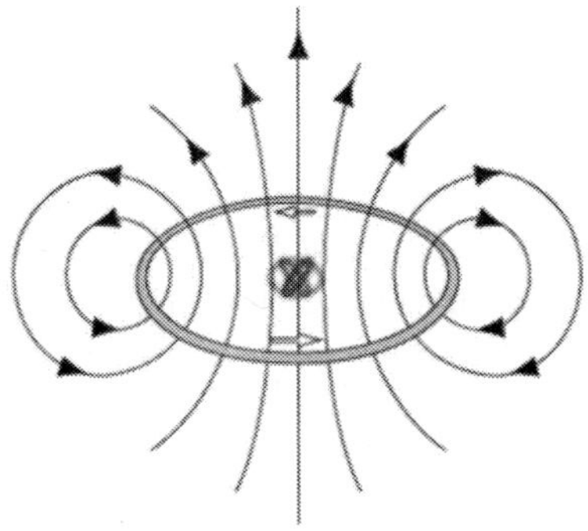

Figure 11. Illustration of the stabilized electron orbit. The gray ring represents the electron's zitterbewegung orbit, the arrows represent magnetic field lines, and the red rings represent the proton's precessing zitterbewegung motion.

precise evaluation of formulas (8)–(11) presented in [27].

Reference [18] reports 2.3 pm inter-nuclei distance measurement in dense states of hydrogen. As discussed in [10], such bond distance is obtained when the zitterbewegung orbits of electrons around neighboring nuclei are phase coherent, and thereby provides further experimental support to our concept of localized zitterbewegung orbit.

Regarding nuclei with $Z > 1$, we note that for ^{4}He such state would not be stable because it has no nuclear magnetic moment. In case of lithium, our theoretical estimation gives approximately 150 eV for the required electron kinetic energy level. However, the presence of a zitterbewegung orbit electron around a lithium nucleus would not meaningfully impact its fusion probability.

Table 1. Comparison of the total electron energy level in zitterbewegung orbit state around various nuclei.

Nucleus	Potential energy (keV)	Kinetic energy (keV)	Total electron energy (eV)
^{1}H	–3.756	3.837	81
^{2}H	–3.756	3.791	35
^{7}Li	–11.44	11.59	150

3.2. Transition to close electron–nucleus proximity state

Since the use of heavier elements for catalyzing electron transition into zitterbewegung orbit was documented in [18], we consider the possibility that the initiating electron kinetic energy for the above-described meta-stable state does not need to come from a free electron, but it may be the kinetic energy of a bound electron's orbit. Specifically, a bound electron of some heavier atom may transition into an electron–proton or electron–deuteron zitterbewegung orbit state when its wavefunction overlaps with an approaching proton or deuteron nucleus. The virial theorem states that in a single electron hydrogen atom the electron's kinetic energy is equal to its ionization energy. In multi-electron atoms this relationship may not be exact, but nevertheless we can use the electron ionization energy to estimate its kinetic energy. Table 2 lists the estimated electron ionization energies for nickel's outer electrons. For the outermost N1

Table 2. Listing of electron ionization energies in nickel's outer orbitals.

Orbital	N1	M5	M4	**M3**	M2	M1
Ionization energy (eV)	7.64	15	15.7	**79**	84.2	124.4

electron orbital, we used the available ionization energy data. For the other electrons, we accounted the relative X-ray transition energies between the N1 and other orbitals, and added the ionization energy of the N1 orbital.

Visible light driven transmutation in the presence of a bromine-containing catalyst (Eosin-Y) was reported in [17]. Since nickel and bromine both appear to be initiating catalysts of nuclear reactions, we compare their electron energy levels. The ionisation energy of bromine's electrons in the brominated organic compound can be estimated to be the similar as in the atomic bromine. Comparing data of Tables 2 and 3, the common energy level appears to be at 79–80 eV. Since [17] reports visible light driven transmutation, the electron kinetic energy for zitterbewegung orbit transition is within 1–2 eV from this common 79–80 eV level. We note the very good match between this 80 eV energy level and its theoretic estimation in Section 3.1.

Table 3. Listing of electron ionization energies in bromine's outer orbitals.

Orbital	N3	N2	N1	M5	**M4**	M3	M2	M1
Ionization energy (eV)	12.5	11.8	23.5	78.3	**79.4**	191.2	198.5	264.8

Using the same methodology, we calculated the electron kinetic energies of other elements around nickel. As seen in Fig. 12, there are several elements which have close to 80 eV kinetic electron energy level. However, elements with more than than 80 eV of electron energy, such as Cr or Cu, may not be suitable if their kinetic energy is slightly excessive with respect to the resonant transition energy level. Besides Ni and Br, at least Co and V are expected to have suitable electron energy level.

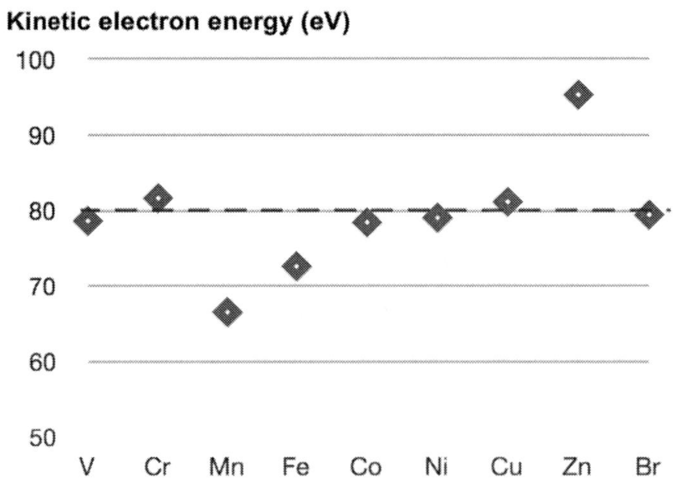

Figure 12. Display of electron energies which are closest to the 80 eV level, shown for elements around Ni.

By comparing Refs. [3] and [5], we observe an important distinction between hydrogen and deuterium fuels: while a nickel-based reactor produces exothermic energy with hydrogen fuel, it only produces exothermic energy with deuterium fuel after applying a palladium surface coating over it. This experimental evidence supports the above identified hydrogen vs. deuterium difference in the electron transition energies. Reference [28] reports nuclear transmutation effects initiated upon passing deuterium through thin CaO layers. These findings were replicated in [29], and its authors also clarified that such transmutations do not occur with with hydrogen gas, but only with deuterium gas. As seen in Table 4, calcium indeed contains approximately 35 eV electron kinetic energy orbitals. According to [30], the M1 orbital's electron binding energy is 0.6 eV higher in CaO than in metallic Ca. Assuming similar energy shift also for the other M-orbitals, in CaO we estimate the M2 and M3 orbitals' kinetic energy levels at 35.6 and 35.1 eV, respectively, in good agreement with our theoretical prediction.

Table 4. Listing of electron ionization energies in calcium's outer orbitals.

Orbital	N1	**M3**	**M2**	M1
Ionization energy (eV)	6.1	**34.6**	**35**	55.5

We make the following proposition about the transition to the close electron–nucleus proximity state: (i) A transition into zitterbewegung orbit around a proton occurs at the 80 eV electron kinetic energy level, and at 35 eV in case of a deuteron; (ii) The electrons of interest are those which have only a small gap with respect to the required energy level – however these are chemically inactive inner electrons. 1–10 eV energy range collisions between atoms can energize such inner electrons of interest to the transition energy level.

The required 1–10 eV collision energy is higher than what can be normally supplied by thermal heating. However, such collision energy may be supplied by a Fermi potential difference between interfaces, by an applied electric field in an electrolysis setup, or by energetic plasmon oscillations. In [21], accelerated uranium decay was observed when aqueous solution of uranium ions was subjected to laser induced plasmons on gold surface. In that study, the measured plasmon energy was over 5 eV. The most plausible explanation involves the above outlined process, resulting in one of gold's or uranium's electrons to be energized to the right kinetic energy level for the transition into a highly localized electron orbit around the H or D nucleus of the water solvent. Some of these resulting quasi-neutron-like particles accelerate uranium's decay by fusing into the uranium nucleus. Such intermediating role of the H or D nucleus is proved by the varying decay rates observed in H_2O vs. D_2O solution.

3.3. Experimental conditions for the transition process

On the basis of the above outlined electron transition process, this section analyzes collisions between atoms in those experimental setups where continuous energy production has been observed. The analysis of Ni–Li phase diagram reveals that Li alloys with Ni up to 10–15 atomic%, above which ratio there are two immiscible phases: an Ni-rich and a Li-rich phase. Similarly, the Cu–Li phase diagram indicates that Cu has a very low alloying capability with Li. Therefore, in our Li–Ni–Cu fueled experiments there is a phase boundary between the molten Li rich and the solid Cu/Ni rich phases. The Fermi level difference between these two phases is estimated to be 5–7 V. Figure 13 illustrates this electronic structure and the influx of cations into the accelerating boundary region during melting. Since Cu and Ni are +2 charged in the metallic environment, the ions crossing over the molten-solid phase boundary during melting process gain 10-14 eV on the average, accelerated by the electric field between these two phases. Their subsequent collision with ions in the molten phase produces the condition allowing the transition of some inner electron into close proximity zitterbewegung orbit. In other words, some fraction of these accelerated ions' kinetic energy provides the missing electron energy for the highly localized transition. Since the Constantan alloy has a continuous melting

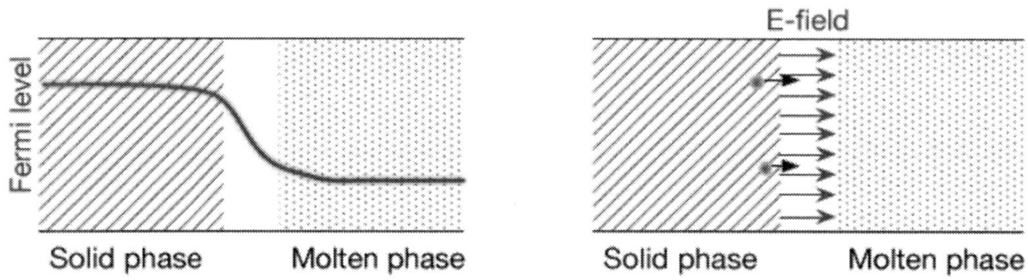

Figure 13. Illustration of the Fermi level difference between solid/molten phases in our experiment (*left*) and the ion flux into the accelerating region during melting (*right*.)

temperature range between 1250 and 1300°C, the continuous acceleration of ions during the heating phase may explain the apparently constant reaction power during the heating phase. There is no similar ion accelerating process during the cooling phase; this difference corresponds to the observed approximately zero reaction power during the cooling phase.

The role of the surface interface has also been reported in hydrogenated/deuterated Ni experiments. In [2], the authors reported the exothermic reaction power to be related to the surface area size of the interface between hydrogenated constantan and borosilicate fiberglass. In [3], the authors reported the exothermic reaction power to require the presence of deuterated nickel – palladium interface. Such interfaces are needed for creating a Fermi level difference between two solid state materials in physical contact. While there is no molten phase in these experiments, H^+ or D^+ ions may be diffusing across such interface, and thereupon being accelerated by the electric field counter-balancing to the Fermi level difference. Indeed, the authors of [2] noted the adsorption of hydrogen on the surface of fiberglass, and the authors of [3] used palladium which has a high hydrogen absorption. These choices of surface layer materials set up the condition for H^+ or D^+ ions diffusing across an interface involving a Fermi level difference. The exothermic reaction power should be therefore proportional to the H^+ or D^+ diffusion rate, which is related to the temperature by the $\exp(-E_a/k_B T)$ factor, where E_a is the activation energy for diffusion. As shown in Fig. 14, the exponential power dependence on the reactor temperature was indeed confirmed in [3].

Altogether, the concept of electron transition to highly localized zitterbewegung orbit appears to be a useful starting point for explaining the experimental observations. In various experiments, the observed reaction dynamics matches the proposed trigger condition of 1–10 eV range excitations. Identifying the details of the consequent electron mediation enabled nuclear processes requires further study. This analysis points to the role of hydrogen in our experimental setup. This suspected role of hydrogen will be validated in future replications of the experiments described in [4], where the fuel containers shall be assembled and sealed under inert gas atmosphere.

3.4. Production of highly localized electrons by braking energetic particles

The rate of p or D involving nuclear fusion reactions has been extensively studied in various materials. The fusion enhancement rate is generally characterized by the screening energy parameter U_e, characterizing the fusion rate enhancement over a range of incoming ion beam energies. Up to now, the observed fusion rate enhancement was generally thought to be a consequence of delocalized electron screening, which is described by the Thomas–Fermi screening model.

Table 5 shows the screening energy parameter in various background materials of interest, from data reported in [12,14,15]. A very strong fusion enhancement is seen in hydrogenated graphite; in this environment the 5.6 MeV

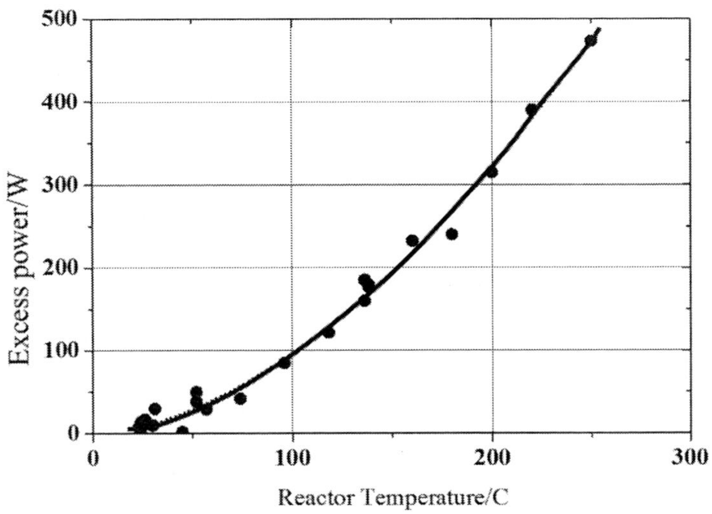

Figure 14. The exponential temperature dependence of excess reaction power in a solid-state deuterated Ni fuel, reproduced from [3].

Table 5. Fusion rate enhancement characterized by the screening energy parameter for the p+^{7}Li (A), D+^{6}Li (B), and p+^{19}F (C) reactions.

	Material	H stoichiometry	U_e
	TiH	1.03	3.9 keV
A	$PdH_{0.2}$	0.21	3.6 keV
	Graphite	0.06	10.3 keV
	LiF		350 eV
B	Solid Li		400 eV
	Liquid Li		700 eV
	Kapton		0 keV
C	Amorphous carbon (hydrogenated)		36 keV
	TiH		73 keV
	Graphite (hydrogenated)		115 keV

electron-assisted nuclear de-excitation has been also observed. Strong fusion rate enhancements are observed for hydrogenated Ti and Pd materials, and by analogy it is reasonable to expect hydrogenated Ni to also have similar screening energy parameter.

A significantly higher fusion rate enhancement is seen in molten lithium than in solid lithium or lithium fluoride. Interestingly, this liquid phase rate enhancement appears to be specific to lithium, and not a general solid-liquid difference in metals. In related experiments [12,13], the authors performed deuterium bombardment of solid/molten indium and ^{6}Li metals respectively. These studies show that at 10 keV bombardment, the reaction rate is about two orders of magnitude higher in solid indium than in liquid indium. In contrast, the reaction rate has been about 40% higher in liquid lithium than in solid lithium.

Altogether, the screening energy data shows several major inconsistencies with the delocalized electron screening model. (i) The fusion rate enhancement is 3–4 orders of magnitude higher in some cases, e.g. in hydrogenated graphite, than the theoretically predicted Thomas–Fermi screening model based screening energy parameter. (ii) The solid–liquid phase transition has strong influence on the screening energy, which is unexpected. (iii) The chart of the fusion rate vs. ion beam energy deviates from the theoretically expected chart. The analysis of this deviation prompted the authors of [20] to suggest a "nuclear reaction resonance" at 105 $\pm$ 15 eV. Perhaps coincidentally, such 105 eV resonant ion energy implies 105/3=35 eV average kinetic energy along each spacial axis during its deceleration, which is matching the 35 eV kinetic electron energy required for the establishment of zitterbewegung orbit around D nucleus.

When the fusion reaction involves p or D, any highly localized electrons around the p or D nuclei increase the probability of fusion between these screened p or D and other nuclei. Therefore, measuring the fusion enhancement rate is also a suitable proxy for measuring the production rate of such highly localized zitterbewegung orbit electrons. The correspondence between the observation of surprisingly strong screening energy parameter in graphite and the observation of electron-assisted nuclear de-excitation indicates that the screening energy parameter may in fact measure the rate of highly localized electron production. At a given proton beam energy, the theoretical rate of p+^{7}Li fusion is much higher than the rate of p +^{19}F fusion due to the Coulomb barrier difference. Since the rate of highly localized electron production in a given material is the same in either case, the apparent screening energy parameter becomes an order of magnitude higher for the p +^{19}F fusion case, as seen in the data of Table 5.

Based on the above discussed transition model, the production of highly localized electrons around a proton may happen via direct excitation of delocalized electrons to 80 eV kinetic energy or via the small excitation of those bound electrons which have already close to 80 eV kinetic energy. In case of a deuteron nucleus, this energy level is 35 eV.

While the 105 $\pm$ 15 eV resonance proposed in [20] is too low value for any conceivable nuclear process, it is indicative of the required electron kinetic energy level for transition into highly localized state, assuming a thermalization type coupling between kinetic ion energy and electron excitation at similar energy. While in metals the electronic excitation lifetime is proportional to the inverse square of the excitation energy, in graphite the electronic excitation lifetime is proportional to the inverse of the excitation energy [24]. This property of graphite may explain the higher screening energy parameter in the graphite environment.

Regarding the excitation of already close to 35 eV inner electrons, Refs. [22,23] provide insightful data. Reference [22] reports 600 eV screening energy for D–D fusion in PdO, in contrast to the 300 eV screening energy for Pd, measured using the same methods. This is again contrary to the Thomas–Fermi screening model, which predicts a higher screening energy parameter for the delocalized electron containing Pd. We estimate the kinetic energies of palladium's orbitals, and surprisingly find that none of them are in the proximity of 35 eV. This points to the possibility that oxygen's L1 orbital may be near 35 eV, at some locations of PdO. According to [30], oxygen's L1 orbital kinetic energy is varying in the wide range of 12 eV, depending on oxygen's oxidation state. Reference [23] shows that in Zr environment the D–D fusion enhancement factor varies with the amount of Zr surface oxidation. This dependence is shown in Fig. 15. For zirconium, we also find that none of its orbitals are in the proximity of 35 eV. Therefore, palladium-oxide and zirconium-oxide involving experiments reveal that oxygen appears to be the common source of fusion enhancement. However, since the fusion enhancement peaks at a certain level of surface oxidation, it is not the fully oxidized state of zirconium surface where oxygen's L1 orbital energy has close to 35 eV kinetic energy. We suggest that there may be an uncommon oxidation state where oxygen's L1 orbital energy is close to 35 eV, and such uncommon oxygen state may be present at the "nuclear active environment" sites. Several researchers of palladium–deuterium reactors suggest the so-called "nuclear active environment" surface sites to be the active spots where nuclear reactions occur.

In summary, measurements of the screening energy parameter show results which are incompatible with the delocalized electron screening concept and appear to be quantifying the production rate of highly localized zitterbewegung orbit electrons by energetic particles. The data shows that the kinetic energy for such meta-stable state initiation may indeed originate either from a delocalized electron energized to 80 eV or 35 eV, or from a similar kinetic energy of

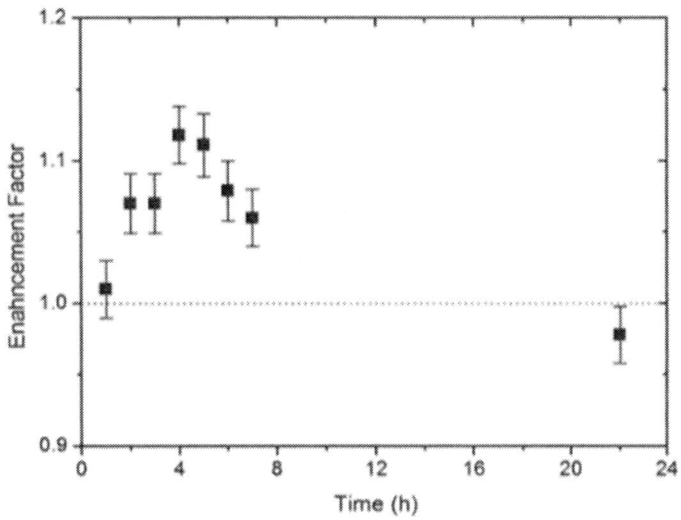

Figure 15. The D–D fusion enhancement factor measured for 15 keV deuteron bombardment of Zr, reproduced from [23]. The chart is showing the enhancement evolution with respect to the initial fusion rate. The Zr surface is oxidation-free at the start of experiment, and oxidizes gradually.

a bound electron's orbit. The previously overlooked role of oxygen in deuterium-fueled experiments is pointed out. We note that Tables 2–4 relate to experiments demonstrating nuclear energy production and transmutation; i.e. the identified energy level is derived from a broad base of diverse experiments.

The possibility of a positive feedback loop between the energetic electrons produced by a nuclear reaction and the production of more highly localized electrons during the braking of these energetic electrons explains the burst-like reaction dynamics, which was experimentally observed.

3.5. Quantitative predictions

If a physical process model is valid, it must be possible to derive refutable predictions from it. Such predictions are specific to the model, and the proposed process may be proven correct by the validation of these predictions. In this section we therefore make a number of quantitative predictions which are unique to the electron mediated nuclear reaction model described herein, and which may be used for its validation:

(1) **Observable signatures at 35 eV or 80 eV electron kinetic energy**. The most important prediction is the above explained role of the 80 eV energy level in hydrogen fueled reactors and 35 eV energy level in deuterium fueled reactors. We predict that various other transition signatures shall be found at these electron kinetic energy levels. Specifically, we predict that the "runaway electron" production of hot fusion reactors is a signature of this phenomenon. It has been known for over 40 years that hot plasmas occasionally enter into a state characterized by emission of energetic electrons and concurrent enhanced neutron emission. Such state tends to appear during the disruption or shutdown of the hot plasma current. The so-called "Dreicer mechanism", which most related publications mention as a proposed explanation for this phenomenon, is contrary to the basic principles of plasma physics and thermodynamics. Figure 16 shows the appearance of a "runaway current

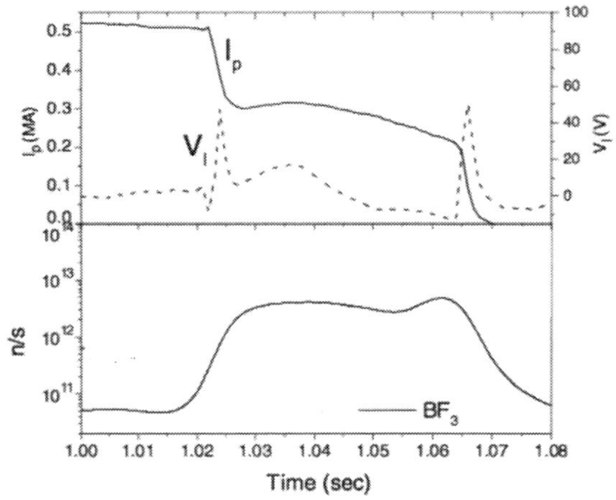

Figure 16. Energetic electron production in a tokamak reactor during a post-disruption current plateau, reproduced from [25]. I_P indicates the plasma current, V_I indicates the tokamak loop voltage, and the bottom part shows the amount of neutrons measured by a BF_3 detector. The plasma disruption takes place at about 1.02 s. We predict approximately 35 eV electron temperature during such current plateau.

plateau", which suddenly appears after a thermal cooling phase, and whose duration coincides with the neutron production process. The appearance of such current plateau is unexpected and unexplained in the context of the "Dreicer mechanism" theory. We predict that the electron temperature in such a "runaway current plateau" shall be eventually measured to be close to 35 eV, and that the nuclear origin of the energetic electrons shall be recognized. In [25], the authors estimate the electron temperature during runaway current plateaus to be 42 eV, based on the current evolution characteristics. This indirect estimation is fairly close to the 35 eV electron temperature predicted by us.

(2) **690 THz radiation signature of proton-electron zitterbewegung orbit establishment and 212 THz radiation signature of deuteron–electron zitterbewegung orbit establishment**. The proposed magnetic stabilization of zitterbewegung orbit states implies nuclear Zeeman level splitting, analogously to the resonance exploited nuclear magnetic resonance imaging devices. For the proton-electron system, we predict visible photon emission near the 690 THz frequency level upon the meta-stable zitterbewegung orbit establishment. This prediction is based on the calculation of magnetic field strength at the center of zitterbewegung orbit [10]. For the deuteron-electron system, we predict 212 THz emission, which is in the near-infrared range. Reference [31] indeed reports unexplained non-thermal near-infrared emission in a deuterium electrolysis based experiment, and mentions several other experiments documenting the same phenomenon. The intensity of the emission was found to be correlated with the measured non-chemical power output, which demonstrates that the zitterbewegung orbit electron state catalyzes nuclear reactions. An interesting point about this prediction is that an accurate measurement of the predicted frequency reveals the exact magnetic field strength at the center of the zitterbewegung orbit. As far as we know, no experiment measured yet the frequency of this photon emission.

(3) **Metastable hydrogen molecule with picometer-range internuclei distance**. Consequent to the zitterbewegung orbit's magnetic moment, two proton–electron systems in zitterbewegung orbit state shall experience magnetic attraction, and shall assume the energetically most favorable stacked orbit orientation. This prediction appears to be validated by the 2.3 pm inter-nuclei distance measurement reported in [18], which corresponds

to the theoretically expected inter-nuclei distance discussed in [10].

(4) **Energetic electron emission with energies up to the nuclear reaction energy**. We predict that the zitterbewegung orbit catalyzed nuclear reactions emit energetic electrons. Specifically, the energy of emitted electrons may reach up to the nuclear reaction energy level, i.e. 5.6 MeV for p-D fusion and 24 MeV for D–D fusion. For p-D fusion, this prediction is compatible with the observation of 5.6 MeV electron energy [7], and for D-D fusion this prediction is compatible with the observation of up to 15 MeV electron energies [26], where the upper energy threshold was limited by the sensitivity of the detecting equipment. It requires further study to understand why such electron emission is apparently high in burst type reactions and low in continuous reactions.

4. Conclusions

We have shown that the exothermic nuclear reaction in nickel-fueled reactors appears to be electron-mediated. We observe the distinct reaction dynamics of a fast burst and a continuous reaction process. Apparently, these are two different nuclear reaction processes. The analysis of the continuous reaction dynamics shows its highly controllable characteristics. This controllable characteristic, its high 30 W/g reaction power rate, and the reliable reaction start-up in our experimental design imply great potential for future energy production. We constructed a theory to explain the experimental observations in nickel-involving nuclear experiments; this theory builds onto previous works of relativistic electron dynamics. This model leads us to propose the presence of highly localized electron states, which are in close nuclear proximity around a proton or deuteron, and related experimental evidence for such close-proximity electron-nucleus configuration has been reviewed. The initiating electron kinetic energy level of 80 eV for hydrogen and 35 eV for deuterium are identified from experimental data, which is in good agreement with our theoretical estimation. The proposed transition process into highly localized electron states matches well the observed reaction data, and we described this process in sufficient detail for guiding future experimental work. This theory may be therefore useful for validation experiments and optimized reactor designs. The theory of zitterbewegung orbit state and its relation to the Dirac equation are described in more detail in a forthcoming book titled: "*Maxwell–Dirac Theory and Occam's Razor: Unified Field, Elementary Particles and Nuclear Interactions*".

Acknowledgments

We thank Serge F. Timashev, Jirohta Kasagi, and Pekka Janhunen for insightful discussions, and Kimmo Lattu for helping with the reactor preparation.

References

[1] J. Kasagi, T. Itoh and Y. Iwamura, Search for γ-ray radiation in NiCuZr nano-metals and H_2 gas system generating large excess heat, *Proc. ICCF-21 Int. Conf. on Condensed Matter Nucl. Sci.*, Fort Collins, USA, 2018.

[2] F. Celani, A. Spallone, B. Ortenzi, S. Pella, E. Purchi, F. Santandrea, S. Fiorilla, A. Nuvoli, M. Nakamura, P. Cirilli, P. Boccanera and L. Notargiacomo, Observation of macroscopic current and thermal anomalies, at high temperature, by hetero-structures in thin and long constantan wires under H_2 gas, *J. Condensed Matter Nucl. Sci.* **19** (2016) 29–45.

[3] T. Mizuno, Observation of excess heat by activated metal and deuterium gas, *J. Condensed Matter Nucl. Sci.* **25** (2017) 1–25.

[4] A. Kovacs, D. Brown and F. Ek, Exothermic reactions in the partially molten Li–Ni–Cu alloy, *J. Condensed Matter Nucl. Sci.* **25** (2017) 159–180.

[5] S. Focardi, V. Gabbani, V. Montalbano, F. Piantelli, S. Veronesi, Evidence of electromagnetic radiation from Ni–H Systems, *Proc. ICCF-11 Int. Conf. on Condensed Matter Nucl. Sci.*, Marseille, France, 2004.

[6] M. Valat, A. Goldwater, R. Greenyer, R. Higgins and R. Hunt, Investigations of the Lugano HotCat reactor, *J. Condensed Matter Nucl. Sci.* **21** (2016) 81–96.

[7] M. Lipoglavsek, S. Markelj, M. Mihovilovič, T. Petrovič, S. Itajner, M. Vencelj, J. Vesič , Observations of electron emission in the nuclear reaction between protons and deuterons, *Phys. Lett.* B **773** (10) (2017) 553–556, DOI:10.1016/j.physletb.2017.09.004.

[8] A. Takahashi, A. Kitamura, R. Seto, Y. Fujita, Taniike, Y. Furuyama, T. Murota and T. Tahara, Anomalous exothermic and endothermic data observed by nano-Ni-composite samples, *J. Condensed Matter Nucl. Sci.* **19** (2016) 23–32.

[9] T. Ohtsuki, K. Ohno, T. Morisato, T. Mitsugashira, K. Hirose, H. Yuki and J. Kasagi, Enhanced electron-capture decay rate of ^{7}Be encapsulated in C_{60} cages, *Phy. Rev. Lett.* **93** (11) (2004).

[10] F. Celani, A. O. Di Tommaso and G. Vassallo, The electron and occam's razor, *J. Condensed Matter Nucl. Sci.* **25** (2017) 76–99.

[11] T.A. Góngora and R.G. Stuart, The charge radius and anapole moment of a free fermion, *Zeitschrift für Physik* C – *Particles and Fields* **55** (1) (1992) 101–105.

[12] J. Kasagi, Screening potential for nuclear reactions in condensed matter, *Proc. ICCF-14 Int. Conf. on Condensed Matter Nucl. Sci.*, Washington, DC, 2008.

[13] J. Kasagi and Y. Honda, Screening energy of the d+d reaction in an electron plasma deduced from cooperative colliding reaction, *J. Condensed Matter Nucl. Sci.* **19** (2016) 127–134.

[14] A. Cvetinovic, M. Lipoglavěk, S. Markelj, J. Vesič, Molecular screening in nuclear reactions, *Phy. Rev.* C **92** (6) (2015).

[15] M. Lipoglavsek, Catalysis of nuclear reactions by electrons, EPJ Web of Conferences, Vol. 165, 2017.

[16] Y. Iwamura, T. Itoh, J. Kasagi, A. Kitamura, A. Takahashi and K. Takahashi, Replication experiments at tohoku university on anomalous heat generation using nickel-based binary nanocomposites and hydrogen isotope gas, *J. Condensed Matter Nucl. Sci.* **24** (2017) 191–201.

[17] G. Lu and W. Zhang, Photocatalytic hydrogen evolution and induced transmutation of potasium to calcium via low energy nuclear reaction (LENR) driven by visible light, *J. Mol. Catalysis* **31** (5) (2017) 401–410.

[18] S. Zeiner-Gundersen, Hydrogen reactor for rydberg matter and ultra dense hydrogen, a replication of Leif Holmlid, presentation at the ICCF-21 conference, Fort-Collins, USA, 2018.

[19] G.H. Miley and J.A. Patterson, Nuclear transmutations in thin-film nickel coatings undergoing electrolysis, *J. New Energy* **1** (3) (1996).

[20] M. Kaczmarski, K. Czerski, D. Weissbach, A. Huke, G. Ruprecht and A. I. Kilic, Threshold resonance contribution to the thick target ^{2}H(d,p)^{3}H reaction yield, *Acta Physica Polonica* B, **48** (3) (2017) 489–493.

[21] A.V. Simakin and G.A. Shafeev, Accelerated alpha decay under laser exposure of metallic nanoparticles in aqueous solutions of uranium salt, *Phys. Wave Phenomena* **19** (1) (2011) 30–38.

[22] J. Kasagi, Low-energy nuclear reactions in metals, *Progr. Theoret. Phys. Supplement* **154** (1) (2004) 365–372.

[23] K. Czerski, D. Weissbach, A.I. Kilic, G. Ruprecht, A. Huke, M. Kaczmarski, N. Targosz-Ślęczka and K. Maass, Screening and resonance enhancements of the ^{2}H(d,p)^{3}H reaction yield in metallic environments, *Europhys. Lett.* **113** (2) (2006).

[24] S. Xu, J. Cao, C.C. Miller, D.A. Mantell, R.J.D. Miller and Y. Gao, Energy dependence of electron lifetime in graphite observed with femtosecond photoemission spectroscopy, *Phy. Rev. Lett.* **76** (3) (1996) 483–486.

[25] J.R. Martin-Solis, B. Esposito, R. Sánchez, F.M. Poli and L. Panaccione, Enhanced production of runaway electrons during a disruptive termination of discharges heated with lower hybrid power in the Frascati Tokamak Upgrade, *Phy. Rev. Lett.* **97** (16) (2006).

[26] C. Paz-Soldan, C.M. Cooper, P. Aleynikov, D.C. Pace, N.W. Eidietis, D.P. Brennan, R.S. Granetz, E.M. Hollmann, C. Liu, A. Lvovskiy and R.A. Moyer, Spatiotemporal evolution of runaway electron momentum distributions in tokamaks, *Phy. Rev. Lett.* **118** (25) (2017).

[27] P. Kálmán and T. Keszthelyi, Solid state internal conversion, *Phy. Rev.* C **69** (3) (2004).

[28] Y. Iwamura, M. Sakano and T. Itoh, Elemental analysis of Pd complexes: effects of D_2 gas permeation, *Japanese J. Appl. Phys.* **41** (7R) (2002).

[29] T. Hioki, N. Takahashi, S. Kosaka, T. Nishi, H. Azuma, S. Hibi, Y. Higuchi, A. Murase and T. Motohiro, Inductively coupled plasma mass spectrometry study on the increase in the amount of Pratoms for Cs-ion-implanted Pd/CaO Multilayer complex with deuterium permeation, *Japanese J. Appl. Phys.* **52** (10R) (2013).

[30] NIST X-ray photoelectron spectroscopy database, https://srdata.nist.gov/xps.
[31] M. Swartz, G. Verner and A. Weinberg, Non-thermal near-IR emission from high impedance and codeposition LANR devices, *Proc. ICCF-14 Int. Conf. on Condensed Matter Nucl. Sci.*, Washington, DC, 2008.

J. Condensed Matter Nucl. Sci. 29 (2019) 440–452

Research Article

Resonant Surface Capture Model

Xingzhong Li*, Zhanmin Dong, Changlin Liang and Guisong Huang

Department of Physics, Tsinghua University, Beijing, China

Abstract

A resonant surface capture model is proposed to explain the various phenomena: the temperature dependence of excess heat, nuclear fusion cross-section data from beam–target experiments and from condensed matter nuclear reactions. This model is based on Oppenheimer's stripping nuclear reaction, and Bethe's solar energy calculation using the resonance effect to put the incoming projectile at the edge of target nucleus without forming compound nucleus. This is a new kind of resonance at low energy: the width of which is proportional to $E_0^{3/2}$ instead of $\exp\left[-\frac{C}{\sqrt{E_0}}\right]$, when the resonance energy $E_0 \to 0$.

Keywords: Lithium–hydrogen fusion reaction, 3-Parameter formula for fusion cross-section, Resonant surface capture model, Temperature dependence of excess heat, Width of resonance

1. Introduction – The Concept of Resonant Surface Capture

In 2016, Storms [1] published an important discovery: the temperature dependence of excess heat is a straight line in the semi-logarithmic plot (Fig. 1, blue dotted line), and the slope of this straight line is close to the activation energy (E_a) of the diffusion coefficient. At a first glance, it was a big surprise: the excess heat is an inelastic nuclear scattering process, but the diffusion is an elastic nuclear scattering process. How can be related to each other? What is hidden behind this experimental phenomenon?

This phenomenon reveals an important role of the nuclear resonance. During the elastic process, the resonance would be fully developed without any damping; then, the resonance would put the peak of the wave function at the edge between the nuclear potential well and the Coulomb barrier (Fig. 2, red line at $r = a$). The peak implies that the nucleon in the target nucleus would have a chance to direct interact with the nucleon in the projectile, because the nucleons are sticking together and inside the range of nuclear interaction due to resonance. This is a surface capture process that occurs without forming a compound nucleus. The reaction energy would be carried away by the charged nuclear products, and be transferred to the surrounding electrons as excess heat. Thus an inelastic nuclear scattering process follows an elastic nuclear scattering due to the resonance. Two different processes are thus connected together by the resonance effect. We may call this model as resonant surface capture model.

*Corresponding author. E-mail: lxz-dmp@tsinghua.edu.cn.

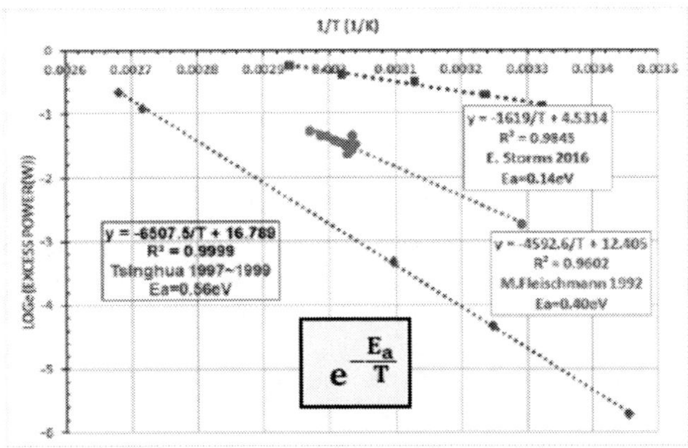

Figure 1. The temperature dependence of excess is a straight line in semi-logarithmic plot.

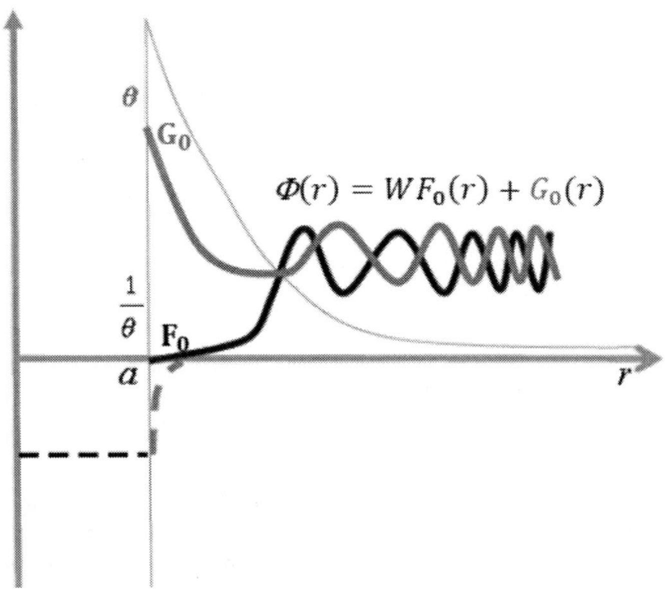

Figure 2. The resonance makes the wave function peaked at the edge of nuclear potential well (*red line*).

This direct nuclear interaction is very similar to the nuclear stripping interaction proposed by Oppenheimer in 1935 [2], and justified by the experimental data. The only new feature here is the resonance which put the projectile at the edge of the target nucleus. This process does not excite the target to a highly excited nuclear state; therefore, there would be no strong Gamma radiation accompanying this excess heat.

2. Equation – Resonance appears when $|W| \leq 1$ Instead of $W = 0$

The concept of resonant surface capture may be expressed by a peaked wave function inside a Coulomb field which is connected to the nuclear potential well (thin blue line in Fig. 2). In quantum mechanics any wave function, $\Phi(r)$, inside a Coulomb field may be expressed by a linear combination of two independent solutions of Schrodinger equation as:

$$\Phi(r) = WF_0(r) + G_0(r). \tag{1}$$

Here, $\Phi(r)$ is the reduced radial wave function of the relative motion between injected projectile and the target nucleus; $F_0(r)$ and $G_0(r)$ are the regular and irregular Coulomb wave functions for zero orbital angular momentum, respectively, because we are only interested in the low energy case. When $r \to a$, $F_0(r)$ decreases rapidly to a very small number in order of $\frac{1}{\theta}$ (lower black solid line in Fig. 2) and $G_0(r)$ increases rapidly to a very large number in order of θ (upper red solid line in Fig. 2). θ depends on energy exponentially as:

$$\theta = \sqrt{\frac{e^{2\pi\eta} - 1}{2\pi}}, \tag{2}$$

θ^2 is the Coulomb barrier factor which is similar to the reciprocal of Gamow factor; and $\eta \equiv \frac{1}{k\,a_c}$ – the Coulomb variable;

$$a_c \equiv \frac{4\pi\epsilon_0\hbar^2}{\mu Z_1 Z_2 e^2},$$

the Coulomb unit of length and k is the momentum of relative motion between projectile and target nucleus

$$k \equiv \sqrt{\frac{2\mu E}{\hbar^2}}$$

and E is the energy of this relative motion in the center of mass system. Z_1e and Z_2e are their electrical charges, μ is the reduced mass, ϵ_0 the dielectric constant of vacuum, $\hbar$ is the Planck constant divided by 2π. It is interesting to notice that: Although $|F_0(r)||_{r=a} \ll |G_0(r)||_{r=a}$, the logarithmic derivative of $F_0(r)$ and $G_0(r)$ at $r = a$ are similar in magnitude [3], i.e.

$$\left|\frac{1}{F_0(r)}\frac{\partial F_0(r)}{\partial r}\right|_{r=a} \approx \left|\frac{1}{G_0(r)}\frac{\partial G_0(r)}{\partial r}\right|_{r=a}, \quad \text{when } k \ll \sqrt{\frac{2}{a\,a_c}}. \tag{3}$$

This feature of $F_0(r)$ and $G_0(r)$ at $r = a$ makes the coefficient of the linear combination, W, extremely large in most cases because the wave function has to be connected smoothly to the wave function inside the nuclear potential well. The logarithmic derivative of wave function $\Phi(r)$ at $r = a$ must equal to D_L, the logarithmic derivative of wave function inside the nuclear potential well. It makes W very large at low energy, because

$$W = -\frac{G_0(a)}{F_0(a)}\left[\frac{D_L - \frac{1}{G_0(r)}\frac{\partial G_0(r)}{\partial r}}{D_L - \frac{1}{F_0(r)}\frac{\partial F_0(r)}{\partial r}}\right]_{r=a} = -\theta^2\left[\left(\frac{a_c}{a}\right)\frac{D_L - \frac{1}{G_0(r)}\frac{\partial G_0(r)}{\partial r}}{D_L - \frac{1}{F_0(r)}\frac{\partial F_0(r)}{\partial r}}\right]_{r=a}, \tag{4}$$

and θ^2 is a very large number at low energy. While θ^2 is known, the factor in square bracket in Eq. (4) is unknown, because it involves unknown nuclear property, D_L .

Figure 2 is plotted for the case of W=1. Usually, we think that when the factor in square bracket in Eq. (4) makes $W = 0$; then, there will be only $G_0(r)$ in Eq. (1), and the wave function, $\Phi(r)$, would be peaked at the boundary between Coulomb field and nuclear potential (Fig. 2 upper red solid line). So the resonant phenomenon would appear at $W = 0$. However, even if $W \neq 0$, provided that $|W| \leq 1$, we may still have a wave function, $\Phi(r)$, peaked at the boundary between Coulomb field and nuclear potential as shown in Fig. 2. Therefore, in order to study the resonant scattering, we should find the energy dependence of W, and search the pair of nuclei which makes $|W| \leq 1$.

3. A Tool for Finding Energy Dependence of W – 3-Parameter Formula for Fusion Cross-section

In quantum mechanics, W is directly related to low energy fusion cross-section as [4–10]:

$$\sigma_{\text{Fusion}}(E) = \frac{\pi}{k^2} \frac{(-4W_\text{i})}{(W_\text{r})^2 + (1 - W_\text{i})^2}. \tag{5}$$

Here W_r and W_i are the real and imaginary parts of W. Equation (5) gives a tool to determine the W from the experimental data of fusion cross-section, $\sigma_{\text{Fusion}}(E)$. Equation (4) has shown that W might be separated into two factors: the Coulomb field factor, θ^2, which is known as a real number by Eq. (2), and the remaining complex nuclear factor. We may rewrite the W as

$$W \equiv \theta^2(w_\text{r} + w_\text{i}). \tag{6}$$

The Coulomb field factor θ^2, varies rapidly with energy; and the unknown nuclear factor, $(w_\text{r} + w_\text{i})$, might be found through low energy fusion cross-section data. From the d + T $\rightarrow$ n + ^{4}He fusion cross-section, we have found that w_r varies slowly with energy, and w_i is almost a constant at low energy. The d + T fusion cross-section may thus be expressed by a 3-parameter formula very well:

$$W_\text{r} = \theta^2(C_1 + C_2E), \quad W_\text{i} = \theta^2 C_3, \tag{7}$$

$$\sigma_{\text{Fusion}}(E) = \frac{\pi}{k^2} \frac{(-4C_3)}{\theta^2[(C_1 + C_2E)^2 + (\frac{1}{\theta^2} - C_3)^2]}. \tag{8}$$

Figure 3 gives the comparison between this 3-parameter formula and the experimental data. Here the black circles are experimental data from the National Nuclear Data Center (ENDF/B-VIII.0) [11]. The red lines are from 3-parameter formula, Eq. (8). In order to show the good fit, both logarithmic scale (left axis) and linear scale (right axis) are plotted to show the fit from 200 eV to 200 keV [12].

Indeed, Eq. (8) has been justified by more than 15 pairs of fusion cross-section data. It gives two important features of W: (i) W_r and W_i are the products of θ^2 and a constant at very low energy, respectively. (ii) At very low energy both W_r and W_i are extremely rapidly decreasing when energy is increasing, because θ^2 is an extremely rapidly decreasing function when energy is increasing as expressed by Eqs. (2)–(4). When $k \rightarrow 0$, $\eta \rightarrow \infty$; then, $\theta^2 \rightarrow \infty$ exponentially. On the other hand, when $k \rightarrow \infty$, $\eta \rightarrow 0$; then, $\theta^2 \rightarrow \eta \rightarrow 0$. These two features imply that there must be an energy, E_0, above which both W_r and W_i are less than 1. Moreover, if this resonance energy, E_0, is low enough; then, we may show that the resonant *elastic* nuclear scattering cross-section would be a *step-wise* function of energy E due to the rapid variation of θ at low energy.

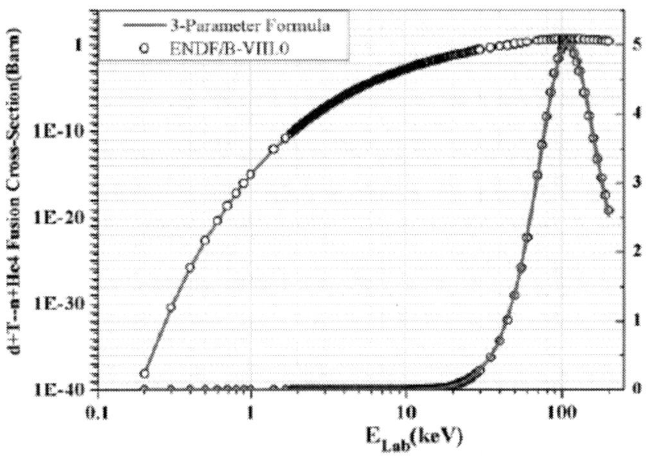

Figure 3. d+T→n+ ^{4}He fusion cross-section data reveal energy dependence of *W*.

4. Low-energy Resonant Elastic Nuclear Scattering Cross-Section – A Step-wise Function of Energy E

In quantum mechanics, the elastic scattering cross-section is expressed by W as well [13]

$$\sigma_{\text{Elastic}}(E) = \frac{\pi}{k^2} \frac{4}{(W_\text{r})^2 + (1 - W_\text{i})^2}. \tag{9}$$

When energy is greater than the resonant energy. E_0 , both $W_r$and$W_i$are less than 1; then

$$\sigma_{\text{Elastic}}(E) = \frac{\pi}{k^2} \frac{4}{(W_\text{r})^2 + (1 - W_\text{i})^2} \to \frac{4\pi}{k^2}. \tag{10}$$

When energy is less than the resonant energy. E_0, both $W_\text{r}$and$W_\text{i}$are rapidly increasing to be greater than 1; then

$$\sigma_{Elastic}(E) = \frac{\pi}{k^2} \frac{4}{(W_r)^2 + (1 - W_i)^2} \propto \frac{4\pi}{k^2\theta^4} \to 0. \tag{11}$$

We may illustrate this behaviour by assuming two resonance energy, $E_{0\text{r}}$ and $E_{0\text{i}}$ for W_r and W_i, respectively:

$$W_\text{r} = \frac{\theta(E)^2}{\theta(E_{0\text{r}})^2}\left(1 - \frac{E}{E_{0\text{r}}}\right) \text{ and } W_\text{i} = \frac{\theta(E)^2}{\theta(E_{0\text{i}})^2}. \tag{12}$$

This assumption gives:

$$W_\text{r} = 0 \quad \text{at} \quad E = E_{0\text{r}}, \tag{13}$$

$$W_\text{r}^2 \gg 1, \quad \text{when } E \ll E_{0\text{r}} \text{ and } W_\text{r}^2 \ll 1, \quad \text{when } E \gg E_{0\text{r}}, \tag{14}$$

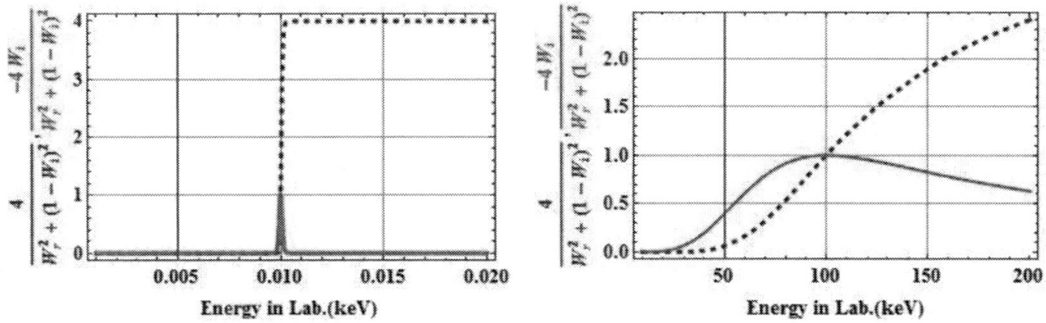

Figure 4. Changing from $E_0 = 100$ keV (*right*) and $E_0 = 10$ eV (left) shows that blue dotted line becomes a step-wise line, and red line becomes a peak-wise line.

$$W_\mathrm{i}^2 \gg 1, \quad \text{when } E \ll E_{0\mathrm{i}} \text{ and } W_\mathrm{i}^2 \ll 1, \quad \text{when } E \gg E_{0\mathrm{i}}. \tag{15}$$

In Fig. 4, the blue dotted line shows the factor

$$\frac{4}{(W_\mathrm{r})^2 + (1 - W_\mathrm{i})^2}$$

as a function of energy for two cases: $E_{0\mathrm{r}} = E_{0\mathrm{i}} = E_0 = 100$ keV, and $E_{0\mathrm{r}} = E_{0\mathrm{i}} = E_0 = 10$ eV. It is evident that the blue dotted line becomes a step-wise function when E_0 is getting smaller. We may expect a much steeper line when E_0 approaches thermal energy (~25 meV).

5. Low-energy Resonant Inelastic Nuclear Scattering Cross-Section – A Peak-wise Function of Energy E

According to resonant surface capture model, the cross-section of nuclear reaction is a transition after this resonant elastic scattering [2,14]; hence,

$$\sigma_{\text{Surface capture}}(E) = \frac{\pi}{k^2} \frac{4}{(W_\mathrm{r})^2 + (1 - W_\mathrm{i})^2} \left[\int \psi_\mathrm{f}^* H_\mathrm{int} \psi_\mathrm{i} \, \mathrm{d}\tau \right]^2, \tag{16}$$

Here,

$$\left[\int \psi_\mathrm{f}^* H_\mathrm{int} \psi_\mathrm{i} \, \mathrm{d}\tau \right]^2$$

is the transition probability from the initial state, ψ_i, to the final state, ψ_f . The initial state, ψ_i, is a plane wave at $r \to \infty$ formed by the incoming projectile and the target nucleus; and the final state, ψ_f, is a halo state formed by the captured nucleon in the halo of target nucleus (purple dotted line in Fig. 2). This transition happen only if the overlapping between ψ_i and ψ_f is not negligible. ψ_f decreases rapidly when $r > a$; hence, the overlapping integration becomes not negligible only if ψ_i is large at the edge $r = a$, i.e. in the case of resonance when G_0 is dominant in the linear combination. Thus $\psi_\mathrm{i} \propto G_0 \propto \theta$. Or

$$\sigma_{\text{Surface capture}}(E) = \frac{\pi}{k^2} \frac{4}{(W_\mathrm{r})^2 + (1 - W_\mathrm{i})^2} \left[\int \psi_\mathrm{f}^* H_\mathrm{int} \psi_\mathrm{i} \, \mathrm{d}\tau \right]^2 \propto \frac{\pi}{k^2} \frac{4}{(W_\mathrm{r})^2 + (1 - W_\mathrm{i})^2} \theta^2, \tag{17}$$

when energy is greater than the resonant energy. E_0, both W_r and W_i are less than 1; then

$$\sigma_{\text{Surface Capture}}(E) = \frac{\pi}{k^2}\frac{4}{(W_\mathrm{r})^2 + (1 - W_\mathrm{i})^2}\theta^2 \to \frac{4\pi}{k^2}\theta^2, \tag{18}$$

when energy is less than the resonant energy. E_0, both W_r and W_i are rapidly increasing to be greater than 1; then

$$\sigma_{\text{Surface Capture}}(E) = \frac{\pi}{k^2}\frac{4}{(W_\mathrm{r})^2 + (1 - W_\mathrm{i})^2}\theta^2 \to \frac{4\pi}{k^2\theta^4}\theta^2 \to 0, \tag{19}$$

In Fig. 4, the red solid line shows the factor

$$\frac{4}{(W_\mathrm{r})^2 + (1 - W_\mathrm{i})^2}\theta^2$$

as a function of energy with E_0 =10 eV (having assumed both W_r and W_i are equal to 1 at same energy E_0). The additional factor, θ^2, in numerator adds a steep decreasing behaviour in the region where $E > E_0$. Thus the resonant surface capture cross-section becomes a peak-wise function with peak at $E \approx E_0$. We may expect much more sharp peak behavior when E_0 approaches thermal energy (~25 meV).

6. Average Over Maxwell Distribution Function

In order to study the temperature dependence of excess heat and diffusion in Fig. 1, we have to average the collision rate over the Maxwell velocity distribution. For the elastic collision rate,

$$\langle\sigma_{\text{Elastic}}\, v\rangle = \left(\frac{\mu}{2\pi k_\mathrm{B}T}\right)^{3/2}\int_0^\infty \left(\frac{\pi}{k^2}\frac{4}{W^2+1}\right) v \exp\left[-\frac{\mu v^2}{2k_\mathrm{B}T}\right] 4\pi v^2\, \mathrm{d}v \propto \int_{E_0}^{\infty} \exp\left[-\frac{E}{T}\right] \mathrm{d}E \propto \exp\left[-\frac{E_0}{T}\right]. \tag{20}$$

The step-wise behaviour of $4/(W^2 + 1)$ just changes the lower limit of integration from 0 to E_0 , it results an exponential factor, $\exp\left[-(E_0/T)\right]$.

For the resonant surface capture rate, this same exponential factor will appear as well:

$$\langle\sigma_{\text{Excess Heat}}\, v\rangle = \left(\frac{\mu}{2\pi k_\mathrm{B}T}\right)^{3/2}\int_0^\infty \left(\frac{\pi}{k^2}\frac{4}{W^2+1}\right)\left|\int \psi_\mathrm{f}^* H_{\text{int}}\Psi_\mathrm{i}\,\mathrm{d}\tau\right|^2 v \exp\left[-\frac{\mu v^2}{2k_\mathrm{B}T}\right] 4\pi v^2\, \mathrm{d}v$$
$$\propto \int_0^\infty \frac{4}{W^2+1}\theta^2 \exp\left[-\frac{E}{T}\right] \mathrm{d}E \propto \exp\left[-\frac{E_0}{T}\right]\Delta E. \tag{21}$$

The peak-wise behavior of

$$\frac{4}{W^2+1}\theta^2$$

just turns the integration into the product of $\exp\left[-E_0/T\right]$, and ΔE, i.e. the peak value of integrand at E_0 times the energy width ΔE.

Now we have shown that both the elastic collision rate and the resonant surface capture rate share the same exponential factor $\exp\left[-E_0/T\right]$ as shown in Fig. 1. Indeed it gives the hint that there must be a low-energy resonance at the energy E_0 which is close to the activation energy of diffusion coefficient, E_a .

7. Comparison with More Experimental Data

In addition to the theoretical derivation of Storms' straight line, we would like to see if there is any more experimental evidences to support this behavior. The first such evidence may be Fleischmann and Pons' famous "Heat after Death" experiment in 1992 [15]. Before the electrolyte reached the boil point, there were 12 data points for excess heat power at certain temperatures. In Fig. 1, the red dotted line goes just along with these 12 red points. Only a few points after the triggering are not on this straight dotted line. However, the slope of this straight line is different from that of Storms' blue dotted line which is thought to be the activation energy of deuteron diffusion coefficient in palladium. Then what is the possible reason for this difference? As we remember, Fleischmann and Pons would like to add LiOD in the electrolyte in order to enhance the conductivity of electrolyte for obtaining greater current density which was believed to be an important condition to have excess heat at those early days. As mentioned by Storms, after long time of electrolysis there was a possibility of forming lithium–palladium alloy in the cathode surface. Therefore, we might think the activation energy of deuteron diffusion in lithium deuteride. The search in literature did not give the activation energy for deuteron diffusion in LiD in the temperature range of interests. The only available activation energy in this temperature range is for lithium diffusion in LiH, i.e. $E_a = 38.5\ \mathrm{kJ/mol} = 4632$ K (~0.399 eV) [16], it is very close to the slope of that red dotted line in Fig. 1 (4593 K~0.396 eV). According to literature [16], the activation energy of D in LiD (or H diffusion in LiH) is expected to be close to $38.5\mathrm{kJ/mol}$ in the temperature range of interests. More experimental data are desirable.

Having been encouraged by this comparison, we re-examined Tsinghua University data from gas-loading experiments performed early in 1996–1999 [17,18]. The purple dotted line in Fig. 1 shows another good fit with experimental data points. It seems that both in gas-loading and in electrolysis experiments, the temperature dependence of excess heat is a straight line in the semi-logarithmic plot, and there should be a low energy resonance between deuteron (or proton) and some target nucleus. We would like to address the possible target nucleus here for electrolysis experiments and shall discuss this gas-loading straight line in another proceedings paper [19].

8. The Evidences of Low-energy Resonance for Nuclear Reaction

From 3-parameter formula for fusion cross-section Eq. (8), it is possible to find the low-energy resonance as well. If there is a low-energy resonance at $E = E_0$ for elastic scattering; then, the rapidly decreasing θ^2 would make $\theta^2[(C_1 + C_2E)^2] \to 0$ in the region $E > E_0$. Since most of fusion cross-section data are taken from beam–target experiments using accelerator, the beam energy is greater than the resonance energy, E_0, which is supposed to be near thermal energy. Thus we may expect that the fusion cross-section would be fit by only one parameter, C_3, as

$$\sigma_{\text{Fusion}}(E) = \frac{\pi}{k^2}\frac{(-4C_3)}{\theta^2[(C_1 + C_2E)^2 + (\frac{1}{\theta^2} - C_3)^2]} \to \frac{\pi}{k^2}\frac{(-4C_3)}{\theta^2[(\frac{1}{\theta^2} - C_3)^2]} \to \frac{\pi}{k^2}\frac{(-4C_3\theta^2)}{(1 - C_3\theta^2)^2}. \tag{22}$$

Figure 5 shows the fusion cross-section for $p + {}^6Li \to {}^3He + {}^4He$. Black points are from ENDF/B-VIII.0, and the red line is from Eq. (22) with only one parameter $C_3 = -5.62$. The left and right axes are for logarithmic and linear scale, respectively, to show the good fit ($R^2 = 0.99999$). Indeed the existence of low energy resonance for $p + {}^6Li \to {}^3He + {}^4He$ is provided by Lipinski's patent as well [20].

Figure 6 shows the fusion cross-section for $d + {}^6Li \to {}^4He + {}^4He$. Black points are from ENDF/B-VIII.0, and the red line is from Eq. (22) with only one parameter $C_3 = -0.1768$. The left and right axes are for logarithmic and linear scale, respectively, to show the good fit ($R^2 = 0.99997$). Indeed the existence of low-energy resonance for $d + {}^6Li \to {}^4He + {}^4He$ is provided by Pons' heavy water electrolysis experiment using enriched ^{6}LiOD as well. After long period of electrolysis using palladium cathode Pons observed a big amount of helium with excess heat [21].

These experiments just explain the slope of red straight line in Fig. 1.

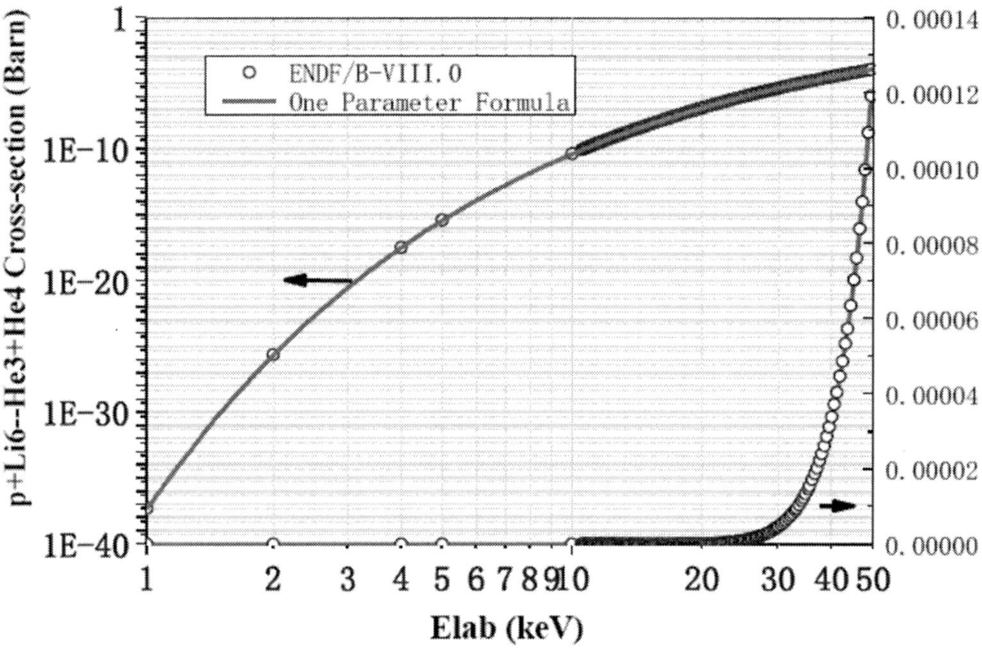

Figure 5. $p + {}^6Li \rightarrow {}^3He + {}^4He$ fusion reaction cross-section at low energy may be fit by 1-parameter formula.

9. Two Kinds of Resonance Widths

The widths of peaks in Fig. 3 (at 100 keV) and in the integrand of Eq. (21) are very different, because the resonance peak in Fig. 3 is determined by $W_r = 0$, and the resonance peak in the integrand of Eq. (21) for a transition after the ***pure*** elastic collision is determined by $W = W_r = 1$.

Since the Coulomb barrier factor

$$\theta = \sqrt{\frac{e^{2\pi\eta} - 1}{2\pi}} \approx \sqrt{\frac{e^{31.4 Z_1 Z_2 \sqrt{\frac{m_1}{E_{\text{lab keV}}}}}}{2\pi}},$$

the changing rate of θ^2 would be much slower when

$$E_{\text{lab keV}} > \left(\frac{31.4 Z_1 Z_2 \sqrt{m_1}}{2}\right)^{2/3}.$$

Here, $E_{\text{lab keV}}$ is the kinetic energy of the injected projectile in laboratory system in unit of keV, m_1 is its mass number (an integer),

$$\left(\frac{31.4Z_1Z_2\sqrt{m_1}}{2}\right)^{2/3} \approx 7.9\ \text{keV}$$

for d+T fusion. This affects the width of resonance peak.

The resonance peak in Fig. 3 for d+T fusion is near 100 keV which is greater than

$$\left(\frac{31.4Z_1Z_2\sqrt{m_1}}{2}\right)^{2/3}.$$

therefore, the variation of $W_\mathrm{r} \equiv \theta^2 w_\mathrm{r}$ around 100 keV is mainly determined by $w_\mathrm{r} = C_1 + C_2E$. The width would be determined by

$$\Delta E_\mathrm{h} \approx \frac{2}{\theta^2 C_2} \quad \text{at} \quad E_{0\mathrm{h}} = -\frac{C_1}{C_2} \quad \text{where} \quad w_\mathrm{r} = 0.$$

This expression may give a wrong impression that the width of resonance would be extremely small as

$$\Delta E_\mathrm{h} \to \left|\frac{2}{\theta^2 C_2}\right|_{E_{\mathrm{lab\ keV}} \to 0} \to 0.$$

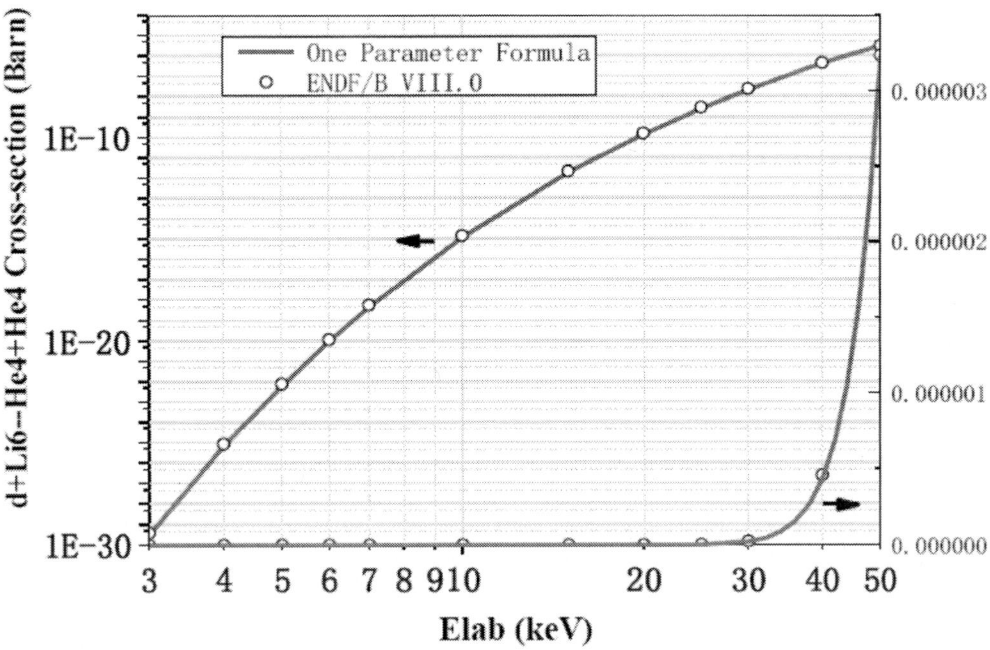

Figure 6. d + ^{6}Li → ^{4}He + ^{4}He fusion reaction cross-section at low energy may be fit by 1-parameter formula.

This is not true, for $E_{\text{lab keV}} \to 0$. Indeed in the integrand of Eq. (21) for a transition after the ***pure*** elastic collision (where a resonance energy is supposed to be $E_{\text{0lab keV}} < 0.01\text{keV}$), $E_{\text{0lab keV}}$ is much smaller than

$$\left(\frac{31.4Z_1Z_2\sqrt{m_1}}{2}\right)^{2/3} \quad \left(\text{for p} +^6 \text{Li fusion} \quad \left(\frac{31.4Z_1Z_2\sqrt{m_1}}{2}\right)^{2/3} \approx 13\,\text{keV}\right)$$

therefore, the variation of $W_\mathrm{r} \equiv \theta^2 w_\mathrm{r}$ around 0.01 keV in the integrand of Eq. (21) is mainly determined by θ^2 instead of w_r (because at this low energy, $w_\mathrm{r} \to C_1 = \text{constatnt}$). The width of resonance peak would be determined by the peak-wise behavior of $\frac{4}{W^2+1}\theta^2$. The step-wise denominator determines the up-slop behavior of the peak at $E_{0\mathrm{c}}$ with

$$w_\mathrm{r} = \frac{1}{\theta(E_{0\mathrm{c}})^2} \quad (\text{i.e. } W = W_\mathrm{r} = 1),$$

and the θ^2 in numerator determines the down-slop of peak with a width of

$$\Delta E_{0\mathrm{c}} \approx \frac{2\,\mathrm{Log}\frac{1}{2}}{31.4Z_1Z_2\sqrt{m_1}}(E_{\text{0Labk eV}})^{3/2} \quad \left(\text{i.e. } \theta[E_{0\mathrm{c}} + \Delta E_{0\mathrm{c}}]^2 = \frac{1}{2}\theta[E_{0\mathrm{c}}]^2\right).$$

Although the energy width $\Delta E_{0\mathrm{c}}$ still goes to zero, when $E_{\text{0Lab keV}} \to 0$, it is much more slower than $\frac{1}{\theta^2}$. This resonance peak at low energy will have enough contribution to the integration in Eq. (21).

10. Conclusion

The essence of resonance is to put the wave peak at the edge of the nuclear boundary; then, Coulomb repulsion becomes Coulomb *attraction*. Once the injected projectile sticks on the surface of target nucleus, the nucleons near the surface are in the range of the strong interaction even if no compound nucleus is formed.

This effect was justified by Oppenheimer (1935) [2] in terms of the angular distribution of nuclear products with energetic projectile (which are approximately several MeV). This resonant surface capture model only makes use of the resonance to put the projectile on the surface of the target nucleus. Indeed Bethe applied a similar calculation for the weak interaction $\mathrm{p} + \mathrm{p} \to \mathrm{D} + \mathrm{e}^+ + \text{neutrino}$ in the solar energy model [14]. Bethe included the resonance effect partly in his calculation, but the approximation in his simplification prevented him from fully considering the resonance effect. However, the overlapping of the initial wave function with the final wave function provided a good estimate of the transition probability both in Oppenheimer's stripping reaction and in Bethe's solar energy calculation.

Our new finding is a feature of low-energy resonance in condensed matter which is quite different from the resonance in beam–target experiments. The resonance energy, E_0, is no longer determined by the $W_\mathrm{r} = 0$ in the denominator of Eq. (5), instead it is determined by $W_\mathrm{r}^2 + W_\mathrm{i}^2 \leq 1$. Therefore, the width of resonance is no longer decreasing as $1/\theta^2 \to 0$ at low energy, instead it is decreasing as $(E_{0\mathrm{c}})^{3/2}$ which is much slower than $1/\theta^2$.

The slope of straight lines in Storms' semi-logarithmic plot and the data from experiments (Beam–target data from the National Nuclear Data Center, Lipinski patent for "Low energy Hydrogen–Lithium Fusion Reactor", and Pons' electrolysis with enriched lithium (^{6}LiOD)) all point to ^{6}Li as a possible fuel candidate in Condensed Matter Nuclear Science.

The two-step resonant surface capture model explains not only the low energy nuclear reaction without Coulomb repulsion, but also the releasing of nuclear energy without strong Gamma and neutron radiation, because there were no highly excited nuclei involved in surface capture reaction (no compound nucleus formed).

We have to further find the collective mechanism in condensed matter which makes the resonant surface capture reaction happen on a large scale with a self-sustaining mode. Then a safe and clean nuclear energy source would be available to the whole world.

Acknowledgements

Many thanks to T. Passell for his guiding to lithium stripping reaction [22], to E. Storm for his discovery of temperature dependence of excess heat, to M. Miles for his thermodynamic derivation using Eyring theory [23], to A. Nicolas for his information about activation energy of H(D) diffusion in metals. This work is supported by The Ministry of Education (#20091770437), The Ministry of Science and Technology (Fundamental Research Division, #2009CB226113), Natural Science Foundation of China (#10475045 & #21153003) and Tsinghua University (Basic Research Fund (985-III)).

References

[1] E. Storms, How basic behavior of LENR can guide a search for an explanation, *J. Condensed Matter Nucl. Sci.* **20** (2016) 100–138.

[2] J.R. Oppenheimer and M. Phillips, Note on the transmutation function for deuterons, *Phys. Rev.* **48** (1935) 500–502. S.T. Butler, Direct nuclear reactions, *Phys. Rev.* **106** (1957), 272–286.

[3] M. Abramowitz and I.A. Stegun, *Handbook of Mathematical Functions*, 10th Printing, Dover, New York, p. 542 (14.6.9), December, 1972.

[4] X.Z. Li, Q.M. Wei and B. Liu, A new simple formula for fusion cross-sections of light nuclei, *Nucl. Fusion* **48** (2008) 125003.

[5] X.Z. Li, A new approach towards fusion energy with no strong nuclear radiation, *Nucl Fusion and Plasma Phys* **16** (2) (1996) 1–8 (in Chinese), see also *J. New Energy* **1** (4) (1996) 44–54 (in English).

[6] X.Z. Li, J. Tian, M.Y. Mei and C.X. Li, Sub-barrier fusion and selective resonant tunneling, *Phys. Rev.* C **61** (2000) 024610.

[7] X.Z. Li, Nuclear physics for nuclear fusion, *Fusion Sci. Technol.* **41** (2002) 63.

[8] X.Z. Li, B. Liu, S. Chen, Q.M. Wei and H. Hora, Fusion cross sections for inertial fusion energy, *Laser Part. Beams* **22** (2004) 469–477.

[9] X.Z. Li, Z.M. Dong and C.L. Liang, Studies on $p+{}^{6}Li$ fusion reaction use 3-parameter model, *J. Fusion Energy* **31** (2012) 432–436.

[10] M. Kikuchi, *Frontiers in Fusion Research – Physics and Fusion*, Springer, London, 2011, p. 31.

[11] National Nuclear Data Center, Brookhaven National Laboratory, ENDF/B-VIII.0 (2018) is available on Internet http://www.nndc.bnl.gov.

[12] C.L. Liang, Z.M. Dong and X.Z. Li, Selective resonant tunneling – turning hydrogen-storage material into energetic material *Current Sci.* **108** (4) (2015) 519.

[13] L.D. Landau and E.M. Lifshitz *Quantum Mechanics*, 3rd. Edn.,revised and enlarged Reprinted 1991 (with corrections) p. 507.

[14] H.A. Bethe and C.L. Critchfield, The formation of deuterons by proton combination, *Phys. Rev.* **54** (1938) 248.

[15] M. Fleischmann, and S. Pons, Calorimetry of the Pd–D_2O system: from simplicity via complications to simplicity, *Phys. Lett.* A **176** (1993) 118.

[16] Hj. Matzke and V.V. Rondinella, Diffusion in carbides, nitrides, hydrides and borides 5.3 Diffusion in hydrides, Landolt –Börnstein, New Series III/33B1, pp. 5–49.

[17] Xing Z. Li, Bin Liu, Xian Z. Ren, Jian Tian, Wei Z. Yu, Dong X. Cao, Shi Chen, Guan H. Pan and Shu X. Zheng, Pumping effect – Reproducible excess heat in a gas-loading D/Pd system, in *The 9th Int. Conf. on Cold Fusion, Condensed Matter Nuclear Science*, 2002, Tsinghua University, Beijing, China, Tsinghua Univ. Press.

[18] Zhan M. Dong, Chang L. Liang, Bin Liu, Qing M. Wei, Jian Tian, Shu X. Zheng, Jin Z. Yu and Xing Z. Li, Studies on anomalous phenomena of D/Pd systems using a gas-loading process – a stride towards neutrino detection, *J. Condensed Matter Nucl. Sci.* **4** (2011) 119–131.

[19] Zhan M. Dong, Shu X. Zheng, Chang L. Liang and Xing Z. Li, Temperature dependence of excess heat in gas-loading experiments, appear in ICCF-21 proceedings paper.

[20] S.A. Lipinski and H.M. Lipinski, *Hydrogen–Lithium Fusion Device*, WO2014/189799 A9 (27 Nov. 2014).

[21] Private Communication (by J.-P. Biberian).

[22] T. Passell, The case for deuteron stripping with metal nuclei as the source of the Fleischmann–Pons excess heat effect, *J. Condensed Matter Nucl. Sci.* **15** (2014) 1–7.

[23] Melvin H. Miles and Iraj Parchamazad, The Eyring rate theory applied to cold fusion, presentation in SSICCF-20, Xiamen, China, Sept. 28–30, 2016 and in ICCF-20, Sendai, Japan, 2–7 October 2016.

J. Condensed Matter Nucl. Sci. 29 (2019) 453–471

Research Article

Theoretical basis for Nuclear-waste Remediation with Femto-atoms and Femto-molecules

Andrew Meulenberg*
Science for Humanity Trust Inc., USA

Jean-Luc Paillet
University of Aix-Marseille, 13007 Marseille, France

Abstract

The relativistic quantum mechanics equations for atoms predict deep electron orbits with radii in the femto-meter range in addition to the known atomic orbitals. In prior papers, we have explored a model for the deep-orbit solutions of these relativistic equations and of the resulting hydrogen and helium femto-atoms (and even femto-molecules). One prediction of this model, based on observations from successful cold fusion (CF) results and previously mentioned, is that of hard-radiation-free transmutation. An extension of this important feature is that of the relativistic long-range electromagnetic forces of the deep-orbit electrons that can draw a femto-atom or molecule through a lattice to an excited or unstable nucleus. The earlier papers on this topic assumed the deep-orbit electrons to have kinetic energies in the 1–2 MeV range. Our recent work has replaced and/or augmented the low-MeV range with ∼100 MeV values. These highly relativistic electrons create basically the same remediation characteristics as those at 1–2 MeV. The selective attraction of mobile femto-atoms or molecules to radionuclides means that, not only their transmutation products but, all radioactive materials in the vicinity are preferentially made to decay by multi-particle, but fast, processes. This ability to so neutralize such materials explains some of the outstanding questions about low-energy nuclear reaction (LENR) results, such as why known characteristic decay products of observed neutron-activated transmutations are not seen. Presently, the model is only a possible explanation of observations. This paper is not a "how-to" document. It seeks to consolidate, update, and expand our prior theoretical material on selective transmutation and nuclear-waste remediation that could lead to suggestions for experimental testing and possible confirmation of the proposed model.

Keywords: Cold fusion, Deep-orbit electrons, Liénard–Wiechert potentials, Relativistic equations, Selective transmutation

1. Introduction

The s-orbits of atomic electrons pass through the atoms' nuclear region in which the kinetic energy gained from the Coulomb potential of the nuclear and electron charges briefly makes such electrons relativistic. The relativistic

*Corresponding author. E-mail: Mules333@gmail.com.

Schrödinger (Klein–Gordon or K–G) and Dirac equations predict deep electron orbits with radii in the femto-meter range In such orbits, the electrons are continuously relativistic; but, no standard models predict how electrons can get there or what happens when/if they do so. In prior papers, we have explored the nature of deep-orbit solutions of the relativistic equations [1–4] and of the resulting femto-atoms (and even femto-molecules) [5–7]. Since the existence of the deep-orbits has not been experimentally observed and the mathematics of the relativistic-quantum-mechanics equations leading to the deep orbits has been contested, this whole topic could be considered speculative. Nevertheless, the "normal" solutions to these equations are fundamental to quantum mechanics (QM) because they have been validated experimentally. Analysis shows the deep orbits, if populated, to be both self-consistent and consistent with experimental observations in Cold Fusion (CF)

One prediction of our electron deep-orbit (EDO) model, based on observations from successful cold fusion results and mentioned in several of the above references, is that of hard-radiation-free transmutation. An extension of this important feature is that of the relativistic long-range electromagnetic (EM) forces of the deep-orbit electrons [8,9], which can draw a femto-atom or molecule through a lattice to an excited or unstable nucleus. This attraction provides the basis for selective remediation of radioactive waste.

The selective attraction of mobile femto-atoms or molecules to radionuclides means that, not only transmutation products but, all radioactive materials near the source of neutral femto-atoms are preferentially made to decay by multi-particle, but fast, processes.[a] This ability to so neutralize such materials explains some of the outstanding questions about low-energy nuclear reaction (LENR) results, such as why characteristic decay products of known transmutations (e.g., those by neutron activation) are not seen. It is also a means of further validating the electron deep-orbit model of Cold Fusion. Presently, the model is only a possible explanation of observations. But, confirmation of transmutation in biological systems [10] introduces other possibilities that may, or may not [11], be consistent with the EDO model.

If the substrate in a CF-active system is intentionally doped (with specific isotopes and radioactive elements), it should be possible to direct, and thereby determine and quantify details: of the CF processes, of the subsequent transmutation pathways and of hard-radiation mitigation. This paper[b] seeks to consolidate and expand our prior survey [12], which includes references to our work on selective transmutation and nuclear-waste remediation and to introduce into the topic a recent finding of much higher kinetic-energy deep-orbit electrons [3] (and the references therein) than had been considered in prior works on the topic.

Section 2 provides some further background as to why early cold fusion results have led to thoughts of a different process for transmutation. Section 3 (and its subsections) provides details on the specific mechanisms involved with a model of cold fusion that invokes the existence of deep-orbit electrons and resulting femto-atoms as the principal actors in the different fusion process, the high mobility of an active agent, the transmutation of lattice nuclei without expected radiation, and the greater attraction of the active agent to radionuclides within the lattice. This is followed by a brief discussion of the heuristics and limitations of quantum mechanical models to address these mechanisms and a summary list of the key points of the paper.

[a] Multi-particle interactions are generally low probability because of the timing requirement that all interacting particles must be together at the same place and time (or, at least, within a narrow window). The EDO, when populated, provides constant proximity of multiple bodies and therefore fulfils this simultaneity requirement 100% of the time. Furthermore, it will provide additional reaction pathways and will often increase reaction rates by reducing or eliminating barriers in the normal interactions. Often, the model used here is a semi-classical picture of a point-like particle. Thus, an $l = 0$ electron will transit the nuclear region; but, over time, this trajectory (which has a frequency dependent on electron velocity and orbit and which has no angular dependence) will trace out the isotropic distribution function, with a peak at $r = 0$, generally pictured for the s-orbital.

[b] Work presented at ICCF-21, The 21st International Conference for Condensed Matter Nuclear Science, 3–8 June, 2018, Fort Collins, CO, USA.

2. Basis for CMNS Nuclear-waste Remediation

We have proposed that relativistic deep-orbit electrons are the basis for all CF results. The ability of such electrons to initiate and participate in nuclear fusion and radiative decay processes, generally without the normally-observed processes of conventional nuclear physics, leads us to the topic of this paper. The path to this proposal begins with the Pons and Fleischman electrochemical work with deuterium [13] that led to the observation of heat in excess of that possible from chemical reactions. In subsequent experiments, it was found that, associated with this excess heat, the expected product ratio of fusion of two deuterons was not the same as was well-known for the D+D reaction. Instead of the normal neutron to proton ratio, n/p = ∼50/50, results for D+D $\rightarrow$ ^{3}He + n and ^{3}H + p, the neutron decay path was suppressed by many orders of magnitude (6–8) relative to the proton decay path. Furthermore, even the tritium results of the proton decay path of D+D were orders of magnitude lower than would be expected for the excess heat measured. Thus was demonstrated the first examples of altered-decay processes associated with transmutation via Cold Fusion. Almost from the beginning of Cold Fusion, the topic was publicly rejected by those who "knew" what could happen at the atomic, the molecular, and the nuclear levels. Their major arguments have now been countered on the theoretical level (e.g., [14]) and by thousands of experiments. The theoretical models and experimental results open up a new approach to the transition region between atomic and nuclear physics. Their understanding can now be extended into other areas.

In the years of experimental efforts to reliably reproduce the Cold Fusion heat effect and to understand its operative mechanism, it became clear that other transmutations were also taking place [15]. Again, the expected gamma-decay radiations from transmutation products (i.e., those known from neutron activation) were not observed. Initially, this lack of observables was used by skeptics to deny that transmutation had occurred or to claim that the "strange" isotopes found in CF experiments were due to contamination. It was attributed by some to the low rates of cold fusion. This latter attribution was despite the many comments (generally to debunk the claims of measured excess heat) about radiation killing everybody in the laboratory, if the measured heat was really from transmutation. With the accumulated evidence for ^{4}He being a dominant product of cold fusion and having excess heat produced being proportional to the determined production of ^{4}He [16], the standard model for DD fusion had to be recognized as limited in scope. Thus, DD fusion (the transmutation of D+D $\rightarrow ^4$He) became a key source of information about both the fusion/fission processes and the suppression of radiation and nuclear-fragmentation processes in CF-induced transmutation. Moving beyond the "anomaly" of ^{4}He production, isotopic profiles of cathodes from excess-heat-producing experiments showed both the decrease in certain "natural" isotopes of the substrate elements and the increase in others that would be considered logical transmutation products [15]. This led to the thought that, associated with heat production in cold fusion, nuclear remediation might be possible. *The further observations (particularly in H+H $\rightarrow$ D cold fusion results with NiH) that the "substrate isotopic distribution did not show proportionate changes" pointed to selective transmutation.* Was this a result of differences in capture cross sections for some "product" of Cold Fusion? What mechanism(s) can be considered to be responsible for such selectivity?

3. Deep-orbit Electrons and their Interactions

The relativistic QM equations give an electron deep orbit (EDO) with a binding energy for hydrogen at BE = –509 to –507 keV [17] when an electron with or without spin and a point nucleus is assumed. However, the equations do not provide specific information on the electron energies (kinetic and potential). This must be found based on other considerations. Therefore, the proposed deep-orbit electrons are relativistic with model-dependent kinetic energies of $1 < \mathrm{KE} \leq \sim 100$ MeV. The lower limit is established by the minimum requirements for D + D $\rightarrow$ ^{4}He with reduced or zero fragmentation. It is what had been assumed in our prior work. The upper limit is established by recent work with the assumption that the equality in the Heisenberg Uncertainty Relation (HUR) is obeyed [1,18]. Values of KE = ∼100 MeV are possible based on both the virial theorem for a $1/r$ Coulomb potential and on the inequality in the

HUR; but, for our present purposes, the 100 MeV value is representative of a high-KE deep-orbit electron. The failure of the low-MeV KE values (representative of a low-KE deep-orbit electron) to obey the HUR is to be addressed in detail elsewhere.

Our prior work on nuclear remediation was based on the lower limit value of KE. In the present work, we extend that and include the high-KE option of the EDOs to see if it is also consistent with the prior work. The recent work looks at additional effects that are only 2nd and 3rd order for atomic electrons. However, for femto-atoms, they could dominate the orbits. A deep-orbit electron's proximity to the nucleus greatly lowers the electric-dipole moment (but not the spin effects? See below and in [18]) of the electron/nucleus pair. The small orbital-dipole moments reduce both the far-field magnetic effects and the electric-dipole effects of these electrons relative to those from atomic electrons.

If nuclear spin is included in the relativistic model, then there are additional attractive and a repulsive potentials to consider [18]. An additional attractive potential draws the electron closer to the center of the nucleus, but may raise the actual potential at that location since any repulsive centrifugal potential must be overcome to shrink the EDO toward the center. Both effects alter the virial theorem for stable orbits.

In the present model, we assume that spin requires a finite size "body" and, as a test point (a relativistic charged body with spin) is moved close to the center of that body, its effects are diminished from those based on the standard radial dependence of the spin–spin interaction. The "overlap" of two such spinning bodies will more rapidly diminish the spin–spin, S–S, interaction as $d > 0$. The centrifugal force on a deep-orbit electron, a non-zero perturbation of the $l = 0$ atomic quantum number, also diminishes with overlap of the orbiting body with the center of rotation. Nevertheless, this latter is only a single-body effect and therefore does not decrease as rapidly as does the two-body S–S effect.

The high-energy option [18] of the model would allow the deep-orbit electrons to be energetic enough to fit the Heisenberg Uncertainty Relation (HUR). Since the relativistic equations do not specify the KE and PE of the deep-orbit electrons, but only the binding energy, BE, it is possible to have a relatively low BE (e.g., ~ -500 keV) , when both KE and |PE| are large (e.g., ~100 MeV). This condition is also consistent with the relativistic virial theorem [19] that specifies, for stable Coulomb orbits, a condition KE $\to$ PE as $v \to c$. Thus, the extreme energies might exist within the same orbit (i.e. one with the same binding energy, but not necessarily the same orbital geometry and average radius) and still satisfy the relativistic equations.

The relativistic virial theorem for a pure $1/r$ Coulomb potential gives a relationship between its orbital kinetic and potential energies of KE = $-(\gamma/(\gamma+1))$PE. For an electron with KE = ~100 MeV and γ = ~200, then PE = $-$KE/$(\gamma/(\gamma+1))$ = $-$(201/200)KE. Thus, |BE| = |PE| $-$ KE =~ 0.5 MeV. For an electron with only KE = ~1 MeV, then γ = ~3 and |PE| = KE$/(3/(3+1))$ = 4KE/3. Thus, |BE| = |PE| $-$ KE <~ 0.35 MeV. This is not consistent with the virial theorem for a Coulomb potential. However, for an orbit within the multi-fermi range of a nucleus, there are too many additional forces to consider before assuming a 1/r potential in the near-nuclear region.

The interactions of the deep-orbit electrons with the charged nucleon components have never been significantly explored because this deep-orbit has not been accepted since early rejection of the neutron model as a "proton plus electron". That rejection preceded, by many decades, more-elaborate models for the internal structure of the nucleon. Nevertheless, if quarks (and their components?) are charged, then it would be impossible for the deep-orbit electrons not to interact strongly with them. However, if the vibration or orbital frequency[c] of these electrons is so high that it

[c]In the K–G deep orbits, the $l = 0$ orbital electron transits the nuclear region and thus may be said to vibrate through the nucleus rather than orbit it. In the Dirac solutions, EDO electrons never fully transit the nucleus. Nevertheless, it might be said to vibrate, as it reflects between the attractive and repulsive potentials about the nucleus, even as it moves about it in a rosette rather than in an elliptical orbit. The vibrational frequency can be higher than the rotational frequency and it could be very much higher than the deBroglie frequency. This is the basis for considering lower kinetic energy EDOs (e.g., KE < 2 MeV) that might violate the HUR based on $\hbar$. The deBroglie frequency would only be a perturbation on this non-HUR deep orbit.

might greatly exceed that of quarks, then it might not be "seen" by them. In this case, it is unlikely that any effective dynamic interaction would be with the nucleon or even its quarks rather than with any charged sub-components. Without knowledge of the frequency nature of spin and its source, we cannot say that the deep-orbit electron's S–S interactions with the nucleons or quarks can be dominant, diminished, or ruled out. Thus, speculation as to the nature of sub-nuclear components, based on coincidences encountered in trying to explain observed CF phenomena, may not be out of place.

The orbital or vibrational frequencies of the deep-orbit electron level(s) could be comparable to that of the nucleons, quarks, and/or subcomponents. Thus, they would interact with the nucleons and components to explain the transfer of energy from the nucleus to the deep-orbit electron. This transfer of energy from the potential energies of a nucleon to the kinetic energy of the electron is basic to the change in mass of the deuterons that is needed to model the new pathway to ^{4}He from D–D Cold Fusion [8]. *It also explains the mechanism of energy transfer from excited nucleons to the lattice and the mechanism of transmutation without emission of energetic radiation or particles* [6]. The following sections provide details of how this transfer may take place and why it is a unique property of the deep-orbit electrons.

3.1. Deep-orbit electrons and their interactions outside of the nucleus

Relativistic deep-orbit (EDO or deep Dirac level, DDL) electrons have strong long-range electromagnetic (EM) forces and fields [8,9]. We look at the maximum-strength **E**-fields based on some assumptions[d] for application of the Liénard–Wiechert potentials [20] and for calculating the maximum values of the fields. One assumption is that the static potential and fields of the nucleus are small relative to those of the relativistic electron's high velocity and acceleration. Therefore, we can generally ignore the proton's essentially static charge and the dipole nature of the electron/nucleus pair. The general expression for the electric field is:

$$\vec{E}(\vec{x},t) = q\left(\frac{\vec{n}-\vec{\beta}}{\gamma^2(1-\vec{\beta}\cdot\vec{n})^3R^2}\right)_{\text{ret}} + \frac{q}{c}\left(\frac{\vec{n}\times[(\vec{n}-\vec{\beta})\times\vec{\dot{\beta}}]}{(1-\vec{\beta}\cdot\vec{n})^3R}\right)_{\text{net}}. \quad (1)$$

The maximum value for the first term on the right-hand side (RHS) depends strongly on the velocity relative to c ($\beta = \mathbf{v}/c$) and on the direction the electron is moving relative to the unit vector $\boldsymbol{n}$ pointing from the "retarded" source toward the maximum field. For highly relativistic orbital motion under these assumptions, which include: the velocity being parallel to $\boldsymbol{n}$; $v \to c$; $\boldsymbol{\beta} = \mathbf{v}/c \to \boldsymbol{n}$; and $\boldsymbol{\beta}\cdot\mathbf{n} \to 1$. Since $\gamma^2 = 1/(1-\beta^2) = 1/(1-\beta)(1+\beta)$, the first term reduces to $q(1+\beta)/(1-\beta)R^2$, where R is the distance from the charge to the test point when the energy was emitted.

The last term depends on both the velocity *and* on the acceleration of the electron as vectors. [e]

The maximum value for the second term becomes $2^{1/2}q\,\dot{\boldsymbol{\beta}}/c(1-\boldsymbol{\beta}\cdot\mathbf{n})^{3/2}R$. Thus this term increases, relative to the velocity term, as the velocity, the acceleration, and the decrease in R. We need not concern ourselves with the retarded (ret) values of the two terms because we have assumed circular orbits, only maximum values, and R is much greater than either the dipole moment or the orbital radius r. Therefore, neither $\boldsymbol{\beta}\cdot\boldsymbol{n}$ nor R change much with the retarded-time correction. Since the following description of fields provides only a crude approximation to the real world at these dimensions, these assumptions are also loose. In particular for near-field values, the dipole moment established by the electron orbit may be comparable to the distance from the dipole center to the test point and thus the

[d] As we use the semi-classical model of the electron for the orbital picture, we use classical electrodynamics here for the description of electric fields generated by moving and accelerating point-like relativistic electrons.

[e] With the same assumptions as above and also assuming the (circular-orbital) acceleration $\dot{\boldsymbol{\beta}}$ to be at right angles to the velocity (and thus to $\boldsymbol{n}$), then the second term becomes $(\boldsymbol{n}\times\boldsymbol{\beta})q\,\dot{\boldsymbol{\beta}}/c(1-\boldsymbol{\beta})\cdot\boldsymbol{n})^2R$. At small angles between $\mathbf{n}$ and $\boldsymbol{\beta}$ and $\boldsymbol{\beta}\to\mathbf{n}$, then $(\mathbf{n}\ \boldsymbol{\beta})/(1-\boldsymbol{\beta}\cdot\mathbf{n}) \to\sim (2/(1-\boldsymbol{\beta}\cdot\mathbf{n})^{1/2}$ (calculated) and $(\mathbf{n}\times\boldsymbol{\beta})/(1-\boldsymbol{\beta}\cdot\mathbf{n}) \to\sim (2/(1-\boldsymbol{\beta}\cdot\boldsymbol{n}))^{1/2}\ (\boldsymbol{n}\times\boldsymbol{\beta}/(1-\boldsymbol{\beta}\cdot\boldsymbol{n})^2 \to 2^{1/2}/(1-\boldsymbol{\beta}\cdot\boldsymbol{n})^{3/2}$.

distances and directions from the electron to the test point are not properly defined as assumed for $R \gg r$. Just as the applicability of the $1/r$ Coulomb potential must become suspect at nuclear distances, the size of the electron becomes important and must also enter consideration, particularly when $R \leq \sim R_c$ (the classical-electron radius at ∼2.8 fm).[f]

The first RHS term in Eq. (1) (giving the static Coulomb term when $v \ll c$) has a $1/R^2$ dependence and the second (including the radiation fields) has a longer-range $1/R$ dependence. Thus, the second term dominates for the far-field (large-R) effects from near-nuclear deep-orbit electrons with their near-c velocities and femto-meter-range (high-acceleration) orbits. The normal interpretation of the radiation far-field is where the **E**-field has become independent of the source (i.e., as a photon). However, we now make a different assumption. Since, in the source region of the maximum $1/R$ **E**-field, the electron has a acceleration component in the $\boldsymbol{n}$ direction, the radial decay of field strength is not strictly $1/R^2$, it is faster. Thus, this far-field energy is actually still bound to the charge [21]. This assumption is supported by the fact that no (photonic) radiation is observed from ground states (where there is not enough angular momentum in any decay to form a photon) or states with only filled states below them. Electrons in such states have a strong dynamic (relativistic?) EM field about them that is in addition to the static-charge field. This bound-field energy is included in the effective mass of the electron [22]. Decay, via photonic radiation of this dynamic energy, from excited states has a probabilistic nature, not a continuous one as the equations and classical electrodynamics might imply. (However, significant change in field strength is continuous during electron transits of the nuclear region and periods of transition between states.) Again, this bound dynamic field, represented by the $1/R$ term, while an energy source, is not generally radiant. When there is no radiation loss from this field, then, it must form a standing wave (evanescent wave) with an EM return wave to balance the outgoing wave and the second term must be considered as a source of far-field energy. Again, the field energy is still bound to the source charge unless/until it is absorbed by a nearby charge or radiated away as a photon. We will thus use this term (variously called the EM, the radiation, and the far-field term) to determine, below, the intense **E**-fields from relativistic deep-orbit electrons.

To gain some perspective on the relative electric-field strengths of the two terms, we can look at the approximate ratios for different orbits and particles. By comparing the two terms of eq. 1 for various conditions, thus using the extremes (Coulomb vs. radiation, velocity vs. acceleration, or near-field vs. far-field terms), it is possible to get a feel for why some things can be ignored when doing atomic physics; but, they become important for the deep-orbit electrons. For our approximations here, we will assume: that near-nuclear orbits are circular; that we can ignore non-Coulomb forces; that the "hidden" γ's cancel in the "$\dot{\beta} = \mathbf{a}/c = \gamma\, \mathrm{F}/\gamma m_0 c = (e^2/R^2)(1/m_0 c)$" term; and that the classical-electron radius $R_\mathrm{c} = e^2/m_0 c^2$. Thus, $\dot{\beta} = cR_\mathrm{c}R^2$. For example, in comparing the two terms for atomic-electron **E**-fields in the atomic or Angstrom range, R_a, comparable to the Bohr orbit, $R_\mathbf{B}$,we have (ignoring factors of 2 and with $v \ll c$):

$$\frac{\text{Atomic Coulomb}}{\text{Atomic EM near-field}} = \frac{\sim q/\gamma^2(1-\beta)^2 R_\mathrm{a}^2}{2^{1/2} q\dot{\beta}/c(1-\boldsymbol{\beta}\cdot\boldsymbol{n})^{3/2} R_\mathrm{a}} = \frac{\sim R_\mathrm{a}}{R_\mathrm{c}} = \frac{\sim 10^{-10}}{3 \times 10{-}15} = \sim 3 \times 10^4. \quad (2)$$

For an atomic electron, the large dominance of the Coulomb field over the non-relativistic radiation field at atomic/lattice distances (e.g. ∼100 pm), with the ratio of the two terms mathematically reducing to the distance

[f] Defining electrons as being point-like may become a problem when its infinite-range electric fields are being considered. However, if we consider 99% of its rest mass to being within the classical radius and the **E**-field amplitude being reduced to ∼1% at the Compton radius of the stationary electron, then we have some idea of the range of validity for our calculations. If the electron is within a Compton wavelength of a proton, the joint **E**-field is already distorted and a dipole representation may be useful. If the deep electron orbit is only a classical-electron radius from a proton, then the sum of Coulomb potentials drops quickly and there is relatively little mono-pole **E**-field left outside the 10 fm range. For most purposes, in the far field, the combination appears neutral. How the acceleration term, in this extreme near-field location, fits reality (and the Liénard–Wiechert potentials) must be left for others to determine. It even raises the question of "when does a pair of leptons become a boson?"

of the test point from the charge relative to the classical electron radius, or R_a/R_c, indicates why the latter term can generally be ignored in the lattice.

A similar comparison, but one for the **E**-field of a deep-orbit electron (in the deep-Dirac level, DDL) with a near-nuclear test point, indicates that the Coulomb and EM terms are nearly of the same order of magnitude in the near field. In this very rough approximation, we have: assumed a near-circular orbit, $R_c/2 < R_{DDL} <\sim R_c$, and used a gamma of 3–4 (as in the earlier publications). Thus, from Eq. (2), modified for the deep orbit (and still cancelling the velocity terms and using $\dot{\beta} = cR_c/R^2$ with $R = R_c$:

$$\frac{\text{DDL Coulomb}}{\text{DDL EM near-field}}(\gamma = 3) = \frac{\sim q/\gamma^2(1-\beta)^2 R_c^2}{q\dot{\beta}/c(1-\beta)^{3/2}R_c} = \frac{\sim c}{\gamma^2(1-\beta)^{1/2}\dot{\beta}R_c} = \frac{\sim 1}{3^2(1-0.94)^{1/2}} = \sim 0.5. \quad (3a)$$

For the deep-orbit electron, the two terms are of the same order-of-magnitude. Updating the results from our more recent work [18], which gives an average gamma of ~200 to make the deep-orbit electrons obey the HUR, complicates things because the actual orbit of the electron is unknown and, if not circular, the velocity (and thus gamma) changes cyclically, rapidly, and greatly. Nevertheless, with the same assumption of a circular orbit as above, the ratio of near-fields decreases by nearly two orders-of-magnitude. From (3a), with $\gamma = 200$:

$$\frac{\text{DDL Coulomb}}{\text{DDL EM near-field}}(\gamma = 200) = \frac{\sim c}{\gamma^2(1-\beta)^{1/2}\dot{\beta}R_c} = \frac{\sim 1}{200^2(1-0.9999876)^{1/2}} = \sim 7\times 10^{-3}. \quad (3b)$$

On the other hand, the effective Coulomb potential for the highly relativistic DDL electron is increased relative to the nearly static potential of (3a) by ~70 times, while the acceleration changes very little.[g] Whatever the details, it is clear that, in the near-field case for deep-orbit electrons, the acceleration term nearly equals or exceeds the velocity term. At large R, the $1/R$ EM field is correspondingly much greater than the $1/R^2$ Coulomb field. *This "long-reach" of the DDL-electron electric field is critical information when trying to explain both the energy transfer from the nuclear region to the lattice and the mechanism for selective remediation.*

How does the DDL far-field compare with atomic electron near-fields? The far-field vs. near-field regime is determined by the distance relation to the orbit. Thus, the atomic near-field, at $R_a =\sim 10^{-10}$ m, is in the deep-orbit far-field. The classical-electron radius ($R_c < 3$ F) is still the DDL near field. The atomic electrons for light nuclei are non-relativistic and therefore are controlled by Coulomb's law in both near- and far-field. The DDL-electron far-field is dominated by the $1/r$ acceleration term. If this is primarily dependent on the Coulomb potential, there are still two things to consider. For a relativistic orbit and assuming $\mathbf{a} = \mathbf{f}/\gamma m_0$ (with the acceleration normal to the velocity vector), then the force can be from the relativistic Coulomb potential, $V_{eff} = \gamma V + V^2/2m_0c^2$, and the effective, or Lorentz, mass (γm_0) is also increased. However, the relativistic effect on force and effective mass would be strongly dependent on the shape of the orbit and the other potentials within this region. With $R_a =\sim R_B$ and $R_{ddl} = \sim R_c = \sim R_B/137^2$:

$$\frac{\text{DDL EM far-field}}{\text{Atomic Coulomb}} = \frac{q\dot{\beta}/c(1-\beta)^{3/2}R_a}{q/R_a^2} = \frac{cR_c}{R_{DDL}^2}\frac{R_a}{c(1-0.9)^{3/2}} = \frac{\sim 137^2}{0.1^{3/2}} = \sim 6\times 10^5. \quad (4)$$

Thus, at lattice dimensions ($>R_a$), the DDL-electron influence is nearly a million times stronger than that of the atomic-electron Coulomb potential. Furthermore, at its ultra-high frequencies (at nuclear levels), this E-field strength cannot be as easily screened by bound lattice electrons as at atomic frequencies. Therefore, while one can generally

[g] Since this factor of ~70 is close to the difference in γ's between the two cases, it is possible that one of the relativistic conversions (e.g., $\dot{\beta} = a/c = f/mc$) needed one more or less γ than used in the above calculations.

ignore atomic-electron effects beyond their nearest neighbors, *the influence of a DDL electron can be seen for a significant volume of the lattice about it.* For the case of a DDL electron obeying the HUR, e.g., for a highly relativistic DDL electron in the near-nuclear region, it is necessary to consider the effective Coulomb potential, $V_{\rm eff} = \gamma V + V^2/2mc^2 =\sim \gamma V_{\rm cb}$, for $\gamma =\sim 200$ [1]. For this large gamma, $v/c =\sim 0.99999$ and the $1/(1-\beta)^{3/2}$ term increases from that of Eq. (4) (of ~32) to over 30 million. Assuming the gammas cancel in the acceleration term, it does not change much (perhaps $R_{\rm ddl}$ reduces to $\sim R_{\rm c}/2$) and the ratio of the two fields of Eq. (4) becomes $\sim 10^{12}$.

If calculations based on these equations and assumptions are valid, the maximum DDL **E**-field from $\sim 1 <$ KE < 2 MeV and from 100 MeV electrons are, respectively, very large and extremely so. Both are at such a high frequency that they will only cause a dither in the lattice-electron orbitals [8]; thus, there is generally no net energy transfer. Nevertheless, if timed and oriented correctly, such a dither in a lattice atomic-electron orbit passing through its nuclear region could shift an electron to a much different orbital. This huge **E**-field could excite, or even strip from their nuclei, a small portion of s-orbit atomic electrons throughout a macroscopic lattice. Such a scenario would account for the large-volume dissipation of excited-nuclear energy needed to explain some of the observed cold-fusion experimental results (such as the low-energy X-ray distribution). If, in a much smaller local lattice volume, the s-orbital electrons are continually (but briefly) removed from their orbits, then state inversion and consequential lasing is possible. Such lasing has been observed in cold-fusion experiments.

Because the DDL orbital frequencies are so high, such electrons would interact with nucleons (directly or via their sub-components) more readily than with lattice electrons. However, because of the mass differences between the DDL electron and nucleons, the energy transfer at each interaction is not great unless resonant conditions exist between the nucleons and/or their sub-components and the DDL electrons. The many and diverse nuclear forces and interactions may permit conditions that produce such resonances. The various shapes of these nuclear potentials alter the probability of significant non-photonic energy transfer [9]. *Such resonances become the basis for the transmutation and selective-remediation models.*

The interaction of DDL electrons with charged and spinning relativistic sub-nucleon components (i.e., quarks and their parts) would perhaps be the greatest interaction of the deep-orbit electrons with both the parent nucleus and the lattice nuclei. This interaction introduces a new perspective on nuclear physics and must be examined closely in terms of what is known from decades of experimental results. It may require an adjustment of, or modification to, the present models of nuclear forces and interactions. A major consideration must be that of the nature of the nucleon charge and its response to a deep-orbit electron. Polarization of the nuclear charge by the proximate electron produces a modified dipole, rather than the monopole fields calculated above. This greatly reduces (perhaps by orders of magnitude) both the amplitude and range of the fields described. The actual amount of reduction cannot be calculated based on present information for the nucleons. Nevertheless, *the residual peak-field intensities in the lattice from the DDL electrons, relative to those normally experienced from atomic electrons, should still be immense* and are the basis of this paper.

3.2. Neutral femto-atoms and molecules

The proposed neutral femto-atoms and -molecules can explain the different processes of cold fusion in both the palladium (Pd) and nickel (Ni) systems. Their deep-orbit electrons provide the means for allowing hydrogen femto-atoms and paired femto-D 'molecules' to penetrate both the electron cloud and the Coulomb barrier about lattice atoms and nuclei, respectively. Nevertheless, there are both similarities and differences in the production and activity of the proton- and deuteron-based femto-bodies [6,7].

A single femto-atom can be formed if the atomic electron of a hydrogen atom (H or D) can tunnel[h] or decay into

[h]Tunneling may be a misnomer here. There is neither physical-potential nor centrifugal barrier between the atomic s-orbital levels and the DDLs. However, there is a barrier of sorts in the inability of the atomic ground state's (an $l = 0$, $k = 1$) s-orbital to provide the angular momentum required

the deep orbit about its nucleus. Such femto-atoms can either fuse with lattice nuclei or combine in a new chemistry to form neutral femto-molecules or positive femto-molecular ions. It is also possible for femto-atoms to fuse with atomic nuclei or perhaps to combine with normal atoms to form mixed, neutral molecules. Thus, hydrogen-based femto-bodies can be femto-atoms H# and D#, femto-molecules H##2 and D##2, or the femto-molecular ions H#p+, H#d+, D#p+, and D#d+, where each # indicates a deep-orbit electron in the atom, ion, or molecule. The mixed molecules, H#2, D#2, HD#, and H#D, contain femto-atoms, but are atomic-scale quasi-molecules because the femto-atoms share an atomic electron in addition to their single deep-orbit electron. These mixed molecules could be the simplest of the halo nuclei that have been observed and measured for several decades. Deuterium could be considered to have the lightest halo nucleus since the two nucleons spend more time outside the nuclear potential than inside.

HD# and H#D are the same mixed-molecule because both the atomic and deep-orbit electrons are shared equally between the two nuclei. They will act, and be measured, as ^{3}H unless very accurate (e.g., spectroscopic) measurements are made. The neutral femto-atoms or femto-molecules will act as stable, but strong, transmutants. They would move through the lattice and into a nucleus as easily as would free neutrons or poly-neutron bodies. *It is this high mobility of the femto-particles that would make them so important in nuclear-waste remediation.* However, there are differences between femto-atoms or molecules and neutrons or poly-neutrons in both their lattice transport and in their interaction with nuclei.

The deep-orbit electron model of CF for D–D fusion proposes that deuterium atoms either come together as femto-deuterides, D# + D#, or form these femto-atoms in the fusion process. Either way, formation of the helium femto-molecule D#D# is a probable mechanism. The question here is whether and how the DDL electrons are shared. If they are not paired by spin or another mechanism, then the positive femto-molecular ion, D#D+, with the deuterons sharing a single DDL electron is the probable configuration. While this ion would attract an electron to form the mixed molecule D#2, neither ion nor molecule is a transmutant without sufficient kinetic energy to initiate the process. The ion is mobile in a lattice, but can't overcome the Coulomb barrier of another nucleus. The mixed molecule is neither mobile nor able to fuse with a nucleus.

Since D–D cold fusion produces transmutation, but the expected mixed molecules are not transmutants, some other entity or mechanism must be involved. The femto-D atom *is* a strong transmutant, so it would act as the femto-H atoms do, until it enters another nucleus. However, because of its greater mass, the mobility and attraction of femto-D to other nuclei is less than that of femto-H. Thus, it will not have the range about the source of femto-atoms that femto-H has and it will have a higher probability of fusing with a like femto-atom in the process. The D#D# entity could be a short-lived femto-molecule and/or some form of diatomic femto-structure. The yet-to-be-discovered nature of the binding and orbit(s) of the paired deep-molecular-electrons as the paired deuterons sink to the lowest potential (the ^{4}He nuclear ground state) would be the determining factor. Until one of the deep-orbit electrons leaves the D#D# entity (a femto-He atom?), it is a short-lived, strong, local transmutant. Thus, d*ifferences between D#D# and H#H# in the lattice are in their lifetime and the type of transmutations observed.*

3.3. Neutral femto-atoms and femto-molecules in the lattice

Selective transmutation begins in the motion of, and forces on, a femto-atom or -molecule in the lattice. As neutral near-nuclear-size bodies, there might not appear to be any major differences in their transport through the lattice

---

to form a photon in the transition to the (near-zero-angular-momentum, $k = -1$, $l = 1$) DDL region. (The change in quantum number, Δk, representing the change in total angular momentum from atomic to DDL, would be 2. If this translates into an E2, or electric-quadrupole, transition then its probability is greatly reduced relative to that of the normal E1, or electric dipole.) Were the deep-level electron KE to approach 100 MeV, then the $k = -2$ ($l = 2$), DDL might be available for photonic decay, via E3; but, again at a further large reduction in transition probability. The possibility of an atomic electron entering the relativistic Coulomb potential regime required to reach these high kinetic energies, via photonic interaction, has not yet been validated or rejected.

compared to that of neutrons (both have high mobility). However, when considering the deep-electron orbits and the internal structure of femto-atoms or -molecule and neutrons, it appears that there may be considerable differences in their ability to interact with lattice electrons and nuclei.

The primary neutral femto-atoms (femto-hydrogen or femto-helium) include one or two protons and zero, one, or two neutrons. Thus, there are common structures between them and neutrons (if one considers the quark model of the nucleus). However, the femto-atoms also contain one or two DDL electrons that make a major difference. Along with the relativistic fields from the deep-orbit electrons, a comparison of the quarks (and their possible sub-components) with deep-orbit and atomic electrons may be key to understanding the selective interaction with the lattice.

Atomic electrons are primarily non-relativistic; but, they have large orbits and therefore may have large dipole moments. The large orbits, with low-electron acceleration and the charge "cloud" they represent, both screen the nuclear charges and prevent close interaction between lattice nuclei. The electron bonds, with and between, nuclei also prevent a ready motion of atoms through the lattice. This restriction of movement negates much of the influence of the large dipole moments that the atomic electron's orbit could provide.

The deep-orbit electrons are relativistic, perhaps highly so; but, they have small orbits and therefore small dipole moments (the spin dipole moments are a separate issue). Their near-nuclear orbits, with high acceleration and the tightly bound charge "cloud" they represent, completely screen the nuclear charges down to the multi-femtometer level and make the femto-atoms and -molecules act as neutrons (neutral bodies) with their ready motion (high mobility) through the lattice. Only when a femto-atom gets to the multi-femtometer distance from a nucleus does the complete screening break down and allow the deep-orbit electron and the nucleus to act as separate charges. However, the tight-binding energy of a deep-orbit electron to its nucleus ($|\mathrm{BE}| < 511$ keV) allows the formation of femto-molecules (or halo nuclei?) and allows such nuclei to get close enough for their nuclear forces to overcome any nuclear-Coulomb repulsion.

As shown above, the relativistic velocities and extreme acceleration of the deep-orbit electrons make the last term in Eq. (1) become dominant in their interaction with lattice electrons and nuclei. Their interaction with lattice electrons are phase related, therefore stochastic (random), and strong interactions are statistical, just as are radioactive decay or photoemission. Their interaction with lattice nuclei are frequency related and strong interactions therefore occur primarily for resonant conditions. The ability of a DDL electron to alter nuclear energy states is dependent on the shape of the nuclear Coulomb-potential well. Since these electrons do not normally have sufficient angular momentum to form a photon and, with the near-field interactions, transverse photons are not able to form in the limited volume between the nucleus and its DDL electron [9] to act as the energy-transfer mechanism. Therefore, energy transfer between an excited nucleus and the lattice is through direct, near-field, EM interaction (or longitudinal photons?) between the nucleus and its DDL electron and, then, between the DDL electron and the lattice electrons (via far-field, EM interaction). However, these same EM fields can also interact with lattice nuclei to cause an attraction of femto-atoms to them.

The strong fields of a DDL electron can shift and split degenerate nuclear and sub-nuclear energy levels without changing any nuclear states. Even a small change in the energy of levels means that a potential can be altered and thereby a force generated. Since the relativistic-electron far-fields are so strong, they can affect nuclei over a large volume in the lattice. The nuclei most affected by these fields will generate the greatest force. All things being equal, the nearest nuclei will have the greatest attraction for the deep-orbit electron and the femto-atoms. However, radioactive nuclei are likely to have the largest changes and, if resonant, might even have state changes induced by the relativistic electrons. *This greater attraction of the DDL electrons for radioisotopes, compared to stable nuclei, is a basis for selective transmutation, a first step in nuclear-waste remediation.*

3.4. Deep-orbit electron as nuclear attractor

As an intermediary between a lattice nucleus and a nucleus with deep-orbit electron, the electron itself may be attracted to both nuclei. If either atom is mobile within the lattice, the electron will draw them together, even over many lattice sites, without any static-charge effects that can be easily shielded by the lattice electrons. The neutral femto-atom may be drawn to any nucleus; but, the lattice, as an isotropic sea of nuclei may not give a particular direction for this motion induced by long-range forces. It is also drawn by any lattice s-orbital electrons that can respond inelastically to the bound EM radiation of the DDL electron. However, as a stochastic process, this effect is also isotropic. So, while there is a preferred motion toward nearby nuclei, the effect is not strong unless there is something special about the interaction.

It is proposed that radioactive nuclei, not being in a lowest-energy configuration and having various modes of getting there, are more susceptible to absorbing and emitting energy in response to the strong fields of the DDL electron. A major reason for this susceptibility is the near-nuclear frequencies of the DDL electron and its fields. If these frequencies are resonant with decay modes of an excited nucleus, then, induced emission is one means by which radioactive decay paths are altered ($F = -\mathrm{d}V/\mathrm{d}x$ from reduction of system potential energy.) Nevertheless, this process is not likely to be one that would create an attraction between femto-atom and radioactive nucleus for long enough to bring two nuclei together. Radioactive nuclei in the lattice are not generally in excited states unless they have already begun their decay process. The more likely process is one of an unstable nucleus having one or more nucleons in less tightly bound states. Such nucleons will respond more strongly to the DDL electric-fields, particularly if they are near resonance. *Any energy transfer between the DDL electron and these weakly bound nucleons will draw the femto-atom toward, and/or induce decay of, the radionuclide.*

In another strong EM process, the case of a fusing pair of deuterons with one or more DDL electrons present, the transfer of nuclear energy to the lattice from the decay to ^{4}He is the change in total energy that defines the attractive force. In the case of a fusing pair of protons with one or more DDL electrons, the transfer of nuclear energy to the lattice is from the weak interaction changing a proton and electron into a neutron and forming the deuteron. The excitation of atomic s-orbital electrons in the lattice would be nearly isotropic; however, that of nucleons in randomly distributed radioactive nuclides would be a directional attractor. It is these states that will more readily respond strongly to the DDL electron(s) and therefore become more attractive to the femto-atoms.

3.5. Neutral femto-atoms and molecules in their interaction with nuclei

If the femto-atoms/molecules were only drawn to radioactive nuclides and then fused in the manner of neutron activation, then this would only be a lower form of nuclear-remediation. Long-term radioactivity would be reduced; but, radioactivity would be increased in the short-term. However, this radioactivity is not what has been observed in cold fusion. Why not?

The deep-orbit electron model is one explanation for fusion with few or no energetic particles/photons being observed. This model was initially created as an explanation for how D–D fusion could proceed to ^{4}He by a pathway never observed (or recognized, if observed) previously in such fusion [23]. When applied to H–H fusion, a variant of the model leads to the formation of femto-H atoms and molecules [5] and also to deuterium (via an accelerated version of the weak-nuclear reaction). Together, these processes further provide an explanation and understanding of radiation-free transmutation that differs significantly from transmutation via neutron activation. Both processes depend on the deep-orbit electron concept and on the fact that the interaction is multi-body [6,7] as in beta decay, rather than two body (see Appendix A) as in neutron activation. The relativistic DDL electron is able to absorb nuclear energy and transmit it much more efficiently to the lattice than can the heavier nuclear components themselves. The continued presence of a third body increases the possible pathways to a lower-energy state. This increases the potential for finding a faster way to "ground" and for a nucleus decaying with particle-energy distributions (such as in beta decay) rather than the observed spectral lines of gamma decay for most neutron activation processes.

The H–H and D–D fusion processes, the simplest of transmutations, often begin with the creation of one or more femto-atoms. Models for this creation are described elsewhere [24–26] and will be assumed here. We begin here with a brief review of the femto-atom, H# or D#, entering a H or D nucleus. This entrance into a nucleus is where differences begin. For H+H# to fuse into a stable deuterium atom, it is necessary for a weak interaction to convert a proton into a neutron. This is normally a low-probability interaction since the two protons normally do not spend much time together. The Coulomb repulsion almost always forces the interaction to be one of scattering rather than of fusion. The deep-orbit electron of cold fusion allows the formation of a meta-stable femto-H molecule or molecular ion, which keeps the protons within femtometers of each other until fusion or some other interaction occurs. The proximity of the energetic electron to the internal charges of the nucleon alters the energy levels of the nucleons and their components via electromagnetic interactions. But, the short-range nuclear force between protons is not normally adequate to bind them against their centrifugal and longer-range Coulomb repulsion. However, the deep-orbit electron reduces this repulsion (actually makes the charge interaction attractive up to a point) [5] and therefore gives time and proximity for the weak interaction needed for the H+H $\rightarrow$ (H + n) $\rightarrow$ D transition to occur. Further details of this nuclear interaction for H+H and for that of D+D $\rightarrow {}^4$He are to be described in a later paper.

For D+D# to fuse, the additional strong nuclear interaction provided by the presence and proximity of neutrons means that fusion is rapid and limited only by the strong centrifugal forces when the deuterons get close enough together.[i] Since there is no weak interaction required in this fusion process, this centrifugal barrier to fusion is not much of a limitation in the presence of nucleon probability overlap, and the femto-molecular state does not last long before it transitions to the fully-fused state. Because no stable nuclear-energy level exists below the ^{3}He + p fragmentation level, the transition from D–D# femto-molecule to excited ^{4}He# to ^{4}He ground state is a continual (non-photonic, but still radiative?) process rather than a quantum "jump" associated with single photons. Thus, in this model of the Cold Fusion process, the initial binding of the two deuterons is by femto-molecule formation mediated by one or two deep-orbit electrons. *The proximity of the energetic DDL electron(s) and the nucleon charges alters the energy levels of the nucleons and their charged sub-components via electromagnetic interactions.* The nuclear forces, strongly enhanced by those EM interactions, greatly speed the fusion process. Nevertheless, *it is the reduction in nuclear mass energy, associated with the presence of the deep-orbit electrons and their influence on (excitation of) nuclear components, that is a likely cause of nuclear remediation.*

This reduction in nuclear mass energy is a result of the transfer of the mass deficit of the nucleus of the incident femto atom or molecule to the target nucleus. This energy comes from the change in potential energy (increase in absolute value of perhaps up to 100 MeV) brought about by the presence of the intense fields of the relativistic deep-orbit electron as it moves from the femto-atom DDL to the new nucleus. These fields alter all energy levels within the nucleus in a random manner. This rattling of the nuclear "cage" until the energies equilibrate and dissipate makes and breaks internal nuclear bonds and allows fusion or fission of the nucleus to a lower energy state. The relativistic deep-bound electron acts as a stirring rod and energy-transfer mechanism at the same time. While the nucleus settles down to a generally lowest state, it transfers energy to the electron which can, in turn, transfer it to the lattice. Thus, this chaotically bound electron can dissipate most or all of the excess energy of the radionuclide, including that which

[i] While the quantum number, $l = 0$, is required to obtain a near-nuclear orbit, this does not necessarily mean that the electron has no angular momentum. For that to happen, the electron must transit the origin exactly and that has a low relative probability. However, when considering all possible $l = 0$ paths through the nuclear region, as many may pass on one side of the nucleus as the other. Thus, centrifugal forces would not be a limitation to rapid fusion. On the other hand, the perturbation theory of QM allows there to be a net offset from this $l = 0$ balance that would provide a centrifugal barrier; albeit one that is much smaller than the h of $l = 1$. Again, the multipath model allows for such transits and barriers within the nuclear region even when the primary orbit does not. This would provide for a "porous" barrier (and tunneling) that decreases as the perturbation increases. Are there resonances within the $l = 0$, $\Delta l \geq 1/100$, picture (i.e., angular momentum $\geq h/100$) that would allow a metastable state that violates the HUR?

would normally be stored in excited levels pending departure as gamma radiation. If the electron ultimately leaves (as it slows down), it takes with it the remains of the excited nucleus energy minus the initial binding energy with which the femto-particle entered. Thus, *the femto-particles not only reduce the radioactivity of impurities (native or introduced) in the lattice, they convert the excess energy of these radionuclides into potentially useful heat.*

Mark Davidson, in his papers on variable mass, e.g., [27] and the references therein, has described the mass changes in a quantum mechanical model. His use of Fock–Stueckelberg theories of off-mass–shell interactions is the only other attempt to explain CF and its consequences in this context. However, that model does not include the concept of deep-orbit electrons, which provides a classical basis for such mass changes. The existence of these electrons should be considered as a constant interaction with a nucleus and therefore it both limits and extends the QM model.

It is precisely at this point that classical physics may provide information that QM does not yet "see". Non-relativistic QM, through the HUR, allows violation of the conservation of energy for brief periods of time during an interaction. Classical physics does not allow that violation; but, through relativity and deep-orbit electrons, it can provide a reason for why such apparent events might occur. While a change in nuclear-mass energy of a few MeV would have a major impact on the decay mode of the excited state of ^{4}He in the D–D fusion process, *the change of ~100 MeV, associated with a HUR-compliant deep-orbit electron, would have a major impact on fusion processes and products of any nucleus.*

4. Discussion

In prior papers, we have described how the nuclear charges, in an excited state of ^{4}He, transfer excess energy to the relativistic deep-Dirac-level electrons with $3 < \gamma < 4$ [28]. For the same process, but one modified to be consistent with the Heisenberg Uncertainty Relation (giving γ =~200) [3], the transfer of such energy from DDL to lattice electrons is much faster because the fields generated are much more intense. The energy transfer from nucleons to DDL electron is not a strong function of deep-electron velocity (being nearly a constant at $v \sim c$); but, it would still be dominated by the near-field term of Eq. (1). All of the energy transferred from the nucleons to the DDL electron is rapidly dissipated to the lattice primarily via the far-field term of Eq. (1), which does not contain a gamma in the denominator. Because of their high KE (~100 MeV), the DDL electrons may not change their relative energy levels much in the process. This is somewhat different from the scenario proposed earlier [28–30] for $\gamma < 4$ (where the DDL electron, with KE=~1.5 MeV, may change its relative energy level and orbital frequency dramatically) and, if valid, the higher-γ DDL model could result in a harder continuum X-ray spectrum generated within the lattice. Nevertheless, the process is the same for both electron-energy ranges and applies to other nuclei during transmutations when a femto-atom fuses with a lattice (or lattice contaminant) nucleus. The DDL electrons, being close to (both inside and outside of) a nucleus, can accept the very-strong, but short-ranged, nuclear-generated fields and can convert their energy into strong fields in the distance. *This ready transfer of nuclear energies to deep-orbit electrons, with the DDL frequencies being on the order of nucleon-component frequencies, is also a basis for DDL-electron-mediated internuclear interactions.*

With this background, it is possible to picture what happens in general when a femto-atom enters a lattice nucleus. The DDL-electron binding energy, on the order of BE = 0.5 MeV (even if its KE = ~100 MeV), is much less than that binding the nucleons (>5 MeV). Therefore, on entering the nucleus, a femto-H will split and the deep-orbit electron now becomes "shared" with all of the nucleons. This sharing, whether the electron is inside or outside the nucleus, distributes the local Coulomb field. The electron's binding energy and potential energy, with respect to the initial proton, are correspondingly reduced in magnitude. The proton that initially bound a deep-orbit electron will thus regain most of the energy (mass) that it had previously shared only with the electron.

In the process of redistributing the new and old energies of the transmuting nucleus, both the incoming proton and the deep-orbit electron have multiple, and competing, choices for the transition. Many, most, or even all, of the paths would not be available to a single proton or neutron entering the nucleus. *These new decay paths are provided by the deep-orbit electron, its effects on nuclear mass, and its ability to mediate energy transfer within the nucleus and from the nucleus to the lattice.* The actual transition will be a result of statistical (competing) processes. Possible mechanisms and examples of the new pathways are taught in the examination of cold-fusion results for D–D fusion, H–H fusion, and transmutations [6,7,29].

In almost all nuclear processes, multiple paths are possible. The observed paths are often just the most probable; but, the differences in probability can be nearly zero or many orders of magnitude. Over the years, many techniques have been discovered and developed to explore, explain, and alter these probabilities. The known energy-levels and decay probabilities of ^{4}He are instructive in explaining how D–D fusion results can be so greatly altered by the cold-fusion process [12,29]. The pathway from the excited state of helium (^{4}He*), just after D–D fusion, to the ^{4}He ground state (via >20 MeV gamma emission) exists; but, it has low probability in the hot-fusion process because fragmentation is so much more probable. Cold fusion results (with ^{4}He#), which show very little fragmentation, might be thought to indicate a suppression of that mode of decay. This would leave energetic gamma decay as the dominant path to ground. However, since such radiation is not observed:

(1) it also must be suppressed by some process *or*,
(2) another faster pathway, perhaps even faster than fragmentation, must be found or,
(3) an entirely different fusion mode must exist (e.g., [31,32] and the references therein).

The present model of deep-orbit electrons satisfies (1) and/or (2). Thus, this mechanism for cold fusion in D–D interactions can be applied to the CF-transmutation results, which, as in the D–D cold-fusion case, do not show the well-known results of hot fusion. The two different fusion models in item (3) of Storms and of Takahashi, both still evolving as is the present model, also contain the concept of confining electrons between pairs (one or more) of resonant hydrogen atoms. Therefore, they have been selected (from the many CF models proposed over the years) to represent different, yet likely, contenders.

Even a comparison of three different theoretical QM models for the hydrogen atom has provided important information as to what different approximations mean in predicted results. The simple QM model, the Schrödinger equation without inclusion of electron spin or relativity, does not show any deep (nor negative-energy) orbits. The Dirac model, with both spin and relativity does show the deep orbits. The Klein-Gordon model (a relativistic version of the simple model, but still without spin) also shows the deep-orbits, but at slightly different binding energies. So the key to the deep-orbits is relativity [3]. Nevertheless, relativistic effects have not been fully explored and exploited because no evidence of deep-electron orbits has been accepted and therefore no results, from experiments specifically designed to seek deep-orbit-electron characteristics, have been provided for any models to work with. While high-atomic-number elements have atomic-electron orbits that are relativistic and some effects have been calculated and confirmed by models, details of the transient nuclear interactions with these electrons are seldom explored [33].

All three QM models generally assume a point-source, infinite-mass, positive charge as the center of attraction. Deviation from these assumptions is generally in the perturbation model, since the integrated effects of nuclear interaction with atomic electrons is small relative to the Coulomb energies involved. While corrections for the finite nuclear mass are made by use of a "reduced" mass, *changes* in the nuclear masses with electron proximity are not considered. Other than as an explanation for cold fusion results, we have not seen where this known effect has been proposed even when physics has been trying for years to understand the difference between calculated proton radii from atomic-electron and muonic-atom experiments.

Arguments have been made [34] that challenge the several presently 'accepted' versions of the internal structure and nature of a nucleus, its nucleons, and their constituent parts. It is difficult to go beyond a number of assumptions when making predictions or calculations regarding nuclear interactions. Introduction of a relativistic electron into the near-nuclear region complicates things even further. However, if relativistic DDL electrons exist, their dynamic EM fields at the nucleus would be much stronger than those of any nucleon, or externally applied field, and major effects would be expected. These electron-based fields could have a greater influence during the transmutation process than those of a femto-atom's nucleus. Introduction of a DDL electron reversibly alters nuclear (nucleon and quark) energy levels, masses, and processes even before fusion begins. Furthermore, relative to the motion of such an electron, *the fusion process (a rearrangement of nucleons) is not instantaneous; it is a transition (generally the source of gammas). The final product is a result of statistical (competing) processes mediated by the fields of the DDL electron.*

5. Summary of Deep-orbit Electron Effects on Selective Radioactivity Remediation

We have covered a lot of territory to provide a basis for this topic. We will try to organize the major points here. The key points are in italics.

(1) Because of their relativistic velocities and very-strong acceleration toward the center of charge, the DDL EM fields, if present, *dominate the fields* about an atom (Section 3.1).
(2) In cold fusion, *direct transfer* of nuclear energy to lattice s-orbit electrons is via DDL electrons (Section 3.1: "Thus, this huge **E**-field could excite ... a small portion of s-orbit atomic electrons throughout a macroscopic lattice").
(3) *Neutral femto-atoms* and molecules allow protons to avoid a nuclear-Coulomb barrier (Sections 3.2 and 3.3). This charge neutrality and small size permits their unobstructed mobility through a lattice and into a nucleus.
(4) *Near-field attraction* (induced-dipole to induced-dipole) of femto-atoms and molecules to lattice atoms and nuclei is strong (Sections 3.2 and 3.3).
(5) Even *attraction of femto-atoms and molecules to distant excited lattice nuclei, or to radionuclide* impurities, is possible via far-field forces (Section 3.2 and 3.3).
(6) Deep-orbit electrons inside a nuclear region can produce *neutrons* via "fast" weak interactions (electron capture in Sections 3.4 and 3.5).
(7) *Multiple possible paths* exist for femto-atom transmutation (Section 4).
(8) Deep-orbit electrons give *m multiple decay paths* for excited transmutation products (Section 4).
(9) *Multi-body interaction* may eliminate gamma "spectral" lines (Section 3.5 and Appendix A).
(10) *The relativistic deep-orbit electron* is a basis for all the above.

Appendix A. Three-body Vs. Two-body Atomic-electron Energy Transfer

This appendix is an attempt (by A. Meulenberg) to provide a physical description of energy transfer that is not expressed in the mathematical formulations of the process. Energy transfer mechanisms are considered critical to cold fusion processes, to the observed transmutations, and to the selectivity of such transmutations. Such an interpretation should not change the mathematics and, it is hoped that, it might give some insight into mechanisms that must otherwise just be accepted on faith. It is not expected to be a complete picture, fully proven. However, it is thought to be self-consistent and with sufficient references to be considered plausible.

The normal process for atomic-energy transfer between two atoms (with both electron excitation and orbital decay) is radiative, via a photon. This can be considered to be a three-body process, two electrons with the third body, a photon, considered to be the exchange medium. An important function of this medium is that of reversing a change in frequency during a transition. This is equivalent to reversing rotation direction in a mechanical system with an idler wheel [35]. In the energy exchange between the electron and the photon, even the process of forming or absorbing a photon is a three-body process, with a third "body", the bound electromagnetic (EM) field[j] considered to be a buffering

[j]The bound EM field is a charge's static electric field plus its velocity- and acceleration-dependent distortions, the measurable **E**- and **B**-fields of an

exchange medium that is coupled to and encompasses both. Why do these processes work, when a simple two-body interaction does not? The following is offered as a possible explanation.

The atomic-energy absorption process can be considered to be a low-energy, resonant (or harmonic), version of Compton (inelastic) scattering [36]. Using a model of the photoelectric effect [37] and a classical view of the atomic electron as being a bound, charged, particle, we see a resonant interaction between the photon and electron. We look at the development of the EM field and the photon in a non-relativistic, atomic-orbital, decay process to see the resonances between electron and bound field and between bound field and forming photon [38]. As an atomic electron decays from one orbit to a greater binding-energy state (i.e., to one with a higher absolute value, | BE|), both its kinetic energy and orbital frequency (not its spectral frequency, which is related to energy differences) increase. This means that its bound EM field increases in amplitude. Furthermore, being driven by the electron's motion, the dynamic EM-field frequency increases as well. This process is critical to the energy conservation of the system and to the formation of a photon [38].

Similarly, when a photon is absorbed by an atomic electron, it is initially coupled (quasi-bound) to it through the electrostatic charge and the phase-related, bound, dynamic-EM field. Cycle-by-cycle it raises the electron to a higher-orbit with a lower binding energy (i.e., |BE| is lower) and with a higher potential energy (V_2, where $V_1 < V < V_2 < 0$). The point in each case is that, in a *decay* process, the nascent photon (a subset of the bound dynamic-EM field) increases frequency (and energy) along with the orbital electron. In an *excitation* process, the quasi-bound-photon frequency decreases and it loses energy as the electron gains energy (reduces its binding energy |BE| in the Coulomb field). This process preserves energy and momentum conservation in a continuous transition instead of in a "magical" and instantaneous quantum jump that depends on "off-shell", non-conservative, components of an interaction to define the process.

The photon and electron frequencies do not change at the same rate and the difference is tied up in the remainder of the bound EM field of the local interaction between nucleus, electron, and bound photon. When the nascent- or quasi-bound-photon and electron orbital frequencies (or their harmonics) are resonant, a photon can be respectively released from, or absorbed into, this bound EM field.[k] Thus, this field may be considered a buffering body, just as a bound electron is a body inside an atom. Without mention of this physical buffering field, the photo-electric process is just a quantum-mechanical description (grey box model [39]) of an observable effect.[l]

Just as the free (or bound) electron transition to (deeper) atomic orbitals requires an intermediary, the transfer of energy from an excited atomic electron to another bound electron (in the same or another atom) requires a third body. This process cannot happen as readily by a two-body mechanism because of frequency mismatches. If an excited bound electron tries to transfer energy to another charged body, via its electric or magnetic fields, it will be poorly coupled (non-resonant) if the frequencies do not match. If they do match and energy transfer begins, then the "sending" bound electron will go to higher orbital frequencies as it decays. Similarly, a "receiving" bound electron

electron in motion. Unless/until, the dynamic portion of this EM field forms into a photon and leaves, it is bound to the source and acts as part of its distributed mass energy. Nevertheless, in some of this discussion, it can be considered a separate (but coupled) body. In the present standard model, virtual photons are considered to be the force-exchange medium of charge. When equated to evanescent waves or the energy associated with **E**-field lines, they are also bound energy with effective mass and this perhaps becomes a many-body problem. Thus, in this regard, the present model differs little from this aspect of the standard model.

[k] The standard model would say that a virtual photon has become "real".

[l] The connection between the continuous process described here and the quantum mechanical description of a quantum "jump" is that of two coupled oscillators. QM only compares the initial and final states, nothing in between. Determining the eigenvalue of the transition is equivalent to finding the oscillation frequency between the two states from the diagonalization of a matrix representing the classical simultaneous equations for the two oscillators. This Heisenberg representation of QM is thus equivalent to the classical problem with the additional insight of identifying the mutual-oscillation frequency with the difference in energies between the oscillators and its relationship with the photon that is emitted or absorbed in the process. While QM greatly simplifies the description of the process, it loses information on the transition process and the function and importance of the bound EM field as a third "body".

will go to lower frequencies. Thus, an initial frequency match will rapidly lose resonance and coupling strength. This process eliminates a net-energy transfer between two electrons in the same energy state even though there will be a cyclic (reversible) energy exchange between coupled oscillators.

On the other hand, if an excited bound electron tries to directly transfer energy to another atom's electron with a higher orbital frequency (i.e., more tightly bound), via its electric and/or magnetic fields, it will be poorly coupled (non-resonant) initially. However, it will slowly move toward resonance and a more rapid transfer of energy as its frequency increases and that of the receiving electron decreases. Nevertheless, there is no net energy transfer at the mutual resonance point where their frequencies are the same. This is because both electrons are radiating the same amount of energy and they simply act as coupled oscillators changing their energy levels at a frequency that is lower than that of either of their orbits. The probability of a net exchange of energy is very low because bound electrons are unlike simple harmonic oscillators with fixed frequencies as their kinetic energy changes. In a Coulomb potential, before reaching the end point of a decay to a stable orbit from an excited state, the lower level of the sending electron has a higher frequency than that of the receiving electron and will not be able to stimulate it to even lower frequencies. In the standard model, the two states will simply exchange energy back and forth via "virtual" photons.

The photon, as a third body, is able to maintain near-resonance conditions (frequencies) between two different bound-electron states throughout the exchange of energies (but not necessarily simultaneously). Even if the bound-electron states of different atoms have the same energies and no net-energy transfer takes place, there is an ongoing interaction that can raise and/or lower the system energy. This is perhaps why virtual photon exchange (evanescent waves [39]) is considered to be a model for the charge interactions of the Coulomb potential. Based on a standing-wave model of the photon and electron [41,42] and a more detailed knowledge of the nature of the internal fields (i.e., gradients of potentials) of the photon [43], a mechanism for the attractive and repulsive forces induced by these virtual photons is proposed elsewhere.

The deep-orbit electron, as a third body, is also able to maintain near-resonance conditions between two different nuclei throughout an exchange of energies. In addition, it has the capability to exchange energy between a nucleus and an atomic electron. While it does this by creating photons from energy it receives from an excited nucleus or by direct excitation of lattice-atom s-orbital electrons, it can also cause a physical attraction between nuclei and between nuclei and nearby atoms (via their bound electrons).

Most interactions of nuclei with atomic electrons, via deep-orbit-electron fields, are not strong because of the differences in orbital frequencies (e.g., $> 10^{19}$ vs. $\sim 10^{10}$ Hz). The deep-orbit-electron frequencies become comparable to those of the lattice electrons only when (and as), the nuclear-decay-energy transfer exceeds the deep-orbit-electron's ability to transfer energy to the lattice. With this condition, the deep-orbit electron will be raised in its potential well and its orbital frequency will generally decrease as it is forced toward ionization.[m]

Direct excitation/ionization of lattice electrons by intense deep-orbit-electron near-fields is a relatively short-range effect because of its rapid drop in field intensity with distance. The initially lower-intensity far-fields of deep-orbit-electrons can, at greater distances, still interact with nuclei (and their components) because the frequencies are comparable. This interaction can produce long-range forces between nuclei that may be similar to that of the Coulomb potential. However, because of their disproportionately high frequency, they may be only weakly screened by the atomic electrons of the lattice.

The point of this appendix is to indicate that simultaneous multi-body interactions have characteristics different from two-body interactions, which are generally the cause of spectral lines rather than of a spectral continuum. Thus,

[m]An important property of the DDL electron is its ability to absorb energy and rise out of the potential well by decreasing its average orbital radius as well as by increasing it. These are not photonic transitions. They are non-quantized transients, related to a transition process itself, and therefore not addressed by QM accept as virtual photons. This ability, normally ignored for atomic electrons, is important in its interactions with the nuclear potential [9], which has varying slopes at different energies and distances.

any nuclear-reaction products expected, based on experience with two-body reactions, will be only second-order effects in the presence of deep-orbit electrons.

Acknowledgement

This work is supported in part by HiPi Consulting, Windsor, VA, USA and by the Science for Humanity Trust Inc., Tucker, GA, USA.

References

[1] J.-L. Paillet and A. Meulenberg, Advance on electron deep orbits of the hydrogen atom for LENR, *J. Condensed Matter Nucl. Sci.* **24** (2017) 258–277.

[2] A. Meulenberg and J.-L. Paillet, Implications of the electron deep orbits for cold fusion and physics, *J. Condensed Matter Nucl. Sci.* **24** (2017) 214–229.

[3] J.-L. Paillet and A. Meulenberg, Special relativity: the source of electron deep orbits, *Found. Phys.* **47**(2) (2017) 256–264.

[4] A. Meulenberg and J.-L. Paillet, Nature of the deep-Dirac levels, *19th Int. Conf. on Condensed Matter Nucl. Sci.*, Padua, Italy, 15/05/2015, *J. Condensed Matter Nucl. Sci.* **19** (2016) 192–201.

[5] A. Meulenberg and J.-L. Paillet, Basis for femto-molecules and -ions created from femto-atoms, *ICCF-19, 19th Int. Conf. on Condensed Matter Nucl. Sci.*, Padua, Italy, 15/05/2015, *J. Condensed Matter Nucl. Sci.* **19** (2016) 202–209.

[6] A. Meulenberg, Femto-helium and PdD transmutation, *ICCF-18, 18th Int. Conf. on Condensed Matter Nucl. Sci.*, Columbia, Missouri, 25/07/13, *J. Condensed Matter Nucl. Sci.* **15** (2015) 106–117.

[7] A. Meulenberg, Femto-atoms and transmutation, *ICCF-17*, Daejeon, South Korea, 12–17 August, 2012, *J. Condensed Matter Nucl. Sci.* **13** (2014) 346–357.

[8] A. Meulenberg and K.P. Sinha, Deep-orbit-electron radiation emission in decay from $^4H^{*\#}$ to ^4He, *ICCF-17*, Daejeon, South Korea, 12/8/12, *J. Condensed Matter Nucl. Sci.* **13** (2014) 357–368.

[9] A. Meulenberg, Radiation coupling: nuclear protons to deep-orbit-electrons, then to the lattice, *J. Condensed Matter Nucl. Sci.* **15** (2015) 125–136, http://coldfusioncommunity.net/pdf/jcmns/v15/125_JCMNS-Vol15.pdf.

[10] J.-P. Biberian, Biological transmutations, *J. Condensed Matter Nucl. Sci.* **28** (2019) 21–27, http://coldfusioncommunity.net/pdf/jcmns/v28/21_JCMNS-Vol28.pd.

[11] Hideo Kozima, Nuclear transmutations and stabilization of unstable nuclei in the cold fusion phenomenon, *J. Condensed Matter Nucl. Sci.* **28** (2019) 28–49, http://coldfusioncommunity.net/pdf/jcmns/v28/28_JCMNS-Vol28.pd.

[12] A. Meulenberg, Extensions to physics: what cold fusion teaches, Special Section: *Low Energy Nuclear Reactions*, M. Srinivasan and A. Meulenberg (Eds.), *Current Sci.* **108** (4) (2015) 499–506. http://www.currentscience.ac.in/cs/Volumes/108/04/0499.pdf

[13] M. Fleischmann and S. Pons, Electrochemically induced nuclear fusion of deuterium, *J. Electroanal. Chem* . **261** (2A) (1989) 301–308, doi: 10.1016/0022-0728(89)80006-3.

[14] A. Meulenberg and K.P. Sinha, New visions of physics through the microscope of cold fusion invited paper, *17th Int. Conf. on Condensed Matter Nucl. Sci.*, Daejeon, South Korea, 12–17 August, 2012, *J. Condensed Matter Nucl. Sci.* **13** (2014) 378–390.

[15] M. Srinivasan, G, Miley, E. Storms, *Low-Energy Nuclear Reactions: Transmutations*, Ch. 43, in *Nuclear Energy Encyclopedia: Science, Technology, and Applications*, Wiley, New York, 2011, https://doi.org/10.1002/9781118043493.ch43.

[16] A. Lomax, Replicable cold fusion experiment: heat/helium ratio, Special Section: Low Energy Nuclear Reactions, *Current Sci.* **108**(4) (2015) 574–577, http://www.currentscience.ac.in/Volumes/108/04/0574.pdf.

[17] J.A. Maly, J. Vávra, Electron transitions on deep Dirac levels I, *Fusion Technol.* **24** (1993) 30.

[18] J.-L. Paillet and A. Meulenberg, On highly relativistic deep electrons, *ICCF-21, 21st Int. Conf. for Condensed Matter Nucl. Sci.*, 3–8 June, 2018, Fort Collins, CO, USA, to ppear in JCMNS, 2019, https://www.youtube.com/watch?v=SxPrXqfNS5Q&feature=youtube, http://viXra.org/abs/1902.0398.

[19] https://en.wikipedia.org/wiki/Virial_theorem#In_special_relativity.

[20] J.D. Jackson, Classical Electrodynamics, Wiley, New York, 1962, p. 467, and https://en.wikipedia.org/wiki/Li%C3%A9nard%E2%80%93Wiechert_potential#Corresponding_values_of_electric_and_magnetic_fields.

[21] A.L. Kholmetskii, O.V. Missevitch, R. Smirnov-Rueda and T. Yarman, Propagation of Electromagnetic Fields in Near and Far Zones: Actualized Approach with Non-Zero Trace Electro-Magnetic Energy-Momentum Tensor, Progress in Electromagnetics Research M 22:57-72, January 2012, DOI: 10.1016/j.aop.2017.06.022

[22] C.A. Mead, *Collective Electrodynamics: Quantum Foundations of Electromagnetism*, The MIT Press, Cambridge, MA, 2000.

[23] A. Meulenberg and K.P. Sinha, Tunneling beneath the ^{4}He$*$ fragmentation energy, at *239th ACS National Meeting*, San Francisco, CA, March 2010, and *J. Condensed Matter Nucl. Sci.* **4** (2011) 241–255.

[24] K.P. Sinha and A. Meulenberg, Lochon catalyzed D–D fusion in deuterated palladium in the solid state, *National Academy Sci. (India) Lett.* **30** (7,8) (2007) arXiv:0705.0595v1.

[25] A. Meulenberg and K.P. Sinha, Composite model for LENR in linear defects of a lattice, *ICCF-18, 18th Int. Conf. on Condensed Matter Nucl. Sci.*, Columbia, Missouri, 25/07/2013, http://hdl.handle.net/10355/36818.

[26] A. Meulenberg, Pictorial description for LENR in linear defects of a lattice, *ICCF-18, 18th Int. Conf. on Condensed Matter Nucl. Sci.*, Columbia, Missouri, 25/07/2013, *J. Condensed Matter Nucl. Sci.* **15** (2015) 117–124, http://hdl.handle.net/10355/36535.

[27] M. Davidson, Off-mass-shell Particles and LENR, *J. Condensed Matter Nucl. Sci.* **19** (2016) 4–55. https://www.lenr-canr.org/acrobat/BiberianJPjcondensedr.pdf#page=54.

[28] A. Meulenberg, From the naught orbit to ^{4}He ground state, *16th Int. Conf. on Condensed Matter Nucl. Sci.*, Chennai, February 6–11, 2011, *J. Condensed Matter Nucl. Sci.* **10** (2013) 15–29.

[29] A. Meulenberg, Femto-atom and femto-molecule models of cold fusion, *Infinite Energy Magazine* Issue 112 (2013) 41–45.

[30] A. Meulenberg and K.P. Sinha, Lochon and Extended-lochon models for LENR in a lattice, *Infinite Energy Magazine* Issue 112 (2013) 29–32.

[31] E. Storms, How the explanation of LENR can be made consistent with observed behaviour and natural laws, special section: low energy nuclear reactions, *Current Sci.* **108** (2015) 531–534.

[32] A. Takahashi, Development status of condensed cluster fusion theory, special section: low energy nuclear reactions, *Current Sci.* **108** (2015) 514–515.

[33] J.-L. Paillet and A. Meulenberg, Deepening questions about electron deep orbits of the hydrogen atom, *Proc. 12th Int. Workshop on Hydrogen Loaded Metals*, Asti, Italy, June 5–9 2017, *J. Condensed Matter Nucl. Sci.* **26** (2018) 54–68.

[34] N.D. Cook, *Models of the Atomic Nucleus, Introduction*, 2nd Edn., 2010, Springer, Berlin.

[35] https://en.wikipedia.org/wiki/Idler-wheel#Applications.

[36] https://en.wikipedia.org/wiki/Compton_scattering.

[37] https://en.wikipedia.org/wiki/Photoelectric_effect#Emission_mechanism.

[38] A. Meulenberg, Virtual and real photons, In the appendix of Paper 8121-38, *Proc. SPIE Optics + Photonics 2011*, Conference 8121, The Nature of Light: What are Photons? IV, 21–25 August 2011, San Diego, CA, USA.

[39] https://en.wikipedia.org/wiki/Grey_box_model.

[40] A.A. Stahlhofenand and G. Nimtz, Evanescent modes are virtual photons, *Europhys. Lett.* **76**(2) (2006) 189–195, DOI: 10.1209/epl/i2006-10271.

[41] J.G. Williamson, The nature of the photon and the electron, In: The Nature of Light: What are Photons? VI, San Diego, CA, USA, 9 Aug. 2015, p. 957015, ISBN 9781628417364 (doi:10.1117/12.2188259.

[42] A. Meulenberg, W.R. Hudgins and R. F. Penland, The photon to electron/positron-pair transition, *Proc. SPIE. 9570*, The Nature of Light: What are Photons? VI, 95700S, September 10, 2015, doi: 10.1117/12.2187489.

[43] A. Meulenberg, W.R. Hudgins and R.F. Penland, The photon: EM fields, electrical potentials, and AC charge, Proc. SPIE, 9570, The Nature of Light: What are Photons? VI, 95700C, September 10, 2015; doi:10.1117/12.2187509.

J. Condensed Matter Nucl. Sci. 29 (2019) 472–492

Research Article

On Highly Relativistic Deep Electrons

Jean-Luc Paillet*

University of Aix-Marseille, 13007 Marseille, France

Andrew Meulenberg

Science for Humanity Trust Inc., USA

Abstract

We address a number of questions relating to the progress of our study on the relativistic-electron deep orbits (EDOs): (1) How to combine different EM potentials having two possible versions (attractive and repulsive), while rejecting unrealistic energies? (2) What about the angular momentum of the deep electrons? How is the Heisenberg Uncertainty Relation satisfied in these EDOs? (3) From where is extracted the high kinetic energy (of order 100 MeV) of the deep-orbit electrons? (4) What is the behavior of the effective potential V_{eff} as a function of distance to the nucleus? (5) What is the order of magnitude of the radiative corrections for the EDO's? (6) What is the relation between EDO solutions of the Dirac equation and the high energy resonances (with high binding energies) corresponding to a semi-classical local minimum of energy?

Keywords: Confinement, Deep electron levels, Heisenberg uncertainty relation, Highly relativistic bound electrons, LENR

1. Introduction

The observed generation of heat, in excess of that possible from chemical reactions, from deuterium-loaded palladium substrates at room temperature [1] led to a field of endeavor called "cold fusion". It was proposed that, as in the well-known hot ($\sim 10^7$°C) fusion processes of the sun, the embedded deuterium atoms were somehow able to come together close enough to fuse and liberate significant nuclear energy. This incredible news was initially welcomed because of the need for cheaper energy sources, the known fusion reactions, and the many decades of research supporting them. However, the extensive base of well-known and accepted nuclear physics soon became a reason that cold fusion was rejected by those knowledgeable in the field. The results of cold fusion did not agree at all with what was known from hot-fusion research. Either the cold fusion research was faulty or something new was happening. Assuming the data was correct, what was new?

Was the solid state or crystalline environment of the palladium substrate somehow able to help two deuterons overcome their Coulomb repulsion? If so, how? Many models for this mechanism were proposed, and, correctly

*Corresponding author. E-mail: jean-luc.paillet@club-internet.fr.

or incorrectly, subsequently rejected. Many of these relied on placing an electron between the deuterons for a much greater period than that of the bound atomic electrons. It was known that this was possible with a heavy electron (e.g., the muon, with $>200\times$ the electron mass); but, the known fusion of deuterium in the presence of muons resulted in the same products predicted by hot fusion, but not in cold fusion. Thus, this effect, alone, could not be the appropriate mechanism for cold fusion. Was there a mechanism that produced electron placement in such a manner as to alter the fusion process itself? Fortunately, the energy levels of the ^{4}He nucleus were well-known and could provide an answer that both depended on an electron spending more time between the deuterons and altered the products of the resulting d–d fusion reaction [2]. This answer did not explain how an electron could spend more time between deuterons; but, it is consistent with prior [3–5], and more recent [6–8], models that did so.

Classical physics allows an electron to orbit close to the nucleus for short periods, during which time the nuclear Coulomb barrier is strongly shielded. This was an approach explored for a time until it was noticed that quantum mechanics (QM), in the form of the Relativistic Klein–Gordon and Dirac equations, had predicted the existence of electron deep-electron orbits (EDOs) many decades ago. These models did not depend on spin for this prediction. The introduction of relativity into the QM equations made the difference between *only* the atomic orbitals and those *plus* the deep orbits. Unfortunately, since no deep orbits had ever been seen and the concept of the neutron as a proton plus tightly bound electron had been rejected, interest in this solution of the relativistic equations was greatly reduced. It periodically reappeared (almost every 10 years); however, because the deep-orbit solution had a singular point (at $r = 0$) when a strict Coulomb potential and point charges were considered, this portion of these important equations were regularly rejected on a mathematical basis alone.

In trying to understand a physical basis for the deep orbits, relativistic effects were examined in greater detail [9,10] and the several arguments against the EDOs were successfully refuted [11,12]. However, the greatest argument for the reality of these deep orbits is in the predictions of physical effects based on such orbits [13–20]. Once it was established that the EDOs could explain most or all of the observed cold fusion effects, it became important to look at other relativistic effects [21,22] to improve the model. Inclusion of actual nuclear details have been made as successive approximations and not all "branch" points have been explored yet.

This paper is a continuation, with a brief overview, of that study. It starts with the assumption that the Heisenberg Uncertainty Relation (HUR) applies to the nuclear range as well. From this it is possible to calculate a limit for the relativistic coefficient, γ, and, then, to look at the deep-orbit-electron's relativistic interactions with its orbit and the nuclear components (e.g., spin and charge). The greatly enhanced forces and potentials from relativity and proximity create very large energies (100 s of MeV) relative to the static calculations and yet the binding energy of the deep-orbit electron is still relatively small ($|\mathrm{BE}| < 0.511$ MeV) because it is a difference between kinetic and potential energies. Since the observables are small differences between large numbers, many assumptions and relations, made in historical work for different ranges, may no longer be valid. Thus, continued refinement, testing, and examination of assumptions and premises are required for this transition to the highly relativistic nuclear regime. An exploration of QED in this context brings another tool into the effort.

2. Relativistic Confinement Energies and the Relativistic Coefficient γ

In previous studies on magnetic interactions [22,23], we showed magnetic potentials to have very high energy near the nucleus; as a consequence, we could expect the HUR to be respected in this zone. Next, we adopted a *new strategy*: to directly address the HUR as a starting point, while considering an electron confined in a sphere of radius r.

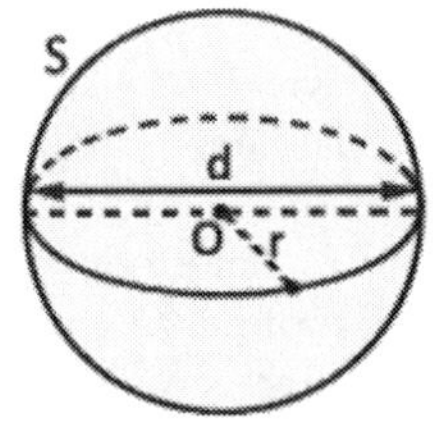

The HUR can be expressed by the inequality $\Delta p\,\Delta x \geq \hbar/2$, where p is the norm of the momentum of the considered particle and x is its radial location. The delta indicates the uncertainty in these parameters. As in most QM textbooks (see e.g. [24,25]) we can put $p \sim \hbar/r$, where the "2" has been removed, to take a reasonable average value for p based on this relation. Then, we consider the relativistic expression of momentum, $p = \gamma mv$, where m is the rest mass of the electron, v its velocity, and γ the relativistic coefficient defined by $\gamma = (1\text{-}\ v^2/c^2)^{-1/2}$. We can deduce $\gamma\ v = \hbar/mr$.

Now, by using the expression of γ, one has $(\gamma\ v)^2 = (cv)^2/(c^2-v^2) = (\hbar/mr)^2$. From $(cv)^2/(c^2-v^2) = (\hbar/mr)^2$ and, by simple algebraic transformations[a], we obtain $\gamma^2 \sim 1+\hbar^2/(mrc)^2$. Recognizing the reduced Compton wavelength of the electron as a constant, $\lambda\!\!\!^{-}_{c} = \hbar/mc$, we have relation (1), expressing γ as a function of the confinement radius r:

$$\gamma^2 \sim 1 + (\lambda\!\!\!^{-}_{c}/r)^2, \quad \text{i.e.} \quad \gamma \sim [1 + (\lambda\!\!\!^{-}_{c}/r)^2]^{(1/2)}. \tag{1}$$

Moreover, as $\lambda\!\!\!^{-}_{c} \sim 386$ F for an electron, and for the EDOs, r is of order a few F, one has $r \ll \lambda\!\!\!^{-}_{c}$, and thus $(\lambda\!\!\!^{-}_{c}/r)^2 \gg 1$. In this case, the previous relation expression can be reduced to a very simple one:

$$\gamma \sim \lambda\!\!\!^{-}_{c}/r. \tag{2}$$

Note that for highly relativistic velocities $v \to c$ and $\gamma^2 \sim \hbar^2/(mcr)^2$, and we have relation (2) as well. The kinetic energy, $\text{KE} = (\gamma\text{-}1)\,mc^2$, becomes $\sim ((\lambda\!\!\!^{-}_{c}/r) - 1)\,mc^2$ and, expressed as a function of the de Broglie wavelength λ_{dB} assuming the quantum condition for the lowest closed orbit of $\lambda_{\text{dB}} = 2\pi r = 2\pi\hbar/p = h/p$, $\text{KE} \approx (\lambda_c\ /\lambda_{\text{dB}}\ \text{-}1)\,mc^2$ with $\gamma \approx \lambda_c\ /\lambda_{\text{dB}}$, where λ_c is the Compton wavelength under non-reduced form.

2.1. Examples of confinement energies at some ends of the energy scale

(1) For the case of an atomic electron, we consider r equal to the Bohr radius $r_{\text{B}} = \hbar/\alpha cm$, where α is the coupling constant of electro-magnetism. The kinetic energy corresponding to the confinement of an electron in a radial region corresponding to r_{B} is equal to $E \approx p^2/2m$, a non-relativistic expression, because we know the electron is not very relativistic (see its γ in Section 2.2). While putting $p = \hbar/r_{\text{B}}$, we obtain $E \approx (\hbar/r_{\text{B}})^2/2m = mc^2\alpha^2/2$, which is the expression of the classical Rydberg energy, of numerical value $\sim$13.6 eV, a well-known result associated with the ground state of the hydrogen atom and the HUR.

(2) Consider now the case where $r = 2$ F, where an electron is very close to the nucleus, as in the EDO.

We can expect the electron to have relativistic speed, requiring the relativistic expression of the energy, i.e. $E = (p^2c^2 + m^2c^4)^{1/2}$. A numerical computation shows that $pc \gg mc^2$, and so, for this near-nuclear orbit, $E \sim pc \sim 98.6$ MeV.

[a] With $s = h/mr$, $(cv)^2 = (cs)^2\text{-}(vs)^2$, we deduce $v^2(c^2+s^2) = (cs)^2$, $(v/c)^2 = s^2/(c^2+s^2)$, $\gamma^2 = 1/(1-(v/c)^2) = 1/[1\text{-}s^2/(c^2+s^2)] = 1+s^2/c^2$.

2.2. Examples of relativistic coefficients, for the confinement radii indicated in Section 2.1

(1) We consider again $r = r_B$, as in Section 2.1 (1). Then we compute γ in two different ways.

(a) First, we know the kinetic energy is the Rydberg energy $E_{Ryd} = mc^2\alpha^2/2$. But the kinetic energy KE is also given by the expression KE = (γ-1) mc^2. From KE = E_{Ryd}, we can deduce $\gamma = 1+\alpha^2/2$. From this, we have $\beta^2 = (v/c)^2 = 1\text{-}1/\gamma^2 = (1\text{-}1/(1+\alpha^2/2)^2) \sim ((1+\alpha^2/2)^2\text{-}1) \sim \alpha^2$, so $v \sim \alpha c \sim \mathbf{c}/137$, a well-know result calculated by Arnold Sommerfeld.

(b) Now, we compute γ by means of the results deduced from HUR. As one has $r = r_B$, the condition $r \ll \lambda\!\!\!^-_c$ is not satisfied and we use the full expression for γ (1): $\gamma \sim (1 + (\lambda\!\!\!^-_c/r_B)^2)^{(1/2)}$. With $\lambda\!\!\!^-_c = \hbar/mc$ and $r_B = \hbar/\alpha cm$, we obtain $\gamma \sim (1+\alpha^2)^{(1/2)} \sim 1+\alpha^2/2$, while neglecting the terms of order α^n for $n \geq 4$. So we still have a very good approximation of the result of Sommerfeld.

(2) Consider the case $r = 2$ F. As $r \ll \lambda\!\!\!^-_c$, we can use expression (2) to compute $\gamma \sim \lambda\!\!\!^-_c/r = 386/2 = 193$, and $\beta = 0.99998\ldots$ The electron is ultra-relativistic. The question is therefore the following: can a Coulomb potential confine such an ultra-relativistic electron? We show how to solve this question positively in Section 3.

3. The Relativistic Effective Potential Energy V_{eff} is Strong Enough to Confine Electrons in Deep Orbits

Because of the high level of the relativistic coefficient γ, it is interesting to consider the effects of the relativistic correction to the static Coulomb potential, as indicated in [26,27], under the resulting form of an effective dynamical potential noted V_{eff}, and already considered in [28,10]. The general form Eq. (3) of V_{eff}, comes from the development of relativistic quantum equations (Dirac, Klein–Gordon) with the expression of the relativistic energy of a particle in a central field for a Coulomb potential energy V:

$$V_{eff} = V(E/mc^2) - V^2/2mc^2. \tag{3}$$

This transformation is little known since, in the atomic cases and for light elements, we have $E \sim mc^2$ and $V \ll mc^2$ that leads to $V_{eff} \sim V$. On the other hand, while considering the relativistic expression of the E of an electron in the potential V, i.e. $E = V+(\mathbf{p}^2c^2 + m^2c^4)^{1/2}$, we can deduce the following form Eq. (4), including the coefficient γ:

$$V_{eff} = \gamma V + V^2/2mc^2. \tag{4}$$

Now, we put the full expression of γ (1), $\gamma \sim [1 + (\lambda\!\!\!^-_c/r)^2]^{(1/2)}$, into Eq. (4), to obtain V_{eff} as a function of r:

$$\begin{aligned} V_{eff} &= -(\alpha c\hbar/r)([1 + (\lambda\!\!\!^-_c)^2/r^2]^{(1/2)} - \alpha\hbar/2mcr) \\ &= -(\alpha c\hbar/r)([1 + (\lambda\!\!\!^-_c)^2/r^2]^{(1/2)} - \alpha\lambda\!\!\!^-_c/2r). \end{aligned} \tag{5}$$

For r of order a few F and by reduction of γ, this expression can be simplified into:

$$V_{eff} \sim (\alpha c\hbar\lambda\!\!\!^-_c/r^2)(1 - \alpha/2). \tag{6}$$

One can also write the following approximate (equivalent) forms, while neglecting the term $\alpha/2$ in the second parenthesis of Eq. (6):

$$V_{eff} \sim \gamma V \sim (\lambda\!\!\!^-_c/r)V \sim \lambda\!\!\!^-_c e^2/r^2 \sim \alpha c\hbar\lambda\!\!\!^-_c/r^2 \sim -\alpha\hbar^2/mr^2. \tag{7}$$

When looking at formula (4), we can see the first term of the sum, equal to γV, which expresses a strengthening of the attractive potential V, since γ is always ≥ 1. But the second term of the sum in Eq. (4), $V^2/2mc^2$, has a positive sign, that means a repulsive action. The question is therefore whether V_{eff} is always a reinforcement of V.

- For $r \ll \lambda\!\!\!^{-}_{\mathrm{c}}$, expression (6) allows us to answer yes.
- If we release this condition, we have to use expression (5), and this one does not allow to answer the question easily in a purely algebraic way. Previously [28], we could show V_{eff} is always a reinforcement of V, but only for quasi-circular orbits.

Now, by using a numerical method, we can show without any hypothesis on the shape of the orbit, the following result: $V_{\mathrm{eff}} < V < 0$ when $r < 52.91741577$ pm, to compare with the Bohr radius $r_{\mathrm{B}} \sim 52.9177210$ pm. Such precision may seem ridiculous, as the computation of γ from the HUR is based on an approximate principle and the s-orbit is nearly linear. Nevertheless, we can give the following results for V_{eff}.

With only the condition $r <\sim r_{\mathrm{B}}$, i.e. for any energy level under the ground state and independent of any of the HUR analysis above, we have:

(1) V_{eff} is always attractive,
(2) $|V_{\mathrm{eff}}| > |V|$, i.e. V_{eff} is always a strengthening over the static Coulomb potential,
(3) Moreover, expressions (6) and (7) show that: when r decreases sufficiently and $\to 0$, V_{eff} has a behavior in K/r^2, with $K \sim 8.9 \times 10^{-41}$ SI units, i.e. Jm^2.

Finally, and most importantly for the EDO's, if computing V_{eff} near the nucleus, e.g. for r ~2 F, we have $\gamma \sim 193$, as indicated in Section 2.2, and we obtain $V_{\mathrm{eff}} \sim -139$ MeV, whereas the kinetic energy KE $= (\gamma - 1)\, mc^2 \sim 192 \times 511$ keV ~ 98 MeV. With such a high value, V_{eff} can easily confine an electron in this region.

In [10], we showed that Special Relativity is the source of the EDO's. Now, we show that the HUR, which seemed an impediment for the EDO's, provides its proper resolution thanks to relativity.

4. Looking for a Resonance Near the Nucleus

The deep orbit electrons have the following features:

- They are highly relativistic.
- They are subjected to several electro-magnetic interactions of high intensity, some of which are not involved in the Dirac equation used until now for determining the EDOs for a single particle.
- Note also that, in the "nuclear zone", the deep-orbit electrons are certainly subject to fairly high radiative corrections. But the Coulomb electric field, strengthened by a relativistic effect corresponding to V_{eff}, seems sufficient by itself to retain an electron in the nuclear zone.

Under these conditions, the question of EDO stability seems a very difficult problem to solve. Nevertheless, to have a first estimate of a possible stable resonance, we can use a well-known semi-classical approximation, which consists of seeking a local minimum of energy (LME), in an approximate way similar to that used for finding the ground state, the Bohr level. One can find this in most Quantum Mechanics textbooks, e.g. [24].

In fact, we combine two approaches for doing this:

- To attempt to determine which interactions have the greatest role(s) in generating a resonance.
- To compute a total energy, while respecting the HUR.

For the latter point, we consider the relativistic expression of energy, in which the norm of momentum $|\mathbf{p}|$ is replaced by $\hbar/r$, that gives the following expression, denoted E_{H} (H for Heisenberg):

$$E_{\mathrm{H}} = \sqrt{\frac{\hbar^2 c^2}{r^2} + m^2 c^4}. \tag{8}$$

With relation (1), $\gamma \sim [1 + (\lambda_c/r)^2]^{(1/2)} = [1 + \hbar^2/(mcr)^2]^{(1/2)}$, one easily verifies that the kinetic energy KE = (γ-1) mc^2 is also equal to $E_H - mc^2$. Indeed

$$\text{KE} = mc^2[1 + \hbar^2/(mcr)^2]^{(1/2)} - mc^2 = [m^2c^4 + \hbar^2c^2/r^2]^{(1/2)} - mc^2 = E_H - mc^2. \tag{9}$$

In principle, we add to E_H, a term V representing a potential energy, where V is a function of the radius. Thus we obtain the total energy E, represented by the following relation: $E = E_H + V$. Then, we look for an LME for various combinations of potentials included in the term V and we determine the radius of this local minimum. Of course, V systematically includes the dynamical effective potential V_{eff}, given its essential role for the existence of EDOs indicated in Section 3. But, before developing the question of the combination of potentials included in V, it is interesting to look at the properties of the "minimal combination", by putting $V = V_{eff}$ and thus $E = E_H + V_{eff}$.

4.1. Study of the case, where the only potential taken into account is the effective potential V_{eff}

In this section, we have PE = V_{eff} and $l = 0$.

First, we plot below (Fig. 1) three curves: $|V_{Cb}|$, $|V_{eff}|$ and KE = $E_H - mc^2$, in loglogplot. The static Coulomb potential is denoted by V_{Cb}, to avoid confusion with the combination of potentials V. To make it easier to read the calculation results, we put $r = \rho \times 10^{-12}$ and we compute with ρ.

In this figure, we can observe the following:

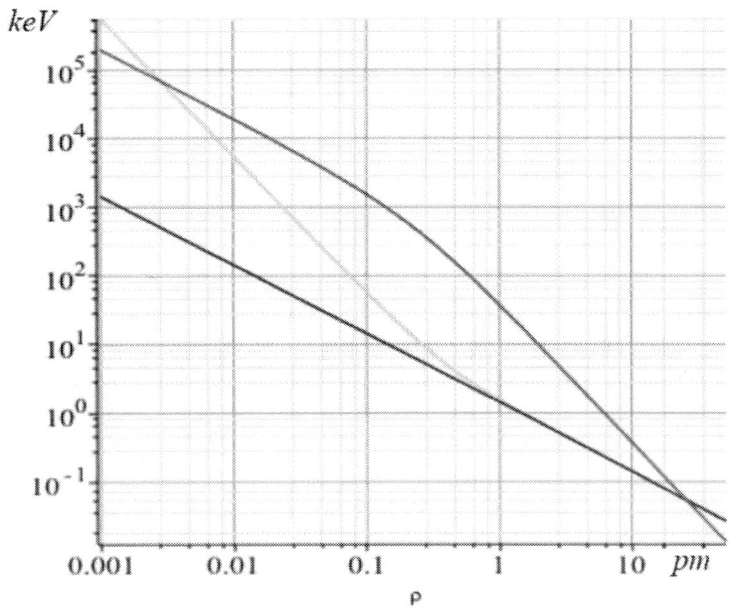

Figure 1. Loglogplot of energies (in keV) for radius denoted by ρ in pm, where 1 F< ρ < 53 pm. (*Blue*) norm of Coulomb potential $|V_{Cb}|$, (*Green*) norm of the effective potential $|V_{eff}|$, and (*Red*) kinetic energy KE.

(1) $V_{\rm eff}$ is always a strengthening of $V_{\rm Cb}$, as indicated in Section 3. In the figure, because of the extent of the scale in loglogplot, $V_{\rm eff}$ is indistinguishable from $V_{\rm Cb}$ at high ρ values and only separates when ρ decreases and reaches 1 pm.
(2) $|V_{\rm eff}|$ and KE intersect at two points, near the ends of the radius scale in Fig. 1. The numerical solution of the equation $|V_{\rm eff}| = $ KE gives these points: ~26.45 pm and ~2.828 F, approximated in the figure. For a non-relativistic orbit, the virial theorem states KE $= |\text{PE}|/2$. Thus, the Bohr radius at this stable point is ~53 pm. For relativistic orbits, KE $= |\text{PE}|\gamma/(\gamma+1)$ and KE $\to |\text{PE}|$ as $v \to c$, and 2.83 F is near the classical electron radius (~2.82 F). This is not a simple coincidence.[b]

Most important is the presence of both crossing points of the curves of KE and $|V_{\rm eff}|$, 2.82 F and 26.5 pm, indicating the possibility of resonance in two regions, where $|V_{\rm eff}|$ becomes stronger than KE:

- either for $\rho > 26.5$ pm: in this area, there are the well-know atomic energy levels, whose lowest is the ground level (Bohr) at ~53 pm, where $V_{\rm eff} \sim V_{\rm Cb} = -2$ KE ~ -26 eV,
- or for $\rho < 2.8$ F, where we might expect resonance of type "EDO".

Of course, when taking into account further EM interactions near the nucleus, this limit $\rho < 2.8$ F could move slightly. We say "slightly", because the energies of the interactions, considered further, are relatively small compared to $V_{\rm eff}$ and KE, where we set aside huge, physically "unreasonable," interactions for the orbits of present interest.

Another important, but difficult, question concerns the possibility of transition between the "atomic" zone and the "EDO" zone. We will give some reflections on this further question. Nevertheless, while considering the $E = E_{\rm H}+V_{\rm eff}$ and the derivative $\mathrm{d}E/\mathrm{d}\rho$ to find possible local extrema of E by solving $\mathrm{d}E/\mathrm{d}\rho = 0$, we find as follows.

(1) An obvious local minimum at $\rho = r_{\rm B}$ ~52.9 pm, for the atomic ground state.
(2) A local maximum at ρ ~5.6 F, where KE + $V_{\rm eff}$ ~17 MeV, that represents a high "pseudo-barrier" for a transition from atomic zone to EDO zone. We can call it the "Heisenberg barrier", since it is due to the very-high kinetic energy required by HUR. In fact, at this radius, we have: KE ~34.6 MeV, while $V_{\rm eff} \sim -17.6$ MeV. Thus, the barrier is 34.6–17.6 MeV ~17 MeV. On the other hand, below the EDO zone, E becomes negative and continuously decreases towards $-\infty$. Under these conditions, the existence of further repulsive interactions is necessary to generate a resonance. We represent, in Fig. 2, the plotted curve of KE + $V_{\rm eff}$ with a radius scale adequate to observe the maximum and the behavior for ρ very small.
Remark: KE+$V_{\rm eff}$ has the same behavior as E and extrema have the same location, since $E = $ KE+$V_{\rm eff}+mc^2$.

4.2. What combination of potentials to consider, for finding a deep LME?

From our previous works on magnetic interactions [22,23], including the study of the Barut–Vigier model and related works, we are led to some conclusions about combinations of potentials capable of producing a LME near the nucleus.

(1) In particular, it seems that we have to rule out the possibility of a *spin–orbit interaction* for an electron in the EDO zone: indeed, the energy $E_{\rm SO}$ associated with the spin–orbit interaction is given by the following expression

$$E_{\rm SO} = \frac{\mu_0 e^2}{8\pi m^2}\frac{1}{r^3}L{\cdot}S = \xi(r)L{\cdot}S, \tag{10}$$

[b]Let us calculate for which value of radius r we have the maximum possible value of |BE|, i.e. BE $= -mc^2$, *while supposing we are in a resonance case.* With BE = KE + PE, and the approximate expression (7) for $V_{\rm eff}$, we have $-mc^2 \sim (\gamma-1)mc^2 + \gamma V_{\rm Cb} \to V_{\rm Cb} \sim -mc^2$, i.e. $\alpha ch/r \sim mc^2$, $r \sim \alpha ch/mc^2$. This is the expression for the electron classical radius $r_{\rm e}$. Somehow $r_{\rm e}$ is a limiting radius for a resonance under the hypothesis PE $= V_{\rm eff}$.

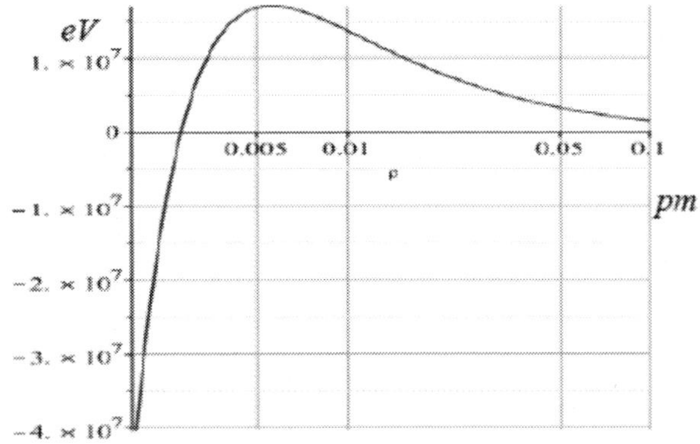

Figure 2. Semilogplot of KE+V_{eff} (in eV) with ρ in pm, for 2 F < ρ <0.1 pm.

where, for any value of the quantum angular momentum $l \neq 0$,

- the attractive case corresponds with $L{\cdot}S = -\ (\hbar^2/2)\ (l+1)$,
- while the repulsive one, with $L{\cdot}S = +\ (\hbar^2/2)\ l$.

But, previous computations of the energy E_{SO} associated with the spin–orbit interaction in attractive mode, while supposing l =1 and applying the usual quantization rules, give $E_{\text{SO}} \sim -13$ GeV for an electron at a distance r ~2 F, i.e. in the expected region of the EDOs. Such a huge value does not seem physically reasonable and this term would prevent a bound state. If considering the repulsive case, with l =1, the formulas above show we obtain one half of the previous value, i.e. 6.5 GeV, which is still physically unreasonable. Under these conditions, one has to consider the angular momentum to be $l = 0$.

(2) Moreover, if $l = 0$, there is no *centrifugal (repulsive) potential*, since this is proportional to $l(l+1)$ in the quantum formulations (e.g. the relativistic Schrödinger equation, [29]) and we can think the orbit is essentially linear through, or (if a hard core exists) at least toward, the nucleus.

(3) As a consequence, to balance the attractive potential energy V_{eff} and to expect a resonance near the nucleus, we have to consider the *repulsive version* of the *magnetic spin–spin interaction* (i.e. the triplet state), denoted by V_{RSS}. Here, we recall that the spin–spin interaction can be expressed by the following general formula (see e.g. [30,31]):

$$H_{\text{SS}} = -\frac{\mu_0}{4\pi}\left[\frac{1}{r^3}\left(3\left(M_{\text{p}}\cdot\hat{r}\right)\left(M_{\text{e}}\cdot\hat{r}\right) - M_{\text{p}}\cdot M_{\text{e}}\right) + \frac{8\pi}{3}M_{\text{p}}\cdot M_{\text{e}}\delta\left(r\right)\right], \tag{11}$$

where M_{p} and M_{e} are the magnetic moments of the proton and the electron, respectively and $\hat{r}$ denotes a radial unit vector. The magnetic moments are related to the respective electron and nuclear (i.e. proton, here) spins S_{e} and S_{p} by the following formulas:

$$M_{\text{e}} = (e/2m_{\text{e}})S_{\text{e}} \quad \text{and} \quad M_{\text{p}} = (2.79|e|/m_{\text{p}})S_{\text{p}}. \tag{12}$$

By introducing the total spin $S = S_{\text{e}} + S_{\text{p}}$ and discounting anyrelativistic effects for the moment, one can write the following relation:

$$S_{\text{e}}{\cdot}S_{\text{p}} = (1/2)(S^2 - (S_{\text{e}})^2 - (S_{\text{p}})^2) = (\hbar^2/2)[s(s+1) - 3/2]. \tag{13}$$

The only possible values of s are $s = 0$ ("singlet" state) and $s = 1$ ("triplet" state), which gives two cases:

- $s = 0 \Rightarrow S_e \cdot S_p = -(3/4)\hbar^2$, i.e. "attractive case",
- $s = 1 \Rightarrow S_e \cdot S_p = +(1/4)\hbar^2$, i.e. "repulsive case".

While extrapolating the known values of spin–spin interaction energy computed at the Bohr radius, to a general expression of the spin–spin interaction energy for any radius r. We found [23], for the repulsive spin–spin (RSS) version noted V_{RSS}, $V_{RSS} \sim 3.4 \times 10^{-56}/r^3$ SI, i.e. J/m^3. For example, for $r \sim 5$ F, we have $V_{RSS} \sim 2.7 \times 10^{-13}$ J $\sim$1.7 MeV.

Note that, for a particle in a relativistic regime, the spin tends to lean in the direction of the motion of the particle [32] and *we could think that it leads to a weakening of the effect on* V_{RSS}.

(4) We have also to take into account a further interaction, always present and repulsive. We recall this term is caused by an interaction between the magnetic moment of the electron with the charge of the proton and involves the squared norm of the magnetic vector potential of the electron,

$$\mathbf{A}(r) = \frac{\mu_0}{4\pi} \frac{\mathbf{m} \times \hat{r}}{|\mathbf{r}|^2},$$

where $\mathbf{m}$ is the magnetic moment of the electron. The complete energy term associated with $\mathbf{A}^2$ has the form $e^2\mathbf{A}^2/2m$ and is considered [33,34] to be expressing a *diamagnetic* energy with a behavior in $1/r^4$. The energy of this diamagnetic interaction, noted V_{dia} and although very weak compared to V_{RSS}, has to be included in the combination of potentials. We found [22,23]

$$V_{dia} = K_4/r^4, \quad \text{with } K_4 = \left(\frac{\mu_0}{4\pi}\right)^2 \frac{\mathbf{e}^4\hbar^2}{4m_e^2 m_p},$$

where m_e is the electron mass, m_p the proton mass, and $K_4 \sim 1.3 \times 10^{-71}$ SI units, i.e. J m^4. *Again, relativistic effects on the spin vector orientation could reduce the values obtained for* $\mathbf{m} \times \mathbf{r}$.

In Appendix 1, we give elements of discussion about magnetic interactions for relativistic electrons.

4.3. Effects of radiative corrections

We have to take into account the effects of radiative interactions, which are strong in the nuclear area.

On one hand, some EM interactions become very strong when the radius decreases because of behavior in inverse powers of r, mostly in powers -2, -3 and -4 for V_{eff}, V_{RSS} and V_{dia}, respectively. On the other hand, *radiative corrections*, which are specific EM interactions deduced from the quantization of the EM field, have to be taken into account, as they can modify the intensity of the EM interactions considered so far.

4.3.1. *Some rudiments about radiative corrections*

The radiative corrections are determined in the framework of the Quantum Electrodynamics (QED), the first theory where QM and Special Relativity are combined in the most suitable manner, i.e. in a covariant way (see e.g. [35]). Moreover, on account of relativity, QED implements an additional quantization of a new kind, the mis-named "2nd quantization", where it is applied to the numbers of considered particles, as particles can be created or destroyed (because of the matter–energy equivalence). Moreover, it implements also the quantization of the EM field, in a way similar to a system of independent harmonic oscillators, with ladder operators of destruction/creation. QED is a fully achieved theory, as it has undergone intense development for decades, mainly in the analysis of interactions involving free leptons and photons, occurring during scattering experiments. In principle, QED includes all relativistic effects, since any fermion quantum field is based on Dirac theory.

In brief, an electromagnetic scattering is associated with the transitions between an initial composite state to a final one, where a transition is caused by various interactions between the initial leptons and possibly photons. It is completely determined by computing the scattering matrix (or S-matrix), whose elements are the probability amplitudes for transition from an elementary initial state to a final one. The various interactions are usefully represented by Feynman diagrams, as they constitute a true algebraic coding for computing and combining all the various operators and propagators involved in the S-matrix elements. At each vertex, the sum of all the momenta is null, to express momentum conservation. Note that, generally, real particles are virtually destroyed in inputs of a transition, while various real particles can be virtually created at the outputs, and intermediate energy propagation is done by means of virtual photons.

Most important elements for the success of QED are that any transition can be decomposed into a series, thanks to a Dyson expansion of the general evolution operator (or "S operator"), and the Wick's theorem applied to the Dyson expansion that permits a finite, fully covariant, expression at any order of expansion. This makes QED a perturbation theory, i.e. it obtains successively more accurate descriptions by computing the S-matrix at an increasing order of the perturbation (of course, at the expense of a quick increase of the computation complexity). Finally, mathematical and computational complexity arises at higher orders, principally, because of loops in the Feynman diagrams: indeed, in a loop, the momentum cannot be determined by the conservation law. So, one has to consider all possible values by summing from 0 to ∞. This implies the presence of diverging integrals and led to the difficult, but efficient, methods of renormalization to overcome this obstacle.

As mentioned above, QED has been principally developed in the framework of scattering theory. It has been much less often applied to bound states, like atomic states; moreover, it seems harder to apply (less suitable?) for these states. We can cite, in the case of bound states, the approximate methods of the "Theory of the external field", particularly developed in [36] and derived from works of Furry (e.g. [37]). Summarily, in the concept of an "external field", the nucleus does not directly participate to scattering as a particle, but only the EM field generated by the nucleus is taken into account, with exchanges of (possibly virtual) photons. As famous examples of problems solved by QED in bound states, that contributed to its huge success, we can cite: the computation of the "anomalous magnetic momentum of the electron", "the Lamb shift in the hydrogen-like atoms", "the radiative transitions between bound states", as well as the analysis of "Bremsstrahlung". While these are small effects, relative to the masses involved, and therefore fit into the perturbative regime of QED that may not be valid for the nuclear interactions, their exploration may be instructive as we move toward the nucleus where the effects would be large.

4.3.2. *The Lamb shift*

For the subject of our work, i.e. the study of possible resonance near the nucleus, we are mainly concerned with the question of the Lamb shift (see e.g. [35], Section 9.6.2). Historically [38], the Lamb shift is a small energy splitting, observed between the $2S_{1/2}(n=2, l=0, j=1/2)$ and the $2P_{1/2}(n=2, l=1, j=1/2)$ orbitals of H atom. However, these two different levels are degenerate for the Dirac equation, i.e. have the same energy as solutions of the equation, since their energy depends only on n and j, and not on l. Nevertheless, observation gives an extra energy of order 4.4 μeV for the 2s, compared to the 2p, orbital. This is due to the fact that an S orbital enters the nuclear domain where the electric field is very high; but, because of the centrifugal potential (associated with angular momentum $l=1$), a P orbital does not penetrate to that region. A similar energy shift exists for the 1S orbital, between the energy computed according the Dirac equation and the observed energy level, but with a greater value $\sim$35 μeV.

Remark: the extra energy associated with the Lamb shift corresponds in fact to *a decrease of the binding energy*:

Lamb shift has a global repulsive effect on the bound electron.

Hans Bethe was the first [39] to give an approximate, non-relativistic, derivation of the Lamb shift from the concept of self-energy of the bound electron (see Section 4.3.3). The computations of Lamb shift are very complex, requiring use of the whole arsenal of QED in the difficult case of a bound state. In principle, the computations are divided into two parts:

- one part using the approximate methods of the "Theory of the external field", or "free interaction picture",
- another part with a bound representation, the so-called "Bound Interaction Picture" (BIP).

There are very few complete demonstrations in Quantum Field Theory textbooks or courses: we can cite

- a rigorous treatment in [36] including some non-relativistic approximation,
- a derivation announced as relativistic in the course of Dyson [40] but, with a non-relativistic dipole approximation, the BIP part is not covariant.

The difficulties of Lamb shift computations are due, in particular, to the fact they involve Feynman diagrams of higher order, i.e. including loops.

4.3.3. *Feynman diagrams for Lamb shift at lowest order*

Here, we show diagrams including only one loop.

There are two kinds of phenomenon involved in the Lamb shift:

- "Electron Self-Energy"(SE), corresponding with the diagram in Fig. 3. One can show (and observe) that it has a repulsive effect on the bound electron.
- "Vacuum Polarization"(VP), sometimes called also "photon self-energy", corresponding to the diagram in Fig. 4. It has an attractive effect on the bound electron. VP leads to a shell of pairs (e^-, e^+) around the bound electron, which leads to a screening effect. From the remark above, one can already deduce the effects of SE are stronger than those of VP.

Of course, such diagrams represent algebraic QED terms occurring in the Dyson expansion, after applying Wick's theorem. Moreover, the self-energy is associated with a *mass-renormalization*, whereas the vacuum polarization with a *charge renormalization*. Here, we do not write QED terms and we do not develop renormalization methods, because it is far beyond this paper.

Both phenomena of the Lamb shift are completely determined from a mathematical point of view, in the QED theory. Nevertheless it is almost impossible to find, in the literature, simple and understandable physical interpretations.

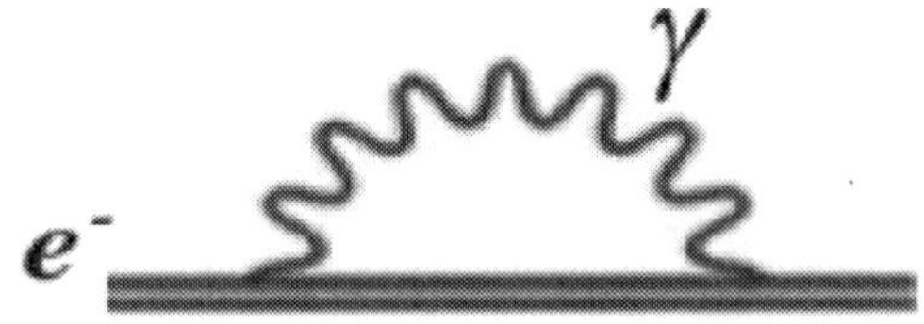

Figure 3. Electron Self-energy. e^- represents an electron, γ represents a (virtual) photon, emitted and reabsorbed by the electron. The thick line represents the fact that the electron is bound.

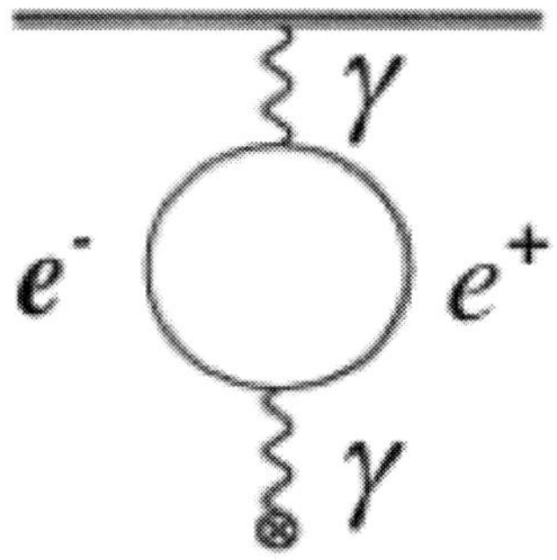

Figure 4. Vacuum polarization. e^- represents an electron, e^+ represents a positron, a photon (represented by γ) is emitted, at the bottom, by the nucleus (indicated by a cross), it produces a virtual lepton pair (e^-, e^+), this pair recombines by emitting a photon, which reaches a bound electron (not indicated).

At least for SE, we propose the following interpretation, maybe a bit simplistic, but corresponding to a well-known and clear physical concept, *the action–reaction* principle: any electrically charged particle acts on its environment by creating an EM field around itself and then, reciprocally, the so-modified environment reacts on this EM field of the particle. The acceleration of a bound particle increases this bound EM (evanescent) field that, as a standing wave, has an outgoing and a return component. The development of a magnetic field (bound to the charge) when moving is the additional energy. Likewise, any distortion of the electrostatic field (the least-energy distribution of a charge at rest) is an increase in the bound-field energy of moving charge.

For VP, a possible interpretation could be more complex, as the environment considered is the (sometimes polemical) "quantum vacuum", including energy fluctuations with creation/destruction of virtual pairs of particle–antiparticle. These energy fluctuations composed of oscillating electric fields interact with the steady-state charge fields of electrons and nuclei. The steady-state fields will polarize the oscillating fields resulting in transient charge separation and a reduction in the steady-state far-field. This effect is observed in the special case of polarization of energetic photons [41] in the creation of real electron-positron pairs where the energy and separation of the charge fields is permanent. Again, if the conditions are not "correct" for complete and stable separation, the polarized waves (virtual pairs) simply recombine as part of a reversible process.

4.4. How can we express the effect of Lamb shift on EDO's binding energy at resonance?

Despite the numerous studies of lamb shift on hydrogen-like ions with more and more precise results, this question is very difficult for several reasons:

- There is no fully analytical formulation of the two effects (SE, VP) of the Lamb shift: we indicate, below, a well-known expression of SE for the fundamental level. It includes a multiplicative factor $F(Za)$, which is given only by means of tables (see e.g. [42,43]): $\Delta E_{\mathrm{SE}} = (\alpha/\pi)(Z\alpha)^4 \mathrm{F}(Z\alpha) mc^2$. For VP, there are mainly asymptotic formulations (short/long ranges) based on Uehling potential [44].
- The computations suppose energies, which are not, or only slightly, relativistic (case of heavy elements), whereas EDO's are highly relativistic.
- From tables [45], the ratio of relative effects $|\mathrm{VP/SE}| < 1$, but it is variable: it seems to increase with Z, certainly because the s-orbital electrons are spending more time closer to the nucleus. For example: for H, $|\mathrm{VP/SE}| \sim 0.025$, whereas for ion U^{+91} one has $|\mathrm{VP/SE}| \sim 0.27$.

While continuing our study with comparisons and extrapolations from tables, and awaiting more suitable results in the future, we present here our approximate results of an LME computation.

According to our discussion on magnetic interactions (Appendix 1), we recall that we take into account two possibilities concerning the effect of magnetic interactions for highly relativistic electrons:

(a) Either, the magnetic moment of the electron is unchanged or little changed by Special Relativity and, in this case, the LME computation is performed by combining potentials indicated in Section 4.2, where the Lamb shift effects are simulated by weakening the near–nuclear interactions.
(b) Or, the spin magnetic moment is considerably weakened for highly relativistic electrons and, in this case, we compute a LME by considering the attractive potential energy V_{eff} and a repulsive quasi-potential energy deduced from the Lamb shift.

4.4.1. *Computations and results in the first case (a) with magnetic interactions*

Without reporting tedious details of our calculations, we can indicate the following:

- For the repulsive effect of SE, we simulate a linear weakening $V_{\text{Cbw}}(r)$ of the static (attractive) Coulomb potential $V_{\text{Cb}}(r)$, by a coefficient K when approaching the nucleus, i.e. at a radius $r_1 > r_0$, where r_0 is the charge radius of the nucleus ~0.84 F. Next we deduce the dynamical effective potential V_{effw} from V_{Cbw}.
- For the attractive effect of VP, we simulate a weakening of the repulsive magnetic potentials $V_{\text{RSS}}$and V_{dia}, by putting $E = E_{\text{H}} + V_{\text{effw}} + V_{\text{RSS}}/C + V_{\text{dia}}/D$, where C and D are constants >1.

Note the choice of only the repulsive version, V_{RSS}, of the spin–spin interaction, has been made from computing experience. Moreover, while seeking a LME for resonance, we try to check an important question not yet evaluated in previous works: *is the binding energy* (BE) *of an electron caught in this resonance consistent with that predicted by relativistic equations?* To satisfy this condition, we must have at least -511 keV< BE<0. Moreover, for an EDO, one can expect a "rather high" value of |BE|.

Numerous calculations, based on the expression of the total energy E indicated above, i.e. $E = E_{\text{H}} + V_{\text{w}}$, where $V_{\text{w}} = V_{\text{effw}} + V_{\text{RSS}}/C + V_{\text{dia}}/\text{D}$, and with coefficients of weakening still arbitrary, confirm the possible existence of LME close to the nucleus. Moreover, we can also verify that the BE of an electron in the LME, has a "suitable" value, i.e. of the order of magnitude close to the solutions of the Dirac equation.

In Fig. 5, we display an example of the curve BE $= E - mc^2 = E_{\text{H}} - mc^2 + V_{\text{w}} =$ KE$+ V_{\text{w}}$ to clarify the value of electron BE in the LME. For the same purpose, we restrict the scale of abscissa for the radius, denoted by ρ in fermi, while the energy values are in keV. The LME corresponds to $\rho \sim$1.63 F, where BE ~ -470 keV, KE ~120.58... MeV and PE $= V_{\text{w}} \sim -121.05...$ MeV. Note the relativistic virial theorem [46] is satisfied: with precise computed values, we have KE/|PE| $= 0.996\ldots$, $\gamma = 236.96\ldots$ and $\gamma/(\gamma+1) = 0.9958\ldots \sim$ KE/|PE|.

4.4.2. *Computations and results in the second case (**b**), without magnetic interactions*

We indicate in Appendix 2 some elements of principle, about the determination of a repulsive quasi-potential V_{LS} that we deduced from the Lamb shift for EDOs in a recent unpublished study. Here we report only results of the computation where we considered the total energy $E = E_{\text{H}} + V$, with $V = V_{\text{eff}} + V_{\text{LS}}$ and $V_{\text{LS}} \sim 6.23 \times 10^{-56}/r^3$ SI. Several computations, made while varying values of decimals, show the existence of an LME, where the binding energy BE of the electron satisfies -511 keV $<$ BE $<$ 0 and can have value of the order of magnitude close to EDO solutions of relativistic equations.

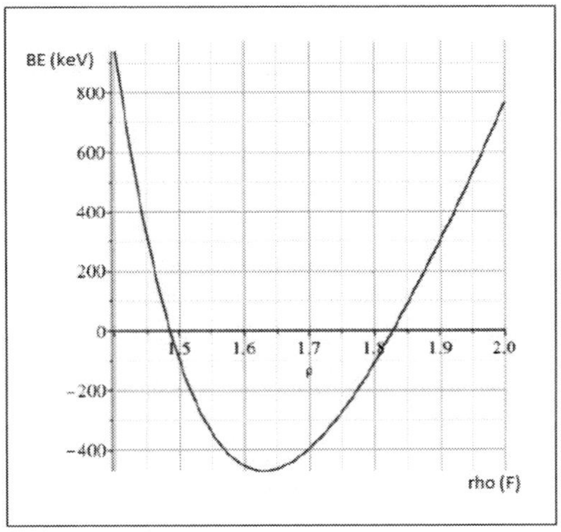

Figure 5. Plot of electron BE for 1.4 F< ρ < 2 F.

In Fig. 6, we plot an example of the curve BE, where the LME corresponds to ρ ~1.41 F, BE ~ − 487 keV, KE ~ 140.03... MeV, PE = V ~ -140.52... MeV, γ = 275.00... We also verify that the relativistic virial theorem is satisfied.

5. Summary and Discussion of Results

(1) At this point, we can relate the results obtained for deep LME, even if only approximate, with the EDO solutions of the Dirac equation.

- In the situation considered in Section 4.4.1. while varying attenuation parameters of the EM interactions, we obtained LME locations between 1.1 F and 2 F and, on the other hand, the mean radii of EDO's obtained by the Dirac equation [21] are from 1.2 to 1.6 F, except for the value obtained when the radial quantum number $n' = 1$. Similarly, in the situation Section 4.4.2, while varying the precision of formulas, we obtain similar LME locations. Finally, in both alternatives considered from our discussion on magnetic interactions, the results have comparable orders of magnitude.
- Seeking an LME, with fixed potentials, provides only one value corresponding to an energy well, whereas the Dirac equation provides an infinity of EDOs levels. But, it is the same when one seeks the size order of electron LME in the simple Coulomb potential of a proton, as e.g. in [24]: one finds the fundamental Bohr level at 53 pm, whereas the Schrödinger equation provides an infinity of energy levels, including the fundamental level plus excited levels corresponding to additional resonances.
- In Section 4.2. (3) on *spin–spin interactions*, we noted that in a relativistic regime, the spin tends to lean in the direction of the motion of the particle [32], i.e., it is getting closer to the helicity, and we wondered if this does not lead to a weakening of the interaction energy. But, in fact, this interpretation assumes the movement of the electron becomes more perpendicular to the direction of the spin of the nucleus, as in the classical image of an orbit on the ecliptic of a planetary system. Of course, there is no reason for

the moment of the electron to have any preferred direction with respect to the orientation of the nuclear spin: one could think that spins tend to align by magnetic coupling, for reaching a lower energy state. Nevertheless, because of quantum mechanics rules on energy transfers, things are more complicated. The evolution of a system of two spins in interaction is precisely described in [47]: under the effect of the coupling, both spins process about their resultant J, with an angular velocity proportional to |J| and the coupling factor.

(2) For our calculations in the previous section, we have taken into account combinations of high-energy, $1/r^n$, potentials that are partially converted into actual kinetic energy for deep electrons. So, we have energies of order 100 – 200 MeV, while we have systematically eliminated potentials that give unrealistic energies (of order of GeV) due to angular momentum effects (spin–orbit interaction and "centrifugal potential"). Nevertheless, we can legitimately ask where do these high energies come from? The most plausible answer is that this energy is taken from the rest mass of the proton, which is of order 1 GeV. Different hypotheses exist about what constitutes the mass of a proton. Most known, from experiment in the LHC [48] and from electron–proton inelastic scattering (e.g. [49]), is that a proton is actually a "soup" of quarks, antiquarks, and gluons in a perpetual shuffling (creation/destruction of pairs) and in highly relativistic movement, but including two up-quarks (with "base" mass ~2.4 MeV, [50]) and one down-quark (with "base" mass ~5 MeV), named "valence" quarks. In this vision, the mass of a proton would come mainly from the relativistic energy of its constituents. In another point of view, based on the "Constituent Quark Model" [51]: in the low-energy limit of Quantum Chromodynamics (QCD), which concerns the current atomic nuclei, the constituent quarks appear like "dressed" current quarks, i.e. current quarks surrounded by a cloud of virtual quarks and gluons. This cloud underlies the large constituent-quark masses, of order 336–340 MeV. In this model, energy could still

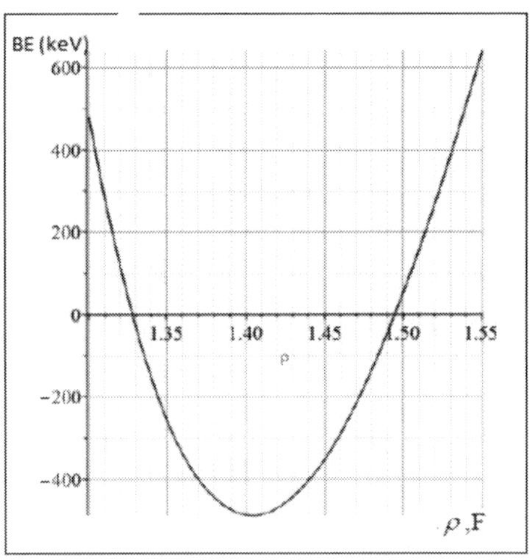

Figure 6. Plot of electron BE for 1.3 F< ρ < 1.55 F.

be taken from the clouds surrounding the current quarks. Of course, the process of energy transfer from the proton-reservoir to the relativistic deep electron, with its proximate intense fields, should be the subject of a detailed study.

(3) In this paper, seeking an LME with fixed potentials is a preliminary study, carried out in a semi-classical way and with a rather coarse view of the effects of radiative corrections (essentially Lamb shift) involved near the nucleus. We are currently pursuing a study to express in a more detailed and semi-analytical manner the involvement of the Lamb shift in the computation of the resonance.

Here, we do not claim to prove rigorously that there is an LME, but we show that the existence of an LME is possible if some conditions are met (related to the combinations of potential energies). And, what is encouraging is that many calculations lead to this possibility, with a BE consistent with that predicted by relativistic equations [21]. Moreover, the resonance is confirmed by the fact that KE and PE satisfy the relativistic virial theorem.

Appendix A. Discussion about Magnetic Interactions for Relativistic Electrons

The magnetic interactions, such as spin-spin and the diamagnetic term (Section 4.2), involve the magnetic moment of the electron associated with its spin. For a relativistic electron, one knows [32] that, when the velocity increases, the direction of the spin tends to become parallel to the momentum **p**, like helicity, which can lead to a weakening of the magnetic interactions. But recently, a referee caught our attention on the fact that the magnetic moment of the electron decreases in inverse proportion of its energy. Also, and although very few documents deal with this question, we ended up actually finding papers [52,53], where the magnetic moment is given (or equivalent forms) by the relation $\mathbf{m} = e\hbar c/2E(\mathbf{p})$, where $E(\mathbf{p})$ is the total energy of an electron having momentum **p**. We note this result is deduced by reasoning of Dirac spinors in a simplified situation, i.e. for a free electron, or a electron moving in a constant magnetic field. What can be deduced for a electron bound in central potential V, a combination of several fields including a nuclear Coulomb field?

(a) On one hand, the total energy is given by $E(\mathbf{p}) = (m^2c^4 + p^2c^2)^{(1/2)} + V$. But for deep electrons, we expect a high binding energy |BE| (in absolute value) of several hundred keV [9–11,21], near the rest mass m, thus $E(\mathbf{p}) \ll mc^2$. As a consequence, we would have $\mathbf{m} \gg \mathbf{e}\hbar/2mc$ and maybe a major strengthening of the magnetic interactions. However, while the effective **m** may be greatly increased because of relativity, the reorientation of the electron spin axis greatly decreases the $\mathbf{S}\cdot\mathbf{S}$ relation (as well as $\mathbf{L}\cdot\mathbf{S}$, for $l = 1$). It may compensate exactly and thus using $\mathbf{m} = e\hbar/2mc$, without correcting for the $\mathbf{S}\cdot\mathbf{S}$ decrease, might be a good assumption.

(b) On the other hand, we can consider the following equivalent form of the expression $\mathbf{m} = e\hbar c/2\mathrm{E}(\mathbf{p})$: $\mathbf{m} = e\hbar/2\gamma mc$, where γ is the usual relativistic coefficient. Can we consider that we have essentially to take into account the role of γ for determining the value of **m**, for a free electron *as well as a bound electron*? As a consequence, since deep electrons are highly relativistic, **m** would be very small and the magnetic interactions would be negligible compared to V_{eff}.

For the present paper, as it is very difficult to know what is the correct alternative for deep electrons, we consider both alternatives. More precisely, we are taking into account an additional effect, that of the radiative corrections, very important near the nucleus, for which we give some indispensable elements of explanation in Section 4.3.

(1) Then, while dealing with alternative (a), we keep the usual expression $\mathbf{m} = e\hbar/2mc$, and we put the total energy E equal to $E = E_{\mathrm{H}} + V$, where V will be a combination of three potential energies, V_{eff}, V_{RSS}, and V_{dia}, taking into account the effect of radiative corrections.

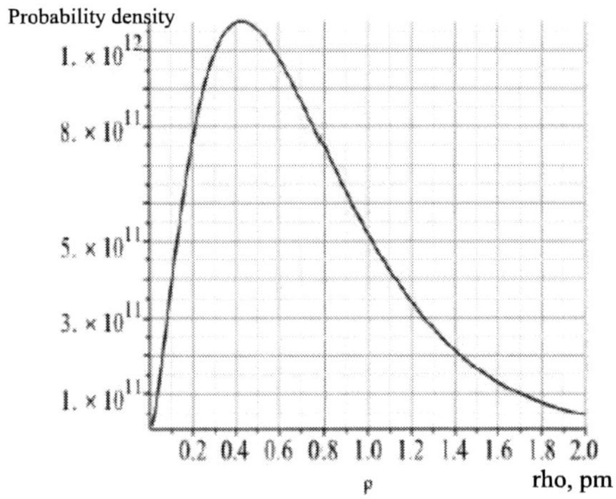

Figure 7. Plot of the normalized radial probability density for 1s orbital of uranium, for $\rho < 2$ pm. The maximum corresponds to $\rho \sim 0.43$ pm.

(2) In the case of alternative (b), we will give an overview of a possible treatment where the magnetic interactions, being considered negligible compared to the attractive interaction V_{eff}, are not taken into account. But, the potential well needed for the existence of deep electrons can still be achieved thanks to a repulsive quasi-potential deduced from the Lamb shift and combined with V_{eff}.

Appendix A.1. Determination of a repulsive quasi-potential associated with Lamb shift for EDOs

Here we give only a schematic view of a work started on the occasion of a presentation at the 13th International Workshop on Anomalies in Hydrogen Loaded Metals (5–9 Oct. 2018, Grecchio, Italy). We build a repulsive potential as a function of the electron orbit radius from calculations on Self-energy (SE), which has a repulsive effect, while considering Vacuum Polarization (VP) calculations lead to a weakening of SE, since VP, which has an attractive effect, has to be subtracted from SE.

Appendix A.1.1. *The starting point of our method*

Our method consists of analyzing the progressions of SE and VP for atomic electrons for nuclei with increasing Z and, for comparisons and extrapolations, we consider the radius corresponding to the maximum probability density, noted *rmx*. Then, we consider the "Z increases" are equivalent to the "radius r decreases": Z increasing $\Leftrightarrow$ the coupling force increases $\Leftrightarrow$ the radius *rmx* decreases.

Nevertheless, as a deep electron is very far from an atomic electron, we use an *intermediate step* in terms of strength of the Coulomb electric field: the hydrogen-like uranium ion. This allows us to section the "distance" between an atomic case and a deep orbit, and above all, to take advantage of many calculations and experiments on the Lamb shift (LS) for heavy elements. Nevertheless, we have to be careful, because, for heavy elements, LS often includes terms that are not radiative corrections: for example, the nuclear size (NS), which has a significant effect. So, to calculate the LS for deep electrons, we discard any effect other than SE and VP.

Note: from now on, we consider the LS only for 1s electrons, for any chemical elements.

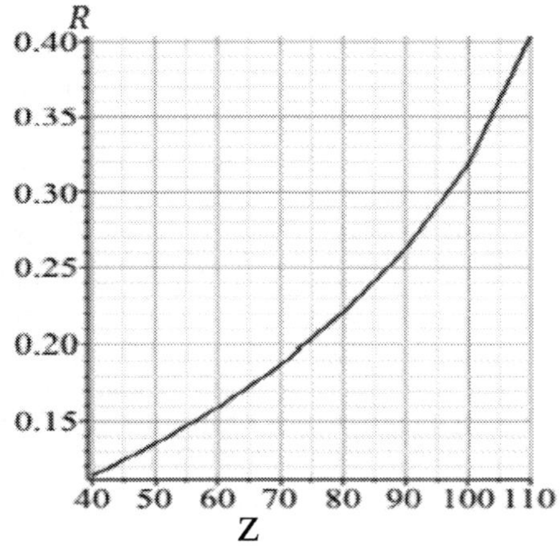

Figure 8. Smooth plot of ratio $R = |\text{VP}|/\text{SE}$, as function of Z, from data given in [45] and for $40 \leq Z \leq 110$.

In Fig. 7 we plot the normalized radial probability density for the 1s orbital of uranium, for $\rho < 2$ pm. For uranium, the radius corresponding to the maximum density probability, noted *rmx*(U) is equal to ~427.3 F.

As indicated in Section 4.3 on radiative corrections, the energy shift due to SE can be expressed by the following formula: $\Delta E_{\text{LS}} = (\alpha/\pi)(Z\alpha)^4 \text{F}(Z\alpha)\, mc^2$. This formula has two factors depending on Z. The first, $(Z\alpha)^4$, quickly increases with the electric Coulomb field and can lead to a possible analytic formulation as function of the radius r.

From data tables, one can see the second factor, $F(Z\alpha)$, slowly decreases with Z. Moreover, again from a table of results, the ratio |VP|/SE increases with Z.

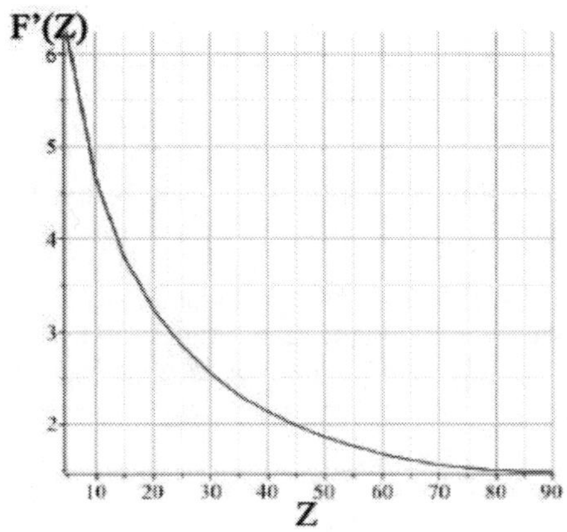

Figure 9. Smooth plot of factor $F'(Z)$, as function of Z, from data given in [42] and for $5 \leq Z \leq 90$.

Appendix A.1.2. *The form of the analytic expression of SE, and extrapolations from data tables*

Elements of comparison between SE values for the hydrogen atom and uranium lead us to take SE $=$ C/r^3 as more "reasonable" than C/r^4. And, we can expect the reducing of SE to the power 3 can be "absorbed" thanks to the progression of |VP|/SE. We give in Fig. 8, a smooth plot of the ratio $R =$ |VP|/SE, as function of Z, from [45]

With the data table, from $Z = 40$, we can recognize and extrapolate a slow geometric progression of the form $R_n = R_0 q^n$, where $q \sim 1.2$, $R_0 = R(Z = 40)$ and $n = (Z - 40)/10$. Thus it can compensate the reduction of SE indicated above, without having to subtract |VP| from SE. Next, we consider the factor $F(Z\alpha)$, expressed more simply as $F'(Z)$. In Fig. 9, we give a smooth plot of $F'(Z)$, built from [42], for $5 \leq Z \leq 90$.

From the data table, we can recognize and extrapolate a quasi-exponential decreasing progression of the factor $F'(Z)$, while considering the progression of Z multiplied by successive integer powers of 2, from $Z = 5$: in fact, at each step, $F'(Z)$ is divided by a number k which is very slowly increasing. To sum up, while considering the ratio of *rmx*(U) /r_{edo}, where r_{edo} is an expected value of the LME for an EDO, fixed to $\sim$1.4 F, the factor $F'(92) \sim 1.49$ for uranium, and SE $=$ 355 eV for uranium, we obtain an approximate value of F' for an EDO, noted $F'_{\mathrm{edo}} \sim 0.021$. From F'_{edo}, we deduce $V_{\mathrm{LS}} \sim 6.2 \times 10^{-56}/r^3$ in SI units or $V_{\mathrm{LS}} \sim 3.8 \times 10^{-37}/r^3$ in eV (with *r* in meters).

Acknowledgements

The authors wish to thank the reviewer for pointing out the problems associated with the spin magnetic moment of a relativistic electron. This work is supported in part by HiPi Consulting, Windsor, VA, USA and by the Science for Humanity Trust Inc., Tucker, GA, USA.

References

[1] M. Fleishmann and S. Pons, Electrochemically induced nuclear fusion of deuterium, *J. Electroanal. Chem.* **261**(1989) 301.

[2] A. Meulenberg and K.P. Sinha, Tunneling beneath the ^{4}He$*$ fragmentation energy, presented at 239th ACS National Meeting, San Francisco, CA March 2010, *J. Condensed Matter Nucl. Sci.* **4** (2011) 241–255.

[3] K.P. Sinha and A. Meulenberg, Lochon Catalyzed D–D fusion in deuterated palladium in the solid state, *National Academy of Sci. (India) Lett.* **30** (7,8) (2007) arXiv:0705.0595v1.

[4] K.P. Sinha and A. Meulenberg, A model for enhanced fusion reaction in a solid matrix of metal deuterides, *Proc. of the 14th Int. Conf. on Cold Fusion (ICCF-14)*, 10-15 Aug. 2008, Washington DC, p. 633.

[5] K.P. Sinha, Model of low energy nuclear reactions in a solid matrix with defects, PDF Special Section: Low Energy Nuclear Reactions, Section editors: M. Srinivasan and A. Meulenberg, *Current Sci.* **108** (4) (2015) 516, available at http://www.currentscience.ac.in/cs/php/feat.php?feature=Special%20Section:%20Low%20Energy%20Nuclear%20Reactions&featid=

[6] A. Meulenberg and K.P. Sinha, Composite model for LENR in linear defects of a lattice, ICCF-18, *18th Int. Conf. on Condensed Matter Nucl. Sci.*, Columbia, Missouri, 25/07/2013, http://hdl.handle.net/10355/36818.

[7] E. Storms, How basic behavior of LENR can guide. A Search for an explanation, *J. Condensed Matter Nucl. Sci.* **20** (2016) 100, www.iscmns.org/CMNS/JCMNS-Vol20.pdf and the references therein.

[8] A. Takahashi, Physics of cold fusion by TSC theory, *J. Condensed Matter Nucl. Sci.* **12** (2013) and https://www.researchgate.net/publication/311614659_Physics_of_Cold_Fusion_by_TSC_Theory_full_paper.

[9] J.L. Paillet and A. Meulenberg, The basis for electron deep orbits of the hydrogen atom, *ICCF-19, 19th Int. Conf. on Cond. Matter Nuclear Science*, Padua, Italy, 15/05/2015, *J. Condensed Matter Nucl. Sci.* **19**(2016) 230–243.

[10] J.L. Paillet and A. Meulenberg, Special relativity: the source of electron deep orbits, *Foundations of Physics*, Vol. 47, Issue 2, Springer, Heidelberg, Feb. 2017, pp. 256–264.

[11] J.L. Paillet and A. Meulenberg, Arguments for the anomalous solutions of the Dirac equations, *J. Condensed Matter Nucl. Sci.* **18** (2016) 50–75, http://viXra.org/abs/1506.0177.

[12] A. Meulenberg and J.L. Paillet, Physical reasons for rejecting arguments against deep-Dirac levels, *J. Condensed Matter Nucl. Sci.* **24** (2017) 230–235.

[13] A. Meulenberg, Femto-atoms and transmutation, *17th Int. Conf. on Condensed Matter Nucl. Sci.*, Daejeon, South Korea, 12–17 August, 2012, *J. Condensed Matter Nucl. Sci.* **13** (2014) 346–357.

[14] A. Meulenberg and K.P. Sinha, Deep-orbit-electron radiation emission in decay from 4He*# to 4He, *17th Int. Conf. on Condensed Matter Nucl. Sci.*, Daejeon, South Korea, 12–17 August, 2012, *J. Condensed Matter Nucl. Sci.* **13** (2014) 357–368.

[15] A. Meulenberg and K.P. Sinha, New visions of physics through the microscope of cold fusion invited paper, *17th Int. Conf. on Condensed Matter Nucl. Sci.*, Daejeon, South Korea, 12–17 August, 2012, *J. Condensed Matter Nucl. Sci.* **13** (2014) 378–390.

[16] A. Meulenberg and K.P. Sinha, Deep-electron orbits in Cold Fusion, *17th Int. Conf. on Condensed Matter Nucl. Sci.*, Daejeon, South Korea, 12–17 August, 2012, *J. Condensed Matter Nucl. Sci.* **13** (2014) 368–377.

[17] A. Meulenberg, Extensions to physics: what cold fusion teaches, *Special Section: Low Energy Nuclear Reactions,* Section, M. Srinivasan and A. Meulenberg (Eds.), *Current Science,* Vol. 108, No. 4, 25 Feb. 2015. http://www.currentscience.ac.in/cs/php/feat.php?feature=Special%20Section:%20Low%20Energy%20Nuclear%20Reactions&featid=100

[18] A. Meulenberg, Femto-helium and PdD transmutation, *ICCF-18, 18th Int. Conf. on Condensed Matter Nucl. Sci.*, Columbia, Missouri, 25/07/2013, *J. Condensed Matter Nucl. Sci.* **15** (2015) 106–117.

[19] A. Meulenberg and J.L. Paillet, Basis for femto-molecules and -ions created from femto-atoms, *ICCF-19, 19th Int. Conf. on Condensed Matter Nucl. Sci.*, Padua, Italy, 15/05/2015, *J. Condensed Matter Nucl. Sci.* **19** (2016) 202–209.

[20] A. Meulenberg and J.L. Paillet, Implications of the electron deep orbits for cold fusion and physics, *J. Condensed Matter Nucl. Sci.* **24** (2017) 214–229.

[21] J.L. Paillet and A. Meulenberg, Electron deep orbits of the hydrogen atom, *Proc. 11th Int. Workshop on Hydrogen Loaded Metals*, Airbus Toulouse, France, 15–16 Oct. 2015, *J. Condensed Matter Nucl. Sci.* **23** (2017) 62–84.

[22] J.L. Paillet and A. Meulenberg, Advance on electron deep orbits of the hydrogen atom for LENR, *Proc. ICCF-20, 20th Int. Conf. on Condensed Matter Nucl. Sci.*, Sendai, Japan, October 2–7 2016, *J. Condensed Matter Nucl. Sci.* **24** (2017) 258–277.

[23] J.L. Paillet and A. Meulenberg, Deepening questions about electron deep orbits of the hydrogen atom, *Proc. 12th Int. Workshop on Hydrogen Loaded Metals*, Asti, Italy, June 5–9 2017, *J. Condensed Matter Nucl. Sci.* **26** (2018) 54–68.

[24] A. Messiah, *Quantum Mechanics*, Vol. 1, North-Holland, Amsterdam, 1967.

[25] Cl. Cohen-Tannoudji, B. Diu and Fr. Laloë, *Quantum Mechanics*, Vol. 1, C_I, Wiley-VCH, New York, London, Sydney, Toronto, 1977.

[26] S.V. Adamenko and V.I. Vysotskii, Mechanism of synthesis of superheavy nuclei via the process of controlled electron-nuclear collapse, *Foundations Phys. Lett.* **17** (3) (2004) 203–233.

[27] S.V. Adamenko and V.I. Vysotskii, Evolution of annular self-controlled electron–nucleus collapse in condensed targets, Foundations Phys. **34** (11) (2004) 1801–1831.

[28] [28] J.L. Paillet and A. Meulenberg, Relativity and Electron deep orbits of the hydrogen atom, *Proc. of the 1st French Symp. RNBE-2016 on Condensed Matter Nucl. Sci.*, Avignon, 18–20 March 16, *J. Condensed Matter Nucl. Sci.* **21** (2016) 40–58.

[29] L.I. Schiff, *Quantum Mechanics*, 3rd Edn., McGraw-Hill, New York, 1968.

[30] Cl. Cohen-Tannoudji, B. Diu and Fr. Laloë, *Quantum Mechanics*, Vol. 2, Ch. XII B, Wiley-VCH, Hoboken, New-Jersey, 1977.

[31] H. Murayama, *Quantum Mechanics*, 221A Lecture Notes, Fall, 2006, Berkeley, http://hitoshi.berkeley.edu/221a/hyperfine.pdf, last access: Sep. 2018.

[32] R. Klauber, *Student Friendly Quantum Field Theory*, 2nd. Edn., Sandtrove Press, Fairfield, Iowa, 2013.

[33] W.H.E. Schwarz, An Introduction to Relativistic Quantum Chemistry, *Relativistic Methods for Chemists*, Chapter 1, *Challenges and Advances in Comp. Chem. and Phys. Series*, J. Leszczynski (Ed.), Vol. 10, Ed. By M. Barysz, Yasuyuki Ishikawa, Springer, Berlin, 2010.

[34] Cl. Cohen-Tannoudji, B. Diu and Fr. Laloë, *Quantum Mechanics*, Vol. 2, Ch. VII, D_{VII}, Wiley-VCH, Hoboken, New-Jersey, 1977.

[35] F. Mandl and G. Shaw, *Quantum Field Theory*, 2nd. Edn., Wiley, Chichester, UK, 2010.

[36] J.M. Jauch and F. Rohrlich, *The Theory of Photons and Electrons*, Addison-Wesley, Reading, USA, 2nd., Printing, June 1959.

[37] W.H. Furry, On bound states and scattering in positron theory, *Phys. Rev.* **81** (1951) 115.

[38] W. Lamb and R. Retherford, Fine structure of the hydrogen atom by a microwave method, *Phys. Rev.* **72**(3) (1947) 241–243.
[39] H.A. Bethe, The electromagnetic shift of energy levels, *Phys. Rev.* **72** (1947) 337.
[40] F.J. Dyson, *Advanced Quantum Mechanics*, 2nd. Edn., arXiv:quant-ph/0608140v1 18 Aug. 2006.
[41] A. Meulenberg, W.R. Hudgins and R.F. Penland, The photon to electron/positron-pair transition, *Proc. SPIE.* 9570, The Nature of Light: What are Photons? VI, 95700S, September 10, 2015, doi: 10.1117/12.2187489.
[42] P.J. Mohr, Self-energy correction to one-electron energy levels in a strong Coulomb field, *Phys. Rev.* **46** (7)(1992)4421–4424.
[43] P.J. Mohr, Quantum electrodynamics calculations in few-electron systems, *Phys. Scripta* **46**(1993) 44–51.
[44] A.M. Frolov and D. Wardlaw, Analytic formula for the Uehling potential, arXiv:1110.3433v4 [nucl-th] (2012).
[45] W.R. Johnson and G. Soff, The Lamb shift in hydrogen-like atoms, $1 \leq Z \leq 110$, *Atomic Data and Nuclear Data Tables*, Vol. 33, Academic Press, New York, 1985, pp. 405–446.
[46] Virial Theorem. https://en.wikipedia.org/wiki/Virial_theorem, 3 Sept. 2018.
[47] Cl. Cohen-Tannoudji, B. Diu and Fr. Laloë, *Quantum Mechanics*, Vol. 2, Ch. X, Fx, Wiley-VCH, New York, 1977.
[48] Strassler Matt., Checking What's Inside a Proton, in Blog Of Particular Significance. https://profmattstrassler.com/articles-and-posts/largehadroncolliderfaq/whats-a-proton-anyway/checking-whats-inside-a-proton/, last access: Sept. 2018.
[49] [49]J. Martin Victoria, Probing the Proton: Electron–Proton Scattering, Lect. 8, Collider Physics, University of Edinburgh, results from HERA collider DESY in Hamburg, 2012. https://www2.ph.ed.ac.uk/~vjm/Lectures/SHParticlePhysics2012_files/PPNotes3.pdf, last access: Sept 2018.
[50] J. Beringer (*Particle Data Group*) et al. 2012, *PDGLive Particle Summary 'Quarks'*, PR **D86**, 010001 (2012), http://pdg.lbl.gov, last access: Sept. 2018.
[51] D. Diakonov, Foundations of the constituent quark model, *Prog. Part. Nucl. Phys.* **36** (1996) 1–18, arXiv:nucl-th/9603023v1, 17 Mar. 1996.
[52] W. Zawadski, Spin magnetic moment of a free relativistic electron, *Phys. Rev.* D **3** (8) (1971) 1728.
[53] S. Sasabe, Spin-magnetic moment of dirac electton and role of Zitterbewegung, *J. Modern Phys.* (2014) 534–542.

J. Condensed Matter Nucl. Sci. 29 (2019) 493–502

Research Article

Lattice Confinement of Hydrogen in FCC Metals for Fusion Reactions

Han H. Nee*, Arsen V. Subashiev and Fracsisco M. Prados-Estéves

Target Technology Company LLC, 564 Wald, Irvine, CA 92618, USA

Abstract

Clusters of H isotope atoms segregated to vacancies, divacancies and vacancy–impurity complexes in FCC metals are proposed as fuel for low energy nuclear reactions (LENR). Such clusters combine extremely high H atomic density, large values of screening potential, and as a result, a low LENR ignition energy in eV region. Besides, high average H density can be achieved due to the superabundant vacancy state (SAV) formation. These conclusions are made based on the density functional theory (DFT) modeling of these clusters, estimations of the nuclear reaction rates using experimental data for the nuclear reaction cross sections and a wide set of experimental studies of charging and recharging of various Ni samples with H and D atoms.The results were analyzed using temperature programmed desorption. The experiments confirmed extremely high loading of Ni samples with H isotopes. We discuss the problem of suitable ignition mechanisms and sustainability conditions.

Keywords: DFT, Divacancies, Fusion reaction rate, Ignition energy, Lattice confinement, Nickel, Screening potential

1. Introduction

The experimental studies of excess heat produced by thermal, electrical, laser and other types of excitation of metals charged with hydrogen isotopes revealed a high energy gain in many observations [1]. In a number of cases the released energy cannot be interpreted in terms of traditional condensed matter physics [2]. It suggests an involvement of reactions with nuclear transformations, with much higher energy release in a single reaction event. It is well understood now that the implementation of stable nuclear reactions with low excitation energy (LENR) requires: proper fuel (with high H isotope concentration), ignition (or triggering with the required energy) and achieving conditions of sustainability (which requires H isotope supply and the energy escape from the reaction zone).

The former experimental studies reached the consensus that Pd and other FCC metals can be the best candidates for the fuel. Moreover, both former and recent studies [1–3] imply an important role of structural defects for the charging of metals with hydrogen. The experimental findings were summarized in a number of reviews which stress

*Corresponding author. E-mail: hannee@targettechnology.com.

the requirement for the creation of optimal nuclear active sites (or "nuclear active environments" [4]) for the energy production.

We have explored the possibility of LENR implementation in Ni as a more practical alternative to Pd especially in the case of a large scale application. We have found that the clusters of H isotopes in Ni, as well as in Pd, segregated to vacancies and divacancies, and vacancy–impurity complexes have unique properties that can make LENR possible. First, the average concentration of H isotope ions segregated in vacancies (and divacancies) is extremely high. Second, the screening potential for H atoms in a divacancy can reach anomalously large values. Strong screening makes the required ignition energy for LENR shift down to the eV range.

The last conclusion is based on presented estimations of the reactivity and reaction rates in the H isotope clusters using experimental data for the screening potential for the nuclear fusion cross sections. At energies $\leq$10 keV a number of face-centered cubic (FCC) metals manifest huge enhancements and indicate large values of the screening potential. We suppose that these data are obtained for the samples with structural defects.

The paper is organized as follows. First, we present results of *ab-initio* modeling of hydrogen clusters bound to vacancies, divacancies and vacancy–impurity defects. Further, we discuss the screening potential for these clusters and indicate configurations for which the anomalously high values of the screening potential can be found.

We also estimate the rate of the nuclear reaction in mH–V clusters in Ni and Pd. The main remaining problem which was recognized long before is the possible mechanisms of ignition of the nuclear reaction. The required energy is in the eV range, far above that available from thermal excitation. Our findings are summarized in the conclusions.

2. Vacancies, Divacancies and Vacancy–mpurity Complexes in FCC Metals

We used Quantum Espresso Density Functional software to calculate binding energies for various structural defects, such as vacancies, interstitial H and D atoms, divacancies, vacancy–impurity complexes and clusters of H atoms segregated to vacancies and divacancies. These calculations are based on the supercell approach when a FCC lattice supercell of 32 atoms is used with periodic boundary conditions on the supercell boundaries. Technical details, such as the validation of the approach, the accuracy estimation, the choice of the pseudopotentials and so on can be found in publication [5]. To eliminate the errors due to inter-cell interactions, the results were checked for some of the structures using computations with a larger supercell of 128 atoms. Much larger computer time requirements eliminated the possibility to use a large supercell for all complexes.

The structure of the supercells with a single vacancy and a divacancy formed by two vacancies in the nearest neighbor positions are shown in Fig. 1 with an indication of the possible positions of H isotopes

segregated to the adjacent octahedral sites of the lattice. While for a vacancy all segregation sites are equivalent, for a divacancy there are four O-c sites (shown in Fig. 1 by triangles) at the corner positions, two O-s -sites (open squares) at "shared" sites near the divacancy center in the divacancy atomic plane (shaded by hatching) and four t-sites (closed squares) at the top in the upper (3-d) and lower (first) layers. Dashed lines are guides for the eyes and indicate octahedrons of O-sites). Filling of all octahedral sites suggests that the maximum occupancy is 10. Shared sites are found to have much higher segregation energy (0.4 eV) than the rest of the sites (0.3 eV). These results are in line with the previously published results for these clusters [6,7] which, however, were restricted to only a limited number of configurations for the populating of available segregation sites. The formation energy for a number of various mH–V_2 clusters for $m = 0, 1, \ldots, 10$ is shown in Fig. 2. Although the total number of clusters with different configurations is quite high (2^{10}) the formation energies of different clusters with the same m varies within 0.2 eV, so that the slope of the dashed line in Fig. 2 is close to the average segregation energy $E_{\text{seg}} = 0.3$ eV. At $m = 10$ the formation energy becomes negative, which corresponds to the state of superabundant vacancy (SAV) state formation, proposed by Fukai [8]. According to our calculations, the SAV state for Ni is different from that for Pd: the binding of two single vacancies in the divacancy (with binding energy 30 meV/vac) and the binding energy in mH–V_2 cluster are higher than for Pd. Therefore concentration of mH–V_2 clusters in SAV phase should be comparable to that of mH–V clusters.

Note that the formation of SAV state in thermal equilibrium should take a long time due to the slow diffusion process of vacancies from the surface. It is accelerated at elevated temperature and pressure. However in the case

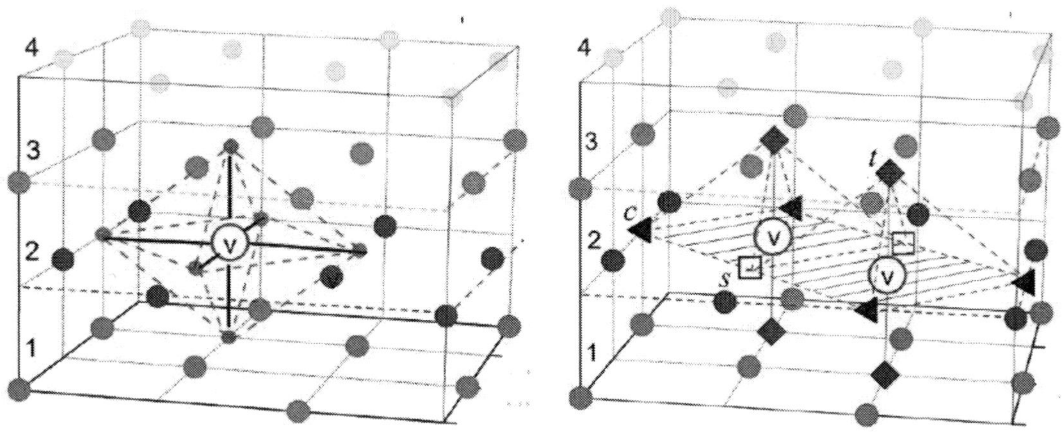

Figure 1. Preferred positions of H atoms at octahedral sites in a single vacancy(left panel) and a divacancy formed by emptying two nearest neighbor Ni sites (*right panel*). Shown is a 32 atom supercell with four atomic layers marked by solid dots of different colors and vacancies shown by hollow dots.

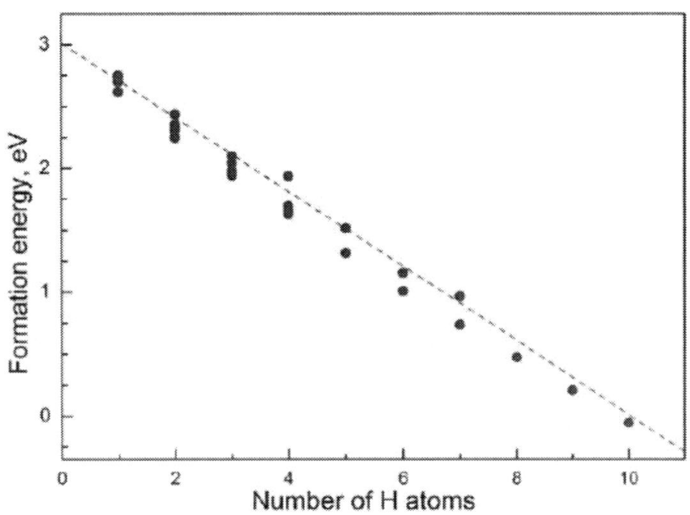

Figure 2. Formation energies for various mH–V_2 clusters with $m = 0, 1, 2, \ldots, 10$.

Table 1. Mass and particle density for different H phases as compared to that of 6H–V and 10H–V_2 clusters in metallic Ni and Pd.

Phase	Mass density (g/m^3)	H density (10^{22} at/cm^3)	Ref.
H (mol liquid)	0.07	4.2	[9]
H (mol. Solid)	0.076	4.54	[9]
DT ice	0.25	5.6	[10]
Metallic H	0.7	41.9	[11]
6H–V cluster in Pd	0.73	43.8	This publ.
10-V_2 cluster in Pd	1.06	63.5	This publ.
6H-V cluster in Ni	1.01	60	[5]
10-V_2 cluster in Ni	1.45	87	[5]
H in the Sun, avg.	1.4	83.8	[12]

of an initial structure with high vacancy concentration or when the hydrogen charging process includes creation of vacancies, as shown by Staker in Appendixes A and B of his paper, in this volume [16], it can be predominant.

We have also calculated the defect formation volume, i.e. the volume into which H is segregated, and further, the average atomic density for a 6H–V and a 10H–V_2 clusters of segregated H atoms. The data on the hydrogen concentration in various phases and also in the hydrogen clusters are summarized in Table 1.

For a divacancy in Ni this density appears to be very high, $n = 8.7 \times 10^{23}$ at/cm^3, which is two times higher than that for a calculated density of metallic hydrogen and close to the average atomic density in the Sun. Somewhat smaller density, $n = 6 \times 10^{23}$ at/cm^3, is found for a 6H–V vacancy cluster. This fact is important for the possibility of a nuclear reaction in such clusters, since the energies of atoms in various positions inside the clusters are quite close. The density of H isotopes in similar clusters in Pd is at least 1.37 times smaller due to a larger lattice constant and its substantial increase in H charging. Calculated binding energy for the subsequent segregation in Ni varies from ≈ 0.27 eV for mH–V clusters to 0.4 eV for mH–V_2 clusters and shows a sufficient stability of these clusters. For Pd these energies are $E_v = 0.2$ eV and $E_{v2} = 0.26$ eV. In both cases the SAV formation seems to be more probable in metallic Ni, than in Pd. However, in Pd it takes place in a hydride phase, rather than in the metallic phase.

Modeling of the clusters in Ni composed of a single vacancy with an impurity atom at the nearest neighbor site was also performed. We considered Al, Li and Mo impurities that can be present in catalytic Ni for technological reasons. We have found that among other impurities Li favorably increases H binding energy to a vacancy, Al does not have much effect, and Mo can have a negative effect on the SAV phase formation. Thus, the average H concentration can be further increased by doping impurities (e.g. Li) with enlarged binding energy of mH–V–I clusters.

We have also calculated the electronic charge distribution in mH–V and mH–V_2 clusters. These calculations showed that the positions of H atoms are close to the "apparent internal surface" where the drop of the metal electron density is maximal. For better understanding of the screening of the hydrogen charge it is useful to consider the redistribution of the electronic density $\delta\rho = \rho_{\mathrm{Ni+H}}$-$\rho_{\mathrm{Ni}}$-$\rho_{\mathrm{H}}$ (the change in electronic density in Ni and near H atoms due to segregation) for a cluster with segregated H atoms. It is depicted in Fig. 3 for a 10H–V_2 cluster and an interstitial H at the tetragonal site of Ni lattice.

Charge density contours are drawn from –0.1 to 0.1 with the interval of 0.001 electrons/Å^3. The redistribution results in the depletion of density of the orbitals of the nearest Ni atoms and accumulation of the excessive electronic charges on H atoms. The total accumulated charge is of about 0.1 of electron charge, and it is distributed almost spherically inside a radius of about 10 nm. This redistribution cannot be interpreted as the manifestation of anomalously high screening potential.

However, the modeling showed the possibility of hydrogen segregation into anomalous pair states with distances smaller than the 25 nm (the distance between the nearest octahedral sites). The density distribution for two of the possible configurations is presented in Fig. 4. Paired H states showed smaller distances between H atoms in divacancy

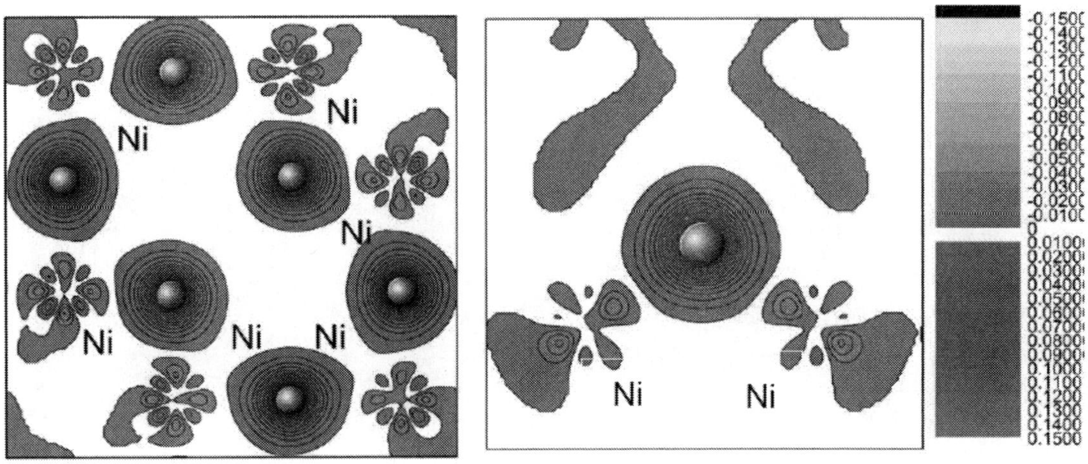

Figure 3. Redistribution of electron density in 10H–V_2 cluster in the divacancy (100) plane (*left panel*) as compared to that for a H interstitial atom at a tetragonal site (*right panel*). Blue and gray contours show depletion regions, pink and red indicate accumulation of electronic clouds around gray balls of H atoms in Ni lattice.

clusters. It can be interpreted as the reduced repulsion between the H ions at the inner surface of a divacancy. We consider it as a manifestation of a dipole–dipole, but not Coulomb, repulsion between the paired ions.

The phenomenon is known as a "supercapacitor effect", equivalent to an anomalously large screening potential experimentally found for many metallic surfaces [13]. It was found in the last decade that the capacitance of the extremely thin dielectric layers with complicated geometry is much larger than that predicted by the Helmholtz (1853) formula $C_{\mathrm{H}} = \varepsilon\varepsilon_0 S/d$, where $\varepsilon\varepsilon_0$ is the permittivity of the capacitor, S is the surface area and d is the distance from the charged plane to the metal surface.

The origin of the effect is the discrete nature of the charges. In deposition of a charge at the metal surface every charge initiates the charge redistribution in a metal known as an appearance of an image charge, which forms a dipole at the surface. The repulsion between the two dipoles is much weaker than the Coulomb repulsion between charges on the homogeneously charged plane. The standard estimation of the screening potential in the dense homogeneous plasma gives the value $U = Ze^2/(\varepsilon_0 a)$, where a is the average distance between the charges. At the surface this distance can be several times smaller due to the dipole formation. Note that the quantitative estimation of this effect requires more detailed calculations.

To summarize, one may consider the formation of the SAV phase in FCC metals with high average H concentration and extremely large local H concentration in mH–V and mH–V_2 clusters as a lattice confinement.

3. Experimental Studies of Charging of Ni Samples with H and D Isotopes

We have studied the charging of nickel and nickel catalyst powders and also samples pressed into cylinders for further LENR experiments with Hydrogen and further recharging it with deuterium. At 83 atmospheres and 250°C the charging times were at least 24 h, but typically several days. We have studied the results with the Horiba EMGA 821 hydrogen analyzer, which enabled us to obtain the temperature programmed desorption (TPD) spectra. With the linear ramp of temperature the measured desorption rate exhibits a number of peaks which are interpreted as various bound states of hydrogen. For FCC metals the most important are three peaks associated with (i) H bound at the surface; (ii)

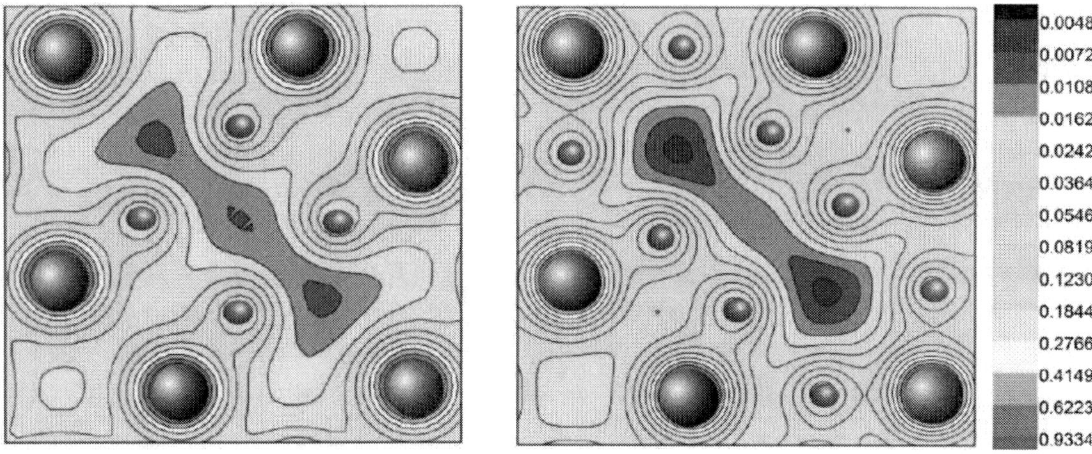

Figure 4. Electron charge distribution in (001) plane of mH-V_2 cluster in Ni with 4 and 8 segregated H atoms in anomalous *(pair)* positions with reduced distance between H atoms. Color panel shows density values in atomic units (1/au = 6.748 e/Å^3). Preferable positions are at the density ≅ 0.22 e/Å^3.

H segregated to single vacancies, and (iii) H segregated to divacancies. These spectra for Ni samples are presented in the Fukaibook [8] (see Fig. 5.64 in [8]). Note that the low-temperature desorption line due to surface hydrogen depends on the initial state of the surface and the heat treatment of the sample and is frequently missing if the sample is kept for some time at room temperature in Argon atmosphere.

Typical experimental desorption spectra for catalytic Ni are shown in Fig. 5 for H and D charged samples. Two characteristic peaks at 200°C and 400°C are attributed to H (D) desorption from the surface and the single vacancy clusters, correspondingly. Their position is close to that reported by Fukai [8] and others. Some shift of the peaks is attributed to a difference in the temperature ramps. The peak positions correlate with binding energies determined in *ab-initio* calculations. The intensity of the peaks in the desorption spectra can account for the concentration of the different clusters. The exact numbers vary from sample to sample. The total concentration of H(D) in the samples is very high, up to 8 at.%. If attributed to clusters with maximal occupation of octahedral sites, it corresponds to the cluster concentration of about 1 at.% (also very high).

These findings were confirmed by a large series of experiments on Ni samples and Ni samples with Li and Al impurities loaded with hydrogen with subsequent thermal desorption. Achieved average H concentration in Ni was ≈8 at.%, which corresponds to ≈1 at.% of the mH–V and mH–V_2 cluster concentration.

We believe that the SAV state (also known as the Fukai Phase for Pd) plays a crucial role in many excess heat experiments. In experiments with Pd electrolytic charging with D [1,3] a long charging and incubation time was found to be necessary to start excess heat production. This time was especially long at 30°C loading, but shorter at 100°C. The SAV state formation by deuterium segregated to vacancies and divacancies in the electrolytic process is a slow process of vacancy diffusion into metal from the surface and creation of divacancies. The diffusion of H isotopes goes much faster and does not add much to the incubation period.

In other cases [14,15] the initial material or the sample preparation technique includes the creation of multiple vacancies that are further stabilized into the SAV state when charged with the hydrogen isotope. For example, in the experiments of Moiser-Boss et al. [15], a Pd film was electro-deposited from PdCl solution at an extremely high electric current density. This process creates a dendritic film structure with vacancies in it. Again, D segregation is a

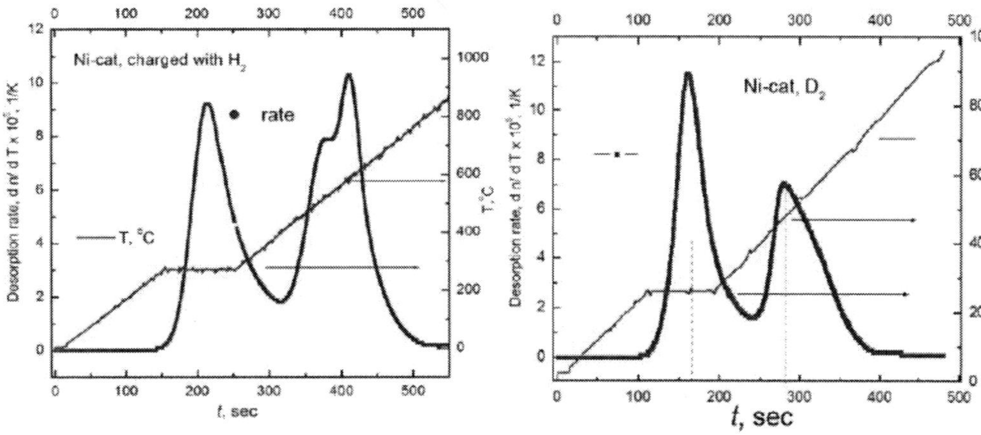

Figure 5. TPD spectra of Ni-catalyst charged with H_2 (*left panel*), to mass concentration 0.13 wt.%; right panel shows TPD spectra for Ni-catalyst sample recharged with D_2. Mass ratio is 0.24 wt.%.

rapid process that does not slow the SAV phase formation. So, no incubation time is required. Similar conditions are fulfilled for catalytic Ni, where due to the production process the initial concentration of vacancies is very high. The crucial role of the Fukai Phase formation for the excess heat generation in electrolytic cell experiments was recently reported by Staker [16].

The formation of the SAV is facilitated by: (i) high-current electrolysis [15], (ii) forced Kirkendall effect [17], (iii) thermal or mechanical treatments of the metal powder, e.g. quenching of metal from the molten state [18] or (iv) by low energy ion (high energy electron) beam bombardment [14] and (v) preprocessing of the metal with plastic deformation as in severe plastic deformation (SPD) or induced by loading-unloading-reloading of hydrogen isotopes with accompanying electromigration current, in addition to the electrolysis current, as detailed by Staker [16]. These techniques should be combined with fast H isotope loading to stabilize SAV phase. Importantly, the presence of SAV phase can be experimentally tested by temperature programmed desorption experiments.

4. Nuclear Reaction Rates

We estimated the cross sections and reactivities for the nuclear reactions between light nuclei in clusters of D segregated to vacancies or divacancies in Ni and Pd [19,20]. We assumed the thermal excitation mechanism (i.e. Maxwell energy distribution of the H isotope atoms in the cluster). The screening model based on the mean field potential of the electron cloud in the metal plasma was employed with the experimental values of the screening potential [21]. For this case the nuclear fusion reaction includes the temperature induced tunneling of ions from the high energy tail of the Maxwell distribution through the Coulomb barrier. Importantly, the effects of screening in the strongly correlated dense plasma are known to be much larger than that predicted by Debye or Thomas–Fermi models [22].

The reactivity of the reaction is defined as $\langle\sigma(v)v\rangle$ where σ is the reaction cross section and v is the relative velocity between the nuclei that will fuse in the collision. For a case of high density plasma (produced in the process of ignition) the reactivity in the cluster will be strongly enhanced due to the screening by the factor F =exp(Γ). The factor was scrupulously studied in a number of papers on cold fusion in astrophysical objects (see [23] for the references). At high temperatures the tunneling occurs at distances much smaller than the screening radius or average distance between the H isotope ions. For this case $\Gamma = U/(k_{\mathrm{B}}T)$. It depends only on the ratio of the value of average screening potential U

and the average particle energy reflected by their temperature (the local screening approximation). The enhancement factor grows exponentially at low temperatures. However, at lower temperatures and metallic densities the turning point of tunneling is shifted to larger distances between the ions, where the screening potential is reduced. Therefore the local screening approximation fails. The exponential increase is modified but still holds true until $\Gamma \leq 170$. Importantly, the corresponding maximal value of the enhancement factor can be really huge $F_{\max} = 10^{74}$. At lowered temperatures (energies) the growth of enhancement is slowed down and is given by [23]

$$F = \exp[\Gamma(1 - c\Gamma^{4/3})]. \tag{1}$$

Here c is a coefficient that can be estimated in a homogeneous dense Coulomb plasma model [19,23], which gave $c = 1.8 \times 10^{-4}$ for Pd and $c = 2.1 \times 10^{-4}$ for H clusters in Ni. We have calculated reactivities for D–D reactions for the deuterium clusters in Pd and Ni, using experimental values of the screening potential [22]: $U_{\mathrm{Pd}} = 800$ eV and $U_{\mathrm{Ni}} = 394$ eV. These reactivities with screening factors calculated using Eq. (1) are presented in log scale in Fig. 6 (solid lines). The reactivities in the local screening approximation, and also without screening are shown by dashed lines. Vertical dashed line indicates the applicability limit of the results. Both the screening effects and the non-local corrections result in the exponentially large variations of the reactivity values. One can further estimate the optimal excitation energy as $k_{\mathrm{B}}T_0 = U/170$, which gives the values $k_{\mathrm{B}}T_0 = 2.317$ eV for Ni, and $k_{\mathrm{B}}T_0 = 4.706$ eV for Pd. From Eq. (1) it follows that the reduced values of screening factors at the optimal energy are $F_{\mathrm{R,Pd}} = \exp(152)$, $F_{\mathrm{R,Ni}} = \exp(147)$. The corresponding reactivities are $\langle\sigma(v)v\rangle_{\mathrm{Ni}} = 1.7 \times 10^{-10}$ cm^3/s and $\langle\sigma(v)v\rangle_{\mathrm{Pd}} = 1.2 \times 10^{-9}$ cm^3/s.

We can estimate the nuclear fusion reaction rates in 6D–V and 10D–V_2 clusters in Pd and Ni, taking into account the density values of segregated H isotopes in monovacancies and divacancies from Table 1. Then, for a single 6D–V cluster in Ni the reaction rate is $R = 3 \times 10^{13}$ 1/s and $R = 4.7 \times 10^{13}$ 1/s for Pd. The advantage of Pd over Ni is not so large because of the higher density of segregated atoms in Ni. The high burning rate only indicates that the limiting stage of the energy production will be the H isotope diffusion and segregation to the vacancies and divacancies. These reaction rates would correspond to a very high energy production for a gram of fuel, far above any chemical sources of energy. The ignition of the nuclear reaction requires the energies in D nuclei at least several eV, which is far above what can be achieved in the thermal heating experiments. Due to the high reaction rates the vacancy clusters with segregated H are most probably the Nuclear Active Environment sites [24] for the LENR.

In the case of reaching the ignition energy, the LENR energy dissipation can be provided by the plasmon emission by the charged product particles along the escape tracks. The rate of the nuclear burning will depend on the H isotope diffusion to the vacancy cluster from the bulk and can be controlled by the ambient H pressure. Therefore, the ignited reaction could be self-sustained.

The huge remaining problem is finding a suitable ignition mechanism, since our calculation ruled out thermal excitation. In the electrolytic cell experiments, inhomogeneities of the surface structure could be a reason for the large fluctuations of local electric field. The mechanism of the energy transfer to the ions in this region remains uncertain. The other candidate is the plasmon mechanism of ignition that could provide a suitable energy range (the surface plasmon energies are $\hbar\omega_{\mathrm{p}} \cong 7$ eV). Available mechanisms of plasmon excitation are also well known (e.g. high energy electron beams [14]). However, the conversion of the plasmon energy into the energy of colliding ions remains problematic.

5. Conclusions

Our theoretical and experimental studies of H isotope clusters in Ni showed that mH–V, mH–V_2, and mH–V–I complexes give high H loading and an extremely high local concentration of H for the SAV metal phase. Large experimental

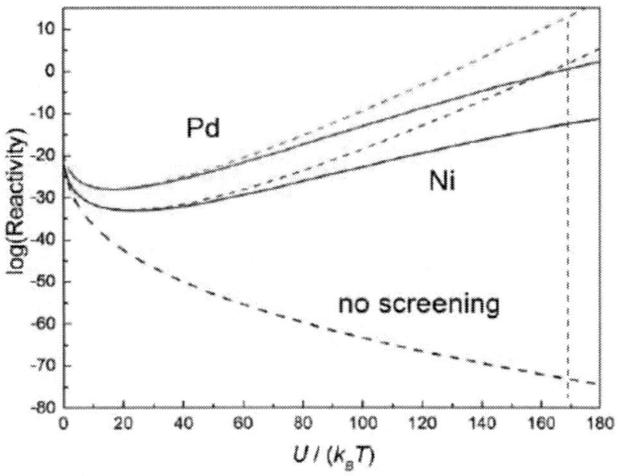

Figure 6. Reactivity for D–D reactions in Pd and Ni (in cm^3/s units, log scale) enhanced by the screening factor $\exp(\Gamma)$, see Eq. (2), as a function of ratio U/k_BT; also shown is the reactivity without screening and reactivities in the local screening approximation (*dashed lines*).

values of the screening potential can be attributed to an anomalously high screening effect at the inner surface of the metal in these clusters. Our estimation of the reactivity in the conditions of optimum excitation that maximizes the screening enhancement showed that the optimal excitation energy (for the thermally enhanced tunneling through the Coulomb barrier) is in the range of 2–5 eV, making LENR ignition more plausible. Therefore, the vacancy clusters with segregated H are most probably the Nuclear Active Environment sites. The H isotope clusters can provide lattice confinement which makes these materials suitable as a LENR fuel. Finally, in our view, LENR can and should be understood and implemented within the boundaries imposed by solid state and nuclear physics.

References

[1] M.C. H.McKubre, F.L. Tanzella, P.L. Hagelstein, K. Mullican and M. Trevithick, The need for triggering in cold fusion reactions, in *10th Int. Conf. on Cold Fusion*, P. L.Hagelstein (Ed.), Condensed Matter Nuclear Science, Word Scientific, USA, 2005. pp. 199–212.

[2] M.R. Swartz, Survey of the observed excess energy and emissions in lattice assisted nuclear reactions, *J. Scientific Exploration* **23** (4) (2009) 419–436.

[3] M.C.H. McKubre, Cold Fusion – CMNS – LENR; past, present and projected future status, *J. Condensed Matter Nucl. Sci.* **19** (2016) 183-191.

[4] E. Storms, An explanation of low-energy nuclear reactions (Cold Fusion), *J. Condensed Matter Nucl. Sci.* **9** (2012) 86–107.

[5] A. Subashiev and H. Nee, Hydrogen trapping at divacancies and impurity–vacancy complexes in nickel: first principles study, *J. Nucl. Materials* **487** (2017) 135–142.

[6] D. Tanguy, Y. Wang and D. Connetable, Stability of vacancy-hydrogen clusters in nickel from the first-principles calculations, *Acta Materialia* **78** (2014) 135–143.

[7] R. Nazarov, T. Hickel and J. Neugebauer, *Ab-initio* study of H-vacancy interactions in FCC metals: implications for the formation of superabundant vacancies, *Phys. Rev.* B **89** (2014) 144108–1–17.

[8] Y. Fukai, *The Metal-hydrogen System: Basic Bulk Properties*, 2nd edition, Springer, Berlin, 2005, pp. 147–158.

[9] J. Zhang, T.S. Fisher, P.V. Ramachandran, J.P. Gore and I. Mudawar, A review of heat transfer issues in hydrogen storage

technologies, *J. Heat Transfer* **127** (2005) 1391–1399.

[10] D.C. Wilson, P.A. Bradley, N.M. Hoffman, F.J. Swenson, D.P. Smitherman, R.E. Chrien, R.W. Margevicius, D.J. Thoma, L.R. Foreman, J.K. Hoffer, S.R. Goldman, S.E. Caldwell, T.R. Dittrich, S.W. Haan, M.M. Marinak, S.M. Pollaine and J.J. Sanchez, The development and advantages of beryllium capsules for the national ignition facility, *Phys. Plasmas* **5** (1998) 1953–1959.

[11] W.J. Nellis, Dynamic compression of materials: metallization of fluid hydrogen at high pressures, *Rep. Prog. Phys.* **69** (2006) 1479–1492.

[12] J.-L. Basdevant, J. Rich and M. Spiro, *Fundamentals of Nuclear Physics*, Springer Science and Business Media, NY, 2004, pp. 229–304.

[13] B. Skinner, M.S. Loth and B.I. Shklovskii, Capacitance of the double layer formed at the metal/ionic-conductor interface: how large can it be? *Phys. Rev. Lett.* **104** (12) (2010) 128302–128306.

[14] M. Swartz and P.L. Hagelstein, Transient vacancy phase states in palladium after high dose-rate electron beam irradiation, *J. Condensed Matter Nucl. Sci.* **14** (2014) 50–60.

[15] P.A. Moiser-Boss, S.R. Chub, M. Fleischman, M.A. Imam, M.H. Miles and S. Szpak, *Thermal and Nuclear Aspects of the Pd/D_2O System*, S. Szpak and P.A. Moiser-Boss (Eds.), Vol. 1, *A decade of Research at Navy Laboratories: SPAWAR Systems Center*, San Diego, US Navy, 2002.

[16] M.R. Staker, Coupled calorimetry and resistivity measurements, in conjunction with an emended and more complete phase diagram of the palladium - isotopic hydrogen system, *J. Condensed Matter Nucl. Sci.* (2019), this volume.

[17] H.J. Fan, M. Knez, R. Sholz et al., Influence of surface diffusion on the void formation induced by the Kirkendall effect: the basic concept, *Nano Lett.* **7** (4) (2007) 993–997.

[18] W. Wycisk and M. Feller-Kniepmeier, Quenching experiments on high-purity nickel, *Phys. Stat. Sol.* (a) **37** (1) (1976) 183–191.

[19] F. Prados-Estévez, A. Subashiev and H. Nee, Nuclear fusion by lattice confinement, *J. Phys. Soc. Jpn.* **86** (7) (2017) 074201–074211.

[20] F. Prados-Estévez, A. Subashiev and H. Nee, Strong screening by lattice confinement and resultant fusion reaction rates in fcc metals, *Nucl. Instrum. Methods Phys. Res.* B **407** (2017) 67–72.

[21] F. Raiola, B. Burchard, Fülöp, Gy. Gyürky, S. Zeng, J. Cruz, A. Di Leva, B. Limata, M. Fonseca, H. Luis, M. Aliotta, H.W. Becker, C. Broggini, A. D'Onofrio, L. Gialanella, G. Imbriani, A.P. Jesus, M. Junker, J.P. Ribeiro, V. Roca, C. Rolfs, M. Romano, E. Somorjai, F. Strieder, and F. Terrasi, Enhanced electron screening in d(d,p)t for deuterated metals, *Eur. Phys. J.* A **19** (2004) 283–287, *ibid.* Enhanced d(d,p)t fusion reaction in metals, **27** (2006) 79–82.

[22] A. Huke, K. Czerski, P. Heide, G. Rupreht, N. Targosz and W. Zebrowski, Enhancement of deuteron-fusion reactions in metals and experimental implications, *Phys. Rev.* C **78** (2008) 015803–1–20.

[23] A.I. Chugunov, H.E. DeWitt and D.G. Yakovlev, Coulomb tunneling for fusion reactions in dense matter: Path integral Monte Carlo versus mean field, *Phys. Rev.* D **76** (2) (2007) 025028.

[24] E. Storms, How basic behavior of LENR can guide. a search for an explanation, *J. Condensed Matter Nucl. Sci.* **20** (2016) 100–120.

J. Condensed Matter Nucl. Sci. 29 (2019) 503–511

Research Article

A Possible Signature of Neutron Quarks – Leptons via Gluon Interaction in Solids

V.G. Plekhanov*

Fonoriton Science Laboratory, Garon Ltd., Lasnamae 22- 3, Tallinn 11413, Estonia

Abstract

The experimental evidence for a macroscopic manifestation of the residual strong interaction in the optical spectra of solids (luminescence and reflection) which differ by term of one neutron from each other (using LiD crystals instead LiH ones) is presented. As far as the gravitation, electromagnetic and weak interactions are the same in both of kind crystals, this only changes the residual strong interaction. Therefore, we conclude that the renormalization of the energy of electromagnetic excitations (electrons, excitons, and phonons) is carried out by the residual strong nuclear interaction. The necessity to take into account the more close relation between quantum chromodynamics and quantum electrodynamics is underlined. In the first step quantum electrodynamics should take into account the residual strong interaction at the description of the dynamics of elementary excitations (electrons, excitons, and phonons) dynamics.

Keywords: Excitons, Leptons, Quantum electrodynamics and chromodynamics, Quarks, Strong interaction

1. Introduction

Recently, the direct dependence of the energy of interband transition E_g in solids (e.g. LiH_xD_{1-x} crystals) on the strong nuclear interaction was demonstrated [1]. The present report is devoted to advance a description of the experimental results demonstrated by the above dependence. According to contemporary physics, the Universe is made up of matter fields, whose quanta are fermions and force fields (whose quanta, in turn, are bosons). Basically, fermions can be classified into two groups: elementary and composite fermions. Elementary fermions are leptons (electron, electron neutrino, muon, muon neutrino, tau, and tau neutrino and quarks (up, down, top, bottom, strange, and charm)). Hadrons (neutrons and protons) containing an odd number of quarks, and nuclei made of an odd number of nucleons (e.g. $^{13}_{6}C$ nuclei contain six protons and seven neutrons) are considered to be composite fermions. Elementary fermions are the fundamental building blocks of matter and antimatter [2,3]. Bosons are identical particles having zero or integer spins. Unlike fermions, bosons do not obey the Pauli Exclusion Principle. The SM [4] only consists of five

*E-mail: vgplekhanov@gmail.com.

elementary bosons (see Fig. 11 in [5]). They are namely the Higgs boson, gluon, Z and $W^{\pm}$ bosons. The Higgs boson have zero electric charge and zero spin, and is the only scalar boson.

The discovery of the neutron by Chadwick in 1932 [2] may be viewed as the birth of the strong interaction: it indicated that the nuclei consists of protons and neutrons and hence the presence of a force that holds them together, strong enough to counteract the electromagnetic repulsion. In 1935, Yukawa [6] pointed out that the nuclear force could be generated by the exchange of a hypothetical spinless particle, provided its mass is intermediate between the masses of proton and electron – a meson. Yukawa predicted the pion [2,7]. The strong forces do not act on leptons (electrons, positrons, muons, and neutrinos), but only on protons and neutrons (more generally, on baryons and mesons – this is the reason for the collective name hadrons). It holds protons and neutrons together to form nuclei, and is insignificant at distances greater than 10^{-15} m ([6] see below). Its macroscopic manifestations have been restricted up to now to radioactivity and the release of nuclear energy [3,4]. Quantum chromodynamics (QCD) is the theory of the strong interaction, responsible for binding quarks through the exchange of gluons to form hadrons (baryons and mesons).

Our present knowledge of physical phenomena suggests that there exist four types of forces between physical bodies (see, e.g. [8,11]): (1) gravitational, (2) electromagnetic, (3) strong, and (4) weak (see, e.g. Table 8 in [5]).

Both the gravitational and the electromagnetic forces vary in strength as the inverse square of the distance and so are able to influence the state of an object even at very large distances. When nuclear physics developed, two new short-range forces joined the ranks. These are the nuclear forces, which act between nucleons (proton, neutron, etc.) and the weak force, which manifest itself in nuclear β-decay (see, e.g. [7). The nuclear force is a result of the residual strong force binding quarks to form protons and neutrons. Subatomic physics deals with all entities smaller than the atom. The modern quantum mechanical view of the three fundamental forces (all except gravity) is that particles of matter (fermions neutrons, protons, and electrons) do not directly interact with each other, but rather carry a charge, and exchange virtual particles (gauge bosons photons, gluons, and gravitons) which are the interaction carriers or force mediators. As can be seen from Table 8 in [5], photons are the mediators of the interaction of electric charges (protons, electrons, and positrons); and gluons are the mediators of the interaction of color charges (quarks). In the present day, the accepted view is that all matters are made of quarks and leptons. As can be seen, of the three pairs of quarks and leptons, one pair of each – the quark u and d and the leptons e^- and ν^e (electrons neutrino) – are necessary to make up the everyday world, and a world which contained only these would seem to be quite possible.

These facts, and a summary of modern nuclear and subatomic physics (see, e.g. [2,7]) allow us to draw several conclusions in regard to nuclear forces, most notably that the binding energy of a nucleus is proportional to the number of nucleons and that the density of nuclear matter is approximately constant. This leads to the conclusion that nuclear forces have a "saturation property" [2,7]. It seems from the last conclusion that it is enough to change the number of neutrons in the nucleus to change the strength of the nuclear force. This constitutes the main ideas of the isotope effect [8].

2. Experimental

The apparatus used in our experiments has been described in several previous publications [9–11]. For clarity, we should mention here that a home-made immersion helium cryostat and two identical double-prism monochromators were used. One monochromator was used for the excitation, and the other, which was placed at right-angle to the first, for analyzing the luminescence and scattering of light. In our experiments we investigated two kinds of crystals (LiH and LiD) which differ by one neutron. The single crystals of LiH and LiD were grown from the melt by a modified Bridgeman–Stockbarger method (see [18,5] and references quoted therein). The crystals were synthesized from 7Li metal and hydrogen of 99.7% purity and deuterium of 99.5% purity. Virgin crystals had a slightly blue–green color, which can be attributed to a nonstoichiometric excess of lithium present during the grown cycle. On annealing for several days (up to 20) at 500°C under $\sim$3 atm of hydrogen or deuterium, this color was almost completely eliminated.

Because of the high reactivity and high hygroscopy of the investigated crystals an efficient protection against the atmosphere was necessary. Taking into account this circumstance, we have developed special equipment which allows us to prepare samples with a clean surface cleaving them in a bath of helium cryostat with normal or superfluid liquid helium [5].

3. Results

We should briefly review electronic excitations in solids. According to the modern concept, the excitons can be considered [12] as the excited N-particles system: An electron from the valence band (see Fig. 1 in [11]) is excited into the conduction band. The attractive Coulomb potential between the missing electron in the valence band, which can be regarded as a positively charged hole, and the electron in the conduction band gives a hydrogen-like spectrum with an infinite number of bound state and ionization continuum (Fig. 71 in [13]). Below we will briefly describe the results of the optical spectroscopy of isotope-mixed solids. In our experiments we have investigated the low-temperature optical spectra (Fig. 1 reflection and Fig. 2 luminescence) of LiH_xD_{1-x} crystals ($0 \leq x \leq 1$) which differ by one neutron from each other. The mirror reflection spectra of mixed and pure LiH and LiD crystals cleaved in liquid helium are presented in Fig. 1. All spectra have been measured with the same apparatus under the same conditions. As the deuterium concentration increases, the long-wave maximum (n = 1S excitons [12]) broadens and shifts towards the shorter wavelengths.

As demonstrated early (see, e.g. review [13]) most low-energy electron excitation in LiH crystals are the large-radius excitons [12]. Exciton luminescence is observed when LiH (LiD) crystals are excited in the midst of fundamental absorption. The spectrum of exciton photoluminescence of LiH crystals cleaved in liquid (superfluid) helium consists of a narrow (in the best crystals, its half-width is $\Delta E \leq 10$ meV) phononless emission line and its broader phonon repetitions, which arise due to radiated annihilation of excitons with the production of one to five longitudinal optical (LO) phonons (see Fig. 2).

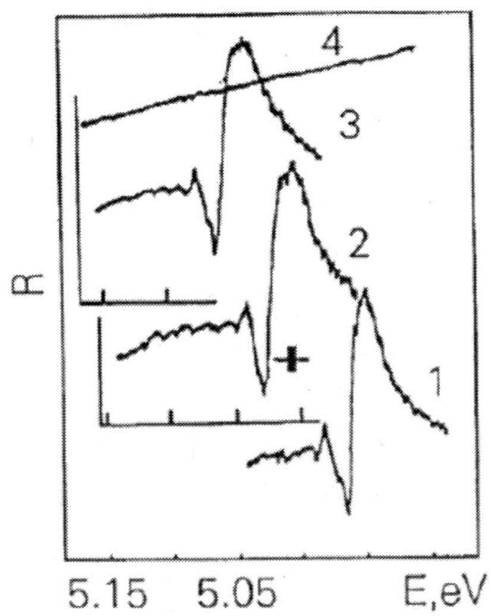

Figure 1. Mirror reflection spectra of crystals: (1) LiH, (2) LiH_xD_{1-x}, (3) LiD at 2 K, and (4) light source without crystals.

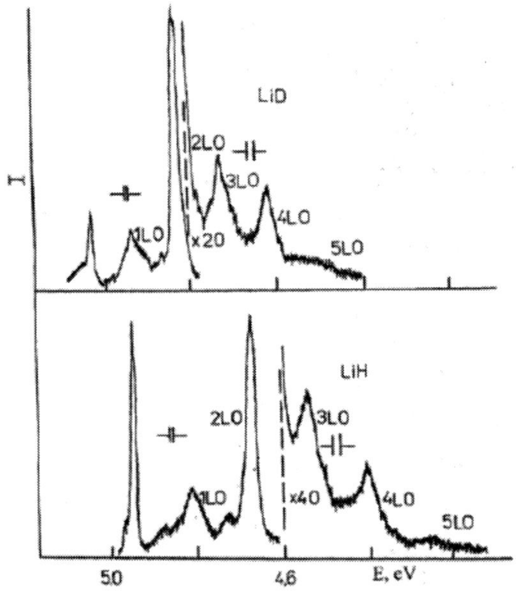

Figure 2. Photoluminescence spectra of free excitons at 2 K in LiH and LiD crystals cleaved in superfluid helium.

The phononless emission line coincides in an almost resonant way with the reflection line of the exciton ground state, which is indication of the direct electron transition X1–X4 of the first Brillouin zone [14]. The lines of phonon replicas form an equidistant series biased toward lower energies from the resonance emission line of excitons.

The energy difference between these lines in LiH crystals is about 140 meV, which is very close to the calculated energy of the LO phonon in the middle of the Brillouin zone [15,16] and which has been measured (see, e.g. [17,18] and references quoted therein). The isotopic shift of the zero-phonon emission line of LiH crystals equals 103 meV. As we can see from Fig. 2 the photoluminescence spectrum of LiD crystals is largely similar to the spectrum of intrinsic luminescence of LiH crystals. There are, however, some related distinctions. First, the zero-phonon emission line of free excitons in LiD crystals shifts to the short-wavelength side on 103 meV. These results directly show the violation of the strong conclusion (see, e.g. [2,7]) that the strong force does not act on leptons. The second difference concludes in less value of the LO phonon energy, which is equal to 104 meV. The simplest approximation, in which crystals of mixed isotopic composition are treated as crystals of identical atoms having the average isotopic mass, is referred to as virtual crystal approximation (VCA) [19]. Going beyond the VCA, in isotopically mixed crystals one would expect local fluctuations in local isotopic composition within some effective volume, such as that of an exciton. As follows from Fig. 1, excitons in LiH_xD_{1-x} crystals display a unimodal character, which facilitates the interpretation of their concentration dependence. Figure 3 shows the concentration dependence of the power of strong nuclear interaction, i.e. dependence on the neutron concentration. In the first approximation the mechanism of isotope shift will be connect with the neutron magnetic field of the deuterium nucleus (neutron) (to be published separately). As can be seen from Fig. 3, the VCA method cannot describe the observed experimental results. As was shown early [18] this deviation from linear low (VCA approximation) is connected with isotope-induced disorder in isotope mixed crystals LiH_xD_{1-x}. According to Lifshitz [20] the isotopic disordering ought to be classified as site disordering of the crystal lattice.

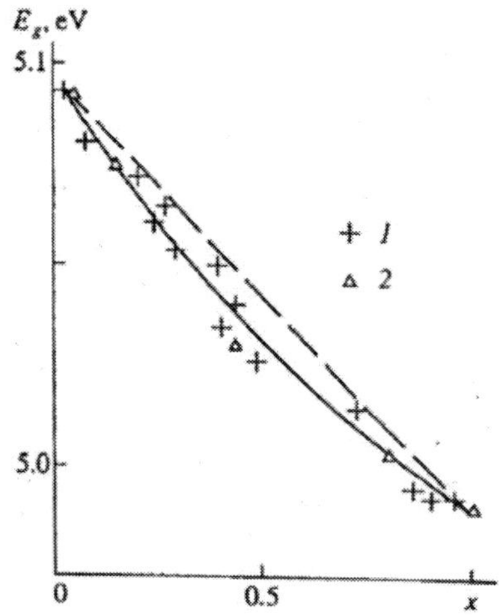

Figure 3. Dependence of the interband transition (strong interaction) energy E_g in mixed crystals on the concentration x (number of neutrons N). The straight dashed line is the linear dependence of coupling constant strong interaction $\alpha_s = f(N)(E_g = f(x))$ in the virtual model. The solid line corresponds to the calculation using the polynomial of second degree $E_g = E_g(\text{LiD}) + \{E_g(\text{LiH}) - E_g(\text{LiD}) - b\}x - bx^2$, where $b = 0.046$ eV is curvature parameter [18]. Points derived from the reflection spectra indicated by crosses, and those from luminescence spectra by triangles.

A comparison of the experimental results with luminescence and reflection in the crystals which differ by one neutron allows the following conclusions:

(1) Adding one neutron (using LiD crystals instead LiH ones) is caused in the increase of exciton energy to 103 meV.
(2) At the addition of one neutron, the energy of LO phonons is decreased on the 36 meV, that is directly seen from luminescence and scattering spectra.

Both characteristics are macroscopic. These experimental results can open an avenue for new nuclear physics. Moreover, our results demonstrate very important information that high energy physics could obtain via the rather simple and inexpensive equipment of experimental physics.

A comparison of the dependence of the residual strong nuclear force in different substances shows that the residual strong nuclear force has a nonlinear character on the number of neutrons (Fig. 4). We should stress that in all experimental results we have, the case where the perturbation theory works very well – the isotope shift 0.103 eV is much smaller than interband transition energy E_g and more less than nuclear energy (results which will be published separately).

4. Discussion

The nucleus is the central part of an atom consisting of A – nucleons, Z – protons, and N – neutrons. The atomic mass of the nucleus is equal to $Z + N$(see Fig. 1 in [5]). A given element can have many different isotopes, which differ from one another by the number of neutrons contained in the nuclei [2,7,21]. Modern physics distinguishes three fundamental properties of atomic nuclei: mass, spin (and related magnetic moment), and volume (surrounding field strength) which are the sources of the isotopic effect (see, also [18]).

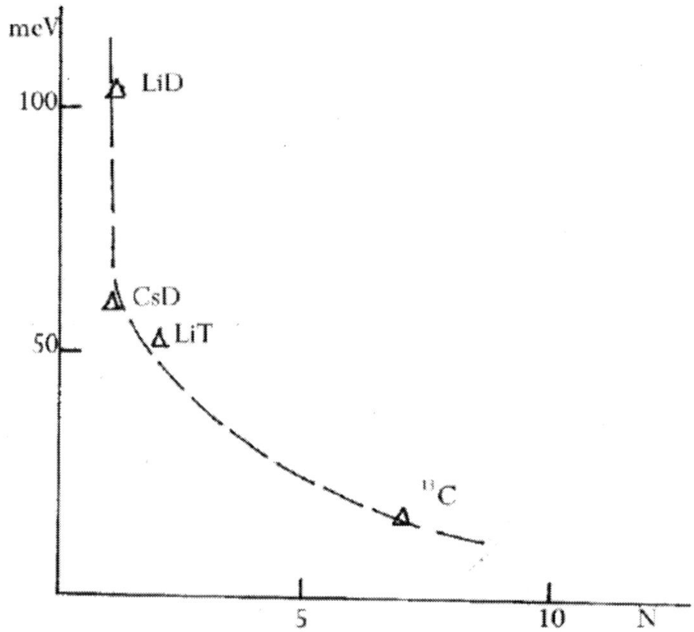

Figure 4. The dependence of the strong force on the number of neutrons in different substances.

LiH (LiD) crystals with a lattice of NaCl type, whose parameters are close to cubic crystals, are dielectrics with a band gap of $E_g = 4.992$ eV ($E_g = 5.090$ eV for LiD) at 2 K [22]. These crystals have an identical electronic structure. The energy band structure of these substances is also identical [13,23]. All three kinds of forces – gravitational, electromagnetic, and weak are also the same for the above compounds. The difference between these substances consists out at one neutron in the nucleus of deuteron. Below we should briefly consider some peculiarities of the physics of deuterons. Nucleons can combine to make four different few-nucleon systems, the deuteron (p + n), the triton (p + 2n), the helion (2p + n), and α-particle (2p + 2n). These particles are grouped together because they are all stable (apart from the triton which has a half-life of about 12 years and so may be treated as a stable entity for most practical purposes). They have no bound excited states (except the α-particle which has two excited states at about 20 and 22 MeV [7]), and they are frequently used as projectile in nuclear reactions. Few-nucleon systems provide the simplest systems to study nuclear structure (see, e.g. [24]). The deuteron provides important information about the nucleon –nucleon interaction. As was noted, the deuteron consists of a proton and a neutron and is the only bound state of two nucleons. Its binding energy is 2.2245 MeV and its total angular momentum J and parity are 1^+ [7]. Since the intrinsic parities of the neutron and the proton are positive parity of the deuteron implies that the relative orbital angular momentum of the neutron and the proton must be even. If the orbital angular momentum L is a good quantum number, states with lower orbital angular momentum generally have lower energy than states with higher angular momentum, and so we expect the ground state of the deuteron to have orbital angular momentum $L = 0$, so that it is in an S state. Then, if the spins of the proton and the neutron in the deuteron are parallel, we expect the magnetic moment of the deuteron to be approximately the sum of the magnetic moments of the proton and neutron, namely $\mu_p + \mu_n = (2.793 - 1.913)\ \mu_N = 0.880\ \mu_N$ ($\mu_N = e\ \hbar/2m_p$) [7]. If, however, the spins are anti-parallel , we expect it to be $(2.793 + 1.913)\mu_N = 4.706\ \mu_N$. Experimentally it is $0.857\ \mu_N$ [7,25] so the spins of the proton and neutron are parallel and so the total spin S of the deuteron is one, since $J = L + S$, $J = 1$. The small but definite difference between $\mu_d = 0.857\ \mu_N$ and $\mu_p + \mu_n = 0.880\ \mu_N$ is due, as will show below, to tensor character of strong forces in deuteron. We thus conclude that the ground state of deuteron is a triplet S state. However, this cannot be the whole story because S states are spherically symmetrical and thus have no quadrupole moment. This is contradicted by experiments. Experimentally the deuteron has a positive quadrupole moment of 0.29 fm^2 [25]. The deviation of the actual deuterium moment from the S state moment can be explained if it is assumed that the deuteron ground state is a superposition of S and D states. Part of the time, the deuteron has orbital angular momentum $L = 2$. Independent evidence for this fact comes from the observation that, as was shown above, the deuteron has a small, but finite, quadrupole moment (see, also [26]). As is well-known, the electric quadrupole moment measures the deviation of a charge distribution from sphericity [8]. The quadrupole moment of a disk shaped (oblate) nucleus Q is negative. A positive quadrupole moment of $Q = 0.29$ fm^2 according experiment indicates that the deuteron is slightly elongated on the z-axis, like an olive (prolate). Quantum mechanical definition of quadrupole moment for a single proton [25] is described by:

$$eQ = e\int (3z^2 - r^2)\,\mathrm{d}t. \tag{1}$$

Thus, if the quadrupole moment is not equal to zero then the eigenfunction of the ground state of the deuteron assigns a probability of 0.04 to finding a ^{3}D1 state and a probability of a 0.96 to finding a ^{3}S1 state. The last one points to the tensor character of the nucleon–nucleon interaction (for more details see, e.g. [2,7]). Nuclear magnetic dipole and electric quadrupole have a similar importance in helping us to interpret the deuteron structure (see, also [23]).

The motion of the electrons produces a magnetic field B_I at the nucleus, which interacts with the electron magnetic moment μ (see, e.g. [27]):

$$E = \mu \cdot B_I. \tag{2}$$

Typical energy differences of hyperfine multiplets are only about 10^{-7} eV (in case of the deuteron it is 3.16×10^{-7} eV (see also [28])). This value is more than seven orders less than we observe in experiments: the isotopic shift of the n = 1s excitons is equal to 0.103 eV.

The short range character of the strong interaction does not possess a direct mechanism for elementary excitation energy renormalization, which was observed in the experiments. Nevertheless, our results were very close of the isotope shift exciton energy in the case $^{12}C_x\ ^{13}C_{1-x}$ diamond crystals to the indicated value above in LiH crystals. Indeed, in such experiments we have isotope shift in $^{12}C_x\ ^{13}C_{1-x}$ diamond crystals approximately 15 meV (see, e.g. [8] and references therein) per one neutron and on seven neutrons we get $15 \cdot 7 = 105$ meV. This value is very close to the observed one (103 meV) in LiH crystals.

However, at present time, we can distinguish the following mechanisms of this renormalization:

(1) Electric field of the neutron's quarks – this mechanism is limited by the boundary of the neutron.
(2) The possible new structure of the quarks and leptons – the so-called preons [29–34].
(3) The most likely mechanism is connected to the magnetic-like field of the neutron quarks.

For the solution of this new task we need more experimental as well as theoretical investigations.

We should underline that the experimental observation of the manifestation of residual strong nuclear interaction in the optical spectra of solids opens an avenue to nuclear and elementary particle physics.

5. Conclusion

The artificial activation of the strong nuclear interaction by adding one (or two, or more) neutrons to atomic nuclei leads it to the direct observation of the strong interaction in low-temperature optical spectra of solids. This conclusion opens a new avenue in the investigation of the strong nuclear interaction by means the condensed matter alike traditional methods (including accelerating technique). Experimental observation of the renormalization of the elementary excitation energy of solids by the strong nuclear interaction stimulates its count in the process of describing the elementary excitations dynamics in quantum electrodynamics. The present article continues to develop the connection between nuclear and condensed matter physics.

Acknowledgements

I thank Profs. J.G. Buitrago for very enlightening discussion and Professor Jed Rothwell and G.A. Plehhanov for improving my English.

References

[1] V.G. Plekhanov, Deposit in VINITI, N202– B2012 (in Russian).
[2] B. Povh, K. Rith, Ch. Scholz and F. Zetsche, *Particles and Nuclei,* Springer, Berlin, 2006.
[3] D.H. Perkins, *Introduction to High Energy Physics*, Cambridge University Press, Cambridge, 2000.
[4] W.N. Cottingham and D.A. Greenwood, *An Introduction to the Standard Model of Particle Physics,* Cambridge University Press, Cambridge, 2007.
[5] V.G. Plekhanov, *Isotope Effect – Macroscopical Manifestation of Strong Interaction,* LAMBERT Academic Publishing, Saarbrücken, Germany, 2017 (in Russian).
[6] Y. Yukawa, *Proc. Phys. Math. Soc.* Jap. **17** (1935) 48.
[7] E.M. Henley and A. Garcia, *Subatomic Physics,* World Scientific, Singapore, 2007.
[8] V.G. Plekhanov, *Isotopes in Condensed Matter,* Springer, Heidelberg, 2013.
[9] V.G. Plekhanov, *Phys. Rev.* B **53** (1996) 9558.
[10] V.G. Plekhanov, *Prog. Mat. Sci.* **51** (2006) 287.
[11] V.G. Plekhanov, *J. Astrophys. and Aerospace Technol.* **5** (2017) 46.
[12] R.S. Knox, *Theory of Excitons,* Academic Press, New York, London, 1963.
[13] V.G. Plekhanov, *Phys. Rep.* **410** (2005) 1.
[14] S. Baroni, G. Pastori Parravicini and G. Pezzica, *Phys. Rev.* B **32** (1985) 4077.
[15] J.L. Verble, J.L. Warren and J.L. Yarnell, *ibid* **168** (1968) 980.
[16] H. Dammak, E. Antoshchenkova, M. Hayoum and F. Finocchi, *J. Phys.: Condensed Matter* **24** (2012) 435402.
[17] V.G. Plekhanov and A.A. O'Konnel, Bronin (Moscow) *JETP Lett.* **27** (1978) 413.
[18] V.G. Plekhanov, *Giant Isotope Effect in Solids,* Stefan University Press, La Jola, CA, 2004.
[19] I. Nordheim, *Ann. Phys.* (Leipzig) **401** (1931) 641.
[20] I.M. Lifshitz, *Selected Works,* Science, Moscow, 1987 (in Russian).
[21] F.E. Close, *An Introduction to Quarks and Partons,* Academic Press, London, New York, 1979.
[22] V.G. Plekhanov, *Phys.* – Uspekhi **40** (1997) 553.
[23] V.G. Plekhanov, *Modern View of the Origin of Isotope Effect,* LAP, LAMBERT Academic Publishing, Germany, 2018.
[24] J. Carlson and R. Schivalia, *Rev. Mod. Phys.* **70** (1998) 743.
[25] A.R. Edmonds, *Angular Momentum in Quantum Mechanics,* Princeton University Press, Princeton, New Jersey, 1957.
[26] H. Kopferman, *Kernomomente,* Acedemische, Verl - Ges., 1956.
[27] V.G. Plekhanov, *Phys. Sci. Intern. J.* **18** (2018) 1.
[28] V.G. Plekhanov, in *Proc. ISINN*, Dubna, Russia, 2018, p. 49.
[29] O.W. Greenberg, A new level of structure, *Phys. Today* 38 (1985) 22–30.

[30] B. Robson, *The Generation Model of Particle Physics*, in *Particle Physics*, E. Kennedy (Ed.), In Tech. Open Access Publisher, Rijeka, Croatia, 2012.
[31] M. Aghirescu, *The Cold Genesis of Matter and Fields*, Science Publishing Group, New York, 2015.
[32] C.S. Kalman and I. D'Souza, *Preons: Models of Leptons, Quarks and Gauge Bosons as Composite Particles,* World Scientific, Singapore, 1992.
[33] C.S. Kalman, Why quarks cannot be fundamental particles, *Nuclear Phys.* B: *Proceedings Supplements* **142** (2005) 235–237.
[34] O. Consa, The helicon: a new preon model, *Progr. Phys.* **14** (2019) 215–219.

J. Condensed Matter Nucl. Sci. 29 (2019) 512–524

Research Article

Transmutations Involving the Di-neutron in Condensed Matter

Cheryl D. Stevenson*

Department of Chemistry, Illinois State University, Normal, IL 61790-4160, USA

John P. Davis

PragmaChem LLC, 222 Prospect Place, Danville, IL 61832-1863, USA

Abstract

It has been recently revealed how a lattice bound proton, in an electrochemical cathode, can absorb a relativistic electron yielding a neutron. There is no longer much doubt that the analogous electron capture by a deuteron, in condensed matter, ($e^-_{real} + {}^1p_{\uparrow 0}n_\uparrow \rightarrow {}_0n_{\downarrow 0}n_\uparrow + \nu_e$) is even more efficient, where the arrows indicate nucleon spins. The life time of the newly formed di-neutron may be vanishingly short, but in the palladium deuteride matrix, we show that it is certainly sufficiently long to undergo transmutation with constituents of the matrix as well as added "impurities" to yield both isotopic and isotonic products: i.e. (${}_0n_{\downarrow 0}n_\uparrow + {}^{106}Pd \rightarrow {}^{108}Pd$). When the agent capturing the di-neutron is ^1H, tritium is the result. When it is ^2H, ^4He and heat are the products (the Fleishmann Pons Heat Effect). Consistent with the tendency of di-neutrons to cluster, multiple di-neutrons are occasionally captured yielding neutron heavy nuclei. This is followed by beta decay and the respective isotonic products. We show that, within the PdD matrix, these di-neutron reactions involve phonon enforced quantum tunneling. The di-neutron-phonon mechanism fully accounts for the following: (1) the evolved heat commensurate with ^4He production, (2) the production of tritium when small amounts of ^1H are present, (3) the release of feeble neutrons, (4) the formation of Pr when Cs is present, (5) the ostensive intermittent reproducibility and erratic (chaotic) heat out-puts, and (6) the seemingly major problem of obfuscating the Coulombic repulsion issue. Indeed, the antics of the di-neutron can fully explain the important aspects of normal temperature LENR heat effects and transmutations. Further, the di-neutron paradigm (mechanism) falls within the normal bounds of the Standard Model, and it is without the necessity of exotic forms of matter.

Keywords: Chaos, Di-neutron, Electron Capture, ^3H, ^{4m}H, Palladium-128, Transmutation

1. Introduction

Although not the most probable path to the lightest heavy isotope, it took only about three minutes after the Big Bang for proton–neutron collisions to begin forming deuterons ($H^+ + {}_0n \rightarrow {}^2H^+ + \gamma$). The first atoms, to be formed in our universe, resulted from the capture of electrons by protons in free space (FS) to yield atomic hydrogen. That action

*E-mail: cdsteve@ilstu.edu.

© 2019 ISCMNS. All rights reserved. ISSN 2227-3123

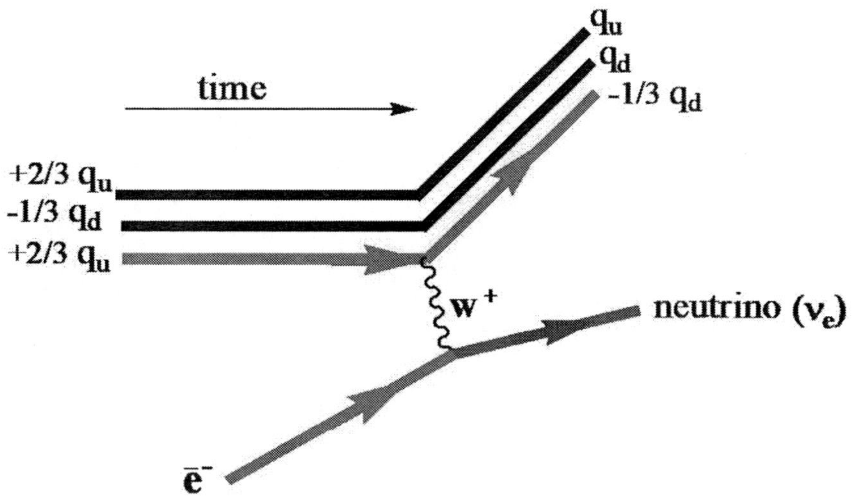

Figure 1. The Feynman diagram for the capture of a relativistic electron by an up quark which, in turn, becomes a down quark. In this process, a very short lived W^+ boson is emitted by the up quark that annihilates the electron to yield an electron neutrino.

originated about 380,000 years later. This seemingly simple process (reaction 1) is actually more fascinating than first meets the eye. If the electron comes in "heavy" (with relativistic energy: $E = (pc)^2 + (m_0c^2)^2$), it can bore in closer to the nucleus than the Bohr radius of 529 pm. In this range, the wave function for the relativistic electron ($\bar{e}^-$) can overlap with weak force fields carried by the W and Z^0 bosons. The W boson interaction results in the merging of the electron (e^-) with the proton and consequent formation of a neutron ($_0n$) and a neutrino. Since the weak force is involved, reaction 2 is best illustrated by a Feynman diagram, Fig. 1.

$$^1H^+ + e^- \rightarrow H^{\cdot}, \tag{1}$$

$$^1H^+ + \bar{e}^- \rightarrow {}_0n + \nu_e. \tag{2}$$

Interestingly, if the incoming electron interacts with the Z^0 weak force boson, it is repelled by the proton ($^1p = {}^1H^+$) and is scattered off via the weak hypercharge [1]. Since the weak force does recognize parity, the scattering is different depending upon how the electrons are polarized (Fig. 2) [1]. The weak force is the fundamental force that recognizes parity.

The mutual capture of the protons and neutrons (reaction 3) gives rise to isotopes, and as indicated the deuteron is a spin triplet. At first, this seems contradictory to the Fermi–Dirac rules of degeneracy pressure: no two fermions, comprising a single entity, can be in the same quantum state. The strong nuclear force, which binds the two baryons in

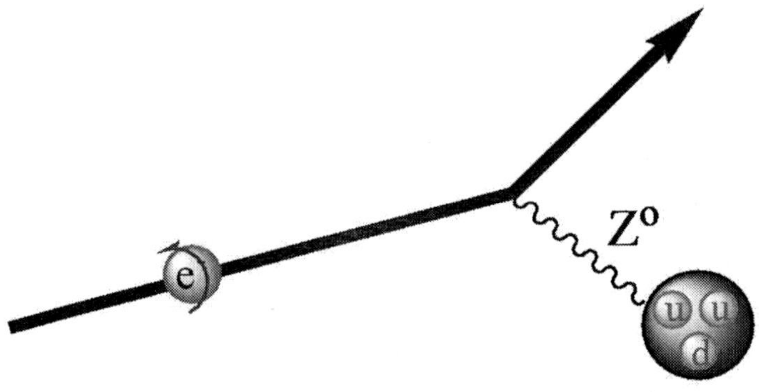

Figure 2. A cartoonish diagram of the scattering of a left-handed electron by a proton (two up quarks and one down quark). The scattering is due to weak neutral currents (the Z^0 boson). Parity violation in this weak interaction yields a scattering asymmetry (different scattering for left and right-handed electrons). The measurement is carried out by firing polarized beams of electrons through an unpolarized liquid H_2 target. The proton's weak charge has a value of 0.0719 ± 0.0045 [1].

reaction 3, cannot feel charge differences or the very small mass difference between ${}_0n$ and ${}^1p$. Consequently, another quantum Eigen value is required to explain the deuteron. Isospin was originally invented by Werner Heisenberg to distinguish between the proton and neutron, and explain the stability of the deuteron (${}^1p_0n$), which is an isospin singlet. The detailed mathematical model for isospin was worked out five years later, in 1937, by Wigner [2].

$$ {}^1H^+ + {}_0n \rightarrow {}^1p_{\uparrow 0}n_{\uparrow}. \quad (3)$$

The two fermions occupying the helium-2 nucleus (the di-proton (${}^1p^1p$)) must be also anti-symmetric under exchange and obey Fermi–Dirac statistics. It would clearly be an isospin triplet and also suffers from Coulombic repulsion pressure. For these reasons, the nucleus of ${}^2He$ has never been detected. The remaining di-nucleon is the di-neutron (${}_0n_0n$) [3], and it has major chemical significance in the condensed phase. Writing in *Chem. Rev.* [4], Pekka Pyykkö claimed that a bound di-neutron (${}_0n_0n$) would be "the ultimate noble-gas molecule." He also correctly predicted the effects of relativistic electrons in gold, cesium, and other heavy elements [5,6]. Unlike ${}^1p^1p$, ${}_0n_0n$ would exist in the absence of Coulombic repulsion.

Forty years prior to Pyykkö's review article [4], calculations carried out by Migdal [7,8] suggested that the di-neutron is, however, unbound, but these early calculations did not the halt the empirical search for the bound di-neutron [9]. Interest in finding the ${}_0n_0n$ was further discouraged just five years later when effective range considerations again suggested an unbound state [10]. Hence, isospin and other theoretical considerations seem to favor an unbound free space di-neutron (FS_{nn}). On the other hand, Migdal suggested that the di-neutron could exist in the vicinity of a neutron heavy nucleus. Consistent with Migdal's predictions [11,12], di-neutrons were shown to exist as semi-isolated entities in neutron heavy exotic nuclei like ${}^5H$, ${}^6H$, and ${}^8He$ [13]. Disbelief in the bound ${}_0n_0n$ grew even as confirming empirical observations were periodically reported. For example, in 1993, Bertulani et al. found that the energy needed to remove a pair of neutrons from ^{11}Li is 200 keV, while that needed to remove a single neutron is 1 MeV [14]! A possible empirical suggestion of $_0$n$_0$n first appeared in the same year that the Pyykkö paper (2012) appeared. At the National Super conducting Cyclotron laboratory, ^{16}Be was generated via beam–target collisions. The ^{16}Be was found to immediately decay by di-neutron emission to ^{14}Be, and the two neutrons "fly off" as a pair [15,16]. If isospin were to have caused the sputtering apart of the two neutrons, they should have flown away in opposite directions. The authors of the ^{16}Be decay paper state: "The di-neutron character of the decay is evidenced by a small emission angle between the two neutrons." However, it should be mentioned that this paper has been discussed in the literature over 60 times, and the vast majority of these articles conclude, via theoretical arguments only, that the two neutrons observed in ref. 15 were not bound. On the other end of the argument, it has been suggested that the existence of a bound FS_{nn} is consistent with Big Bang nucleosynthesis [17]. In 2014, Hammer [3] used Lagrangian calculations to resurrect some theoretical possibility of a $^1S_0$ free space di-neutron (FS_{nn}) [3]; in this paper [3] he states that $_0n_\downarrow {}_0n_\uparrow$ " cannot be excluded to next-to-leading order in pionless EFT." Hammer's statement seems in contrast to the vast majority of quantum mechanical conclusions concerning the FS_{nn}, but his work is too important to ignore.

Witała and Glöckle incorrectly stated that even a slightly bound di-neutron might solve some open problems involving break-up reactions (e.g. reaction 4). Their computations turned out to be incorrect and their reaction will probably remain only a hypothetical "Gedanken-reaction" in free space [18–20]. However, we will provide evidence that an analogous "ship in a bottle" type synthesis (reaction 5) really results in a $^1S_0$ condensed phase di-neutron with a life time sufficiently long to allow its involvement in transmutation and "Fleishman Pons Heat Effect" reactions.

$$_0\mathrm{n} + {}^1\mathrm{p}_\uparrow {}_0\mathrm{n}_\uparrow \rightarrow {}_0\mathrm{n}_\downarrow {}_0\mathrm{n}_\uparrow + {}^1\mathrm{p}, \tag{4}$$

$$\overline{\mathrm{e}}^- + {}^1\mathrm{p}_\uparrow {}_0\mathrm{n}_\uparrow \rightarrow {}_0\mathrm{n}_\downarrow {}_0\mathrm{n}_\uparrow + \nu_\mathrm{e}. \tag{5}$$

History, astonishingly, reveals a multitude of speculations and calculations on even more complicated assemblies of neutrons: clusters of di-neutrons [21]. A number of empirical observations [21,22] of the breakup of ^{14}Be suggest that a multi-neutron cluster is liberated. Most probably it arises from the reaction: $^{14}\mathrm{Be} \rightarrow {}^{10}\mathrm{Be} + ({}_0\mathrm{n}_0\mathrm{n})_2$ [21]. According to Bertulani (the winner of a number of awards in nuclear physics including the John Simon Guggenheim Memorial Foundation Fellowship), the tetra-neutron can exist as a kind of "di-neutron–di-neutron molecule," in which the Hamiltonian would be:

$$H = -(\hbar^2/2m_{0\mathrm{n}}) \sum_{i=1}^{i=4} \Delta_i + V,$$

where V is the neutron–neutron potential [21]. Confinement in condensed matter would hinder escape and render $_0$n$_0$n–$_0$n$_0$n, as a resonant state, much more tenable than it would be in free space. The di-neutron cluster is not essential to LENR, but the formation of at least a transient associated pair of neutrons, in condensed matter, is.

2. Results and Discussion

The electron has a gyromagnetic ratio that is more than two orders of magnitude greater than that of the proton. Hence, deuteron electron capture (reaction 5) results in symmetry breaking and consequent flipping (scrambling) of the resulting neutron spin. When the two spins land in the singlet state, the di-neutron is trapped intact ("ship in a bottle"), and is trapped in the absence of degeneracy pressure as a $^1S_0$ condensed matter entity (Fig. 3).

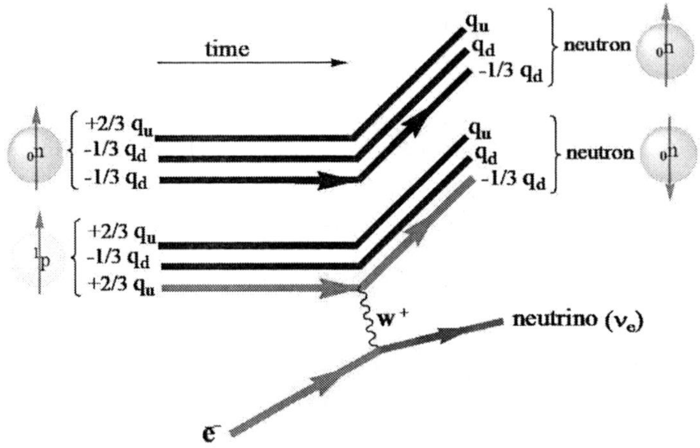

Figure 3. A Feynman diagram for the capture of a relativistic electron by a proton (two up quarks and one down quark) and decaying to a neutron (two down quarks and one up quark) and an electron neutrino. The breaking of spin (purple arrow) symmetry by the proton spin flip allows the newly formed neutron to become one member of a ($n_\downarrow n_\uparrow$) of singlet spin in condensed matter.

Although a free space reaction 4 is not reasonable [18,19], a viable reaction that is stoichiometrically similar, reaction 5 appears to take place in condensed matter. Widom and Larson have provided a mechanism that is free of Coulombic repulsion, for an electro-chemical electron capture processes ($e^- + {}_zA \rightarrow {}_{z-1}A + \nu_e$) in condensed matter [23]. This yields low momentum neutrons when ${}_zA = {}^1H$. Due to the large mass energy ($E = m_0c^2$) of the muon (μ^-), the analogous reaction involving the muon (μ^- in reaction 6) occurs spontaneously after simply mixing muons into a hydrogen system [24]. Reaction 6 continuously takes place in Earth's upper atmosphere. However in contrast, shooting muons through a thin film into deuterium produces a pair of non-interacting neutrons, reaction 7 [25], and these empirical results are consistent with a calculated (LO perturbation theory) negative scattering length (a_{nn}) of $-22.9\pm$ 4.1 fm [26]. This is evidence for the non-existence of a bound FS_{nn}. Also, Using a neutron beam, ${}_0n$ capture breakup experiments produced results calculation-ably compatible with an a_{nn} value of $-18.7\pm$ 0.7 fm for the ${}^1S_0$ FS_{nn} described in reaction 4 [27].

$$\mu^- + {}^1p \rightarrow {}_0n + \nu_\mu, \tag{6}$$

$$\mu^- + {}^1p_\uparrow{}_0n_\uparrow \rightarrow {}_0n + {}_0n + \nu_\mu. \tag{7}$$

The muon is much heavier than is the electron, however, in condensed matter, the electron energy can be augmented by local electromagnetic field fluctuations due to vibrating electric and magnetic multi-pole moments [23,28,29]. Such fluctuations ("condensed matter Zitterbewegung") [29] add velocities to the electrons (e.g. emitted from the anode) in directions other then x translation. They also can lead to greatly enhanced quantum tunneling via phonon interaction [30]. Writing in Science, Ben Powell (in a perspective of Hassan et al. [31]) explains how low energy excitations in condensed-matter can be very different than the electrons, neutrons, and protons that make up the material because the periodic array of atoms breaks translational and rotational symmetries. Consequently, "heavy" relativistic electrons, when injected in an appropriate lattice, can be absorbed by lattice deuterons resulting in viable low momentum di-neutrons, as described in Fig. 3. To find the consequences of such di-neutrons, we should look for anomalous isotopic augmentations ($e^- + {}^2H^+ + {}^mA \rightarrow {}^{m+2}A$) observed after electron injection into lattice material containing deuterons. Actually, such experiments have been carried out [32].

In a much over looked article published in *Analytical Chemistry* in 1991, D. Rolison and W. O'Grady revealed empirical confirmation of nuclear transmutation resulting from palladium foil electrolysis in D_2O [32]. They utilized time of flight mass spectroscopy (TOF-SIMS) to analyze samples before and after electrolysis. Their results were thought to be a result of isotopic anomalies. The puzzling aspects of their data arise from the ostensibly "simple" addition of a di-neutron to ^{106}Pd to yield ^{108}Pd. The natural abundance of palladium-106 is 27.33% while that of palladium-108 is 26.46%. However, during the electro-reduction of D_2O using a palladium deuteride cathode, the ratio of these two isotopes began to merge; and after 10 min of electrolysis, the abundance on the cathode surface of ^{108}Pd (atomic wt. 107.9038917) was found to be greater than that of ^{106}Pd (atomic wt. 105.9034857). The corresponding transmutation (reaction 8), assisted by phonon enforced quantum tunneling effects, represents a gain of 2.000406 mass units. This is extremely strong evidence for, at least, a short lived associated pair of neutrons in the condensed phase.

$$e^- + {}^1p_{\uparrow 0}n_{\uparrow} \rightarrow {}_0n_{\downarrow 0}n_{\uparrow} + {}^{106}\mathrm{Pd} \rightarrow {}^{108}\mathrm{Pd}. \tag{8}$$

Early on, the Rolison–O'Grady results did not garner the deserved recognition for two basic reasons: (1) The di-neutron was then thought to not exist and (2) Isotopic augmentation was perceived as very controversial, because nuclear effects observed in ambient temperature chemical systems were unknown. In fact, a Widom Larson type electron capture followed by a condensed matter fusion is the only reasonable mechanism for their observed ^{106}Pd to ^{108}Pd transmutation. In contrast to the early lack of recognition and as testimony to the scientific rigor and accuracy of their empirical observations [32], Debra R. Rolison won the prestigious William H. Nichols Medal for original research in 2018.

More recent experiments carried out by Jean-Paul Biberian revealed similar transmutations with the isotopes of silver [33]. Following electrolysis of a thin film of silver impregnated palladium deuteride on silicon, a mass spectral SIMS surface analysis revealed an isotopic ratio of ^{109}Ag/^{107}Ag to be 10, while the natural abundance ratio is 1.06 (see reaction 9). Since both the Pd and Ag isotope studies were carried out with the inclusion of all proper control experiments, including analysis before and after electrolysis, error seems unlikely. It is always possible that contamination spoiled only the post-electrolysis sample, but this is extremely unlikely in both studies. Further, contamination with an unnatural isotope distribution seems absurd, especially since we are dealing with highly regarded and even award winning scientists.

$${}^{107}\mathrm{Ag} + {}_0n_0n \rightarrow {}^{109}\mathrm{Ag}. \tag{9}$$

Reactions 8 and 9 demonstrate that the di-neutron serves as a transmutation reactant. Since di-neutrons have a tendency to cluster [21,22], analogous reactions (transmutations) involving the tetra-neutron might be anticipated. The fusion of $({}_0n_0n)_2$ (or just two di-neutrons) with a stable nucleus would most likely result in an unstable nuclear product with too many neutrons relative to protons. The result of this would be beta (β^-) decay and a stepping up in atomic number. An irrefutable display of this phenomenon was reported by Iwamura and his team at the Mitsubishi Labs [34]. Iwamura et al. took advantage of the ability of cesium metal to inject electrons into a reducible lattice. Specifically, they synthesized a multi layer sandwich complex with bulk Pd on the bottom, alternating CaO and Pd layers in the middle, and a Pd thin film on top. They then deposited Cs metal (from the vapor) onto the top layer [34].

Cs is an ideal selection, as ^{133}Cs has a natural abundance of 100%, and Cs has the lowest ionization potential of the stable elements. Deuterium gas was then passed through the sandwich complex. Given the experimental procedures that lead to reactions 8 and 9, this arrangement should produce a relatively large burst of di-neutrons due to the flow of D_2 directly through the electron encrusted palladium manifold. Post-reaction analysis of the top layer yielded the unmistakable presence of praseodymium-141! ^{141}Pr was, of course, not present on the pre-reaction sandwich. Anyway, Pr would be a most improbable contaminant. Apparently, ^{133}Cs fused with four di-neutrons, or possibly two of Bertaluni's tetra-neutrons, to yield ^{141}Cs. This isotope is unstable and undergoes β^- decay ($t_{1/2} = 25$ s), ultimately yielding ^{141}Pr (reactions 10–12). The presence of ^{141}Pr is apparently a result of the tendency of di-neutrons to cluster. Other transmutation products via ostensible fusion with di-neutron clusters were observed by the Mitsubishi group: for example the formation of samarium from barium [34]. It is suspected that an analogous birth is possible for a number of rare earth elements.

$$2\,^{133}\mathrm{Cs} + 2\,^{1}\mathrm{p}_0\mathrm{n} \rightarrow (_0\mathrm{n}_0\mathrm{n})_2 + 2\,^{133}\mathrm{Cs}^+, \tag{10}$$

$$2\,(_0\mathrm{n}_0\mathrm{n})_2 + {}^{133}\mathrm{Cs}^+ \rightarrow {}^{141}\mathrm{Cs}, \tag{11}$$

$$^{141}\mathrm{Cs} \rightarrow {}^{141}\mathrm{Pr} + 2\beta^-. \tag{12}$$

3. Conclusions

For the case of the W boson mediated $e^- - {}^1p$ interaction, a high energy neutron is usually the result (Fig. 1). However, when the proton is a member of a deuteron, a di-neutron is formed, which in the condensed phase can have very low momentum and participate in low energy transmutations. It has been shown here that the di-neutron, which exists as an isospin triplet and a spin singlet (Fig. 3), can be found in deuteriated cathodes during electrolysis. Di-neutron activity is suspected from the otherwise unexpected presence of unnatural heavy isotopic materials as revealed in [15,17] (see reactions 8 and 9). When such systems are very neutron heavy, β^- decay can lead to rare isotonic products as reported by Iwamura et al. (see reactions 10–12) [34]. It may be too early to consider possible synthetic proposals involving these transmutation reactions, but the search for otherwise unexpected isotopes and isotones having resulted from di-neutron fusion should be considered.

Simple electrochemical experiments, as described in reactions 8–10, followed up with radiological analysis should demonstrate the occurrence and perhaps usefulness of proposed syntheses. For example, $^{238}UD_3$ is a well studied and useful compound [35] and treatment of ^{238}U with $_0n_0n$ via the electrolysis of $^{238}UD_3$ impregnated PdD would lead to uranium-240 ($^{238}\mathrm{U} + {}_0\mathrm{n}_0\mathrm{n} \rightarrow {}^{240}\mathrm{U}$). Uranium-240 has a half life of 14.1 h and decays via β^- emission to ^{240}Np. ^{240}Np, in turn, undergoes further β^- emission ($t_{1/2} = 61.9$ days) to produce ^{240}Pu. The radiological signature of this process would be relatively easy to identify and not require work up and high resolution mass spectral analysis to prove out the viability of the process. A plethora of analogous schemes involving actinides and lanthanides, which are in high demand, are not difficult to imagine.

Finally, it should be mentioned that, in palladium hydride, deuterons or deuterium can readily absorb relativistic electrons to form di-neutrons, but the deuterium must first be freed from bonding to the Pd. This situation is readily met, as there is a plethora of electrons from metals with low ionization potentials (i.e. Cs) or from an external anode. Solid state crystals of polyaromatic hydrocarbons, and their electron rich anion radicals serve as explicit examples of this added electron enforced bond destabilization [36]; the electron preferentially destabilizes the isotopically heavy bond. As a result, warming of anion radical crystalline salts, where the added electrons are now in a conduction band of the solid material, results in bond rupture and hydrogen evolution [37,38]; added electrons attenuate the barrier to bond rupture [39]. The added electrons, whether from an anode or Cs metal, protrude into anti-bonding bands, and relative Pd–D (H) bonding decreases due to zero point energy differences (D relative to H) [36–39].

Nobel Laureate Roald Hoffmann wrote [40]: "In general, the interactions on a solid surface resemble those in molecules," and: "shifts of electron density around the Fermi level, have bonding consequences." This means that the electron count has an important bonding influence on the solid lattice surface that is analogous to what has been observed for molecules in solution [41]. The addition of a sufficient number of electrons to the PdH or PdD system requires that the added electrons go into anti-bonding orbitals (solution) or bands. Electron addition destabilizes the individual bonds, more so for the isotopically heavy (deuterium) systems. This bond rupture, followed by electron capture, di-neutron formation, and transmutation is illustrated in Fig. 4.

It is actually very unlikely that a given electron will be able to approach the nucleon sufficiently close in order to quantum mechanically overlap with the weak field of the nucleon, but it does happen. And, when it does, reaction 5

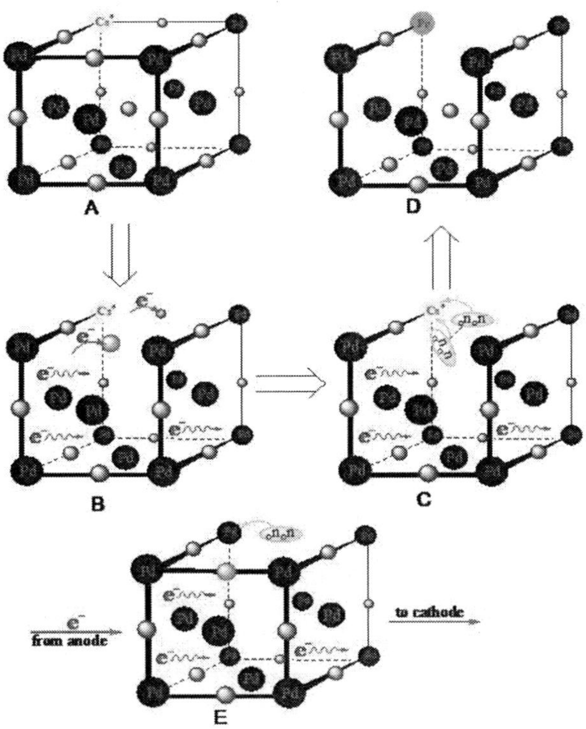

Figure 4. (A) Representation of a single surface cell of Cs impregnated PdD lattice with one face centered Pd atom missing. (B) Electron infusion from the Cs changes the electron count, and Pd–D bond breakage takes place followed by deuteron capture of the Zitterbewegung-ing electrons. (C) Newly formed low momentum di-neutrons may be attached, and the tetra-neutron fuses with a nearby cesium. (D) The heavy cesium becomes Pr (via ^{141}Cs beta decay (reaction 10). (E) In the case of the Rolison O'Grady experiment [32], the electrons come from an attached anode and the di-neutrons fuse with a ^{106}Pd to form its heavier isotope: ^{108}Pd.

takes place followed by considerable local heating. The local heating makes it more probable for a similar nearby event, which produces even more heat; hence, a cascade of events proceeds. This also explains the observation of transmutation products clumped together in "hot spots" on the post-electrolysis cathode surface [42]; phonon enforced quantum tunneling is clearly involved [30].

Quantum tunneling is sufficiently important to be dubbed the "The Third Reactivity Paradigm" of chemical reactions [42]. We have observed how the involvement in tunneling can change the quantum nature of subatomic particles [43,44]. For example the EPR spectrum of a tunneling electron renders all proton hyperfine interactions invisible and shifts the resonance field by nearly a gauss [44]. The observables of di-neutrons in condensed matter are likewise perturbed, as they are always close enough to transmutation active species and surrounded by phonon activity to be involved in active quantum tunneling. As mentioned above, it has been shown how surface phonon waves can drive disparate quantum systems [30]. Hence, both the deuteron electron capture (Fig. 3) and the di-neutron addition (Fig. 4) are surface reactions that involve phonon assisted quantum tunneling to yield the transmutation products, "hot spots," etc. [45]. Phonons are one of the two cornerstones of condensed matter (the other being the electron) [46].

This same phonon tunneling ${}_0n_0n$ transmutation activity, also leads to the formation of heat, tritium, and ${}^4He$ during observation of the FP heat effect [47,48]. Electrolysis of a PdD/H system causes the preferential release of

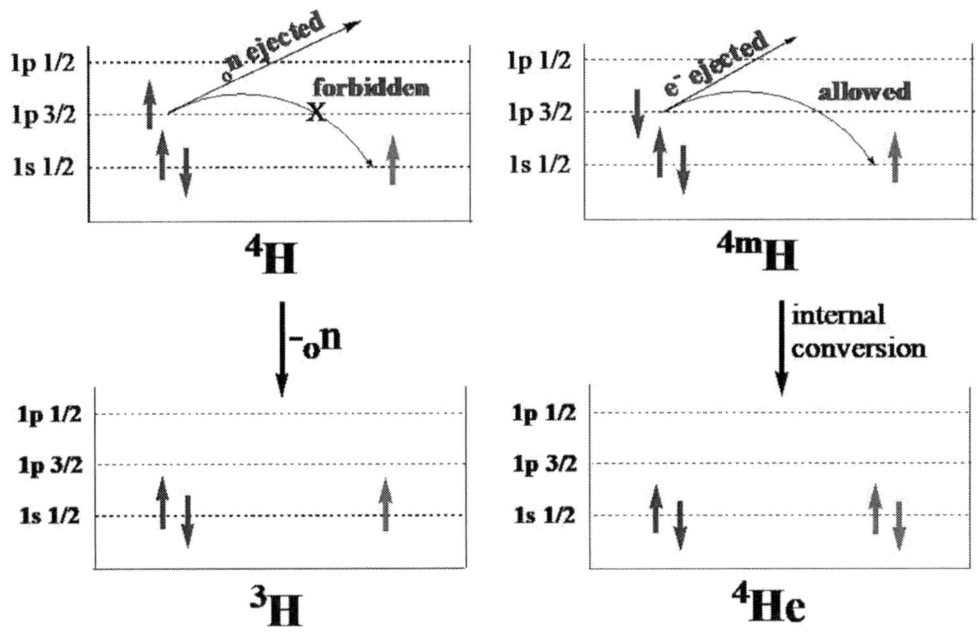

Figure 5. (*Left*) An energy level diagram showing the ejection of a (*blue*) neutron by ${}^4H$ to form the tritium nucleus. (*Right*) the relaxation of ${}^{4m}H$ (formed by the fusion of a di-neutron with a deuteron) to ${}^4He$. This process is spin allowed. Note that the (*blue*) neutron converts to the (*red*) proton in the absence of a spin flip.

D_2 [36]. Electron capture ($_0n^1p + e^-$) yields the low momentum $_0n_0n$, which is confined in the cathode. The di-neutron does fuse with a proton to yield tritium ($_0n_0n + {}^1p \rightarrow {}^3H$) or with a deuteron to yield helium-4 and excess heat ($_0n_0n + {}^1p_0n \rightarrow {}^{4m}H \rightarrow {}^4He$ + heat) [36].

This is the *Fleishmann Pons Heat Effect*! The ^{4m}H is a very short lived intermediate and is probably a meta-stable nuclear spin state of hydrogen-4 (Fig. 5). The mechanistic details of these LENR reactions, involving the di-neutron, have been recently explained in the *Int. J. Hydrogen Energy* [36].

As expressed by Mark Twain,"The reports of my death are greatly exaggerated;" so it is for the di-neutron. Not only that, but there is evidence that it tends to join with others forming neutron clusters [49,50], and this tendency would be augmented in a PdD matrix. We recognize this as a nuclear metathesis shown in Fig. 6. It is analogous to the Nobel Prize winning chemical metathesis [51].

The observed effects of FPHE, LENR, and low energy transmutations nicely fit into a paradigm that includes at least a transient di-neutron and/or its mutual clusters. Further, this paradigm lacks inconsistencies with the parameters of the standard model and is without the evocation of exotic and unusual forms of matter (e.g. high temperature Fermi or Bose condensates, and quantum entanglements). The di-neutron,under "cold" conditions, fuses with many isotopes(^1H, ^2H, ^{106}Pd, ^{107}Ag, ^{133}Cs, etc.) without having to overcome the e^2/r^2 term in the Lagrangian. It may be time to resurrect the original semantics describing Prof. Fleischmann's discovery as "COLD FUSION."

The fact that the nuclear physics may be different for the di-neutron when constrained in condensed matter is not at all surprising, as this is true even for the ephemeral neutrino. Neutrino flavor oscillations are different (perturbed)when the oscillations occur in condensed matter. This is apparently due to a condensed phase resonance that enhances flavor mixing [52]. Even the rate constant for β-decay, which was once thought to tick as an imperturbable clock, is perturbed when ^7Be is softly constrained in an organic fullerene [53]. Likewise, symmetry breaking resonances in condensed matter renders the quantum field of $_0n_0n$ perturbed.

Finally, the exact quantum physics of a number of empirical observations, for example: (1) the condensed matter

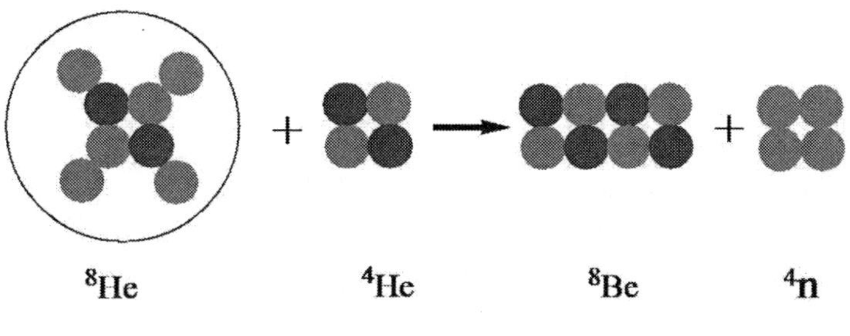

Figure 6. A nuclear metathesis of ^8He with ^4He producing ^8Be and a tetra-neutron, which is probably a resonate state involving (nn + nn) and ($^3{}_0n$ + n). The green spots represent neutrons and the blue spots represent protons.

perturbations upon neutrino oscillation[52], (2) the relativistic electrons in heavy metals [4,6,54], (3) the condensed matter Widom–Larson absorption of electrons [36] and (4) the fundamental nature of the di-neutron [3], has yet to be quantum mechanically resolved. In particular, Hagelstein published some important and interesting quantum inconsistencies in the WL electron capture process (e.g. missing radiation) [55]. However, as the complications involved in computer assisted quantum calculations become more and more untangled, consistency with fundamental quantum theory will undoubtedly be revealed. In the mean time, contemporary attacks on any of the listed (1)–(4) empirically based conclusions have (so far) been based only upon approximation methods of computation.

References

[1] The Jefferson Lab Qweak Collaboration, A search for parity violating new physics at the tev scale by measurement of the proton's weak charge, *Nature* **557** (2018) 207–211.

[2] E. Wigner, On the consequences of the symmetry of the nuclear Hamiltonian on the spectroscopy of nuclei, *Phy. Rev.* **51** (1937) 106–119.

[3] H.-W. Hammer and S. König, Constraints on a possible dineutron state from pionless EFT, *Phys. Lett.* B **736** (2014) 208–213.

[4] P. Pyykkö, The physics behind chemistry and the Periodic Table, *Chem. Rev.* **112** (2012) 371–384.

[5] E. Scerri, Cracks in the Periodic Table, *Scientific American* June (2013) 69–73.

[6] P. Pyykkö, Relativistic effects in structural chemistry, *Chem. Rev.* **88** (1988) 563–594.

[7] A.B. Migdal, Two interacting particles in a potential well, *Sov. J. Nucl. Phys.* **16** (1973) 273–280.

[8] A.B. Migdal, Two interacting particles in a potential hole, *Yadern. Fiz.* **16** (1972) 427–434.

[9] W.S. Lyon and H.H. Ross, *Nucleonics. Anal. Chem.* **56** (1984) 83R–88R.

[10] V.M. Suslov, M.A. Braun, I. Filikhin, I. Slaus and B. Vlahovic, Neutron–neutron effective range parameters, *Phys. Reports* **173** (1989) 257–300.

[11] K. Seth and B. Parker, Evidence for dineutrons in extremely neutron-rich nuclei, *Phys. Rev. Lett.* **66** (1991) 2448–2451.

[12] C.A. Bertulani, L.F. Canto and M.S. Hussein, The structure and reactions of neutron-rich nuclei, *Phys. Reports* **226** (1983) 281–376.

[13] K. Seth and B. Parker, Evidence for dineutrons in extremely neutron-rich nuclei, *Phys. Rev. Lett.* **66** (1991) 2448–2451.

[14] C.A. Bertulani, L.F. Canto and M.S. Hussein, The structure and reactions of neutron-rich nuclei, *Phys. Reports* **226** (1983) 281–376.

[15] A. Spyrou, Z. Kohley, T. Baumann, D. Bazin, B.A. Brown, G. Christian, P.A. DeYoung, J.E. Finck, N. Frank, E. Lunderberg, E.S. Mosby, W.A. Peters, A. Schiller, J.K. Smith, J. Snyder, M.J. Strongman, M. Thoennessen and A. Volya, First observation of ground state dineutron decay: ^{16}Be, *Phys. Rev. Lett.* **108** (2012) 102501.

[16] F.M. Marques, Detection of neutron clusters, *Phys. Rev.* C **65** (2002) 044006.

[17] J.P. Kneller and G.C. McLaughlin, Effect of bound di-neutrons upon big bang nucleosynthesis, *Phys. Rev.* D **70** (2004) 043512.

[18] H. Witała and W. Glöckle, Di-neutron and the three-nucleon continuum observables, *Phys. Rev.* C **85** (2012) 064003.

[19] H. Witała and W. Glöckle, The nn quasifree nD breakup cross section: discrepancies with theory and implications for the $^1S_0$ nn force, *Phys. Rev.* C **83** (2011) 034004.

[20] W. Tornow, H. Witala and R.T. Braun, Determinations of the neutron–neutron scattering length ann from kinematically incomplete neutron–deuteron breakup data revisited, *Few-Body Systems* **21** (1996) 97-–130.

[21] C.A. Bertulani and V. Zelevinsky, Is the tetraneutron a bound dineutron–dineutron molecule? *J. Phys.* G **29** (2003) 2431–2437.

[22] LENPIC Collaboration, Few-nucleon and many-nucleon systems with semilocal coordinate-space regularized chiral nucleon–nucleon forces, *Phys. Rev.* C **98** (2018) 014002.

[23] A. Widom and L. Larsen, Ultra low momentum neutron catalyzed nuclear reactions on metallic hydride surfaces, *Euro. Phys. J.* C **46** (2006) 107–111.

[24] S. Ando, F. Myhrer and K. Kubodera, Capture rate and neutron helicity asymmetry for ordinary muon capture on hydrogen, *Phys. Rev.* C **63** (2000) 015203.

[25] L.E. Marcucci and R. Machleidt, Muon capture on the deuteron and the neutron–neutron scattering length, *Phys. Rev.* C **90** (2014) 054001.

[26] J. Kirschner and D.R. Phillips, Constraining the neutron–neutron scattering length using the effective field theory without explicit pions, *Phys. Rev.* C **84** (2011) 054004.

[27] D.E. Gonzalez Trotter, F. Salinas Meneses, W. Tornow, C.R. Howell, Q. Chen, A.S. Crowell, C.D. Roper, R.L. Walter, D. Schmidt, H. Witała, W. Glöckle, H. Tang, Z. Zhou and I. Šlaus, Neutron–deuteron breakup experiment at $E_n = 13$ Mev: determination of the $^1S_0$ neutron–neutron scattering length, *Ann. Phys. Rev.* C **73** (2006) 034001.

[28] M. Davidson, Theories of variable mass particles and low energy nuclear phenomena, *Found. Phys.* **44** (2014) 144–174.

[29] I. Stepanov, M. Ersfeld, A.V. Poshakinskiy, M. Lepsa, E.L. Ivchenko, S.A. Tarasenko and B. Beschote, Coherent electron Zitterbewegung, arXiv:1612.06190 (2016).

[30] K.J. Satzinger, Y.P. Zhong, H.-S. Chang, G.A. Peairs, A. Bienfait, M. Chou, A.Y. Cleland, C.R. Conner, É. Dumur, J. Grebel, I. Gutierrez, B.H. November, R.G. Povey, S.J. Whiteley, D.D. Awschalom, D.I. Schuster and A.N. Cleland, Quantum control of surface acoustic-wave phonons, *Nature* **563** (2018) 661–665.

[31] B.J. Powell, The expanding materials multiverse, *Science* **360** (2018) 1073–1074.

[32] D.R. Rolison and W.E. O'Grady, Observation of elemental anomalies at the surface of palladium after electrochemical loading of deuterium or hydrogen, *Anal. Chem.* 63 (1991) 1697–1702.

[33] Jean-Paul Biberian, Anomalous isotopic composition of silver in a palladium electrode, *Int. Conf. on Condensed Matter Nucl. Sci.*, Fort Collins, CO, June 3–8, 2018.

[34] Y. Iwamura, T. Itoh, N. Yamazaki, H. Yonemura, K. Fukutani and D. Sekiba, Recent advances in deuterium permeation transmutation experiments, J. Condensed Matter Nucl. Sci. **10** (2013) 63–71.

[35] P.F. Souter, B.P. Kushto, L. Andrews and M. Neurock, Experimental and theoretical evidence for the formation of several uranium hydride molecules, *J. Am. Chem. Soc.* **119** (1997) 1682–1687.

[36] C.D. Stevenson and J.P. Davis, Hydrogen and deuterium isotope effects beyond the electromagnetic force, *Int. J. Hydrogen Energy* **42** (2018) 20011–20021.

[37] K. Ballard, R.C. Reiter and C.D. Stevenson, Calorimetrically measurable enthalpic isotope effect, *J. Phys. Chem.* A **110** (2006) 14050–14053.

[38] C. D. Stevenson, C.V. Rice, P.M. Garland and B.K. Clark, Thermal and laser pyrolysis of hydrocarbon anion radicals, *J. Org. Chem.* **62** (1997) 2193–2197.

[39] D.A. Hrovat, J.H. Hammons, C.D. Stevenson and W.T. Borden, Calculations of the equilibrium isotope effects on the reductions of benzene-d_6 and cyclooctatetraene-d_8, *J. Am. Chem. Soc.* **119** (1997) 9523–9526.

[40] R. Hoffmann, A chemical and theoretical way to look at bonding on surfaces, *Rev. Mod. Phys.* **60** (1988) 601.

[41] R. Hoffmann, How chemistry and physics meet in the solid state, *Angew. Chem. Int. Ed.* **26** (1987) 846–878.

[42] P.R. Schreiner, Tunneling Control of chemical reactions: the third reactivity paradigm, *J. Am. Chem. Soc.* **139** (2017) 15276–15283.

[43] C.D. Stevenson, L.J. Hienle, J.P. Davis and R.C. Reiter, Tunneling and sterically induced ring puckering in a substituted [8]annulene anion radical, *J. Am. Chem. Soc.* **124** (2002) 2704–2708.

[44] R. Rathore, S.H. Abdelwahed, M.K. Kiesewetter, R.C. Reiter and C.D. Stevenson, Intramolecular electron transfer in Co-facially π-stacked fluorenes: evidence of tunneling, *J. Phys. Chem.* B **110** (2006) 1536–1540.

[45] f. scholkmann, and d.j. nagel, is the abundance of elements in earth's crust correlated with LENR transmutation rates? *J. Condensed Matter Nucl. Sci.* **19** (2016) 281–286.

[46] L. Lou, *Introduction to Phonons and Electrons*, Vol. 2, World Scientific, New Jersey, 2003, pp. 102–114.

[47] M. Fleischmann, S. Pons, M. Anderson, Lian Li and M. Hawkins, Calorimetry of the palladium–deuterium-heavy water system, *J. Electroanal. Chem.* **287** (1990) 293–351.

[48] M.H. Miles, R. Hollins, B.F. Bush, J. Lagowski and R. Miles, Correlation of excess power and helium production D_2O and H_2O during electrolysis using palladium electrodes, *J. Electroanal. Chem.* **346** (1993) 99—117.

[49] K. Kisamori S. Shimoura, H. Miya, S. Michimasa, S. Ota, M. Assie, H. Baba, T. Baba, D. Beaumel, M. Dozono, T. Fujii, N. Fukuda, S. Go, F. Hammache, E. Ideguchi, N. Inabe, M. Itoh, D. Kameda, S. Kawase, T. Kawabata, M. Kobayashi, Y. Kondo,

T. Kubo, Y. Kubota, M. Kurata-Nishimura, C.S. Lee, Y. Maeda, H. Matsubara, K. Miki, T. Nishi, S. Noji, S. Sakaguchi, H. Sakai, Y. Sasamoto, M. Sasano, H. Sato, Y. Shimizu, A. Stolz, H. Suzuki, M. Takaki, H. Takeda, S. Takeuchi, A. Tamii, L. Tang, H. Tokieda, M. Tsumura, T. Uesaka, K. Yako, Y. Yanagisawa, R. Yokoyama and K. Yoshida, Candidate resonant tetraneutron state populated by the ^{4}He(8He;8Be) reaction, *Phys. Rev. Lett.* **116** (2016) 052501.

[50] F.M. Marqués, N.A. Orr1, H. Al Faloul, G. Normand and N.M. Clarke, On the possible detection of 4n events in the breakup of ^{14}Be, arXiv:nucl-ex/0504009v1 (2005).

[51] T.J. Bannin, P.P. Datta, E.T. Kiesewetter and M.K. Kiesewetter, Synthesizing stilbene by olefin metathesis reaction using guided inquiry to compare and contrast wittig and metathesis methodologies, *J. Chem. Educ.* **96** (2019) 143–147.

[52] T.K. Kuo and James Pantaleone, Neutrino oscillations in matter, *Rev. Mod. Phys.* **61** (1989) 937–979.

[53] A. Ray, P. Das, S.K. Saha, S.K. Das, J.J. Das, N. Madhavan, S. Nath, P. Sugathan, P.V. M. Rao and A. Jhingan, Change of ^{7}Be decay rate in exohedral and endohedral C fullerene compounds and its implications, *Phys. Rev.* C **73** (2006) 034323.

[54] P. Ball, On the edge of the Periodic Table, *Nature* **565** (2019) 552–554.

[55] P.L. Hagelstein, Electron mass enhancement and the Widom–Larsen model, *J. Condensed Matter Nucl. Sci.* **12** (2013) 18–40.

J. Condensed Matter Nucl. Sci. 29 (2019) 525–547

Research Article

Electron Structure, Ultra-dense Hydrogen and Low Energy Nuclear Reactions

Antonino Oscar Di Tommaso and Giorgio Vassallo*,†
Università degli Studi di Palermo, Dipartimento di Ingegneria (DI), Viale delle Scienze, 90128 Palermo, Italy

Abstract

In this paper, a simple Zitterbewegung electron model, proposed in a previous work, is presented from a different perspective based on the principle of mass−frequency equivalence. A geometric−electromagnetic interpretation of mass, relativistic mass, De Broglie wavelength, Proca, Klein−Gordon, Dirac and Aharonov−Bohm equations in agreement with the model is proposed. A non-relativistic, Zitterbewegung interpretation of the 3.7 keV deep hydrogen level found by J. Naudts is presented. According to this perspective, ultra-dense hydrogen can be conceived as a coherent chain of bosonic electrons with protons or deuterons located in the center of their Zitterbewegung orbits. This approach suggests a possible role of ultra-dense hydrogen in some aneutronic and many-body low energy nuclear reactions.

Keywords: Aharonov–Bohm equations, Aneutronic and many-body low energy nuclear reactions, Compact structures, De Broglie wavelength, Electron structure, Dirac equation, ESR, Heisenberg's uncertainty principle, Klein–Gordon equation, Josephson constant, LENR, natural units, Proca equation, relativistic mass, Ultra-dense hydrogen, Zitterbewegung

Nomenclature

$\gamma_x^2 = \gamma_y^2 = \gamma_z^2 = -\gamma_t^2 = 1$ were $\{\gamma_x, \gamma_y, \gamma_z, \gamma_t\}$ are the four basis vectors of $Cl_{3,1}(\mathbb{R})$ Clifford algebra, isomorphic to Majorana matrices algebra [1]

$\gamma_i\gamma_j = -\gamma_j\gamma_i$ with $i \neq j$ and $i, j \in \{x, y, z, t\}$;

$\partial = \gamma_x \frac{\partial}{\partial x} + \gamma_y \frac{\partial}{\partial y} + \gamma_z \frac{\partial}{\partial z} + \gamma_t \frac{1}{c} \frac{\partial}{\partial t}$

$I = \gamma_x\gamma_y\gamma_z\gamma_t$

$I_\Delta = \gamma_x\gamma_y\gamma_z$

1. Introduction

According to Carver Mead, mainstream physics literature has a long history of hindering fundamental conceptual reasoning, often *"involving assumptions that are not clearly stated"* [2]. One of these is the unrealistic assumption of

*Corresponding author. E-mail: giorgio.vassallo@unipa.it.
†Also at: International Society for Condensed Matter Nuclear Science (ISCMNS).

Nomenclature

Symbol	Name	SI units	Natural units (NU)
$\boldsymbol{A}_\square$	Electromagnetic four-potential	$\mathrm{V\ s\ m^{-1}}$	eV
$\boldsymbol{A}_\triangle$	Electromagnetic vector potential	$\mathrm{V\ s\ m^{-1}}$	eV
A_t	Time component of electromagnetic four potential	$\mathrm{V\ s\ m^{-1}}$	eV
A	Electromagnetic vector potential module	$\mathrm{V\ s\ m^{-1}}$	eV
m	Mass	kg	eV
$\boldsymbol{F}$	Electromagnetic field bivector	$\mathrm{V\ s\ m^{-2}}$	$\mathrm{eV^2}$
$\boldsymbol{B}$	Flux density field	$\mathrm{V\ s\ m^{-2}} = T$	$\mathrm{eV^2}$
$\boldsymbol{E}$	Electric field	$\mathrm{V\ m^{-1}}$	$\mathrm{eV^2}$
V	Potential energy	$\mathrm{J{=}kg\ m^2\ s^{-2}}$	eV
$\boldsymbol{J}_\square$	Four current density field	$\mathrm{A\ m^{-2}}$	$\mathrm{eV^3}$
$\boldsymbol{J}_\triangle$	Current density field	$\mathrm{A\ m^{-2}}$	$\mathrm{eV^3}$
ρ	Charge density	$\mathrm{A\ s\ m^{-3} = C\ m^{-3}}$	$\mathrm{eV^3}$
x, y, z	Space coordinates	m*	$\mathrm{eV^{-1}}$
t	Time variable	s†	$\mathrm{eV^{-1}}$
c	Light speed in vacuum	$2.997\,924\,58 \times 10^8\ \mathrm{m\ s^{-1}}$	1
$\hbar$	Reduced Planck constant	$1.054\,571\,726 \times 10^{-34}\ \mathrm{J\ s}$	1
μ_0	Permeability of vacuum	$4\pi \times 10^{-7}\ \mathrm{V\ s\ A^{-1}\ m^{-1}}$	4π
ϵ_0	Dielectric constant of vacuum	$8.854\,187\,817 \times 10^{-12}\ \mathrm{A\ s\ V^{-1}\ m^{-1}}$	$\frac{1}{4\pi}$
e	Electron charge	$1.602\,176\,565 \times 10^{-19}\ \mathrm{A\ s}$	$0.085\,424\,546$
α	Fine structure constant	$7.2973525664 \times 10^{-3}$	$7.2973525664 \times 10^{-3}$
m_e	Electron rest mass	$9.10938356 \times 10^{-31}\mathrm{kg}$	$0.5109989461 \times 10^6\ \mathrm{eV}$
λ_c	Electron Compton wavelength	$2.426\,310\,2389 \times 10^{-12}\ \mathrm{m}$	$1.229\,588\,259 \times 10^{-5}\ \mathrm{eV^{-1}}$
K_J	Josephson constant	$0.4835978525 \times 10^{15}\ \mathrm{Hz\ V^{-1}}$	$2.71914766 \times 10^{-2}$
r_e	Reduced Compton electron wavelength (Compton radius)	$r_\mathrm{e} = \frac{\lambda_\mathrm{c}}{2\pi}$	
r_c	Electron charge radius	$r_\mathrm{c} = \alpha r_\mathrm{e}$	
T_e	Zitterbewegung period	$T_\mathrm{e} = \frac{2\pi r_\mathrm{e}}{c}$	

*$1.9732705 \times 10^{-7}\ \mathrm{m} \simeq 1\ \mathrm{eV^{-1}}$.
† $6.5821220 \times 10^{-16}\ \mathrm{s} \simeq 1\ \mathrm{eV^{-1}}$.

point-like shaped elementary particles with *intrinsic* properties as mass, charge, angular momentum, magnetic moment and spin. According to the laws of mechanics and electromagnetism, a point-like particle cannot have an "intrinsic angular momentum". Moreover, a magnetic moment must necessarily be generated by a current loop, that cannot exist in a point-like particle. Furthermore, the electric field generated by a point-like charged particle should have an infinite energy. Therefore, an alternative realistic approach that fully addresses these *very basic* problems is indispensable. A possibility is given by a *Zitterbewegung* interpretation of quantum mechanics, according to which charged elementary particles can be modeled by a current ring generated by a massless charge distribution rotating at light speed along a circumference whose length is equal to particle Compton wavelength [3,4]. As a consequence, every elementary charge is always associated with a magnetic flux quantum and every charge is coupled to all other charges on its light cone by time-symmetric interactions [2]. The aim of this paper is to present a gentle introduction to an electron Zitterbewegung model together with some observations that deems to reinforce its plausibility.

The present paper is structured in the following way. In Section 2 the deep connection between some basic concept as space, time, energy, mass, frequency, and information is exposed. In Section 3 an introduction to a Zitterbewegung electron model is presented, together with a geometric-electromagnetic interpretation of Proca, Klein–Gordon, Dirac and Aharonov–Bohm equations. In Section 4 a simple geometric interpretation of relativistic mass and De Broglie wavelength is proposed. In Section 5 the relation of Electronic Spin Resonance (ESR) frequency with Larmor precession frequency of the Zitterbewegung orbit is presented. Finally, in Section 6 some hypotheses on the structure of ultra-dense hydrogen are formulated, whereas Section 7 deals with the possible role of ultra-dense hydrogen in low energy nuclear reactions.

N.B. In this paper all equations enclosed in square brackets with subscript "NU" have dimensions expressed in natural units. The mathematical notation used in Sections 3.3–3.5, based on real Clifford algebra $Cl_{3,1}(\mathbb{R})$, is introduced in [1].

2. Energy, Mass, Frequency and Information

The concept of measurement plays a fundamental role in all scientific disciplines based on experimental evidence. The most used measurement units (such as the international system, SI) are based mainly on human conventions not directly related to fundamental constants. To simplify the conceptual understanding of certain physical quantities it is convenient to adopt in some cases a measurement system based on universal constants, such as the speed of light c and the Planck's quantum $\hbar$.

Considering that a measure is an event localized in space and time, the quantum of action can be seen, in some cases, as an objective entity in some respects analogous to a bit of information located in the space-time continuum. In accordance with Heisenberg's uncertainty principle, the result of the measurement of some values (such as angular momentum) cannot have an accuracy less than half a single Planck's quantum. Therefore, to simplify the interpretation of physical quantities, it may be useful to adopt a system in which both the speed of light and the quantum of action are dimensionless quantities (pure numbers) having a unit value, i.e.: $c = 1$ and $\hbar = 1$. In this system, the constancy of light speed makes possible to use a single measurement unit for space and time, simplifying, in many cases, the conceptual interpretation of physical quantities. The energy of a photon, a "particle of light", is equal to Planck's quantum multiplied by the photon angular frequency. By using the symbol T to indicate the period of a single complete oscillation and λ the relative wavelength, it is, therefore, possible to write

$$E = \hbar\omega = \frac{2\pi\hbar}{T} = \frac{2\pi\hbar c}{\lambda}. \tag{1}$$

By using natural units, period and wavelength coincide and the above expression is simplified in

$$\left[E = \omega = \frac{2\pi}{T} = \frac{2\pi}{\lambda}\right]_{\text{NU}}. \tag{2}$$

The subscript NU highlights the use of natural units for expressions contained within square brackets. This equation indissolubly links some fundamental concepts, as space, time, energy and mass, giving the possibility to express an energy value simply as a frequency or as the inverse of a time, or even as the inverse of a length. Vice versa, it allows to use as a measurement unit of both space and time a value equal to the inverse of a particular energy value as the electron-volt. Therefore, to compute photon wavelength in vacuum with natural units it is sufficient to divide the constant 2π by its energy. This value will correspond exactly to the period of a complete oscillation. Hence, in natural units the inverse of an eV can be used as a measurement unit for space and time:

$$L_{(1\,\mathrm{eV})} = 1\ \mathrm{eV}^{-1} \approx 1.9732705\ \times\ 10^{-7}\ \mathrm{m} \approx 0.2\ \mu\mathrm{m},$$

$$T_{(1\,\mathrm{eV})} = 1\ \mathrm{eV}^{-1} \approx 6.582122\ \times\ 10^{-16}\ \mathrm{s} \approx 0.66\ \mathrm{fs}.$$

Consequently, an angular frequency can be measured in electron volts:

$$1\ \mathrm{eV} \approx 1.519268\ \times\ 10^{15}\ \mathrm{rad\,s^{-1}}.$$

Following these concepts, it is possible to define a link between fundamental concepts of information, space, time, frequency and energy. A "quantum of information" carried by a single photon will have a "necessary reading time" and a "spatial dimension" inversely proportional to its energy. A simple example is given by radio antennas (dipoles), whose length is proportional to the received (or transmitted) "radio photons" wavelength and inversely proportional to their frequency and to the number of bits that can be received in a unit of time. In this perspective, the concept of energy is closely linked to the "density" of information in space and in time.

3. Electron Structure

The famous Einstein's formula $E = mc^2$ becomes particularly explanatory if expressed in natural units:

$$[E = m]_{\mathrm{NU}}\,.$$

Mass is energy and it is, therefore, possible to associate a precise amount of energy to a particle having a given mass. Taking up the considerations made on the deep bond existing between the concepts of space, time, frequency and energy, it is interesting trying to associate the electron rest mass m_{e} to an angular frequency ω_{e}, a length r_{e} and a time T_{e}. In fact Einstein's formula can be expressed as

$$E_{\mathrm{e}} = m_{\mathrm{e}}c^2 = \hbar\omega_{\mathrm{e}} = \frac{\hbar c}{r_{\mathrm{e}}} = \frac{h}{T_{\mathrm{e}}} \tag{3}$$

or adopting natural units

$$\left[E_{\mathrm{e}} = m_{\mathrm{e}} = \omega_{\mathrm{e}} = \frac{1}{r_{\mathrm{e}}} = \frac{2\pi}{T_{\mathrm{e}}}\right]_{\mathrm{NU}}. \tag{4}$$

These constants have a simple and clear interpretation if one accepts a particular electron model consisting of a current ring generated by a massless charge rotating at the speed of light along a circumference whose radius is equal to the electron reduced Compton wavelength, defined as $r_{\mathrm{e}} = \frac{\lambda_c}{2\pi} \approx 0.38616\ \times\ 10^{-12}$ m [3–6]. According to the model described in [4] the charge is not a point-like entity, but it is distributed on a spherical surface whose radius is equal to the electron classical radius $r_c \approx 2.8179 \times 10^{-15}$ m . In Eq. (4) ω_e is the angular frequency of the rotating charge, r_e is its orbit radius and T_e its period. The current loop is associated with a quantized magnetic flux Φ_M equal to Planck's constant ($h = 2\pi\hbar$) divided by the elementary charge e (see Eq. (34) p. 84 [4])

$$\Phi_M = h/e$$

or in natural units

$$[\Phi_M = 2\pi/e]_{NU} \, .$$

The rotation is caused by the centripetal Lorentz force due to the magnetic field associated with the current loop generated by the elementary rotating charge (Eq. (36)).The value of this elementary charge, in natural units, is a pure number and is equal to the square root of the ratio between the charge radius r_c and the the orbit radius r_e (see Eqs. (39) and (40) p. 85 [4]:

$$\left[e = \sqrt{\frac{r_c}{r_e}} = \sqrt{\alpha} \approx 0.0854245\right]_{NU} . \tag{5}$$

Similar models, based on the concept of "current loop", have been proposed by many authors, but have often been ignored for their incompatibility with the most widespread interpretations of Quantum Mechanics [3,5–10]. It is interesting to remember how, already in his Nobel lecture of 1933, P.A.M. Dirac referred to an internal high-frequency oscillation of the electron: "*It is found that an electron which seems to us to be moving slowly, must actually have a very high frequency oscillatory motion of small amplitude superposed on the regular motion which appears to us. As a result of this oscillatory motion, the velocity of the electron at any time equals the velocity of light. This is a prediction which cannot be directly verified by experiment, since the frequency of the oscillatory motion is so high and its amplitude is so small*". In the scientific literature, the German word *Zitterbewegung* (ZBW) is often used to indicate this rapid oscillation/rotation of the electron charge. The rotating charge is characterized by a momentum p_c of purely electromagnetic nature:

$$p_c = eA = e\frac{\Phi_M}{2\pi r_e} = \frac{\hbar\omega_e}{c} = \frac{\hbar}{r_e} = m_e c.$$

In this formula the variable $A = \hbar/er_e$ indicates the vector potential seen by the rotating charge (see Eq. (25), p. 82 [4]. Multiplying the charge momentum p_c by the radius r_e we obtain the "intrinsic" angular momentum $\hbar$ of the electron:

$$p_c r_e = \hbar. \tag{6}$$

Using natural units the momentum p_c has the dimension of energy and it is exactly equal to the electron mass–energy at rest m_e:

$$\left[p_c = eA = E_e = \frac{1}{r_e} = m_e = \omega_e\right]_{NU} .$$

3.1. Aharonov–Bohm equations and Zitterbewegung model

The magnetic Aharonov–Bohm effect is described by a quantum law that gives the phase variation φ of the "electron wave function" starting from the integral of the vector potential $\boldsymbol{A}_\Delta$ along a path [11], i.e.

$$\varphi = \frac{e}{\hbar}\int \boldsymbol{A}_\Delta \cdot \mathrm{d}\boldsymbol{l}. \tag{7}$$

In the proposed Zitterbewegung model, the electron "wave function phase" has a precise geometric meaning: the charge rotation phase. By using (7), a possible counter-test consists in verifying that the phase shift φ along the circumference of the Zitterbewegung orbit is equal exactly to 2π radians. In fact

$$\varphi = \frac{e}{\hbar}\oint \boldsymbol{A}_\Delta \cdot \mathrm{d}\boldsymbol{l} = \frac{e}{\hbar}\int_0^{2\pi r_\mathrm{e}} A\,\mathrm{d}l = \frac{e}{\hbar}\int_0^{2\pi r_\mathrm{e}} \frac{\hbar}{e r_\mathrm{e}}\,\mathrm{d}l = \frac{e}{\hbar}\frac{\hbar}{e r_\mathrm{e}} 2\pi r_\mathrm{e} = 2\pi, \tag{8}$$

because vectors $\boldsymbol{A}_\Delta$ and $\mathrm{d}\boldsymbol{l}$ have the same direction tangent to the elementary charge trajectory. This result is also consistent with the prediction of the *electric* Aharonov–Bohm effect, a quantum phenomenon that establishes the variation of phase φ as a function of the integral of electric potential V in a time interval T, i.e.:

$$\varphi = \frac{e}{\hbar}\int_T V\,\mathrm{d}t. \tag{9}$$

Applying the electric Aharonov–Bohm effect formula to compute the phase shift φ within a time interval $T_\mathrm{e} = \frac{2\pi}{\omega_\mathrm{e}}$ equal to a Zitterbewegung period we obtain the expected result, *i.e.* $\varphi = 2\pi$. In fact, the electric potential of the electron rotating charge can be expressed as

$$V = \frac{e}{4\pi\varepsilon_0 r_\mathrm{c}} = \left[\frac{e}{r_\mathrm{c}}\right]_\mathrm{NU}$$

and its period as

$$T_\mathrm{e} = \frac{2\pi r_\mathrm{e}}{c} = [2\pi r_\mathrm{e}]_\mathrm{NU}\,.$$

A simple calculation, applying (9) and (5), yields the same results:

$$\varphi = \frac{e}{\hbar}\int_0^{T_\mathrm{e}} V\,\mathrm{d}t = \frac{e}{\hbar}VT_\mathrm{e} = \frac{e}{\hbar}V\frac{2\pi r_\mathrm{e}}{c} = \left[\frac{e^2}{r_\mathrm{c}}2\pi r_\mathrm{e}\right]_\mathrm{NU} = 2\pi. \tag{10}$$

Now, by equating the fifth term of (8) and the fourth term of (10) it is possible to demonstrate that

$$A_t = \frac{V}{c} = A = |\boldsymbol{A}_\Delta|\,,$$

$$[A_t = V = A = |\boldsymbol{A}_\Delta|]_\mathrm{NU}\,,$$

$$\boldsymbol{A}_\square^2 = (\boldsymbol{A}_\Delta + \gamma_t A_t)^2 = \boldsymbol{A}_\Delta^2 - A_t^2 = 0. \tag{11}$$

By introducing the differential form of (9) we obtain

$$\mathrm{d}\varphi = \frac{e}{\hbar} V \,\mathrm{d}t$$

and this yields the phase speed

$$\frac{\mathrm{d}\varphi}{\mathrm{d}t} = \omega_{\mathrm{e}} = \frac{e}{\hbar} V = \frac{e^2}{4\pi\varepsilon_0 \hbar r_{\mathrm{c}}} = \frac{c\alpha}{r_{\mathrm{c}}} = \frac{c}{r_{\mathrm{e}}} = \frac{m_{\mathrm{e}} c^2}{\hbar} = \frac{ce}{\hbar} A,$$

$$\left[\frac{\mathrm{d}\varphi}{\mathrm{d}t} = \omega_{\mathrm{e}} = m_{\mathrm{e}} = eV = eA\right]_{\mathrm{NU}}. \tag{12}$$

3.2. Proca equation and Zitterbewegung electron model

A deep connection of Maxwell's equations (see Eq. (97), p. 121 [1])

$$\boldsymbol{\partial} \left(\boldsymbol{\partial} \wedge \boldsymbol{A}_{\square}\right) + \mu_0 \boldsymbol{J}_{\square} = 0 \tag{13}$$

with Proca equation for a particle of mass m

$$\boldsymbol{\partial} \left(\boldsymbol{\partial} \wedge \boldsymbol{A}_{\square}\right) + \left(\frac{mc}{\hbar}\right)^2 \boldsymbol{A}_{\square} = 0, \tag{14}$$

$$\left[\boldsymbol{\partial} \left(\boldsymbol{\partial} \wedge \boldsymbol{A}_{\square}\right) + m^2 \boldsymbol{A}_{\square} = 0\right]_{\mathrm{NU}} \tag{15}$$

emerges if we prove that equation $\left[\mu_0 \boldsymbol{J}_{\square} = m^2 \boldsymbol{A}_{\square}\right]_{\mathrm{NU}}$ can be applied to the electron Zitterbewegung model introduced in [4]. In this model the electron's charge orbit delimits a disc-shaped volume with radius r_{e} and height $2r_{\mathrm{c}}$. Inside this volume the *average* Zitterbewegung current density $\bar{\boldsymbol{J}}_{\mathrm{e}}$ can be computed dividing the Zitterbewegung current by one half the disc vertical section $\mathcal{A}$:

$$\bar{\boldsymbol{J}}_{\mathrm{e}} = \frac{\boldsymbol{I}_{\mathrm{e}}}{\mathcal{A}},$$

where

$$\mathcal{A} = 2 r_{\mathrm{e}} r_{\mathrm{c}} = 2\alpha r_{\mathrm{e}}^2,$$

$$\bar{\boldsymbol{J}}_{\mathrm{e}} = \frac{\boldsymbol{I}_{\mathrm{e}}}{\mathcal{A}} = \frac{\boldsymbol{I}_{\mathrm{e}}}{2\alpha r_{\mathrm{e}}^2}. \tag{16}$$

From [4], p. 82, we have that

$$\left[\boldsymbol{I}_{\mathrm{e}} = \frac{\alpha \boldsymbol{A}_{\triangle}}{2\pi}\right]_{\mathrm{NU}}$$

and substituting it in (16) we get

$$\left[\bar{\boldsymbol{J}}_{\mathrm{e}} = \frac{\boldsymbol{A}_{\Delta}}{4\pi r_{\mathrm{e}}^{2}}\right]_{\mathrm{NU}},$$

$$\left[\mu_0 \bar{\boldsymbol{J}}_{\mathrm{e}} = 4\pi \bar{\boldsymbol{J}}_{\mathrm{e}} = \frac{\boldsymbol{A}_{\Delta}}{r_{\mathrm{e}}^{2}} = \omega_{\mathrm{e}}^{2} \boldsymbol{A}_{\Delta} = m_{\mathrm{e}}^{2} \boldsymbol{A}_{\Delta}\right]_{\mathrm{NU}}.$$

Remembering that the electron's electromagnetic four potential $\boldsymbol{A}_{\square} = \boldsymbol{A}_{\Delta} + \gamma_t A_t$ associated to the rotating charge is a light-like vector (i.e. $\boldsymbol{A}_{\square}^2 = 0$, see Eq. (11)) we can write the following relations:

$$\left[\mu_0 J_{et} = \frac{A_t}{r_{\mathrm{e}}^{2}} = \omega_{\mathrm{e}}^{2} A_t = m_{\mathrm{e}}^{2} A_t\right]_{\mathrm{NU}},$$

$$\left[\mu_0 \bar{\boldsymbol{J}}_{e\square} = \mu_0 \left(\bar{\boldsymbol{J}}_{\mathrm{e}} + \gamma_t J_{et}\right) = m_{\mathrm{e}}^{2}\left(\boldsymbol{A}_{\Delta} + \gamma_t A_t\right) = m_{\mathrm{e}}^{2} \boldsymbol{A}_{\square}\right]_{\mathrm{NU}},$$

and consequently (QED):

$$\left[\mu_0 \bar{\boldsymbol{J}}_{e\square} = m_{\mathrm{e}}^{2} \boldsymbol{A}_{\square}\right]_{\mathrm{NU}}. \tag{17}$$

3.3. Proca and electromagnetic Klein–Gordon equations

In this paragraph and in the next one we will use only natural units, omitting the subscript NU. The aim is to show the connection of Proca equation with an "electromagnetic version" of Klein–Gordon equation. By applying the operator $\boldsymbol{\partial}\wedge$ to Proca equation

$$\boldsymbol{\partial}\left(\boldsymbol{\partial} \wedge \boldsymbol{A}_{\square}\right) + m^2 \boldsymbol{A}_{\square} = 0, \tag{18}$$

$$\boldsymbol{\partial F} + m^2 \boldsymbol{A}_{\square} = 0,$$

we get

$$\boldsymbol{\partial} \wedge \boldsymbol{\partial F} + m^2 \boldsymbol{\partial} \wedge \boldsymbol{A}_{\square} = 0,$$

$$\boldsymbol{\partial} \wedge \boldsymbol{\partial F} + m^2 \boldsymbol{F} = 0.$$

Now, by writing Maxwell's equations considering an averaged four-current vector density

$$\boldsymbol{\partial F} = -4\pi \bar{\boldsymbol{J}}_{\square}, \tag{19}$$

and by applying to both members the operator $\boldsymbol{\partial}\cdot$ we obtain the following expression

$$\boldsymbol{\partial} \cdot \boldsymbol{\partial F} = -4\pi\, \boldsymbol{\partial} \cdot \bar{\boldsymbol{J}}_{\square} = 0,$$

that is equal to zero as a consequence of the charge–current conservation law. For this reason, the term $\boldsymbol{\partial} \wedge \boldsymbol{\partial F}$ can be safely substituted by the term $\boldsymbol{\partial}^2 \boldsymbol{F}$:

$$\boldsymbol{\partial} \wedge \boldsymbol{\partial F} = \boldsymbol{\partial}^2 \boldsymbol{F} - \boldsymbol{\partial} \cdot \boldsymbol{\partial F} = \boldsymbol{\partial}^2 \boldsymbol{F}.$$

As a result we obtain a Klein–Gordon-like equation where the electromagnetic bivector $\boldsymbol{F}$ substitutes the "wavefunction" ψ:

$$\boldsymbol{\partial}^2 \boldsymbol{F} + m^2 \boldsymbol{F} = 0. \tag{20}$$

A similar equation for the electromagnetic four potential can be obtained simply by applying the Lorenz gauge condition $\boldsymbol{\partial A}_{\square} = \boldsymbol{\partial} \wedge \boldsymbol{A}_{\square}$ to Proca equation:

$$\boldsymbol{\partial}^2 \boldsymbol{A}_{\square} + m^2 \boldsymbol{A}_{\square} = 0 \tag{21}$$

or

$$\boldsymbol{\partial}^2 \boldsymbol{A}_{\square} + \omega^2 \boldsymbol{A}_{\square} = 0. \tag{22}$$

It is important to note that the Lorenz gauge condition can been applied to Maxwell's equations (19) only when *an averaged* four current density vector value is used. In this case the electromagnetic four potential is also an averaged value and no more an harmonic function of space–time [1].

3.4. The electromagnetic Dirac equation

By following the same conceptual pattern of the previous paragraph, an electromagnetic–geometric version of the Dirac equation (23),

$$i\partial\!\!\!/\psi - m\psi = 0 \tag{23}$$

should have the form

$$\boldsymbol{\partial F} - \boldsymbol{mF} = 0. \tag{24}$$

Here $\boldsymbol{m}$ cannot be a scalar, being $\boldsymbol{\partial F}$ a vector and $\boldsymbol{F}$ a bivector, respectively, but rather a space-like vector with module m. A possible candidate for $\boldsymbol{m}$ is a vector that has the same direction of the Zitterbewegung radius $\boldsymbol{r}$ and a module $m = \frac{1}{r} = \omega$. Calling $\boldsymbol{r}_u$ a unit vector in the same direction of $\boldsymbol{r}$, Eq. (24) becomes

$$\boldsymbol{\partial F} - \omega \boldsymbol{r}_u \boldsymbol{F} = 0, \tag{25}$$

where the operator $\boldsymbol{\partial}$ of $Cl_{3,1}(\mathbb{R})$ substitutes the Dirac operator $i\partial\!\!\!/$, the Zitterbewegung angular frequency ω the electron mass and the electromagnetic bivector $\boldsymbol{F}$ the wave function ψ. The unit vector $\boldsymbol{r}_u$ is always orthogonal to the vector potential and therefore:

$$\boldsymbol{r}_u^2 = 1,$$

$$\omega \boldsymbol{r} = \boldsymbol{r}_u,$$

$$\boldsymbol{r} \cdot \boldsymbol{A}_{\square} = 0.$$

By applying (19) to (25) we can write

$$4\pi \boldsymbol{J}_{\square} + \omega \boldsymbol{r}_u \boldsymbol{F} = 0 \tag{26}$$

whereas, by applying (17) to (26) and remembering that $\boldsymbol{F} = \boldsymbol{\partial} \boldsymbol{A}_{\square}$, we obtain:

$$\omega^2 \boldsymbol{A}_{\square} + \omega \boldsymbol{r}_u \boldsymbol{\partial} \boldsymbol{A}_{\square} = 0,$$

that can be written as

$$\boldsymbol{r}_u \boldsymbol{\partial} \boldsymbol{A}_{\square} + \omega \boldsymbol{A}_{\square} = 0.$$

Now, by left multiplying the last equation for the unit vector $\boldsymbol{r}_u$ we obtain a Dirac-like equation for the electromagnetic four potential

$$\boldsymbol{\partial} \boldsymbol{A}_{\square} + \omega \boldsymbol{r}_u \boldsymbol{A}_{\square} = 0. \tag{27}$$

Multiplying for the elementary charge e (27) becomes

$$e \boldsymbol{\partial} \boldsymbol{A}_{\square} + e\omega \boldsymbol{r}_u \boldsymbol{A}_{\square} = 0. \tag{28}$$

Moreover, by multiplying the electromagnetic four-potential for the ratio $\frac{e}{\omega}$, we obtain a light-like vector that can be interpreted as the charge four-velocity $\boldsymbol{c} - \gamma_t$ (see Eq. (60) of [1] and Eq. (12))

$$\frac{e}{\omega} \boldsymbol{A}_{\square} = \boldsymbol{c} - \gamma_t, \tag{29}$$

that left multiplying by $\boldsymbol{r}_u$ becomes

$$\frac{e}{\omega} \boldsymbol{r}_u \boldsymbol{A}_{\square} = \boldsymbol{r}_u \boldsymbol{c} - \boldsymbol{r}_u \gamma_t. \tag{30}$$

Now, by applying (30) to (28) and remembering that $\boldsymbol{\partial} \boldsymbol{A}_{\square} = \boldsymbol{F}$, (28) becomes

$$e\boldsymbol{F} = -\omega^2 \left(\boldsymbol{r}_u \boldsymbol{c} - \boldsymbol{r}_u \gamma_t \right). \tag{31}$$

Applying the identity $\boldsymbol{F} = (\boldsymbol{E} + I\boldsymbol{B})\, \gamma_t$ (see Eq. (73) of [1], Eq. (31) becomes

$$e \left(\boldsymbol{E} + I\boldsymbol{B} \right) \gamma_t = -\omega^2 \left(\boldsymbol{r}_u \boldsymbol{c} - \boldsymbol{r}_u \gamma_t \right). \tag{32}$$

This last equation can be split in two equations. The first one deals with the electric field $\boldsymbol{E}$:

$$e\boldsymbol{E}\gamma_t = \omega^2 \boldsymbol{r}_u \gamma_t.$$

Applying the identity $eA = \omega$, the square ω^2 can be written as $eA\omega$, namely a term that is equal to the module of the force generated on an elementary electric charge by the time derivative of a rotating vector potential:

$$e\boldsymbol{E} = eA\omega\mathbf{r}_{\boldsymbol{u}} = -e\frac{\mathrm{d}\boldsymbol{A}_{\vartriangle}}{\mathrm{d}t}. \tag{33}$$

This electric force has the same value of the centrifugal force acting on a mass m rotating with angular frequency ω at distance r from its orbit center:

$$\boldsymbol{r}_u = \omega\boldsymbol{r} = m\boldsymbol{r},$$

$$e\boldsymbol{E} = m\omega^2\boldsymbol{r}.$$

The second part of (32) deals with the magnetic flux density field $\boldsymbol{B}$:

$$eI\boldsymbol{B}\gamma_t = -\omega^2\boldsymbol{r}_u\boldsymbol{c},$$

$$eI_{\vartriangle}\boldsymbol{B} = -\omega^2\boldsymbol{r}_u\boldsymbol{c}$$

that right multiplying for $\boldsymbol{c}$ becomes:

$$eI_{\vartriangle}\boldsymbol{B}\boldsymbol{c} = -\omega^2\boldsymbol{r}_u.$$

As $\boldsymbol{B}$ and $\boldsymbol{c}$ are orthogonal vectors in the Zitterbewegung model, it is possible to write also:

$$eI_{\vartriangle}\boldsymbol{B}\wedge\boldsymbol{c} = -\omega^2\boldsymbol{r}_u$$

that, using ordinary vector algebra, becomes:

$$e\boldsymbol{c}\times\boldsymbol{B} = -\omega^2\boldsymbol{r}_u. \tag{34}$$

Finally, merging (33) with (34) we obtain an equation that tell us that the mass-less rotating charge, with momentum $\boldsymbol{p} = e\boldsymbol{A}_{\vartriangle}$, is subjected to a centripetal magnetic force $-\omega^2\boldsymbol{r}_u$:

$$e\boldsymbol{c}\times\boldsymbol{B} = e\frac{\mathrm{d}\boldsymbol{A}_{\vartriangle}}{\mathrm{d}t} = \frac{\mathrm{d}\boldsymbol{p}}{\mathrm{d}t}, \tag{35}$$

$$e\boldsymbol{c}\times\boldsymbol{B} = -m\omega^2\boldsymbol{r} = -\omega^2\boldsymbol{r}_u. \tag{36}$$

These easy to interpret equations confirm the correctness of the original choice of $\omega\boldsymbol{r}_u$ for the vector $\boldsymbol{m}$ in the electromagnetic version of Dirac equation (24).

3.5. Proca equation, electric charge quantization and Josephson constant

An interesting consequence of Eq. (22) is the magnetic flux and electric charge quantization. In this paragraph we call "wave amplitude" the module A of vector potential $\boldsymbol{A}_{\vartriangle}$ in Eq. (22)

$$\boldsymbol{A}_{\square} = \boldsymbol{A}_{\vartriangle} + \gamma_t A_t, \qquad A = |\boldsymbol{A}_{\vartriangle}| = A_t$$

Substituting ω with eA in Eq. (22) we obtain *a non-linear wave equation for the electromagnetic four potential, where the wave angular frequency is proportional to the wave amplitude and the proportionality coefficient is the "electric charge quantum", i.e. the elementary charge e.*

$$\left[\partial^2 \boldsymbol{A}_{\square} + e^2 A^2 \boldsymbol{A}_{\square} = 0\right]_{\mathrm{NU}}, \tag{37}$$

$$\left[\partial^2 \boldsymbol{A}_{\square} + \alpha A^2 \boldsymbol{A}_{\square} = 0\right]_{\mathrm{NU}}. \tag{38}$$

In this equation the ratio frequency/amplitude, ν/A, expressed in natural units is a pure number equal to half the value of Josephson constant K_J :

$$\left[\frac{v}{A} = \frac{1}{2}K_\mathrm{J}\right]_{\mathrm{NU}}.$$

The product of wave amplitude and wave period T is equal to another constant exactly equal to a magnetic flux Φ_M, a value two times the magnetic flux quantum Φ_o (see Fig. 1). It is a reasonable conjecture to consider (37) also valid for other charged elementary particles. In natural units we have

$$\left[AT = \frac{\omega T}{e} = \frac{h}{e} = \Phi_\mathrm{M} = 2K_\mathrm{J}^{-1}\right]_{\mathrm{NU}},$$

where

$$\Phi_\mathrm{M} = 2\Phi_\mathrm{o} = 4.13566766 \times 10^{-15}\ \mathrm{V\,s},$$

$$[\Phi_\mathrm{M} = 73.55246018]_{\mathrm{NU}}.$$

4. Geometric Interpretation of Relativistic Electron Mass and De Broglie Wavelength

If an electron moves along an axis z orthogonal to its charge rotation plane, it will describe an helical trajectory whose length is $L = c\Delta t$ and whose z-axis length is $l = v_z\Delta t$. The electron mass is exactly equal to the inverse of the helix radius r if expressed in NU, i.e. $m = r^{-1}$. An acceleration along z, implies a smaller radius and, hence, a mass increase. Using the Pythagorean theorem it is possible to write the value of the radius r as a function of v_z [4,5]:

$$r = r_\mathrm{e}\sqrt{1 - \frac{v_z^2}{c^2}}$$

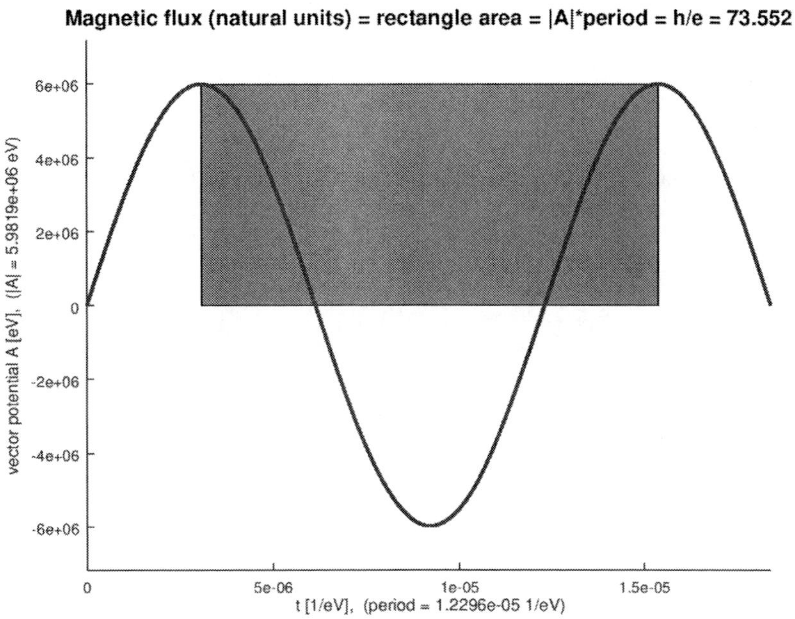

Figure 1. A possible explanation of magnetic flux and electric charge quantization: in electromagnetic Klein–Gordon/Proca equation vector potential amplitude time wave period is a constant $\Phi_m = h/e$.

and the related mass variation

$$m = \frac{\hbar\omega}{c^2} = \frac{m_e}{\sqrt{1 - \frac{v^2}{c^2}}}.$$

The charge momentum is proportional to the angular frequency and it has a direction tangent to the helical path. The relativistic momentum of charge is, then,

$$p_c = eA = \frac{\hbar\omega}{c} = \frac{\hbar}{r} \tag{39}$$

or, using natural units,

$$\left[p_c = \omega = \frac{1}{r} = m\right]_{\mathrm{NU}}.$$

Equation (39) suggests a particular interpretation of the Heisenberg uncertainty principle: an electron, whose charge has a momentum p_c, cannot be confined within a spherical space of radius R less than r. This means that it must be

$$R > r = \frac{\hbar}{p_c}.$$

Now, the charge momentum vector $\boldsymbol{p}_{\rm c} = e\boldsymbol{A}_{\vartriangle}$ can be decomposed into two components: $\boldsymbol{p}_{\perp}$, that is orthogonal to electron velocity and another one, $\boldsymbol{p}_{\|}$, that is parallel, i.e. in the z-direction. Therefore the charge momentum can be expressed as

$$\boldsymbol{p}_{\rm c} = \boldsymbol{p}_{\perp} + \boldsymbol{p}_{\|}.$$

The magnitude of component $\boldsymbol{p}_{\perp}$ is a constant, independent from velocity v_z, and is proportional to the charge angular speed $\omega_{\rm e}$ in the xy-plane [12]. Therefore,

$$p_{\perp} = \frac{\hbar\omega_{\rm e}}{c} = m_{\rm e}c$$

or in natural units

$$[p_{\perp} = \omega_{\rm e} = m_{\rm e}]_{\rm NU}\,,$$

whereas the component $p_{\|}$ is the momentum of the electron and is proportional to the *instantaneous* angular frequency $\omega_z = v_z/r$

$$p_{\|} = \frac{\hbar\omega_z}{c} = \frac{\hbar v_z}{cr} = \frac{\hbar\omega}{c^2}v_z = mv_z$$

or in natural units

$$\left[p_{\|} = \omega_z = \frac{v_z}{r} = mv_z\right]_{\rm NU}.$$

Using again the Pythagorean theorem it is possible to write the following equations

$$\omega_{\rm e} = \frac{v_{\perp}}{r} = \frac{\sqrt{c^2 - v_z^2}}{r} = \frac{\sqrt{c^2 - v_z^2}}{r_{\rm e}\sqrt{1 - \frac{v_z^2}{c^2}}} = \frac{c}{r_{\rm e}}. \tag{40}$$

and, as a consequence of (40), also

$$\omega = \frac{c}{r}.$$

But

$$\omega_z = \frac{v_z}{r}$$

and, therefore, the sum of squares of the angular frequencies yields the following relations

$$\omega^2 = \omega_{\rm e}^2 + \omega_z^2, \qquad p_{\rm c}^2 = p_{\perp}^2 + p_{\|}^2,$$

and, finally,

$$m^2c^2 = m_{\rm e}^2c^2 + m^2v_z^2. \tag{41}$$

For the sake of simplicity we will use the symbol p to indicate the electron momentum $p_{\|}$

$$p = p_{\|} = mv_z.$$

According to De Broglie hypothesis, ω_z is the instantaneous angular frequency associated to a particle with rest mass m_{e}, relativistic mass m and velocity $v_z = \omega_z r$. As a consequence

$$p = mv_z = \frac{\hbar\omega}{c^2} v_z = \frac{\hbar}{cr} v_z = \frac{\hbar\omega_z}{c} = \hbar\frac{2\pi}{\lambda} = \hbar k \tag{42}$$

or

$$\left[p = mv_z = \omega v_z = \frac{v_z}{r} = \omega_z = \frac{2\pi}{\lambda} = k\right]_{\mathrm{NU}}.$$

Equation (42) yields

$$\frac{p}{k} = p\frac{\lambda}{2\pi} = \hbar. \tag{43}$$

where the term $k = 2\pi/\lambda$ is the wave number of the electron and λ the related De Broglie wavelength. Of course, if we observe the electron at a spatial scale much larger than its Compton wavelength and at a time scale much higher than the very short period $T \approx 8.1 \times 10^{-21}$ s of the Zitterbewegung rotation period, for a constant speed v_z, the electron can be approximated to a point particle, provided with "mass" and charge, which moves with a uniform motion along the z-axis of the helix. Particularly, Fig. 2 represents the helical trajectories of electrons moving at different speeds.

5. ESR, NMR, Spin and *Intrinsic* Angular Momentum

As shown in the previous paragraph, in the proposed model, the electron has an angular momentum $\boldsymbol{\hbar}$ and a magnetic moment $\boldsymbol{\mu}_{\mathrm{B}}$, equal to Bohr magneton. It is, therefore, reasonable to assume that, in presence of an external magnetic field, the electron is subjected, as a small gyroscope, to a torque $\boldsymbol{\tau}$ and to a Larmor precession with frequency ω_{p}. The only difference with a classical gyroscope is the quantization of the $\hbar_{\|}$ component of the angular momentum $\boldsymbol{\hbar}$ along the external flux density field $\boldsymbol{B}_{\mathrm{E}}$. This component can take only two possible spin values, namely $\hbar_{\|} = \pm\frac{1}{2}\hbar$ (see [4], p. 83). The two spin values will correspond to two possible values for the angle θ formed between the angular momentum vector and the external magnetic field vector: $\theta \in \left\{\frac{\pi}{3}, \frac{2\pi}{3}\right\}$:

$$\hbar_{\|}^2 + \hbar_{\perp}^2 = \hbar^2, \quad \hbar_{\|} = \pm\frac{1}{2}\hbar.$$

The torque exerted by the external flux density field $\boldsymbol{B}_{\mathrm{E}}$ is

$$\tau = |\boldsymbol{\mu}_{\mathrm{B}} \times \boldsymbol{B}_{\mathrm{E}}| = \mu_{\mathrm{B}} B_{\mathrm{E}} \sin(\theta)$$

and the related Larmor precession angular frequency is

$$\omega_{\mathrm{p}} = \frac{B_{\mathrm{E}}\mu_{\mathrm{B}}}{\hbar}. \tag{44}$$

The precession angular frequency will correspond to two possible energy levels:

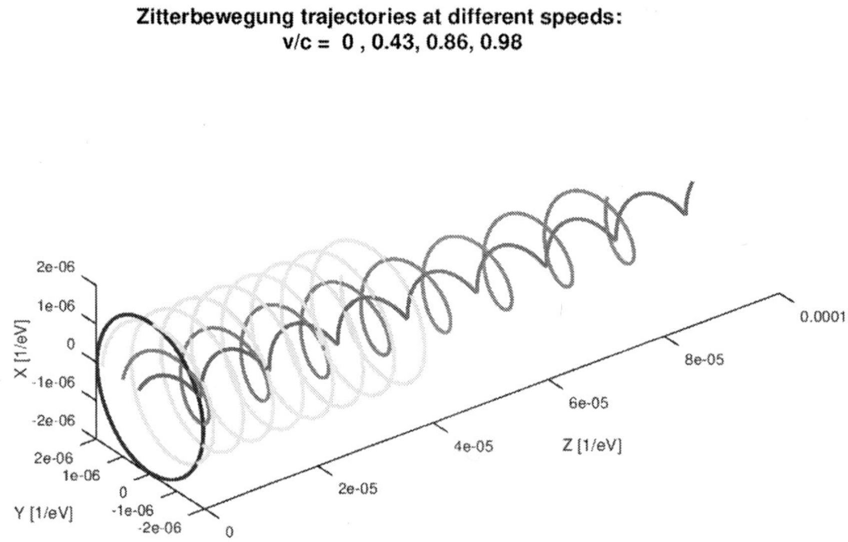

Figure 2. Zitterbewegung trajectories for different speeds.

$$E_{\mathrm{H}} = \hbar\omega_{\mathrm{p}} \quad \text{if} \quad \theta = \frac{2\pi}{3}$$

and

$$E_{\mathrm{L}} = -\hbar\omega_{\mathrm{p}} \quad \text{if} \quad \theta = \frac{\pi}{3}.$$

The difference of energy levels corresponds to the Spin Electronic Resonance (ESR) frequency ν_{ESR}:

$$\Delta E = E_{\mathrm{H}} - E_{\mathrm{L}} = 2\hbar\omega_{\mathrm{p}} = \hbar\omega_{\mathrm{ESR}} = h\nu_{\mathrm{ESR}}. \tag{45}$$

From (44) and (45) it is possible to determine the ESR frequency as

$$\nu_{\mathrm{ESR}} = 2\frac{B_{\mathrm{E}}\mu_{\mathrm{B}}}{h}. \tag{46}$$

For instance, an external magnetic flux density field equal to $B_{\mathrm{E}} = 1.5$ T yields a frequency $\nu_{\mathrm{ESR}} \approx 42$ GHz. By calling s the spin value and μ the nuclear magnetic moment we can also generalize (46) for particles other than the electron. In this case the term used is Nuclear Magnetic Resonance (NMR) frequency, which is equal to

$$\nu_{\mathrm{NMR}} \approx \frac{B_{\mathrm{E}}\mu}{hs}. \tag{47}$$

For instance, for isotope $^{7}_{3}\mathrm{Li}$, with $s = 3/2$, $\mu \approx 1.645 \times 10^{-26}$ and $B_{\mathrm{E}} = 1.5\,\mathrm{T}$, the NMR frequency is $\nu_{\mathrm{NMR}} \approx 24.8$ MHz, whereas for isotope $^{11}_{5}\mathrm{B}$ we have $s = 3/2$, $\mu \approx 1.36 \times 10^{-26}$ $\mathrm{J\,T^{-1}}$ and NMR frequency is $\nu_{\mathrm{NMR}} \approx 20.5$ MHz. Another example deals with isotope $^{87}_{38}\mathrm{Sr}$ with $s = 9/2$ and $\mu \approx 5.52 \times 10^{-27}$ $\mathrm{J\,T^{-1}}$. In this case NMR frequency is $\nu_{\mathrm{NMR}} \approx 278$ kHz for $B_{\mathrm{E}} = 0.15$ T with a Larmor frequency $\frac{\omega_{\mathrm{p}}}{2\pi} = \frac{1}{2}\nu_{\mathrm{NMR}} \approx 139$ kHz.

5.1. Electron spin and coherent systems

In the proposed model, the electron, in presence of an external magnetic field, is subjected to Larmor precession and its spin value $\pm\hbar/2$ is interpreted as the intrinsic angular momentum component parallel to the magnetic field. It is interesting to note that a hypothetical technology, able to align the intrinsic angular momentum of a sufficient number of electrons, could favor the formation of a coherent superconducting and super-fluid condensate state. In this state, the electrons would behave as particles with whole spin $\hbar$ and would no longer be subject to the Fermi–Dirac statistic. The compression effect (pinch) of an electrical discharge, accurately localized in a very small "capillary" volume, inside which a very rapid and uniform variation of the electric potential occurs, could favor the formation of a superconducting plasma. The conjecture is based on the possibility that, as a consequence of Aharonov–Bohm effect, a rapid, collective and simultaneous variation of the Zitterbewegung phase catalyzes the creation of coherent systems like those described by K. Shoulders and H. Puthoff [13]: *"Laboratory observation of high-density filamentation or clustering of electronic charge suggests that under certain conditions strong coulomb repulsion can be overcome by cohesive forces as yet imprecisely defined"*.

6. Hypotheses on the Structure Of Ultra-dense Hydrogen

In relativistic quantum mechanics, the Klein-Gordon equation describes a charge density distribution in space and time. In this equation a term $m^2c^2/\hbar^2$ appears, whose interpretation becomes simple and intuitive if one uses natural units and the principle of mass–energy–frequency equivalence. In particular, it is possible to recognize this term as the square of the Zitterbewegung angular frequency ω:

$$\left[\frac{m^2c^2}{\hbar^2} = m^2 = \omega^2\right]_{\mathrm{NU}}.$$

In the paper "*On the hydrino state of the relativistic hydrogen atom*" [14], the author, by applying the Klein–Gordon equation to the hydrogen atom, finds a possible deep energetic level of $E_0 \approx 3.7$ keV (see Eqs. (16) and (17)) at a distance r_0 from the nucleus. In particular Naudts demonstrates that

$$E_0 \approx m_{\mathrm{e}}c^2\alpha \approx 3.7\,\mathrm{keV}$$

at a distance from nucleus equal to

$$r_0 \approx \frac{\hbar}{m_{\mathrm{e}}c} \approx 0.39 \times 10^{-12}\ \mathrm{m}.$$

According to the author, the E_0 level corresponds to the hypothetical state of a relativistic electron: "*The other set of solutions contains one eigenstate which describes a very relativistic particle with a binding energy which is a large fraction of the rest mass energy*". It is possible to formulate an alternative hypothesis according to which the radius r_0 is simply the radius r_e of the Zitterbewegung orbit, in the center of which the proton is located. Consequently the energy, E_0, can be interpreted as the electrostatic potential energy between the electron charge and the proton:

$$E_0 = \frac{1}{4\pi\epsilon_0}\frac{e^2}{r_\mathrm{e}} = \frac{\hbar}{r_\mathrm{e}}\alpha c = m_\mathrm{e}c^2\alpha, \qquad \left[E_0 = \frac{e^2}{r_\mathrm{e}} = \frac{\alpha}{r_\mathrm{e}} = \omega_\mathrm{e}e^2 = m_\mathrm{e}\alpha\right]_{\mathrm{NU}}.$$

A series of numerous experiments conducted by Leif Holmlid of the University of Gothenburg, recently replicated by Sindre Zeiner–Gundersen [15], seems to demonstrate the existence of a very compact form of deuterium [16–18]. Starting from the kinetic energy (about 630 eV) of the nuclei emitted in some experiments, achieved by irradiating this particular form of ultra-dense deuterium with a small laser, a distance between deuterium nuclei of about 2.3×10^{-12} m has been computed, a value much smaller than the distance of about 74×10^{-12} m that separates the nuclei of a normal deuterium molecule. Therefore, it is possible to advance an hypothesis on the structure of ultra-dense hydrogen (UDH) starting from the electron Zitterbewegung model. The proton is considerably smaller than Zitterbewegung orbit radius r_e, consequently an hypothetical structure formed by an electron with a proton (or a deuterium nucleus) in its center would have a potential energy of

$$\left[\frac{-e^2}{r_\mathrm{e}} \approx -3.7\ \mathrm{keV}\right]_{\mathrm{NU}},$$

a value corresponding to the energy in the X-ray range with a wavelength of about 3.3×10^{-10} m. The distance between the deuterium nuclei in the Holmlid experiment could be explained by an ordered linear sequence of ultra-dense particles in which the rotation planes of the electron charges are parallel and equidistant. In these hypothetical aggregates, the Zitterbewegung phases of two neighboring electrons differ by π radians and the distance d_c between the charges of the two electrons is equal to the distance traveled by light in a time equal to a rotation period T. This distance amounts to $d_\mathrm{c} = cT = \lambda_\mathrm{c} \approx 2.42 \times 10^{-12}$ m. In this case, the distance between the nuclei d_i can be obtained by applying the Pythagorean theorem, as shown in Fig. 3, yielding the value

$$d_i = \sqrt{\lambda_\mathrm{c}^2 - \left(\frac{\lambda_\mathrm{c}}{\pi}\right)^2} \approx 2.3 \times 10^{-12}\ \mathrm{m}.$$

This UDH model is in agreement with the third assumption of Carver Mead "Alternate World View": *"every element of matter is coupled to all other charges on its light cone by time-symmetric interactions"* [2].

6.1. Ultra-dense hydrogen and anomalous heat generation in metal–hydrogen systems

The combustion of a mole of hydrogen (about two grams) generates an energy of 286 kJ (or 240 kJ if we do not take into account the latent heat of vaporization of water), a value that corresponds to an energy of 1.48 eV per atom. The formation of an ultra-dense hydrogen atom would release an energy of 3.7 keV per atom, a value 2500 times higher. The conversion of only two grams of hydrogen into ultra-dense hydrogen would then be able to generate an energy of 715 MJ ≈ 198 kWh. Consequently, the hypothesis, according to which in some experiments the development of anomalous heat is partially or totally due to the formation of ultra-dense hydrogen, cannot be excluded. Following *an alternative* hypothesis, the $\alpha m_\mathrm{e}c^2 \approx 3.7$ keV energy is not emitted as an X-ray photon but is stored in the electron mass–frequency–energy, with a consequent small Zitterbewegung orbit radius reduction. By defining m_eu and r_eu the mass and the radius, respectively, in this new state we have:

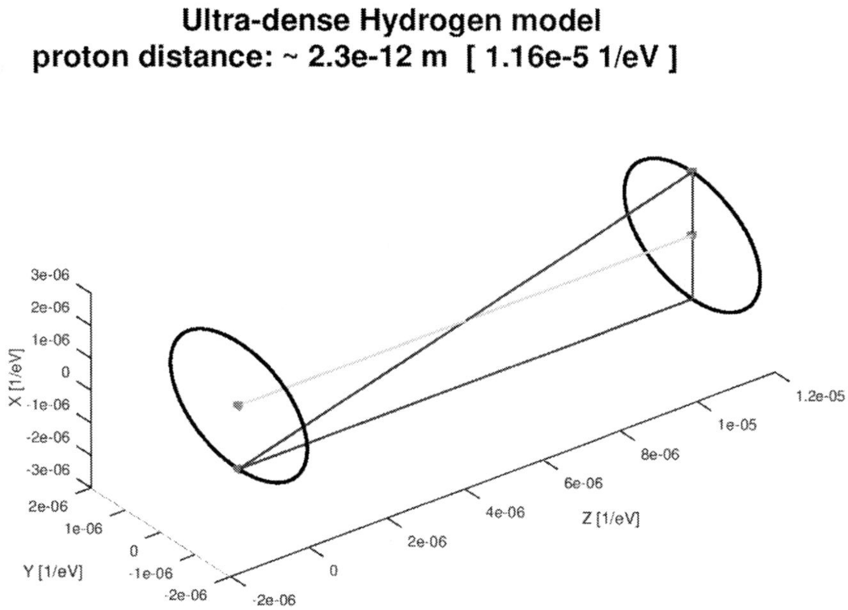

Figure 3. Ultra-dense hydrogen model..

$$m_{\rm eu}c^2 = m_{\rm e}c^2 + \alpha m_{\rm e}c^2 \approx 514.728\,{\rm keV}. \tag{48}$$

The mass increase implies a Zitterbewegung radius reduction. In fact

$$m_{\rm e}c^2 = \hbar\omega_{\rm e} = \frac{\hbar c}{r_{\rm e}},$$

$$m_{\rm eu}c^2 = m_{\rm e}\left(1+\alpha\right)c^2 = \hbar\omega_{\rm eu} = \frac{\hbar c}{r_{\rm eu}},$$

and therefore

$$r_{\rm eu} = \frac{\hbar}{m_{\rm e}\left(1+\alpha\right)c} = \frac{r_{\rm e}}{1+\alpha}.$$

This radius reduction generates a potential energy decrease:

$$\triangle E_{\rm p} = \frac{e^2}{4\pi\varepsilon_0}\left(\frac{1}{r_{\rm e}} - \frac{1}{r_{\rm eu}}\right) = \frac{e^2\alpha}{4\pi\varepsilon_0 r_{\rm e}} = \left[\frac{\alpha^2}{r_{\rm e}}\right]_{\rm NU} \approx 27.2\,{\rm eV}.$$

Following the Carver Mead "transactional" interpretation of photons, the eventual (or necessary?) emission of the ultraviolet 27.2 eV photon may be favored by a "Mills catalyst" [19,2].

Another Zitterbewegung model for deep electron states has been recently presented by A. Kovacs et al., aimed at explaining their impressive experimental results [20].

7. Ultra-dense Hydrogen and Low-energy Nuclear Reactions

In the proposed model the particles of hydrogen or ultra-dense deuterium are electrically neutral but have a magnetic moment almost equal to electron's one. This is a value 960 times higher than the neutron magnetic moment. A particle with magnetic moment $\boldsymbol{\mu}$ is subjected, in presence of a magnetic field $\boldsymbol{B}$, to a force $\boldsymbol{f}$ proportional to the gradient of $\boldsymbol{B}$

$$\boldsymbol{f} = \nabla\left(\boldsymbol{B} \cdot \boldsymbol{\mu}\right).$$

Therefore, the magnetic field $\boldsymbol{B}$ generated by a nucleus could exert a considerable "remote action" on the particles of ultra-dense hydrogen. This force could be the source of the "long range potential" mentioned in a theoretical work of Gullström and Rossi, "Nucleon polarizability and long range strong force from $\sigma I = 2$ meson exchange potential" [21]:

"A less probable alternative to the long range potential is if the e-N coupling in the special EM field environment would create a strong enough binding to compare an electron with a full nuclide. In this hypothesis, no constraints on the target nuclide are set, and nucleon transition to excited states in the target nuclide should be possible. In other words these two views deals with the electrons role, one is as a carrier of the nucleon and the other is as a trigger for a long range potential of the nucleon".

Hence, it is possible that, according to this scenario, electrons would have a fundamental dual role as catalysts of low-energy nuclear reactions (LENR): the first as neutralization-masking effect of the positive charge of hydrogen or deuterium nuclei, a necessary condition to overcome the Coulomb barrier, the second as the source of a relatively long-range magnetic force.

By using the Holmlid notation "H(0)" to indicate ultra-dense hydrogen particles, it is possible to hypothesize a LENR reaction involving the ${}^{7}_{3}\mathrm{Li}$, an isotope that constitutes more than 92% of the natural Lithium

$${}^{7}_{3}\mathrm{Li} + \mathrm{H}\left(0\right) \rightarrow 2\,{}^{4}_{2}\mathrm{He} + \mathrm{e}. \tag{49}$$

This reaction would produce an energy of about 17.34 MeV mainly in the form of kinetic energy of helium nuclei, without emission of neutrons or penetrating gamma rays. A similar reaction, able to release about 8.67 MeV, could be hypothesized for the isotope ${}^{11}_{5}\mathrm{B}$

$${}^{11}_{5}\mathrm{B} + \mathrm{H}\left(0\right) \rightarrow 3\,{}^{4}_{2}\mathrm{He} + \mathrm{e}. \tag{50}$$

Emissions in the X-ray range would still be present in the form of braking radiation (*Bremsstrahlung*) generated by the deceleration caused by impacts of helium nuclei with other atomic nuclei.

The three "miracles" required by the low-energy nuclear reactions could therefore find, for example, in the reaction (49) a possible explanation:

(1) Overcoming the Coulomb barrier: the ultra-dense hydrogen particles are electrically neutral.
(2) No neutrons are emitted: the reactions products of (49) and (50) consist exclusively of helium nuclei and an electron.

(3) Absence of penetrating gamma radiation: the energy produced is mainly manifested as kinetic energy of the reaction products and as X-ray emission from *bremsstrahlung*. However a probability for gamma radiation from excited intermediate products and from secondary interaction of high energy alpha particles could not be completely dismissed.

The mechanical energy of the alpha particles produced by the reactions could be converted with a reasonable yield directly into electrical energy or into usable mechanical energy [22], avoiding the need for an intermediate conversion into thermal energy. None of the three miracles is required to justify the production of abnormal heat due to ultra-dense hydrogen formation.

In the Iwamura experiment the low-energy nuclear transmutation of elements deposited on a system formed by alternating thin layers of palladium (Pd) and calcium oxide (CaO) was observed. The transmutation occurs when the system is crossed by a flow of deuterium. The CaO layer, essential for the transmutation, is hundreds of atomic layers far from the surface where the atoms to transmute are deposited or implanted. It is, therefore, necessary to find a mechanism that explains the remote action, the role of the CaO and the overcoming of the Coulomb barrier by deuterium nuclei. An interesting hypothesis could derive from considering the formation of ultra-dense deuterium (UDD) at the interface between calcium oxide and palladium, an area in which the high difference in the work function between Pd and CaO favors the formation of a layer with high electron density (Swimming Electron Layer or SEL) [23]. The ultra-dense deuterium could subsequently migrate to the area where the atoms to transmute are present. Therefore, aggregates of neutral charged ultra-dense deuterium would be, according to this hypothesis, the probable responsible for the transmutation of Cs into Pr and Sr into Mo. It is possible that strontium oxide, with its very low work function, substitutes the calcium oxide role in Celani's experiments [24]. By using again the Holmlid notation "D(0)" to indicate "atoms" of ultra-dense deuterium, the hypothesized *many-body reactions* in Iwamura experiments [25] would be very simple:

$$^{133}_{55}\mathrm{Cs} + 4\mathrm{D}(0) \rightarrow {}^{141}_{59}\mathrm{Pr} + 4\mathrm{e},$$

$$^{88}_{38}\mathrm{Sr} + 4\mathrm{D}(0) \rightarrow {}^{96}_{42}\mathrm{Mo} + 4\mathrm{e},$$

$$^{138}_{56}\mathrm{Ba} + 6\mathrm{D}(0) \rightarrow {}^{150}_{62}\mathrm{Sm} + 6\mathrm{e}.$$

In the above equations the symbols $4\mathrm{D}(0)$ and $6\mathrm{D}(0)$ represent picometric, coherent chains of ultra-dense deuterium particles. The short distance between deuterons in such hypothetical structures may favor these otherwise difficult to explain many-body nuclear transmutation. In this context, the electrons would have the precise role of deuterium nucleus vectors within the nucleus to be transmuted.

8. Conclusions

In this paper a simple Zitterbewegung electron model has been introduced, where the concepts of mass-energy, momentum, magnetic momentum and spin naturally emerge from its geometric and electromagnetic parameters, thus avoiding the obscure concept of "intrinsic property" of a "point-like" particle. An intuitive geometric interpretation of relativistic mass and De Broglie wavelength has been presented. Using only electromagnetic and geometric concepts an interpretation of Proca, Dirac, Klein–Gordon and Aharonov–Bohm equations based on this particular electron model has been presented. A non linear equation for electromagnetic four potential has been introduced that directly implies electric charge and magnetic flux quantization.

Electronic Spin Resonance (ESR) frequency has been computed starting from a spin model based on the Larmor precession frequency of Zitterbewegung rotation plane. A very simple model for ultra-dense hydrogen, where electron has only spin angular momentum, has been proposed, highlighting its possible role in many-body and aneutronic low energy nuclear reactions.

Acknowledgements

Authors wish to thank Francesco Celani and Giuliano Bettini for helpful discussions and suggestions.

References

[1] F. Celani and A.O. Di Tommaso and G. Vassallo, Maxwell's equations and Occam's razor, *J. Condensed Matter Nucl. Sci.* **25** (2017) 100–128.

[2] Carver Mead, The nature of light: what are photons? *Proc. SPIE*, 8832: 8832 – 8832 – 7, 2013.

[3] David Hestenes, Quantum mechanics from self-interaction, *Found. Phys.* 15(1) (1985) 63–87.

[4] F. Celani, A.O. Di Tommaso and G. Vassallo, The electron and Occam's razor, *J. Condensed Matter Nucl. Sci.* **25** (2017) 76–99.

[5] Oliver Consa, Helical model of the electron, *General Sci. J.* (2014) 1–14. https://www.gsjournal.net/Science-Journals/Research

[6] J. Paul Wesley and David L. Bergman, Spinning charged ring model of electron yielding anomalous magnetic moment, *Galilean Electrodynamics* **1** (1990) 63–67.

[7] A.L. Parson and Smithsonian Institution, *A Magneton Theory of the Structure of the Atom (with Two Plates)*, Vol. 65, in Publication (Smithsonian Institution), Smithsonian Institution, 1915.

[8] D.L. Bergman and C.W. Lucas, Credibility of common sense science, *Found. Sci.* (2003) 1–17. Reprinted by permission of *Galilean Electrodynamics* **1** (1990) 63—67. http://citeseerx.ist.psu.edu/viewdoc/download?doi=10.1.1.490.4984&rep=rep1&type=pdf.

[9] David Hestenes, The zitterbewegung interpretation of quantum mechanics, *Found. Phys.* **20**(10) (1990) 1213–1232.

[10] Richard Gauthier, The electron is a charged photon, Vol. 60, *APS April Meeting 2015*, 2015. http://meetings.aps.org/link/BAPS.2015.APR.Y16.4.

[11] Y. Aharonov and D. Bohm, Significance of electromagnetic potentials in the quantum theory, *Phys. Rev.* **115** (1959 485–491.

[12] G. Bettini, The Moebius Strip: a Biology of Elementary Particles, *viXra.org*, *Quantum Physics*, 2010. http://vixra.org/pdf/1004.0035v4.pdf.

[13] H.E. Puthoff and M.A. Piestrup, Charge confinement by casimir forces, arXiv:physics/0408114, 2004.

[14] J. Naudts, On the hydrino state of the relativistic hydrogen atom, *ArXiv Physics e-prints*, July 2005.

[15] S. Zeiner-Gundersen and S. Olafsson, Hydrogen reactor for Rydberg matter and ultra dense hydrogen, a replication of Leif Holmlid, *Int. Conf. Condensed Matter Nucl. Sci.*, ICCF-21, Fort Collins, USA, 2018.

[16] Shahriar Badiei, Patrik U. Andersson and Leif Holmlid, High-energy Coulomb explosions in ultra-dense deuterium: Time-of-flight-mass spectrometry with variable energy and flight length. *Int. J. Mass Spectrometry* **282**(1–2) (2009) 70–76.

[17] Leif Holmlid. Excitation levels in ultra-dense hydrogen p(-1) and d(-1) clusters: structure of spin-based Rydberg matter, *Int. J. Mass Spectrometry* **352** (2013) 1–8.

[18] Leif Holmlid and Sveinn Olafsson, Spontaneous ejection of high-energy particles from ultra-dense deuterium D(0), *Int. J. Hydrogen Energy* **40**(33) (2015) 10559–10567.

[19] R.L. Mills, J.J. Farrell and W.R. Good, *Unification of Spacetime, the Forces, Matter, and Energy*, Science Press, Ephrata, PA 17522, 1992.

[20] Andras Kovacs, Dawei Wang and Pavel N. Ivanov, Investigation of electron mediated nuclear reactions, *J. Condensed Matter Nucl. Sci.* **29** (2019).

[21] Carl-Oscar Gullström and Andrea Rossi, Nucleon polarizability and long range strong force from $\sigma_{I=2}$ meson exchange potential, arXiv 1703.05249, 2017.

[22] Alfonso Tarditi, Aneutronic fusion spacecraft architecture, 2012, NASA-NIAC 2001 PHASE I RESEARCH GRANT, Final

Research Activity Report (SEPTEMBER 2012), pp. 1–33,
https://www.nasa.gov/sites/default/files/atoms/files/niac_2011_phasei_tardittianeutronicfusionspacecraftarchitecture_tagged.pdf.

[23] H. Hora, G.H. Miley, J.C. Kelly and F. Osman, Shrinking of hydrogen atoms in host metals by dielectric effects and inglis-teller depression of ionization potentials, *Proc. 9th Int. Conf. on Cold Fusion, (ICCF9)*, 2002, pp. 1–6.

[24] F. Celani, C. Lorenzetti, G. Vassallo, E. Purchi, S. Fiorilla, S. Cupellini, M. Nakamura, P. Boccanera, R. Burri, B. Ortenzi, L. Notargiacomo and A. Spallone, Steps to identify main parameters for ahe generation in sub-micrometric materials: measurements by isoperibolic and air-flow calorimetry, *ICCF21, Int. Conf. on Cold Fusion*, 3–8 June 2018, pp. 1–19, DOI: 10.13140/RG.2.2.16425.29287.

[25] Y. Iwamura, T. Itoh and N. Yamazaki, H. Yonemura, K. Fukutani and D. Sekiba, Recent advances in deuterium permeation transmutation experiments. *J. Condensed Matter Nucl. Sci.* **10** (2013) 76–99.

International Society for Condensed Matter Nuclear Science (ISCMNS)
The Willows, Hobro,
Wolverley, Kidderminster
Worcs DY11 5ST
England

ISBN 978-1-5108-9472-3

CURRAN ASSOCIATES INC.
proceedings.com